D0231342

Macmillan Foundations

A new series of introductory texts across a wide range of subject areas to meet the needs of today's lecturers and students

Foundations texts provide complete yet concise coverage of core topics and skills based on detailed research of course requirements suitable for both independent study and class use – *the firm foundations for future study.*

Published

Biology
Chemistry
Politics

Forthcoming

Economics
History of English Literature
Mathematics for Science and Engineering
Modern British History
Modern European History
Physics
Psychology

Biology

JULIAN SUTTON

First published 1998 by
MACMILLAN PRESS LTD
Houndmills, Basingstoke, Hampshire RG21 6XS
and London
Companies and representatives
throughout the world

ISBN 0–333–65860–4 paperback

A catalogue record for this book is available from the British Library.

This book is printed on paper suitable for recycling and made from fully managed and sustained forest sources.

10 9 8 7 6 5 4 3 2
07 06 05 04 03 02 01 00 99 98

Typeset by Footnote Graphics, Warminster, Wilts
Printed in Malaysia

Contents

PART 4

Genes and life cycles

PART 5

The Working Organism

PART 6

Organisms in Context

PART 7

Applied Topics

Preface

IN ORDER to use this book effectively, you ought to know why I have written it this way. My aim is to introduce the important ideas of Biology clearly, and in enough detail for you to be able to move straight on to more specialized literature written for undergraduates. The emphasis is on concepts rather than facts. There are plenty of facts here, but they are included to help explain the concepts, as needed. I have used scientific terms where they help, but never for their own sake. Mathematics is kept to a minimum. You will find little here about practical techniques, or the history of scientific ideas. All these things are important, but there are other ways to learn about them.

Arranging topics in sequence poses a difficult problem. We study Biology at many levels, from molecule to ecosystem. Should we start with small scale processes and then use them to explain things on a larger scale, or should we begin at the big end of the spectrum because these things are more familiar to us from our everyday experience? My compromise is to start big with an overview of the diversity of life, how we group organisms and give them names. Next comes natural selection, because we can explain very little in Biology without it. Then we go to the other end of the size scale with molecular and cell biology, and get bigger by way of genetics, reproduction, physiology, ecology and behaviour. The last two units are rather different. They deal with applications of biological principles in disease and biotechnology. Here, detailed examples are what matters. In order to pack enough detail into a reasonable space, these units consist of a series of information boxes, each dealing with an example or application.

Few people will sit down and read this book from beginning to end. Many readers will need to refer to a particular unit immediately. The 'Connections' boxes at the start of each unit help you to know which other units give useful background information. The index directs you to the key sections about each topic, not simply every page which mentions the word.

Biology is not only about special organisms in special places. The same principles apply to living things everywhere. The organisms which live around you are, if anything, more interesting than exotic ones, because they are there to be studied. The more you watch, think about and experiment with a plant or animal, the more interesting (and often more surprising) its life becomes. Throughout the book I relate biological ideas to the organisms which are familiar to me, around my home. I live on the edge of a country town in south west England. Agricultural grasslands are all around, grazed by sheep and cattle. The high moorland of Dartmoor is a few kilometres away on one side; the rocky shores of the English Channel are on the other. Do not be put off because my familiar organisms are not yours. The challenge is to get out and see the living things in your own backyard, and apply the same principles to them. I am passionate about my subject. I have written this book to encourage you to make it your own.

Julian Sutton

Acknowledgements

I am grateful to Sarah Sutton and Liz Mower for uninhibited comments on large parts of this book; to Richard Bennett-Lovesey, John Burston, Mark Edwards, Richard Graham, Chris Holt, Richard Jones and Tom Maclennan for comments on particular units from a range of viewpoints; and to five anonymous reviewers for a detailed appraisal of a draft version. Frances Arnold, Liz Jones and others at both Macmillan Press and Footnote Graphics have taken my stack of words and sketchy diagrams, and made them into a book.

It is traditional for authors to thank their partners and children for their support, tolerance, encouragement and so on. I used to think this was a polite convention. Now I know better, and gratefully follow tradition.

Grateful acknowledgement is made to the following for permission to use copyright material:

American Association for the Advancement of Science for Figure 27.2 for data from. J. H. Connell, 'Tropical rain forests and coral reefs as open, non-equilibrium systems', *Population Dynamics* (1979), editors R. M. Anderson et al., Blackwell Scientific Publications, Figure 7.3; American Chemical Society for Box 10.1 from A. Bennet and L. Bogorad, 'Properties of subunits and aggregates of blue-green algal biliproteins', *Biochemistry*, **10** (1971), Figure 1; Blackwell Science Ltd for Box 26.2 and Figure 26.4 for data from A. R. Watkinson, 'Interference in pure and mixed populations of *Agrostemma githago*', *Journal of Applied Ecology*, **18** (1981), Figures 1, 2, 3, 4 and 5; Cambridge University Press for material on page 408 from *How Animals Move* (1953) by Sir James Gray, Plate XII; Field Studies Council Publications for Figure 26.2 from R. A. D. Cameron, 'Strand structure, species composition and succession in some Shropshire woods', *Field Studies*, **5** (1980), Figure 3; Peter Goldblatt for Box 1.2 from P. Goldblatt, 'The genus *Watsonia*, a system monograph', *Annals of Kirstenbosch Botanic Gardens*, **19**, Table 4 and Figure 8; Nature for data in Figure 25.9 from D. Bradley et al., 'Control of inflorescence architecture in Antirrhinum', *Nature*, **379** (1996), Copyright © 1996 Macmillan Magazines Ltd; Munksgaard for data in Figure 26.1 from N. J. Collins, 'Growth and population dynamics of the moss *Polytrichum alpestre* in the maritime Antarctic', *Oikos*, **27** (1976) Figure 9.3, Copyright © 1976 Oikos; and John Wiley & Sons Ltd for data in Figure 27.1 from A. J. Davy and R. L. Jefferies, 'Approaches to the monitoring of rare plant populations', *The Biological Aspects of Rare Plant Conservation* (1981), editor H. Synge, Figure 1.

Photographs:
Animal Photography, p. 119; Ardea London, pp. 61, 100, 245, 249, 305, 339, 399, 469; Frances Arnold (by kind permission of Fuller, Smith & Turner P.L.C.), p. 149; Camera Press, pp. 190, 285, 344, 455; J Allan Cash, p. 319; Dr Jack Cohen, University of Warwick, p. 223; Bruce Coleman Collection, p. 262; Imperial War Museum, p. 346; LEO Electron Microscopy, p. 65; Andrew Masson, p. 450; Richard T. Mills, p. 9; Barry Marsh, University of Southampton, p. 20; National Medical Slide Bank, pp. 260, 350, 369, 390, 503; Not US Ltd, p. 381; Oxford Scientific Films, p. 105, 155, 183, 209; Planet Earth Pictures, p. 283; RSPB Images, p. 344; Martyn Rix, p. 438; Patrick Salvadori, pp. 415, 462; Science Photo Library, p. 77; Steve Redwood, pp. 175, 329, 404, 461, 491; Julian Sutton, pp. 92, 371, 422.

Every effort has been made to contact all the copyright-holders but if any have been inadvertently overlooked the publishers will be pleased to make the necessary arrangements at the first opportunity.

Introduction

Science and the nature of explanation

We are an inquisitive species. We want to know what is around the next bend. We climb mountains because they are there. As children we are tempted to take clocks apart to find out how they work, or to flush playing cards down the lavatory to see what happens. Science is a natural extension of this curiosity. Our ambitious aim is to understand how the world works. Biology is a blanket term for all those areas of science which have to do with living things.

As we investigate the world, we do several things. We observe, compare, look for patterns and relationships, speculate and make hypotheses, formulate ideas about how things might be. We also experiment. Experiments are what makes science distinctive. An experiment involves changing the world and seeing what happens, setting up particular conditions and measuring the effect. Experiments allow us to test hypotheses. Simple observation may suggest several possible explanations. Carefully designed experiments help us to distinguish good explanations from bad. We might be interested in the distribution of *Hyacinthoides non-scriptus*, the plant known in England as the bluebell. It is a common woodland plant in many parts of England. Why does it grow there, but not in other habitats such as meadows or moorland? Perhaps it grows best in shade. This is a perfectly sensible hypothesis, but only experiments can show whether or not it is correct. If we grow bluebells in a garden, under controlled conditions so that light is the only factor which varies, they grow better in direct sunlight than in shade. The hypothesis was wrong. It turns out that competition from other plants, such as grasses, excludes bluebells from other habitats.

This is just one example of how science advances, which shows the importance of experiments, but there are other ways. This book is not about the process of scientific discovery. That is best learnt by direct experience, working with others in the trade. All scientific education, from primary school to postdoctoral research fellowship, involves an induction into the ways of science. This book is about what science tells us about the living world, the explanations that it has, and has not, provided.

We can look at nature on many different scales. I write this sitting on the ground, leaning against a birch tree on Dartmoor. The tree, a single organism, is made up of organs: stems, roots and leaves. Each is made up of several tissues, which in turn are made up of cells. Cells consist of many compartments, each with distinctive chemistry. All are made up of interacting molecules. Returning to a larger scale, my tree is not alone. I can see other birch trees in the area, some larger and older than others. They form a population. The birch population interacts with populations of gorse, heather and other plants, forming a community. All the living and non-living things around here constitute a still higher level of organization, the ecosystem. As I watch banks of cloud rolling in from the Atlantic, I am reminded that some processes can only be understood on a global scale.

Science tries to explain observations at one level in terms of processes going on at a lower level. This is called reductionism. It does not imply that the whole is just the sum of the parts. A body is not simply a collection of cells; it is a collection of interacting cells. Many complex properties of the body as a whole can be explained in terms of the interactions between its cells.

We do not seek to account for every detail of a system in a reductionist explanation. Nature is vast, our minds are small. We aim to make models, pictures of nature which are simple enough to fit into our minds, yet retain important features of the real thing. We may never understand every factor which affects the distribution of bluebells; we may never want to. If we can draw out a few factors which have a big effect from the sea of factors which have some effect, and use them to explain important features of bluebell distribution, then we have made a useful model. Only then are we fulfilling our ambition as scientists. We are going beyond description, to explanation.

PART 1

The Diversity of Organisms

Describing the Diversity of Life

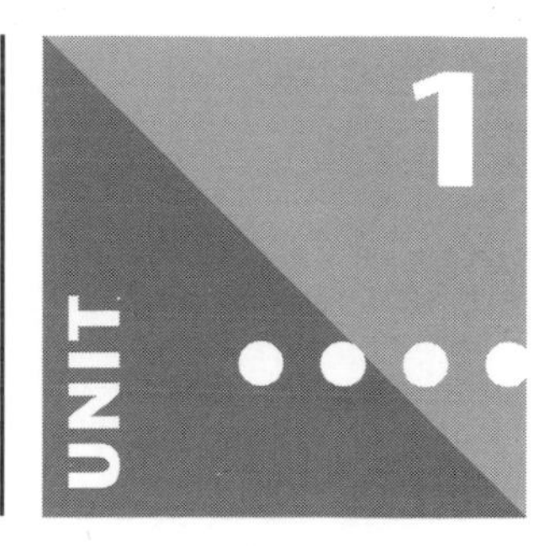

Connections

► This unit includes an overview of the major groups of living things, as well as covering how classifications are made. You may not want to study the mass of information as a whole, but refer to it while reading other parts of the book.

Contents

1.1 Why classify?

Living things come in all shapes and sizes. There are lots of them. Our minds cannot easily handle this much diversity. We need a map to find our way around it, a framework in which each type of organism has its place. Any sort of classification would do this job, but a scientific classification tries to do much more. The aim is to make classification 'natural', that is to reflect evolution. Closely related organisms are grouped together, more distantly related organisms in other, related groups. Also, the system of scientific names is tied in to classification, so names say at least something about where a species fits into the overall scheme. This is a tall order. It is no surprise that **taxonomy**, the science of classification, continues to develop. Classifications change as new organisms are discovered and as taxonomists re-evaluate the relationships between the ones we already know about. Names sometimes change to take account of this.

Some of us tend to put taxonomists down. 'They only describe,' we might say, 'they don't explain'. We moan when familiar names change. However, we cannot function as biologists without classification. If we want a good classification, we must accept changes as it improves. Further, many taxonomists have much wider interests: they may be geneticists, gardeners, evolutionary biologists, naturalists, ecologists or several of these at once. They are interested in what taxonomy has to say to their science, and in what it can do for taxonomy. If we want to make sense of the diversity of life, we must value taxonomy and taxonomists.

1.2 Boxes within boxes: the taxonomic hierarchy and scientific names

The system of classification we have used for several hundred years works like this. Organisms belong to one of a few major groups called kingdoms. We recognize five

Taxonomy: The science of classification.

Table 1.1 How two animal species are placed in the taxonomic hierarchy

	2 spot ladybird	**human**
Kingdom	Animalia	Animalia
Phylum	Arthropoda	Chordata
Class	Insecta	Mammalia
Order	Coleoptera	Primates
Family	Coccinellidae	Hominidae
Genus	*Adalia*	*Homo*
Species	*Adalia bipunctata*	*Homo sapiens*

or six kingdoms at present (Section 1.4). Each kingdom is divided into several phyla (singular: phylum). Each phylum is further divided into classes, then orders, then families, then genera (singular: genus). Each genus contains one or more species. So, each species belongs to several groups at different levels. It is always grouped with similar species. At the lowest level, the genus, the species are very similar: different sorts of buttercup, for example. At the highest levels, the similarities are broader: for example, all types of plant belong to one kingdom, Plantae. This is a hierarchical scheme, a system of boxes within boxes.

Taxonomists sometimes recognize distinct forms of a species. Species are then divided into even smaller groups. Subspecies are distinct in several ways; forms and varieties probably differ by only one or a few genes. The cultivated forms of plants are distinct varieties. They belong to a special type of group, the cultivar.

The system of scientific names is based on two levels of the hierarchy: the genus and the species. Each species is given a two part name, or binomial. The first is the name of the genus, the generic name. The second is unique to that species, the specific name. For example, the two spot ladybird is called *Adalia bipunctata*. The generic name always gets a capital letter, the specific does not. The name is in Latin, or at least a latinized form of words from other languages, so it is printed in italics, or underlined in handwriting. Names are latinized for historical reasons. The specific name has some relevance to the species: for example *bipunctata* refers to two spots. Groups below the species level are included in the name if relevant. Groups above the generic level are never included.

With scientists in different times and places describing species, some are bound to be named twice. The first name to be published is the one we use, once somebody has sorted out the mess. Sometimes the same name gets applied to species defined in slightly different ways. To distinguish these, the name of the person who applied the name is sometimes included after the specific name. *Adalia bipunctata* becomes *Adalia bipunctata* L. The 'L.' stands for Linnaeus (Carl von Linne) who devised the binomial system and named lots of familiar European plants and animals. Table 1.1 gives some examples of how species fit into the taxonomic hierarchy, and their names.

1.3 How classifications are made

Making a classification involves human judgement. No one can write down a definition of exactly how different two species must be before they are put into different genera, for example. It comes down to the taxonomist's opinion, based on experience of how other people have set up genera in the past.

The taxonomist must look at a wide range of characters. Some are more significant than others. Differences which could be down to one or a few genes, such as flower colour or plant height, can evolve quickly. They tell us little about how closely related two species are. Other differences which must involve many genes reflect major changes taking a long time. The different body plans of, say, a beetle, a haddock and a slug suggest that these animals should be widely separated in a classification. A difference in just one character is not enough to place two organisms in different groups. The taxonomist is looking for differences in several characters at once. After the classification has been made, it is sometimes possible to pick out a single character which can reliably be used to distinguish the two groups in practice.

This traditional way of classifying organisms is still in use today. However, two other techniques are available: numerical taxonomy and cladistics.

Numerical taxonomy is an attempt to make classification objective, to take unreliable human judgement out of the process. Very many characters are coded numerically, in all the species to be grouped. Computer software is used to calculate the degree of similarity between every possible pair of species. Then they are arranged in a sequence which puts the most similar species together (Fig. 1.1). This is not an evolutionary tree. The branching points simply show the percentage similarity between the species at the tips of those branches. The taxonomist then decides how similar two species should be in order to belong to the same genus, and so on. In this example, 90% is the cut off. Species J and E belong to one genus, species G to another. If 80% is the cut off for families, J, E and G all belong to one family. The other five species belong to two genera in a second family. All belong to one order.

Numerical taxonomy does not remove the taxonomist's judgement, however. A computer cannot choose reliable characters by itself. The taxonomist must still select the characters to be coded, rejecting the ones that would be rejected in a traditional classification.

The main use of numerical taxonomy is in classifying microorganisms. Bacteria are classified mainly on biochemical and genetic characters. Protein sequences and DNA sequences (Units 3 and 7) are easy to code and cluster by computer, but are hard to handle in more traditional ways.

Traditional and numerical classifications are based on the **phenotype**, an

Phenotype: An organism's characteristics.

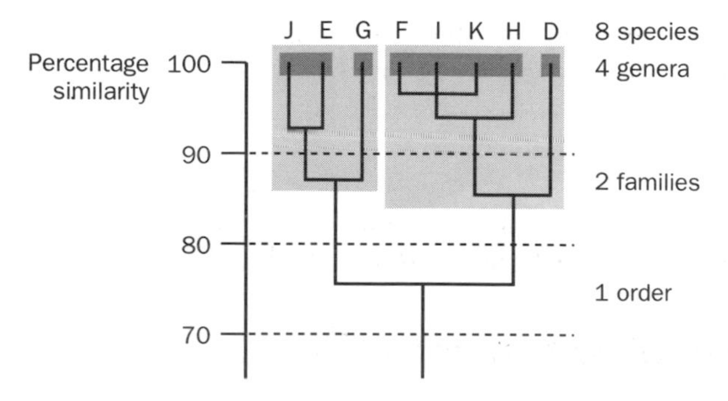

Fig. 1.1 An example of how a group of eight species (D–K) might be classified, using numerical taxonomy.

Phenetic classifications: Classification based on organisms' characteristics.

Phylogeny: The evolutionary history of a group.

Phyletic classifications: Classification based on the evolutionary history of a group.

Cladogram: A tree-like diagram, showing when related groups diverged from one another.

Cladistics: A method of making classifications, based on cladograms.

organism's characteristics. They produce a **phenetic classification**. The aim of cladistics is to make a classification based on **phylogeny**, the evolutionary history of a group. This is called a **phyletic classification**.

The **cladogram** is the basis of **cladistics**. This is a tree-like diagram, in which branching points are the times when two groups diverged, as in Fig. 1.2. All five groups shared an ancestor at the bottom of this tree. Genes flow up the page, from parent to offspring, through countless generations. The first branching point is when the evolutionary lines leading to the five modern groups separated. The ancestors may have looked very similar to one another, and nothing like the modern groups. The branching point is simply the time when the flow of genes split into two streams. We cannot use fossils on their own to make cladograms; the fossil record is far too patchy. Instead, the taxonomist must try to work out which *primitive* characteristics the shared ancestors had, and which *derived* characteristics have appeared more recently. Crudely, characteristics which are scattered among the species without following any pattern are probably primitive. When several characteristics are shared by a group of species, but are not found in other species, the characteristics are probably derived. Groups are defined by shared derived characteristics. For example, body hair is a derived characteristic which sets the mammals apart from the rest of the vertebrates. In Fig. 1.2, species A, B and C would be grouped together because they share some derived characteristics which D and E do not. A and B would have further derived characteristics which C does not, and so on. Sometimes a character may evolve twice in different groups: parallel evolution. How can parallel derived characters be distinguished from primitive characters? In the end, it comes back to the taxonomist's judgement: we cannot get away from it.

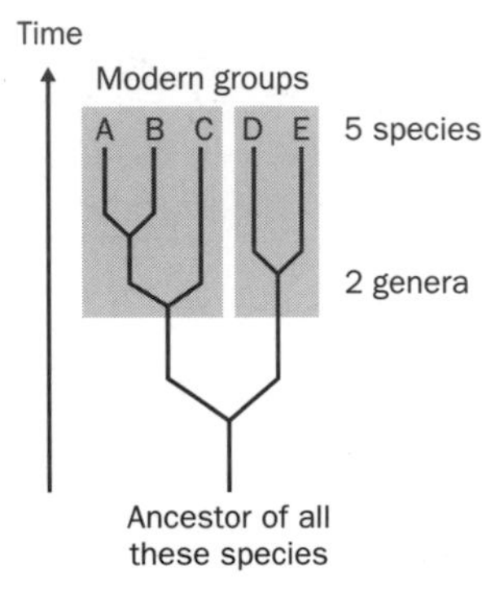

Fig. 1.2 A cladogram, and how it can be used to group species.

Cladistic classification was first used in insect taxonomy and has spread to other fields. Box 1.1 gives a famous example of the debate which followed. Box 1.2 shows a real example of cladistic analysis, from plant taxonomy. It is increasingly accepted as a useful taxonomic tool, if used intelligently.

Box 1.1

A classic cladistic controversy

How should we classify the crocodiles and their relatives? Should they be placed with the rest of the reptiles, in the traditional way, or do they belong with the birds?

A cladogram suggests that they should be placed with the birds. However, the ancestors of these modern groups, just after they diverged, must have been very similar to one another: still essentially reptiles. The birds have evolved in a very distinctive way since branching. The other groups have diverged much less. A phenetic classification would recognize this, and place the crocodiles and other reptiles together. In the end, the taxonomist must choose which classification is most useful.

Cladistic analysis, like any other tool, should be the slave of the taxonomist, not the other way round.

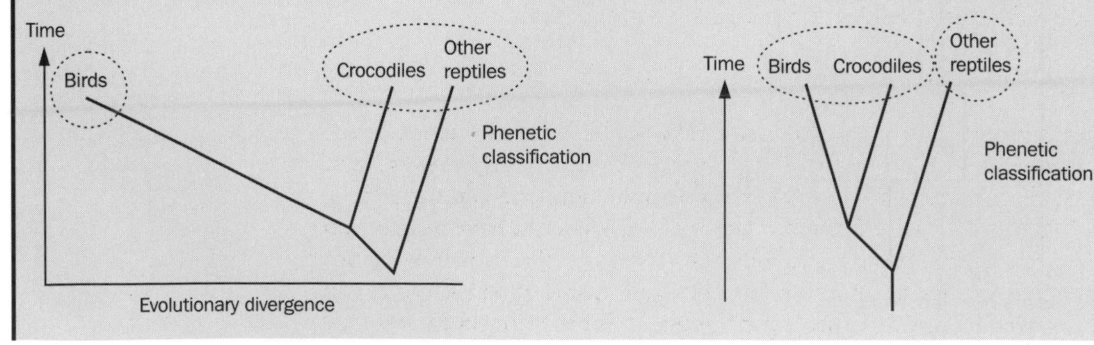

Box 1.2

A cladistic classification

Here is part of a modern classification which uses cladistic analysis. *Watsonia* is a genus of flowering plants in the iris family (Iridaceae), native to southern Africa. Here, the genus is divided into a number of Sections and Subsections, each containing one or several related species.

First, the taxonomist's detailed knowledge of the genus led to identification of 13 characters, each with derived and primitive states. Primitive states are shown in brackets in the table. (Don't worry if a few of the botanical terms are unfamiliar: concentrate on how the information is used.) So, taking character 6, he believes that ancestors of the genus had pink or purple flowers. Orange and red flowers appeared later in its history.

Then groups were set up which share derived characteristics. These are shown in the cladogram. There seems to have been parallel evolution in four characters, shown by double lines in the cladogram.

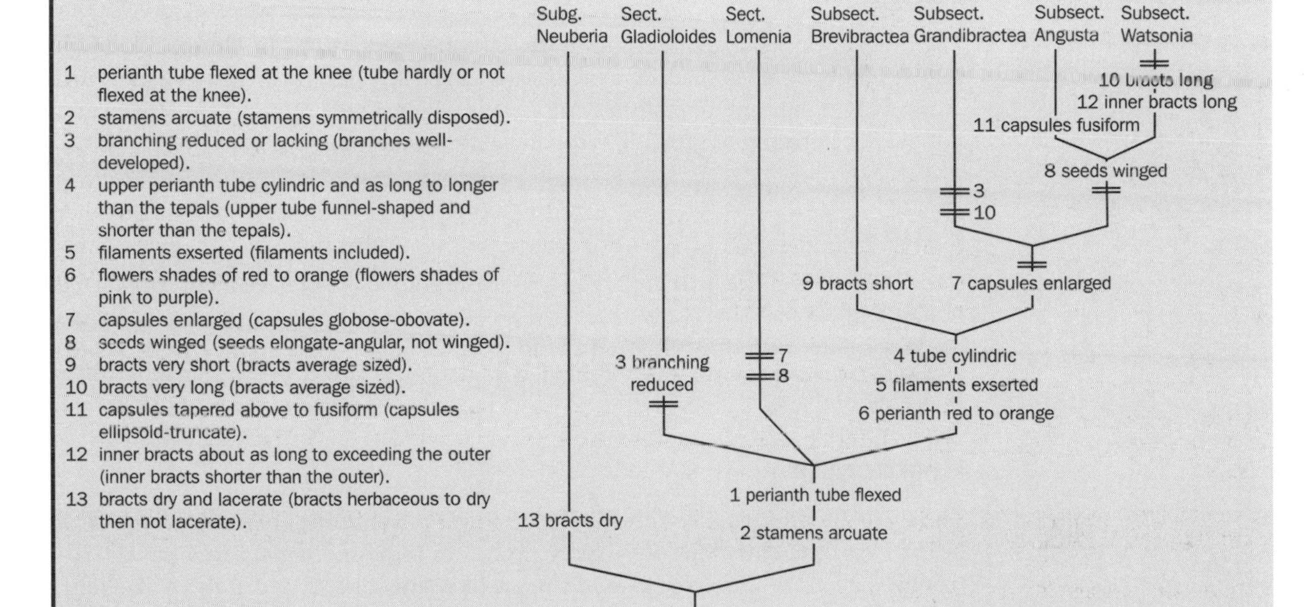

1.4 Five kingdoms, or even six

Organisms are grouped into five kingdoms, arguably six.

Prokaryotae

These are the prokaryotes: the bacteria and their relatives. They have a simple structure (Section 4.3) with no nucleus and few of the other internal compartments of eukaryotic cells. They include the true bacteria, the mycoplasmas, and the photosynthetic cyanobacteria (blue-green algae). The many phyla of this kingdom have been set up largely on biochemical characteristics, and will not be given a detailed treatment in this book. One group, the Archaea, are quite distinct genetically from both prokaryotes and eukaryotes, although their cell structure is prokaryotic. It is becoming clear that they should belong to a sixth kingdom. Most of these strange organisms live in extreme environments, such as hydrothermal vents in deep ocean trenches.

Animalia

The animals are eukaryotes. Their bodies are made up of many cooperating cells, forming different types of tissues. The cells have no cell wall. Most animals have distinct organs. Most have a more or less centralized control system, involving nerves and hormones. They feed on organic matter that has already been made by other organisms: they are heterotrophs.

Fungi

The fungi are also eukaryotes. They have cell walls made strong by chitin. The body, called a mycelium, is made up of threads called hyphae. These have a cell wall around them, but only sometimes cross-walls which divide them into distinct cells. Fungi are heterotrophs, growing through whatever they feed on.

Plantae

The plants are multicellular eukaryotes. Like the animals, they have highly organized tissues and organs. Their cells have rigid walls based on cellulose. Their movements are very limited and control systems tend to be localized. They are nearly all photosynthetic: they make their own food using light energy and small molecules.

Protoctista

These are the simple eukaryotes. Some are single celled, some are multicellular with little or no differentiation into tissues. Some are photosynthetic, some are heterotrophs. There are protoctistans with plant-like, animal-like and fungus-like cells, whilst others have intermediate characteristics. The other kingdoms are very distinct from one another, but it is not easy to define where the Protoctista should end and where the other eukaryotic kingdoms should begin.

Oddities

Two groups do not fit comfortably into this classification. The viruses are genetic parasites which are not made up of cells (Boxes 30.1–30.6). Arguably, they are not living things at all. They certainly do not belong to any of these kingdoms. It is generally agreed that they cannot be part of the classification system at all. They are probably best thought of as groups of genes which have become semi-independent, returning to cells as parasites.

The lichens are associations between fungi and single-celled photosynthetic organisms, either prokaryotic or eukaryotic. Both benefit from the relationship, and they form distinctive bodies which even reproduce as one. The whole is certainly more than the sum of the parts. They are sometimes put into a phylum, Mycophycophyta, but this has to sit outside the five kingdoms.

All five kingdoms are here: heron (Animalia) perched on tree (Plantae) with lichens (Fungi and Protoctista), whilst every surface in sight is covered in bacteria (Prokaryotae).

1.5 A trip through the eukaryotic kingdoms

Here are some of the major phyla from each kingdom. It is not a complete catalogue, which could become quite bewildering, but the larger and more familiar phyla are here. For each, important features are briefly described and some examples are given. Few characteristics are shared by *every* member of a group. There are exceptions to most rules, but to keep things simple I have not written 'usually', 'mostly' and so on.

Kingdom Protoctista

The members of this group which photosynthesize are traditionally called algae. All the rest, except the Oomycota, are called protozoa.

Phylum Chlorophyta

The green algae are the most plant-like protoctists. They have cellulose in their walls, they store starch, and they photosynthesize using similar pigments to plants: chlorophylls a and b, carotenes and xanthophylls. However, each cell has one big chloroplast rather than many little ones. Their body forms include single cells which swim using flagella, spherical or disc-shaped colonies, filaments and sheet-like seaweeds. They live in fresh water, the sea, soil and on moist surfaces.

Phylum Rhodophyta

The red algae have plant-like cells. They contain chlorophylls a and b, carotenes, xanthophylls and phycobilins. A few live in fresh water, most in the sea. Some form filaments, others sheets of cells. The coralline red algae are coated in hard calcium carbonate.

Phylum Phaeophyta

The brown algae are nearly all seaweeds. A few form filaments but most have large, multicellular bodies with some tissue differentiation. They have chlorophylls a and c, carotenes, and a brown xanthophyll pigment, fucoxanthin.

Phylum Bacillariophyta

The diatoms are photosynthetic, unicellular organisms which sometimes form small colonies. Their unusual cell walls include silica as well as cellulose. The walls come in two halves called valves. One fits over the other like a lid on a box. They have chlorophylls a and c, carotenes and fucoxanthin. Diatoms move in a mysterious, gliding way which seems to involve the secretion of mucus. They live in fresh water and the sea.

Phylum Dinophyta

The dinoflagellates are unicells which swim using flagella. Most are photosynthetic, but some are heterotrophic instead, or as well. Most have cell walls which often have a distinctive sculptured appearance. One of the two flagella often lies in a groove in the wall. Most are marine plankton, that is they live in the upper layers of the sea.

Phylum Euglenophyta

Another phylum of unicells with flagella. The flagella arise from a hollow in the cell surface. Some have unusual chloroplasts surrounded by three membranes. Others are heterotrophic. The cell is protected by a tough, flexible extracellular matrix, the pellicle, but there is no cellulose wall. Most live in fresh water.

Phylum Zoomastigina

These are heterotrophic, flagellate unicells without a cell wall or pellicle. Many are parasites of animals or plants, for example *Trypanosoma*, which causes sleeping sickness.

Phylum Apicomplexa

The sporozoans are heterotrophic unicells, with a pellicle but no flagella. They are parasites with complex life cycles. Diploid and haploid stages alternate, as in plants and some other protoctists, such as the Rhodophyta. At least one stage has a complex of organelles at the cell apex (hence the name), which the cell uses to attach to and enter the host cell. The malaria parasites are famous examples.

Phylum Ciliophora

The ciliates are heterotrophic unicells. A pellicle keeps their shape constant. They are covered in rows of cilia, used in movement and sometimes for catching food. Each cell has two nuclei. The diploid micronucleus is a 'normal' nucleus, whose genes are passed on to the next generation. The macronucleus has many copies of each chromosome, and controls synthesis of proteins in the large amounts needed by these big cells. Some of the largest protoctists are ciliates; exceptional species may be up to 5 cm long. Most are free-living in fresh water and the sea, but parasites, forms which live in colonies, and species which live attached to surfaces by stalks, are all known.

Phylum Rhizopoda

These heterotrophic unicells are amoebae: cells which can change shape and crawl by extending the cell in a particular direction. Some species secrete a hard, open 'pot' around the cell, called a test. This is made of chitin, sometimes with solid particles from the environment. They live in fresh water, the sea, soil, and sometimes as parasites.

Phylum Foraminifera

Another group of amoeboid unicells which secrete tests. Sometimes the test is studded with particles from the environment, sometimes it has calcium carbonate deposits. Fine extensions of the cell form nets which extend beyond the test, catching protoctists and small animals for food. The whole net may be centimetres across. They live in the sea, either floating in the plankton or on the sea bed.

Phylum Eumycetozoa

The slime moulds have a weird, unique life cycle. Amoeboid cells live and feed in the soil or on dead organic matter. Then they get together in one of two ways. In some, two cells fuse, then divide to form a large, slimy mass containing many nuclei but not made up of distinct cells: a plasmodium. In others, many amoebae group together to form a multicellular colony. Cells move within the colony, which crawls along as a unit: it looks just like a tiny slug. The plasmodium or 'slug' produces fruiting bodies, often on stalks. They make spores which disperse and develop into amoebae.

Phylum Oomycota

The oomycetes are the most fungus-like protoctists. The body is a mycelium, a network of thread-like hyphae. It grows through living or dead plant tissues, digesting them as it goes. Cell walls are based on cellulose fibres, not chitin. Asexual spores have flagella, which are never seen in fungi. The downy mildew diseases of plants, and the disease known as potato blight, are caused by oomycetes.

Kingdom Fungi

Phylum Zygomycota

The zygomycetes are a group of rather simple moulds, including the bread mould *Mucor*. The short-lived mycelium has no cross-walls. They produce asexual spores in pin-like structures which stick up from the mycelium. Sexual reproduction involves hyphae from two mycelia joining: two nuclei fuse, one from each, forming a zygospore. This then divides by meiosis, and goes on to form a sporangium.

Phylum Ascomycota

The ascomycetes are diverse. Most have a mycelium of hyphae with cross-walls, although yeasts are unicellular. They have in common the ascus, a type of spore sac. Two nuclei from different individuals fuse to make a diploid nucleus. It divides by meiosis, normally followed by a round of mitosis, to form a line of haploid spores enclosed by a wall. This is the ascus. The ascomycetes include powdery mildews, which are parasites of plants, some moulds including *Penicillium*, and some longer-lived organisms such as truffles and morels.

Phylum Basidiomycota

The basidiomycetes have hyphae with cross-walls. The mycelia are often long lived, but produce more short-lived fruiting bodies such as toadstools. A toadstool is made up of many closely packed hyphae. They carry basidia: swollen diploid cells at the tips of hyphae in the gills. Each basidium divides by meiosis to form four spores. The spores are thrown violently into the air spaces between the gills, and float away. The basidiomycetes include mushrooms and toadstools, bracket fungi and puffballs.

Kingdom Animalia

Phylum Porifera

The sponges are simple animals showing little organization of cells into tissues. They have no nervous system. In the simpler sponges, the body is one or several jar-like colonies of cells. The wall of the jar is made up of two layers of cells, and has pores in it. Water flows in through the pores, and out through the opening at the top. Cells with beating flagella drive the water through. They also have a ring of microvilli which filters out food particles. More complex sponges have a labyrinth of chambers and passages for water, but in every case water is driven through the system. Sponges have an internal skeleton of spicules. These are spikes made of various mixtures of protein, silica and calcium carbonate. Sponges live in water, mostly the sea.

Phylum Cnidaria

This phylum includes the jellyfish, sea anemones, hydrozoans and corals. The animals have radial symmetry, and are bags with an opening at just one end. This acts as a mouth and anus. Digestion takes place in the inside of the bag, the enteron. A ring of tentacles surrounds the mouth. The body wall is made up of two layers of cells (it is diploblastic). All the remaining animal phyla have a third layer (they are triploblastic). There is little or no organ development. Cells with stinging organelles are common. These are highly organized infoldings of the plasma membrane. They can shoot out very rapidly to form a sharp, stinging hair. The nervous system is a net of nerve cells with no central organization.

There are two body forms: the medusa and the polyp. The familiar sea anemone body is a big polyp; jellyfish are medusae. Corals are colonies of polyps, secreting calcium carbonate spicules. Many cnidarians have both body forms in their life cycle.

Phylum Platyhelminthes

These soft-bodied animals have bilateral (left–right) symmetry, like all the following phyla. The body has three layers of tissue (it is triploblastic). Their mouths open into blind-ended guts. Unlike the cnidarians, they have organs for controlling the salt content of the body.

The flatworms (class Turbellaria) live in fresh water, the sea and the soil. A carpet of cilia on the underside allows the smaller species to creep or swim. Some have a crude head, with eyes, chemoreceptors for detecting food, and the mouth.

The flukes (class Trematoda) are parasites in animals' bodies. They tend to be leaf-shaped, rather than worm-like. Specialized organs attach the fluke to its host. They are usually in a ring around the fluke's mouth. Many have extraordinary, complex life cycles with several distinct stages.

The tapeworms (class Cestoda) are gut parasites of vertebrates. They have attachment organs at one end, but no gut. Their food has already been digested by the host, and is absorbed through the body wall.

Phylum Nematoda

The nematodes, or roundworms, are a huge and successful group. Like the insects, there is reason to believe that most species have not yet been described. Many live in soil or sediments. Some are parasites. Most nematodes are only a few millimetres long. The body is a pointed cylinder. The gut runs from the mouth at one end to the anus near the other end. Nematodes are not made up of repeating segments, unlike the annelid worms. The body has a cavity in which the internal organs are found (Fig. 1.3). This is quite different to the gut cavity, which is a narrow tunnel of 'outside world', stretching through the body. The body cavity itself is completely enclosed. This type of body cavity is derived from the space inside the hollow early embryo. It is called a pseudocoel.

Phylum Rotifera

The rotifers are tiny animals. They are typically cone-shaped, surrounded by a tough cuticle. They have a gut and pseudocoel, as in the nematodes. There are two rings of cilia at the wide end of the cone. Many species use these to swim. Rotifers are found in fresh water, or sometimes in the sea, wet soil and sand. They are strikingly similar to the larvae of several other groups, including molluscs and annelids.

Phylum Annelida

The segmented worms have a body plan based on repeating units. The body is elongated; the gut has a mouth and an anus. The outer surface is covered by a thin,

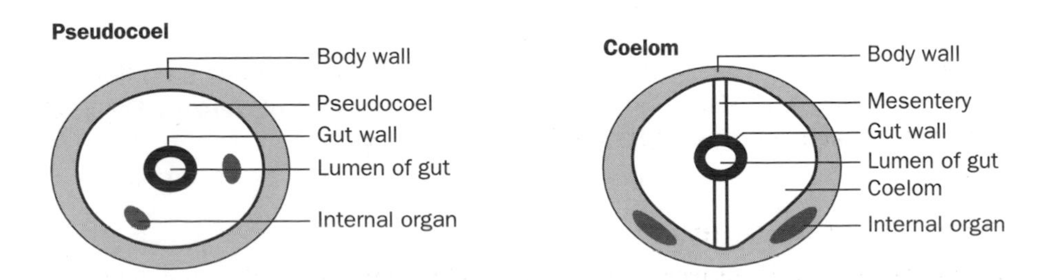

Fig. 1.3 Two types of body cavity in animals: the pseudocoel and the coelom.

flexible cuticle, with chitin bristles called chaetae. The body cavity develops in the body wall of the embryo, not from the central space: it is a coelom, not a pseudocoel (Fig. 1.3). A coelom can develop in several ways, but the gut always ends up suspended in the cavity by strands of issue called mesenteries. All the following animal phyla have a coelom.

Members of class Oligochaeta mostly live in soil or fresh water. They include the earthworms. Oligochaetes have chaetae attached singly or in pairs, and have no distinct head. Class Polychaeta includes many marine worms such as the lugworm and ragworm. The head is more distinct, and chaetae are in large groups on either side of each segment. The leeches (class Hirudinea) are carnivores and blood-sucking parasites with suckers at both ends of the body, and no chaetae.

Phylum Mollusca

The molluscs are a large, diverse group. The body plan includes a rather squat body with a more or less well-developed head, a shell of protein and calcium carbonate, a rough tongue-like radula used for scraping away food particles, and gills inside a cavity under the shell.

In class Gastropoda, the shell is in one piece, usually coiled. This class includes the limpets, winkles, snails and slugs (snails with reduced shells). Land snails have an air-filled lung in the mantle cavity, instead of gills.

The bivalves (class Pelycopoda) have a two-part shell. They include cockles, mussels and oysters. The radula and eyes are absent. Most are filter feeders.

Class Cephalopoda includes squid, octopus and cuttlefish. The shell is internal or absent. The well-developed head has forward-facing tentacles for catching food. A radula and tough beak break up the food. Cephalopods chase their prey, using jet propulsion (Section 24.8). The largest invertebrates are giant squid, sometimes over 20 m long.

Phylum Bryozoa

The bryozoans are colonies of tiny animals, zooids. The colony has a modular construction, as in plants. Each zooid leads a relatively independent life, and new ones can be added by asexual reproduction.

The bottle-shaped zooid body sits snugly in an open tube made of chitin or calcium carbonate. The top of the 'bottle' sticks out, with a ring of feeding tentacles round the mouth. The gut runs down into the body, and loops back up to an anus near the mouth. Bryozoans are filter feeders, found mostly in the sea.

Phylum Arthropoda

The arthropods are extraordinarily diverse. Most invertebrates which live on land are arthropods. They have a rigid exoskeleton based on chitin, made up of jointed plates moved by muscles. Jointed mouthparts and limbs are characteristic. The head is well developed and includes a brain.

Using a crude, rather old-fashioned classification, four classes are very familiar: several of these groups are often given phylum status.

The insects (class Insecta) have a body divided into head, thorax and abdomen. They have compound eyes and a pair of antennae on the head. The thorax has three pairs of legs and two pairs of wings. They are basically a terrestrial group, using a system of internal air pipes for gas exchange (Section 20.4). Many have a complex life cycle involving larvae. The insects include bees, flies, crickets, beetles, moths, fleas, aphids and many other familiar groups.

The spiders and mites (class Arachnida) are also terrestrial. They have four pairs of legs. The body is divided into just two parts, the prosoma and the opisthosoma. The prosoma is a sort of head without antennae, and is very different from other arthropod heads. They have simple eyes only.

The crustaceans (class or superclass Crustacea) are mainly aquatic. The head is more or less distinct, with two pairs of antennae. Several body segments each have a pair of appendages, many of which are leg-like. They include shrimps, lobsters, crabs, barnacles and, on land, woodlice.

Class Chilopoda includes the centipedes while class Diplopoda includes the millipedes. Both are terrestrial, with distinct heads and one pair of antennae. The centipedes are carnivores, and have one pair of legs on each segment. The herbivorous millipedes have two.

Phylum Echinodermata

These marine animals have an unusual symmetry, based on five rays with no distinct head. They usually appear to have true radial symmetry, but are sometimes bilateral, especially the larvae. They have an internal skeleton beneath the body surface, of calcium carbonate spikes or plates. An internal network of water-filled pipes acts as a hydraulic system. This operates 'tube feet' used for movement, or feeding tentacles. The echinoderms include starfish, sea urchins and sea cucumbers.

Phylum Chordata

Most, but not quite all chordates are vertebrates. All chordates have a common set of features at some stage of their development. These include a hollow nerve cord running along the dorsal (back) side of the body (in vertebrates this is the spinal cord); a stiff rod of tissue running alongside the nerve cord, at least in the embryo, called the notochord; and a tail which extends beyond the anus. The vertebrates also have an internal skeleton of bone or cartilage, including a vertebral column (backbone).

Class Osteichthyes

The bony fish have a bony skeleton. They live in water and have no limbs. The skin is covered in overlapping scales; fins supported by spikes of bone have a role in swimming. Gills are used for gas exchange. They include mackerel, trout and eels.

Class Chondrichthyes

The cartilaginous fish differ from the bony fish in several ways. Most obviously, the skeleton is made of cartilage, not bone. The fins are fleshy. This group includes sharks, dogfish and rays.

Class Amphibia

The amphibians include toads, frogs, salamanders and newts. They have two pairs of limbs, a bony skeleton and skin without scales. The larva, a tadpole, lives in water. It has gills and a prominent tail. The adult may live in water or on land, but uses lungs for gas exchange. Fertilization is external, along with many fish but no other vertebrates.

Class Reptilia

The reptiles include snakes, lizards, turtles and crocodiles. They have two pairs of limbs, a bony skeleton and a scaly skin. The limbs are modified for walking or swimming; snakes have none. Lungs are used for gas exchange. Fertilization is internal; the eggs have a leathery shell. In some species, the eggs are incubated and hatch inside the female.

Class Aves

The birds differ from the reptiles in having feathers, forelimbs modified as wings, and eggs with hard shells. They have no teeth, but have a beak around the mouth. Fertilization is internal, but eggs are incubated outside the body.

Class Mammalia

The mammals differ from the reptiles in having hairy skin and mammary glands which provide milk for the young. Most belong to one of two groups, the eutherian (placental) mammals and the marsupials. In the eutherian mammals, such as lions, porcupines and apes, the embryo develops much longer inside the female, nourished by the placenta. The marsupials, including kangaroos and wombats, do not have this.

Kingdom Plantae

The major plant phyla are introduced here. Many of their distinctive features relate to their life cycles and reproduction. They are discussed in more detail in Unit 14. It is important to realize that most plants have two distinct bodies in a single life cycle: a gametophyte made up of haploid cells, and a diploid sporophyte (Unit 11).

Phylum Bryophyta

In most mosses and liverworts, the gametophyte consists of tiny leaves, one or a few cells thick, attached to stems anchored to the ground by rhizoids. These simple organs have little tissue differentiation. Their structures are unlike the organs of other plants. In the thallose liverworts, the gametophyte is a thick, branching plate of tissue, often forming a crust on the ground. The sporophytes of mosses and liverworts are short-lived spore capsules on stalks. Bryophytes have no xylem or phloem, unlike the remaining phyla, known as vascular plants.

Phylum Filicinophyta

The sporophyte is the dominant body of the ferns. They have true roots, stems and leaves. The leaves are often large and finely divided. Spores are made by sporangia in groups on the backs of the leaves. In most ferns there is only one type of spore. The gametophyte is small, short lived and rather like a thallose liverwort.

Phylum Lycopodophyta

The clubmosses also have a dominant sporophyte, with small leaves. Sporangia are carried in groups at the ends of some stems. Some clubmosses make two sorts of spore: large ones which produce female spores and smaller male ones.

Phylum Coniferophyta

The conifers have no free-living gametophyte. The sporophyte is a tree or shrub. The stems are woody, and continue to thicken throughout their lives. The leaves are needle like. Conifers produce seeds: an embryonic plant in a case, which develops from a fertilized ovule.

Phylum Angiospermophyta

The flowering plants also make seeds, but they are inside a fruit which develops from an ovary (Section 14.3). The ovary is, ultimately, what defines a flower. Some flowering plants have woody stems, some do not. There are two ancient groups within the phylum, which differ in many details of structure. The monocotyledons (Class Monocotyledonae) include grass, orchid, palm and iris families. The dicotyledons (Class Dicotyledonae) include rose, maple, fig and pea families.

Summary

- The system of classification is a framework for understanding the diversity of life.
- Organisms are classified using a hierarchical system of groups within groups.
- The main levels within the taxonomic hierarchy are as follows (highest first): kingdom, phylum, class, order, family, genus, species.
- The scientific name of a species has two parts, the generic and specific names.
- Phyletic classifications are based on the evolutionary history of a group.
- Cladistic analysis can be used in making a phyletic classification.
- Phenetic classification is based only on the characteristics of species as we see them now.
- Real classifications may combine some elements of both. It is impossible to classify organisms without some human judgement.
- Organisms are grouped into five kingdoms: the Prokaryotae, Protoctista, Fungi, Animalia and Plantae.

Exercises

1.1. The foxglove of north western Europe is a flowering plant. It is a dicotyledon, belonging to the family Scrophulariaceae, genus *Digitalis*, species *purpurea*. Its name was first given by Linnaeus.
(i) Which kingdom, phylum and class does the foxglove belong to?
(ii) Accurately write out the scientific name of the foxglove.

1.2. The natterjack toad is an amphibian, belonging to the order Salientia, family Bufonidae, genus *Bufo*, species *calamita*. It was described by the 18th Century biologist Laurenti.
(i) Which kingdom, phylum and class does this toad belong to?
(ii) Accurately write out the scientific name of the natterjack toad.

1.3. Look at the classification of *Watsonia* in Box 1.2. Character 3, reduced branching, is thought to have evolved twice within the group: parallel evolution. Using only the evidence in this cladogram, suggest why the taxonomist may have come to this conclusion.

1.4. Go out of your door and look around. Classify the organisms you see into kingdoms, and further if possible. This unit will probably not help you identify unfamiliar organisms, but it should help you place familiar ones in the taxonomic hierarchy.

Further Reading

Berenbaum, M. *Bugs in the System: Insects and their Impact on Human Affairs* (Reading MA: Addison-Wesley, 1995). A fun read, and very stimulating.

Gow, N.A.R. and Gadd, G.M. *The Growing Fungus* (London: Chapman & Hall, 1995). A thorough introduction to the structure and biology of fungi. It is arranged according to areas of Biology, not taxonomic groups.

Gullan, P.J. and Cranston, P.S. *The Insects: an Outline of Entomology*. (London: Chapman & Hall, 1994). A thorough, accessible survey of this important animal group.

Ingold, C.T. and Hudson, H.J. *The Biology of Fungi* (6th ed.) (London: Chapman & Hall, 1993). Shorter and lighter than Gow & Gadd, part-devoted to a group by group treatment.

Ingrouille, M. *Diversity and Evolution of Land Plants* (London: Chapman & Hall, 1992). The major plant phyla are covered in this text.

Margulis, L., Schwartz, K.V. and Dolan, M. *The Illustrated Five Kingdoms* (New York: Harper-Collins, 1994). Examples of organisms from over 100 phyla, in their habitats. Full of information, a book to browse through and gain a better overview of the diversity of life.

McNeill, Alexander R. *Animals* (Cambridge: Cambridge University Press, 1990). A group by group approach, which includes selected aspects of biology and physiology. The physics of movement is dealt with well: not surprising, since it is written by an authority on the subject.

Pechenick, J.A. *Biology of the Invertebrates* (3rd ed.) (Dubuque IA: Wm C. Brown, 1996). A group by group overview. Not over-detailed, it is easy for the beginner to see the whole picture.

Prescott, L.M., Harley, J.P. and Klein, D.A. *Microbiology* (3rd ed.) (Dubuque IA: Wm C. Brown, 1996). A comprehensive textbook including good sections on medical and industrial microbiology. See other references in Unit 30.

South, G.R. and Whittick, A. *Introduction to Phycology* (Oxford: Blackwell Science, 1987). A book on the algae, organized according to areas of biology rather than taxonomic groups.

Young, J.Z. *The Life of Vertebrates* (Oxford: Oxford University Press, 1950). By no means a modern book, this is exceptional for its clarity and insight into many aspects of vertebrate biology.

Explaining the Diversity of Life: Natural Selection

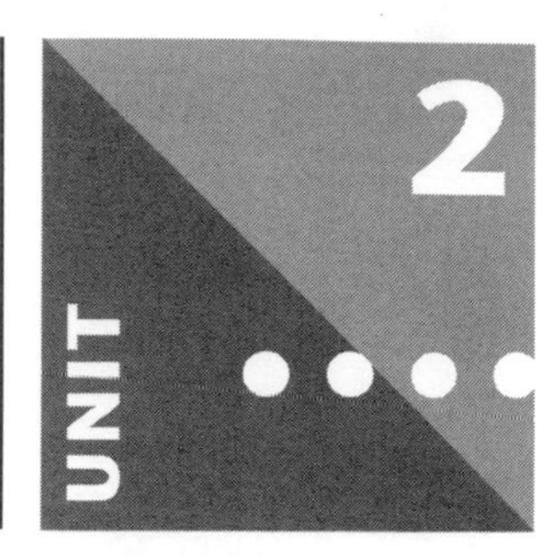

Connections

▶ Evolutionary theory is a central part of Biology. The basic ideas of evolution and natural selection are an important introduction to the rest of this book. However, this unit might be heavy-going if you have not studied Biology before. Return to the tricky bits later, especially once you have a better understanding of genetics and life cycles (Units 7, 11 and 12).

Contents

2.1 The theory of evolution

Biology is an historical science. We are the way we are because our parents were like that too. Adult porcupines have baby porcupines, not toads, toadstools or cherry trees. Something is passed on from generation to generation. We now call that something 'genes' (Units 7 and 12). However, genes are shuffled during sexual reproduction (Section 11.3) and offspring may not be quite like their parents. A baby porcupine is a unique individual, but still recognizably a porcupine. To understand the way things are now, we need to understand their history, how they got to be that way.

Our thinking about Biology (and another historical science, Geology) went through some major changes in the 19th Century. In western thought, we had tended to assume that species were put here by God, more or less the way they are today; that the Earth was not as old as we now believe, a matter of thousands of years; and that if there had been changes in the past, they were caused by catastrophes like the Great Flood. As these assumptions were challenged, a new framework of ideas emerged. Firstly, the Earth is old, thousands of millions of years old. Next, much of the landscape and geology we see can be explained by processes which can still be seen happening, very slowly, over thousands and millions of years. Further, things that lived in the past were not all the same as those alive today. Many have become extinct, including some very large ones that are unlikely to be hiding somewhere. Many species, and entire groups of organisms, seem not to have lived before a certain time in the past.

Perhaps small differences between offspring accumulate over many generations, leading to noticeable changes. Today's organisms are the descendants of different organisms alive in the past. This is the theory of **evolution**. Fossils support the theory, but there are other lines of evidence which, taken together, are extremely persuasive. These are summarized in Box 2.1.

Finally, the simple and elegant idea of natural selection occurred to at least two people, Darwin and Wallace. Natural selection is an inescapable mechanism which can lead to evolution (Section 2.2). The issue became whether natural selection could explain most or all evolutionary change, or just trivial parts of it. To the present day, it is the only really plausible mechanism which could drive evolution. Inputs from genetics, animal behaviour, ecology, molecular and cell biology in the 20th Century have built on, but not changed the basic ideas. The theory of evolution remains an essential part of our scientific framework for understanding the world.

Evolution: Inherited changes in a population, accumulated over a number of generations.

Box 2.1

Evidence for evolution

Fossils are almost the only evidence we have about which organisms lived in the past, and when, However, the record is far from complete.

There is a great deal of evidence that evolution has somehow taken place. There is no earth-shattering proof. Rather, there are lots of lines of evidence which, when repeated in group after group of living things, form a body of evidence which is hard to interpret any other way.
Here is an outline of some of these lines of evidence.

Fossils

Fossils are the remains of organisms (or at least the hard parts of them) preserved in rocks. Rocks which form from sediments may contain fossils of organisms which died while the sediment was forming. Different aged rocks contain different fossils. Some fossils are very like organisms which are alive today. Others are not, suggesting that whole groups of organisms such as ammonites and dinosaurs lived in the past, but have become extinct. Some groups are found as fossils in rocks formed only after a certain time. This suggests that new groups have appeared during the history of the Earth.

Comparative anatomy

Some species have similarities to one another. This is the basis of classification. We can distinguish the animals from the fungi, the mammals from the reptiles, and so on. Within each group, smaller and more similar groups can be distinguished. This pattern can easily be explained if evolution has occurred. The more similarities two organisms have, the more recent the ancestor they share. If a whole group shares an ancestor, it is likely to share characteristics. Mammals as diverse as bats, apes and whales have the same pattern of bones in their front limbs. Plants in the pea family share a unique flower structure. None of this makes sense unless evolution has happened.

Comparative biochemistry

Similar proteins, and similar genes coding for them, are found in related species. Similarities in protein and DNA sequences (Unit 3) broadly follow similarities in other characteristics. Chemical similarities support the idea of evolution for the same reasons as structural similarities.

Artificial selection

People breed plants and animals for agriculture, horticulture and sport. Very different breeds of cat, cabbage and cow have been selected over tens, hundreds or thousands of years. The basic method is to select individuals with desirable characteristics and breed from them, then repeat the process on the offspring, and so on. If striking differences can be selected artificially this quickly, surely the same thing could happen naturally, given the much longer times available.

Geographical distribution

Species do not live everywhere. They are restricted to particular habitats and, usually, particular parts of the world. Again and again we find that species are not spread around the world at random. Groups of related species are centred on particular places. The lemurs have a centre of diversity on the island of Madagascar, the gum trees in Australia. This only makes sense if related species evolved from shared ancestors.

2.2 Natural selection

Natural selection is a mechanism that can cause evolution. An easy argument based on some simple observations shows what natural selection is.

Organisms vary enormously (Sections 11.3, 12.5). Even in a single population, each individual has a different set of characteristics. Some of this variation is inherited, caused by differences in the genes passed from parent to offspring. Next, organisms could reproduce much more than they actually do, if resources were not limited. Further, not all offspring survive to maturity. Individuals are competing for resources: the more they get, the more they can reproduce. If individuals vary, it follows that some will be able to get more resources than others. These will have more offspring. They are said to be fitter. The others will have less offspring. They are less fit. If this variation is controlled by genes, then a gene which increases **fitness** will be more common in the next generation. It has to be true! Individuals with the gene have more offspring, which also have the gene. Individuals without the gene have less offspring, which also do not have the gene.

So, genes which increase fitness become more common. This is natural selection. Fig. 2.1 shows a theoretical model of the different ways selection can act on a particular characteristic.

Natural selection and adaptation

The direction that natural selection takes depends on what variation is available, and on the environment. Natural selection can only change tail length in a monkey population if there is some variation. If every monkey has the same length tail, there will be no natural selection until some new genetic variation appears. This could happen by mutation (random changes to genes, Sections 7.5, 11.3) or by immigration of monkeys from another population. Natural selection works in different directions in

Fitness: A measure of how many offspring a particular variant leaves.

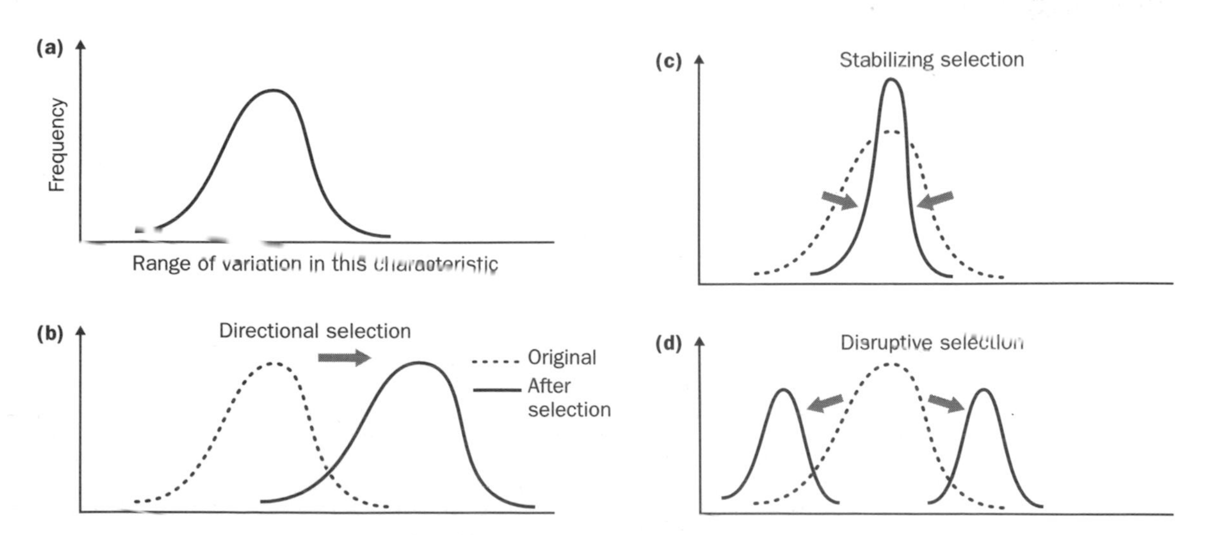

Fig. 2.1 How selection can act on a characteristic. (a) the range of variation in a population, before selection; (b) directional selection: selection favours one extreme of variation; (c) stabilizing selection: selection favours the commonest form, reducing variation; (d) disruptive selection: both extremes are favoured at the expense of the commoner forms.

different environments. A characteristic which increases fitness in one environment may not increase it elsewhere.

Populations of organisms become better suited to their environments. They are said to be adapted. **Adaptation** is the result of natural selection and is inherited. It should not be confused with the ways in which individuals become adjusted to their environments as they develop: these are not inherited. Natural selection does work in particular directions, but not because it is an active process or because something is controlling it. The environment acts as a template. Selection in only some directions will increase fitness.

Some of the best evidence that natural selection is important comes from studying small-scale changes, which have taken place over tens or hundreds of years. Boxes 2.2 and 2.3 show two classic examples. It is much harder to show that natural selection has caused large scale, long-term evolutionary change. We simply cannot make direct observations spanning millions of years. We can only say that the selection we see at work over tens of years *could* lead to big changes over a longer time, and that it is the only plausible explanation we have.

When we see characteristics in an organism, we tend to tell ourselves evolutionary stories about them. Why, for example, are polar bears white? Camouflage, you might argue. Polar bears live and hunt in snowy, icy, white places. Imagine a bear population with variation in coat colour. The paler coloured bears are less obvious to their prey, so they stand a better chance of catching them. They get more food, so reproduce more. Genes leading to white coats get more common. Genes for dark coats become rare. This is very plausible, and makes a good working hypothesis. The danger is that this sort of story may be treated as fact, without being tested, and good evidence is often hard to find. Box 2.4 shows just two pitfalls of interpreting every characteristic in this way.

Adaptations: Characteristics of organisms which suit them to an environment or way of life.

Given enough time, one might expect all populations to become as well adapted as they could be. Then, natural selection would stop and genetic variation would dis-

Box 2.2

Selection at work: I

The peppered moth, *Biston betularia*, has two colour forms. The typical form has pale grey wings flecked with brown. The melanic form has black wings. In 1950s England, the melanic form was more common in sooty, industrial cities; in the countryside, typical moths were more common. Could this be the result of natural selection? Are melanic forms better camouflaged on sooty trees, while typical forms are hard to spot on lichen-covered branches? If so, predation by birds could be what drives natural selection in this case. Several lines of evidence support this idea.

- Old moth collections are preserved in museums. There are no records of melanic forms before 1850. Pollution in cities became more serious around that time. From then on, the melanic form became increasingly common in cities.
- Birds do eat the peppered moth. They find melanic moths more easily on lichen-covered trees, and typical moths more easily on sooty trees.
- This characteristic is inherited. A dominant allele (Section 12.2) controls melanism.
- Typical and melanic moths were released in the industrial West Midlands. Moths were marked so they could be recognized later. Some days later, moths were recaptured in the same area. The percentage of the marked moths which was recaptured was calculated for each type. The experiment was repeated in rural Dorset. As predicted, melanics survived better in the town, typical moths in the country.

This story is repeated in many other moth species.

Box 2.3

Selection at work: II

Heavy metal ions, like copper, are toxic to plants. Metal mines produce spoil heaps of rock and soil rich in metals. Few plants can grow on them. Sometimes, however, populations of plants which seem to be metal-tolerant are found on spoil heaps. Has natural selection favoured metal-tolerant forms? This question was thoroughly investigated in a classic study of the grass *Agrostis capillaris* growing on an ancient copper mine in Wales.

- These individuals are more tolerant of copper ions in soil than 'normal' *Agrostis* plants, when grown together in the laboratory.
- This variation is inherited, but several genes seem to be involved.
- Non-tolerant populations include rare copper-tolerant individuals: there is variation for natural selection to work on.
- Pollen from non-tolerant individuals reaches plants on the mine each year. Natural selection is intense enough to maintain the tolerant population, even though genes are constantly flowing in.
- Pollen from tolerant plants reaches plants on normal soils. Selection works against tolerance here, because tolerant plants are slower growing and compete poorly.
- Artificial selection can produce metal-tolerant populations of *Agrostis* and other grass species.

In the same way, herbicide-resistant populations of some weeds are appearing. Human activity can be a powerful stimulus for evolutionary change.

Box 2.4

It's obvious . . . isn't it?

It is easy to tell evolutionary stories, using natural selection to explain an organism's features. However, not all stories are true. Here are two pitfalls.

Everything is an adaptation

This need not be so. Some structures are there because they were favoured by selection in the past, even though they have no function now. This is called redundancy. Other features may simply be inescapable consequences of the body plan. Why do humans have an appendix? This small, blind-ended branch of the gut seems to have no function. Those of us who have had ours removed get on perfectly well. Perhaps it is a redundant structure. Certainly, some mammals use a larger appendix as a fermenter for culturing microbes which can digest plant cell walls (Section 21.4). Perhaps we had an ancestor which did this. On the other hand, it may be that some part of the genetic controls on development, which mammals share, leads to this branch. In some mammals, a function has led to selection for larger size. In others, it just sits there.

Why do men have nipples? Men share a common development plan with women. A pair of structures is produced in both sexes. In women, they grow and develop useful features. In men, they just sit there.

Why do we have five toes on each foot and five fingers on each hand (all right, four fingers and a thumb)? It is hard to see any adaptive reason for five, rather than four or six, but there is no need to find one. Mammals share a body plan with a particular bone structure in the limbs. For whatever reason, the first mammals had limbs with five fingers or toes. The rest of us are stuck with that. The basic plan may be modified, and in those species we may need an adaptive explanation for the modification.

What happens now happened then

There is a temptation to assume that an organ or feature with a particular function was selected for that purpose. This need not always be true.

Insects normally use their wings for flight. However, there is an evolutionary problem. Wings are useless for flight unless they are big enough. It is hard to see how large wings could evolve in one step, but small steps on the way would have no selective advantage; perhaps wings evolved for some other purpose. Butterflies use their wings as sun traps, to warm up the body before flight. Perhaps wings were at first an adaptation for temperature control, or even for gas exchange. Small wings would be valuable. Bigger wings would be more valuable. Perhaps the use of wings in flight came later.

appear. Only the fittest combinations of genes would survive. Two observations show that this has not happened. Firstly, many genes show **polymorphism** within populations, that is two or more forms of the gene (alleles) can be found. Box 2.5 discusses some reasons for this. Secondly, the fossil record does not suggest that evolution has become slower as time goes on. So why does evolution continue?

The answer must be that the environment keeps changing. The physical environment certainly changes on an evolutionary time-scale. Continents move, mountain ranges form and erode away, big meteorites hit the Earth, ice ages come and go. However, the biological environment may change even more. Every population interacts with other populations: predators, herbivores, competitors, parasites. As these evolve and become better predators or parasites, so the species they interact with evolve better defences, and so on. Interacting species are locked into a never-ending evolutionary struggle. This is known as the Red Queen hypothesis: read Lewis Caroll's *'Through the Looking Glass'* and you will find out why!

Polymorphism: The situation in which more than one version of a gene exists in a population.

Many developments in evolutionary theory are made with the help of mathe-

Box 2.5

Causes of polymorphism

Polymorphism is the situation where two or more distinct forms of a species are found together, in the same population. If it is a result of genetic polymorphism, there must be two or more forms (alleles) of a gene in the population. If one form is fitter than the other, we would expect it to become more common. The polymorphism would break down. So why do polymorphisms exist? The first three explanations dodge the problem in some way. The others meet it head on.

Mutation New copies of an allele appear occasionally by mutation (Section 11.3). Even if natural selection works against that allele, mutation means that it will always be there, at low levels. Many alleles which lead to rare genetic diseases, like cystic fibrosis, are maintained in this way.

Selection has not finished Natural selection is in the process of eliminating an allele, but has not finished. This situation can only be short-lived.

The alleles are selectively neutral The two forms are equally fit. Any changes in allele frequency are due to random changes: genetic drift (Section 2.5) rather than selection.

They need each other Sometimes, each form of a species is fitter if the other is present too. Sexes are an obvious example. Males cannot reproduce without females, and vice versa: the polymorphism continues. The same argument applies to different forms of flowers, and self-incompatibility genes which minimize inbreeding in plants (Box 14.2).

Other times, other places In a changing environment, one form may be fitter some of the time, the other form the rest of the time. The environment may change faster than selection can operate. The result will be a polymorphism. The same could happen if the environment is patchy. One form is fitter in some patches, the other elsewhere. This probably explains a classic example of polymorphism: the colour and banding patterns of a European snail, *Cepaea nemoralis*. Different forms are better camouflaged from birds in different parts of the habitat.

Heterozygous advantage Most familiar organisms are diploid: they have two copies of each gene in their cells (Section 11.1). A heterozygote is an individual which has two different alleles of a given gene (Section 12.2). Sometimes heterozygotes are fitter than either homozygote, so both alleles survive. For example, sickle-cell anaemia is caused by a mutant allele of one of the haemoglobin genes (Section 12.2). People with two copies of this allele carry oxygen in their blood inefficiently. However, heterozygotes are more resistant to malaria than people with only the normal allele. In regions where malaria is common, heterozygotes have a higher fitness. The mutant allele survives, and some homozygotes are inevitably born.

matical models of natural selection. This is not the place for a detailed treatment, but Boxes 2.6 and 2.7 introduce this field.

Natural selection acts on individuals and on genes

Natural selection favours individuals which have a higher fitness. In the same way, it can be argued that natural selection favours the forms of genes which increase fitness. The individuals with that form of a gene have more offspring, so it becomes more common.

Selection affects populations because populations are made up of individuals. Most of the time, it is not easy to show that selection only acts on individuals, because what is good for the individual is also good for the population. However, there are situations where the fitness of a characteristic varies according to how common it is. Then, selection sometimes works against the good of the population. This is particularly apparent in animal behaviour. A simple, very extreme example shows how.

Imagine an animal population which is severely limited by a predator species. An effective deterrent would make the population much more successful. Now imagine a gene which causes a very extreme reaction to a predator. An animal with this gene mounts a frenzied, uninhibited attack on the predator. It is almost always fatal to the animal, but also badly injures the predator. An imaginary population made up of these 'suicide bombers' might be very successful. Predators are deterred and move to other, less hostile prey. Any one individual is very unlikely to have to pay the

Box 2.6

The Hardy–Weinberg model

Mathematical models of the effect of selection on genes in populations begin here. The Hardy–Weinberg model shows what would happen *without* selection. The model applies to populations where the generations do not overlap. It makes a lot of big assumptions: no natural selection; random mating; no emigration or immigration, at least not of one type of individual more than another; no mutation. A Hardy–Weinberg world would be rather boring, and this is the point. If the Hardy–Weinberg model fits a situation, natural selection is probably not happening. If it does not fit, something interesting may be going on.

We need the idea of a gene pool: all the copies of particular gene in the whole population. Imagine a gene pool in which there are just two alleles of a gene, **A** and **a**. There are lots of copies of each. The proportion of the total which are **A** is p. The proportion of the total which are **a** is q. Because all genes are either **A** or **a**, $p + q = 1$.

We can model random mating by randomly selecting two genes from the gene pool for each individual in the next generation. The chance of selecting **A** is p. The chance of selecting a second **A** is also p. So, the probability of an individual being **AA** is $p \times p = p^2$. Similarly, the probability of being **aa** is q^2. The probability of being **Aa** is $p \times q + q \times p = 2pq$ (remember that either p then q or q then p would give **Aa**).

genotype (allele combination)	**AA**	**Aa**	**aa**	Total
frequency in the population	p^2	$2pq$	q^2	1

This can be used to test for selection in a population. If the frequency of the genotypes does not balance in this way, one of the assumptions of the model is probably being broken. Selection may be happening, or another of the original assumptions may not be true. Even more important, the model is an important foundation stone for mathematical theories of natural selection.

Box 2.7

A simple model of selection

Imagine a population which fits all the assumptions of the Hardy–Weinberg model (Box 2.6) except one: natural selection is happening. The **AA** individuals have a higher fitness than the others: a higher proportion survive to breed. We set the fitness of **aa** and **Aa** at 1.0. **AA** has a higher fitness, 1 + x.

p_n is the frequency of the **A** allele in generation n. q_n is the frequency of the **a** allele in generation n.

Allele combination	AA	Aa	aa
Fitness	$1+x$	1	1
Frequency among zygotes (fertilized eggs)	p_n^2	$2p_nq_n$	q_n^2
Relative numbers of adults	$p_n^2(1+x)$	$2p_nq_n$	q_n^2
Total adults	$p_n^2(1+x)+2p_nq_n+q_n^2=1+xp_n^2$		

So what is the frequency of each allele in the next generation, n+1? All the copies of the gene in **AA** individuals are **A**, but only half the copies in **Aa** individuals are **A**.

$$p_{n+1} = \frac{p_n^2(1+x)+\frac{1}{2}(2p_nq_n)}{1+xp_n^2} = \frac{p_n^2(1+x)+p_nq_n}{1+xp_n^2}$$

In the same way:

$$q_{n+1} = \frac{q_n^2+p_nq_n}{1+xp_n^2}$$

The changing frequencies can be modelled by putting the new allele frequencies through the model again and again. For example, this graph shows the increase in **A** when it starts out rare (p = 0.01) but gives a very high fitness (x = 4).

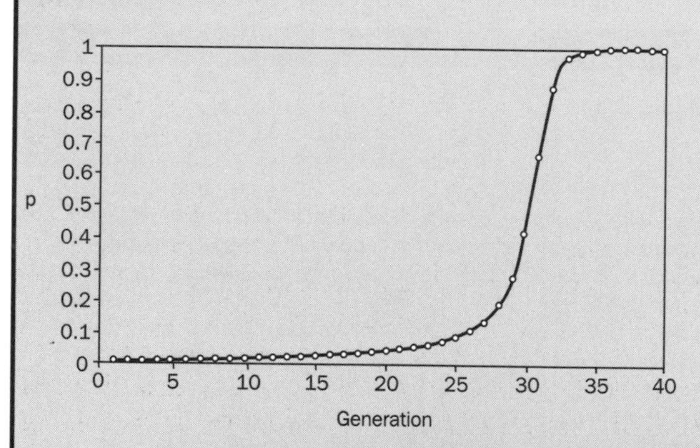

If you can handle this (easy!) maths, evolutionary theory may be for you!

ultimate price during its lifetime, and the whole population benefits. Next, a mutant form of the gene appears. Individuals with this gene do not carry out suicide attacks. In a population dominated by suicide bombers, these 'pacifists' will be protected by the bombers, but no pacifists will die in suicide attacks. This means that in a mixed population, the pacifists have a higher fitness, so the pacifist gene invades the population. At the other extreme, suicide bombers cannot invade a pacifist population. They bear the cost of attacks, but get none of the benefits. So, even though a pure population of suicide bombers would be more successful, it is not favoured by selection. Pacifism is, here, an evolutionarily stable strategy (ESS).

Any form of altruistic behaviour, in which an individual sacrifices itself for the good of the population, is subject to the same problem. Altruistic behaviour is rare, not surprisingly. However, social behaviour is seen in some animals, even though it depends on cooperation and **altruism**. Why?

2.3 Evolution of altruism

Some cases of cooperation are easy to explain. Both individuals get more out of cooperating than not, and it is obvious to the other individual if one is 'cheating'. Lionesses cooperate to hunt because the cost of sharing the meal is less than the benefit from extra kills.

However, true altruism is harder to explain. Here, one individual helps another, but suffers as a result. The other benefits. In these cases, cost and benefit are measured in terms of fitness: the numbers of offspring raised successfully.

The theory of kin selection helps explain most examples of altruism. Organisms get their genes from their parents. This means that relatives have some genes in common. Natural selection could favour a gene that makes an individual help its relatives. There is a real chance that they also carry the gene, so helping them may help the gene to be passed on. There is nothing too surprising about this. We accept that animals use resources in looking after their offspring, because their offspring carry their genes. Why not help any other relation that shares these genes?

The idea of **relatedness** lets us compare the benefits of helping different relations. Relatedness is the probability that two individuals will share a particular gene *as a result of being descended from a shared ancestor* (as opposed to by chance). Some simple examples show how relatedness is calculated in diploid organisms, which include most animals.

We start with parent and child. My son's genes came from two places. Half came from my wife, half from me. In the same way, only half my genes were passed on to him. Parent and child have a relatedness of 0.5.

How about full sisters or brothers, which share the same mother and father? We need to trace their relatedness through both shared ancestors, the mother and the father. One child has half the mother's genes. The other child has a different, randomly selected half of the mother's genes. The chance of the sisters sharing a given gene by this route is only $0.5 \times 0.5 = 0.25$. However, they also share genes from the father. Their relatedness by him is also 0.25, so overall relatedness is $0.25 + 0.25 = 0.5$ (Fig. 2.2). In the same way, we can work out the relatedness of cousins as 0.125, through their shared ancestors, the grandparents.

What does this tell us about the evolution of altruism? Crudely, helping two brothers or eight cousins has the same value as helping yourself. Returning to our 'suicide bomber' behaviour, if a suicidal act saves the lives of more than two brothers or sisters, more than eight cousins, and so on, a gene causing that behaviour will spread through the population. Not all altruistic behaviour is so extreme, but similar arguments are involved.

The most extreme examples of altruistic behaviour involve tightly-knit colonies of relatives. Social insects like bees and ants are good examples (Box 13.1). In fact, worker bees are one of the few real examples of the suicide bomber. Their barbed stings get stuck in the victim, so the bee is ripped to death as it pulls away.

Social animals also go in for division of labour. This is possible since the entire

Altruism: Behaviour which benefits others, rather than oneself.

Relatedness: The probability that two individuals will share a particular gene as a result of sharing ancestors.

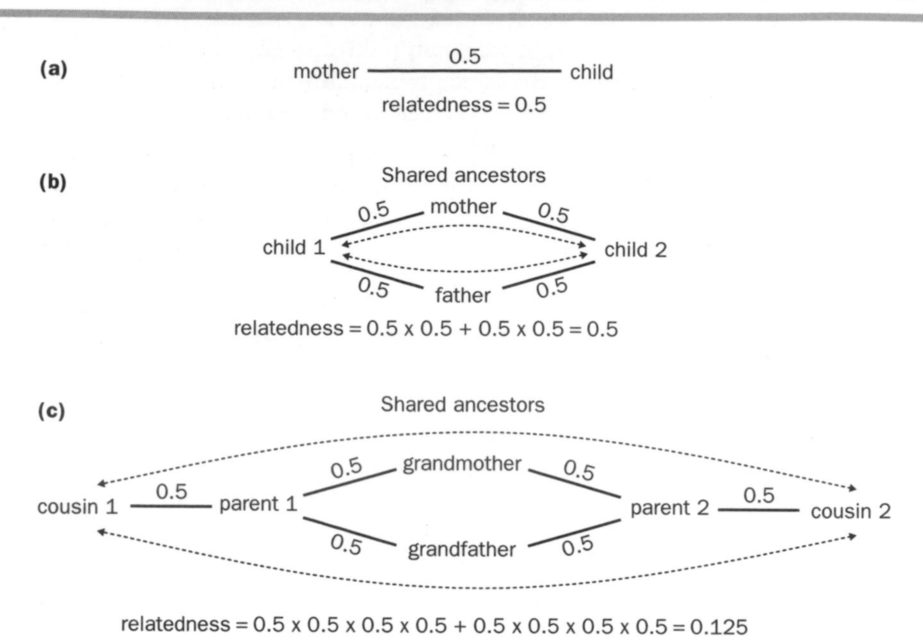

Fig. 2.2 Relatedness of (a) mother and child, (b) brothers or sisters and (c) cousins.

colony is cooperating to pass on its shared genes. Social insects have specialized castes with particular jobs, such as defence, collecting food and reproduction. Castes have even been discovered in a weird social mammal, the naked mole rat.

2.4 Sexual selection

Males make many small gametes (sperm in animals). Females make fewer, bigger gametes (eggs in animals). Also, there are good reasons why males and females are equally common (Section 13.9). These three observations suggest that males and females may have different priorities when mating. This may lead to natural selection working in surprising directions.

Males can father far more offspring than females can bear. They can increase their fitness by mating more. This implies that males will compete for a limited resource, females. Observations on many species support this. (Interestingly, female seahorses compete for males. Male seahorses look after the young, so have a bigger investment in each offspring.)

Females are generally limited by the number of gametes they can make, or the number of young they can raise. This suggests that females will tend to choose high quality mates rather than going for quantity. Observations of behaviour in many species support this. More concrete evidence – genes which influence the type of mate a female chooses – have been found in a species of ladybird.

These two types of sexual selection can have interesting consequences. Contests between males may lead to the evolution of large body size, eye-catching displays or weapons. The large antlers of stags can be explained by an evolutionary arms race. Stags with bigger antlers win more matings, so genes for big antlers spread. Once the whole population has big antlers, genetic mutations which lead to even bigger antlers will be favoured, and so on. The same argument could lead to larger body size in males, but not females. This size difference is commonly seen in animals; our own species, for example.

The difficulty with female choice is that male display characteristics may not always be good indicators that this is a high-quality mate. For example, it is hard to explain the evolution of the peacock's huge, colourful tail feathers, except by female choice. It need not follow, however, that the male with the prettiest tail will be the best father. There are several other ideas about why female choice should evolve. They lead down a steep path into mathematical modelling. This is not the place for a mathematical treatment of evolutionary theory, but it is a satisfying field for anyone who believes in mathematics as a language for discussing nature.

2.5 Where do species come from . . .?

Species are groups of populations which can interbreed to give fertile offspring. We know that the human races belong to one species because in regions where two races historically met, mixed marriages lead to normal, fertile children. We know that chimpanzees, gorillas and orang utans belong to other species; matings between humans and other apes are probably extremely rare, and there are good reasons to believe that matings like this would not produce offspring, let alone fertile offspring.

Different pairs of species are isolated in different ways. Isolation mechanisms can be classified as geographical, prezygotic and postzygotic.

Sometimes we recognize two groups of organisms as distinct species even though the only thing that stops them interbreeding is that they live a long distance apart. This is called geographical isolation.

Prezygotic mechanisms prevent hybrids being formed. In plants, they include ecological isolation by living in different habitat types, isolation in time by flowering at different seasons, the tendency for pollinating animals to visit the same type of flower again and again, and chemical incompatibility, where pollen grains cannot germinate (Section 14.2) on the styles of different species. In animals, signals for attracting and recognizing mates are much more important. They may be physical (colours or shapes, for example) or behavioural. They tend to be unique to a particular species.

Postzygotic mechanisms keep species apart even though hybrid embryos may be formed. Sometimes the embryos last longer than others. In many cases, embryos fail to develop properly and abort. In others, hybrids reach maturity but cannot reproduce. The mule, a horse–donkey hybrid, is an example of this. Rarely, hybrids are fertile, but their offspring, the grandchildren of the original pair, fail to develop or are sterile.

How are new species formed? Occasionally, new plant species are formed by doubling the chromosome number of a sterile hybrid (Box 11.5). This instant **speciation**, in the presence of both parents, is not common. Most speciation probably involves two populations of the same species becoming isolated geographically. Perhaps the species becomes extinct in the middle part of its range, or perhaps it colonizes a new area. Either way, the isolated populations evolve independently for a long time. Later, another change may bring the populations together again. They are now too different to interbreed. Some isolation mechanism has been set up, quite by chance. Postzygotic mechanisms could be the result of changes to the genetic

Species: Groups of populations which can interbreed in nature to produce fertile offspring.

Speciation: The formation of new species.

development plans of the two new species, which are now incompatible. Conflicting instructions cause failure of development. Alternatively, changes in chromosome structure and number could lead to failure when cells divide by meiosis (Section 11.2), because chromosomes cannot pair properly.

A prezygotic mechanism could also have been set up like this. If the two new species had come to rely on different signals for recognition, they would not interbreed. However, prezygotic mechanisms could also evolve after the two species came together again. If hybrids were less fit than the parents, it would be an advantage to both species to avoid hybrid matings. Natural selection could reinforce small differences in recognition signals, leading to complete isolation.

Theory and observation both suggest that speciation can occur without geographical isolation. It seems to be rare.

Small, isolated populations may be important in speciation. Isolation can certainly lead to speciation, but there is some evidence that small size can help. Imagine a small population, made up of individuals picked randomly from a larger population. By chance, a rare form of a gene may be much more common, or even more scarce in the new population. For example, imagine a population of 10 000 diploid organisms (each has two copies of each gene). 1 in 5000 copies (0.0002) of a particular gene is a rare form. Now select 10 individuals at random to start a new population. If one individual has just one copy of the gene, 1 in 20 copies (0.05) will be that form. It has become 250 times more common, instantly, without natural selection. Alternatively, none with that form of the gene may be picked out, and its frequency would fall to zero. Either way, with thousands of genes to consider, a small offshoot population will probably be genetically distinct. This is called **the founder effect.**

Something else may lead to change without selection in small populations. When genes are passed from parent to child, some random events are involved. Which copy of a gene an egg or sperm receives, and which sperm fertilizes which egg are both determined randomly. If, by chance, sperm carrying a particular gene happen to be more successful, that gene will become more common, even though it did not cause the sperms' success. Random effects are far more noticeable in small samples. In larger populations, they are averaged out more effectively. (Try tossing a set of coins again and again until, by chance, you throw all heads. With only three coins it happens quite often: I succeeded after eight throws. With 10 coins it happens less often: I got bored after 50 failed attempts. With 10 000 coins I would have little hope of ever succeeding, even if I could afford to try.) So, random changes in gene frequency are more significant in smaller populations. This is called **genetic drift.**

It is not at all clear how important genetic drift and the founder effect are in speciation.

The founder effect: Small populations, set up as offshoots of larger ones, may by chance have very different allele frequencies.

Genetic drift: Random changes in allele frequencies in a population.

Extinction: The death of an entire species.

2.6 ... and why do species disappear?

Extinction is as real as speciation, yet we know much less about it. Part of the problem is that to investigate past extinctions we must study something that is no longer here to be studied. To investigate extinctions as they happen, we are likely to have to look at organisms which have already become very rare. There are only a few cases where we know exactly why a species became extinct. These are where humans attacked a conspicuous species, for example the dodo, the passenger pigeon and the

Human activity can bring a species to the brink of extinction. The spread of agriculture in the Great Plains of North America was bad news for the American bison. This stack contains about 25 000 skulls.

smallpox virus. In general, we can only study the reasons why species become rare. These include increased predation; introduction of another competing species; an increase in a disease-causing organism; increased human exploitation; a change in the climate; or any habitat change, which could itself be driven by one of these changes. We must assume that factors which have made a species rare could push it over the edge, to extinction. Any of these changes could result from human activity. People have pushed many species into rarity, and some to extinction. It seems inevitable that more will go. The difficulty is predicting how many, and which ones.

Extinction has not always happened at the same rate. The fossil record shows phases where the total number of species has grown, followed by phases of mass extinction. Some of these mass extinctions were bigger than others, and affected some groups more than others. A mass extinction of marine invertebrates in the late Permian period (around 230 million years ago) corresponds with the time when the continents merged to form just one, reducing the area of shallow sea. Another mass extinction, which included most dinosaur species, was at the end of the Cretaceous period (about 64 million years ago). This coincided with a huge meteorite hitting the Earth, which would have thrown dust into the upper atmosphere, causing climatic change. Coincidences are not necessarily causes, however, and the search for the causes of an historical event is particularly hard.

2.7 A drive for complexity?

It is common to see evolution as a process which produces more complex organisms from less complex ones. There are elements of truth in this idea, but pitfalls too.

There is a temptation to look at the organisms alive today, and to put them into a tree or sequence, with some higher up than others, each group having evolved from the group below it. The human species tends to end up at the top of schemes like this! However, today's organisms did not evolve from other organisms alive today. Living bacteria are as modern as today's flowering plants, mammals and insects. These groups share common ancestors. Some have changed from those ancestors more than others, but all have changed. We carry our history in our genes, in our body plans, but that is all. We are not surrounded by our history: every generation faces a brave new world.

On the other hand, more complex organisms exist now than existed early in the history of life. There have been several major increases in complexity, particularly the evolution of the eukaryotes and of the vertebrates. Further, within a single group we often see specialists evolving from generalists. In the mammals, for example, the highly specialized carnivore groups such as the dog and cat families had ancestors with more generalist body plans. However, a specialist need not be more complex.

Returning to the title of this unit, does natural selection explain the diversity and complexity of life? Perhaps not, but it is clear that we cannot understand life in all its diversity without natural selection.

2.8 The origin of life

How did life first appear? This is one of the biggest and hardest historical questions in Biology. We will make two reasonable assumptions: that the Earth could not support life when it first formed; and that life was somehow organized from non-living materials sometime between the formation of the Earth (about 5 billion years ago) and the appearance of the first fossils (about 3.5 billion years ago). If we see life as a chemical phenomenon (Section 4.1), then we must look for a chemical origin of life.

Laboratory simulations suggest that natural processes on the early Earth could have produced a range of organic molecules which we associate with life. In these experiments, electrical discharges (like lightning) or ultraviolet light were passed through a mixture of gases which may have been present in the early atmosphere: carbon dioxide, methane, ammonia and hydrogen. Amino acids, sugars, purine and pyrimidine bases (Unit 3) have all been made in this way. Some clay particles can act as catalysts, and they might have organized the first polymers.

Life is self-replicating. Parents produce offspring. At the molecular level, DNA indirectly brings about its own copying (Section 7.3). This relies on a great deal of molecular machinery, involving RNAs. Most people agree that the first self-replicating molecule was probably an RNA. RNAs can act as information stores and as catalysts (Section 5.4). Moreover, self-replicating RNAs can be made. We can imagine a world populated by self-replicating RNA molecules, reproducing, competing for the nucleotides from which they are made, mutating, evolving. Only later would DNA have been used as a specialized information store, and proteins as specialized catalysts: enzymes. Then there is the problem of how self-replicating molecules organized cells around them as machines for living.

Every stage of this story involves breathtaking leaps. The priority is to show that each step *could* have happened, before we can dream of convincing ourselves that it *did* happen this way. So long as we keep a healthy scepticism, this story is a useful framework for thinking about the origin of life. A sketchy story is better than no story at all; a working hypothesis is better than an empty mind.

How would these issues change if we were to discover life elsewhere in the Universe? This is more than idle speculation. As I type this, the current issue of one highly respected scientific journal includes a paper on the discovery of what might prove to be fossil microorganisms in a meteorite which almost certainly came from Mars. Another carries a report of a recent conference on the search for extra-terrestrial life, without apology or any hint of mockery.

Extra-terrestrial life would not much alter our ideas on the origin of life on Earth. If the early Earth had suitable conditions, why not elsewhere in the Universe too?

Since we can only guess what extra-terrestrial life might be like, based on our experiences of life on Earth, we can only guess what conditions could support other life forms. If microbes did, or even still do exist on Mars, it would be important to discover whether they are related to cells on Earth. Do they share a common ancestor? This would imply that life originated on just one planet, and crossed to the other. Little green men in flying saucers need not be involved! Some of the ancient bacteria that live within rocks might, just possibly, survive inside a fragment of rock thrown into space from the surface of one planet, and eventually hitting the other as a meteorite. Even if life on Earth originated on Mars, and we have no reason to think it did, the problem of the origin of life changes little. The same issues are simply moved a few million miles further from the Sun. This does not mean that it would not be an exciting discovery. Suddenly, we would know that our experience of life might be only a small part of the picture. How much of our biological knowledge would hold true beyond the Earth? We have no idea.

The possible fossils in the martian meteorite show the difficulties of this field. They show some, but not all the chemical characteristics which we use to recognize microfossils on Earth. Their shape looks microbial, but they are much smaller than known microfossils, and there is no evidence of any cell structure. Are they fossils of an unfamiliar form of life, or not fossils at all? Mars is the best place to look for more evidence, and it is a long way away. No martian material has ever been brought back to Earth. The truth is out there, but finding it will be hard, and knowing when we have found it will be even harder.

Summary

- Organisms alive today are the descendants of different organisms alive in the past. The changes which happen over many generations are evolution.
- A body of evidence from the fossil record, patterns of similarity in structure and DNA sequence, geographical distribution and artificial selection suggests that evolution takes place.
- Natural selection is the only general mechanism of evolution we know of.
- Populations show genetic variability. Some variants leave more offspring than others; their genes become more common in the next generation. This is natural selection.
- Fitness is a measure of how many offspring a particular variant tends to leave. Fitter variants become more common.
- Populations of organisms become better adapted to their environments.
- Many of the best-known examples of natural selection involve selection for characters which help organisms survive in polluted environments.
- Natural selection acts on individuals and on genes. It does not favour characteristics which help the wider population, unless they also help the organism itself, or its relatives which have many of the same genes.
- Males can generally increase their fitness by mating with more females, because they make many small gametes. This may lead to selection for characteristics used in competition for mates, or display.

- ◆ Females make fewer, bigger gametes. Genes which lead to females selecting high-quality mates will generally be favoured by natural selection.
- ◆ Species are groups of populations which can interbreed to produce fertile off-spring.
- ◆ Species are kept distinct by many factors which prevent either interbreeding, or production of fertile hybrids, or both.
- ◆ New species appear when a barrier to breeding forms between two populations. This usually involves the populations being far apart for many generations.
- ◆ Small populations may change noticeably even without natural selection, because random effects are more noticeable: genetic drift.
- ◆ When a new area is colonized by a few organisms, some forms of genes may be unusually common or rare in the new population. This depends on the genes which the colonizers happened to have: the founder effect.
- ◆ Many factors make species become rare, but we know little about how a rare organism becomes extinct.
- ◆ Human influence has caused some extinctions, and may cause many more.
- ◆ There have been episodes of mass extinction in the distant past.
- ◆ More complex living things have appeared through evolutionary time.
- ◆ Chemical and physical processes on the early Earth probably produced a range of organic molecules. The surfaces of clay particles may have provided a place for long-chain molecules such as RNAs to form.
- ◆ RNAs can act as both information stores and catalysts. They were probably the first self-replicating molecules. DNA, enzymes and cells came later.
- ◆ The possibility of extra-terrestrial life raises exciting issues, but does not change the basic questions about the origin of life.

Exercises

2.1. Biologists tend to tell evolutionary stories, hypotheses about how a characteristic might have evolved. The following story is rooted in an old and discredited idea about how evolution could take place.

'Background: a plant species lives in a habitat where the upper layers of the soil become very dry in summer. The plants have extremely deep roots. The roots grow to the same depth even in a garden where the plants are watered in the summer. Three closely-related species live in more moist habitats, and do not have deep roots. Story: the ancestors of this species invaded the dry habitat. In order to get water in summer, the roots had to grow deeper and deeper into the soil. They passed this characteristic on to their offspring, which now do the same wherever they grow.'

(i) What is the biological flaw in this story?

(ii) Retell the story in terms of natural selection.

2.2. Does natural selection still take place in human populations?

2.3. Naked mole rats, mentioned at the end of Section 2.3, live in communal burrows. They almost always find mates from the same burrow. A distinct caste has been discovered recently. It is rare (perhaps 1 in 2000 individuals) and has some distinctive physical characteristics. Most importantly, however, these animals have a strong urge to leave their own colony and join another.

(i) Why is the naked mole rat a good candidate among mammals for the evolution of castes?
(ii) What advantage might there be in producing this caste?

2.4. The pheasant is a large, mainly ground-living bird. The females are mottled brown in colour. The males are brightly coloured, with a longer tail than the females. In several other pheasant species, the females look similar, while the males are still brightly coloured but in different ways. Suggest why the males and females look so different.

2.5. Signals for recognizing and attracting mates are important in keeping animal species distinct. List the things you find sexually attractive in other people. Classify them as physical or behavioural

2.6. Work out the relatedness of:
(i) you and your great-grandfather (your grandfather's father);
(ii) you and your aunt (your mother's sister);
(iii) your half-sister (you share the same mother but have different fathers).
(If you don't have all these relations, just play along, please . . .)

Further reading

Textbooks to study:

Bulmer, M. *Theoretical Evolutionary Ecology* (Sunderland MA: Sinauer, 1994). A brisk and efficient tour of the field, which assumes that you are not scared of mathematics. Topics include population dynamics, life history evolution, foraging, selection, applications of game theory in evolution, kin selection, sex ratio, sexual selection, the evolution of sex.

Cowen, R. *History of Life* (2nd ed.)(Oxford: Blackwell Science, 1995). An historical approach to evolution ('what was alive 500 million years ago?' rather than 'how did sexual reproduction evolve?'). This readable book integrates ideas from Biology and Earth science, with lots about the fossil record.

Maynard Smith, J. *Evolutionary Genetics* (Oxford: Oxford University Press, 1989). A relatively advanced, but enthusiastic and authoritative guide to the areas of genetics which help to explain evolution. Population genetics and aspects of molecular genetics are included. Some knowledge of mathematics required.

Ridley, M. *Evolution* (2nd ed.) (Oxford: Blackwell Science, 1996). A big undergraduate textbook, clear and up to date.

Books to read through:

Dawkins, R. *The Selfish Gene* (2nd ed.) (Oxford: Oxford University Press, 1989). More than 20 years after the original edition appeared, it still makes a powerful, provocative case for the idea that selection acts on genes, and clearly presents kin selection. If you enjoy his style, several of his more recent books in a similar vein are easy to find.

Dennett, D.C. *Darwin's Dangerous Idea* (London: Penguin, 1995). An intellectually stimulating modern case for the theory of evolution by natural selection. Hard to summarize: read it if you enjoy ideas.

Gould, S.J. *Ever Since Darwin* (London: Penguin, 1977); Gould, S.J. *Hens' Teeth and Horses' Toes* (London:Penguin, 1983). Gould writes stimulating, semi-popular essays on evolutionary topics in an American magazine. Collections are regularly published: these are just two examples. Great to dip into: you don't have to agree with everything he says to enjoy them.

Lewin, R. *Patterns in Evolution: the New Molecular View* (New York: Scientific American Library, 1997). The uses of molecular data in studying evolution: highly illustrated, and using simple language.

Wilson, E.O. *The Diversity of Life* (London: Penguin, 1992). Wilson addresses the causes of, and threats to the diversity of life head-on. Natural selection, speciation, community structure, the human impact and extinction are covered in a beautifully written book, packed with interesting examples.

PART 2

Molecules and Cells

Molecules and Life

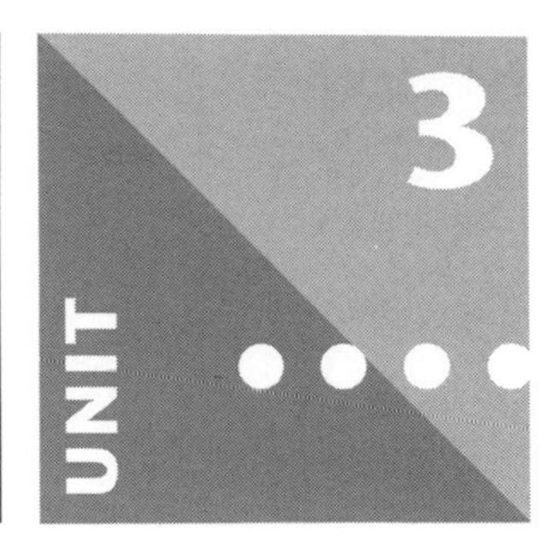

Connections

▶ The first few sections of this unit introduce the idea of molecules to readers who are less confident with basic ideas in Chemistry. The bulk of the unit describes the main groups of molecules which are important in living things. Much of what happens in cells and organisms can be explained in terms of molecules. This unit is an essential introduction to most of this book.

Contents

3.1 Molecules

Atoms rarely go around on their own. If they did, the world would be a rather boring place since there are only a few hundred types of atom, even if one includes very minor variants called isotopes. It is much more common for atoms to be linked more or less intimately in highly structured groups called molecules. Molecules range in size from two to many millions of atoms. They may consist of one or several types of atoms, linked in a particular way by bonds. Molecules are, then, extraordinarily diverse. Living things are made up of molecules. The challenge to the scientist is to explain the nature of organisms and how they work, in terms of the nature of the molecules which make them up, and how they interact.

3.2 Atoms and bonds

Understanding how atoms are linked by bonds requires some knowledge of what atoms are like. A simple model of the atom, dating from the early 20th century, still has a great deal of explaining power. The atom can be thought of as a sphere made up of two regions, one inside the other. The central region, the nucleus, is very small in comparison to the atom as a whole. It is also very dense: most of the atom's mass is found here. The nucleus carries a positive charge. It is made up of two types of particles. All nuclei contain positively charged protons; all except hydrogen atoms also contain uncharged neutrons. The outer part of the atom is mostly empty, but contains electrons which are free to orbit the nucleus. Electrons have a tiny mass, less than one thousandth the mass of a proton or neutron. Each has a negative charge

(a)

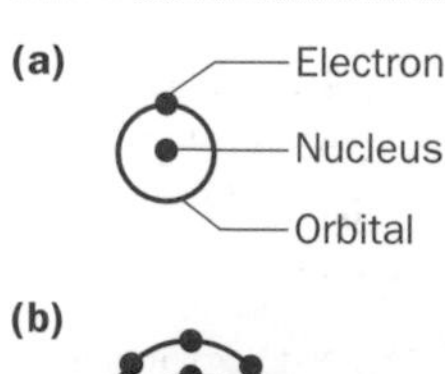

(b)

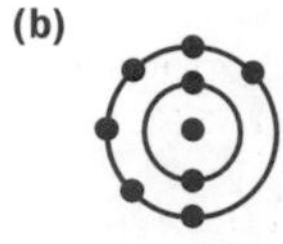

Fig. 3.1 Arrangement of electrons in orbitals. (a) a hydrogen atom, (b) an oxygen atom.

which exactly balances the positive charge of a proton. Atoms have the same number of electrons and protons. Their charges cancel, so atoms have no overall charge. They can become charged by gaining or losing electrons: such charged atoms or molecules are called ions.

The electrons are not spread evenly around the atom. They are much more likely to be found at some distances from the atom than others. One can think of electrons orbiting the nucleus in a series of concentric spheres called orbitals. Each orbital can only hold a certain number of electrons. The innermost orbital can hold up to two, the next up to eight, the next eight again, and so on. The electrons are normally found as near the nucleus as possible: the inner orbitals are filled first. In the hydrogen atom, which consists of just one proton and one electron, the electron is found in the innermost orbital (Fig. 3.1(a)) unless an energy input kicks it, temporarily, into another. In an oxygen atom, with its eight electrons, the inner orbital is filled by two electrons: the other six go into the next orbital (Fig. 3.1(b)).

Bonds

Atoms with orbitals which are only partly filled are generally unstable. They gain stability either by gaining or losing electrons, or by sharing them with another atom. If one could make a mixture of sodium atoms and chlorine atoms, they would rapidly form sodium chloride (Fig. 3.2). Sodium, with just one electron in its outer

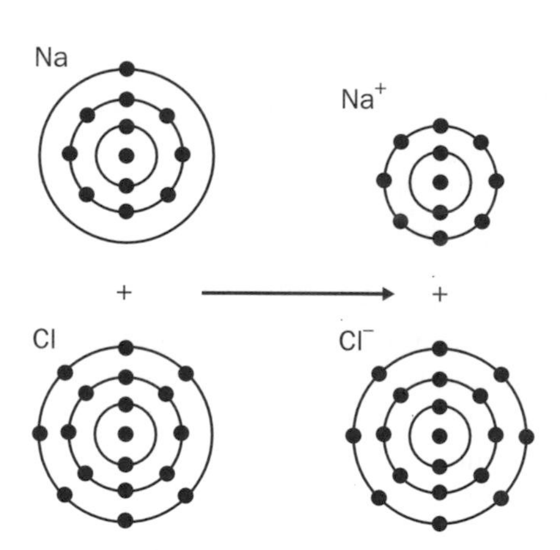

Fig. 3.2 A sodium atom donates an electron to a chlorine atom. Sodium chloride, an ionic compound, is formed.

orbital would give up this electron to chlorine, which is just one electron short of a full orbital. Both gain stability. These atoms have become charged ions: substances like this are called ionic compounds. The ions are held in a regular arrangement within a crystal by the electrical attraction between opposite charges. This is a sort of chemical bond, the ionic bond. If the crystal melts or dissolves, these rather weak bonds break and the ions go their separate ways.

A water molecule is stable because hydrogen and oxygen atoms share electrons (Fig. 3.3). Oxygen needs two more electrons to fill its outer orbital. It shares one of its electrons with a hydrogen atom and in turn receives a share in the hydrogen's electron. Through sharing, the hydrogen atom fills its orbital. The link that results from sharing one pair of electrons is called a covalent bond. The oxygen atom gains its final electron through making a second covalent bond with another hydrogen

O

H H

represented as

O

H H

Fig. 3.3 The water molecule. Hydrogen and oxygen are linked by covalent bonds.

atom. As a result, three atoms gain stability by forming a **molecule**, and the world becomes a wetter place. Molecules are primarily held together by covalent bonds. These are, in general, stronger than other types of chemical bond.

Some molecules have no overall **charge**, but carry small positive and negative charges on different parts of the molecule. This happens when one atom in a covalent bond attracts the shared electrons more strongly than the other. For example, oxygen atoms attract electrons more strongly than hydrogen atoms. In the V-shaped water molecule, the oxygen atom at the tip of the 'V' is slightly negative. The hydrogen end of the molecule is slightly positive (Fig. 3.4): the molecule is said to be **polar**. The positive end of one molecule may be weakly attracted to the negative end of another. Similar attractions may weakly link two different parts of a long, flexible molecule. In biological molecules, local charges are particularly noticeable where there are oxygen–hydrogen, nitrogen–hydrogen, carbon–oxygen or carbon–nitrogen bonds. The weak electrical attractions between local charges are called hydrogen bonds if a hydrogen atom ends up sandwiched between two other, more negatively charged atoms (Fig. 3.5). If hydrogen is not involved, they are simply a type of ionic bond.

Large biological molecules, macromolecules, are held together by covalent bonds. Hydrogen bonds and to a lesser extent ionic bonds hold their covalently-bonded skeletons in particular three-dimensional shapes which give the molecules distinctive properties.

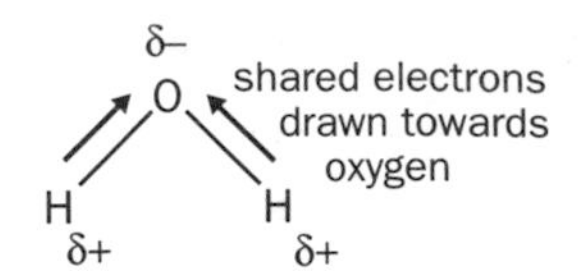

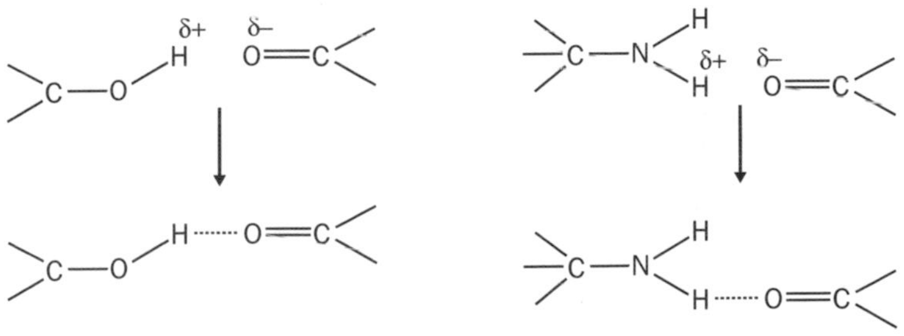

Fig. 3.4 The water molecule is polar. The Greek letter δ (delta) shows that these charges are small and localized.

Fig. 3.5 Two situations in which hydrogen bonds can be formed.

Carbon skeletons, functional groups

Most of the important molecules found in cells are organic molecules, that is molecules based on a skeleton of covalently bonded carbon atoms. Box 3.1 discusses why carbon is so important. The simplest organic molecules are hydrocarbons, made up of just carbon and hydrogen. Only a few of these are at all important in cells (notably ethene, Section 25.7). Most biological molecules have other functional groups attached to the carbon skeleton, replacing hydrogen. Functional groups give the molecule distinctive chemical properties. They also affect its three-dimensional shape and charge distribution. Box 3.2 shows some important functional groups.

Hydrophilic, hydrophobic

Molecules interact with water according to their charge. The water molecule carries local positive and negative charges. One end of the water molecule will be attracted to any part of another molecule which carries a charge, large or small. Molecules

Molecule: A group of atoms linked by covalent bonds.

Charged molecule: A molecule which has extra, or too few electrons.

Polar molecule: A molecule which has a small local charge, but no overall charge.

Box 3.1

Why carbon?

Organic molecules are molecules based on a skeleton of carbon atoms linked by covalent bonds. Most molecules important in Biology are organic.

Carbon is special for two reasons. First, carbon–carbon bonds can be made relatively easily, so carbon chains, rings, multiple rings and other more complicated structures can exist. Other atoms may also be bonded to this skeleton. Second, each carbon atom normally makes four covalent bonds. If these bonds are with four different atoms, the bonds splay out in different directions, pointing out towards the corners of a tetrahedron. These two features, taken together, mean that an enormous diversity of carbon skeletons can exist with distinctive three-dimensional shapes. This is important in Biochemistry, since in many cases shape partly determines function.

Box 3.2

Functional groups

Functional groups attached to the carbon skeleton give organic molecules their distinctive properties. Here are some important functional groups. The free bonds are the ones which link the group to the rest of the molecule.

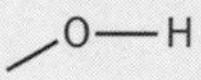

—O—H

Hydroxyl group (–OH). Molecules with one of these, but no other functional groups, are called alcohols.

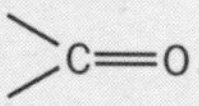

>C=O

Carbonyl group (>C=O). If the oxygen is attached to a carbon at the end of the chain, the molecule is an aldehyde. If it is attached somewhere in the middle of the chain, it is a ketone.

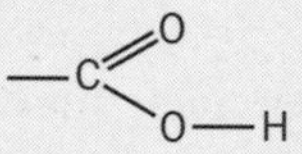

—C(=O)—O—H

Carboxyl group (–COOH). Molecules with one of these are called carboxylic acids. The hydrogen may break off as H^+, leaving a carboxylate group: the group is said to have dissociated.

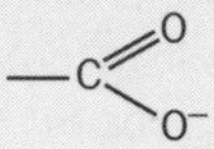

—C(=O)—O^-

C Carboxylate group –COO^-). It carries a negative charge.

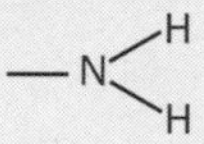

—N(—H)—H

Primary amino group (–NH_2). It can gain H^+, to form –$NH3^+$.

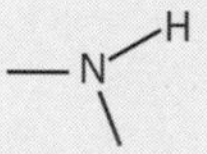

—N(—H)—

Secondary amino group (>NH). The nitrogen forms part of a ring or chain with carbon atoms.

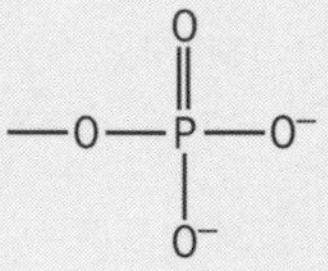

—O—P(=O)(—O^-)—O^-

Phosphoryl group (–OPO_3^{2-}). Sometimes two or three are linked as a chain.

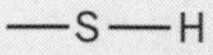

—S—H

Sulphydryl group (–SH). Relatively unusual, but important in the formation of disulphide bridges (–S–S–, Section 3.5).

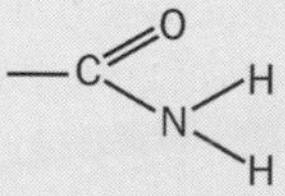

—C(=O)—N(—H)—H

Amide group (–$CONH_2$). Found in some amino acids.

without local charges will not attract water in the same way. In fact, since water molecules will be attracted to one another, non-polar molecules will tend to be left clumped together in clusters which exclude water. Non-polar molecules like this are colourfully described as **hydrophobic** (water-fearing). Molecules, or parts of molecules which are rich in carbon–hydrogen and carbon–carbon bonds, tend to be hydrophobic. The electrons are quite evenly shared in these bonds, so local charges are minute. Polar molecules which interact with water are described as **hydrophilic** (water-loving). Parts of molecules containing carbon–oxygen, oxygen–hydrogen and nitrogen–hydrogen bonds tend to be hydrophilic. Since biological molecules are

Hydrophilic molecules: Attract water molecules.

Hydrophobic molecules: Do not attract water, and clump together.

found in a watery context, whether inside or outside the cell, these simple ideas are important in understanding how biological molecules behave.

3.3 Molecules found in cells

Very many sorts of molecule are found in and around cells. It is not difficult to make sense of all this diversity. Many of these molecules have similar structures and can be grouped into 'families'.

The carbohydrates are based on sugars. The simplest sugars, monosaccharides, include glucose. Other familiar sugars, such as sucrose, are disaccharides: two monosaccharide units bonded together. Polysaccharides consist of many monosaccharides bonded together in chains, sometimes branched. Some are important as energy stores, such as glycogen in the mammalian liver. Others have a structural role, such as cellulose in the plant cell wall.

Proteins are based on another group of simple molecules, amino acids. Like sugars, amino acids can be bonded together into unbranched chains. Short amino-acid chains, peptides, have some important functions. For example, the hormone oxytocin, which stimulates the uterus to contract in labour, is a peptide, nine amino acids long. Longer chains, polypeptides, are even more important. A protein consists of one or more polypeptides, held by hydrogen bonds in a distinctive three dimensional shape. Proteins are important, for example as enzymes, as structural components of cells, in molecular recognition systems and in regulating what passes in and out of cells.

Nucleic acids are also long, unbranched chains, made up of many nucleotides. Deoxyribonucleic acid (DNA) carries the cell's genetic information. It is able to control its own replication, a rare quality which is fundamental to life. Ribonucleic acid (RNA) has more varied roles, mostly related to the way in which information carried in DNA is expressed.

Lipids are the one major group of biological molecules which are insoluble in water. Some include fatty acids as part of their structure. Fatty acids are long-chain hydrocarbons with a carboxyl group at one end. Other lipids have a quite different structure, for example the steroids, which have a multiple ring structure. Lipids also have a range of functions, for example phospholipids form the basis of biological membranes (Section 4.2), steroids have roles in membranes and as hormones, and triglyceride fats act as energy stores.

Some molecules combine elements of more than one of these groups. Glycoproteins, proteoglycans and peptidoglycans all have polysaccharide chains and peptides or polypeptides linked together in some way. Glycolipids are lipids with polysaccharide chains attached.

Not all molecules in cells fall into these groups. Many inorganic ions, such as sodium, chloride, calcium and the various phosphates, are essential. Other organic molecules are important too, many of them as metabolites: intermediates in the metabolic pathways which build up, break down and interconvert biological molecules.

3.4 Carbohydrates

Monosaccharides, the simplest sugars

Structurally, monosaccharides have this in common: one carbon atom is part of an aldehyde or ketone group; each of the others has one hydroxyl group attached. They also share many properties. They are soluble in water, they react chemically in similar ways and they all taste sweet. Most of the important monosaccharides have three, five or six carbon atoms: they are trioses, pentoses and hexoses. There are only two trioses: dihydroxyacetone and glyceraldehyde. They are important as metabolites, steps on the way to and from other sugars rather than useful products in themselves. The most widespread pentoses are ribose and its derivative deoxyribose. They are important because they form parts of nucleotides. Hexoses are common and important. Sometimes they are found on their own, for example glucose is the main energy-storing molecule carried around in mammals' blood. Their greatest importance, however, is when linked together to form pairs or longer chains. Box 3.3 shows the structure of all these, and some other monosaccharides. Fig. 3.6 illustrates two subtly different forms of a monosaccharide, the α and β forms. They interconvert freely when on their own, but not when linked together.

The glycosidic linkage

Monosaccharides can be linked together as pairs (disaccharides) or chains (polysaccharides). The systems of bonds which hold them together are called glycosidic

Box 3.3

Sweet and simple: monosaccharide structure

Trioses (three carbons)

There are just two triose sugars. Look at the position of the carbonyl group: it is on the middle carbon in dihydroxyacetone, and on the end of the chain in glyceraldehyde.

Dihydroxyacetone

Glyceraldehyde

Pentoses (five carbons)

Ribose is an important example. It can exist in an open chain form, showing its similarities with the trioses. However, it usually forms a ring.

usually represented as

Ribose: open chain

Ribose: five-membered (furanose) ring

Look closely to see that this is simply a rearrangement of the atoms. The carbonyl group has reacted with one of the hydroxyl groups. Notice that only four of the carbons form part of the ring. Deoxyribose is an important derivative of ribose. Spot the difference!

Hexoses (you've guessed it – six carbons)

Glucose, like most hexoses, normally exists as a six-membered (pyranose) ring. Fructose is an important exception: it forms a furanose ring. The difference is due to the position of the carbonyl group in the open chain forms.

Glucose

Fructose

In three dimensions

These rings are not flat as we tend to draw them on the page. The ring of glucose is bent along two lines: a 'chair' form. Other pyranose rings have a 'boat' form. Ribose has an 'envelope' form.

Chair

Boat

Envelope

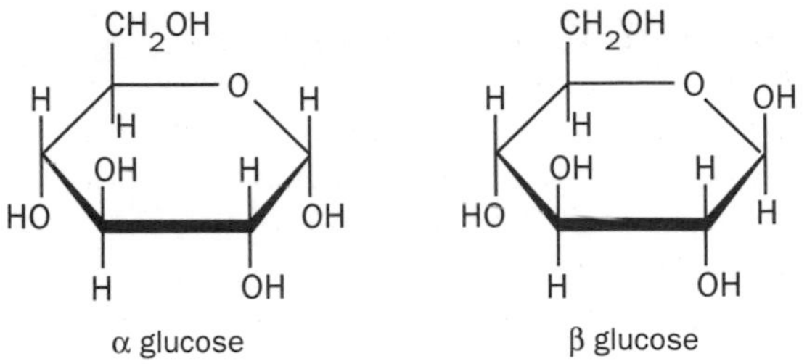

α glucose β glucose

Fig. 3.6 Two forms of a monosaccharide, glucose in this case. In the α form, the hydroxyl group attached to the first carbon in the ring is below the plane of the ring. In the β form it is above the ring.

linkages. Fig. 3.7 shows a typical situation. Two glucose molecules are linked to form a disaccharide, maltose. Carbon 1 of the first glucose is linked via an oxygen to carbon 4 of the second: this is a 1,4 glycosidic linkage. The first carbon is in the α form, and is fixed like this once the linkage is made, so this is properly called an α1,4-glycosidic linkage. Once joined, the two original molecules have become a single molecule: they are now referred to as glucose residues. In making the linkage, a water molecule has been made. Any reaction like this, where water is made when two molecules are joined, is called a condensation reaction. When a glycosidic linkage is broken, water is used: a hydrolysis reaction.

Other carbon atoms may be linked in a similar way: 1,6, 1,3 and 1,2 linkages are found in some polysaccharides. Box 3.4 shows all the disaccharides which are at all familiar. One, sucrose, has a 1,2 linkage.

Polysaccharides

Polysaccharides are long chains of monosaccharides. They differ in which type or types of sugars they are made up of, the type of glycosidic linkages, and whether or

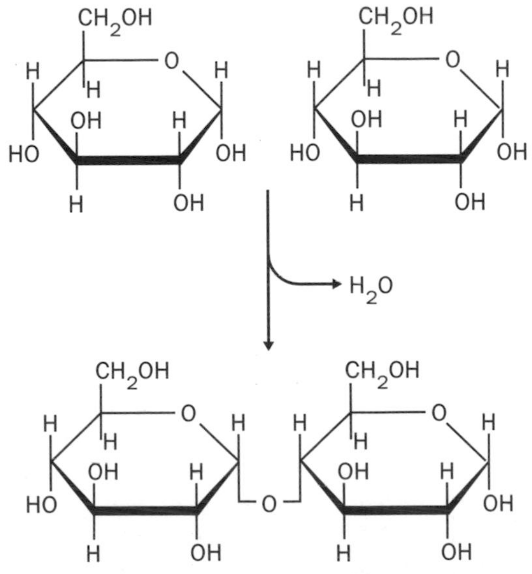

Fig. 3.7 Formation of a glycosidic linkage. Two glucose molecules are linked to form maltose.

Box 3.4

Important disaccharides

Lactose: galactose and glucose, β1,4 linked. Lactose is the sugar found in mammalian milk.

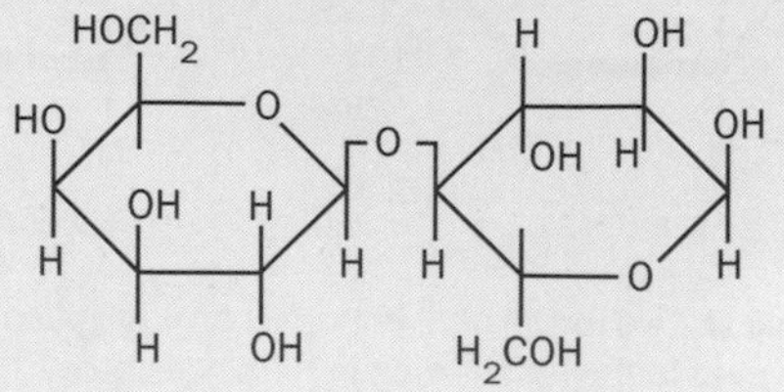

Maltose: two glucose molecules, α1,4 linked. Maltose is important only as a step in the breakdown of starch.

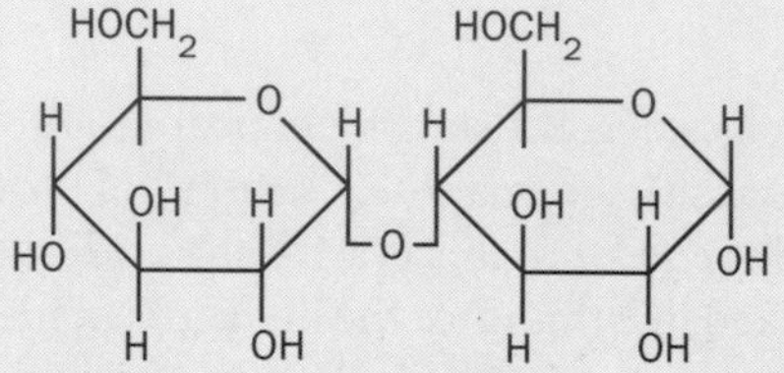

Sucrose: glucose and fructose, α1,2 linked. Sucrose is the sugar which is transported from one part of a plant to another.

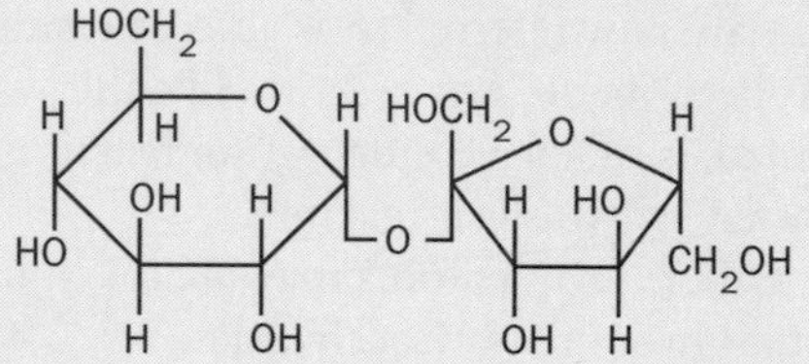

not they are branched. These differences directly affect their function. They control the ways in which the polysaccharide winds around itself to form a three-dimensional shape, or interacts with other chains to form a fibre or mesh. Some examples illustrate this. Starch is used as an energy store in plants. It is found in dense grains in plastids (Section 4.5) within cells. Starch is a mixture of two polysaccharides: amylose and amylopectin. Glycogen is another energy store, found in animal cells. In mammals, most is concentrated in the liver and muscles. Cellulose forms unstretchable fibres in the plant cell wall. Chitin forms similar fibres in fungal cell walls, as well as in the exoskeletons of arthropods, for example crab shells and the hard outer covering of beetles. Alginate is one of a group of polysaccharides found in the brown algae (phylum Phaeophyta), which forms slimy, water-retaining gels.

Amylose, one of the starch polysaccharides, has a very simple structure. It consists of glucose residues with α1,4 glycosidic linkages, and is not branched. The chain coils up to form a hollow helix (Fig. 3.8(a)), held by hydrogen bonds. There is little bonding between different chains, so amylose molecules are simply packed together in a physically weak mass.

Cellulose differs in only one way, but it completely changes the way the molecule behaves. Glucose residues are β1,4 linked in an unbranched chain. The shape of the glucose residues in the β form means that each glucose residue is upside down in relation to the ones either side of it. The chain as a whole is a flat ribbon (Fig. 3.8(b)). Chains lie parallel and are linked by hydrogen bonds. Many chains together form a fibre. This structure is strong because hydrogen bonding is between, rather than within, molecules. In chitin, glucose is replaced by *N*-acetyl glucosamine, which is

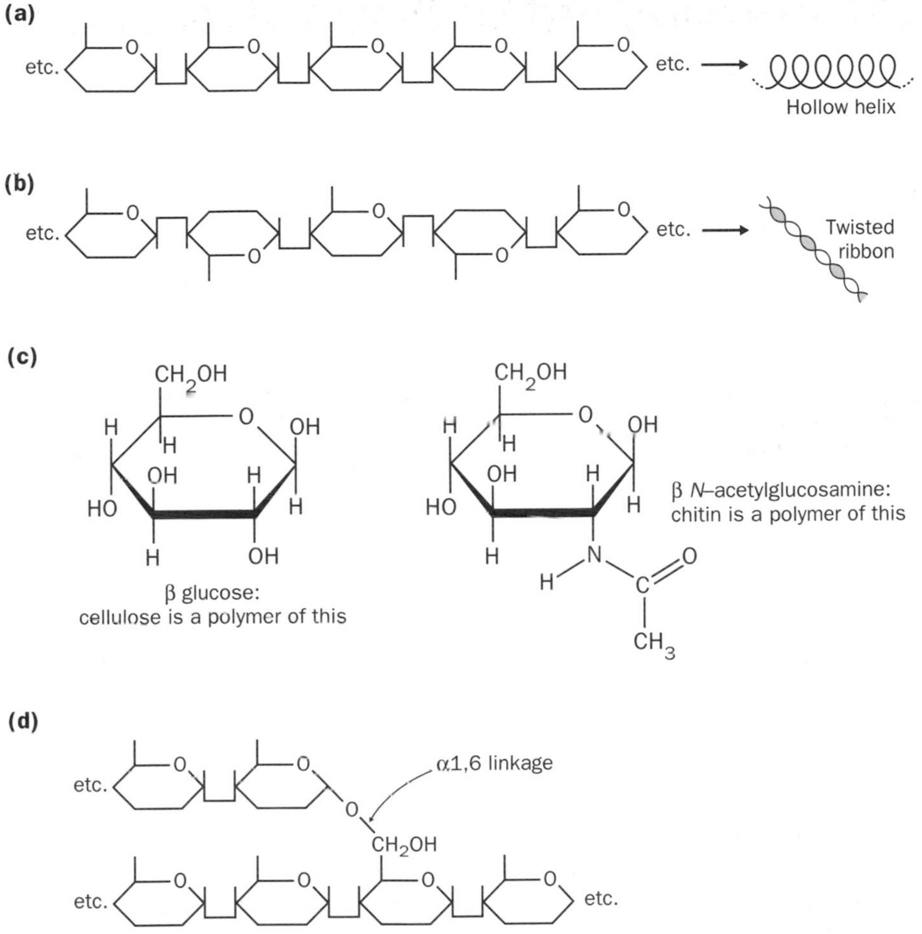

(a) Hollow helix

(b) Twisted ribbon

(c) β glucose: cellulose is a polymer of this

β *N*-acetylglucosamine: chitin is a polymer of this

(d) α1,6 linkage

Fig. 3.8 Polysaccharide structure (see text for explanation). (a) amylose, (b) cellulose, (c) chitin, (d) amylopectin and glycogen.

essentially glucose with a nitrogen-containing side chain instead of one hydroxyl group (Fig. 3.8(c)). Just like cellulose, they are β1,4 linked and form fibres in the same way.

Amylopectin and glycogen are very similar. Both consist of α1,4 linked glucose residues, with occasional α1,6 linkages which give rise to branches (Fig. 3.8(d)). The only difference between them is that glycogen is more branched. Amylopectin has about one α1,6 linkage for every thirty α1,4 linkages. In glycogen, the ratio is more like 1:10.

Alginate is a mixed polymer. The sugar residues are rather obscure: the important point is that some parts of the chain are rich in one type of residue, while the other parts have another type. This means that the polysaccharide is a mosaic of short sections with different properties. One type of section can pair up with the same type on another strand. The other type of section remains unpaired. As a result, many strands are interlocked in an open, three-dimensional mesh (Fig. 3.9). Water, and anything dissolved in it, fills the spaces in the mesh.

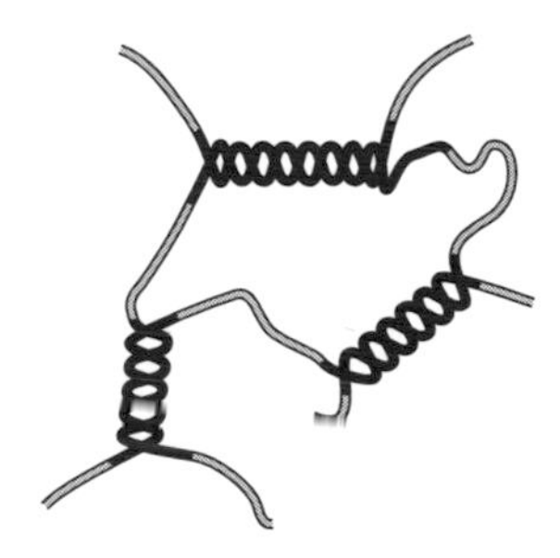

Fig. 3.9 Gel formation in alginate, a mixed polysaccharide. Some parts of each strand (shown dark) form a double helix with another strand. Other parts (light) remain free.

3.5 Proteins

Amino acids

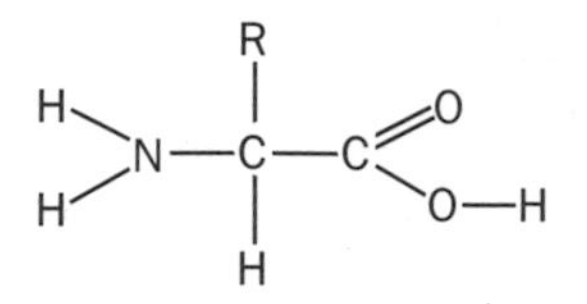

Fig. 3.10 The structure of naturally occurring amino acids. The side chain (R) varies.

An amino acid is an organic molecule which has an amino group as well as a carboxyl group. The amino acids found in cells have much more than this in common (Fig. 3.10). A single carbon atom has the following four things attached to it: an amino group, a carboxyl group, a hydrogen atom and some other group which varies. This last group is called the side chain.

Amino acids are at the same time both acids and bases. Acids tend to increase the concentration of H^+ ions in solution. Carboxyl groups tend to lose an H^+ ion, becoming carboxylate ions ($-COO^-$). This raises the H^+ concentration, so amino acids are acidic. Bases tend to decrease the concentration of H^+ ions in solution. Amino groups do just this, binding H^+ to become $-NH_3^+$. Hence, amino acids are basic as well as acidic. Box 3.5 shows how they exist in different ionic forms at different pH values.

The side chains of amino acids give them their distinctive properties. A basic set of 20 types is found in cells (Box 3.6). A few extras, which are really only modified forms of one of the 20, occur only in rather special situations. The side chains vary in size, and in the extent to which they are locally charged. Side chains may be acidic, basic or neutral, in addition to the acidic and basic groups which all amino acids share.

Amino acids on their own have very few functions within cells. However, they have one property which makes them enormously important: they can be linked to form chains.

Joining amino acids

Amino acids can be linked in three ways. Firstly, the carboxyl group of one may be joined to the amino group of another in a peptide linkage (Fig. 3.11). In this way,

Box 3.5

Amino acids and pH

Both the amino group and the carboxyl group of an amino acid can become charged by gaining or losing H^+. This is influenced by pH.

Under acidic conditions (high H^+ concentration), the amino group tends to gain H^+, becoming positive. The carboxyl group is less likely to lose H^+ when there is already lots in solution.

Under alkaline conditions (low H^+ concentration), the carboxyl group tends to lose H^+, becoming negative. The amino group is less likely to gain H^+ as there is little in solution.

At intermediate pH values, a doubly-charged ion can exist. The zwitterion carries a negative charge on the carboxyl group and a positive charge on the amino group.

zwitterion

acid ⟷ alkaline

Box 3.6

Amino acid diversity . . .

. . . it's the bit on the side which makes them special.

Amino acids share the same basic structure. The side chain gives each one its distinctive properties. Only the side chains are drawn here; the rest of the molecule is just shown as M. Standard three-letter codes for each amino acid are also shown.

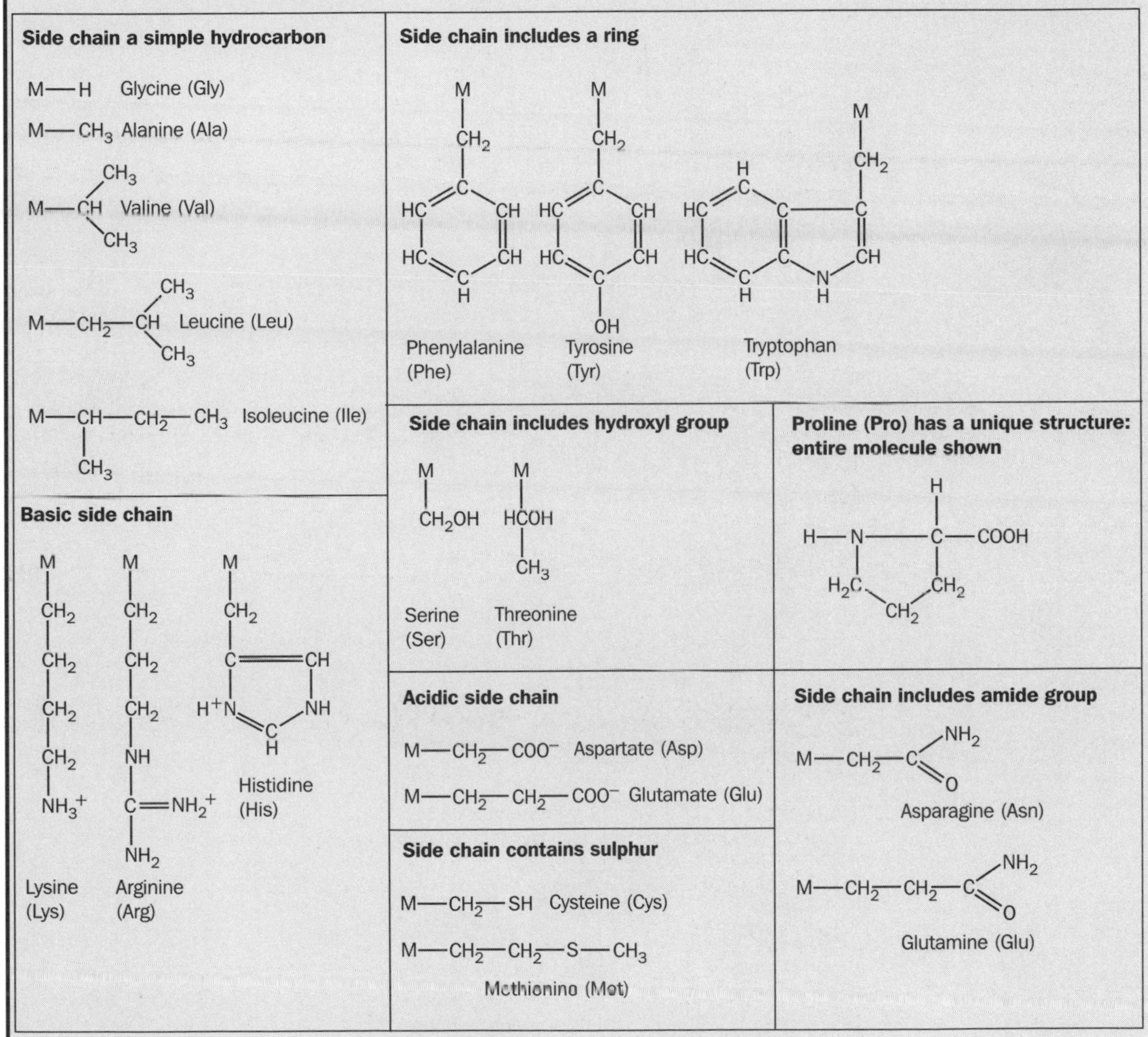

Side chain a simple hydrocarbon

M—H Glycine (Gly)

$M—CH_3$ Alanine (Ala)

$M—CH(CH_3)_2$ Valine (Val)

$M—CH_2—CH(CH_3)_2$ Leucine (Leu)

$M—CH(CH_3)—CH_2—CH_3$ Isoleucine (Ile)

Basic side chain

$M—CH_2—CH_2—CH_2—CH_2—NH_3^+$ Lysine (Lys)

$M—CH_2—CH_2—CH_2—NH—C(=NH_2^+)—NH_2$ Arginine (Arg)

$M—CH_2—C(=CH—NH—CH=N^+H)$ Histidine (His)

Side chain includes a ring

Phenylalanine (Phe)

Tyrosine (Tyr)

Tryptophan (Trp)

Side chain includes hydroxyl group

$M—CH_2OH$ Serine (Ser)

$M—HCOH—CH_3$ Threonine (Thr)

Proline (Pro) has a unique structure: entire molecule shown

Acidic side chain

$M—CH_2—COO^-$ Aspartate (Asp)

$M—CH_2—CH_2—COO^-$ Glutamate (Glu)

Side chain contains sulphur

$M—CH_2—SH$ Cysteine (Cys)

$M—CH_2—CH_2—S—CH_3$ Methionine (Met)

Side chain includes amide group

$M—CH_2—C(=O)NH_2$ Asparagine (Asn)

$M—CH_2—CH_2—C(=O)NH_2$ Glutamine (Glu)

amino acids form unbranched chains, polypeptides. What is left of each molecule in the chain is now called a residue. Secondly, the sulphur-containing side chains of two cysteine residues may be linked in a disulphide bridge (Fig. 3.12). Disulphide bridges may link two polypeptides, or different parts of a single chain to form a loop. Both these linkages are relatively strong, involving covalent bonds. Thirdly, amino acids may be linked more weakly by hydrogen bonds. The pattern of bonding varies

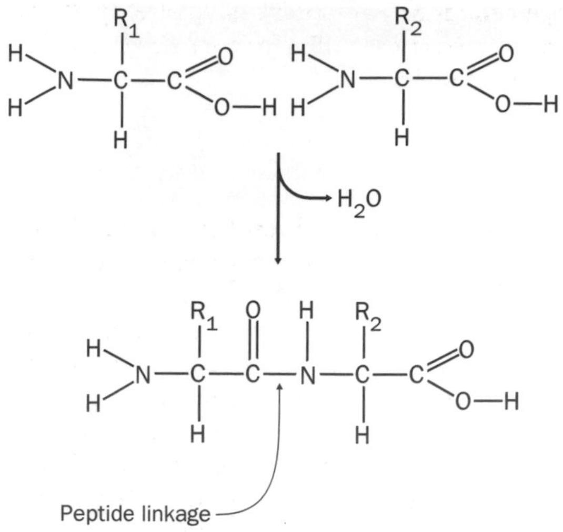

H_2O

Peptide linkage

Fig. 3.11 Formation of a peptide linkage between two amino acids.

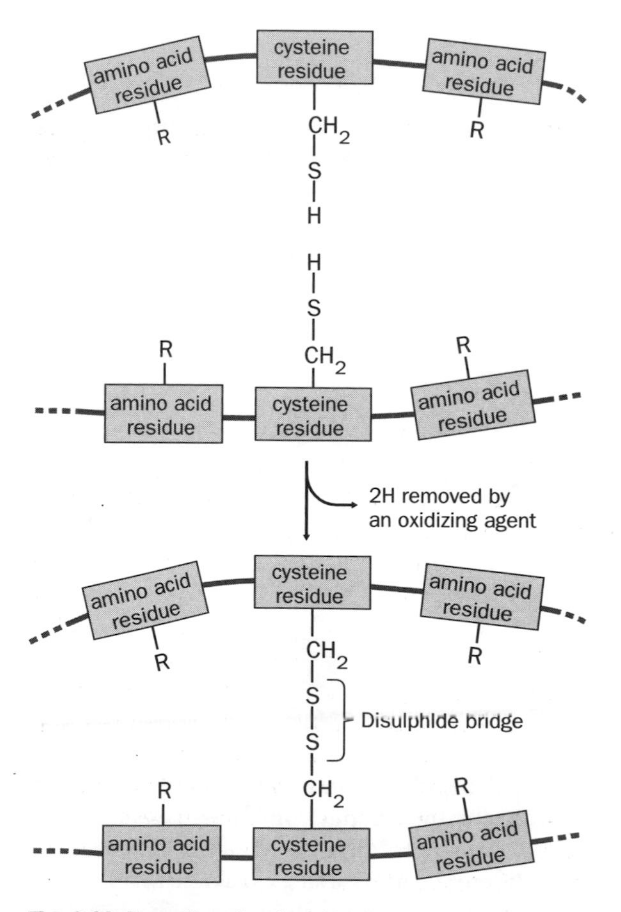

Fig. 3.12 Formation of a disulphide bridge. Cysteine residues in two polypeptides, or different parts of the same polypeptide, are covalently linked.

with the side chains involved. Hydrogen bonds between amino acids already linked in a polypeptide hold the chain in a distinctive three-dimensional shape. At this level, the polypeptide can be described as a protein.

From polypeptide to protein

A protein is one or more polypeptides with a specific three-dimensional shape. Protein structure is traditionally considered at four levels: primary, secondary, tertiary and quaternary structures.

All proteins have a primary structure. This is simply the sequence of amino acids in the polypeptide, plus any disulphide bridges. The primary structure is covalently bonded and comparatively stable. In all proteins, the polypeptides are held in specific shapes by weaker hydrogen bonds, the secondary and tertiary levels of structure.

Secondary structure relates to the ways in which local regions of the polypeptide fall into shape. There are a few well-known examples of regular secondary structures. The α helix is a tight spiral with the amino acid side chains sticking outward (Fig. 3.13(a)). The β pleated sheet consists of parts of one or several polypeptides, lying parallel and linked by hydrogen bonds (Fig. 3.13(b)). Hairpin turns, also called β turns, are groups of four amino acids where the chain turns sharply through 180° (Fig. 3.13(c)). They are especially common where they allow a single polypeptide to form a β pleated sheet (Fig. 3.13(d)).

Tertiary structure is the way in which the polypeptide as a whole forms a shape, each local region still in its secondary structure. It is the tertiary structure that gives proteins their distinctive properties. Box 3.7 shows the three levels in one protein, myoglobin.

Some proteins consist of several polypeptides, each with its own tertiary structure. Haemoglobin, for example, consists of four polypeptides. The way in which these subunits link to form the complete protein is known as the quaternary structure.

Proteins are important

Electrical interactions like hydrogen bonds are central to how proteins do their jobs, as well as to their structure. Proteins can bind other molecules in a very specific way.

Protein fibres form when many similar protein molecules bind together. This is seen in the intermediate filaments which strengthen cells, and in the microtubules and actin filaments which form a 'skeleton' within the cell (Section 4.5: The cytoskeleton). More usually, proteins bind other types of molecule. Enzymes are protein catalysts which speed up particular reactions (Unit 5). They bind the molecules which take part in the reaction, bringing them together in an ideal position to react. Proteins control what crosses biological membranes by acting as channels or carriers (Section 6.5). Much of their specificity lies in electrical interactions, and in binding the molecule which crosses. Channels which are opened or closed by a chemical signal, must also bind the signalling molecule (for example neurotransmitter receptors, Section 17.4). Carrier proteins in blood must bind the molecule they transport: haemoglobin binds oxygen, for example. Antibodies bind specific foreign molecules (Section 23.5). Hormone receptors on the surface of cells bind hormone molecules (Section 16.4).

Proteins bind other molecules, and so have useful functions. Proteins are important.

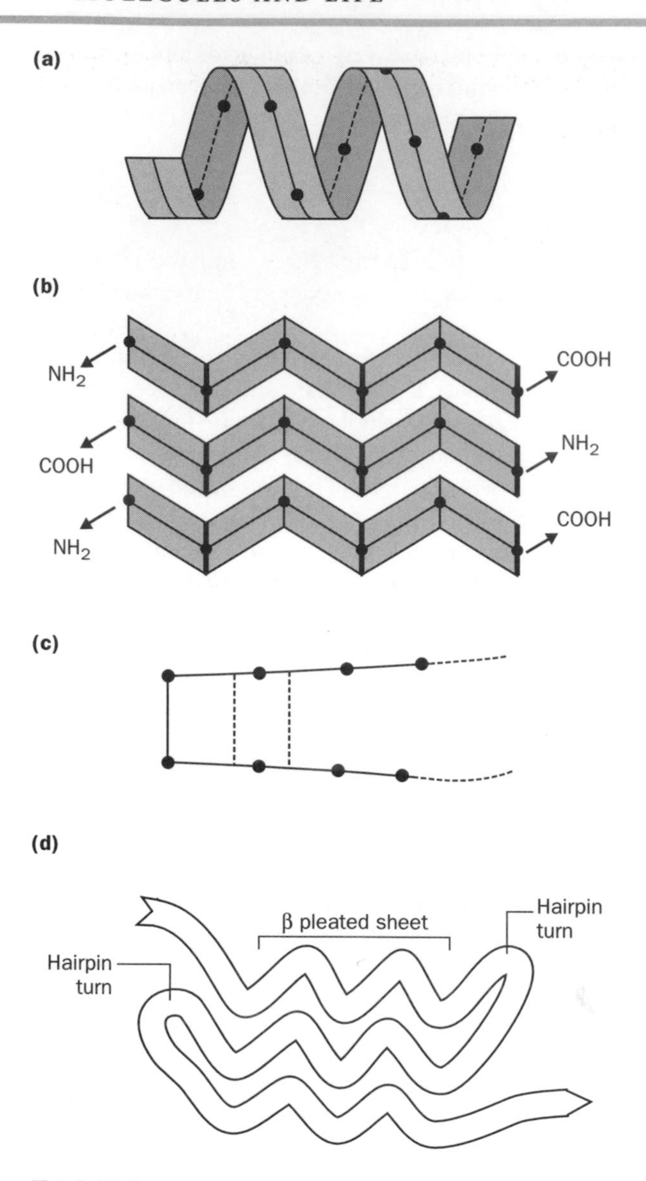

Fig. 3.13 Protein secondary structures (each amino acid residue is shown as a single blob, for clarity). (a) α helix, (b) β pleated sheet, (c) hairpin turns, (d) hairpin turns allow a single polypeptide to form a β pleated sheet.

3.6 Glycoproteins and related molecules

Glycoproteins, proteoglycans and peptidoglycans are a bunch of hybrid molecules which in some way combine chains of amino acids with chains of sugars. The distinctions are slightly blurred.

Glycoproteins are proteins with one or more polysaccharide chains attached. The protein is typically embedded in the plasma membrane of the cell with the polysaccharide chains sticking out (Section 4.6). Many receptors for signalling molecules such as hormones, are glycoproteins. The carbohydrate chains probably help

Box 3.7

Levels of protein structure: myoglobin

Myoglobin is a globular protein which binds oxygen. It is found in mammalian muscles, where it acts as a short-term reserve tank of oxygen (Section 19.2). Sperm whale myoglobin was one of the first proteins to have its three-dimensional structure worked out.

Primary and secondary structure

Myoglobin consists of a single chain of 153 amino acid residues. Its secondary structure is dominated by α helices. About three quarters of the chain is wound into eight separate helices. Here is the amino acid sequence: the boxes show the amino acids involved in each helix.

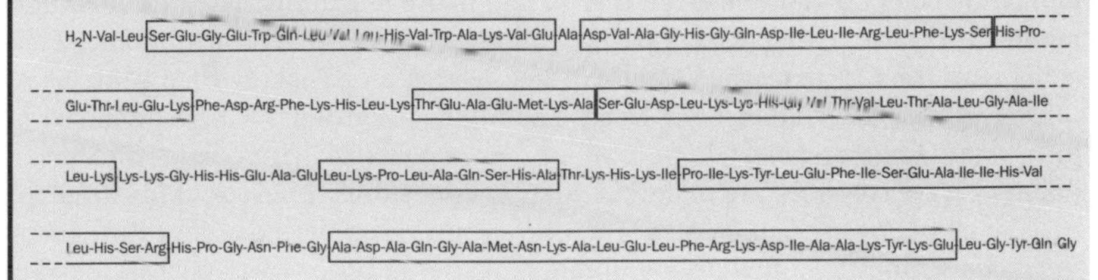

Tertiary structure

The helices are then folded into a very compact, globular structure. A non-protein group, a haem group, is also attached, here shown shaded. Non-proteins intimately linked to proteins are called prosthetic groups.

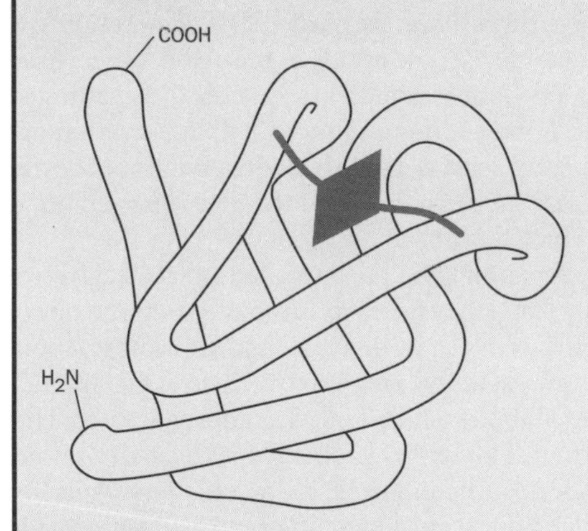

Quaternary structure

This does not apply to myoglobin, since it consists of just one polypeptide.

recognize molecules. The protein component is usually bigger than the carbohydrate portion of the molecule.

Proteoglycans are mostly found outside animal cells, forming part of the extracellular matrix (Section 4.6). Their carbohydrate components are less varied than in glycoproteins. Most of the chains are glycosaminoglycans (GAGs). GAGs are mixed polysaccharides in which alternate sugar residues are amino sugars, often *N*-acetyl glucosamine, as seen in chitin (Fig. 3.8(c)). The carbohydrate content can be much higher than in glycoproteins. The most extreme examples have a polypeptide strand at the core with GAGs sticking off it all the way along, like a piece of tinsel or an angry cat's tail.

The only familiar peptidoglycan is the one which gives strength to bacterial cell walls (Section 4.4). Here, short polypeptides and short chains of sugars are linked in three dimensions to form a tough mesh as big as the wall itself.

3.7 Lipids

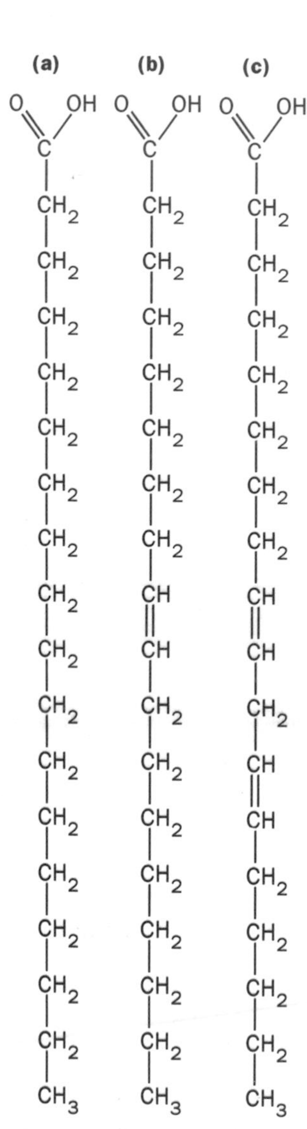

Fig. 3.14 Fatty acids. (a) stearic acid (saturated), (b) oleic acid (monounsaturated), (c) linoleic acid (polyunsaturated).

Lipid molecules are dominated by carbon and hydrogen atoms. They are predominantly non-polar. Either a large part, or all of the molecule is hydrophobic. This, along with the fact that when oxidized they release more energy than proteins or carbohydrates, determines their roles. The following groups will be discussed in turn. Fatty acids are most important as components of triglycerides and phospholipids. Triglycerides are mainly used as energy stores. Phospholipids form the basis of all biological membranes. The next group is the steroids. They have many functions including stiffening membranes, acting as hormones and making skin waterproof. The waxes have much more restricted roles, but they are simple and interesting.

Fatty acids consist of a hydrocarbon chain with a carboxyl group at one end (Fig. 3.14). The hydrocarbon chain is hydrophobic. The chain length varies from 12 to 24 carbons, but fatty acids with 16 or 18 carbons are particularly common. Some fatty acids have one or more carbon–carbon double bonds in the chain, others have only single bonds. Fatty acids with no double bonds are described as saturated (Fig. 3.14(a)). Those with one double bond are monounsaturated; those with more are polyunsaturated (Fig. 3.14(b,c)). Fatty acids with more double bonds or shorter chains have lower melting points, so are more fluid. This difference is passed on to the triglycerides and phospholipids which they form.

Triglycerides consist of three fatty acids, all linked to a smaller molecule, glycerol (Fig. 3.15). Each fatty acid is joined by a reaction between its carboxyl group and one of glycerol's three hydroxyl groups. Water is made, so this is a condensation reaction. The linkage which results is called an ester linkage. Triglycerides are extremely hydrophobic. They clump together as globules which exclude almost all water. This makes them valuable as a way of packing a lot of stored energy into a small volume: they are energy stores. Huge triglyceride droplets almost fill the fat cells which cushion you when you sit down, give some of us elegant body curves, and which provide insurance against the famine which just might be around the next corner.

The simplest phospholipid is phosphatidate (Fig. 3.17(a)). Two fatty acids are linked to glycerol, as in triglycerides. However, a phosphoryl group is attached to the third hydroxyl group of glycerol. This is charged, and makes one end of the molecule hydrophilic. Most other phospholipids are made from phosphatidate by linking a small polar molecule to the phosphoryl group. Three common examples are choline, serine and inositol, resulting in phospholipids called phosphatidylcholine, phosphatidylserine and phosphatidylinositol (Fig. 3.17(b–d)). In all cases, the molecules have two hydrophobic tails and a hydrophilic head (Fig. 3.17(e)). As a result, phospholipid molecules cluster as spheres or sheets, with the tails pointing inwards and the heads pointing out into the surrounding water. These sheets, bilayers, form the basis of all biological membranes.

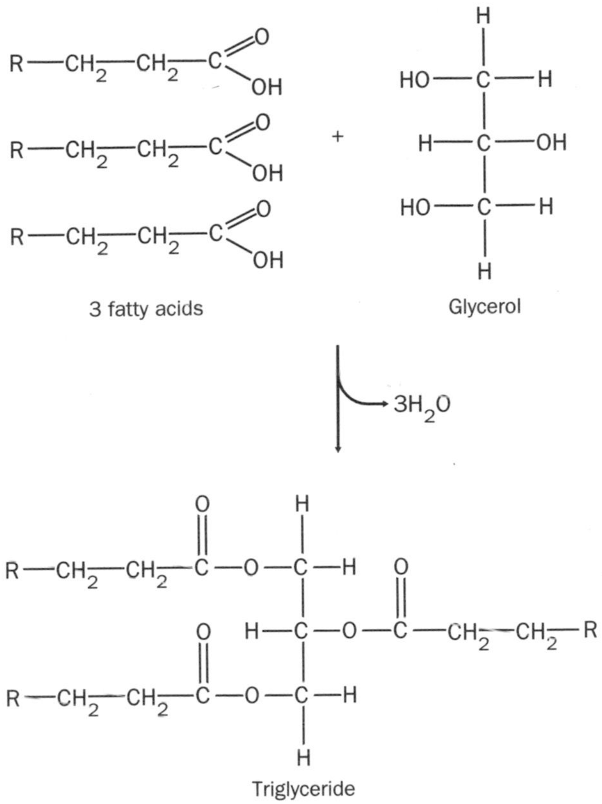

Fig. 3.15 Formation of a triglyceride (=triacylglycerol). R represents the long hydrocarbon chain of a fatty acid.

Steroids are based on a flat, hydrophobic system of rings (Fig. 3.16). Cholesterol is the basic steroid. Others, such as the hormone progesterone, are made by modifying cholesterol. Cholesterol has only one polar group, the hydroxyl group on the first ring. This tiny hydrophilic region allows it to slot into phospholipid bilayers. It stiffens membranes, by preventing the phospholipid tails from flexing too much.

A wax, strictly speaking, is a fatty acid joined to a fatty alcohol by an ester linkage. (A fatty alcohol is a long chain hydrocarbon with a hydroxyl group at one end.) The most familiar is the wax which bees use to build arrays of cells for their larvae: honeycombs. More importantly, they are major energy stores in many marine invertebrates, especially copepod crustaceans. Since these animals are enormously important, waxes have a starring role in marine food chains. Many other lipids have waxy properties: they are solid at room temperature and form water-resistant layers. The waterproof cuticles which coat leaves contain lipids of this type. The upper surface of a holly leaf, for example, is glossy because of its thick cuticle. Some cuticle 'waxes' are hydrocarbons with both carboxyl and hydroxyl groups. They can make ester linkages with one another to form a tough mesh.

Waxes are only found in certain situations. However, the next group of molecules, nucleotides, is essential to the life of all cells.

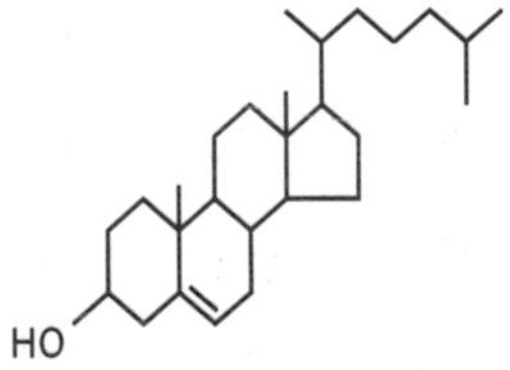

Fig. 3.16 The carbon skeleton of cholesterol, a steroid. The only functional group is the hydroxyl group shown.

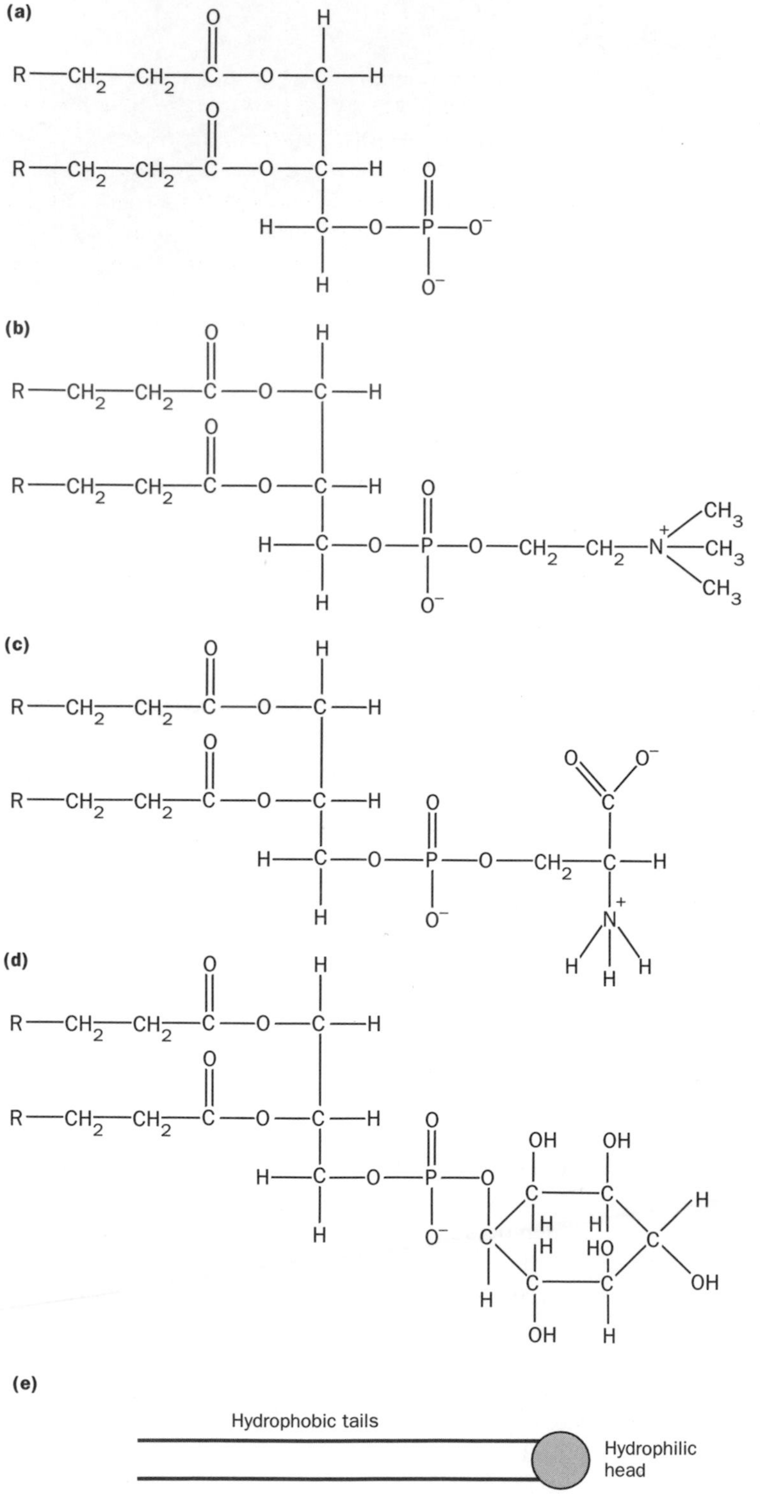

(a)

(b)

(c)

(d)

(e)

Hydrophobic tails

Hydrophilic head

Fig. 3.17 Phospholipids. (a) phosphatidate, (b) phosphatidylcholine, (c) phosphatidylserine, (d) phosphatidylinositol, (e) a standard way of representing any phospholipid.

3.8 Nucleotides

A nucleotide is a hybrid molecule made up of three parts. A pentose sugar is linked to a nitrogen-containing base: together they are called a nucleoside. Then one or more phosphoryl groups are added to carbon 5 of the sugar, making it a nucleotide (Fig. 3.18).

The pentose involved is either ribose or its derivative deoxyribose. Nucleotides made from them are called ribonucleotides and deoxyribonucleotides, respectively. The bases are more diverse (Fig. 3.19). Two groups are particularly important, the purines and the pyrimidines.

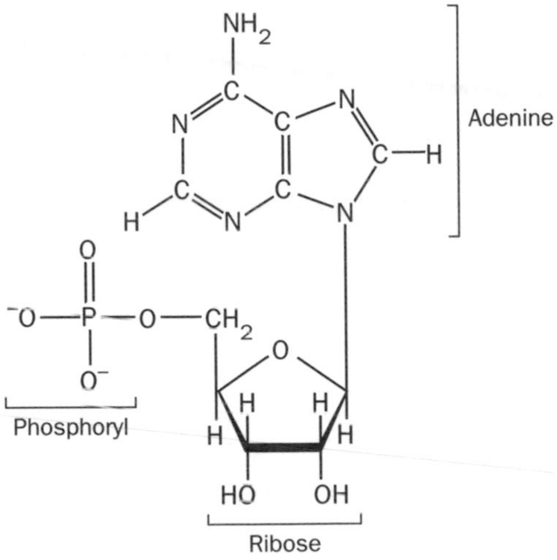

Adenine

Phosphoryl

Ribose

Fig. 3.18 Adenosine monophosphate, a nucleotide.

Pyrimidines

Thymine Cytosine Uracil

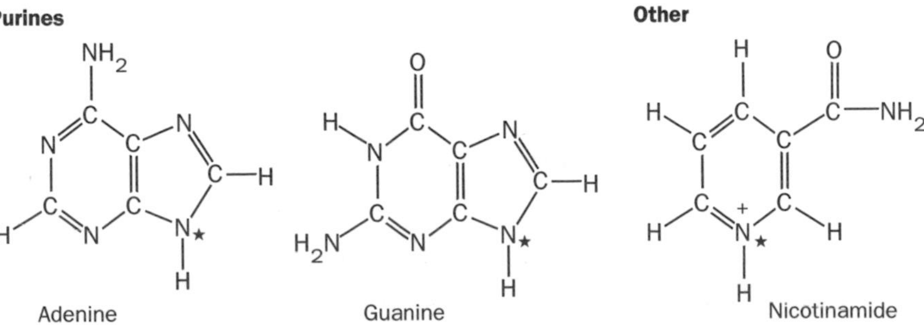

Fig. 3.19 Some organic bases found in nucleotides. They can be joined to pentose sugars at the points labelled *.

Some nucleotides are useful on their own. Adenosine triphosphate (ATP) acts as a multi-purpose, short-term energy store in cells (Section 8.3). Energy is released when either one or two phosphoryl groups are removed, forming adenosine diphosphate (ADP) or adenosine monophosphate (AMP) (Section 8.4). Cyclic AMP, a derivative of AMP, has a completely different use, as a signalling molecule within cells (Section 16.4). Several other nucleotide-based molecules are important as energy carriers or as reducing agents: NAD, NADP and FAD are discussed in detail in Unit 8, along with ATP.

Nucleotides are most commonly found linked into chains as nucleic acids.

Nucleic acids

Nucleic acids are polynucleotides, that is polymers of nucleotides. There are two main types, ribonucleic acid (RNA) and deoxyribonucleic acid (DNA). RNA is a polymer of ribonucleotides. The sugar is ribose in every nucleotide, but the base can be any one of four. It may be a purine: either adenine (A) or guanine (G). Alternatively, it may be a pyrimidine: uracil (U) or cytosine (C). Nucleotides within the chain are linked between the phosphoryl group of one nucleotide to carbon 3 of ribose in the next (Fig. 3.20(a)). The RNA molecule as a whole is a long, single strand of alternate ribose and phosphoryl units, with a base sticking off each ribose (Fig. 3.20(b)). The bases may be in any order, so whilst all RNAs have the same basic structure, they vary enormously in base sequence and length. The two ends of a polynucleotide strand are not the same. The pentose at one end is only linked to the rest of the chain at carbon 3. Carbon 5, with its phosphoryl group, is free. This is the 5′ end, pronounced 'five prime'. At the other end, carbon 3 is free, the 3′ end.

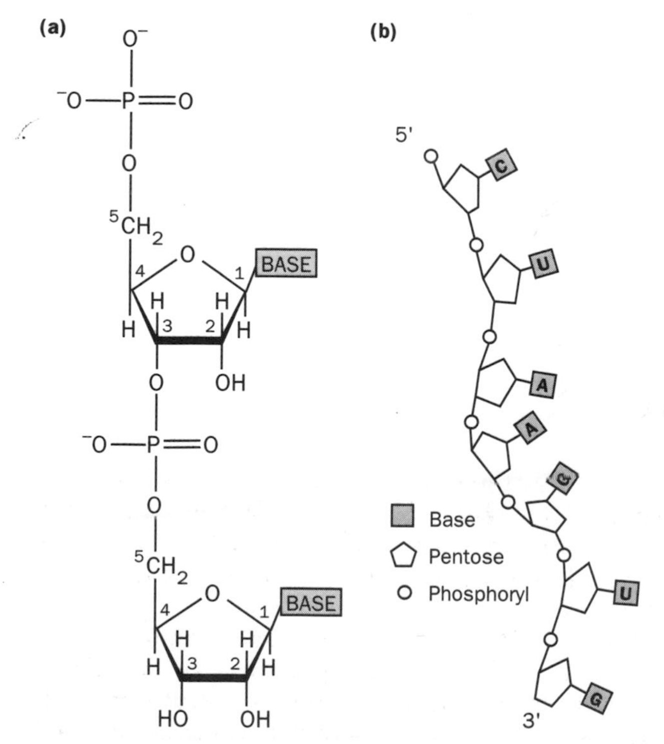

Fig. 3.20 RNA structure. (a) two linked ribonucleotides, (b) a short section of RNA.

DNA is different in several ways. Firstly, the pentose is deoxyribose instead of ribose. Secondly, uracil is replaced by another pyrimidine, thymine (T). The biggest difference is that a DNA molecule is made up of not one but two polynucleotides. The two strands are parallel, but running in opposite directions: they are said to be antiparallel. This is possible because the bases of each pair of nucleotides are held together by two or three hydrogen bonds. These pairings are specific. A purine binds to a pyrimidine, A with T, G with C (Fig. 3.21). This means that if you know the base sequence of one strand, you can work out the other. For example, the sequence 5′-ATTGACCGT-3′ must be paired with 3′-TAACTGGCA-5′ (Fig. 3.22(a)). Drawn flat like this, the molecule looks like a ladder, but in reality the whole structure is twisted into a spiral, both strands turning together in the same direction. This is the famous double helix (Fig. 3.22(b)).

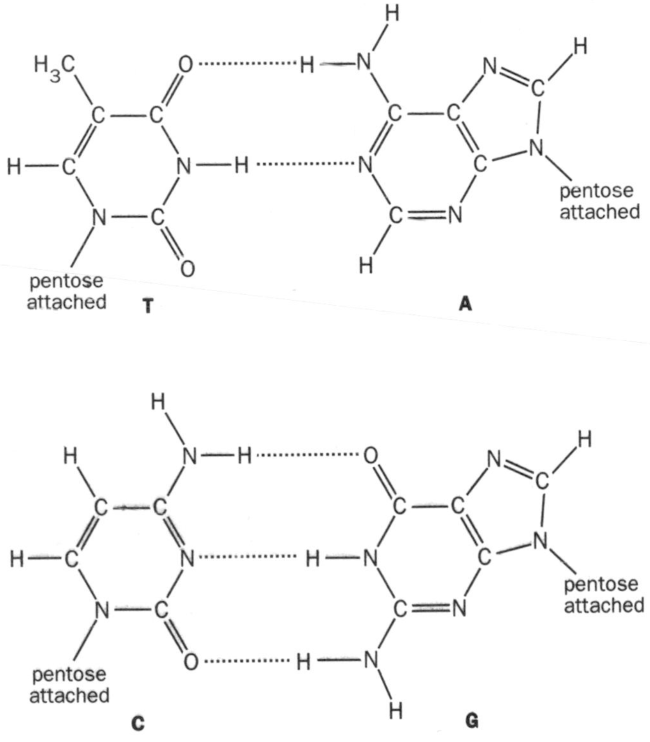

Fig. 3.21 Base pairs in DNA.

DNA acts as an information store. The base sequence codes for the amino acid sequences of every polypeptide that the cell or body can make, plus sequences involved in controlling how this information is released. DNA is the stuff that genes are made of. RNAs have a number of roles, all connected with the flow of information from DNA to polypeptides (Unit 7).

RNA is normally single-stranded, but this does not mean that it cannot form base pairs. Some RNAs form complex three-dimensional shapes in order to do their jobs (Box 7.3). They are held in shape by base pairings between different parts of the same strand. G pairs with C, A with U. Double-stranded RNAs exist, as the genetic material of some viruses (Box 29.2).

(a)

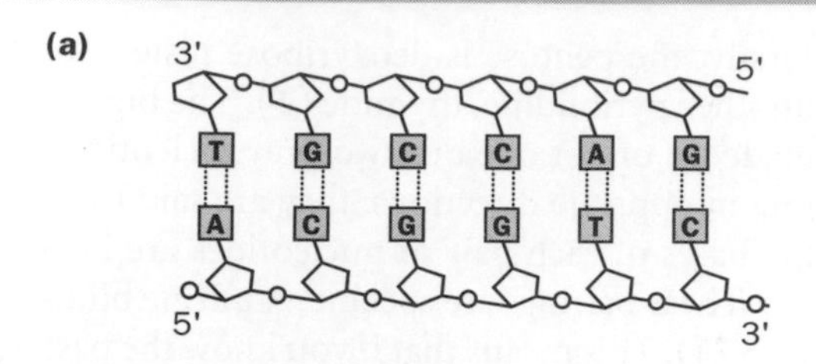

(b)

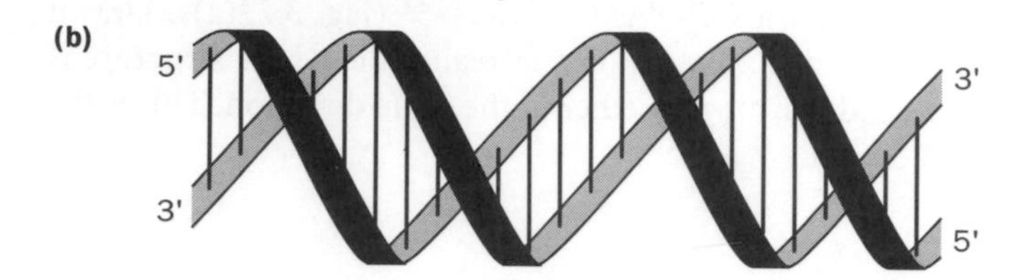

Fig. 3.22 DNA structure. (a) two antiparallel strands linked by hydrogen bonds, (b) the paired strands are twisted into a double helix.

3.9 Simple organic acids

The groups of molecules discussed so far are mostly rather long lived. They have important roles in their own right. Several simple carboxylic acids are abundant in cells as metabolites: steps in the synthesis, breakdown and interconversion of molecules. At the pH values normally found in cells, most of these acids are predominantly in the ionic form, having lost H^+. For example, most of the pyruvic acid in the cell will be in the form of pyruvate. The similarities between this 3 carbon acid and both triose sugars and some amino acids are shown in Fig. 3.23. Fig. 3.24 shows some more examples of important organic acids.

Dihydroxyacetone:
a triose

Pyruvic acid:
a keto acid

Alanine:
an amino acid

Fig. 3.23 A comparison of sugars, keto acids and amino acids.

Succinate	Fumarate	Malate	Oxaloacetate
COO^-	COO^-	COO^-	COO^-
CH_2	CH	$HCOH$	$C{=}O$
CH_2	$\|\| \ CH$	$HCOH$	CH_2
COO^-	COO^-	COO^-	COO^-

Citrate	α-ketoglutarate
COO^-	COO^-
CH_2	CH_2
$HOC—COO^-$	CH_2
CH_2	$C{=}O$
COO^-	COO^-

Fig. 3.24 Some important organic acids.

3.10 The watery environment

Cells contain a good deal of water, typically about 70% of their total mass. Most cells die if they lose much of this. A very few cells, particularly in seeds and in some mosses, can tolerate almost complete drying out. Even these, however, can do nothing until water returns. While desiccated, they are dormant. Cells are usually surrounded by water. Most single-celled organisms spend their lives in the sea, ponds, puddles, liquid films, animals' bodies, the soil or other moist places. Animal cells are bathed in fluid; cell walls act as stiff wet sponges around plant cells, even if the organism lives in a desert or on a cliff face.

Three things make water so important. Firstly, it is liquid at the temperatures of

The chemical reactions of life stop when cells freeze. Only when water is liquid can dissolved molecules diffuse and react.

much of the Earth's surface; secondly, water molecules are polar; thirdly, there is a lot of it on Earth. As a liquid, it can act as a solvent in which molecules can move around. The chemical reactions of life cannot take place if molecules stay still. Being polar means that other polar or charged molecules can dissolve in it. Most large biological molecules are polar. They fall into their correct three-dimensional shapes, held by hydrogen bonds, when water is present. Phospholipids form bilayers as a result of electrical interactions with the solvent, so membranes could not form without water.

Life almost certainly began in water, and living things carry water with them wherever they go, in and around their cells. Without water, the essential things of life stop working. Where there's life, there's water.

Summary

- Molecules are groups of atoms linked by bonds.
- Covalent bonds are strong and hold atoms together in molecules.
- Charged and polar molecules can form weaker ionic and hydrogen bonds.
- Charged and polar molecules attract water molecules: they are hydrophilic. Non-polar molecules are hydrophobic: they repel water.
- Most biological molecules are organic molecules, based on a skeleton of carbon atoms.
- The carbohydrates include sugars (monosaccharides and disaccharides) and polysaccharides: long chains of monosaccharides.
- Some polysaccharides are energy stores; others form strong fibres.
- Amino acids are linked to form unbranched chains: polypeptides.
- One or more polypeptides are held in a particular three-dimensional shape to form a protein.
- Proteins have very many roles, notably as enzymes and in recognizing other molecules.
- Lipids are hydrophobic molecules. Some are energy stores. Others form the basis of biological membranes.
- Nucleotides are linked to form polynucleotides, DNA and RNA.
- DNA is an information store. RNA has roles in the expression of this information.
- Water is present in and around cells. It provides an environment in which these other molecules can carry out their functions.

Exercises

3.1. Identify the functional groups in this molecule.

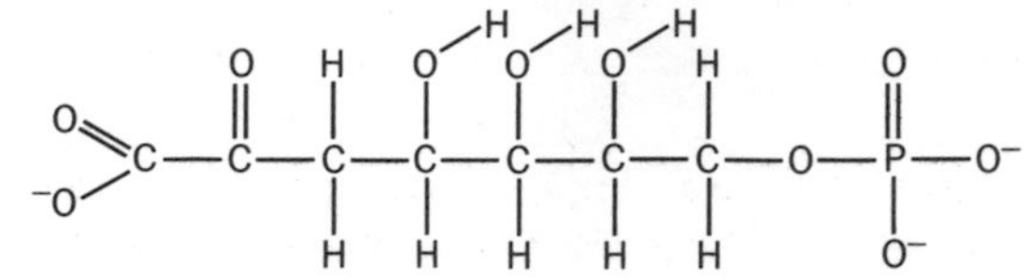

```
                       H    H    H
                      /    /    /
          O    H    O    O    O    H         O
          ‖    |    |    |    |    |         ‖
  O≈C —— C —— C —— C —— C —— C —— C —— O —— P —— O⁻
 ⁻O/           |    |    |    |    |         |
               H    H    H    H    H         O⁻
```

3.2.

(i) Which group does this molecule belong to? Use the information in this unit to name it (you might not carry that much detail in your head). What is its biological significance?

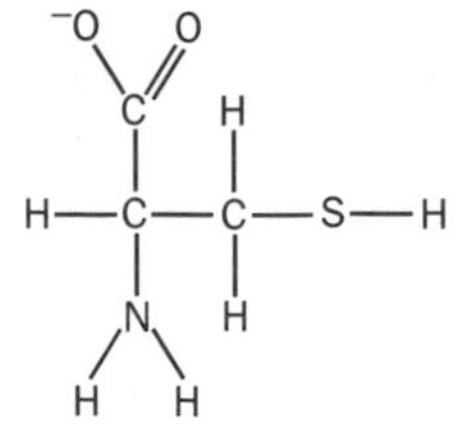

(ii) The molecule in part (i) is drawn in the form it would take at either very high or very low pH. Which?

3.3. Which group does this molecule belong to? Use the correct terms to describe important features of its structure.

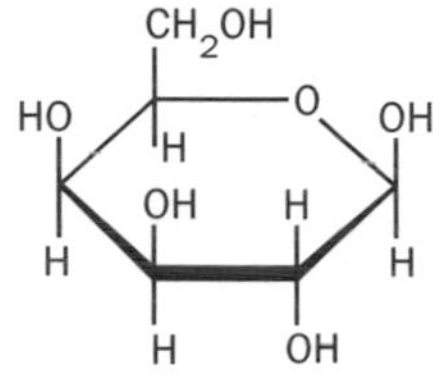

3.4. Which group does this molecule belong to? Be as precise as possible.

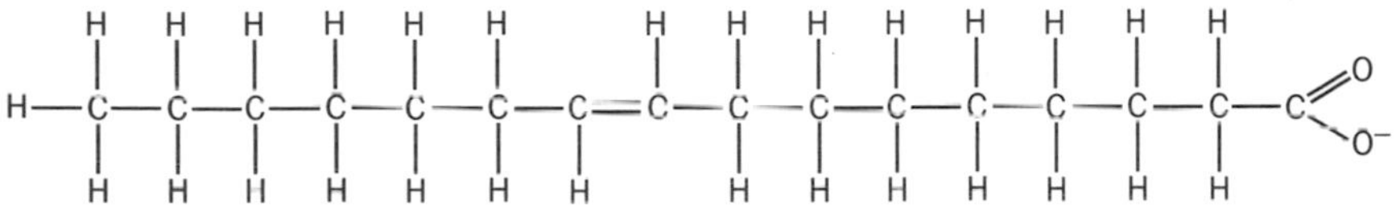

3.5. In which of these are hydrogen bonds important within the structure of the molecule? Amylose, glucose, oleic acid (a fatty acid), DNA, catalase (an enzyme), testosterone (a steroid), cellulose.

3.6. Complete this table comparing DNA and RNA.

	DNA	**RNA**
Which pentose?		
Which bases?		
Single- or double-stranded?		
Roles in cells		

Further reading

Lewis, R and Evans, W. *Chemistry* (Basingstoke: Macmillan Press, 1997). If your understanding of basic chemistry is shaky, this general introduction should help.

Smith, C.A. and Wood, E. *Biological Molecules* (London: Chapman & Hall, 1991). A clear introduction to biochemistry.

Stryer, L. *Biochemistry* (4th ed.)(New York: Freeman, 1995). The classic big undergraduate textbook of biochemistry. There are several broadly similar books on the market, but I would still go for this if looking for clear explanation.

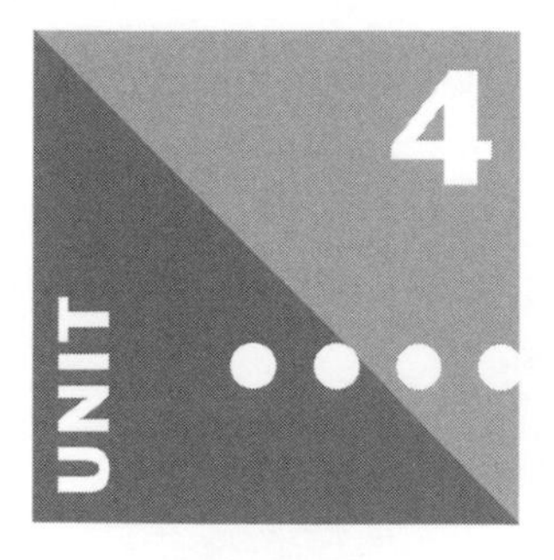

Cell Structure and Function

Contents

Connections

► Organisms are made up of cells, so this unit is an essential introduction to the rest of the book. You should already have a basic knowledge of biological molecules (Unit 3).

4.1 No membrane, no cell

Chemical reactions are what make living things tick. When we try to list the physical things that make life distinctive (reproduction, growth, feeding and respiration, the ability to respond to things in the environment, and so on) we are listing processes to which special molecules and very particular reactions are central. Life is about special chemistry, not completely perhaps, but certainly in part.

Many types of molecules are found only in association with life: proteins, polysaccharides and nucleic acids, for example. Molecules diffuse, and this tends to spread them out more and more evenly. If life is to exist, these special molecules must be imprisoned together in places where the reactions of life can occur. Enzymes, the protein catalysts which both make possible and control most of these reactions, are also trapped here. These places of imprisonment are called cells, and they are enclosed by **plasma membranes**.

Cells influence the chemistry of their environments. Sometimes chemicals are secreted which form permanent structures outside the membrane, for example the cell walls of plants and fungi and the tough material between cells in cartilage.

Some cells live on their own, many bacteria for example. Others live with other cells in bodies, formed by division of a single initial cell. This communal living allows for at least some division of labour. In all cases, however, it is the membrane which allows the distinctive chemistry of the cell to develop. Viruses are not made up of cells, and lack membranes: consequently they also lack most of this special chemistry. They are, however, no more than piratical nucleic acid molecules which have none of the characteristics of life until they find themselves inside a cell, when they can subvert its chemical machinery to their own ends. All truly living things are made up of one or more cells, and so have membranes.

Plasma membrane: The membrane surrounding a cell.

4.2 Membranes

Membranes are impermeable to almost all molecules soluble in water. This allows them to keep the inside of the cell chemically distinct from the outside, and also to enclose distinctive compartments, **organelles**, within the cell. Certain molecules must be imported and exported: more or less specific carrier molecules embedded in the membrane allow this. The membrane acts, then, as a highly selective molecular filter.

Plasma membranes and membranes around organelles share a common structure. They are very thin: **electron micrographs** show a thickness of rather less than 10 nm. They are flexible and have little mechanical strength. The main components of membranes are proteins and lipids, particularly phospholipids.

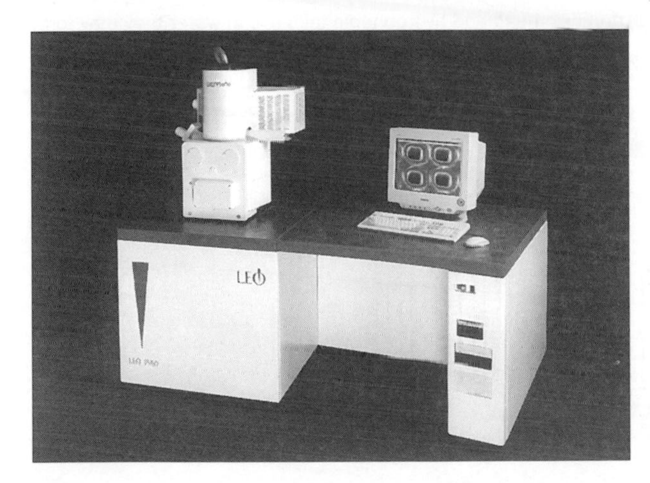

Ordinary microscopes cannot make images of objects or details smaller than 0.5 μm, because they are too small to interact with light waves. Electron microscopes, like this one, can resolve much smaller features, because electrons have a shorter wavelength. Many of the compartments of the cell were discovered in this way.

What happens to pure phospholipids in water gives a clue to the structure of the membrane. The fatty acid chains of a phospholipid are non-polar, so do not attract water molecules: they are hydrophobic. The phospholipid head, however, is charged and interacts with water: it is hydrophilic (Fig. 4.1(a)). In water, the molecules arrange themselves into spheres called micelles (Fig. 4.1(b)) or bilayers (Fig. 4.1(c)). In each case, the fatty acid chains are packed together, excluding water; the heads are in contact with the water and with one another. It seems that a phospholipid bilayer forms the basis of all membranes. Its job is to prevent polar molecules from crossing: the membrane's hydrophobic heart will repel them. Since membranes exist in a watery environment, and as a rule, molecules soluble in water are polar, almost everything is excluded.

The proteins of the membrane are linked to the bilayer in two main ways (Fig. 4.2). Some span the membrane and have regions in both the internal and external environments. Others are attached to one surface of the bilayer. Proteins that span the membrane must have a hydrophobic central region if they are to stay put. This is generally achieved by an α-helix with non-polar amino acid side chains bristling off around it. Hydrogen bonding between the polar parts of the polypeptide inside the helix maximises stability. Sometimes a single α-helix spans the membrane, sometimes there are several (Fig. 4.3). Proteins which are attached to the membrane surface usually have a fatty acid joined to them, which slots into the bilayer; some-

Organelle: Compartment or other structure inside a cell.

Electron micrograph: An image made using an electron microscope, which can resolve much smaller structures than the light microscope.

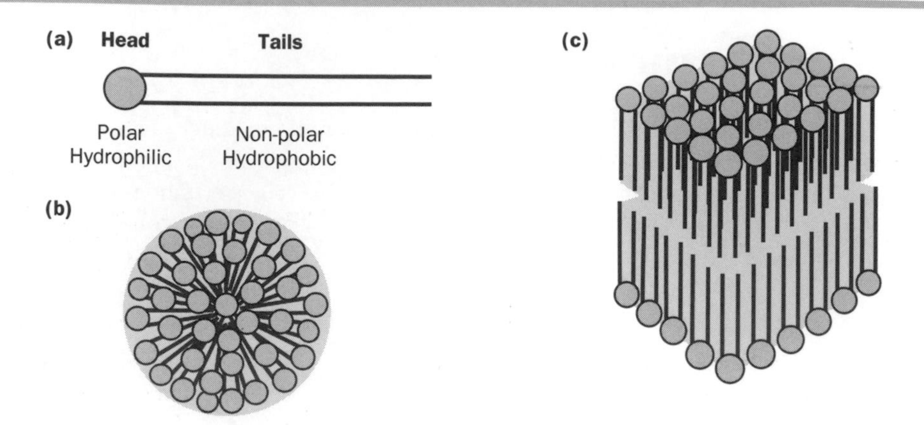

Fig. 4.1 Phospholipids have a hydrophilic head and hydrophobic tails (a). In water they may form micelles (b) or bilayers (c).

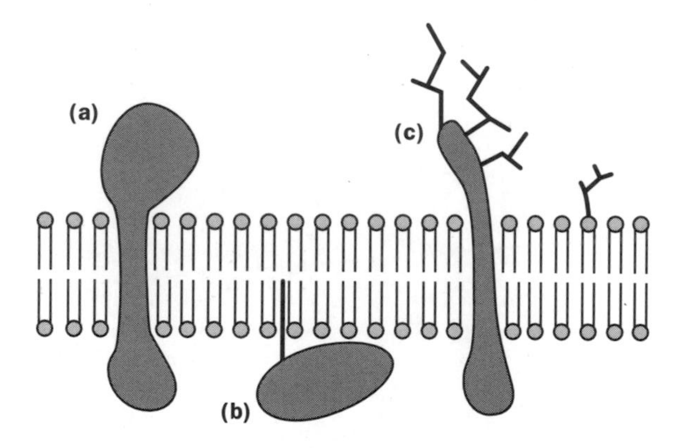

Fig. 4.2 Membrane proteins. (a) a transmembrane protein, (b) a protein anchored to the inner surface by a fatty acid 'anchor', (c) a transmembrane glycoprotein. A glycolipid is also shown.

times an entire phospholipid is attached. Proteins seem never to be embedded in a membrane without spanning its full thickness, probably because this would involve an interruption to the α-helix at the bend, exposing polar groups to a hydrophobic environment.

Many transmembrane proteins act as carriers for specific molecules (Section 6.5): if the lipid bilayer says no, proteins say a qualified yes. In this way, cells and cell compartments can regulate their contents. Other membrane-bound proteins are enzymes. Still others act as receptors: for example, mammalian cells which respond to the hormone insulin have specific receptor proteins in their plasma membranes. These receptor proteins tend to have oligosaccharides or polysaccharides attached to them on the outer side: they are glycoproteins. There are many plasma membrane glycoproteins, and some glycolipids as well. We still know little about most of them: what evidence there is suggests that many may have roles in cell communication and recognition.

The plasma membrane is not a rigid array of molecules. Phospholipids move freely within their own half of the membrane, but 'flip-flop' from one half to the other only rarely, since the head would have to cross the hydrophobic core.

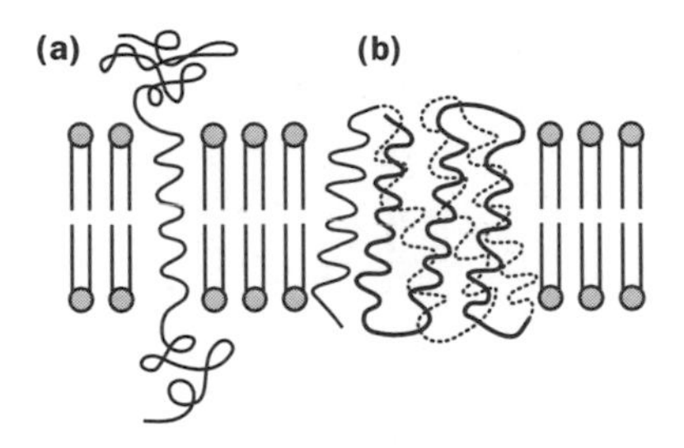

Fig. 4.3 α helices span membranes. Some transmembrane proteins have just one (a), others have several (b), seven in this case.

Individual molecules move in one plane within what remains a regular structure: the bilayer is a liquid crystal. Proteins too can diffuse within the plane of the membrane: 'protein icebergs floating in a sea of lipid' is the classic metaphor. There is no evidence that proteins can ever flip-flop: this means that the two surfaces of the membrane can remain distinct. This modern view of membrane structure is known as the fluid mosaic model. The fluidity of the membrane is enormously important. Bends in the membrane do not become fractures. Proteins released into the membrane at one point can diffuse all over the surface of the cell or organelle.

Just like a liquid, a bilayer may freeze if it gets too cold: the phospholipids cease to diffuse once their motion is not enough to overcome the attraction between adjacent molecules. Its regular structure may also be compromised if it becomes too fluid when hot. The types of lipid in the bilayer affect its fluidity. Unsaturated fatty acid chains (Section 3.7) have kinks in them, which makes it harder for adjacent chains to interact, making the freezing point lower. Steroids such as cholesterol (Section 3.7) are found in membranes and also affect fluidity. Cholesterol binds to phospholipids in a very particular way. It becomes linked to a fatty acid chain, near the head. This stiffens part of the chain, somewhat reducing fluidity. However, in doing so, it prevents adjacent chains coming close together, making freezing more difficult. Cholesterol, then, has an important role in moderating membrane fluidity.

All cells have a plasma membrane. Only some, however, have much in the way of internal membranes. This is just one example of the differences between the two major types of cell: prokaryotic and eukaryotic.

4.3 Prokaryotes and eukaryotes

Microscopic fossils in some of the world's oldest rocks suggest that cells have been around for at least 3500 million years. For at least the first half of that time, all these cells seem to have had the simple structure of present-day bacteria and their relatives: kingdom Prokaryotae. They were prokaryotic cells with a plasma membrane, a type of cell wall, sometimes flagella, but with very little discernible internal structure (Fig. 4.4). Certainly, the cytoplasm contains few or no membranes in prokaryotes. The larger and more complex eukaryotic cells are of more recent origin: the eukaryotes include animals, plants, fungi and other superficially simpler organisms. Some eukaryotic cells have walls, but they are chemically very different to those of prokaryotes. Inside the plasma membrane are many membrane-bound organelles (Fig. 4.5). DNA is confined within the nucleus. The mitochondria are the site of most reactions in aerobic respiration; chloroplasts, the site of photosynthesis in eukaryotes, are members of a wider organelle family, the plastids. The cytoplasm is criss-crossed by a network of

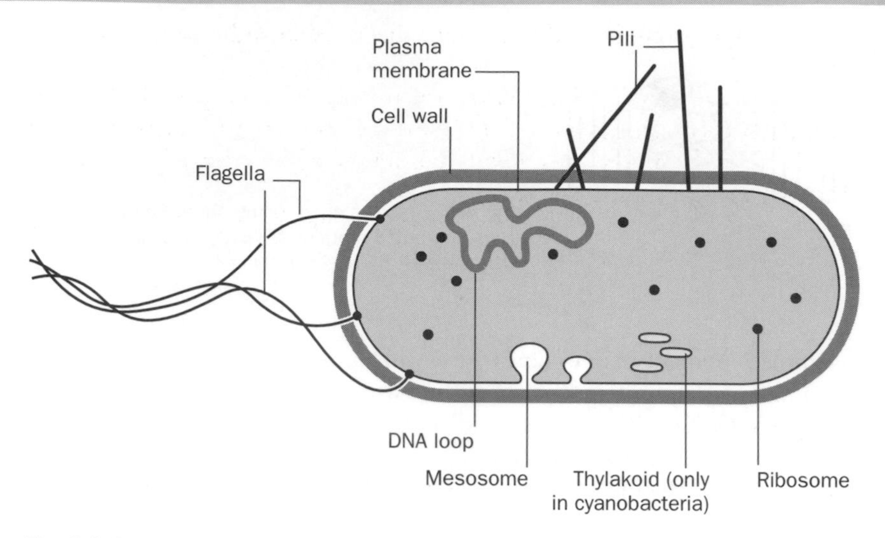

Fig. 4.4 Structure of a generalized bacterial cell.

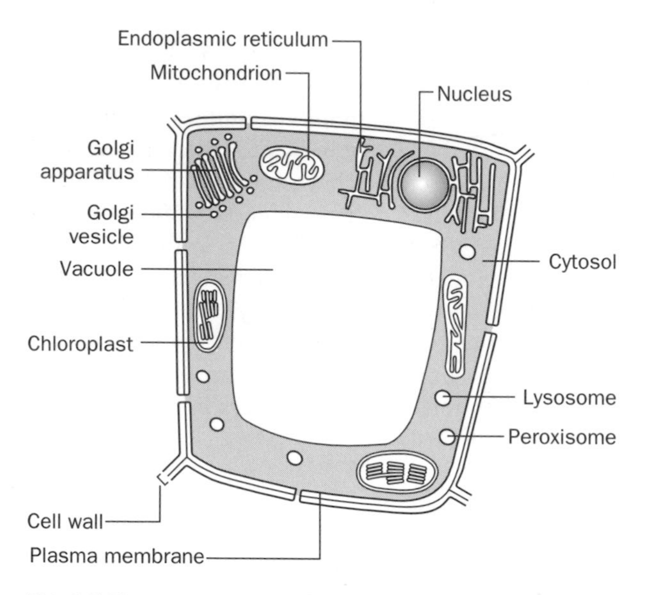

Fig. 4.5 The compartments of the eukaryotic cell, illustrated by a highly diagrammatic plant cell.

membrane pipes and sheets, the endoplasmic reticulum: here, lipids and some proteins are made and transported. Stacks of flattened membrane bags, the Golgi apparatus, process and package many types of molecule to be sent out of the cell or to other organelles. Other small membrane-bound sacs include lysosomes, containing digestive enzymes and peroxisomes, whose enzymes allow them to make and destroy hydrogen peroxide. Plant cells contain large vacuoles: membrane-bound water bags. Their job is essentially to fill space cheaply and to provide a place for storing assorted molecules. The cytoplasm of eukaryotes also contains several types of protein fibre. Together they form the cytoskeleton, a three-dimensional network which gives the cell its shape and is involved in cell movements. With all this going on in the cytoplasm, the term cytosol is used to describe the solution within which all these things are arranged.

One of the few structures found in both prokaryotes and eukaryotes is the ribosome. These precisely organized groups of RNA and protein molecules are the site of protein synthesis. Interestingly, eukaryotic ribosomes are more dense, a useful clue when investigating the evolution of the eukaryotes.

Do not be fooled into thinking that the prokaryotes are dull! They may be structurally simple, but their chemistry is excitingly diverse.

4.4 Prokaryotic cells

The outside story

Almost all prokaryotic cells have a strong cell wall outside the plasma membrane. Its main job is to stop the cell exploding. Most prokaryotes live in environments where the solution outside the cell is less concentrated than the cytoplasm: they will tend to gain water by osmosis (Section 6.2). Prokaryotic cells have no way of getting rid of excess water, so they swell up and would eventually burst, were it not for the strong cell wall which scarcely expands as pressure builds up inside. The strength of the wall is provided by peptidoglycans, hybrid molecules in which polysaccharide chains are linked to one another by short polypeptide chains (Fig. 4.6). The result is an enormous mesh-like molecule which forms the entire wall. Box 4.1 compares the structures of two types of bacterial cell wall.

Only a few prokaryotes have no cell wall, and they live in solutions of the same concentration as their cytoplasm, where osmosis is not a problem. The tiny mycoplasmas are parasites: many species live in the layers of moisture on the mucous membranes of mammals. Their cells are spherical since there is no wall or cytoskeleton to hold them in shape.

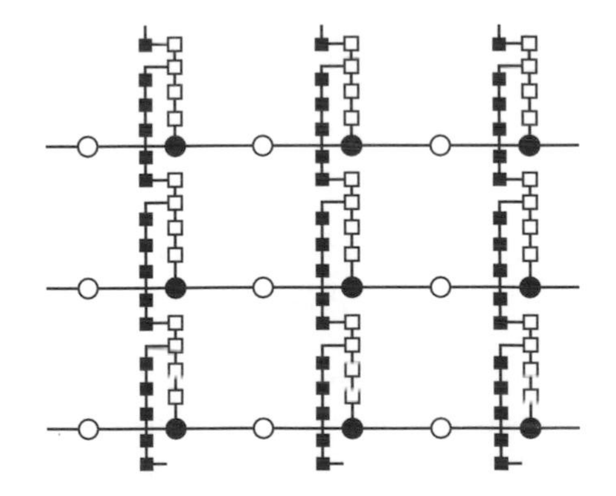

Polysaccharide chains are made up of two different derivatives of amino sugars

● *N*-acetylglucosamine

○ *N*-acetylmuramate

One type of peptide is made up of glycine, another of different amino acids

■ Glycine

□ Other amino acids

Fig. 4.6 A small portion of the bacterial cell wall peptidoglycan.

Box 4.1

Two types of bacterial wall

Bacteria are often classified as either Gram positive or Gram negative. They respond to a particular staining technique in different ways. This reflects a fundamental difference in wall structure. The walls of Gram positive bacteria are basically a rather thick (20–80 nm) peptidoglycan layer. Gram negative bacteria have a very thin (2–3 nm) peptidoglycan layer outside the plasma membrane, then a second membrane around this. The outer membrane has protein pores with large channels. These allow molecules up to the size of sugars or short peptides to pass through: it is still the plasma membrane which has real control over what enters and leaves the cell.

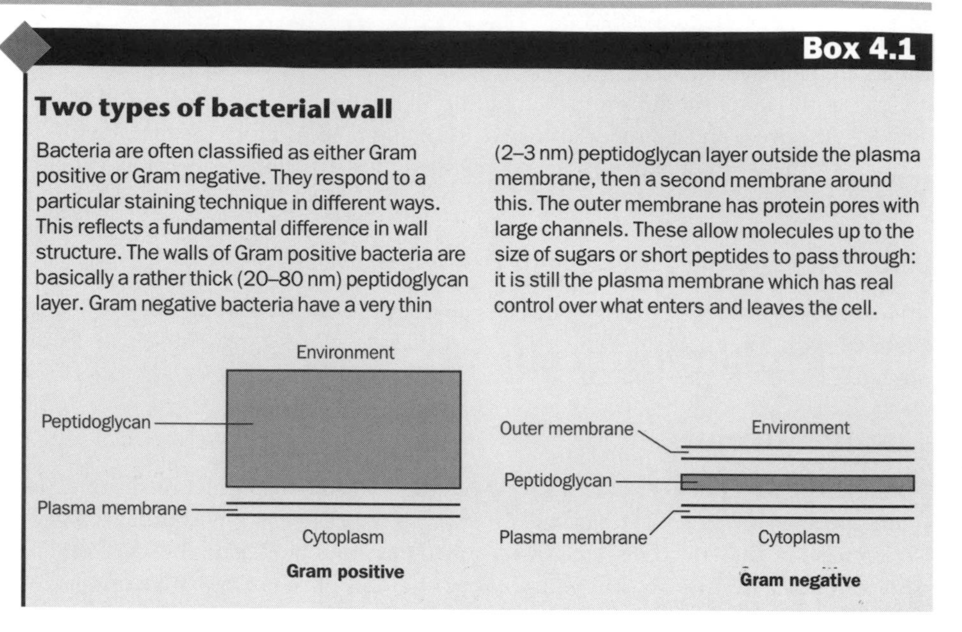

Many bacteria have a capsule or slime layer outside the cell wall. Capsules are polysaccharide gels with little mechanical strength. They cannot be essential for life since some species never have them, and capsule-free mutants of species which do have them can survive in culture. Their function may be to deter predators.

Many bacteria have fine protein strands which trail out into the environment. They are of two quite different types: flagella and pili. Both types are constructed from more or less globular proteins arranged as a hollow spiral. Flagella allow the cell to swim (Section 24.5). They rotate, driven by a protein 'motor' embedded in the plasma membrane. The energy required is supplied by a proton gradient (Section 8.5) across the inner membrane. Flagella have a diameter of 12–18 nm; pili are more variable, between 4 and 35 nm across. Pili cannot move actively. Their role is to stick bacteria together, either when forming film-like communities or for reproduction (Section 11.4).

The inside story

The cytoplasm of prokaryotic cells may not have much in the way of internal membranes, but it is not structureless. The single closed loop of DNA is not enclosed within a nucleus, but stains show that the DNA and its associated proteins are found in a restricted region of the cell. The DNA becomes attached to specific sites in the plasma membrane during cell division, but probably not the rest of the time.

Ribosomes are found in the cytoplasm. They are highly organised groups of proteins and RNA molecules. They provide sites where proteins can be made using the information brought from the genes by messenger RNA (mRNA) (Section 7.3). Each ribosome is made up of two subunits (Fig. 4.7) which only come together when they bind a strand of mRNA. Prokaryotic ribosomes are slightly smaller than those in eukaryotic cells, but they seem to work in a similar way.

The cyanobacteria, a group of photosynthetic prokaryotes, contain flattened membrane bags called thylakoids, much like those found in eukaryotic chloroplasts. Chlorophyll and the other molecules of the photosynthetic machinery are

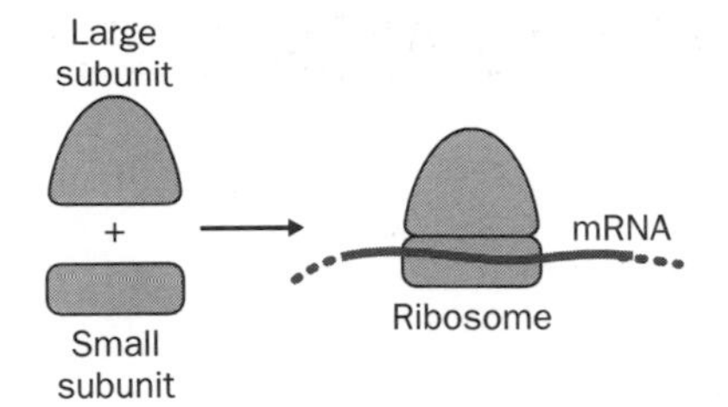

Fig. 4.7 The two subunits of ribosomes come together on a strand of mRNA.

embedded in the thylakoid membrane. In other photosynthetic bacteria they are embedded in the plasma membrane, sometimes in infolded regions called mesosomes. Yet other species have mesosomes associated with respiratory proteins. The difference between a mesosome barely attached to the plasma membrane and a membrane-bound sac free in the cytoplasm is very slight. It is not difficult to imagine how internal membranes could have evolved, and it is in eukaryotic cells that internal membranes are most highly developed.

4.5 Inside the eukaryotic cell

The nucleus

The nucleus is a cage for DNA. Most other molecules can get in and out, but the DNA is stuck inside. Larger structures like microtubules are kept out. Perhaps this prevents the DNA becoming tangled around these other things.

The nucleus is surrounded by not one but two membranes: such double membranes are called envelopes. The inner and outer membranes are continuous. The reason for this can be seen in the structure of the pores which perforate the envelope (Fig. 4.8). A complex of proteins spans both membranes and the space between them, bringing the membranes together round the edge of the pore. Nuclear pores allow molecules up to the size of small proteins to pass freely through a water-filled channel somewhere in their structure. However, certain large proteins, including enzymes needed for copying DNA and for making RNA from a DNA template (Section 7.3) need to get through from the cytoplasm where they are made. The pore complexes seem to include protein carriers which recognize these enzymes and actively transport them into the nucleus (that is, an energy supply is needed).

DNA is not draped haphazardly through the nucleus. Each chromosome is a single DNA molecule with associated proteins; an open chain rather than a closed loop. If they got tangled up, there could be problems when the cell came to division, when the chromosomes 'condense' into short, thick, distinct rods. A few studies on

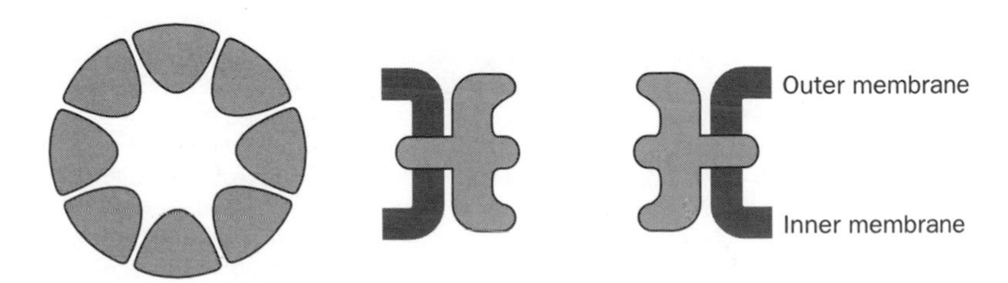

Fig. 4.8 A nuclear pore, viewed from above and in section (see text for details).

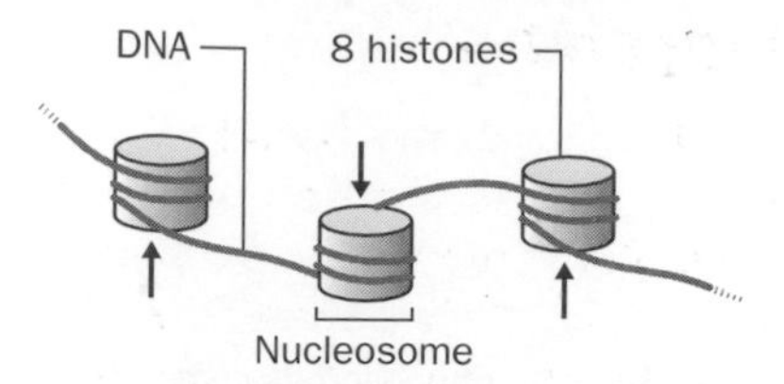

Fig. 4.9 DNA wraps around groups of histone proteins to form nucleosomes.

rather unusual cells have hinted that each chromosome occupies a distinct part of the nucleus when the cell is not dividing. The way in which DNA is effectively shortened by being wound around proteins is much better understood. Histones are the most common DNA-binding proteins and they bind DNA regardless of its sequence (most other DNA-binding proteins are quite specific about the sequence). DNA is wound around groups of histones to form structures called nucleosomes (Fig. 4.9). At the centre is a group of eight histones, two each of four types, H2A, H2B, H3 and H4. Nearly two turns of DNA (146 nucleotide pairs) are wrapped around it; a length of unbound DNA links adjacent nucleosomes. Another histone, H1, binds groups of nucleosomes together, giving even tighter packing. The arrangement is usually a spiral fibre. DNA can stay wound into nucleosomes during translation, when RNA is made from a DNA template, but higher levels of organization must break down first.

The nucleolus is a special part of the nucleus centred on a particular DNA sequence. It is a ribosome factory, and is big enough to be seen under the light microscope. At the centre is a series of genes coding for ribosomal RNA (rRNA) which forms part of ribosomes. The genes are identical copies, grouped in one chromosome. There are so many because, unlike most genes, the RNA is the final product so there is less scope for amplification than in situations where many protein copies can be made from a single RNA strand. The protein components of ribosomes are imported from the cytoplasm, through the nuclear pores. They join rRNA in the outer part of the nucleolus to form the large and small subunits of ribosomes. These are exported back to the cytoplasm through the pores. The nucleolus is, then, a mass of rRNA and ribosomal subunits in various stages of assembly. The faster the cell is making ribosomes, the larger the nucleolus.

Box 4.2 describes some quite different proteins which give the nucleus shape and strength.

Box 4.2

Structural proteins of the nucleus

The nuclear lamina is a protein layer 10–20 nm thick, just inside the envelope and supporting it. It is made of proteins called lamins, which form fibres linked to make a net. During cell division, the nuclear envelope and lamina break down, releasing the DNA from its cage. The lamins are given a phosphoryl group which prevents their linking, so the net breaks down. Another structure is more controversial. When the nucleus is very gently broken down, a three-dimensional mesh of proteins remains in the shape of the nucleus. This has been called the **nuclear matrix**. It appears to be a sort of scaffolding filling the nucleus, through which the DNA is draped: certainly, particular DNA sequences bind to these proteins. However, there is some doubt about whether the matrix exists in the living cell, or whether it is an artefact of the experimental procedure, made up of DNA-binding proteins which are not normally linked in this way.

Endoplasmic reticulum and Golgi apparatus

An interconnecting meshwork of membrane-bound tubes and sheets runs through the cytoplasm. This is the endoplasmic reticulum (ER). Under the electron microscope it is seen in haphazard sections. In three dimensions, however, the ER seems to enclose a single, highly stretched-out compartment. In most cells, most of the ER is studded with ribosomes: this is called rough ER. The function of rough ER is to separate out and transport newly-made proteins which are to be secreted by the cell or incorporated into its membranes. Proteins which are to do their job in the cytosol or the nucleus do not enter the ER. This sorting is possible because the proteins which are to go into the ER have a 'signal' sequence of amino acids at one end. A cunning molecular recognition system allows the ribosome to bind to the ER only if it is making a protein with a signal sequence at the beginning of the chain (Box 7.3). Some ER, called smooth ER, lacks bound ribosomes. It is seen only in some types of cell, and is involved in lipid metabolism. Mammalian liver cells, for example, have a lot of smooth ER which is used to construct **lipoprotein** particles (lipids on their own are not sufficiently hydrophilic to be transported in the blood).

Another small network of tubes which may well be connected to the ER is found in the cytoplasm. This is the calcium sequestering compartment, which actively takes up Ca^{2+}, keeping it at a safely low level in the cytosol. It can release calcium ions rapidly as a 'second messenger', part of the cell's response to external signals (Section 16.4). It is very well developed in muscle cells, where it is called the sarcoplasmic reticulum. Calcium ions are released, triggering contraction, when a nerve impulse reaches the cell (Box 24.5).

The Golgi apparatus consists of round, flat, membrane-bound discs called cisternae. The structure is rather like a pile of hollow dinner plates. These stacks are found in the cytoplasm, usually near the nucleus, and are intimately linked with both the ER and the vesicles (little membrane bags) which carry materials out to the cell surface. Proteins and lipids enter the Golgi stack from the ER at one end. They pass from one cisterna to the next, probably in vesicles which bud off from the first and fuse with the next (Fig. 4.10). As they move through the stack, various modifications occur. The cisternae hold the enzymes needed to build many types of carbohydrate chain. The short, branched chains of glycoproteins and glycolipids are added here, before export to the membranes where they will do their jobs. The long, straight chains bristling from proteoglycans are also constructed here, before vesicles carry them out of the cell to form part of a basal lamina (Section 4.6), the matrix of connective tissue, or as part of a mucus secretion. The large Golgi stacks of plant cells are used to construct the polysaccharide chains of the ground substance: only the cellulose fibres are built up outside the plasma membrane. Material passes through the Golgi bound for other organelles, such as lysosomes, as well as to the plasma membrane and beyond. Some fairly complex sorting must go on here. Every stage of the process involves different sorts of vesicles: how these get to their correct destinations is only beginning to be understood.

Mitochondria

Mitochondria (Fig. 4.11) are strange and important organelles. Although they have a cosy relationship with the rest of the cell, based upon mutual dependence, mitochondria do a lot for themselves. They have their own DNA, multiple copies of a

Lipoprotein: A cluster of lipid molecules bound to a protein.

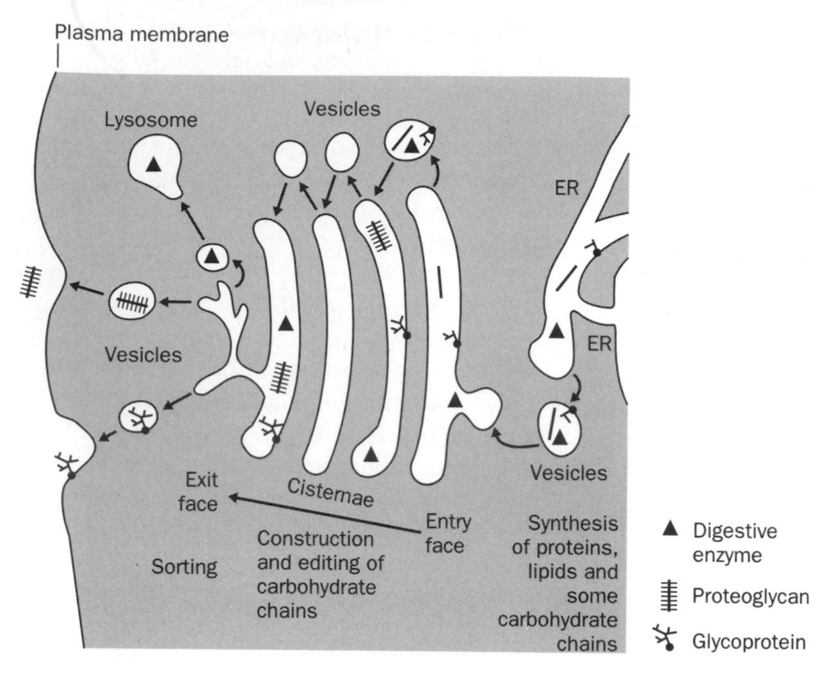

Fig. 4.10 The role of the Golgi apparatus (see text for details).

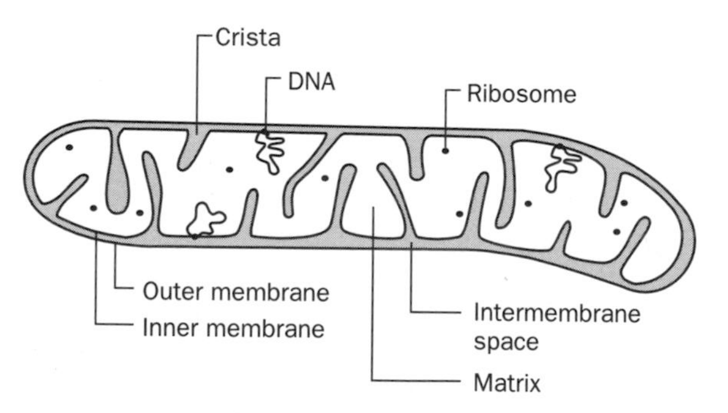

Fig. 4.11 Mitochondrion in longitudinal section.

single closed loop, like a much smaller version of the prokaryotic chromosome. They also have their own ribosomes, which are of the prokaryotic type, even though the rest of the cell has eukaryotic ribosomes. The mitochondrion is able to reproduce independently of the cell and can make some of its own proteins, although others are coded for in the nucleus and imported from the cytosol.

Mitochondria often appear to be roughly sausage-shaped in electron micrographs, although there is evidence that they are highly flexible and constantly change shape in the living cell. They are surrounded by two membranes. The outer membrane has large pores which allow molecules up to the size of small proteins to pass through. It is the inner membrane that controls what enters the mitochondrion (Box 9.3). The inner membrane is deeply folded, giving it an enormous surface area:

each finger-like fold is called a crista. The inner membrane is unusual in having a very large amount of the phospholipid cardiolipin. This makes it especially hard for ions to get through, important since energy released in respiration is transiently stored as a gradient of hydrogen ions (protons) across this membrane (Section 9.2: Oxidative phosphorylation).

The mitochondrion has two internal compartments, the intermembrane space and the compartment inside the inner membrane, the matrix. The intermembrane space is more or less continuous with the cytosol, but lacks enzymes, since the pores in the outer membrane are too small to let them through. The larger matrix is quite distinct, and contains the enzymes needed for the breakdown of fatty acids and, most importantly, the Krebs cycle (Section 9.2). The Krebs cycle is a set of reactions involved in oxidizing the carbon skeletons of all manner of energy-storing molecules. Much of the energy released ends up in the form of NADH (Section 8.5). The chemical machinery of oxidative phosphorylation (Section 9.2) then transfers this energy from NADH to ATP, the energy currency of the cell (Section 8.3). This machinery is found in the inner membrane, which explains the extravagant folding: NADH, made in the matrix, is never far from the place where it is needed.

The major role of the mitochondrion is, then, quite clear: it is an ATP factory.

Plastids

The plastids are a family of organelles found only in plants and plant-like protoctists. Like mitochondria, they have a double membrane, multiple copies of a loop of DNA which codes for some of their proteins, and their own ribosomes, again of the prokaryotic type.

Meristems are the places in a plant where cell division is rapid: shoot and root tips are good examples. The cells of meristems contain small plastids with few internal membranes. They reproduce in order to keep their numbers up as the cells divide. These are proplastids. As a cell differentiates, they develop in different ways according to the needs of the cell (Fig. 4.12). In photosynthetic cells they become chloroplasts (Fig. 4.13) which are the site of photosynthesis. Chloroplasts contain many disc-shaped membrane bags called thylakoids, often in stacks called grana. The thylakoid membranes contain the molecular machinery needed to harness light energy to make ATP and NADPH (Section 10.3). The surrounding solution, the

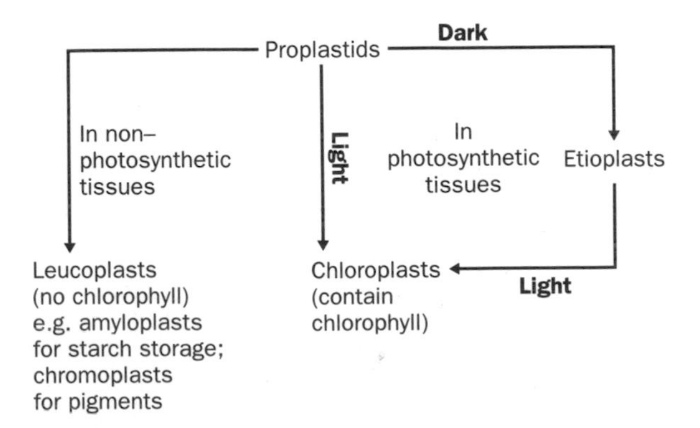

Fig. 4.12 Plastid development.

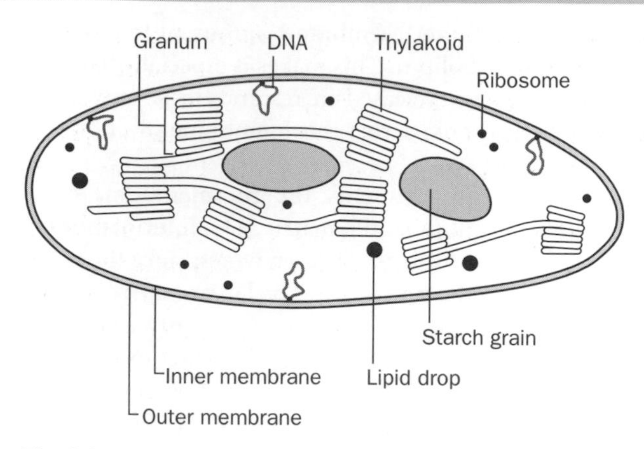

Fig. 4.13 Chloroplast in longitudinal section.

stroma, contains the enzymes needed for the reduction of CO_2 to carbohydrates using NADPH and ATP. Chloroplasts also act as starch stores, with discrete starch grains nestling between the thylakoids. In non-photosynthetic tissues, proplastids develop as chlorophyll-free leucoplasts. Often, these contain large starch grains, acting as a food reserve for the cell: these plastids are amyloplasts. When I eat a baked potato, the starch I digest comes entirely from large amyloplasts in the cells of the tubers. Other specialized plastids include pigment-containing chromoplasts: it is these that make tomato skins red, for example.

Photosynthesis and starch storage may be unique to plants, but plastids, including chloroplasts, have additional functions. Fatty acids and some amino acids are made in plastids, by pathways which occur in other compartments of animal cells.

Peroxisomes, lysosomes and vacuoles

Peroxisomes, lysosomes and vacuoles are surrounded by a single membrane, do not have their own DNA, and tend to appear as variable rounded shapes in electron micrographs. Here the similarities end: their chemistry and functions are very different.

Peroxisomes are centres for all sorts of oxidation reactions that do not take place in the mitochondria, as well as one or two that do happen there. Even though they cannot make their own proteins, new peroxisomes are made by old ones dividing, their proteins being imported from the cytoplasm.

Three crucial types of reaction go on here, all involving hydrogen peroxide (H_2O_2):

$$XH_2 + O_2 \xrightarrow{\text{specific enzymes}} X + H_2O_2$$

$$YH_2 + H_2O_2 \xrightarrow{\text{catalase}} Y + 2H_2O$$

$$2H_2O_2 \xrightarrow{\text{catalase}} 2H_2O + O_2$$

where X and Y represent various molecules that can be oxidized. A good example of the first type of reaction is the oxidation of glycollate to glyoxylate, which occurs in photorespiration, the damage limitation pathway in plants which recycles phosphoglycollate, a wasteful by-product of photosynthesis (Section 10.4). Mammals detoxify alcohol in liver cells, in two different ways. One is an example of the second type of peroxisome reaction. The third type of reaction is a means of disposing of any excess hydrogen peroxide generated by the first type.

In mammals, peroxisomes are especially common in liver and kidney cells, where detoxification is their main role. Two very distinct types are found in plants. Leaf peroxisomes are found next door to chloroplasts and are involved in photorespiration. Glyoxysomes are found in types of germinating seeds which store a lot of energy as lipids. They oxidize fatty acids to succinate, which can then be used to make carbohydrates.

Lysosomes are membrane sacs containing enzymes which catalyse hydrolysis reactions. The major groups of long-chain biological molecules are each built up by condensation reactions and broken down in hydrolysis reactions. These are, then, digestive enzymes. Lysosomes are variable in shape and contain a range of enzyme cocktails. They are a mixed bag of mixed bags. Their main role is digestion within the cell. Material from outside the cell which is to be digested is enclosed within a membrane sac which buds off the plasma membrane and eventually fuses with a lysosome. This exposes the material to enzymes which may include proteases, lipases, glycosidases and nucleases. *Amoeba* and similar protozoa feed like this; phagocytic white blood cells also destroy bacteria this way. Cells can also decommission unwanted organelles in a similar way.

Unlike peroxisomes, lysosomes do not divide by themselves, but are produced by the Golgi apparatus. The inside of a lysosome is quite acidic, at about pH 5. Its

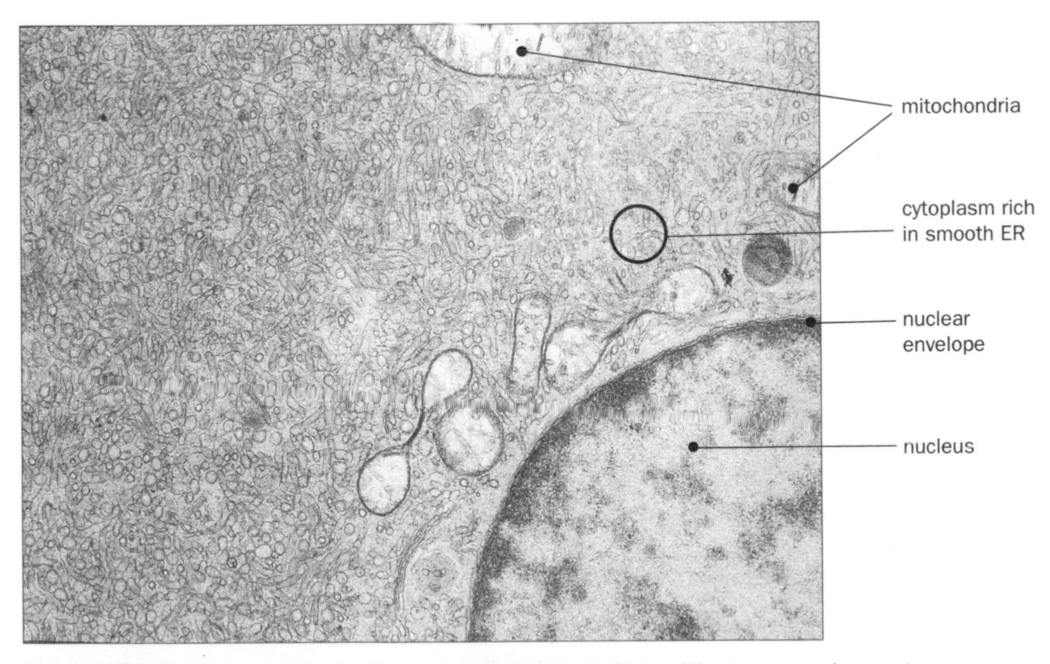

Part of an animal cell, in an electron micrograph. The cytoplasm of this cell has an unusual amount of smooth ER – but every cell type is special in some way.

enzymes work best around this pH, the more neutral cytosol being protected against vandalism by these destructive enzymes, should they accidentally leak out.

Plant and fungal cells contain very large membrane sacs called vacuoles. An adult plant cell generally has just one, which can fill up to 90% of the cell's volume. They have little metabolism of their own, making them an ideal attic for the cell: molecules can be put in there, out of harm's way, and recovered unchanged when required. Their lack of protein filaments, and low levels of enzymes, also make them a cheap way of filling space. They are often used for storing metabolites such as sucrose or malate (Section 10.7: Moist or dry) or as a mineral store. Water follows solutes into the vacuole by osmosis, so it plays a big part in maintaining cell turgor – the stiffness that results from having water at high pressure inside. Vacuoles also contain hydrolytic enzymes, involved in digesting the cell ready for recycling when, for example, a beech leaf dies in autumn. This has led to them sometimes being thought of as big lysosomes.

The cytoskeleton

Cytoplasm is not just a solution, it is criss-crossed by a three-dimensional network of protein fibres. Some are there simply as reinforcements for the animal cell's rather feeble structure: these are the intermediate filaments. The others, actin filaments and microtubules, are dynamic structures, capable of growing, breaking down and attaching to other components of the cell. They are involved in moving things around within the cell, and can push or pull the plasma membrane into shape. Just as carrier proteins in membranes are the machinery of molecular transport, the cytoskeleton forms the apparatus for the movement of organelles and of cells themselves.

Intermediate filaments are polymers of rod-like proteins, coiled around one another to form molecular cables about 10 nm in diameter and indefinitely long: their structure is much like a rope, made up of several coiled strings, each string itself being made up of many shorter, coiled fibres. Examples include the keratins which reinforce epithelial cells (and which are abundant in hair, hooves and nails, which are derived from epithelia), the lamins of the nuclear lamina, and the neurofilament proteins which strengthen the long, vulnerable axons of nerve cells.

Actin filaments are thinner and very flexible. They are also polymers, of the globular protein actin, arranged as a double helix. Actin filaments are concentrated just inside the plasma membrane, in the cell cortex, where they are linked by specific actin-binding proteins to form either gels or bundles. Their major role in many cell types is to maintain and change the shape of the cell. They can push it out, as in the microvilli of cells lining the small intestine; they can also pull it in, as happens when an animal cell divides (Fig. 4.14(a)). In some situations actin filaments are quite unstable and can rapidly lengthen or shorten. Extension of actin filaments is involved when a crawling white blood cell pushes forward the leading edge of the cell (Fig. 4.14(b)).

A group of actin-binding proteins called myosins uses energy from the hydrolysis of ATP to 'walk' along actin filaments. This is the basis of some types of cell movement, and in its most extreme development drives muscle contraction. In muscle cells, myosins are linked to form filaments which run parallel to the actin filaments:

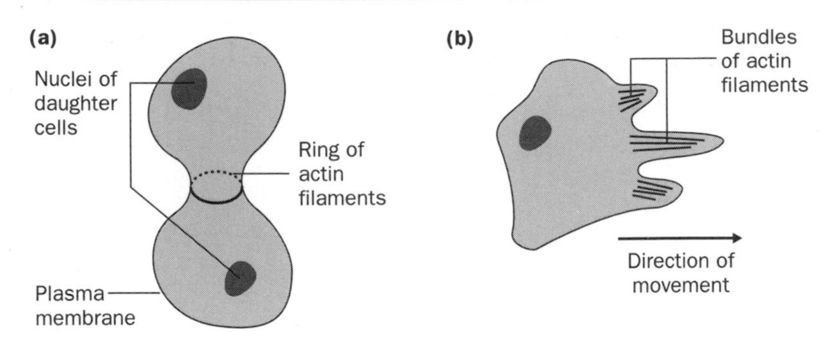

Fig. 4.14 Two roles of actin filaments. (a) in a dividing animal cell, (b) in a crawling white blood cell.

when the myosins start walking, the filaments slide along one another and the whole cell shortens (Section 24.5).

Microtubules are the thickest type of filament (external diameter 25 nm). They are hollow tubes with a shell made up of a spiralling layer of globular tubulin molecules. Microtubules have distinct ends. One end, the plus end, can grow quickly; the minus end grows much more slowly. In a typical cell, microtubules radiate out from a region near the centre of the cell, called the centrosome. This is the cell's most important microtubule organizing centre. The centrosomes of many cells contain centrioles, rather mysterious structures made up of highly ordered microtubules. They seem not to be essential for centrosome function: plant cells, for example, manage perfectly well without them. The plus end of each microtubule grows out from the centrosome, but older tubules wither away again. The result is a constantly changing starburst of microtubules, extending through the cell. These microtubules act as a framework on which organelles can be moved around the cell. Two families of motor proteins use energy from ATP to trundle along microtubules, although it is not known precisely how they do it. Dyneins move along from plus to minus ends, kinesins in the opposite direction. Organelles can be shifted when attached to motor proteins.

When a cell divides, either by mitosis or meiosis (Section 11.2), the microtubules are organized by not one but two centres, situated at opposite ends of the cell. Microtubules run the length of the cell between them, forming the spindle. This is a skeleton along which motor proteins can move chromosomes, separating them into sets for the daughter cells.

A quite different role of microtubules is in cell movement. Eukaryotic flagella and cilia are hair-like extensions of the plasma membrane, the cytoplasm within containing a highly organized group of microtubules running down the filament. Nine pairs of microtubules form a ring, with two more microtubules in the centre. Dyneins attached to each microtubule pair 'walk' along the adjacent pair, causing them to slide. Sliding is organized so that it results in a wave of bending passing down the flagellum – in other words, it beats. This mechanism is discussed in more detail in Box 24.4. The flagellum has its own microtubule-organizing centre, the basal body located in the cytoplasm at the base of the flagellum. Flagella and cilia are the same thing. They are called flagella if there is one or a few, as in sperm, acting as tails for swimming. If there are lots carpeting the cell surface, they are called cilia, as in ciliated protozoa and the epithelial cells lining the bronchi of a mammal.

4.6 Outside the eukaryotic cell

The plasma membrane defines the outer limit of the cell's special chemistry as far as soluble molecules go. Insoluble molecules can stick around, however. The cells of plants and fungi have walls around them. These more or less rigid boxes allow high pressure inside to make the cells stiff without bursting. Animal cells also secrete an extracellular matrix with more varied functions: unlike cell walls, it is only rigid in special situations. This difference has important consequences. Plants do not need a skeleton but cannot move very much. Animals are free to walk, swim and boogie, but require some sort of skeleton unless they are very small. The strength of most skeletons comes, again, from the extracellular matrix of special cells.

Cell walls and the extracellular matrix have one important feature in common: they are composite materials made up of fibres embedded in some ground substance. The fibres resist stretching; the ground substance resists compression and twisting. The same principle is used in artificial materials like reinforced concrete and fibreglass.

In plant cell walls the fibres are cellulose microfibrils made up of many, probably parallel cellulose molecules cross-linked by hydrogen bonds. A typical microfibril might be 5–8 nm in thickness, with tens rather than hundreds of cellulose molecules in any one cross-section. The ground substance includes other polysaccharides usually grouped into two rather diverse categories, hemicelluloses and pectins, as well as proteins and, sometimes, the giant mixed polymer, lignin.

The structure of the cell wall (Fig. 4.15) is best understood by following its development. It first appears as a cell plate which begins to separate the two ends of a dividing cell. Ground substances, especially pectins, are deposited first. They are produced in the Golgi and transported in vesicles which travel to the developing wall along microtubules. The vesicles fuse together, leaving a wall of ground substance with plasma membrane either side. The new plasma membrane started out as the

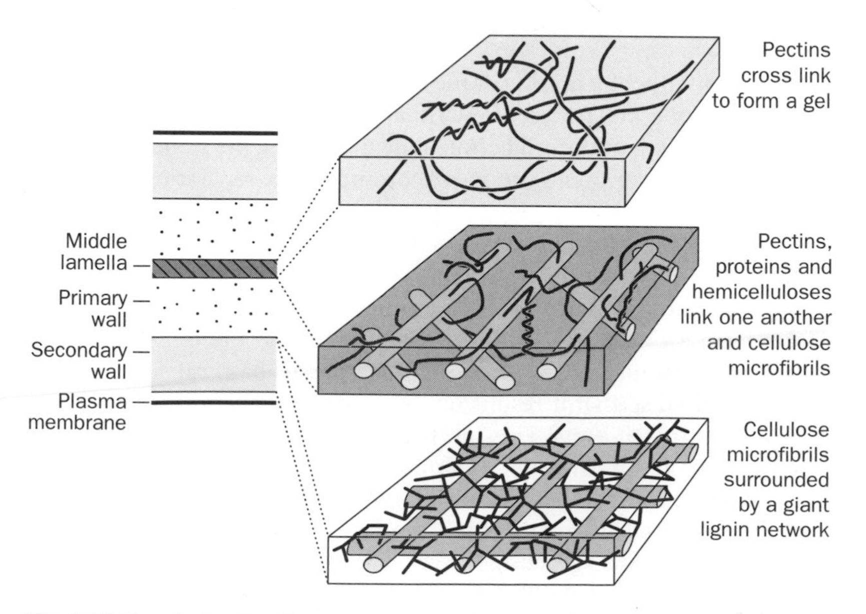

Fig. 4.15 The plant cell wall.

vesicle membranes. In the presence of calcium or magnesium ions, some pectins wrap around one another to form open meshes which trap water: in other words, they form gels. This gel is the thin central layer of the wall, the middle lamella. The rest of the wall will be built either side of it by the two newly distinct cells.

While the cells are still small and young, cellulose microfibrils are synthesized on the outer surface, probably by enzyme complexes embedded in the plasma membrane. Microfibrils are usually laid down parallel to one another, in layers. This gives the new wall strength in some directions but not others, and determines which way the cell can grow, since the wall must stretch if the cell inside is to get larger. Microtubules under the plasma membrane seem to control the direction of microfibril synthesis. How they do this is not known: perhaps they act as tracks along which cellulose synthase complexes move. Pectins, hemicelluloses and proteins continue to arrive in Golgi vesicles and are deposited around the microfibrils. Exactly how these components are arranged is poorly understood, but they certainly pack the spaces between the microfibrils and cross-link them. Both covalent and hydrogen bonds are involved. This composite material is the primary cell wall. Once the cell has stopped expanding, a secondary cell wall is sometimes added. This consists of several layers of microfibrils, typically three, each layer running in a different direction. Hemicelluloses and lignin are the most important components of the secondary wall ground substance. The more lignin that is deposited, the more waterproof the wall becomes. This is particularly important in xylem, the water-carrying tissue of plants.

Animal cells have much more varied extracellular matrices, but their components are selected from a reasonably short menu. Firstly there are structural proteins: collagen which can form either strong inelastic fibres or a tough mesh; elastin which forms stretchy fibres; and laminin, a cross-shaped protein which can form nets. Secondly there are glycosaminoglycans (GAGs), polysaccharides in which alternate sugar residues are amino sugars. GAGs form gels with a high content of trapped water, making them highly resistant to compression. Frequently they are bonded to proteins to form proteoglycans: the largest of these are tinsel-like molecules with GAGs bristling out from a central protein strand. Thirdly there is fibronectin, a protein which helps cells remain attached to the extracellular matrix. Fourthly, minerals may be present, notably hydroxyapatite, found in bone, whose crystals contain calcium, phosphate and hydroxide ions. Box 4.3 shows how different types of extracellular matrix can be made from this short menu.

The role of the basal lamina in keeping tissues distinct shows that it is not enough to think of the extracellular matrix as the dumb, passive product of cells. The cell controls the composition of the matrix, but in many cases the matrix influences the course of development of the cells themselves.

4.7 Connections between cells

Cells within a body have the same genes and cooperate with one another. In this situation, neighbours can become good friends. Several types of connections between cells are known, each of which helps the overall functioning of the tissue. In plants there is only one type: cytoplasmic tunnels through the wall called plasmodesmata. In mammals there are three groups of connections: gap junctions which allow some cytoplasmic contact, desmosomes and adherens junctions which anchor cells to one another, and tight junctions which seal the gap between adjacent cells.

Box 4.3

Extracellular matrix: variations on a theme

Fibrocartilage, found for example in the discs between vertebrae, is extremely tough. Like other connective tissues, it is made of large amounts of extracellular matrix with cells dispersed through it. The matrix is dominated by collagen fibres, within a ground substance largely made up of gigantic GAG proteoglycans.

Elastic cartilage, found for example in the pinna, the visible outer part of the mammalian ear is, predictably, more elastic. It contains more elastin and less collagen.

Bone develops from cartilage. Its important features are that it is enormously strong in compression and is more rigid than cartilage. Tough collagen fibres run through solid, crystalline hydroxyapatite.

Areolar connective tissue forms thin sheets around organs. Collagen and elastin fibres are present as a loose net. The GAG ground substance occupies much of the volume: great physical strength or elasticity are slightly less important here.

Synovial fluid, which fills a cartilage bag in many joints, such as the knee, is there for lubrication. It is a liquid extracellular matrix, dominated by hyaluronate, a GAG which is not linked to a protein and is particularly effective at binding water.

Epithelia are layers of cells which cover surfaces: the linings of the bronchus, the Bowman's capsules in the kidney, and the small intestine are a few examples. Epithelial cells secrete a thin layer of extracellular matrix between them and the connective tissue underlying them. This is the **basal lamina**, sometimes called the basement membrane. The carpet is largely made up of two protein meshes and GAG proteoglycans. The protein meshes are formed by laminin and by collagen in its less usual net form. Basal laminae give the cells molecular clues to where they are. Connective tissue cells do not cross the lamina and epithelial cells only divide so that new cells will be in contact with it. The unusually thick basal laminae in the glomeruli of the kidney act as a molecular sieve, controlling what is filtered out of the blood (Section 22.4). Large proteoglycans define the size of the pores in the sieve.

Plasmodesmata are tunnels through plant cell walls, typically 20–40 nm across (Fig. 4.16). They are lined by a membrane which is continuous with the plasma membrane of the cells on either side of the wall. Smaller molecules can diffuse through in the cytosol, keeping concentrations of ions and metabolites similar in adjacent cells. Almost all the cells in a plant are connected in this way, which might be useful in very local transport of metabolites, water or signalling molecules. Plasmodesmata usually have a tube of smooth endoplasmic reticulum running

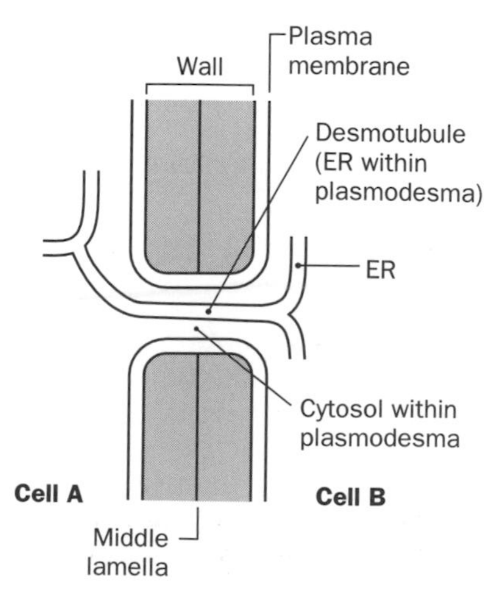

Fig. 4.16 A plasmodesma.

through them. This may or may not have a function, but is certainly a relic of the way in which plasmodesmata form. Cell-wall material is deposited around, but not inside any piece of ER which crosses the developing cell plate, leaving a hole which is perpetuated as the primary and secondary walls are laid down.

Gap junctions between animal cells have a similar function. Small areas of the plasma membrane of adjacent cells are less than 5 nm apart (Fig. 4.17). Groups of transmembrane proteins in each membrane form rings with an aqueous pore in the centre. Pairs of rings link up, resulting in pores running from one cell through to the other. Again, small molecules can pass from cell to cell, although there is some selectivity.

Adjacent cells sometimes need to be tied together to produce barriers which do not open up when they are bent, notably in epithelia. It is not enough for membrane proteins in each cell to link up. Membranes are weak and the junction could rip out, like a button sewn onto worn-out cloth. Instead, the cytoskeletons of the two cells must be linked. Transmembrane proteins bind one another on the outer surface of the membrane and elements of the cytoskeleton on the inside (Fig. 4.18). Desmosomes link intermediate filaments; adherens junctions link actin filaments. Similar junctions link intermediate filaments to the extracellular matrix.

Tight junctions involve quite large bands of the membranes of adjacent cells within an epithelium. They are linked at many points in the band, presumably by transmembrane proteins. The effect is a molecular draught-excluder, limiting the

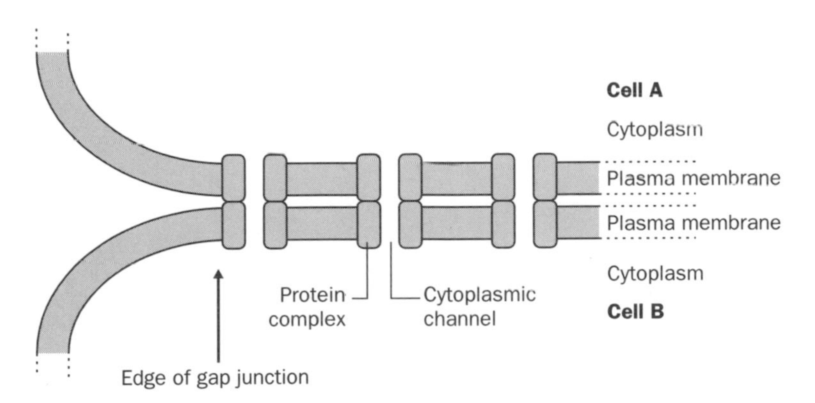

Fig. 4.17 A gap junction.

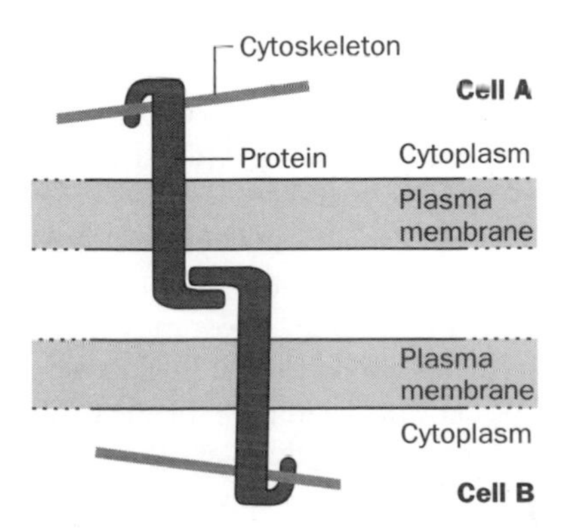

Fig. 4.18 A junction which holds two cells together (desmosome or adherens junction).

diffusion of molecules through the gap between the cells. This gives the cells much more control over what passes through the epithelium. This is of enormous importance whenever the fluids on either side of an epithelium need to be kept distinct. The epithelium of the small intestine, separating tissue fluid and gut contents, and the epithelium of kidney tubules controlling the distinctive composition of urine inside the tubules, are good examples of the need for tight junctions. All tight junctions exclude large molecules like proteins. When it comes to small molecules, however, some tight junctions are tighter than others.

Tight junctions also limit the diffusion of transmembrane proteins in the plane of the membrane. Generally, membrane proteins can diffuse around freely in the plasma membrane. In epithelial cells, however, they cannot cross tight junctions, so the plasma membrane is effectively split into two faces (Fig. 4.19). The cell can then put different transport proteins into the two faces without them mixing later, important if molecules are to be moved across the epithelium in a particular direction.

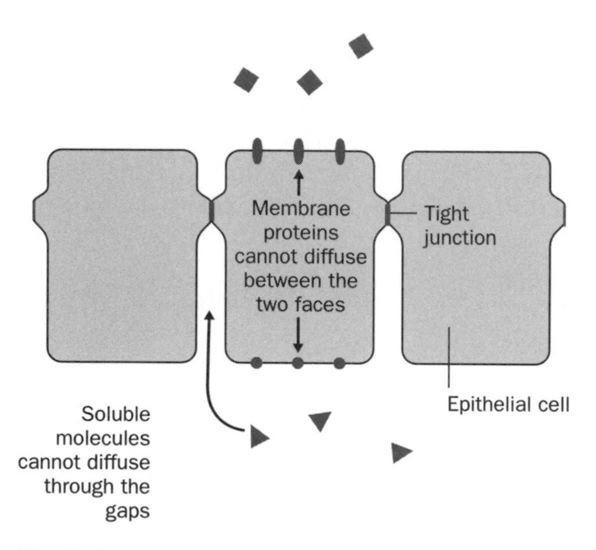

Fig. 4.19 Tight junctions seal gaps between epithelial cells, and keep membrane proteins on one face of the cell.

4.8 From prokaryote to eukaryote

Eukaryotic cells are evolutionary newcomers, but their diversity, complexity and capacity to form relatively big organisms like fleas and trees makes them especially interesting. When looking at their evolution, we need most of all to explain the evolution of organelles, the internal compartments that allow all this chemical complexity.

The evolution of mitochondria and plastids is probably the most bizarre yet least controversial part of the story. Many of their features are distinctly like those of prokaryotes. Their closed DNA loops attached to the membrane, ribosomes of the prokaryotic type, lack of histones and sensitivity to drugs which inhibit prokaryotic protein synthesis, as well as their independent reproduction, all support the idea that they originated as prokaryotic cells living symbiotically within another cell: endosymbiosis (Fig. 4.20(a,b)). Presumably their ancestors were engulfed by the

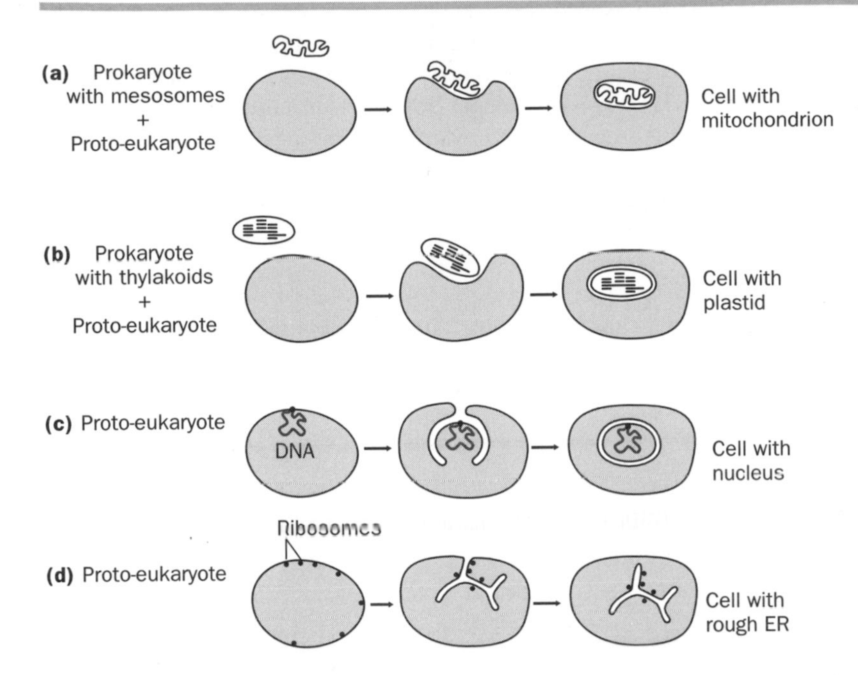

Fig. 4.20 Ideas about the origins of some organelles (see text for details). (a) mitochondrion, (b) plastid, (c) nucleus, (d) rough ER.

plasma membrane of the host cell but not digested. This would explain the presence of the outer membrane.

There has been a search for existing cells resembling the primitive mitochondria and plastids, and the proto-eukaryotes which engulfed them. The giant protozoan, *Pelomyxa palustris*, which lurks in muddy ponds, lacks mitochondria but has symbiotic bacteria which live and reproduce within it. Whilst it has a nucleus – several, in fact – it does not divide by conventional mitosis as do eukaryotes. Enthusiasts see it as a model for the proto-eukaryote. At the very least it shows that endosymbiosis can occur, and that intermediate states between prokaryotes and eukaryotes do exist. A strange photosynthetic prokaryote, *Prochloron*, lives symbiotically with a type of sea squirt. It has an unusual number of chloroplast-like properties, including the right mixture of photosynthetic pigments and thylakoid membranes which are stacked, rather like grana. It makes a good model for the ancestor of plastids in most plants, although some algal groups have very different chloroplasts which may have evolved separately. Good models for the ancestor of mitochondria are harder to find. It must have been capable of aerobic respiration, with the apparatus for electron transport (Section 9.2: Oxidative phosphorylation), perhaps a purple non-sulphur bacterium (Box 10.2) which had lost the ability to photosynthesize.

Enthusiasts have argued for an endosymbiotic origin of microtubules and even the nucleus, but the stories they tell have convinced few others. It seems more likely that organelles such as the ER and nucleus have their origins in infoldings of the plasma membrane. This seems particularly likely when DNA and ribosome attachment are considered. DNA is attached to the plasma membrane of prokaryotes but to the inner membrane of the nucleus in eukaryotes. Some ribosomes are bound to the plasma membrane of prokaryotes, just as some are found on the outside of the eukaryotic ER (Fig. 4.20(c,d)).

These are evolutionary stories, supported at best by circumstantial evidence and analogies drawn from what we find alive today. Unique events in the distant past are not open to experiment. This story-telling does not always feel much like science: the philosopher Karl Popper has argued that it is not. However, for those of us interested in origins, it is the best we can do.

Summary

- Membranes control what can cross them. They keep the chemistry of cells and organelles distinctive.
- A membrane is made up of a phospholipid bilayer.
- Some proteins and glycoproteins span the bilayer. Others are attached to one surface. They act as carriers, receptors and enzymes.
- Lipids and proteins can diffuse only in the plane of the membrane.
- Prokaryotic cells have no nucleus and few organelles. Eukaryotic cells are more complex.
- Most prokaryotes have a peptidoglycan wall to stop them bursting.
- Prokaryotes have a single closed loop of DNA. Like eukaryotes, they have ribosomes for protein synthesis.
- The nucleus is a cage for DNA.
- Eukaryotic DNA is in open strands, wound around histones.
- Ribosomes are made in the nucleolus.
- Rough endoplasmic reticulum transports proteins which are to be secreted or put into membranes. Smooth ER is involved in lipid metabolism.
- The Golgi apparatus processes and packages proteins and glycoproteins.
- Mitochondria and plastids have their own DNA, and reproduce almost independently.
- Mitochondria are ATP factories.
- Plastids develop in several ways: chloroplasts carry out photosynthesis; amyloplasts store starch.
- Peroxisomes carry out diverse oxidations.
- Lysosomes contain digestive enzymes. They help digest unwanted organelles and food particles.
- Plant and fungal cells have big vacuoles. They help maintain turgor, fill space cheaply, and are used for storage.
- The cytoskeleton consists of protein fibres in the cytoplasm: actin filaments, intermediate filaments and microtubules. Their roles include cell movement, controlling cell shape, moving organelles and strengthening cells.
- Cells secrete cell walls or an extracellular matrix around them.
- Cells are connected to their neighbours. Different junctions allow communication between cells or prevent molecules passing through the gap between them.
- Plastids and mitochondria may have evolved from symbiotic prokaryotes.

Exercises

4.1. Group these organelles according to whether they have one, two or no membranes around them:
nucleus, ribosome, vacuole, endoplasmic reticulum, peroxisome, plastid, lysosome, mitochondrion.

4.2. Compare the structure of prokaryotic and eukaryotic cells by completing the following table.

	Prokaryotic	Eukaryotic
Wall present?		
Plasma membrane present?		
Membrane-bound organelles present?		
Is DNA in a nucleus?		
Chromosomes: shape and number		
How do flagella move?		
Relative size of ribosome		
Relative size of cell		

4.3. The role of the β cells in the mammalian pancreas is to make and secrete the hormone insulin. Insulin is a protein. Which organelles would you expect to be especially abundant in these cells, and why?

4.4. In each of these situations, predict which type of junction is found between the cells.

(i) Heart muscle cells are tied together strongly. This maintains the organization of the muscle when the cells contract.

(ii) The Purkinje fibres of the heart are specialized muscle cells which carry impulses rapidly, much like nerve cells. Fast conduction requires that ions can diffuse from one cell to the next quickly.

(iii) Water can move from the cytoplasm of one cell to the next in a plant root, without crossing a plasma membrane.

Further reading

Alberts, B., Bray, D., Lewis, J., Raff, M., Roberts, K. and Watson J.D. *Molecular Biology of the Cell* (3rd ed.) (New York: Garland, 1994). An enormous but excellent textbook which puts the reader in touch with current ideas without too much pain on the way. I know of no better guide to cell biology.

Craigmyle, M.B.L. *A Colour Atlas of Histology* (2nd ed.) (London: Wolfe Medical Publications, 1986). A highly illustrated guide to the ways in which cells are organized to form mammalian tissues.

Enzymes

Contents

Connections

► Enzymes are protein molecules (Unit 3). They are important inside cells in all manner of metabolic processes, including photosynthesis (Unit 9) and respiration (Unit 10). Some cells secrete digestive enzymes (Unit 21). They are also involved in handling genetic information (Unit 7). All aspects of plant and animal physiology (Part 5) require an understanding of enzymes.

5.1 Enzymes are important

Without chemical reactions, there is no life. A lizard dozing on a rock is seething with chemical activity. Enormous numbers of reactions are taking place in every cell. A dead lizard placed beside it looks remarkably similar. In its cells, however, some reactions are no longer taking place, and others are taking place at quite different rates, depending on how long it has been dead, and what killed it. Many of the reactions in living cells take place much more slowly in the test-tube, sometimes so slowly that to all intents and purposes they do not take place at all. What is missing is a **catalyst**.

Catalysts are substances which increase the rate of a chemical reaction without being used up in the reaction (Box 5.1). **Enzymes** are proteins (made by cells) which act as catalysts. They are very specific. Enzymes generally catalyse only one type of reaction, or a group of extremely similar reactions. Most enzymes work in some compartment of the cell that made them. A very few, such as digestive enzymes in animals' guts, are secreted and operate outside the cell.

Enzymes are given names which end in *-ase*. For example, peptidases are enzymes which break polypeptides into amino acids. Not all are named so logically! Box 5.2 gives more examples.

Catalysts: Substances that increase the rate of a chemical reaction without being used up.

Enzymes: Catalytic proteins made by cells.

Substrate: A molecule which takes part in a reaction.

Active site: The part of an enzyme which binds the substrate, and where the reaction takes place.

5.2 How enzymes work

An enzyme binds the **substrate**, the molecule or molecules which take part in the reaction. The substrate binds to one part of the enzyme, the **active site**, which is

Box 5.1

Catalysts

Catalysts increase the rate of a reaction without being used up or permanently changed.

At a very simple level, reactions which release energy in some way are feasible (Section 8.2). Those which take in energy are not feasible: they must be driven by an input of energy.

However, some reactions which release energy take place very slowly. This is because the molecules taking part in the reaction need to gain a certain amount of energy before the reaction can take place, even though all that energy and more is given out again. This initial energy requirement is called the activation energy. Only a small proportion of the reacting molecules will have this much energy at any one time, so the reaction will be slow.

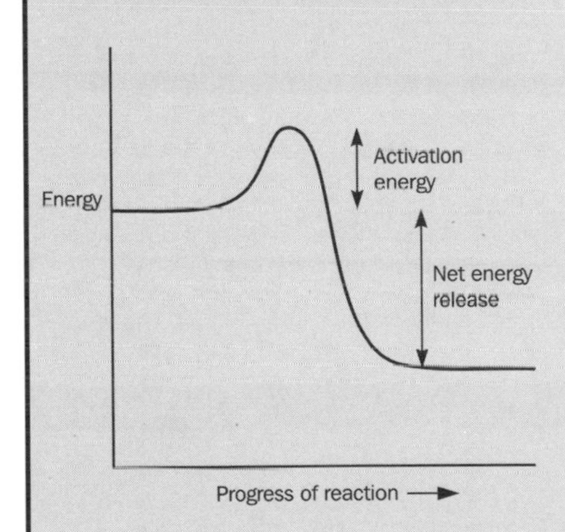

Catalysts reduce activation energy, so the reaction takes place faster.

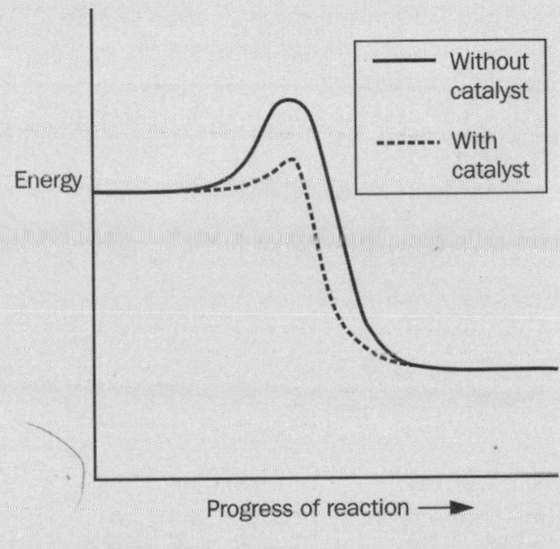

Catalysts have immense economic importance. They make many processes in the chemical industry commercially viable. For example, the industrial synthesis of ammonia: $N_2 + 3H_2 \rightarrow 2NH_3$ is speeded up by an iron catalyst.

Most biological catalysts are protein molecules: enzymes.

usually a groove or cleft in the globular protein molecule. Enzymes are highly selective about what they will bind. This is determined by the precise shape and distribution of electrical charges within the active site. Only a substrate of the right shape and size, with the right charges in the right places, will bind. Some enzymes catalyse a reaction with a single substrate, others with two or more. Box 5.3 illustrates some ideas about how enzymes bind substrates.

Once the substrate has bound, the enzyme allows a particular reaction to occur much more quickly. Reactions are not magical jumps from a substrate to a product. Electrons move, bonds are broken and formed. There are all sorts of short-lived intermediates before the final product is formed. Some are unstable; others need a lot of energy to set them up, even if it is released again afterwards. Enzymes hold the substrates in ideal positions for reaction, stabilize intermediates, and provide an environment where less energy is needed for reactions to take place. As a result, individual molecules are more likely to react. Overall, the rate of reaction increases. Enzymes are not the fairy godmothers of the cell, making the impossible happen. Rather, they are slick middle men, making the right contacts and providing congenial places for deals to be struck.

Box 5.2

Naming enzymes

Most enzymes' names end in *-ase*. Some are named more logically than others. Here are some examples:

Many enzymes are named according to the sort of reaction they catalyse. **Polymerases** catalyse the formation of polymers. For example, DNA polymerase joins nucleotides onto the end of a growing DNA strand. **Dehydrogenases** remove hydrogen atoms from a molecule. The all-important electrons may be passed on to an electron carrier like NAD. For example, alcohol dehydrogenase catalyses the following reaction:

ethanol + $NAD^+ \rightarrow$ ethanal + NADH + H^+

Kinases transfer a phosphoryl group from ATP to another molecule. Phosphofructokinase catalyses the following reaction:

fructose 6-phosphate + ATP $\rightarrow$ fructose 2,6-bisphosphate + ADP

Isomerases interconvert isomers, that is molecules in which the same atoms are arranged in alternative ways. Phosphoglucose isomerase interconverts glucose 6-phosphate and fructose 6-phosphate. **Carboxylases** catalyse the joining of a carbon dioxide molecule to an existing carbon skeleton, forming a carboxyl group. Phosphoenolpyruvate carboxylase catalyses the following reaction:

phosphoenolpyruvate + $CO_2 \rightarrow$ oxaloacetate + P_i

Other names say more about the substrate than the reaction. **Proteases** digest proteins into shorter polypeptides; **peptidases** digest polypeptides into amino acids. **ATPases** catalyse the hydrolysis of ATP, releasing energy. This is always coupled to some other useful reaction or process.

Some names are more obscure, sometimes referring to the biological significance of the enzyme, rather than the reaction or substrate. Many of these are easily accessible or unusually stable enzymes which were discovered early on. **Catalase** catalyses the breakdown of hydrogen peroxide. **Lysozyme** attacks the peptidoglycan of bacterial cell walls, causing the cells to burst (lyse). **Pepsin** is a protease found in the mammalian stomach.

Box 5.3

How does the enzyme bind the substrate?

The 'lock and key' model is a sensible first guess. The substrate fits neatly into the rigid shape of the active site, helped by an appropriate distribution of charges.

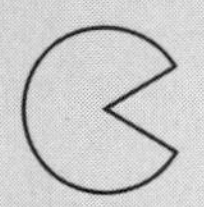

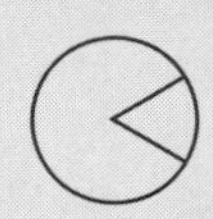

enzyme + substrate → enzyme–substrate complex

Some enzymes really do work like this. A good example is lysozyme, found in tears, saliva and egg white. It digests the cell walls of bacteria, allowing them to burst.

The 'induced fit' model is a bit more subtle. As the substrate binds, the enzyme changes shape.

Carboxypeptidase A is a classic example. This digestive enzyme nibbles away at the carboxyl ends of polypeptides, removing one amino acid at a time.

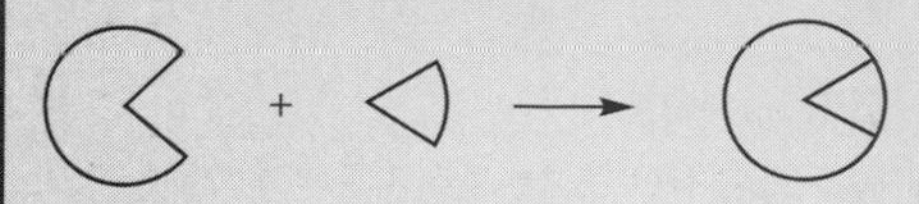

Induced fit is an important feature of some enzymes which bind two substrates. The enzyme has to have **two coupled binding sites**, one for each substrate. When one substrate molecule binds, the enzyme changes shape, making it easier for the second to bind.

Hexokinase is a well-known example. This enzyme catalyses the transfer of a phosphoryl group from ATP to glucose. Both ATP and glucose must bind before the reaction can take place.

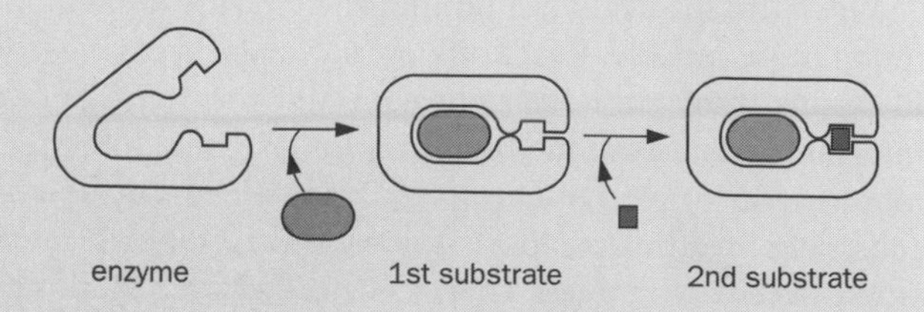

Remember, all these diagrams are just cartoons. They give an impression of what goes on, but do not show what the molecules really look like.

Factors affecting rate of reaction

Enzymes speed up reactions, but by how much? Several factors affect the rate of enzyme-mediated reactions. Temperature, pH, the concentrations of molecules involved in the reaction, the concentration of the enzyme itself, and the presence of other molecules which may activate or inhibit the enzyme, all have an effect on rate. Some affect reactions whether or not an enzyme is involved. Others are specific to enzyme-mediated reactions.

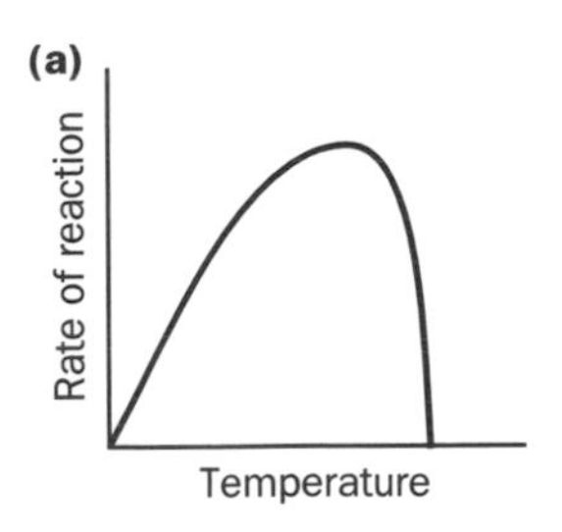

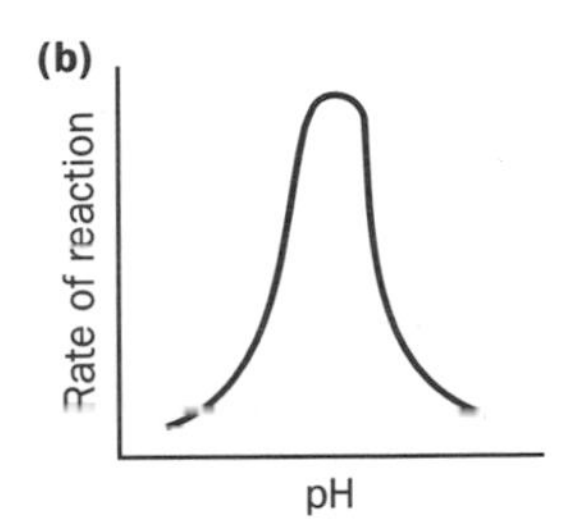

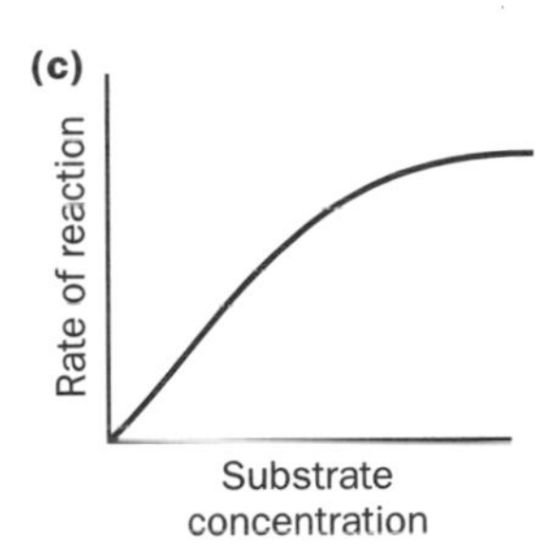

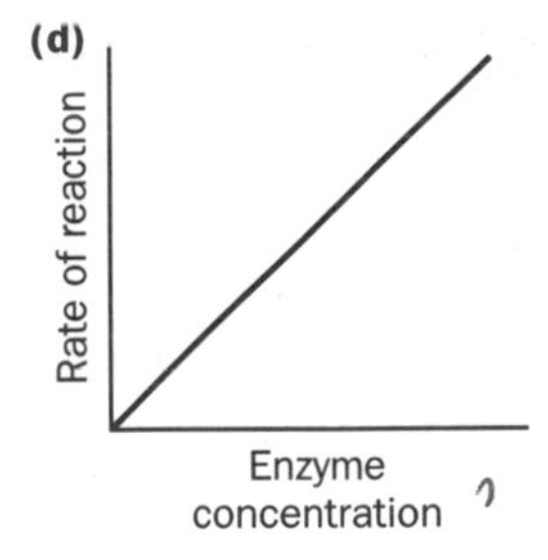

Fig. 5.1 Factors which affect the rate of enzyme-mediated reactions. (a) temperature, (b) pH, (c) substrate concentration, (d) enzyme concentration.

Temperature

Enzymes are inactive in the frozen state. Once the solution has melted, rate of reaction increases with temperature until an optimum temperature is reached. Above this, rate quickly declines (Fig. 5.1(a)). Enzymes work in solution. They are either in solution with the substrate, or bound to a membrane in contact with the solution containing the substrate. If the solution is frozen, the substrate molecules move only by vibrating: they are not able to move from place to place. They cannot collide with the enzyme, so the enzyme has no effect on rate of reaction. Above melting point, molecules move around randomly in solution. The warmer it gets, the faster they move. The faster they move, the more likely they are to collide with enzyme molecules, so reaction rate increases. Temperature also affects the enzyme itself. As well as moving around in solution, the polypeptide chain can twist and flex. If the temperature gets too high, the molecule flexes so vigorously that the hydrogen bonds which hold it in shape are broken. The enzyme loses its distinctive shape and so loses its activity. The enzyme is said to have been **denatured**. This is true of proteins in general, not only enzymes. Not all enzymes have the same optimum temperature. Some are held in shape more strongly than others. These have higher optimum temperatures. Natural selection has led to enzymes having optimum temperatures which reflect their normal working temperatures. Homeothermic ('warm-blooded') animals, such as birds and mammals, tend to have enzymes with an optimum very close to body temperature. They regulate temperature so closely that there is little danger of enzymes denaturing. Arctic plants tend to have enzymes with lower optima than tropical plants. Even the same enzyme from the same plant may have a lower optimum in winter than in summer. This suggests that there may be more than one form of the enzyme. In the test-tube a denatured enzyme is useless for ever more. The process cannot be reversed. In cells, however, there is a team of proteins which restores the shape of denatured proteins, if the damage is not too great.

pH

pH is a measure of the H^+ concentration of a solution. The lower the pH, the more H^+ there is: the solution is more acidic. The higher the pH, the less H^+ there is: the solution is more alkaline. Being charged, H^+ can interfere with hydrogen bonding. It can also alter the pattern of charges on the polypeptide, since both amino and carboxyl groups can gain and lose H^+ (Section 3.5). As a result, an enzyme works best at one particular pH. Activity declines as conditions become more acidic or more alkaline: the enzyme is denatured (Fig. 5.1(b)). In most cells, the cytosol is approximately neutral, with a pH of about 7. Enzymes found in the cytosol generally have their optimum pH in this region. Digestive enzymes in lysosomes have a lower optimum. The contents of a lysosome are at about pH 5. The point is that these

Denature: To inactivate an enzyme by changing its shape.

enzymes work well in the lysosome, but if they accidentally escape into the cytosol, they are less likely to damage the cell. In some special situations, enzymes regularly meet more extreme pH values. Pepsin is a protease which is made by cells in the stomach lining. It is secreted into the horribly acidic stomach contents (Section 21.4) where it digests proteins into shorter peptides. Human pepsin has an optimum pH of between 2 and 3, and is useless at pH 5 or above.

Purified enzymes are used in the food industry. On a smaller scale, we used amylase and pectolytic enzyme to digest starch and pectins, to make this carrot wine crystal clear.

Substrate concentration

The more concentrated the substrate, the quicker the reaction. This is true for chemical reactions in general. In enzyme-mediated reactions, however, there is a second issue. There is a limited number of enzyme molecules available. Each one is occupied for a while each time it binds the substrate and catalyses the reaction. There comes a point when each enzyme molecule is constantly occupied. As soon as one product molecule leaves, another substrate molecule binds. Adding more substrate cannot make the reaction faster. So, overall, as substrate concentration increases, rate of reaction increases at first. As concentration rises further, rate levels off at a maximum value (Fig. 5.1(c)). Box 5.4 introduces a mathematical model which links rate of reaction to substrate concentration.

Enzyme concentration

The more enzyme there is, the faster the reaction (Fig. 5.1(d)). More enzyme molecules means more frequent enzyme–substrate collisions, so more product is made in a given time.

Inhibitors

Inhibitors are molecules which take no part in a reaction, but make the enzyme less effective. They reduce the rate of reaction. Some inhibitors are a normal part of the working cell. Others are potentially dangerous foreign molecules. Inhibitors are classified as either reversible or irreversible. Irreversible inhibitors bind more or less permanently to some part of the enzyme. The enzyme's shape is changed, and it loses its activity. Reversible inhibitors are divided into competitive and non-competitive inhibitors, depending on how they bind. Competitive inhibitors bind reversibly to the enzyme's active site. While the enzyme is bound, the substrate cannot bind. This means that at any one time, a proportion of enzyme molecules is out of action. Rate of reaction is reduced. Competitive inhibitors are normally very much like the substrate in structure. This is why they can bind to the active site. Sometimes the enzyme even catalyses a reaction with the inhibitor as substrate. Non-competitive inhibitors bind to the enzyme somewhere other than the active site. They do not affect substrate binding, but the enzyme becomes a poorer catalyst. Some enzymes affect substrate binding *and* catalytic ability: mixed inhibition. The mathematical model shown in Box 5.4 can help us to distinguish between competitive and non-competitive inhibitors in practice.

Cofactors

Cofactors are a ragbag of molecules which are needed for enzymes to work. Some are permanently bonded to the enzyme, and are in effect a non-protein part of the enzyme. These are called prosthetic groups. Catalase, for example, has an iron-containing haem group, much like that in haemoglobin (Section 19.2). Coenzymes are not attached, and are really extra substrates. They are mostly carriers of some-

Box 5.4

The Michaelis–Menten model

A mathematical model links rate of an enzyme-mediated reaction to substrate concentration.

If a whole set of quite sensible assumptions are made about how the enzyme works, this equation can be derived theoretically:

$$V = V_{max} \frac{[S]}{[S] + K_M}$$

V = rate of reaction; [S] = substrate concentration; V_{max} = the maximum rate of reaction; K_M = the Michaelis constant, specific to the enzyme in question. It is the substrate concentration at which rate is half V_{max}.

As you would hope, this gives a graph just like the one you would expect to get by experiment (Fig. 5.1(c)).

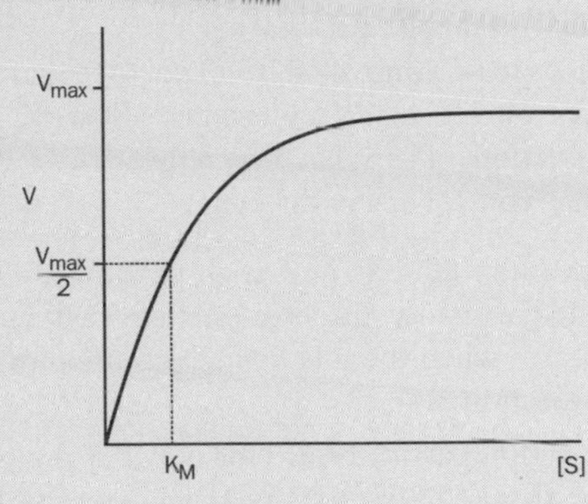

Much more interestingly, it can be shown that 1/V plotted against 1/[S] gives a straight line. This is called a Lineweaver–Burke plot.

This is useful because:

(i) it is much easier to fit a straight line than a particular sort of curve to a set of experimental data;

(ii) different types of inhibitors give distinctive Lineweaver–Burke plots: this helps to distinguish competitive and non-competitive inhibitors.

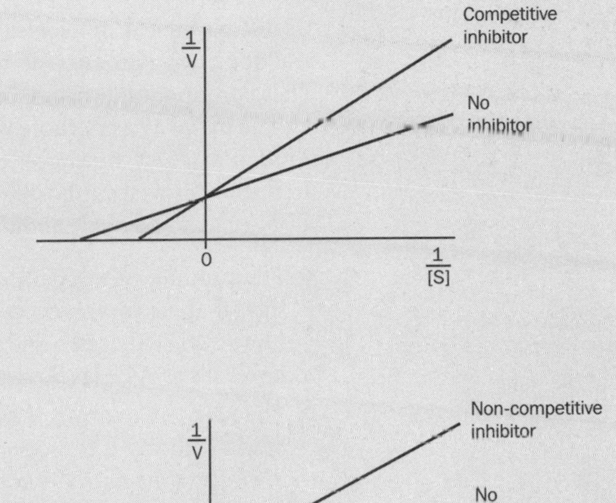

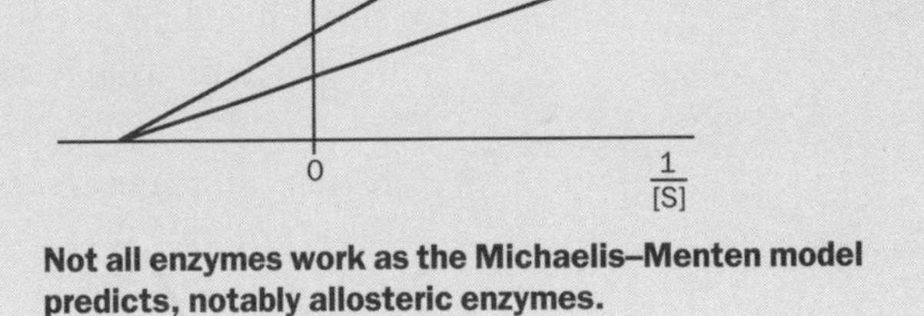

Not all enzymes work as the Michaelis–Menten model predicts, notably allosteric enzymes.

thing useful within the cell. NAD and NADP are electron carriers, ATP acts as an energy carrier (Section 8.4), and coenzyme A is a carrier of 2-carbon groups which can be added on to existing carbon skeletons (Section 9.2: The Krebs cycle). When any of these things is needed for an enzyme-mediated reaction, it is described as a cofactor, even though it has to bind to the active site, along with the substrate.

Phosphorylation

Some enzymes can exist in either an active or an inactive form. The difference is a phosphoryl group attached to a specific amino acid residue. This is in a key position, so that phosphorylation changes the shape of the enzyme. Phosphoryl groups are added by enzymes called protein kinases, and removed by protein phosphatases (Box 16.2). This is an important and rather drastic way in which the cell controls its enzymes.

Activators

Many enzymes are activated by molecules which bind to a site other than the active site. When the activator binds, the enzyme changes shape. This makes it easier for the substrate to bind. Proteins like this which can exist in alternative forms are described as allosteric proteins. There are allosteric inhibitors as well as activators. Activators, inhibitors and phosphorylation are the main ways in which the cell controls enzyme activity. This is best understood in the context of metabolic pathways.

5.3 Metabolic pathways

Useful biological molecules are rarely made by a one-step process. Sequences of reactions, with an enzyme for each step, convert a substrate to the final product. These production lines are called **metabolic pathways**.

The pathway as a whole needs to be controlled, to make sure that there is always enough product, but never too much. This is achieved by allosteric inhibition and activation, as well as phosphorylation, of key enzymes in the pathway. Several generalizations can be made about the control of a pathway:

- The product of the pathway often inhibits it. A high product concentration is a sign that no more needs to be made. Sometimes, pathways are activated by molecules which become scarce when there is lots of product, for example ADP level falls as ATP level rises (Section 8.4).
- The pathway's substrate sometimes activates it. This may happen when the point of the pathway is to store or dispose of the substrate. The more substrate there is, the faster it needs to be converted.
- Early steps in pathways tend to be important control points. Otherwise, quite a lot of substrate could be converted into intermediates in the pathway before all the other reactions grind to a halt. If you want to stop traffic flowing down a road, you put a 'no entry' sign at the entrance, not half-way along.
- Late stages in pathways are sometimes controlled. This is more important in long pathways. If only the first step were controlled, product would be made for some time after the first enzyme was inhibited, as the stocks of intermediates were used up.
- The slowest step in a pathway, the rate-limiting step, makes a logical control point. Controlling its enzyme will certainly affect the rate of the entire pathway. Roadworks slow down traffic flow most of all when they are at a place which is already a bottleneck.
- Most reactions are reversible, at least to some extent. At some balance of substrate and product concentration, the reaction goes neither forward nor backward: a state of equilibrium. Most control points are reactions which are a long way from equilibrium in the normal cell. Activating or inhibiting the enzyme will have an immediate effect on the amount of product being formed. This is related to the previous point: the slowest steps are likely to be far from equilibrium; the fastest steps have a chance to approach equilibrium.

Metabolic pathway: A sequence of reactions in a cell which together make a useful product.

Fig. 5.2 illustrates these points. It shows the most important controls on three related pathways. All three have glucose as substrate. They also share the same first step. This is the phosphorylation of glucose to glucose-6-phosphate. Firstly, glycolysis

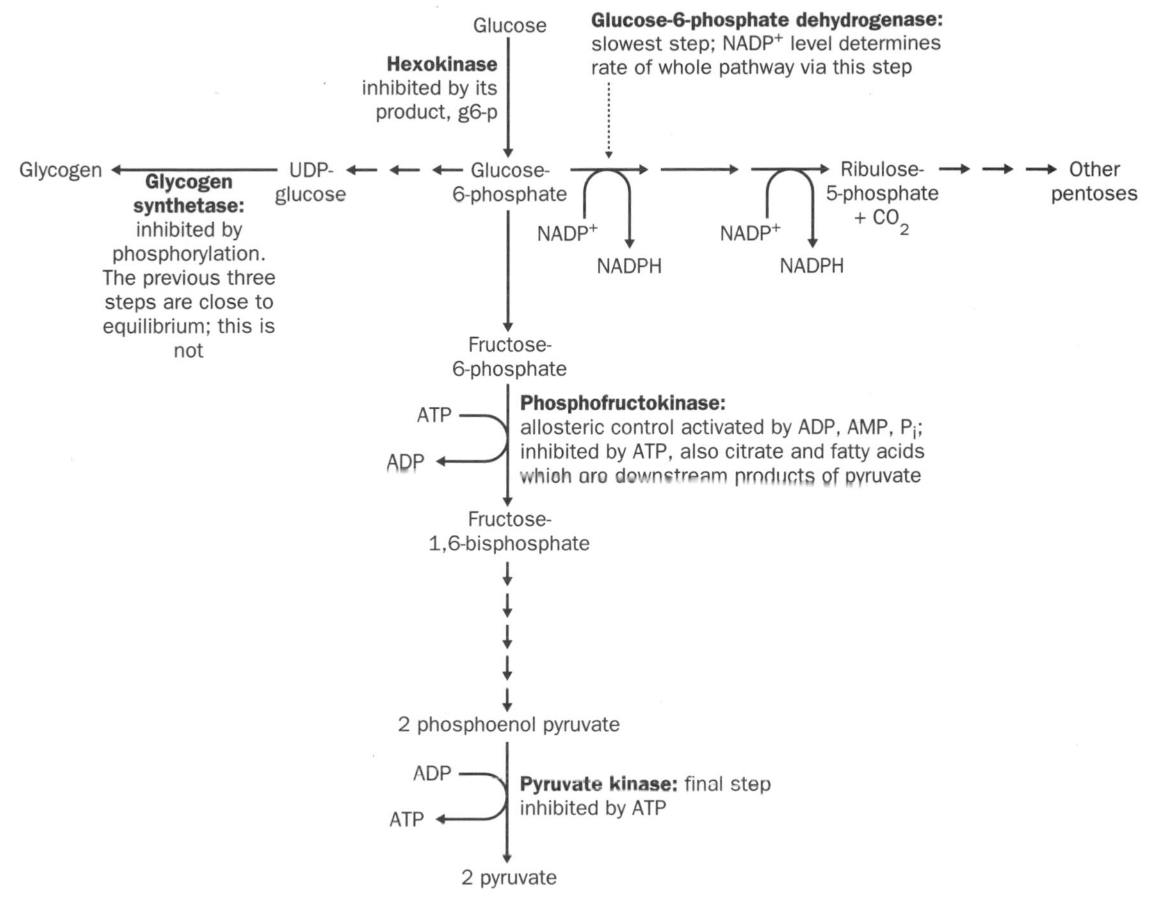

Fig. 5.2 Control points in glycolysis, glycogen synthesis and the pentose phosphate pathway (see text for details).

is a long pathway converting glucose to pyruvate, which feeds into other pathways. The main function of glycolysis and subsequent pathways is to make ATP, an energy store. Secondly, the glycogen synthesis pathway is found in some animal cells. Excess glucose is stored as glycogen, an insoluble polysaccharide (Section 3.4). Finally, the pentose phosphate pathway has two functions. It uses glucose (a hexose sugar) to make pentose sugars. It is also an important way of making NADPH, a reducing agent used in constructing many biological molecules (Section 8.5). Everything is here: allostery, phosphorylation, control of early, late and rate-limiting steps.

5.4 Catalytic RNAs

Enzymes are diverse and highly effective as catalysts. For years, most people assumed that all the catalysts found in cells were proteins. The first concrete evidence that they are not came in the early 1980s when an RNA molecule found in *Tetrahymena*, a ciliate protozoan, was investigated. It was a ribosomal RNA, which forms part of the cell's ribosomes. Like many eukaryotic RNAs, it had sections which needed to be cut out before it could function properly (Section 7.3). This molecular cutting and pasting is usually carried out by complex catalysts called spliceosomes (Box 7.2). Amazingly, the sequence which was to be cut out of the *Tetrahymena* RNA acted as the catalyst for its own removal. No external catalyst was needed. This example is, at the same time mind-blowing and rather trivial. The important point is that catalytic RNA molecules exist. They form complex three-dimensional shapes, with active sites, in the same way as conventional enzymes. These catalytic RNAs are sometimes called ribozymes. Some more or less well understood examples are shown in Box 7.3. They mostly work in close association with proteins. Almost all are involved with the business of making proteins according to the genetic information stored in DNA. So far, none seem to be involved with the cell's general metabolism. At least in this area, enzymes retain their unique position as the cell's catalysts.

Summary

- Catalysts are substances which increase the rate of a chemical reaction without being used up.
- Catalysts reduce the activation energy of a reaction.
- Enzymes are catalytic proteins made by cells.
- Enzymes are highly selective.
- The substrate binds to the active site, a groove or cleft in the enzyme.
- Enzymes have an optimum temperature. Up to this point, raising temperature increases rate of reaction, due to more frequent molecular collisions. Above this point, enzymes are denatured.
- Enzymes have an optimum pH, which varies between enzymes.
- Increasing substrate concentration increases rate of reaction, until the enzyme is saturated.
- Increasing enzyme concentration increases rate of reaction.
- Inhibitors are molecules which make enzymes less effective. Only some are reversible. Competitive inhibitors bind to the active site; many resemble the substrate. Non-competitive inhibitors bind elsewhere on the enzyme.
- Cofactors are diverse substances needed for particular enzymes to work.
- Some enzymes can be activated or inactivated by phosphorylation.
- Some enzymes are controlled by allosteric activators and inhibitors. These molecules bind to sites other than the active site. They change the enzymes' shapes, altering their catalytic activity.
- Metabolic pathways are sequences of reactions, each with its own enzyme, which convert a substrate to the final product. Pathways are tightly controlled.
- Some biological catalysts are RNAs, or RNA–protein complexes.

Exercises

5.1. Can you work out which of the following proteins are enzymes?
(i) haemoglobin, **(ii)** immunoglobulin, **(iii)** phosphoenolpyruvate carboxylase, **(iv)** fibronectin, **(v)** carboxypeptidase A, **(vi)** hexokinase, **(vii)** actin, **(viii)** lactase.

5.2. What do the following observations have in common?
(i) a yeast culture dies if it is boiled;
(ii) heavy metals such as mercury and cadmium are toxic to most cells;
(iii) if the pH of human blood falls below 6.8 or rises above 8.0, the patient normally dies as a result.

5.3. Alcoholic drinks contain ethanol. Do-it-yourself distilling can be dangerous because under some conditions a different alcohol, methanol, can become concentrated in the drink. Methanol is oxidized in liver cells to a toxic product.
One way of treating methanol poisoning is to give the patient a large dose of ethanol. Speculate on how this treatment might work.

5.4. The oxidative branch of the pentose phosphate pathway is a metabolic pathway which produces pentoses from hexoses. It also generates NADPH, needed as a reducing agent in the synthesis of fatty acids, among other things.

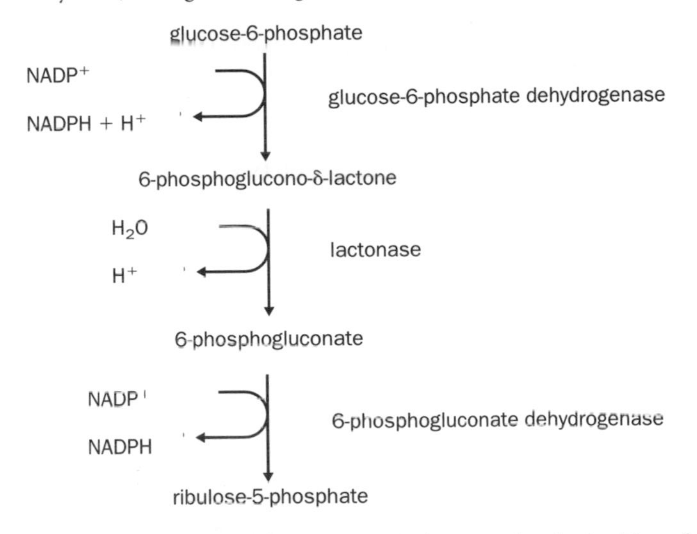

The enzyme glucose-6-phosphate dehydrogenase is the control point in this pathway. $NADP^+$ concentration is the main controlling factor: the higher the concentration, the faster the rate. Explain why these observations make sense, given the function of the pathway.

Further reading

Smith, C.A. and Wood, E.J. *Biological Molecules* (London: Chapman and Hall, 1991). Includes two chapters on enzymes which go a bit beyond this book, but at a gentle pace.

Stryer, L. *Biochemistry* (4th ed.) (New York: Freeman, 1995). Enzymes fall firmly within the scope of biochemistry. This is the classic big undergraduate textbook of biochemistry. There are several broadly similar books on the market, but I would still go for this if looking for clear explanation.

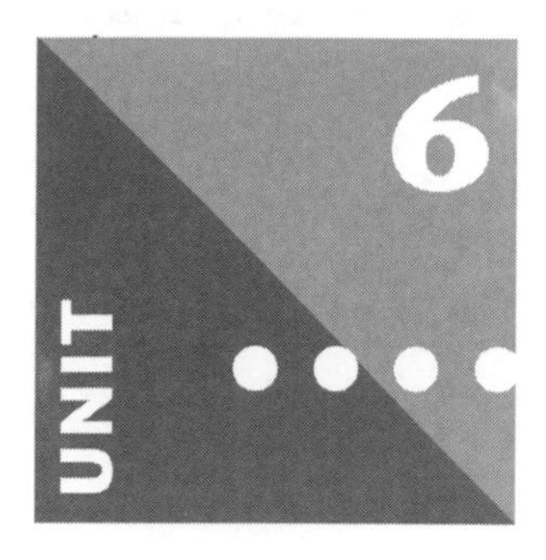

Movement of Molecules

Contents

Connections

▶ Movement of molecules and ions across membranes is central to many aspects of physiology (Part 5), as well as energy conversions in cells (Part 4). This unit is an important introduction to these topics. A basic knowledge of cell structure (Unit 4) and biological molecules (Unit 3) is assumed.

Molecules move

Molecules move constantly. In gases they speed about, occasionally colliding with one another, and bouncing off any solid barriers they meet. In liquids and solutions they also move about, but change direction as a result of collisions much more frequently, since the molecules are packed together much more densely. Even the molecules of a solid move, although this time they are confined to a very small region and vibrate rather than wander. Big, flexible molecules such as polypeptides even show significant internal movement: one part of the molecule can flop, flex or wobble in relation to another.

The higher the temperature, the faster they move. In fact, temperature is defined by the amount of kinetic energy the moving molecules have. Molecules still have plenty of energy at 0°C: this zero point is defined on practical grounds as being the freezing point of water at atmospheric pressure. Water molecules still vibrate within ice crystals. The temperature at which molecules would have no kinetic energy at all is about –273°C, absolute zero. It is probably impossible for matter to reach this temperature, although it can be approached very closely. Biological systems never get near absolute zero in nature, so the strange properties of matter when molecules have all but ceased moving need not concern us here.

It is important for cells that molecules should move. They need useful molecules to move into them from their environment, and waste products to move away. They need signalling molecules and energy-storing molecules to move from one cell to another. Specific molecules must be shifted across membranes to set up and maintain the distinctive chemistry of cells and organelles. Sometimes the natural movement of molecules does the job. Cells frequently select which molecules may cross a membrane, and regulate the rate of flow. Cells can even actively shift molecules across membranes with an input of energy from ATP. Before looking at how cells influence molecular movement, three basic processes will be described: diffusion, osmosis and mass flow. These are all relevant to living systems, but are not unique to them.

6.1 Diffusion

Molecules move in random directions. The effect of this is that they end up more or less evenly spread out, even if they started out in a tight group. Substances diffuse from regions of higher concentration to regions of low concentration. Individual molecules do not all move this way and do not somehow 'know' which way to go: it is the net result of many molecules moving in all directions (Box 6.1). This can happen only if molecules are free to move about. **Diffusion** is obvious in gases, liquids and solutions, but painfully slow or non-existent in solids. The size of the molecule is also important: larger molecules diffuse more slowly.

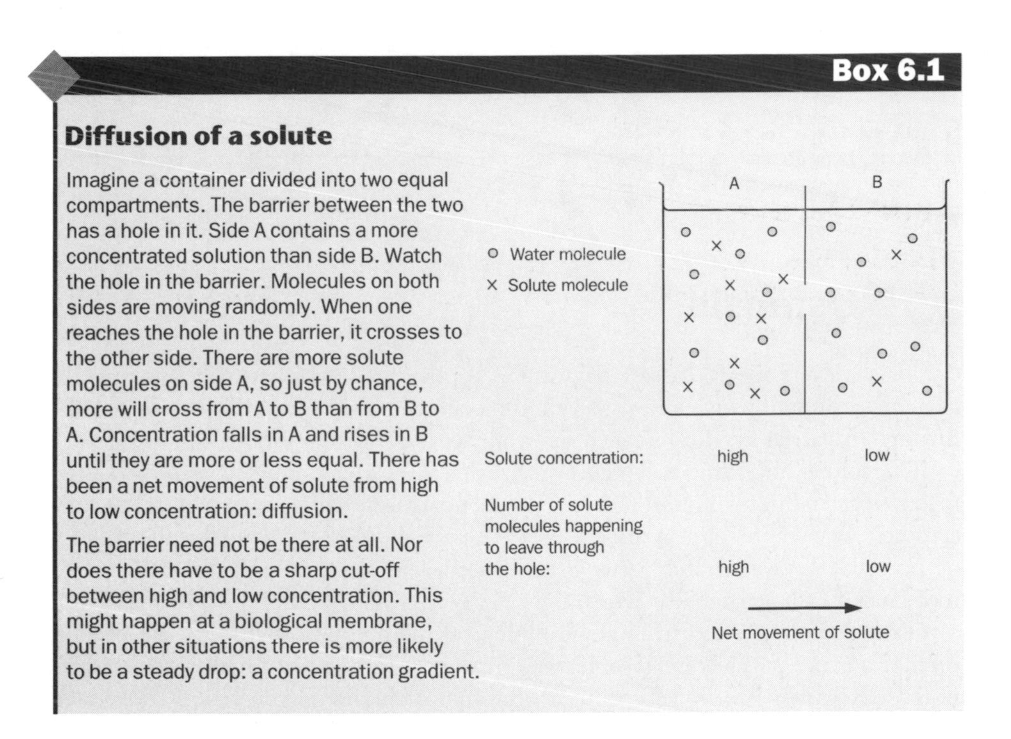

Box 6.1

Diffusion of a solute

Imagine a container divided into two equal compartments. The barrier between the two has a hole in it. Side A contains a more concentrated solution than side B. Watch the hole in the barrier. Molecules on both sides are moving randomly. When one reaches the hole in the barrier, it crosses to the other side. There are more solute molecules on side A, so just by chance, more will cross from A to B than from B to A. Concentration falls in A and rises in B until they are more or less equal. There has been a net movement of solute from high to low concentration: diffusion.

The barrier need not be there at all. Nor does there have to be a sharp cut-off between high and low concentration. This might happen at a biological membrane, but in other situations there is more likely to be a steady drop: a concentration gradient.

Diffusion is extremely important to cells. Consider a typical mammalian cell, constantly supplied with oxygen by the blood. So long as the plasma membrane is permeable to oxygen, it will diffuse in from the higher concentration in the blood to the lower concentration in the cell. On a larger scale, thin things like flatworms and leaves rely on diffusion to exchange gases with the atmosphere. Only thicker organisms such as earthworms and elephants need a blood system, when the distances are to great for diffusion to be quick enough.

Diffusion is one example of the great paradox that a random process can lead inexorably to one particular outcome. That outcome is always disorder.

Diffusion: The net movement of molecules from a region of higher concentration to a region of lower concentration.

Leaves, even big ones, rely on diffusion to bring gases to and from the atmosphere. This is possible only because they are thin.

6.2 Osmosis

Osmosis is a very special case of diffusion. It applies to a solvent, usually water, when a barrier is present which allows solvent molecules, but not solute molecules, to pass through. A barrier like this is called a partially permeable membrane.

If the solute concentration is higher on one side of the membrane, water moves across to that side, tending to dilute it. This is diffusion, not magic. Water molecules are less concentrated when more solute is present, partly because the solute takes up space that could be occupied by water, and partly because some are electrically attracted to solute molecules, forming complexes which are too large to cross the membrane (Box 6.2).

Osmosis is important in biology. Biological membranes, including the plasma membrane, are partially permeable: water can pass through but many solutes cannot. Cells which live in very dilute solutions, including single-celled freshwater organisms and most plant cells, tend to gain water by osmosis. Plant cells can survive this because the rigid cell wall prevents the cell bursting as pressure rises. Less protected animal cells have to take steps to prevent pressure rising. The osmotic problems of cells in concentrated solutions such as seawater are the reverse: how can water loss be restricted?

Osmosis: The net movement of water across a partially permeable membrane, from a region of lower solute concentration to a region of higher solute concentration.

Predicting the movement of water by osmosis becomes tricky when solutions contain different concentrations or different solutes, and when physical pressure may be different on either side of a membrane. The osmotic effect of all the dissolved substances in a solution is summarized as the solute potential. The effects of osmosis and physical pressure are combined in the water potential of the solution (Box 6.3), the most useful tool in predicting water movement.

Box 6.2

Osmosis

Imagine a container divided into two equal halves by a partially permeable membrane. The membrane lets water molecules pass through, but solute molecules cannot. Equal amounts of different solutions are put into the two halves. Side B has a more concentrated solution than side A. The extra solute molecules in side B take up space which could be occupied by water molecules, so there are less water molecules in B. Moreover, molecules which dissolve in water tend to be at least locally charged, so the polar water molecules are attracted to them. At any one time, a proportion of the water molecules is involved in the complexes of solute plus water which form this way. These two factors mean that the concentration of free water molecules is higher on side A, so water diffuses from the more dilute solution in A to the more concentrated solution in B. This is osmosis.

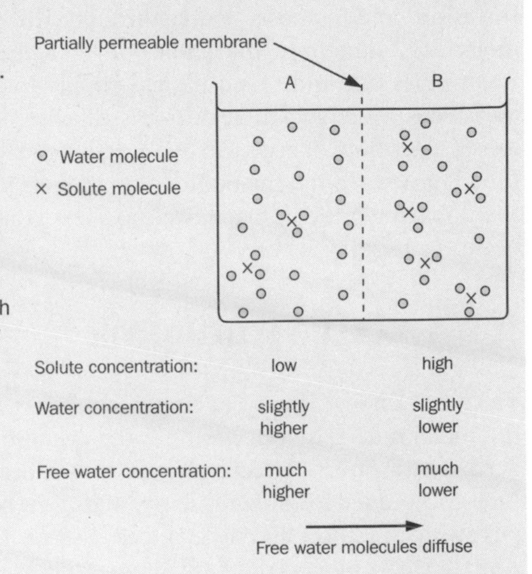

Box 6.3

Water potential

Water potential is the degree to which water tends to leave a particular compartment. Two things influence this. The first is physical pressure. The higher the pressure, the greater the tendency for water to leave. This tendency is called the pressure potential. In practice, this is normally measured by comparison with air pressure, so the pressure potential of water at atmospheric pressure is taken to be zero. The second factor is the concentration of the solution in the compartment. The more concentrated the solution, the more water tends to enter it by osmosis. Water potential is all about the tendency of water to *leave* a compartment, so the solute potential is rather cack-handedly defined as the degree to which water tends to leave by osmosis. Solute potential, then, always has a negative value, unless we are dealing with pure water, which has a solute potential of zero.

Water potential is symbolized by the Greek letter ψ (psi). Since a physical pressure can reverse the effects of osmosis, not only pressure potential, but also solute potential and overall water potential can be measured in pressure units, kilopascals (kPa).

$$\underset{\text{overall water potential}}{\psi_w} = \underset{\text{solute potential}}{\psi_s} + \underset{\text{pressure potential}}{\psi_p}$$

For example, a plant cell might have a pressure potential of 500 kPa and a solute potential of –800 kPa. Its water potential can be calculated as follows:

$$\psi_w = \psi_s + \psi_p = -800 \text{ kPa} + 500 \text{ kPa} = -300 \text{ kPa}$$

Water will only move from a region of higher water potential to a region of lower water potential. This can be used to predict how water will move in any given situation. For example, what will happen if the cell above is put into a beaker of pure water at atmospheric pressure?

In the beaker:

$$\psi_w = \psi_s + \psi_p = 0 \text{ kPa} + 0 \text{ kPa} = 0 \text{ kPa}$$

This is higher (more positive) than the –300kPa of the cell contents, so water will move into the cell. It will only stop when the water potential of the cell has risen to zero, equal to that in the beaker. Water is essentially incompressible, so only a very little needs to enter the cell in order to increase the pressure a great deal. This will scarcely dilute the cell contents, so the change in solute potential will be negligible. Increased pressure potential accounts for more or less all the increase in water potential, so once equilibrium has been reached, the situation is as follows:

WATER	CELL	
	Ψ_s	–300 kPa
	$+\Psi_p$	300 kPa
$\Psi_w = 0$ kPa	$=\Psi_w$	0 kPa

6.3 Mass flow

Diffusion and osmosis work through the random movements of individual molecules. Sometimes, the whole body of a liquid or gas moves along together, as in water pipes, the wind, and the gas supply to my grandma's cooker. This is called **mass flow**. Individual molecules are still moving within the bulk, but this is not what causes mass flow. It is driven by something external, normally a pressure difference. Blood moves around the bodies of mammals by mass flow. The movement of sugar solution from one part of a flowering plant to another, through the phloem, probably occurs in a similar way.

6.4 Active transport

The direction of diffusion or osmosis is governed by concentrations on either side of the membrane. The cell has only the crudest control over this. Frequently, a cell needs to transport molecules across a membrane against a concentration gradient. Energy is needed for this. Similarly, mass flow often requires an energy input. Transport which requires an energy input, usually from ATP, is called **active transport**. Any transport process not requiring an energy input is described as passive.

Cells have specific machinery for driving active transport and for allowing diffusion of only selected molecules. This is considered next.

6.5 Transport across membranes: channels and carriers

Proteins embedded in membranes allow molecules to cross, actively or passively. They can be placed in three groups according to function.

Passive channels for diffusion

These proteins simply provide sites at which molecules can cross the membrane. They do not set up a concentration gradient: a gradient must already be present if molecules are to diffuse. Some **channels** are always open, others can be controlled. Some are large, allowing a variety of molecules to cross. The porins are proteins which form huge pores in the outer membranes of plastids, mitochondria and some bacteria, allowing molecules up to the size of small proteins to pass through. Others are small and specific, such as the voltage-gated potassium channel in the plasma membrane of nerve cells (Fig. 6.1(a)). This allows K^+ to diffuse into the cytosol as part of the sequence of electrical events that is a nerve impulse (Section 17.3). The channel opens and closes according to the voltage across the membrane.

Passive symports and antiports

These **carrier** proteins allow diffusion only if two types of molecule move together. In a **symport**, both cross the membrane in the same direction; in an **antiport**, they go opposite ways. For example, glucose enters animal cells by a sodium/glucose symport (Fig. 6.1(b)). If Na^+ and glucose are both at higher concentrations outside,

Mass flow: Movement of the whole body of a liquid or gas.

Active transport: Transport which requires an energy input.

Channels: Passive protein pores in membranes.

Carriers: Protein 'machines' which transport molecules or ions across membranes.

Symports: Carriers which transport two different molecules in the same direction.

Antiports: Carriers which transport two different molecules in opposite directions.

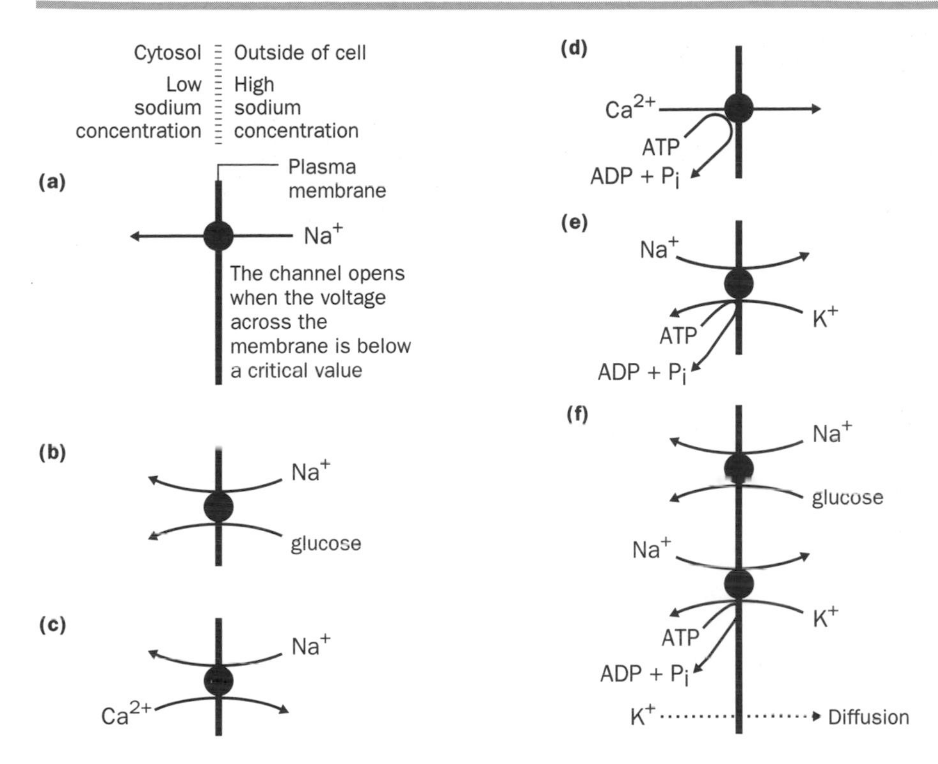

Fig. 6.1 A range of carriers in the plasma membranes of animal cells (see text for details).

they will both diffuse in together. More interestingly, if only Na^+ is more concentrated outside (and it normally is), glucose will still move in with the Na^+: the sodium gradient acts as an energy store, driving glucose uptake. A Na^+/Ca^{2+} antiport in the plasma membrane of animal cells acts in a similar way (Fig. 6.1(c)). Under normal cell conditions, the sodium gradient across the plasma membrane is used to pump Ca^{2+} out of the cell. Ca^{2+} is normally kept at low levels in the cytosol: pulses flood in as a chemical signal that the cell has detected something significant, such as a hormone, in its environment (Section 16.4).

Active pumps and antiports

These carrier proteins use energy from the hydrolysis of ATP to drive movement of molecules across a membrane. They must have active sites for ATPase activity, as well as binding sites for the molecules which are pumped. The plasma membrane Ca^{2+} ATPase of animal cells is a typical pump (Fig. 6.1(d)). This works with the Na^+/Ca^{2+} antiport to keep cytosolic calcium levels low. The plasma membrane Na^+/K^+ ATPase of all animal cells is especially well understood. This active antiport pumps three sodium ions out of the cell for every two it pumps in (Fig. 6.1(e)). Many membranes are quite permeable to K^+, so the major effect is to set up an energy-storing Na^+ gradient across the plasma membrane.This works along with a great variety of sodium-dependent passive symports and antiports, including those described above, to drive much of the active transport in and out of animal cells (Fig. 6.1(f)). Active transport across the plasma membranes of bacteria is organized in an equally economical way. This time, ATPases pump protons (H^+ ions) across the membrane. The proton gradient drives active transport by way of symports and antiports. In the living cell, concentrations of ions and ATP tend to be kept rather constant, so pumps tend to work in one direction only. In experiments, they can be persuaded to work backwards. If K^+ was at low levels and Na^+ at high levels in the cytosol, for example, the Na^+/K^+ ATPase would make, rather than use, ATP.

How do channels and carriers work? Passive channels provide an aqueous pore across the membrane through which ions or molecules can diffuse. Carriers work as molecular machines, binding the molecule which is to be transported, then changing shape, releasing it on the other side (Fig. 6.2).

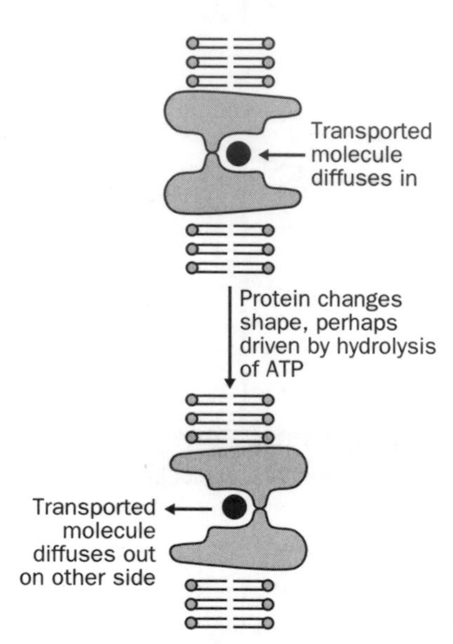

Fig. 6.2 A model of how a carrier protein may transport molecules.

6.6 Endocytosis

Some things are just too big to be transported across the plasma membrane: proteins for example, or even bacterial cells. Such things are taken in by the process of endocytosis, in which vesicles bud off the plasma membrane into the cytoplasm, bringing in other material within them.

A few types of animal cell can take in large pieces of material by endocytosis, which is then called phagocytosis (Fig. 6.3). The large vesicles containing the ingested material fuse with lysosomes. The digestive enzymes from the lysosome then get to work on the material. Some white blood cells can do this, taking in bacteria which have been recognized by antibodies, as well as the body's own dying cells. Some protozoa, such as *Amoeba*, feed like this.

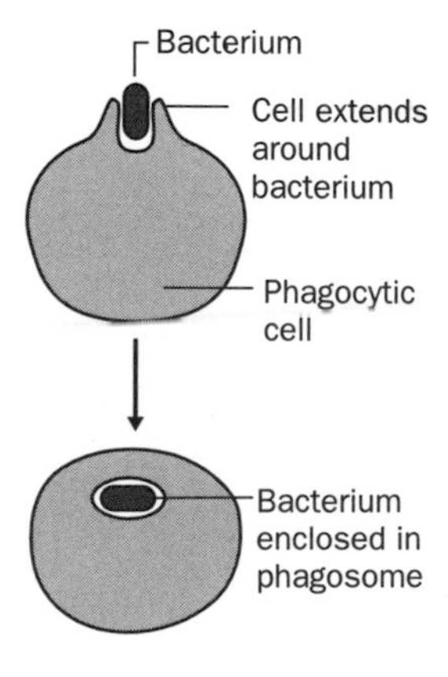

Fig. 6.3 Phagocytosis.

Most animal cells take in large molecules by a small-scale version of the same process: pinocytosis (Fig. 6.4). Small vesicles constantly bud off the plasma membrane. Molecules present in the external solution are likely to be trapped within the vesicles as they form. Some specific molecules are trapped much more efficiently, by receptors in the plasma membrane, which become grouped in the area of membrane which later buds off. Cholesterol, some signalling proteins, and transferrin, a protein which carries iron around the body, are all brought in by receptor-mediated endocytosis. Clathrin is a protein which orchestrates pinocytosis. It binds to the inner surface of the plasma membrane and to other clathrin molecules, forming a web over a small patch of membrane, called a coated pit. Receptors become grouped in the pit which bulges inwards, forming a clathrin-coated vesicle. Once it is completely detached from the plasma membrane, an enzyme dismantles the clathrin network which can then be recycled. The vesicle then fuses with an endosome, a larger membrane sac formed by the fusion of many similar vesicles. Quite what happens next is not fully understood, but some sorting of molecules goes on, some passing to the Golgi vesicles, some to vesicles which fuse with lysosomes, and some types of receptors may return to the plasma membrane in yet another type of vesicle.

Some protozoa, such as *Amoeba*, feed by phagocytosis.

A membrane alone is not enough . . .

The great importance of membranes is that they limit diffusion. They keep cell compartments separate and define their boundaries. Membranes allow special chemistry to go on in particular places, but they do not make that chemistry happen. Something needs to set up the special conditions of each cell compartment which give them their individual characteristics and roles. To a large extent, it is the molecular apparatus of membrane transport which does this. A membrane is only useful once molecules can cross it.

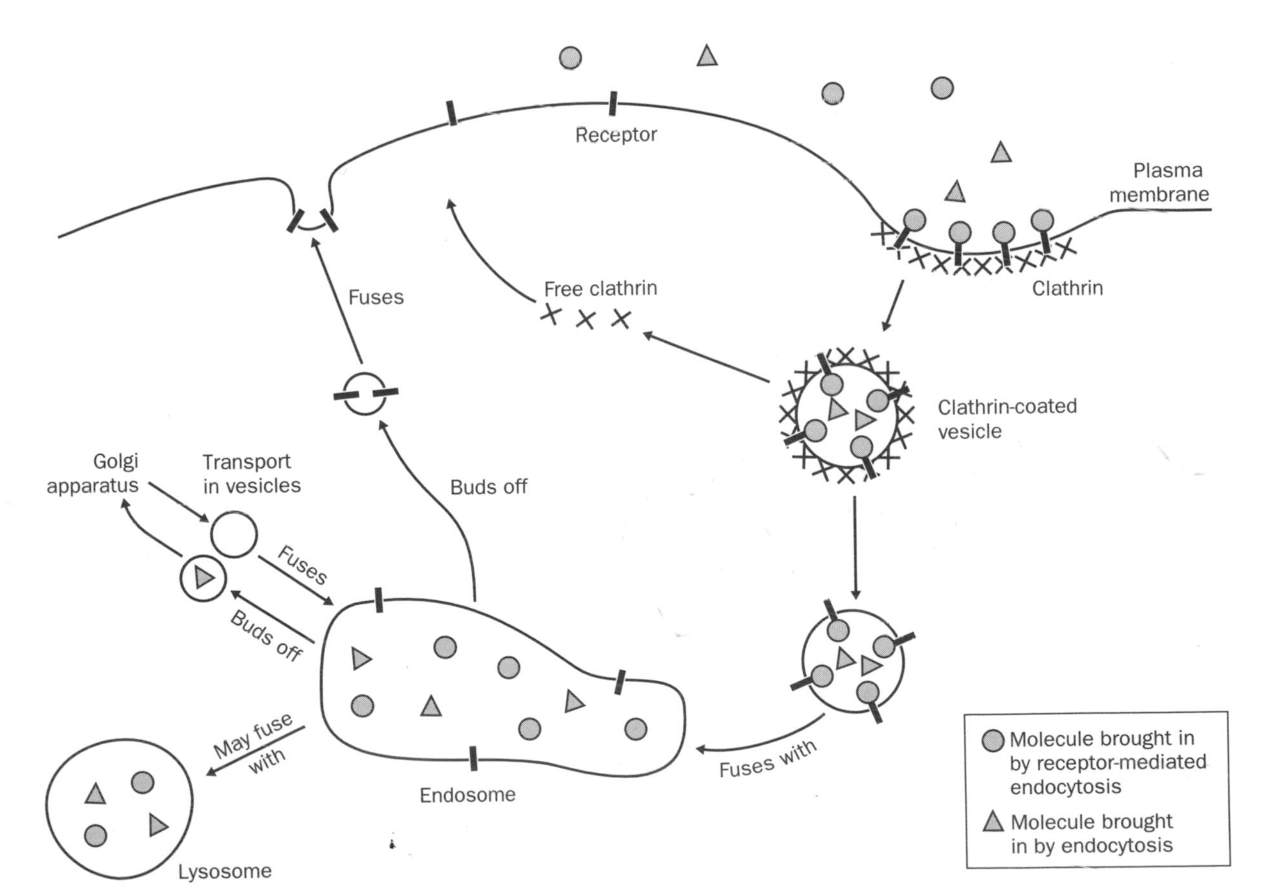

Fig. 6.4 Pinocytosis.

Summary

- Molecules are in constant motion.
- Diffusion is the net movement of molecules from a region of higher concentration to a region of lower concentration.
- Osmosis is the net movement of water across a partially permeable membrane, from a region of lower solute concentration to a region of higher solute concentration.
- Liquids and gases can move along as a body, mass flow.
- Active transport is transport which requires an energy input.
- Proteins embedded in membranes allow molecules to cross, actively or passively.
- Channels are passive protein pores in membranes; some can be opened and closed.
- Carriers are protein 'machines' which bind a molecule or ion, change shape, and release it on the other side of the membrane. Some are active.
- Antiports are carriers which transport two different molecules in opposite directions.
- Symports are carriers which transport two different molecules in the same direction.
- Large molecules and even bigger structures are brought into cells by endocytosis.
- Receptor-mediated endocytosis transports molecules in a highly specific way.
- Clathrin orchestrates endocytosis.

Exercises

6.1. Here are several situations in which substances are moving. For each, decide whether diffusion, active transport, mass flow or endocytosis is involved.

(i) Nitrate ions enter a plant cell, even though nitrate concentration is higher inside the cell than out.

(ii) Antibody proteins are taken up by cells in a baby's gut wall, from her mother's milk.

(iii) Urea moves down the ureter from the kidney to the bladder. The water it is dissolved in, as well as various mineral ions, move along together at the same rate.

(iv) Oxygen enters a cell. The cell is using oxygen in aerobic respiration, so its concentration is much lower inside the cell than out.

6.2. Turn to Box 9.3 which shows various channels and carriers in the membranes of mitochondria. Classify these as channels, symports, antiports, or simple carriers which transport only one thing, and as passive or active.

6.3. A plant cell has solute potential of –1000 kPa, and a pressure potential of 600 kPa.

(i) What is the water potential of this cell?

(ii) If the cell was put into each of the following solutions, would water enter or leave the cell? (Pure water: $\psi = 0$)

Solution A: $\psi = -1000$ kPa
Solution B: $\psi = -600$ kPa
Solution C: $\psi = -400$ kPa
Solution D: $\psi = -200$ kPa

6.4. Plasma membrane is constantly pulled into the cell during endocytosis, yet the total surface area of the cell does not change. Why not?

6.5. Cholesterol is carried in the blood in clusters called low density lipoprotein (LDL). Cells take up LDL by endocytosis. Liver cells regulate the level of cholesterol in the blood, but they can only do this if they can take up LDL (Box 15.2).

A mutant gene prevents uptake of LDL and leads to excessive levels of cholesterol in the blood. Other molecules can still be taken in by endocytosis. Suggest the role of the protein coded for by the normal form of this gene.

Further reading

Alberts, B., Bray, D., Lewis, J., Raff, M., Roberts, K. and Watson, J.D. *Molecular Biology of the Cell* (3rd ed.)(New York: Garland, 1994). To explore movement of molecules across membranes, I suggest the relevant sections of this enormous but excellent textbook which puts the reader in touch with current ideas without too much pain on the way.

Atkins, P. *Physical Chemistry* (5th ed.)(Oxford: Oxford University Press, 1994). For a chemist's view of diffusion and osmosis, try this excellent textbook aimed at chemistry undergraduates.

Salisbury, F.B. and Ross, C.W. *Plant Physiology* (4th ed.) (Belmont CA: Wadsworth, 1992). The water relations of plant cells are the business of plant physiologists. This is a big undergraduate textbook. Any text in this field has to navigate a minefield of conflicting ideas and evidence in so many topics. This book does as well as you could hope in drawing out conclusions without avoiding the arguments.

The Gene at Work

Contents

Connections

► Before studying this unit, you should know about the structure of DNA, RNA and proteins (Unit 3). You should also have a basic understanding of cell structure (Unit 4). This unit explains what genes are, and is an essential background to studying genetic variation and the ways in which genes are passed from one generation to the next (Units 11–14).

7.1 Genes

A seed germinates. The seedling grows up. It is an oak tree, like its parents. My son grows up unmistakeably human, like his parents. And he must get his brains, good looks and charm from somewhere . . . Something is being passed from parents to offspring – genes.

A new individual is formed when two gametes fuse in sexual reproduction. Whether a sperm and an egg are meeting deep inside a female porcupine, or a nucleus from a pollen grain is fusing with a nucleus inside the ovule of a cactus flower, the effect is the same. **Genes** from the two parents are meeting. The new cell has a novel, unique combination of genes. When this cell divides to make a body, the genes must somehow be copied so that each new cell in the body has the same genes. The genes are instructions for the development of each cell. They control the development of distinctive animal cells, forming porcupine muscles or quills, or obviously plant cells in the red petals of a cactus flower. Genes are at work in the cell, shaping the way it works, the way it interacts with other cells, and through this the way in which an entire body is formed. Bodies are only there because of the genes inside their cells. The body is the ultimate product of the genes, a vehicle for their survival.

Genes do not live alone in cells, but in big populations called genomes. The human genome is made up of about 100 000 genes. This huge number of genes is there, copied, in each human cell. Some of these genes are very similar, and appear to be related. Over many generations, the make-up of the genome changes, evolves.

So, genes can be looked at in three ways. They are a set of instructions at work in the living cell; they are the units of inheritance; and they are the components of the genome. This unit is about the gene at work.

Gene: An inherited instruction for cell development; a length of DNA coding for one type of polypeptide.

7.2 Genetic information

Genes are bits of DNA. In prokaryotes, the DNA is found in the cytoplasm as a single loop, a chromosome. In eukaryotes, DNA is shut away in the nucleus as a number of linear chromosomes. In each case, the chromosome is a single, very long DNA molecule. Each chromosome consists of many genes as well as much DNA which does not form part of any gene.

Genes work by holding information on the amino acid sequence of polypeptides. Each gene codes for one type of polypeptide. Human growth hormone is made up of only one polypeptide, so it is coded for by one gene. Insulin, a hormone in mammals which signals high levels of glucose in the blood, is made up of two polypeptides: two genes code for it. Haemoglobin, the protein which carries oxygen in blood, is made up of four polypeptides, two each of two types, so again two genes code for it. Proteins have a vast number of functions (Section 3.5: Proteins are important) and can work in very specific ways. As a result of this the instructions for cell structure, development, function and division are nothing more than instructions for making a very large number of different types of polypeptide.

DNA is able to hold information because it has not one but four types of bases (Section 3.8). The deoxyribose phosphate backbone of the molecule is the same everywhere, but the base attached to any given sugar can be adenine (A), guanine (G), cytosine (C) or thymine (T). The sequence of bases holds information on the sequence of a polypeptide. It is a very simple language, consisting of just four letters.

A DNA molecule consists of two strands linked by paired bases. A on one strand is always linked to T on the other. G and C are linked in the same way. Base pairing makes DNA very stable, important if it is to survive as an information store in cells. It also means that only one of the two strands holds meaningful information. If part of one strand has the sequence GATTTCGGC, the other must be CTAAAGCCG. It would be astonishing if this mirror image also contained useful information. The strand coding for polypeptide sequence is called the sense strand. The other is the antisense strand.

7.3 Transferring genetic information

DNA is an amazing molecule. As well as storing information, it can organize the transfer of this information to other molecules. This goes on in two quite different ways. Firstly, it can control the synthesis of new strands of DNA which are identical to itself. This is **DNA replication**. It happens before cells divide, because the new cells need copies of every gene. Secondly, it can organize protein synthesis using the information stored in the DNA base sequence. This happens in two stages, often with a third added in between (Fig. 7.1). In the first stage, a single strand of RNA is made, using just the sense strand of a single gene (rarely a few neighbouring genes) as a template. This process is called **transcription** (literally 'writing across'). The RNA which is made is called messenger RNA (mRNA). In the next stage, a polypeptide is made, following the sequence coded for by the mRNA strand. This process is called **translation**. In most eukaryotic cells, another stage is needed. In bacteria, the DNA of one gene is a continuous line of information. In most eukaryotes, however, the majority of genes are broken into fragments (**exons**) with lengths of DNA in

DNA replication: The making of an identical copy of a DNA molecule.

Transcription: The formation of an mRNA strand using the DNA of one gene as a template.

Translation: The making of a polypeptide, following the information coded for by mRNA.

Exon: A section of a eukaryotic gene which codes for part of a polypeptide.

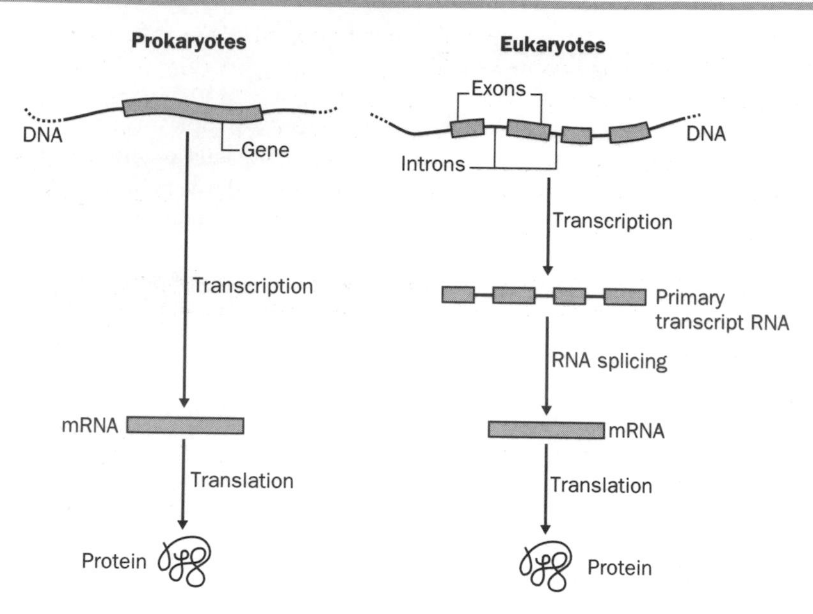

Fig. 7.1 The flow of genetic information from DNA to protein.

Intron: A section of DNA which does not code for polypeptide sequence, between the exons.

RNA splicing: Removes the non-coding introns from the mRNA.

between (**introns**) which do not code for any part of the polypeptide. Introns are transcribed along with the exons, but are edited out of the RNA before it is translated. This is **RNA splicing**.

Transcription, splicing and translation go on constantly in the working cell. A gene can be transcribed many times to make as much of that RNA as is needed. Each mRNA strand can be transcribed many times to make as much of its polypeptide as is needed. All this activity needs to be controlled. Not all proteins are needed all the time, or in every cell of a body. Some are needed in larger amounts than others. The most important controls act on transcription.

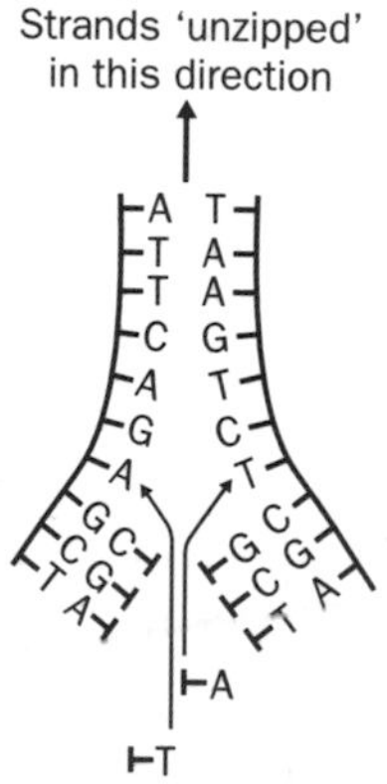

DNA replication

Before a cell divides, all its DNA must be copied. The complimentary base pairing (A–T, G–C) makes this possible. First, the two DNA strands are separated by enzymes. This leaves unpaired bases on both strands (Fig. 7.2). During replication there are lots of free nucleotides within the cell. The deoxyribonucleotides which make up DNA come in four types, one with each sort of base. These pair up with the unpaired bases. No enzyme is needed for this. Random movement, and hydrogen bonding between the 'correct' pairs of bases is all that is needed. An enzyme, DNA polymerase, is needed to link the nucleotides. As a result, there are now two DNA molecules where there was only one before. DNA replication is described as semi-conservative, because each new molecule has one old and one new strand.

Strands 'unzipped' in this direction

Fig. 7.2 During DNA replication, free nucleotides bind to unpaired bases on each strand. They are later joined up by DNA polymerase.

Replication begins at particular places in the DNA: initiation sites. The double helix is separated at this point, and replication proceeds in both directions (Fig. 7.3(a)). The point at which the strands are being separated at any given moment is called the replication fork. Prokaryotic chromosomes have just one initiation site. Replication forks move out around the chromosome from the initiation site. When they meet on the other side of the loop, the two new chromosomes fall apart (Fig. 7.3(b)). The linear chromosomes of eukaryotes have several initiation sites. Each replicating chromosome will have many replication forks at different places (Fig. 7.3(c)).

The principle of replication is very simple. However, the replication fork is a complicated place with a whole set of enzymes needed to solve various technical problems (Box 7.1).

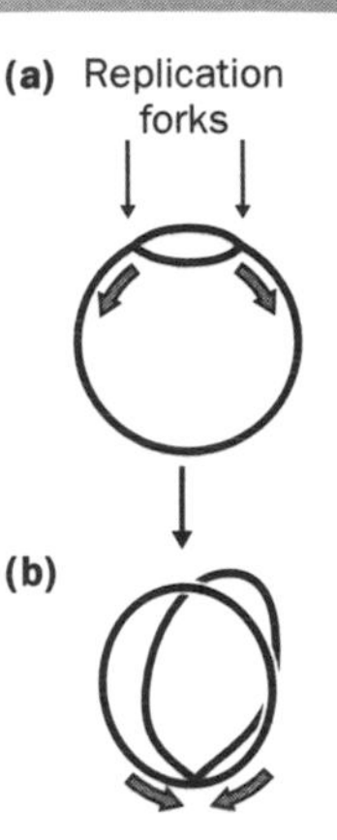

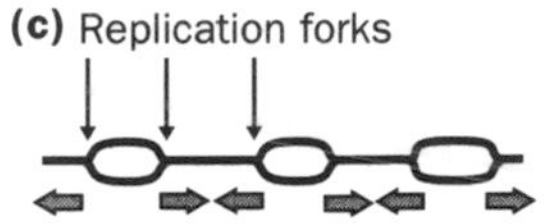

Fig. 7.3 Replication begins at one initiation site in bacterial chromosomes (a) and (b) but at several in eukaryotic chromosomes (c).

Box 7.1

The replication fork

DNA replication proceeds along a chromosome from an initiation site. The point at which the double helix is separated and new DNA strands are formed is called the replication fork. Free nucleotides bind to their complimentary nucleotides in the single strands. DNA polymerase joins them up to form new strands. Sounds simple but it's not . . .

Problem 1

DNA polymerase can only join nucleotides onto the 3′ end of a strand. On the leading strand it trundles along happily, but it can't follow the replication fork on the lagging strand.

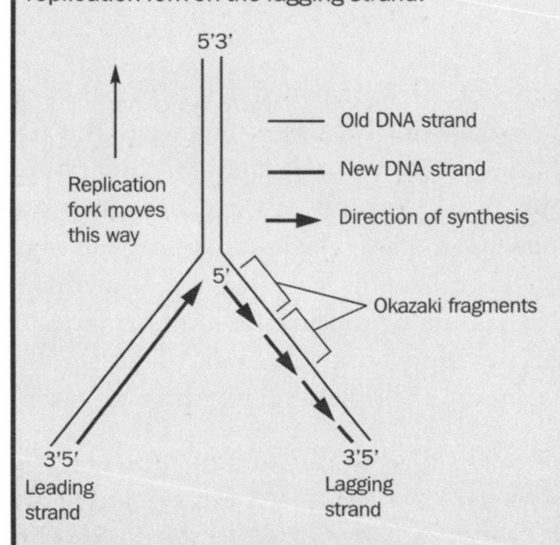

Solution: on the lagging strand, DNA polymerase waits until a length of single-strand DNA is exposed. Then it works away from the replication fork, joining up a short strand of nucleotides; an Okazaki fragment.

Problem 2

DNA polymerase can only join *free* nucleotides to the end of a growing strand. It cannot link the end of one Okazaki fragment to the beginning of the next.

Solution: DNA ligase is an enzyme which does just that.

Problem 3

DNA polymerase can only join nucleotides to the end of an existing chain. It cannot begin an Okazaki fragment from nothing. This 'fussiness' is, in part, what allows it to make so few errors.

Solution: a less fussy enzyme, DNA primase, synthesizes short RNA primers, complimentary to about 10 nucleotides at the start of each fragment. DNA polymerase can then extend these. The primers are removed and replaced by other enzymes.

Problem 4

The double helix needs to be opened in front of DNA polymerase so that nucleotides can come in and pair with the existing strands.

Solution: DNA helicases track along single-stranded DNA, forcing apart the helix ahead of them.

Problem 5

The single strands must be prevented from immediately rejoining.

Solution: helix destabilizing proteins bind to single-stranded DNA yet leave the bases free to pair.

Problem 6

Unwinding the double helix would lead to supercoiling in the double-stranded DNA ahead of the fork. It could easily form tangled dreadlocks because the rest of the chromosome is usually too long to rotate freely. Can't see it? Nor could I at first. Try it with a couple of long bits of string.

Solution: DNA topoisomerases cut, rotate and rejoin the double helix to remove supercoiling.

This marvellous array of co-operating proteins is bound together into a huge complex which moves along with the replication fork.

Transcription

When a cell needs to start making a particular protein, the correct gene needs to be transcribed. The mRNA that is made can then be translated to make the protein. DNA is transcribed one gene at a time. A short base sequence at the start of the gene, the promoter, shows where to begin. It also indicates the sense strand. In eukaryotes, most promoters include a sequence rich in A–T/T–A base pairs, a TATA box.

A group of complex enzymes, RNA polymerases, are central to transcription. An RNA polymerase binds to the promoter and moves along the DNA strand in the $5' \rightarrow 3'$ direction (Fig. 7.4). As the enzyme moves along it unravels the double helix. Free nucleotides bind to the unpaired bases, just as in replication. However, these are ribonucleotides not deoxyribonucleotides, and uracil (U) replaces thymine. RNA polymerase joins the nucleotides along the sense strand to form an RNA chain. As the enzyme passes, the RNA strand peels away from the DNA, and the double helix reforms.

At the end of the gene, another DNA sequence acts as a stop signal. RNA polymerase falls away and the RNA is released. The whole process of transcribing a gene has taken a matter of minutes.

In prokaryotes, transcription makes mRNA ready for translation to form polypeptides. In eukaryotes, the non-coding introns have to be removed in RNA splicing (Box 7.2) before the mRNA can be translated. A few specialized RNAs are useful in themselves and do not need to be translated. Some of these are introduced in Box 7.3 on page 114.

Translation

In translation, a polypeptide is made using the information coded in the mRNA. The big problem is to translate the four-letter language of RNA base sequence to the 20-letter language of amino acid sequence. Two pieces of molecular machinery are involved. The first is transfer RNA (tRNA). This is a family of molecular adaptors which link sequences of three nucleotides (**codons**) to particular amino acids. Table 7.1 shows which amino acid is coded for by each codon. This 'genetic code' is identical in almost all cells. Each tRNA is a single RNA strand wound around itself in places to form a dog-leg. One end of the molecule is covalently bonded to a particular amino acid: a specific enzyme helps bring them together. The other end has three unpaired

Codon: A sequence of three nucleotides which codes for one amino acid.

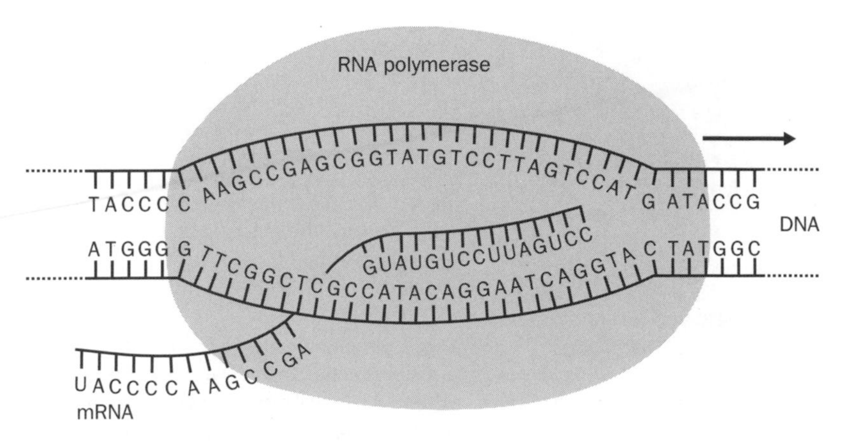

Fig. 7.4 Transcription.

Box 7.2

RNA splicing

Eukaryotes have genes which are interrupted by non-coding sections of DNA: introns. These are transcribed, but are edited out of the primary RNA transcript before translation.

RNA splicing is carried out in the nucleus by spliceosomes. These complex pieces of molecular apparatus are made up of several subunits. Each is made up of a number of proteins and an RNA, in just the same way as a ribosome, but much smaller.

The spliceosome recognizes sequences at each end of the intron and removes it.

In some cases, one type of RNA can be spliced in more than one way. Different versions of the protein are produced as a result. For example, B lymphocytes produce two versions of the antibody molecule (Section 23.5). One is released into the blood when mounting a response to a foreign molecule (antigen). The other is anchored in the plasma membrane. This lets the cell know when an antigen which binds to its own particular type of antibody appears in the body. The membrane-bound version has an extra sequence of hydrophobic amino acids at one end, which become embedded in the membrane. Splicing takes place at a different site when this form is made.

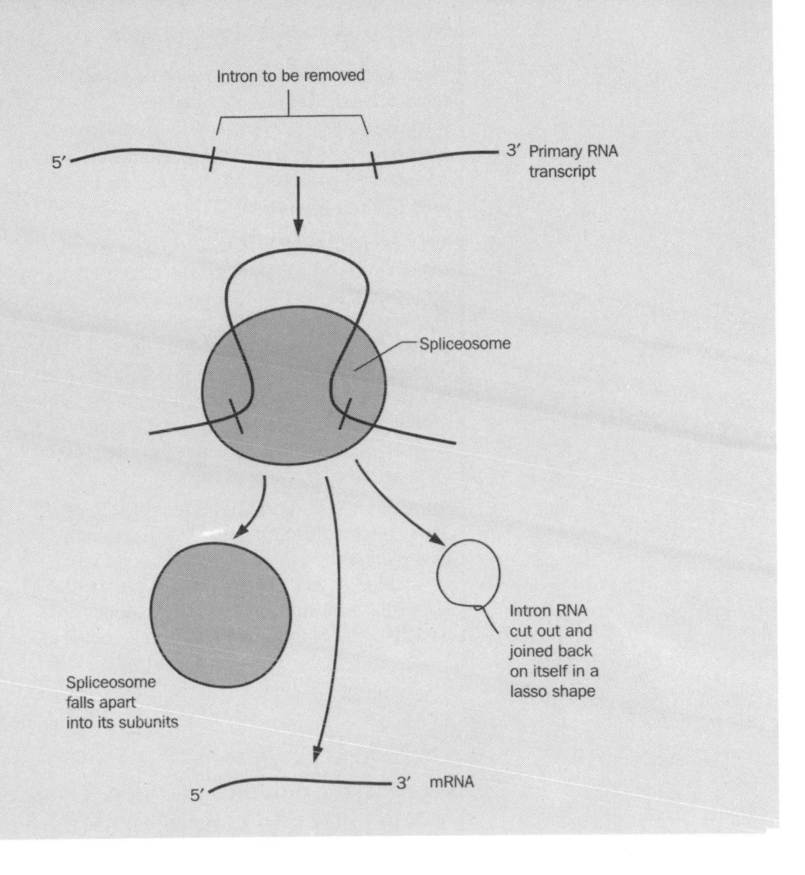

Table 7.1 The genetic code: mRNA codons and the amino acids for which they code

UUC	phenylalanine	UCC	serine	UAC	tyrosine	UGC	cysteine
UUA	leucine	UCA	serine	UAA	STOP	UGA	STOP
UUG	leucine	UCG	serine	UAG	STOP	UGG	tryptophan
CUU	leucine	CCU	proline	CAU	histidine	CGU	arginine
CUC	leucine	CCC	proline	CAC	histidine	CGC	arginine
CUA	leucine	CCA	proline	CAA	glutamine	CGA	arginine
CUG	leucine	CCG	proline	CAG	glutamine	CGG	arginine
AUU	isoleucine	ACU	threonine	AAU	asparagine	AGU	serine
AUC	isoleucine	ACC	threonine	AAC	asparagine	AGC	serine
AUA	isoleucine	ACA	threonine	AAA	lysine	AGA	arginine
AUG	methionine	ACG	threonine	AAG	lysine	AGG	arginine
GUU	valine	GCU	alanine	GAU	aspartic acid	GGU	glycine
GUC	valine	GCC	alanine	GAC	aspartic acid	GGC	glycine
GUA	valine	GCA	alanine	GAA	glutamic acid	GGA	glycine
GUG	valine	GCG	alanine	GAG	glutamic acid	GGG	glycine

Box 7.3

RNA: not just a messenger

Messenger RNA is made when a gene is transcribed, and simply carries information on polypeptide sequences, which is used in translation. Some classes of RNA are never translated, but do useful jobs in their own right.

Transfer RNA (tRNA): tRNAs do the central task in translation: linking base sequence to amino acid sequence (Section 7.3: Translation). Base pairing between different parts of the molecule helps give them characteristic three-dimensional shapes, just like the tertiary structure of a protein. Because of this, tRNAs and various other RNAs can have specific binding and catalytic sites, just like proteins.

Ribosomal RNA (rRNA): Ribosome subunits are ribonucleoproteins: RNAs with proteins arranged around them. No one is quite sure which binding and catalysis tasks are carried out by proteins and which by RNA. Many people suspect that RNA is the major catalyst. Either way, the ribosome would not exist without RNA.

Small nuclear RNA (snRNA): These link with proteins to form the subunits of spliceosomes. Again, the precise role of the RNA component is unclear.

SRP RNA: This forms the core of another ribonucleoprotein, the signal recognition particle (SRP). SRP is found in the cytoplasm and helps bind ribosomes to the endoplasmic reticulum if they are making a polypeptide which needs to enter the ER (Section 4.5). SRP recognizes a signal sequence of amino acids at the start of the relevant polypeptides, as well as binding to the ribosome and a protein in the ER membrane. Again, the precise role of RNA is unclear.

In many viruses RNA is the genetic material (Box 30.2). At least in some plant viruses, some of the RNA seems to have a catalytic role as well, linking coat proteins when the viruses assemble themselves.

bases sticking out, the anticodon. The anticodon binds to the three complimentary bases in mRNA. In some cases, the third base binds only weakly, allowing the same tRNA to bind to more than one codon (e.g. UAC and UAU). This explains why groups of codons differing only in the third base often code for the same amino acid.

The second piece of molecular machinery is the ribosome (Section 4.4). Its role is to bring together the mRNA and tRNAs. It also acts as a catalyst for the joining of amino acids to form the polypeptide. Ribosomes consist of two subunits. Each is made up of ribosomal RNAs (rRNA) and specific proteins.

Ribosomes bind to a start signal near the 5′ end of the mRNA. This is an AUG codon, with other distinctive sequences upstream of it to distinguish it from other AUGs. The ribosome moves along in the 5′ → 3′ direction. As it travels, tRNAs bring in amino acids which are joined to the growing polypeptide. Box 7.4 shows the sequence of events in more detail. As soon as one ribosome is clear of the start codon, another can begin. This means that any one mRNA chain is transcribed by several ribosomes at once (Fig. 7.5). The ribosomes fall away at a stop codon (UAA, UAG or UGA), releasing the finished polypeptides.

In theory, one piece of mRNA could be translated in three different ways, because it can be divided into codons in three different ways (Fig. 7.6). These are called **reading frames**. Normally only one reading frame carries meaningful information, and mRNA is translated only in this one. The position of the start codon determines which reading frame this will be (Fig. 7.6(d)).

Reading frames: The three ways an mRNA sequence can be divided into codons.

Start codons have another interesting consequence. AUG codes for the amino acid methionine. Because of this, all newly made polypeptides begin with methionine.

Box 7.4

On the ribosome

Ribosomes are the sites of translation. They provide a place where tRNA can bind rRNA. They also link amino acids brought in by tRNA to the growing polypeptide chain.

The ribosome can bind two tRNAs. The growing polypeptide chain is linked to one. Another tRNA comes in. Its anticodon is complimentary to the next codon, and it brings the appropriate amino acid with it.

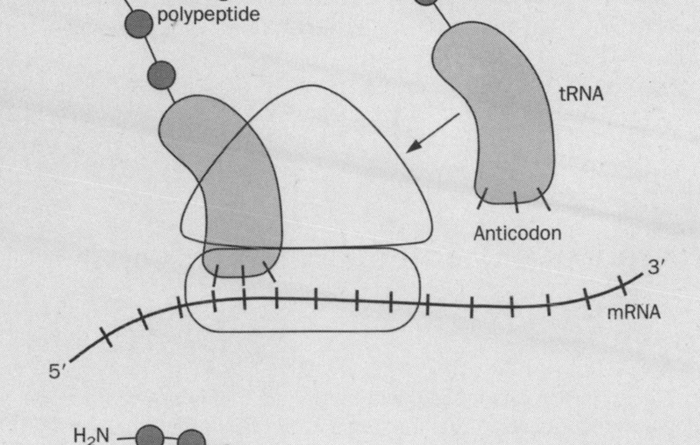

The growing chain is linked to the next amino acid, still attached to its tRNA.

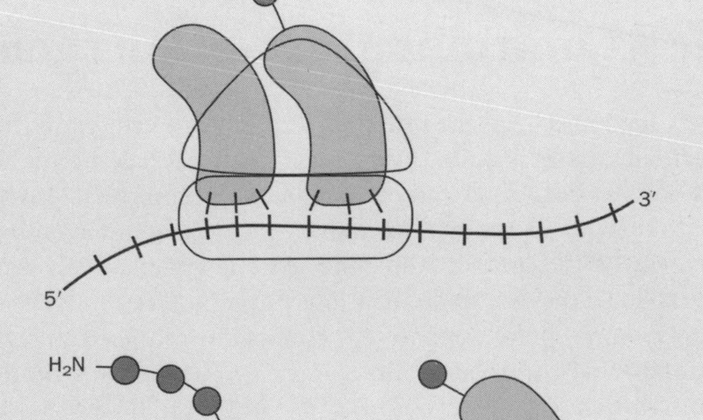

The ribosome slides three nucleotides further down the mRNA. The first tRNA falls away as the second tRNA moves into the first binding site. Now the next tRNA can bind.

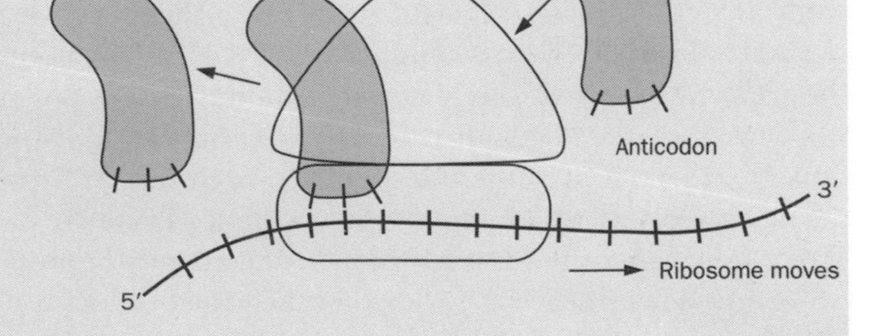

This might affect the function of some completed proteins, so it is removed by an enzyme. It is easy to concentrate on the roles of tRNAs and ribosomes in translation. However, enzymes like this one, and the aminoacyl tRNA synthetases which ensure that each tRNA is loaded with the correct amino acid are also crucially important.

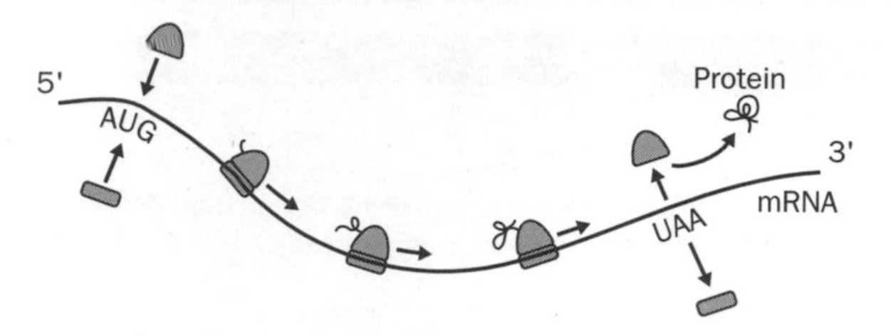

Fig. 7.5 mRNA can be transcribed by several ribosomes at once.

(a) AGU | GAU | GAA | ACU | GUC | AUC | CGC | UCA

(b) A | GUG | AUG | AAA | CUG | UCA | UCC | GCU | CA

(c) AG | UGA | UGA | AAC | UGU | CAU | CCG | CUC | A

(d) AUG | AAA | CUG | UCA | UCC | GCU | CA

Fig. 7.6 An mRNA sequence could be transcribed in any of three reading frames (a–c). The AUG codon determines which reading frame is actually used (d).

7.4 Controlling the working gene

Every human cell has the insulin gene. Only one cell type – the β-cell in the pancreas – actually makes insulin. Every male mammal has the master gene which triggers development of a male rather than a female reproductive system. Yet this gene only needs to make its product for a short time, early in the embryo's development. For the rest of the animal's life, the gene is quite unnecessary. Genes need to be controlled if they are to do their jobs properly. They must be switched on and off. In theory, this could happen by controlling either transcription or translation. Translational controls certainly do exist, but the most important controls are on transcription.

Gene control was first investigated in prokaryotes. It became clear that genes are controlled by specific regulator proteins. These can recognize short, specific sequences in DNA. The protein binds to a DNA sequence upstream of the start of the gene, the operator. This affects the binding of RNA polymerase to the promoter. In some cases, the regulator protein is necessary for polymerase to start transcription. In others, it prevents transcription. Each regulator protein is coded for by a regulator gene elsewhere on the chromosome. These are switched on all the time. This system can control transcription because regulator proteins also bind molecules which provide evidence as to whether the gene product is needed. If the structural gene makes an enzyme to digest some food molecule, transcription should begin when the regulator binds the food molecule. If the gene makes an enzyme involved in manufacturing some essential molecule, the gene should be switched off when the regulator binds that molecule, because it is already present in the cell.

The whole system of structural gene, operator and regulatory gene is shown in Fig. 7.7. Box 7.5 shows two well understood examples in *E. coli*, the gut bacterium.

In eukaryotes, DNA binding proteins also promote and inhibit transcription. The

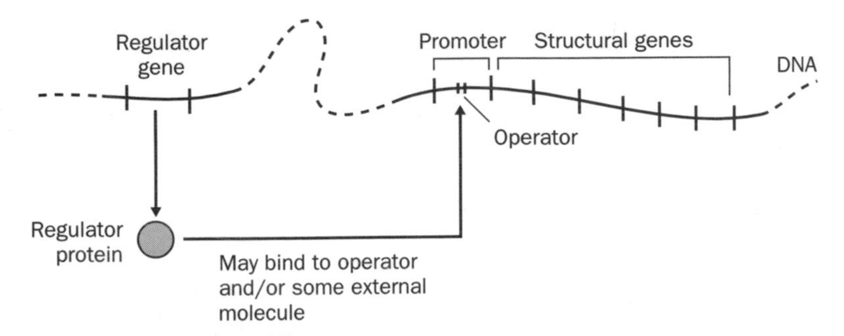

Fig. 7.7 The basis of gene control in prokaryotes (see text for details).

main difference is that many proteins, rather than just one or two, can affect a single gene. There is no way that they could all bind to sites within the promoter: they simply would not all fit. Instead, some bind to sites hundreds or even thousands of nucleotides upstream of the promoter. These also bind to the general transcription factors, proteins which eukaryotic RNA polymerases need to help them recognize promoters. The DNA has to loop back on itself, allowing the regulatory proteins to be in two places at once (Fig. 7.8).

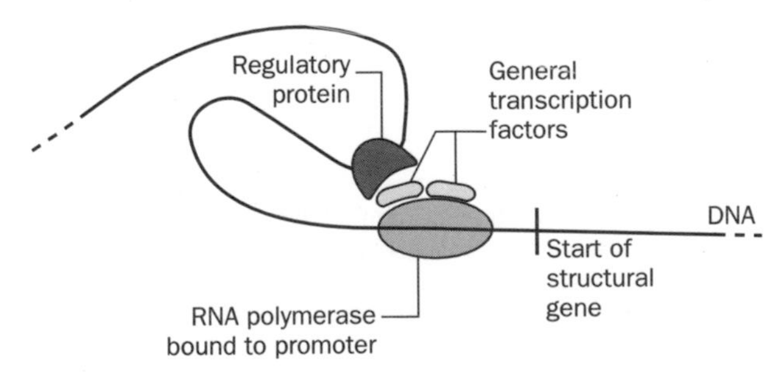

Fig. 7.8 Regulatory proteins can influence transcription even if they bind upstream of the promoter.

Many proteins can control one gene. In the same way, one regulatory protein can control a whole set of eukaryotic genes. This is particularly important in development, when a cell type appears in the embryo for the first time. Many genes need to be turned on or off all at once. One gene, a master gene, makes a regulatory protein which affects, directly or indirectly, all these genes. The developmental problem is now much simpler: control the master gene and everything else follows.

Regulatory proteins do not work in an all-or-nothing way. The rate of transcription can be anything from very rapid to dead slow, depending on the levels of each regulatory protein. In bacteria, the difference in rate is typically 1000-fold. In the bodies of eukaryotes, each cell type has genes which it will never need. These are turned off in more complete, more permanent ways. Some sections of DNA become highly condensed, in a way that is passed on when the cell divides. This prevents transcription. Box 7.6 shows an extreme example of this. In vertebrates, once a gene has been turned off by regulatory proteins, enzymes add a methyl group to cytosine residues. DNA methylation strongly inhibits transcription. It is reversible, but this

Box 7.5

Gene control in bacteria

Transcription of prokaryotic genes is controlled by regulator proteins. Two very different examples show how this helps the gut bacterium, *E. coli*, operate efficiently. In each case, a group of genes with related functions, an operon, is controlled as a unit. An operon is transcribed as a single mRNA strand.

The *trp* operon

The *trp* operon codes for a set of enzymes which make the amino acid tryptophan. The cell needs a supply of tryptophan for protein synthesis. If there is plenty in the cell, no more is needed, so the enzymes are unnecessary. If there is a shortage, the enzymes are required to make more. The genes are only transcribed when tryptophan is at low levels.

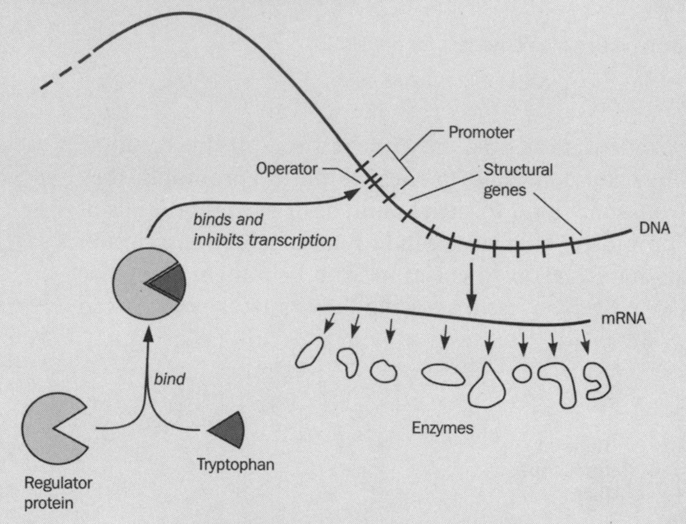

The *lac* operon

The *lac* operon codes for proteins which transport lactose across the plasma membrane, and a digestive enzyme. This breaks lactose into glucose and galactose. Lactose will be present in a mammal's gut only if it is drinking milk. This is normal in infancy but, in most species, unheard of later on. The efficient *E. coli* cell only transcribes the genes if lactose is present. However, if plenty of glucose is present as well, there is no need to bother with lactose. The gene is transcribed only if lactose is present *and* glucose is scarce.

One regulatory protein inhibits transcription and binds *either* to DNA *or* to lactose, not both. A second protein stimulates transcription, but binds to DNA only if it has also bound to cyclic AMP (Section 16.4). In bacteria, cAMP is a hunger signal: glucose is low.

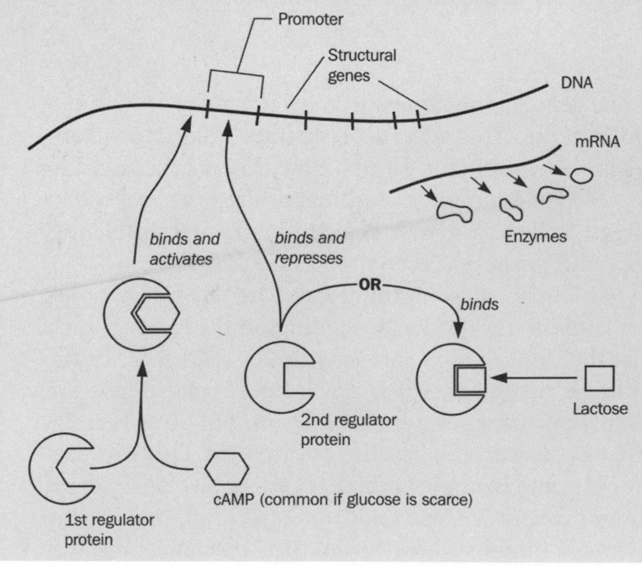

Box 7.C

Inactivating the X chromosome

The sex chromosomes of mammals come in two versions. The long X chromosome has many essential genes. The shorter Y chromosome is non-essential, and carries a gene determining maleness (Section 12.4). Females have two X chromosomes, males have one of each. This could mean that females have more of the proteins coded for on the X chromosome: an unhelpful imbalance. To overcome this, **one entire X chromosome is inactivated in each female cell.**

A female embryo begins life with two active X chromosomes per cell. After a period of cell division, but still quite early in the life of the embryo, one X chromosome is inactivated in each cell. No genes on this chromosome can then be transcribed. Which of the two chromosomes is inactivated seems to be entirely random.

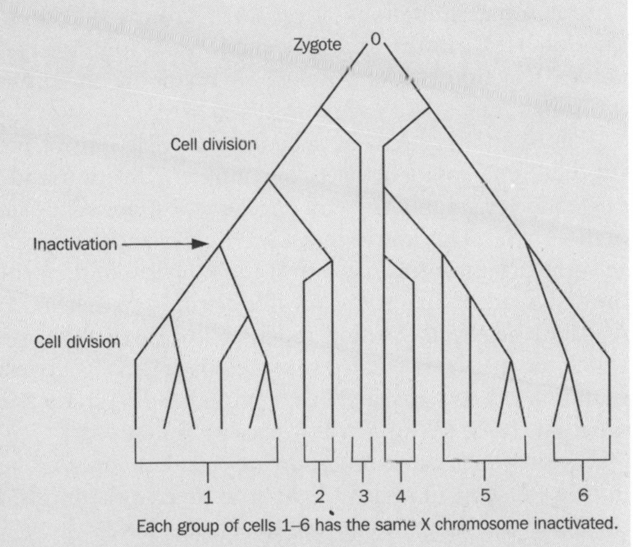

Each group of cells 1–6 has the same X chromosome inactivated.

When the cells divide again, the copies of the inactivated chromosomes remain condensed. This means that the adult body has groups of cells with the same active X chromosome. Tortoiseshell cats show this vividly. Cats carry a fur-colour gene on the X chromosome. Tortoiseshells are always female. One X chromosome carries an allele (version of the gene) which codes for black fur; the other carries an allele for ginger (yellow) fur. Only one is active in each adult cell. The fur has patches of black and ginger. Each patch is derived from a single cell at the stage of development when inactivation took place.

A tortoiseshell cat, keeping itself warm with evidence of X chromosome inactivation.

only happens in special circumstances. These extra controls mean that in eukaryotes, a fully turned-off gene may be transcribed at as little as one millionth the rate of a turned-on gene.

7.5 Mutation: a threat to the working gene

Mutations are accidental changes to the genetic material. Point mutations affect just one nucleotide pair. In doing so they alter the product or control of a single gene. Chromosomal mutations affect large groups of genes, or whole chromosomes (Section 11.3). Both result in genetic variation, and so are the raw material for natural selection (Units 2 and 11). Only point mutations affect the working of an individual gene, so they are considered here.

Point mutations may be substitutions, deletions or insertions (Fig. 7.9). In substitution, one base is replaced by another. Deletion and insertion involve losing a nucleotide or gaining an extra one. Substitution will normally only affect one group of three bases. This means that it will only change one amino acid in the polypeptide. In some positions this may affect the function of the protein, for example if it is in the active site of an enzyme, or if it disrupts an α-helix. In many other positions, it may have no effect. Deletion and insertion are more sinister. They disrupt the way in which the mRNA is divided into codons (Fig. 7.10). Everything downstream of the mutation will be translated in a different reading frame, so many amino acids will be different. These are known as frameshift mutations.

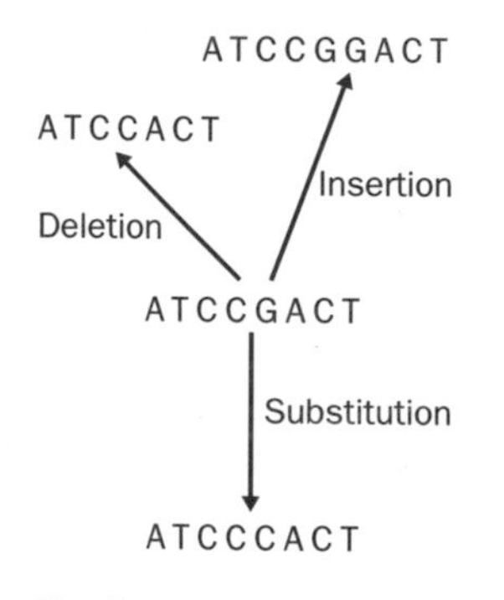

Fig. 7.9 Three types of point mutation.

Factors which cause mutation are called mutagens. These include ionizing radiation like X-rays, ultraviolet light, and diverse chemicals. Box 7.7 shows how some mutagens and defects in the cell can change DNA.

Mutation happens so often that something has to be done about it. Cells have several neat mechanisms for spotting common mutations and repairing the damage. They use teams of enzymes which detect mismatched bases, or bases other than A, T, G and C, and replace them with the correct base. This poses problems. How does the cell know which of a mismatched pair is correct? Which base should replace a non-standard one? Fortunately, some types of substitution are much more common than others. For example, cytosine can change spontaneously to uracil, so a G–U base pair should almost certainly be repaired as G–C. Chemical changes which disrupt the regular structure of DNA can also be corrected. For example, ultraviolet light

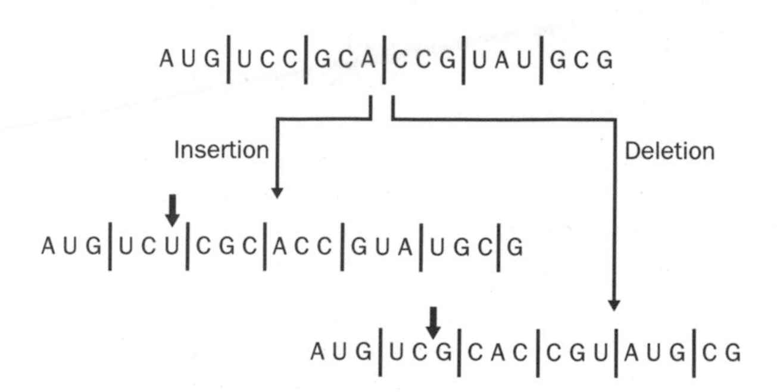

Fig. 7.10 Insertion and deletion cause frameshift.

Box 7.7

How mutations happen

Mutations are changes to the genetic material. Many mutagens and spontaneous events cause specific types of mutation. Here are some examples.

1. Faulty DNA polymerases can result in very high mutation rates. One base pair is replaced by another on replication. Some such faults can give very specific substitutions.
2. Adenine can flip between two forms in which the same atoms are arranged in very slightly different ways. The common form pairs with thymine, but the rare form pairs with cytosine. This results in an A–C base pairing. If not repaired, one of the two new DNA molecules will have the wrong sequence after replication.
3. Nitrous acid deaminates some DNA bases. Adenine is changed to hypoxanthine. This pairs not with thymine but with cytosine. When the DNA repair mechanism spots hypoxanthine, it removes it and replaces it with guanine to match cytosine. Hence, A–T becomes G–C.
4. Molecules with a very similar structure to the normal bases act in a similar way. They are mistaken for one base, and are incorporated into DNA at replication, but then pair with the 'wrong' base.
5. Some flat molecules like acridines slide in between adjacent bases. If this is in a repetitive sequence (e.g. CGCGCGCGCGCG) this can cause the opposite strand to shift, pairing with other bases in the sequence. In the resulting confusion, some bases are left unpaired. Others are added to pair with them, or they are deleted.

promotes bonding between pyrimidine bases which are next to one another on the same strand. These can be cut out and replaced with separate bases.

Mutagens in the environment and the inevitable errors in DNA replication mean that mutation is a constant serious threat to the working gene. DNA repair mechanisms are vital to the cell's survival.

7.6 DNA, RNA and the origin of life

DNA, RNA and proteins are central to the workings of cells. They are linked by extraordinarily complicated processes: transcription, translation and RNA splicing. How did all this complexity evolve? What came first? DNA stores genetic information and organizes its own replication, but cannot catalyse reactions. Proteins can act as effective, specific catalysts, but cannot replicate themselves. Neither on its own could be the basis of anything remotely like life. RNA, however, can do both. It can hold genetic information which can be passed on through complimentary base pairing. It is also increasingly clear that RNAs can act as catalysts, for example as part of the ribosome, and in RNA splicing. Synthetic RNAs have been developed which can bind all manner of molecules very specifically, in a very protein-like fashion. An RNA which could catalyse its own replication would be the basis of a primitive form of life. Synthetic RNAs which can do this, at least in a limited way, have also been developed.

The appearance of self-replicating RNAs in the soup of organic molecules formed by purely chemical means on the early Earth could have marked the beginning of life. Only later did DNA take on the role of a specialized information dump. The information in the second strand allows for repair, in a way that could never happen with RNA. Proteins took on the roles of specialized catalysts, molecular machines and recognition molecules. Before all this, before cells appeared, it is possible to imagine a world populated only by self-replicating RNAs. Reproducing, mutating, competing for the only resource they needed, free nucleotides, they would have been open to natural selection. Life, perhaps, but not as we know it.

Summary

- Genes are passed from parents to offspring.
- Genes control the development and operation of the cell.
- Genes are lengths of DNA.
- Each gene codes for the amino acid sequence of one polypeptide. Information is coded as a base sequence.
- Genetic information is passed on when cells divide because DNA is replicated faithfully.
- The information of one gene is copied as an mRNA strand in transcription.
- In eukaryotes, non-coding introns have to be edited out of the mRNA by RNA splicing.
- Polypeptides are made using the information coded in mRNA in the process of translation.
- In translation, tRNA molecules make the link between groups of three bases and the amino acid they code for.
- Ribosomes are the site of translation.
- Some RNAs, including rRNA and tRNA, are useful in themselves without being translated.
- Genes are controlled by regulatory proteins. They bind upstream of the gene and allow or prevent transcription.
- Mutations are changes to the genetic material.
- Mutagens are factors which cause mutation, including some types of radiation and some chemicals.
- Mutations can render a gene or its polypeptide useless.
- Cells have effective DNA repair mechanisms.
- RNA, not DNA, was probably the basis of the earliest forms of life.

Exercises

7.1. A bacterial gene codes for a short peptide. Study this DNA sequence, then work out:

(i) the base sequence of the mRNA which is made when it is transcribed.

(ii) the amino acid sequence of the peptide made when it is translated.

CCGGTACGAGCGTCCCTGGAGCACGGGTTGATTTATCGGC

(iii) A mutant form of this gene produces a peptide which does not have the last four amino acids. Explain what might have happened.

(iv) Another mutant form of the gene makes an even stranger peptide. The first three amino acids are the same as before, but the next six are different. There are only nine in total, rather than ten. How could this have arisen?

7.2. State the roles of RNA in translation.

7.3. Complete this table, comparing replication, transcription, and translation.

	Replication	Transcription	Translation
Does it happen to whole chromosomes or single genes?			
Where does it happen in a eukaryotic cell?			
What is made?			
Chemical machinery involved			
Site where it begins			

7.4. An experimental technique allows a purified mRNA from a eukaryotic cell to be paired up with the DNA which codes for it. Sequences that match pair up, by complementary base pairing. Under the electron microscope, something like this is seen.

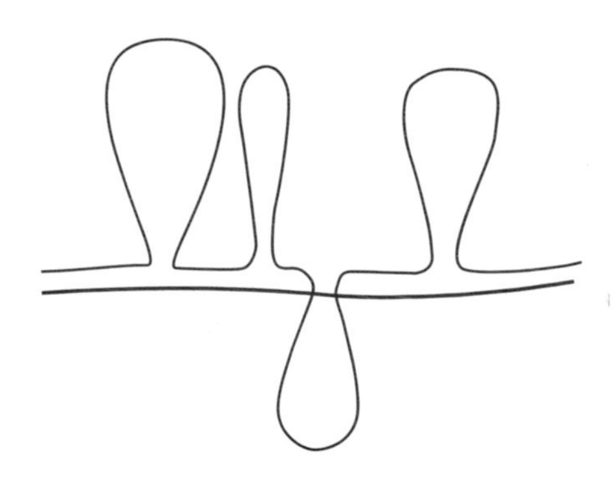

(i) Which strand is DNA, which is RNA, and why are they different lengths?

(ii) What would you expect to see if this was repeated with a prokaryotic gene?

Further reading

Alberts, B., Bray, D., Lewis, J., Raff, M., Roberts, K. and Watson, J.D. *Molecular Biology of the Cell* (3rd ed.) (New York: Garland, 1994). An enormous but excellent textbook which puts the reader in touch with current ideas without too much pain on the way. Molecular genetics is covered in the context of the life of the cell.

Hartl, D.L. *Essential Genetics* (Sudbury MA: Jones & Bartlett, 1996). Bright, clear and well-explained at a simpler level than Lewin. It goes well beyond the scope of this book, without an unbearable acceleration in pace.

Lewin, B. *Genes V*. (Oxford: Oxford University Press, 1994). Massive, up-to-date undergraduate textbook, extending from molecular genetics to molecular aspects of cell biology.

Smith, C.A. and Wood, E.J. *Molecular Biology and Biotechnology* (London: Chapman and Hall, 1991). A clear, straightforward introduction.

A book to read through:

Jones, S. *The Language of the Genes* (London: Harper Collins, 1993). A literate, intelligent explanation of genetics for the general reader. Full of ideas and gentle wit.

Energy

Energy in Living Systems

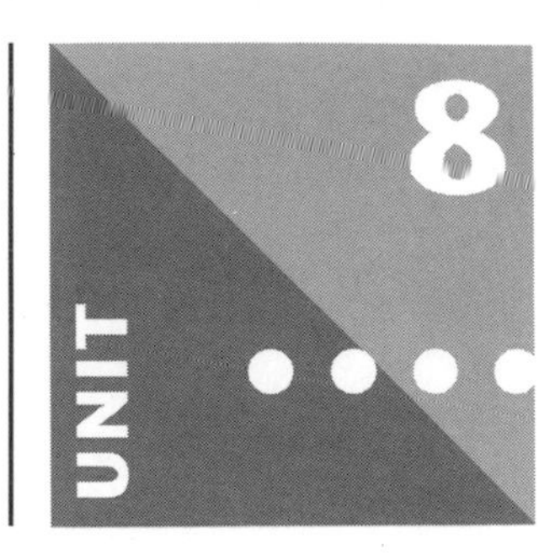

Connections

► This unit is an essential introduction to biological processes which convert energy from one form to another, notably respiration and photosynthesis (Units 9 and 10). Before studying this unit, you should have a basic knowledge of biological molecules, cell structure, enzymes and movement across membranes (Units 3–6). Anyone with a reasonable knowledge of chemistry will find the first few sections basic: almost too basic, perhaps!

Contents

8.1 Energy

The concept of energy is a powerful one. It brings together a wide range of phenomena. A rolling pine cone has energy: the faster it rolls, the more energy it has. The energy of a moving object is called kinetic energy. Kicking the cone to make it go faster gives it more kinetic energy. If the cone stops, its energy must have been passed on elsewhere. Energy is never made or destroyed, it is simply converted from one form to another.

All types of electromagnetic radiation, visible light, infra-red, ultraviolet, microwaves, X-rays and so forth are all forms of energy. An electric current has energy. Hot soup has energy, in fact every object has at least some heat energy: the more it has, the higher its temperature.

As well as these rather dynamic forms, energy can sit still and mind its own business. A compressed spring or a stretched bowstring contain energy, which can stay in that form indefinitely. This sort of stored energy is called potential energy. If I lift a box of eggs onto my head, I have given it potential energy. If my balance is good, that store of energy might remain for minutes or even hours. If the box falls, the energy is released, catastrophically. Most of it ends up as heat.

It is the way in which energy can change from one form to another that links these diverse ideas together. When a publicity seeker throws himself off a high bridge tied to an elastic rope, he falls faster and faster. After a while he slows to a momentary halt on the end of the stretched rope, before being accelerated upward again. He bounces less and less violently until he stops. A single concept, energy, can explain many aspects of this complex chain of events. Energy bounces between two forms, potential and kinetic, slowly being dissipated as heat. The forms change, the energy remains.

Chemical substances have potential energy within the structures of their molecules, some of which may be released as heat during chemical reactions. The combination of the carbon in coal with oxygen has potential energy: heat is released when they react to form carbon dioxide. It is neither possible nor particularly useful to find out how much internal energy each substance possesses. What is both measurable and useful is the change in the internal energy of the system as a whole, estimated indirectly by the amount of heat given out in the reaction.

8.2 Energy and chemical reactions

Only some chemical reactions actually happen. Carbon atoms react with oxygen molecules to make carbon dioxide, but carbon dioxide does not readily break down into carbon and oxygen. How can we predict which reactions are feasible?

When carbon reacts with oxygen, heat is released. This is true for very many reactions and looks as if it could be the basis for a rule. Unfortunately, there are some reactions which happen spontaneously yet take in heat energy: the system cools down.

Entropy is the amount of randomness or disorder in a system. When a crystal dissolves, its regular structure breaks down: there is an increase in entropy. When a complex molecule such as starch is broken down to carbon dioxide and water, there is an increase in entropy. In very many spontaneous reactions, entropy increases, but again there are exceptions, so again this cannot be a general rule.

It turns out that in spontaneous reactions which take in heat, there is a particularly large increase in entropy. In spontaneous reactions in which there is a decrease in entropy, there is a large output of heat. It is something in the balance between heat changes and entropy changes which determines whether or not a reaction will take place.

The free energy (G) of a system is a term that puts together the internal energy of a system, its orderliness (lack of entropy), and energy stored as the pressure of any gases in the system. If a reaction takes place at constant pressure and the free energy decreases, two things could allow this. Either the internal energy has decreased and has been given out as heat, or the orderliness has decreased, so increasing the entropy of the system, or both could have happened together. Either way, the reaction is feasible if free energy decreases.

The change in free energy during a reaction is symbolized by **ΔG**. Δ is the Greek letter delta. It is often used to symbolize a change in something, here a change in G, the free energy. If free energy decreases during a reaction, ΔG will be less than zero, a negative number. A negative value for ΔG shows that a reaction can occur spontaneously. It does not, however, show the rate of reaction which could be anything from extremely rapid to immeasurably slow.

ΔG varies with pressure, temperature, and the concentrations of reactants and products. ΔG may prove only to be negative above a certain temperature, or when a reactant is present at a high enough concentration. Tables of standard ΔG values for many reactions have been published. These are at standard pressure, temperature and concentrations. Conditions under which a reaction might really take place may be very different. Standard values give, at best, only a crude indication of whether a reaction is feasible in the living cell.

Entropy: The amount of disorder in a system.

ΔG: The free energy change for a reaction.

Activation energy and catalysts

Some reactions are feasible yet take place so slowly that, in practice, they do not occur at all. Very many reactions which take place within cells are like this. If the reactants are placed in a suitable solution within a test-tube, no product is made. The reason for this is that many reactions require an input of free energy (the activation energy) before they will take place, even though much more free energy is then released (Box 5.1). If the activation energy is high, only a small proportion of molecules at any one time have enough energy to react, so rate of reaction will be very low.

If the temperature is higher, more molecules will have sufficient energy to react. The presence of a catalyst will also increase reaction rate. Catalysts are molecules which decrease the activation energy of a reaction (Box 5.1). They may take part in the reaction, but remain unchanged after the reaction has taken place. If activation energy is lower, more molecules will have enough energy to react, so rate will increase. Most catalysts within cells are enzymes: proteins (Unit 5). Some rather special RNA molecules also have catalytic activity (Section 5.4).

Oxidation, reduction and energy

At a very simple level, a molecule becomes oxidized when it gains oxygen or loses hydrogen. It becomes reduced when it loses oxygen or gains hydrogen. So when carbon burns in oxygen to form carbon dioxide, the carbon becomes oxidized.

$$C + O_2 \rightarrow CO_2$$

When nitrogen and hydrogen react together in a fertilizer factory, nitrogen is reduced.

$$N_2 + 3H_2 \rightleftharpoons 2NH_3$$

This is a reversible reaction: when ammonia breaks down to nitrogen, hydrogen is lost, so nitrogen is being oxidized.

This is not the whole story, however. Some oxidations and reductions involve neither oxygen nor hydrogen. To understand this, consider what happens when hydrogen and oxygen react to form water.

$$H_2 + \tfrac{1}{2}O_2 \rightarrow H_2O$$

Hydrogen gains oxygen, so becomes oxidized; oxygen gains hydrogen, so becomes reduced. The water molecule involves two covalent bonds between oxygen and hydrogen atoms (Fig. 8.1). Each bond can be thought of as the result of two atoms sharing a pair of electrons (Section 3.2: Bonds). The positively-charged nuclei of both atoms attract the negatively-charged electrons. The atom which attracts the electrons more strongly will gain a greater share of those electrons. In this case, oxygen attracts electrons more strongly than hydrogen, so as it becomes reduced it tends to gain electrons. As hydrogen becomes oxidized, it loses electrons. This leads us to a new, much more general definition of **oxidation** and **reduction**: oxidation is loss of electrons; reduction is gain in electrons.

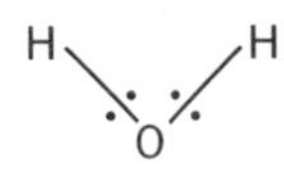

Fig. 8.1 The water molecule: the oxygen atom has a greater share of the electron pair involved in each covalent bond.

We can apply this to situations where oxygen and hydrogen are not involved. If copper atoms lose electrons to become copper(I) ions, they have been oxidized:

$$Cu \rightarrow Cu^+ + e^-$$

Oxidation: Loss of electrons.

Reduction: Gain in electrons.

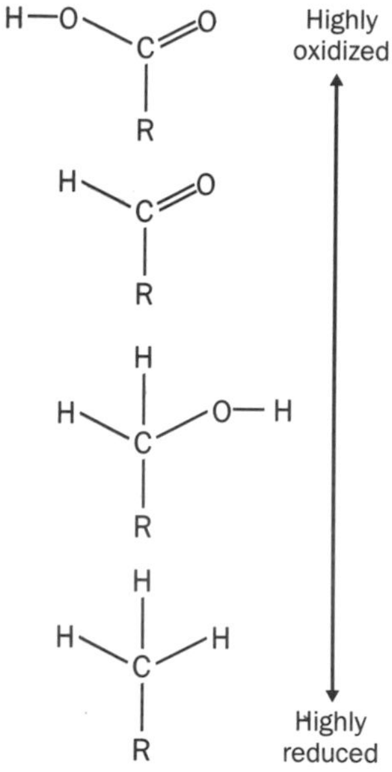

If copper(II) ions gain electrons to become copper(I) ions, they have been reduced;

$$Cu^{2+} + e^{-} \rightarrow Cu^{+}$$

When hydrogen and oxygen react, oxygen is reduced whilst hydrogen is oxidized. Oxidation and reduction always go together. If Cu^{2+} is reduced to Cu^{+}, the electrons needed for this have to come from somewhere, generally from another atom, ion or molecule, which itself becomes oxidized.

Biochemicals are carbon compounds and, in general, the carbon within them is in a reduced state. A carbon atom gets a bigger share of the electrons in a C–H bond than in a C–O bond, so in general the more hydrogen attached to the carbon skeleton, the more reduced it is (Fig. 8.2). The oxidation of most biochemicals is feasible under the conditions within cells. This means that we can think of reduced biochemicals as energy stores. Free energy is released when they are oxidized. Starch, glycogen, fats and proteins all release free energy on oxidation. Each can be used as an energy reserve by cells or organisms. The reverse is also true: an input of free energy is required to build up these reduced molecules. Sometimes the synthesis of the molecule is directly coupled to another, energetically feasible, reaction. In other cases, the reaction involves a potent biological reducing agent such as NADPH (Section 8.5) which itself needs an input of free energy for its synthesis.

Highly oxidized

Highly reduced

Fig. 8.2 Some functional groups, arranged according to the degree of oxidation of the carbon atom.

Phosphorylation and energy

Many biological molecules have phosphoryl groups attached to their carbon skeletons. Some molecules hang on grimly to their phosphoryl groups. Others readily give them up to other molecules: these have a high phosphoryl-group transfer potential. Many reactions in which such a molecule gives up its phosphoryl group are energetically feasible: free energy is released. These molecules can, then, be thought of as energy stores. The most important molecule of this type is adenosine triphosphate (ATP).

8.3 The need for an energy currency

Cells and organisms need energy for three main things: movement, transport of molecules and ions across membranes and the synthesis of biological molecules. Within each of these categories are a huge number of reactions, each requiring an input of free energy.

The cell's sources of energy are much less diverse: all eukaryotic cells, and most prokaryotes too, obtain their energy from the oxidation of reduced biochemicals such as carbohydrates, fats or proteins: this is respiration (Unit 9). Some cells and organisms gain these molecules by feeding; others build them up from simple materials using light as an energy source: photosynthesis (Unit 10).

Cells and organisms can only work if the reactions which require an input of free energy are linked to the oxidation of particular types of energy-storing molecules. Direct coupling of all possible reaction pairs would be unimaginably complicated. What is needed is some molecule which acts as an energy currency, being formed as a result of all sorts of reactions which release energy, and being used to drive diverse energy requiring reactions. **ATP** is such a molecule.

ATP: Adenosine triphosphate; a nucleotide which acts as an energy currency in cells.

The role of ATP in the cell is much like the role of money in society. Bartering is a cumbersome system. The car mechanic can eat bread in exchange for servicing the baker's van, but if the dairy farmer rides around in a pony trap, milk ceases to be part of the mechanic's diet. Money avoids the need for the direct coupling of transactions: it is a universal carrier of value. In the same way, ATP is an all-purpose carrier of free energy.

8.4 ATP

Adenosine triphosphate (ATP) is a nucleotide (Section 3.8). The molecule consists of three units: ribose (a pentose sugar), adenine (a purine base) and a triphosphate group (Fig. 8.3).

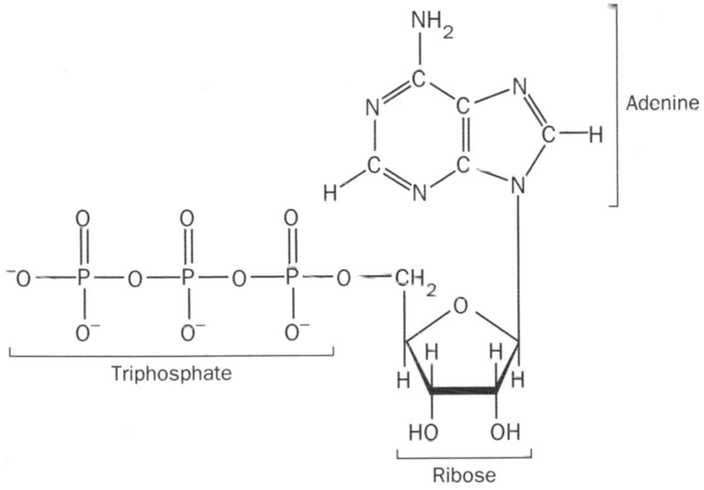

Fig. 8.3 The ATP molecule.

ATP gives up one or two phosphate groups very easily. This is partly the result of repulsion between the four negative charges of the triphosphate group. If the phosphate group on the end of the chain is broken off, adenosine diphosphate (ADP) and the inorganic ion orthophosphate (HPO_3^{2-}) are made (Fig. 8.4). This is a hydrolysis reaction. Biochemists usually use the symbol P_i for orthophosphate.

$$ATP + H_2O \rightarrow ADP + P_i + H^+$$

In the same way, two phosphates can be removed as pyrophosphate, $HP_2O_7^{3-}$ (symbolized as PP_i), leaving adenosine monophosphate (AMP).

$$ATP + H_2O \rightarrow AMP + PP_i + H^+$$

Both these reactions have a similar, highly negative, ΔG value. Under standard conditions, $\Delta G = -30.6$ kJ mol^{-1}. Under typical conditions within a cell, the value would be more like -50kJ mol^{-1}.

ATP is an energy carrier because hydrolysis of ATP releases free energy. The ADP–P_i bond itself does not store energy: the system as a whole releases energy when the reaction takes place. Reforming ATP from ADP requires an equivalent input of free energy.

$$ADP + P_i + H^+ \rightarrow ATP + H_2O$$

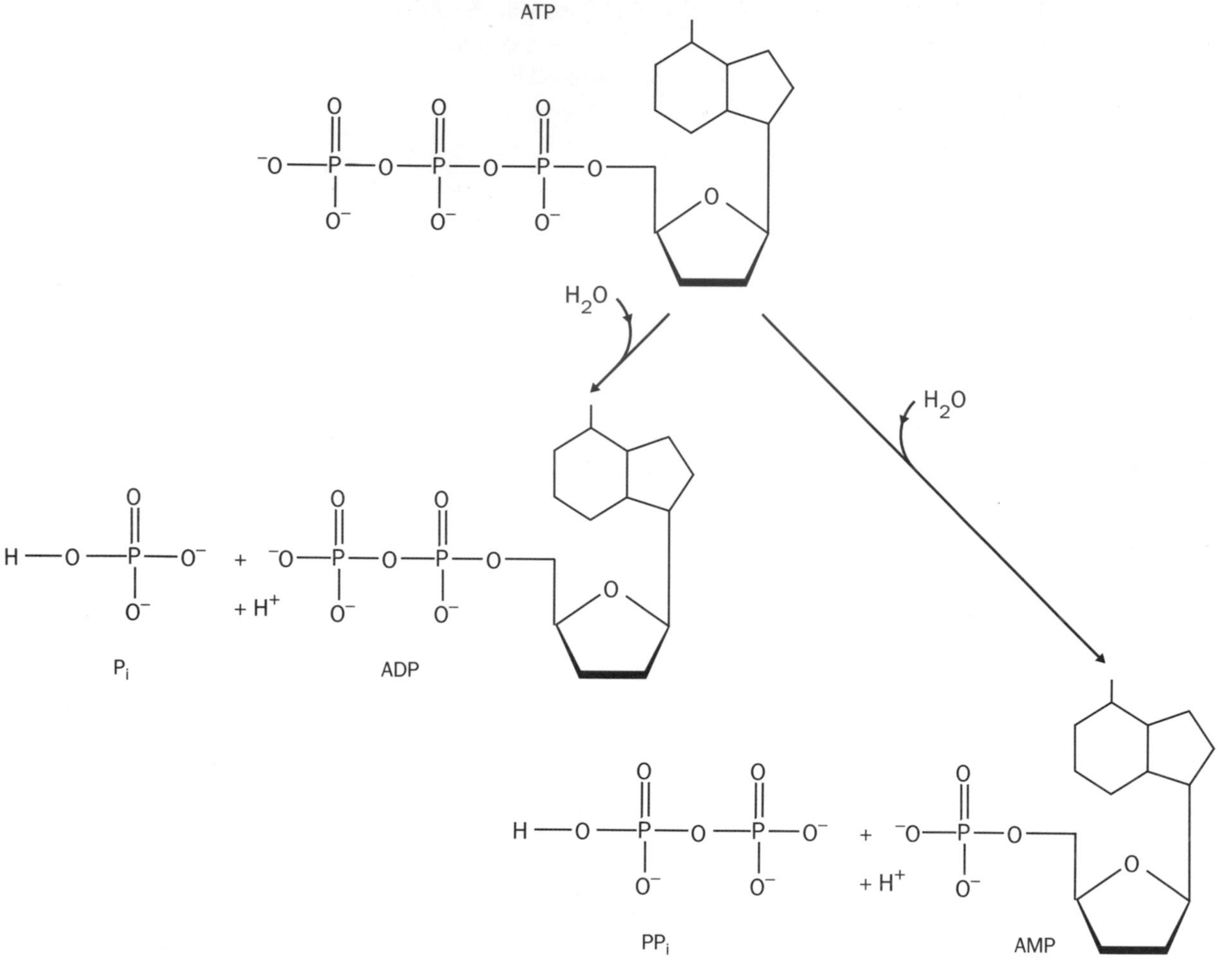

Fig. 8.4 The hydrolysis of ATP to either ADP or AMP.

In respiration, ATP is made by coupling this reaction to oxidation of reduced molecules (Unit 9). In photosynthesis, it is linked to the absorption of light (Unit 10).

ATP is used up when its hydrolysis is coupled to the transport of molecules across membanes, or to movement, or to the synthesis of biological molecules. Enzymes which catalyse a reaction whilst at the same time catalysing the hydrolysis or formation of ATP are called ATPases.

In some situations, ATP is hydrolysed to AMP instead of ADP. ATP cannot be reformed directly from AMP. Instead, another ATP molecule first transfers one phosphate group, leaving two molecules of ADP.

$$\text{AMP} + \text{ATP} \rightarrow 2\text{ADP}$$

ATP can then be formed by phosphorylation of ADP as usual.

ATP is of such importance that its concentration in cells cannot be allowed to fluctuate freely. Just as governments attempt to control the amount of money circulating in the economy, so cells closely regulate the levels of ATP within them. A low ATP concentration stimulates those processes which increase ATP level, and vice versa.

ATP is a relatively short-term energy store. It is significant on the timescale of seconds and minutes rather than days or years.

Other phosphorylated energy stores

Some reactions which build up biochemicals are coupled to the hydrolysis of other nucleotides which are very similar to ATP. Guanosine triphosphate (GTP), for example, differs only in that adenine is replaced by another purine base, guanine. GTP and ATP can be freely interconverted if the appropriate enzyme is present:

$$ATP + GDP \rightleftharpoons ADP + GTP$$

Phosphocreatine (Fig. 8.5) is a quite different molecule which has an even higher phosphate-group transfer potential than ATP. For hydrolysis of phosphocreatine, $\Delta G = -43\ kJ\ mol^{-1}$ under standard conditions, and even more negative under the conditions within the cell. Phosphocreatine is found in vertebrate muscle. When a muscle begins vigorous contraction, ATP is used and its concentration will tend to drop. An increased respiration rate would soon restore the ATP level, but until this happens, ATP concentration must be maintained. Phosphocreatine acts as an energy reserve for this situation. As ATP is used up, phosphocreatine passes its phosphate group to ADP, reforming ATP. During intense muscle activity, phosphocreatine levels fall, but ATP levels stay constant.

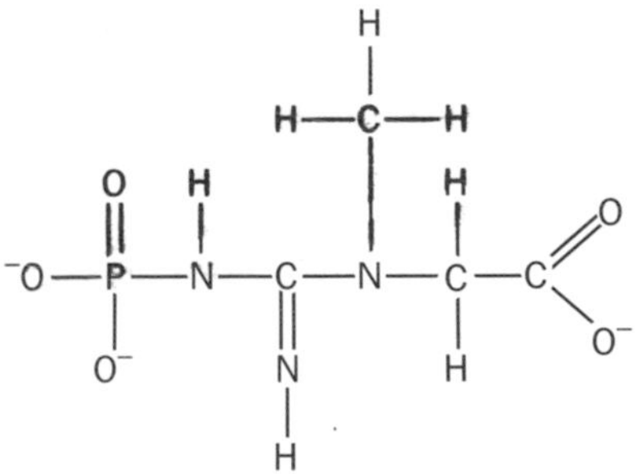

Fig. 8.5 The structure of phosphocreatine.

8.5 Reduced energy stores

In the context of the cell, reduced biochemicals are energy stores. When they are oxidized, free energy is released. The longer-term energy stores of cells and organisms are of this type. Plants most often use starch as an energy reserve, less commonly other polysaccharides (eg. inulin in Jerusalem artichoke tubers), disaccharides (eg. sucrose in sugar beet) or fats (eg. in sunflower seeds). Mammals use glycogen, a polysaccharide, as a medium-term energy store, over a timescale of hours, and fats in the longer term. In each case, when they are oxidized in respiration, the release of free energy is coupled, indirectly, to ATP synthesis. However, before reaching ATP much of this free energy passes to a short-term, reduced energy storing molecule, NADH.

Nicotinamide adenine dinucleotide (NAD) exists in two forms, an oxidized form, NAD^+ and a reduced form, NADH (Fig. 8.6). Two electrons are needed to reduce NAD^+, and H^+ is taken up at the same time:

$$NAD^+ + H^+ + 2e^- \rightarrow NADH$$

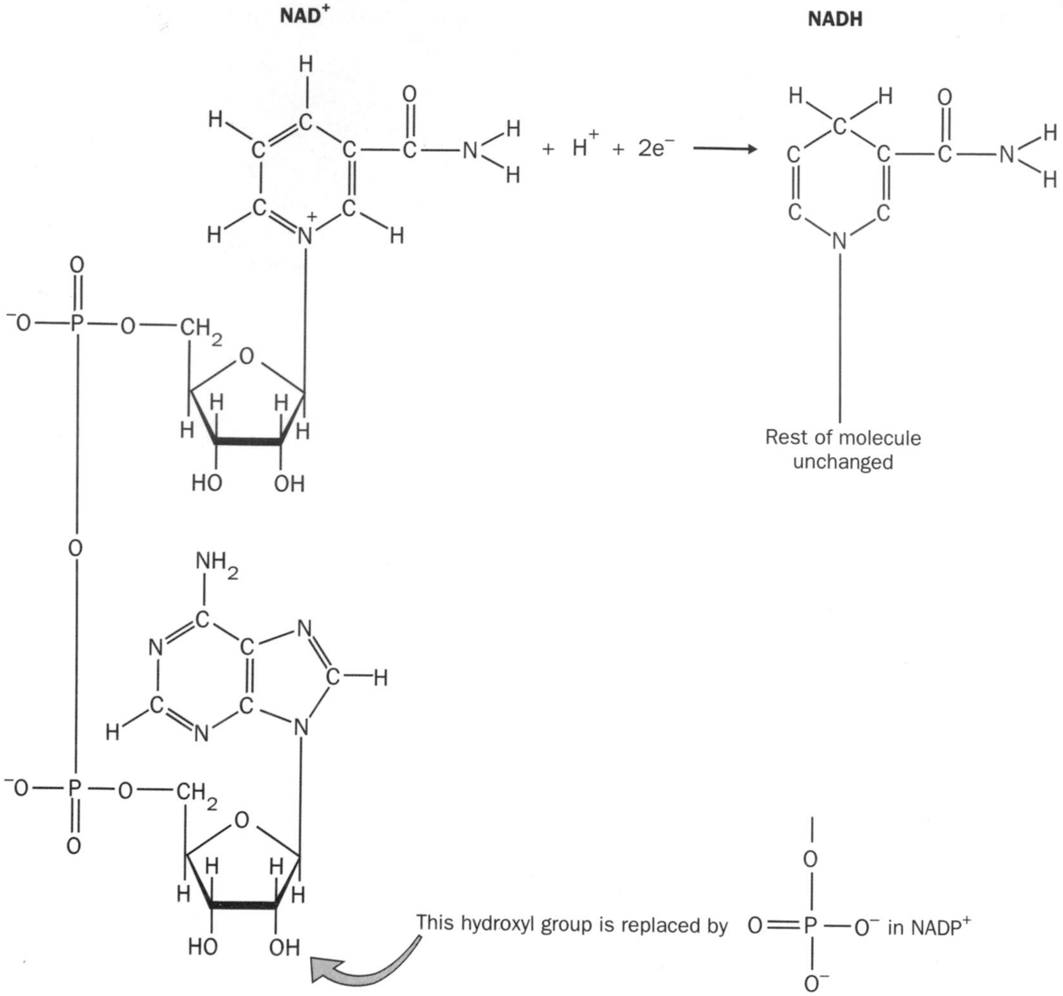

Fig. 8.6 The structure of NAD^+ and its relationships with NADH and $NADP^+$.

NADH can be thought of as a reducing agent and also, therefore, as a carrier of free energy. The energy is passed on when NADH gives up its electrons to other electron carriers. Even though ΔG for oxidation of NADH is highly negative, it does not reduce everything in sight. Specific enzymes are required, allowing its electrons to be channelled, ultimately, to oxygen in aerobic respiration. ATP synthesis is coupled to this.

NAD^+ is usually reduced to NADH in dehydrogenation reactions, where two hydrogen atoms are removed from some other molecule. We can think of two hydrogen atoms as two electrons and two hydrogen ions. NAD^+ is gaining two electrons and one H^+ from the other molecule; the second H^+ goes into the surrounding solution:

$$NAD^+ + RH_2 \rightarrow NADH + R + H^+$$

where R is the other molecule.

NADPH is very similar to NADH, differing only in one phosphate group (Fig. 8.6). Its function is quite different, however. NADPH acts as a reducing agent in the

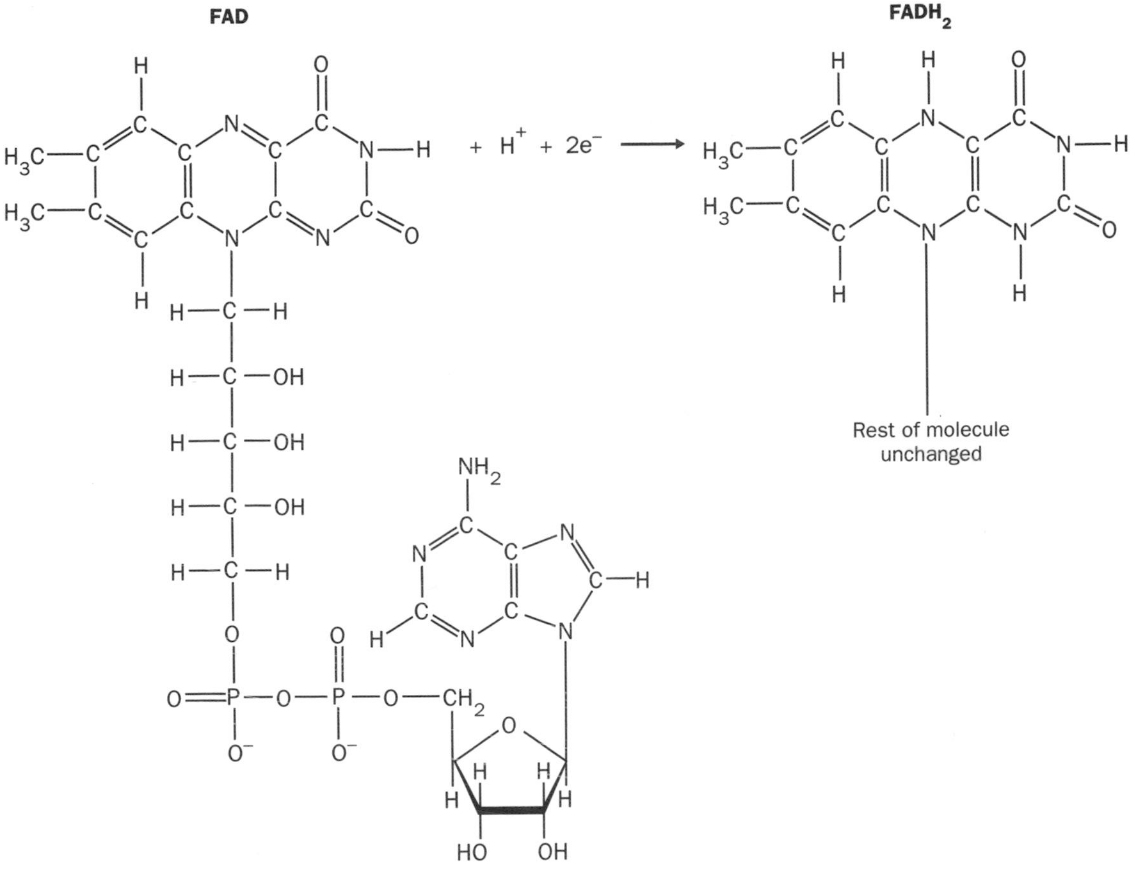

Fig. 8.7 The structure of FAD and its relationship with $FADH_2$.

synthesis of reduced biochemicals, for example in the formation of sugars in photosynthesis, and in the synthesis of fatty acids.

Flavin adenine dinucleotide (FAD) is another electron carrier which has a role in coupling oxidation of storage molecules to ATP synthesis (Section 9.2). FAD can also exist in two forms (Fig. 8.7). This time, two hydrogen ions are carried in addition to the two electrons:

$$FAD + 2H^+ + 2e^- \rightarrow FADH_2$$

Chemical gradients as energy stores

Substances tend to diffuse from regions of high concentration to regions of low concentration. As they do so, orderliness decreases, entropy increases: diffusion is a spontaneous process. On the other hand, concentrating a substance into one region leads to a decrease in entropy: a free energy input is required to build up a gradient. Concentration gradients, then, are energy stores. Gradients of protons (H^+) in particular, but also of ions such as Na^+ are important energy stores in cells on a very short timescale. In aerobic respiration, a proton gradient is used to couple the oxidation of electron carriers to ATP synthesis (Section 9.2: Oxidative phosphorylation). Simply, electron carriers in the inner membrane of the mitochondrion pass electrons

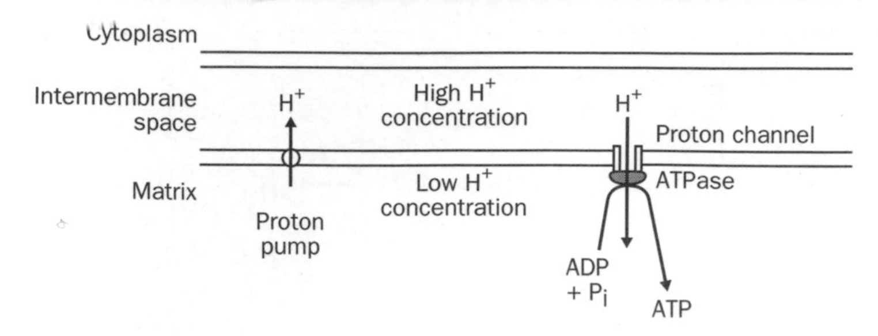

Fig. 8.8 An example of the role of a proton gradient as an energy store: ATP synthesis in the mitochondrion.

from NADH and $FADH_2$ to oxygen. As they do so, the free energy released is used to pump protons out of the matrix of the mitochondrion (Fig. 8.8). The H^+ concentration in the space between the inner and outer membranes rises. Protons will tend to diffuse back into the matrix, but the membrane is impermeable to them. They may only diffuse through specific protein channels which are attached to ATPases. As protons flow back into the matrix, the free energy released is used to phosphorylate ADP to ATP.

Proton gradients couple oxidation of electron carriers to ATP synthesis during photosynthesis (Section 10.3), they drive the transport of potassium ions into guard cells when stomata open (Section 18.3), and provide energy for the rotation of bacterial flagella (Section 4.4). These are just a few examples of their importance.

Gradients of other ions store energy in the same way. A sodium gradient, for example, drives the reabsorption of glucose into the blood during urine formation in the kidney of a mammal (Section 22.4).

Energy stores are interrelated

Energy is transferred from store to store. When a polar bear runs, energy is transferred from long-term energy stores to kinetic energy via many intermediates.

glycogen → glucose → NADH and $FADH_2$ → other e^- carriers → proton gradient → ATP → motion

8.6 Entropy and life

The passage of time is marked by two trends, the tendency for energy to end up as heat, and the tendency for entropy to increase. The universe, if it is a closed system, is destined to end up as a thin soup of diffuse, warm matter. Building complex molecules from simple ones in photosynthesis goes against the trend, as does the growth of cells, the evolution of life itself. Life, it would seem, is going against the flow of energy in the rest of the universe. However, these rules apply to whole systems. A cell is part of a wider system, along with its environment. A decrease in entropy within a cell must be linked to an even greater heat production if ΔG for the reaction is to be negative. Cells constantly release heat to their environments. The system as a whole follows the rules.

Sometimes a pine cone dropped into the river, having raced under the bridge, will be caught in a swirl of water and, for a moment, be carried back towards the bridge before being swept on in its relentless progress downstream. The sight may be magical, the explanation need not be. For all the water moving upstream in the eddy, other water is flowing even faster downstream; taking the flowing water as a whole, the momentum stays the same before, during and after the eddy. From the point of view of the pine cone, this is irrelevant. It really does go against the flow. So it is with life.

Summary

- Energy can be transferred from one form to another, but cannot be made or destroyed.
- Molecules have potential energy within their structures.
- The free energy of a system (G) combines the internal energy of a system and its entropy (disorderliness).
- The free energy change in a reaction (ΔG) determines whether or not a reaction is feasible. If ΔG is negative, the reaction is feasible. If ΔG is positive, the reaction could only take place if coupled to another reaction which is feasible.
- Oxidation is the loss of electrons; reduction is the gain of electrons.
- Oxidation and reduction are always coupled. When something becomes oxidized, something else must become reduced.
- Under the conditions inside cells, the oxidation of most biochemicals is feasible. Reduced biochemicals such as carbohydrates, fats and NADH can act as energy stores.
- Some phosphorylated molecules, such as ATP, can act as energy stores.
- ATP acts as an energy currency within cells, coupling many energy-releasing reactions to many energy-requiring reactions.
- A chemical gradient across a membrane can act as an energy store.

Exercises

8.1. A greyhound is an energy-converting machine. What are its energy inputs and outputs?

8.2. An iron ion undergoes this change:

$Fe^{3+} + e^- \rightarrow Fe^{2+}$

Is it being oxidized or reduced?

8.3. One of the reactions of photosynthesis is as follows:

1,3-diphosphoglycerate + NADPH $\rightarrow$ glyceraldehyde 3-phosphate + $NADP^+$ + P_i

Is 1,3-diphosphoglycerate being oxidized or reduced?

8.4. A protein in the plasma membrane of bacteria pumps hydrogen ions out of the cell. This is coupled to the hydrolysis of ATP. Does transporting protons across this membrane use or release energy?

8.5. The amount of energy released when ATP is hydrolysed is quite modest. Would a molecule which released much more energy in some reaction be more effective as an energy currency?

Further reading

Lewis, R. and Evans, W. *Chemistry* (Basingstoke: Macmillan Press, 1997). This unit glosses over many important ideas about energy in chemistry. If you want to start filling in the gaps, try this sister volume. If your understanding of basic chemistry is shaky, this general introduction should help.

Stryer, L. *Biochemistry* (4th ed.)(New York: Freeman, 1995). Energy stores and energy transformations in cells are dealt with nicely by this classic big undergraduate textbook of biochemistry. There are several broadly similar books on the market, but I would still go for this if looking for clear explanation.

Respiration

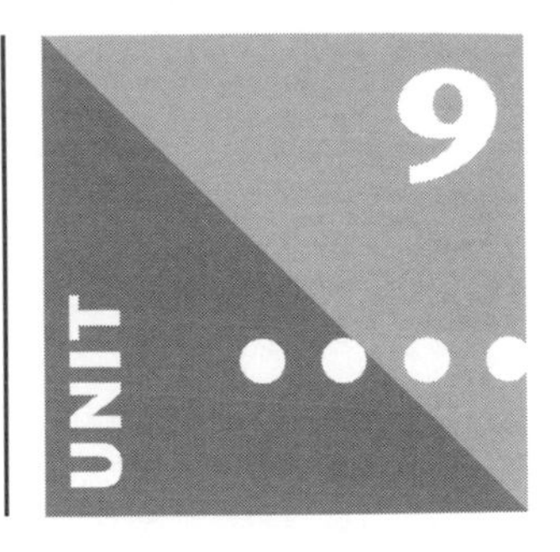

Connections

▶ Before studying this unit, you should have a basic understanding of biological molecules, cell structure, enzymes and movement of molecules across membranes (Units 3–6). Unit 8 is an essential introduction to the concept of energy, which is central to respiration. The chemical machinery of photosynthesis (Unit 10) has much in common with the machinery of respiration. Gas exchange (Unit 20) is made necessary by respiration.

Contents

9.1 Making energy available

Cells have various stores of chemical energy available to them. Photosynthetic plant cells have starch reserves. Other plant cells will have chemical energy transported to them in the form of sucrose. Mammalian cells have a continuous supply of glucose, which is maintained at a more or less constant level in the blood. In some circumstances, fats or proteins are used as energy supplies. The energy in these storage and transport molecules is not immediately available to the cell in order to meet its energy requirements, whether for cell movement, active transport of particular molecules across membranes, or synthesis of biochemicals. These processes require energy from ATP (Section 8.4). Cells need to break down storage and transport molecules in a series of reactions, using the energy released to make ATP. These reactions, taken together, make up the process of **respiration**.

The chemical reactions involved in respiration are diverse. Not only are different substrates broken down in different ways, but there are both **aerobic** (oxygen-requiring) and **anaerobic** (not oxygen-requiring) pathways to consider. Aerobic respiration of glucose is of central importance in many types of cell, so seems a sensible place to start. All that follows refers to eukaryotic cells: plants, animals, fungi and protoctists. Bacteria, as well as lacking mitochondria, respire in various ways strange and alien to us animals.

Respiration: The breakdown, in cells, of energy-rich molecules, with the release of energy. (**Note:** Respiration **is not** breathing or gas exchange.)

Aerobic respiration: Respiration requiring oxygen.

Anaerobic respiration: Respiration not requiring oxygen.

9.2 The chemistry of respiration

Sugar burns quite easily in air, as anyone who has spilt it on the stove knows. Oxygen is used, carbon dioxide and water are made. Reduced carbon in the sugar is oxidized to carbon dioxide; at the same time, oxygen is reduced from molecular oxygen to water.

$$\underset{\text{any monosaccharide}}{C_6H_{12}O_6} + \underset{\text{oxygen}}{6O_2} \rightarrow \underset{\text{carbon dioxide}}{6CO_2} + \underset{\text{water}}{6H_2O}$$

This is a single reaction, and the considerable amount of free energy released is given out, devastatingly, as heat. The equation for aerobic respiration as a whole is identical to this: the process was in fact known as 'physiological combustion' in the 19th century. However, many reactions are involved, and little more than half the free energy released is as heat: the rest is used, directly or indirectly, to make ATP.

Glycolysis

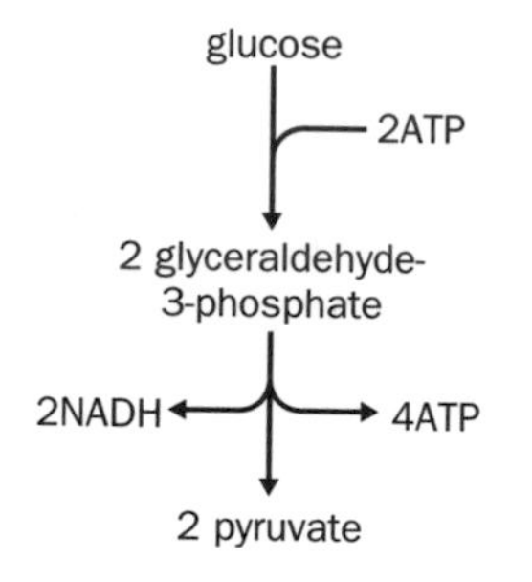

Fig. 9.1 The major steps of glycolysis.

Aerobic respiration of glucose involves three quite separate stages: glycolysis, the Krebs cycle and oxidative phosphorylation. One stage, **glycolysis**, is common to aerobic and anaerobic respiration. Glycolysis takes place in the cytoplasm of the eukaryotic cell. Glucose molecules are first phosphorylated, then broken down to two molecules of pyruvate, each of which contains three carbon atoms. A small amount of ATP is made directly. Glycolysis involves oxidation of glucose: this can be coupled to reduction of NAD^+ to NADH (Section 8.5), which can later be used to make more ATP if oxygen is present. Only a little ATP and NADH are made in glycolysis, which is primarily a preparation for the reactions of the Krebs cycle. Glycolysis is actually a multi-step pathway. One key step is the point at which the 6-carbon skeleton of glucose is divided into two. Two molecules of the 3-carbon sugar phosphate, glyceraldehyde-3-phosphate, are made, which are then converted to pyruvate (Fig. 9.1). Boxes 9.1 and 9.2 show the entire pathway, and how it is regulated.

The Krebs cycle

The **Krebs cycle** takes place in the matrix of the mitochondrion. What happens, essentially, is that the carbon in pyruvate is oxidized to carbon dioxide, as the carbon skeleton is demolished, one carbon atom at a time. A little ATP, and much more NADH is made.

The link between glycolysis and the Krebs cycle is very interesting. A large complex of enzymes does a number of things to pyruvate. Firstly, one carbon atom is removed as carbon dioxide. This is both a decarboxylation and an oxidation. The oxidation of pyruvate is linked, as always, to a reduction: NAD^+ is reduced to NADH. The remaining 2-carbon unit is linked to Coenzyme A to form acetyl–Coenzyme A.

CO_2
pyruvate → acetyl–CoA
CoA NAD^+ $NADH+H^+$

Glycolysis: The pathway which converts glucose to pyruvate.

Krebs cycle: A cycle of reactions which oxidizes pyruvate to carbon dioxide.

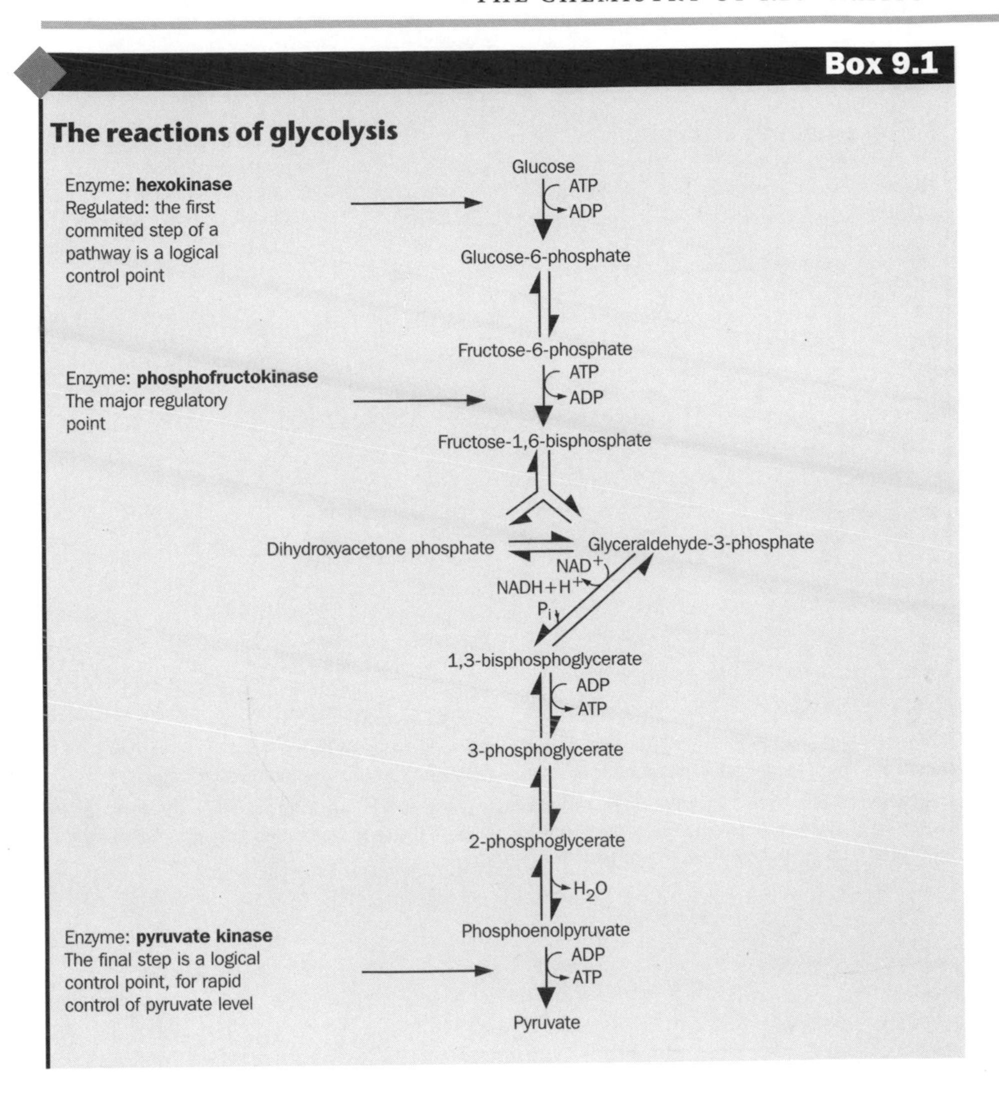

Acetyl–Coenzyme A is an extremely important molecule: the 2-carbon acetyl group is bonded only weakly to the coenzyme, so the group can readily be transferred to another molecule. Wherever the carbon skeletons of biochemicals are built up, two carbons at a time, acetyl–Coenzyme A provides the carbons. Once one acetyl group has been given up, Coenzyme A is free to pick up another. If the acetyl group is a building block, Coenzyme A is the bricklayer.

Acetyl–Coenzyme A provides a way for carbon to enter the Krebs cycle. The acetyl group is added to a 4-carbon acid (oxaloacetate) to make a 6-carbon acid (citrate). Citrate is then oxidized and decarboxylated back to oxaloacetate in a series of steps. The importance of this is that some ATP, a great deal of NADH, and some $FADH_2$ (another, closely-related reducing agent) are made. Both NADH and $FADH_2$ can be used to make ATP later. Carbon dioxide is made as a waste product. Oxaloacetate remains to continue the cycle. Taking the cycle as a whole, the amount of carbon entering as acetyl groups equals the amount lost as carbon dioxide. The key features of the Krebs cycle are shown in Fig. 9.2.

Were we to follow the atoms of a glucose molecule through glycolysis and the

Box 9.2

Some amazing enzymes

Phosphofructokinase (PFK) catalyses the step in glycolysis which is central in controlling the pathway. The reaction is controlled as follows in mammalian liver (muscle is excitingly different):

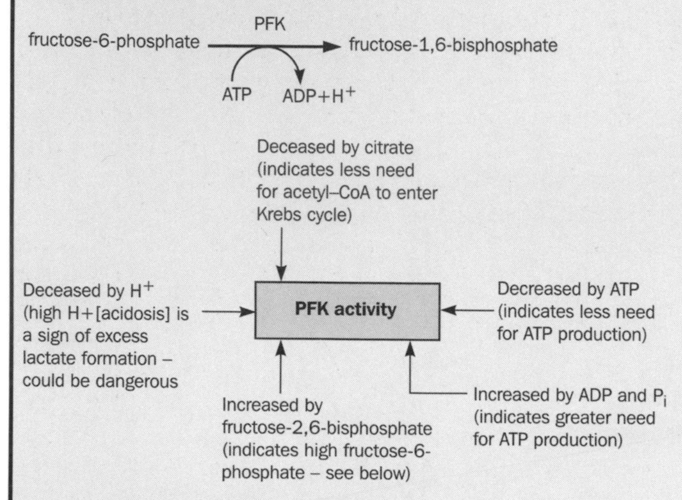

Fructose-2,6-bisphosphate (f-2,6-bp) is made from fructose-6-phosphate (f-6-p) by an enzyme activated by f-6-p. The reverse reaction is catalysed by another enzyme inhibited by f-6-p.

So when f-6-p is abundant, f-2,6-bp is also abundant, and vice versa. This amplifies the variation in f-6-p level, giving more sensitive control. Amazingly, both these antagonistic enzymes are different parts of a single polypeptide chain: a **tandem enzyme.**

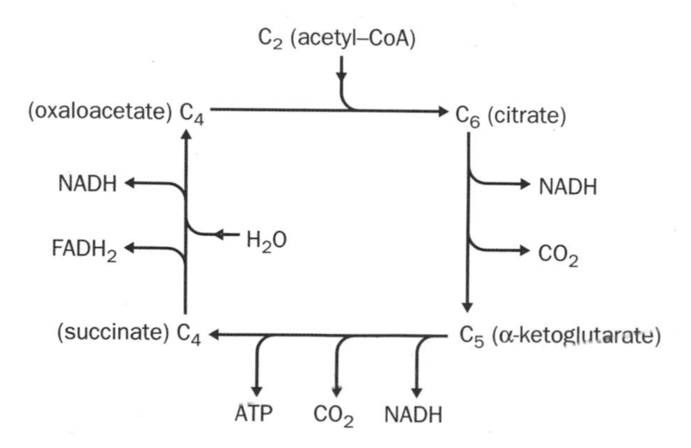

Fig. 9.2 The major steps of the Krebs cycle.

Krebs cycle, we would find that all the carbon and oxygen atoms have been lost as carbon dioxide. The hydrogen atoms are still kicking around, either in NADH or $FADH_2$, or as H^+ released to the solution during the reduction of NAD^+ to NADH. It may come as some surprise that oxygen has not been used, nor has water been made: this happens much later.

Oxidative phosphorylation

Glycolysis and the Krebs cycle produce NADH and $FADH_2$. These molecules are highly reduced, and as usual in biological molecules, reduction goes hand in hand with a high free-energy content. The final stage in aerobic respiration is to oxidize NADH and $FADH_2$, whilst reducing oxygen to water. The free energy which is released is used to make ATP. This process is called **oxidative phosphorylation**.

Oxidative phosphorylation has two stages. The first stage uses the free energy released when NADH and $FADH_2$ are oxidized to set up a gradient of protons (H^+ ions) across the inner membrane of the mitochondrion. This uses a set of chemical apparatus called the respiratory chain, which is embedded in the inner membrane. The proton gradient acts as a very short-term energy store. The second stage is the use of this stored energy to make ATP, by means of an ATPase enzyme.

The respiratory chain consists of three large enzymes which span the membrane, plus two types of smaller molecules which can move freely within the membrane. Electrons are passed, via these enzymes, from NADH to oxygen. To become oxidized is to lose electrons, to become reduced is to gain them (Section 8.2). So, as NADH is oxidized to NAD^+ it gives up two electrons, which are passed on to a specific site (in fact a prosthetic group) in the first enzyme, NADH-Q reductase. This is one of several sites in the enzyme which can gain and lose electrons. The electrons are passed from one site to the next in a specific sequence. Each site is easier to reduce than the next, so the free energy required for the reduction of each site gets progressively lower. Free energy is released, then, whenever electrons are passed between sites. This free energy is used to pump a proton across the membrane, out of the matrix and into the space between the inner and outer membranes (Fig. 9.3). The electrons are then passed on to a small molecule, ubiquinone.

Ubiquinone can move freely and randomly within the membrane, and passes the electrons on to the second enzyme, cytochrome reductase, when they collide (Fig. 9.4). Electrons are passed from site to site within this enzyme, as before. Protons are again pumped, and the electrons are passed on to cytochrome c, a small protein which can diffuse along the outer surface of the membrane. This passes electrons on to the third enzyme, cytochrome oxidase; protons are pumped and the electrons are passed to an oxygen molecule. With the addition of H^+ ions from the solution, water is made. Two molecules of NADH must be oxidized in order to reduce each oxygen molecule, since only two electrons are released from each.

Oxidative phosphorylation: Transfer of energy from NADH and $FADH_2$ to ATP in the respiratory chain.

$$O_2 + 4e^- + 4H^+ \rightarrow 2H_2O$$

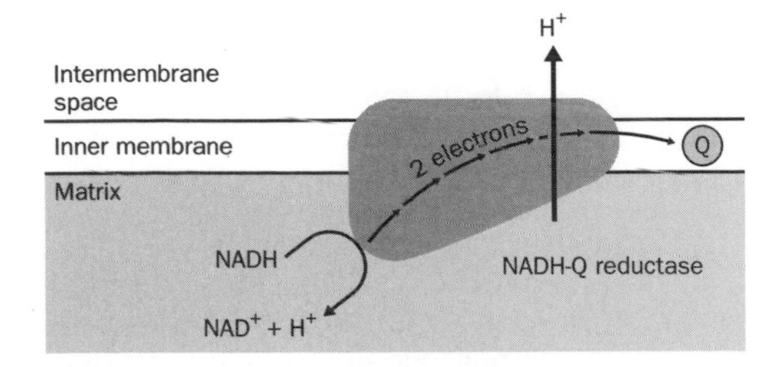

Fig. 9.3 NADH-Q reductase: the first enzyme in the respiratory chain.

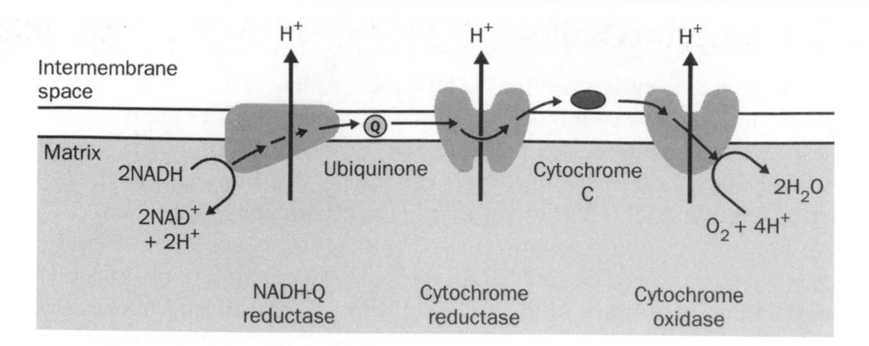

Fig. 9.4 Electron transport in the respiratory chain.

The role of oxygen in respiration at last becomes clear: it is simply a dustbin for electrons. Electrons need to be lost from the last enzyme of the respiratory chain if further electrons are to be passed down the chain. Much of their reducing power has been lost by this stage; only something which accepts electrons very easily will take them. Oxygen is just such a molecule. Moreover, it is abundant in the environment of many cells, it is a small molecule which diffuses readily across membranes, and the product of the reaction, water, is harmless. It is an ideal vehicle for the disposal of electrons. Some bacteria use other molecules to do a similar job, but that is another story.

The result of all this proton pumping is that there is a much higher concentration of protons in the intermembrane space than in the matrix. The intermembrane space will then be more acidic, and more positively charged than the matrix. So, across the membrane there will be both a chemical gradient, which can be measured as a pH difference, and an electrical gradient. Protons will tend to diffuse back to the matrix down the concentration gradient, as well as being attracted by its relatively negative charge. This is prevented, however, because the membrane is impermeable to protons. To pump further protons out of the matrix requires energy, to push them against this electrochemical gradient. The gradient, then, is a store of energy.

This energy store is harnessed to make ATP in a simple and elegant way. An ATPase enzyme, shaped crudely like a mushroom, is embedded in the inner membrane: the stalk of the mushroom is a channel across the membrane, through which protons may pass. The cap of the mushroom, on the matrix side of the membrane, has an active site which can bind ATP and phosphate. To make ATP from ADP and phosphate requires energy only in the presence of water. The active site of the ATPase is a tight cleft which excludes water, so once ADP and phosphate are bound, ATP is formed freely. As protons fall back down the gradient, through the channel, free energy is liberated: this is used to change the three-dimensional structure of the enzyme, so that water is allowed into the active site. The ATP becomes hydrated, and can diffuse back out into the matrix. Box 9.3 shows how ATPase is just one of the transport proteins in the membranes of the mitochondrion.

The number of protons pumped for each electron transported is not known for certain; neither is the number of protons which must flow back through the ATPase in order to release one molecule of ATP. The overall result, however, is quite clear: about 2.5 molecules of ATP are made when one molecule of NADH is oxidized. When $FADH_2$ is oxidized, only about 1.5 ATP molecules are made. This is because $FADH_2$ has a lower reducing power (and so has less stored energy): it can only pass

Box 9.3

The mitochondrial envelope

The matrix of the mitochondrion must maintain its distinctive chemical environment. On the other hand, some substances must move between it and the cytoplasm. Specific carrier proteins in the membranes make this balancing act possible.

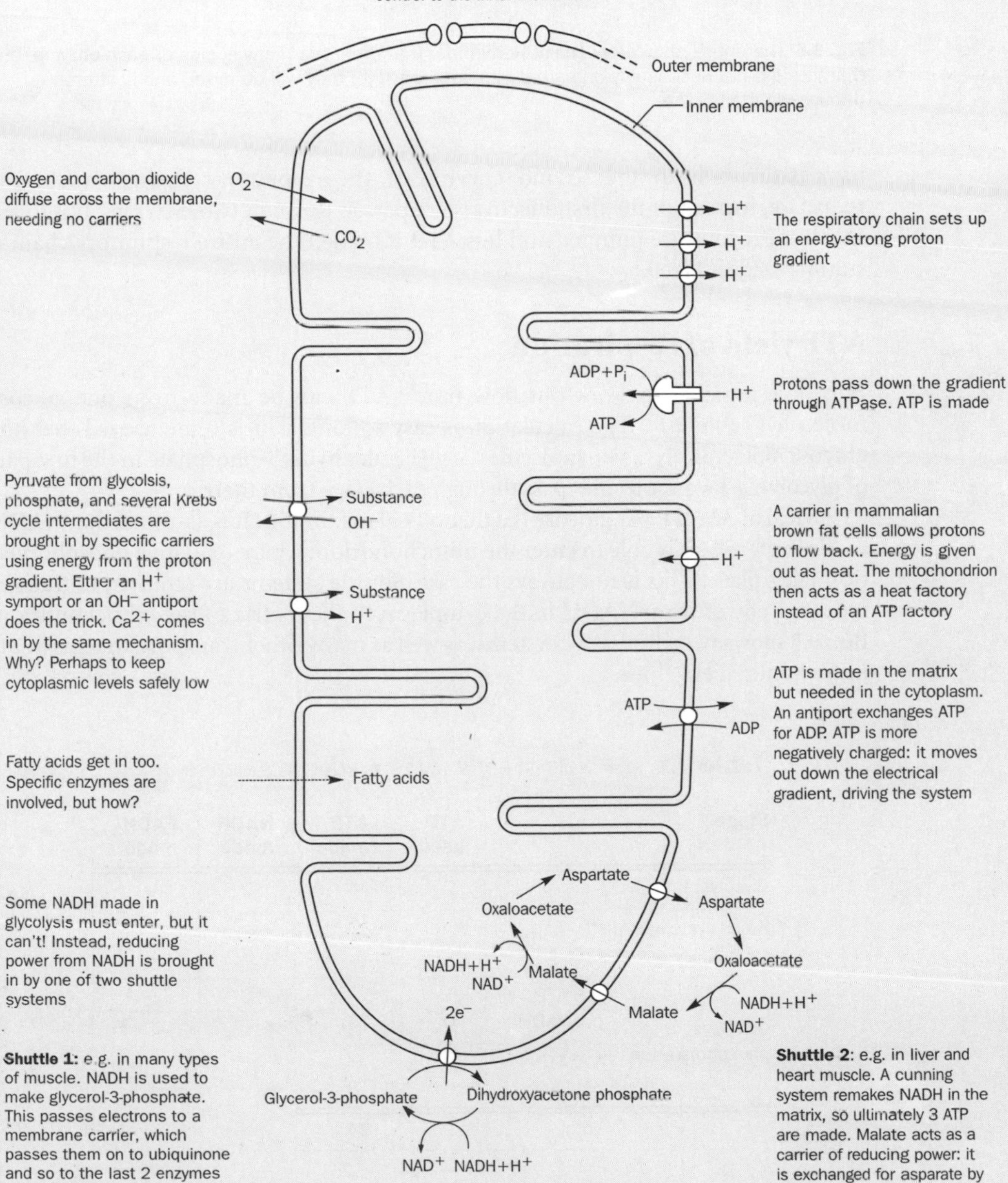

Porin has a large hydrophillic pore. Molecules up to the size of small proteins can pass through freely. This leaves control to the inner membrane

Outer membrane

Inner membrane

Oxygen and carbon dioxide diffuse across the membrane, needing no carriers

O_2

CO_2

H^+
H^+
H^+

The respiratory chain sets up an energy-strong proton gradient

$ADP + P_i$

ATP

H^+

Protons pass down the gradient through ATPase. ATP is made

Pyruvate from glycolsis, phosphate, and several Krebs cycle intermediates are brought in by specific carriers using energy from the proton gradient. Either an H^+ symport or an OH^- antiport does the trick. Ca^{2+} comes in by the same mechanism. Why? Perhaps to keep cytoplasmic levels safely low

Substance
OH^-

Substance
H^+

H^+

A carrier in mammalian brown fat cells allows protons to flow back. Energy is given out as heat. The mitochondrion then acts as a heat factory instead of an ATP factory

ATP

ADP

ATP is made in the matrix but needed in the cytoplasm. An antiport exchanges ATP for ADP. ATP is more negatively charged: it moves out down the electrical gradient, driving the system

Fatty acids get in too. Specific enzymes are involved, but how?

Fatty acids

Some NADH made in glycolysis must enter, but it can't! Instead, reducing power from NADH is brought in by one of two shuttle systems

Aspartate

Oxaloacetate

Aspartate

$NADH + H^+$
NAD^+ Malate

Oxaloacetate

$NADH + H^+$
NAD^+

$2e^-$

Malate

Shuttle 1: e.g. in many types of muscle. NADH is used to make glycerol-3-phosphate. This passes electrons to a membrane carrier, which passes them on to ubiquinone and so to the last 2 enzymes of the respiratory chain. Only 2 ATP molecules are made

Glycerol-3-phosphate

Dihydroxyacetone phosphate

NAD^+ $NADH + H^+$

Shuttle 2: e.g. in liver and heart muscle. A cunning system remakes NADH in the matrix, so ultimately 3 ATP are made. Malate acts as a carrier of reducing power: it is exchanged for asparate by a pair of carriers

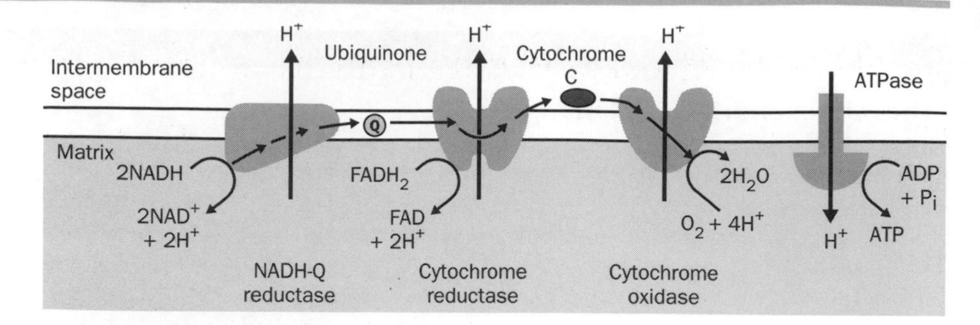

Fig. 9.5 The entire respiratory chain. Remember that there are many copies of each enzyme, and that they are not necessarily organized into sets like this: they will be mixed and scattered through the membrane.

its electrons on to the second enzyme of the respiratory chain, cytochrome reductase, instead of the first. Electrons are passed through two enzymes, not three, so fewer protons are pumped and less ATP is made. The entire respiratory chain is summarized in Fig. 9.5.

ATP yield of respiration

It is now possible to work out how much ATP can be made from one glucose molecule (Table 9.1). The calculation is easy so long as it is remembered that one glucose molecule gives two molecules of glyceraldehyde-3-phosphate in the first part of glycolysis. Two molecules pass through each stage from there on.

A yield of 32 ATP per glucose is a theoretical maximum. It is assumed that NADH made in glycolysis is able to enter the mitochondrion, where oxidative phosphorylation takes place. This is not always the case. Shuttle systems are required to transfer reducing power from NADH in the cytoplasm to the matrix of the mitochondrion. Box 9.3 shows two alternative systems, as well as many other transport systems in the mitochondrial envelope.

Table 9.1 ATP yield of aerobic respiration of glucose

Stage		ATP used	ATP made	NADH made	$FADH_2$ made
Glycolysis		2	4	2	0
Krebs cycle including pyruvate → acetyl–CoA		0	2	8	2
				10	2
Oxidative phosphorylation	2.5 ATP per NADH	0	25		
	1.5 ATP per $FADH_2$	0	3		
	Total	2	34		
			−2		
	Net ATP Yield		32		

9.3 Respiration of other substrates

Glucose is an important substrate in respiration, but it is not the only one. Other carbohydrates, fats and proteins may each be used as energy sources in particular situations (Fig. 9.6). Polysaccharides such as starch, and disaccharides such as sucrose are broken down into monosaccharides or monosaccharide phosphates. These can be interchanged quite freely, and enter the early stages of glycolysis.

Triglyceride fats are less straightforward. First they must be broken down into glycerol and fatty acids. Glycerol, with its three carbon atoms, is converted in three steps to a triose phosphate, glyceraldehyde-3-phosphate, and so enters glycolysis. The long, fatty-acid chains are oxidized, and broken down two carbons at a time in the matrix of the mitochondrion. Each 2-carbon group forms acetyl–coenzyme A, which can enter the Krebs cycle; NADH and $FADH_2$ are also made.

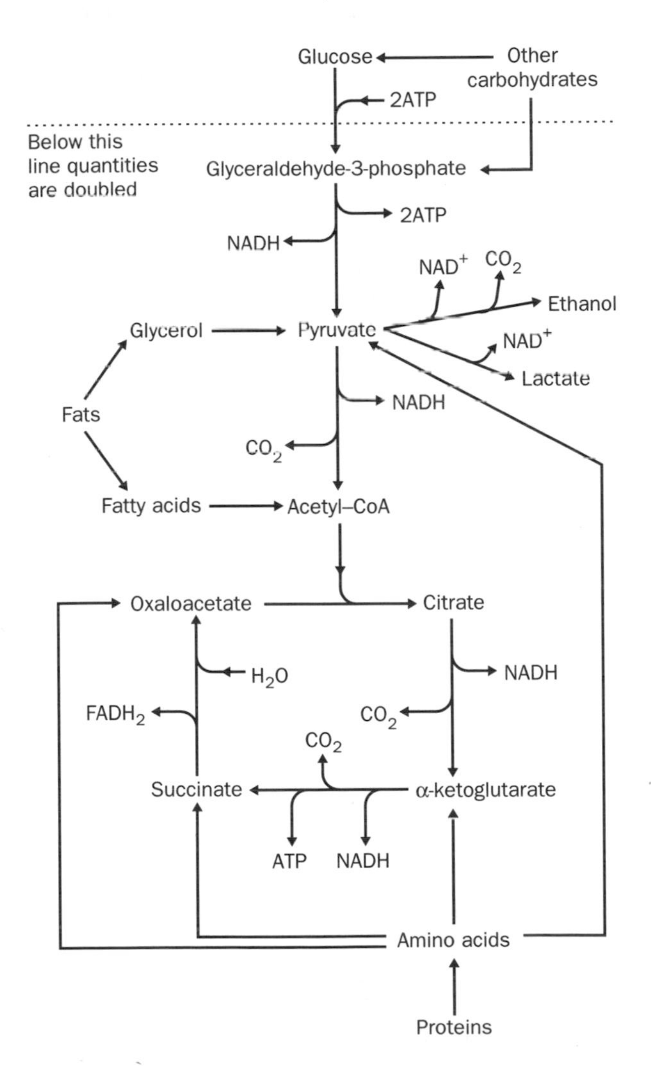

Fig. 9.6 Relationships between glycolysis, the Krebs cycle, and anaerobic pathways, showing the entry of various substrates.

If proteins are to be respired, they must first be broken down into amino acids. Whilst amino acids share a common structure centred on the α-carbon atom, their carbon skeletons are diverse, due to the variation in side chains. This means that the carbon skeletons, once stripped of their amino groups, enter the respiratory pathway at different points. Some enter as pyruvate, others as acetyl–coenzyme A, many as α-ketoglutarate or other Krebs cycle intermediates. The amino groups are removed either to other organic acids (transamination) or are given out as ammonium ions (deamination). Ammonium ions are, indirectly, toxic to animals and must either be excreted rapidly or converted to urea (Section 22.2).

9.4 Anaerobic respiration

Cells do not always have vast amounts of oxygen available to them. The cells of animals living in stagnant mud or as parasites in a sheep's gut, cells near the centre of a large potato, muscle fibres in the fleeing victim of a charging rhinoceros, and root cells of wheat seedlings in a flooded field, are just a few examples. Sometimes there is little oxygen in the organism's environment, sometimes it cannot reach the cells in question fast enough. In each case the cell has energy requirements: oxygen may be too scarce for aerobic respiration alone to meet them. Fortunately, it is possible to break down sugars and make ATP without the use of oxygen. This is called anaerobic respiration.

Oxygen is required only at the end of the respiratory chain. Why should this stop respiration so effectively? Oxygen, remember, is an electron dustbin. If the last electron-carrying site in the chain cannot throw away its electrons to oxygen, it will be permanently reduced. This means that the previous carrier cannot pass on its electrons, and so on until the whole chain grinds to a halt. No NADH is oxidized, no protons are pumped, no ATP is made. The Krebs cycle stops too, because NAD^+ runs out, now that the respiratory chain is no longer recycling it. The oxidation steps in the Krebs cycle require NAD^+ to take away electrons as NADH: if there is no NAD^+, there is no reaction.

Glycolysis might also be expected to stop, since it too uses NAD^+ and makes NADH. However, the pyruvate made in glycolysis can be reduced in a way which uses up NADH and reforms NAD^+. Glycolysis continues, making overall just two molecules of ATP for each glucose molecule used, compared with up to 32 in aerobic respiration.

There are two main ways in which NAD^+ can be remade. The first is alcoholic fermentation, seen in yeast, some other microbes and certain plant cells.

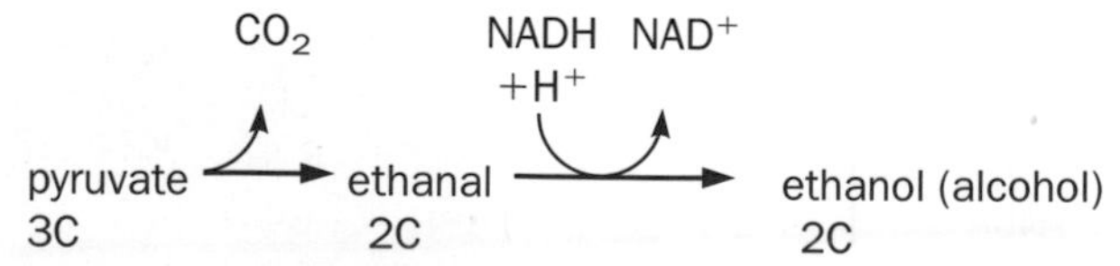

NAD^+ is reformed as ethanol is made, but ethanol is toxic. To a yeast cell, the world is a sewer, and excretion of ethanol solves the problem in the short term. Even so, in some strong, sweet wines, fermentation stops when ethanol concentration rises to a toxic level: the yeast population has poisoned its world. In plant cells, which have to live in harmony with their neighbours within the plant, the problem is more acute, and many plants adapted to waterlogged soils have means of avoiding this self-inflicted injury (Box 18.5).

A lot of anaerobic respiration has gone on here!

The second way of regenerating NAD^+ is to reduce pyruvate to lactate.

$$\text{pyruvate} \xrightarrow{\text{NADH} + \text{H}^+ \;\rightarrow\; \text{NAD}^+} \text{lactate}$$

This is seen in many microbes, such as the bacteria which sour milk. It also occurs in animal cells, for example in muscle during intense exercise. Lactate is a metabolic dead-end. Nothing can be done with it apart from remaking pyruvate. Microbes excrete the lactate. Mammals, which cannot sustain anaerobic respiration for long, re-oxidize it to pyruvate in the liver, where local oxygen levels are likely to be higher than in very active muscle.

In both processes, pyruvate is used and NAD^+ is regenerated. Neither uses oxygen, but alcoholic fermentation produces carbon dioxide.

9.5 Respiration rate of organisms

So far, we have looked at respiration at the cellular level. Cells live together in tissues, organs, organisms and populations. It is perfectly valid to think of not only cells but organisms as respiring, so long as respiration does not become confused in the mind with gas exchange or breathing. It is easy to see how this confusion can arise. Animals exchange carbon dioxide for oxygen, and breathe to ventilate the lungs, only because they respire. Aerobic respiration uses up oxygen, which must be replenished; it produces carbon dioxide which must be disposed of. Gas exchange provides the most accurate way of estimating rate of respiration. Whether one is looking at respiration in a tomcat or a germinating cabbage seed, rate of oxygen consumption is an excellent measure of rate of aerobic respiration. If twice as much glucose is respired, twice as much oxygen will be used up, and so on. Carbon dioxide production could equally well be measured.

The greatest practical problem is that when glucose is the substrate, exactly the same number of oxygen molecules are used as molecules of carbon dioxide are made.

$$C_6H_{12}O_6 + 6O_2 \rightarrow 6CO_2 + 6H_2O$$

Chemists will recall that whatever the gas, the same number of molecules take up the same volume, at any given temperature and pressure. The volume of CO_2 made exactly equals the volume of O_2 used, so direct measurements of volume change are useless. The usual solution is to remove carbon dioxide with a base such as sodium hydroxide: the rate at which gas volume falls then shows the rate of oxygen consumption.

Factors affecting respiration rate

Several factors affect the rate of aerobic respiration:

- *Temperature* Within the range 0°C to approximately 40°C, an increase leads to higher respiration rate. Rate approximately doubles with every 10°C rise. Above the optimum temperature, rate decreases rapidly as enzymes are denatured.
- *Substrate concentration* This increases the respiration rate of single cells to a maximum level. This is harder to investigate in more complex organisms, in which potential substrates are regulated and interconverted.
- *Oxygen concentration* Again, rate is increased to a maximum level, but this is difficult to investigate in cells which are also capable of anaerobic respiration.
- *Specific toxins* For example, cyanide is a potent inhibitor of the respiratory chain.

9.6 Metabolic rate

The rate of respiration of a whole animal is usually called the **metabolic rate**. This varies greatly according to the activity of the animal. **Basal metabolic rate** (BMR) is the metabolic rate when at rest: this is the rate of respiration needed to provide energy for vital functions such as heart-beat, breathing, nerve action, kidney function, cell repair and heat production.

There is much variation in BMR, both within and between species. In humans, BMR varies through life, reaching a peak about one year of age. Individuals of the same age may have different rates, perhaps related to levels of the hormone thyroxine.

A comparison of mammalian species shows that, in general, the smaller the animal the higher the BMR. An obvious explanation is that this is a result of mammals maintaining a constant body temperature in a usually colder world. As body size increases, volume (and hence mass) increases faster than surface area: bigger animals have a lower surface area/volume ratio. Heat is lost only through the surface, so a small animal will lose heat relatively quickly, and will need a higher metabolic rate if it is to keep warm. However, similar trends have been observed in many groups of poikilothermic animals, such as fish. This observation remains something of a mystery.

Metabolic rate: Respiration rate of an entire animal.

Basal metabolic rate: Metabolic rate at rest.

9.7 Respiratory quotient

When one starts to compare respiration rates of two cells or organisms, there is another problem: the type of respiration affects the exchange of gases. When proteins and fats are respired, and when alcoholic fermentation occurs, oxygen used no longer equals carbon dioxide made. Respiratory quotient (RQ) puts a figure on the balance between oxygen consumption and carbon dioxide production:

$$\text{RQ} = \frac{\text{rate of CO}_2\text{ production}}{\text{rate of O}_2\text{ consumption}}$$

(RQ is a ratio, so does not have a unit.)

RQ values indicate particular types of respiration (Table 9.2). RQ for aerobic respiration of glucose or other carbohydrates is 1.0 because the volumes are identical, and any number divided by itself gives 1. Alcoholic fermentation produces CO_2 but uses no oxygen. RQ will be some number divided by zero, which equals infinity. Where oxygen is present at a low level, a mixture of aerobic and anaerobic respiration will give an RQ of between 1.0 and infinity. Aerobic respiration of fats produces more NADH per carbon than carbohydrates. Oxygen is used when NADH is oxidized, so oxygen consumption will be greater than carbon dioxide production: RQ will be less than 1.0, typically about 0.7. Similarly, RQ for a protein substrate will be about 0.9.

Table 9.2 Types of respiration and their effect on RQ

Aerobic/Anaerobic	**Substrate**	**RQ**
Purely anaerobic (alcoholic fermentation)	any	∞
Aerobic	carbohydrate	1.0
Aerobic	protein	0.9
Aerobic	fat	0.7
aerobic (with carbon retention)	any	<1.0, potentially <0.7
Mixed aerobic and anaerobic	any	between 1.0 and ∞
Aerobic	mixed	between 0.7 and 1.0
Anaerobic (lactate formed)	any	—

An intermediate RQ value of, for example, 0.8 could result from various combinations of fat, protein and carbohydrate. Respiration of fat with a very small amount of alcoholic fermentation could also give this result.

An RQ of less than 0.7 would indicate that some carbon is not being released as carbon dioxide. Either carbon skeletons are being siphoned off from the Krebs cycle, perhaps to make amino acids, or carbon dioxide is being used in the cell. Apart from photosynthesis, this could happen when urea is synthesized (Section 22.2), or when cells are building sugars from other energy stores. For example, when blood glucose is low and glycogen is exhausted, the mammalian liver converts fats to sugars. A critical step in this pathway is the carboxylation of pyruvate.

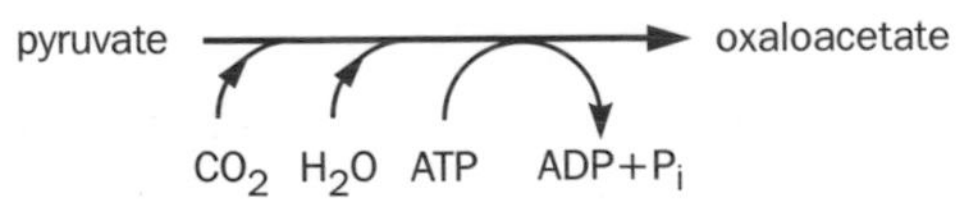

$$\text{pyruvate} \xrightarrow{\;CO_2,\ H_2O,\ ATP\ \rightarrow\ ADP + P_i\;} \text{oxaloacetate}$$

From molecule to organism

There is a direct link between events at the molecular level and observations made on whole organisms. RQ values, measured easily for large chunks of living matter, whether a hamster or a yeast culture, give piercing insights into which pathways are operating within cells. Mammals have lungs simply because the most oxidized form of carbon is a gas, and because the terminal electron acceptor of the respiratory chain is also a gas, oxygen. When we suffocate, we die only because there is nowhere for electrons passed down the chain to go. We are truly the creatures of our own chemistry.

Summary

- Respiration is the breakdown, in cells, of energy-rich molecules, releasing energy.
- Aerobic respiration requires oxygen. Anaerobic respiration does not.
- Aerobic respiration of glucose involves three distinct stages: glycolysis, the Krebs cycle and oxidative phosphorylation.
- Glycolysis converts glucose to pyruvate. Some energy is transferred to ATP and NADH.
- Glycolysis takes place in the cytoplasm.
- The Krebs cycle oxidizes pyruvate to carbon dioxide. Energy is transferred to NADH, $FADH_2$ and ATP.
- The Krebs cycle takes place in the matrix of the mitochondrion.
- Oxidative phosphorylation is the transfer of energy from NADH and $FADH_2$ to ATP.
- It involves a chain of electron carriers in the inner membrane of the mitochondrion. This takes high energy electrons from NADH. As they pass down the chain they lose energy. At the end, they are passed to oxygen.
- The energy released by the electron transport chain is used to set up an energy-storing proton gradient across the inner membrane. This drives ATP synthesis.
- Under anaerobic conditions, NADH cannot be oxidized, so the entire system stops.
- Anaerobic respiration involves reducing pyruvate to ethanol or lactate. This oxidizes some NADH, allowing glycolysis to continue.
- The ATP yield of anaerobic respiration is very low.
- Fatty acids and amino acids can also be used as substrates in respiration.
- Metabolic rate is the respiration rate of an entire organism.
- Basal metabolic rate is the metabolic rate at rest.
- Small animals tend to have higher metabolic rates.
- Respiratory quotients give clues to the substrate and type of respiration going on in an organism.

Exercises

9.1. Dinitrophenol (DNP) is a simple organic compound which interferes with oxidative phosphorylation. It carries protons across the inner membrane of the mitochondrion, allowing the proton gradient to break down. What will be the effect of DNP on the following:

(i) electron transport; **(ii)** ATP synthesis; **(iii)** glycolysis and the Krebs cycle?

In the presence of DNS, what becomes of the energy released in respiration?

9.2. Rotenone inhibits electron transport within NADH-Q reductase. How will this affect the respiratory pathways?

9.3. Some parasitic animals living in the mammalian gut have an unusual mechanism for anaerobic respiration. Glycolysis takes place as normal. Some of the pyruvate is then used to make an organic molecule which acts as an electron acceptor at the end of the respiratory chain, instead of oxygen. This is more efficient than the forms of anaerobic respiration described in this unit. Why?

9.4. Why do children have a higher basal metabolic rate than adults?

9.5. RQ was calculated at intervals during the germination of an oat seed. It began at 1.0, then rose steadily to 5.7. Explain.

Further reading

Salisbury, F.B. and Ross, C.W. *Plant Physiology* (4th ed.) (Belmont CA: Wadsworth, 1992). A big undergraduate textbook, which covers some of the special features of plant respiration.

Schmidt-Nielsen, K. *Animal Physiology: Adaptation and Environment* (4th ed.) (Cambridge: Cambridge University Press, 1990). One of the all-time great textbooks! Exciting, authoritative and easy to read, it puts animal physiology in the context of the environment. Whole body issues are discussed, such as metabolic rate.

Stryer, L. *Biochemistry* (4th ed.) (New York: Freeman, 1995). The classic big undergraduate textbook of biochemistry gives a good treatment of the chemistry of respiration. There are several broadly similar books on the market, but I would still go for this if looking for clear explanation.

Photosynthesis

Contents

Connections

► Before studying this unit, you should have a basic understanding of biological molecules, cell structure, enzymes and movement of molecules across membranes (Units 3–6). Unit 8 is an essential introduction to the concept of energy, which is central to photosynthesis. The chemical machinery of respiration (Unit 9) has much in common with the machinery of photosynthesis. An understanding of photosynthesis is important when dealing with other areas of plant physiology (Unit 18).

10.1 The food problem: construct or consume?

Cells need food. They require a supply of reduced organic molecules such as carbohydrates or fats. These act as energy stores which can be broken down in respiration, releasing usable energy. Some organisms get hold of these food molecules ready made. Animals feed on other organisms which already contain these molecules; fungal hyphae grow through tissues which may or may not still be alive, and digest them as they go. This way of getting food is called heterotrophic nutrition. Cells and organisms which do this can only exist because other cells and organisms have the biochemical apparatus to make food molecules themselves. To do this requires a supply of simple molecules from which to construct these larger molecules, and an energy supply. This is called autotrophic nutrition, and in global terms one type of autotrophic nutrition is enormously important: **photosynthesis**.

In photosynthesis, light is the energy supply for the construction of large food molecules. Almost all photosynthetic cells use the same raw materials: carbon dioxide and water. Carbohydrates are made, oxygen is a waste product. The biochemical apparatus of photosynthesis is extremely similar in all these cells, which include the photosynthetic eukaryotes, both plants and protoctists, as well as a large group of photosynthetic bacteria, the cyanobacteria. Some very different types of photosynthesis occur in a few other groups of bacteria (Box 10.2): whilst these are in-

Photosynthesis: Building up carbohydrates, using carbon dioxide, water and light energy, and making oxygen as a waste product.

significant on a global scale, they give exciting clues to the evolution of the standard photosynthetic apparatus.

This unit is about how plants, photosynthetic protoctists and cyanobacteria make carbohydrates from carbon dioxide and water, using light energy. Many stages are involved, but the overall reaction is given by the following equation:

$$nCO_2 + nH_2O \xrightarrow{\text{light}} (CH_2O)_n + nO_2$$

More specifically, if hexose sugars are the product:

$$6CO_2 + 6H_2O \longrightarrow C_6H_{12}O_6 + 6O_2$$

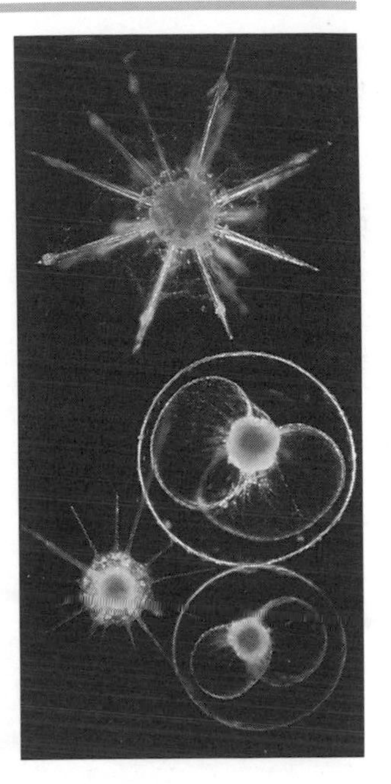

Even the ocean is a hotbed of photosynthesis. These protoctists, phytoplankton, live in the uppermost layer of seawater.

10.2 Photosynthesis

In photosynthesis, carbon is converted from a highly oxidized state in carbon dioxide to a much more reduced state in carbohydrates. Reduction is, at heart, the gaining of electrons (Section 8.2), so a source of electrons is required for photosynthesis. Reduced organic molecules tend to act as energy stores, so photosynthesis also requires an energy supply to drive the reduction of carbon. Water is the source of electrons, and light provides energy, but the process is not a simple one. There are many reactions involved, which in eukaryotes take place in the chloroplasts. These reactions neatly fall into two groups, the light reactions and the carbon-fixing reactions.

In the light reactions, a phosphorylated energy store, ATP, and a potent reducing agent, NADPH are made. Energy comes from light absorbed by chlorophyll, a large organic molecule. Electrons are supplied when water molecules are split, producing oxygen and H^+ ions. The light reactions take place in, and at the surfaces of, the thylakoid membranes inside the chloroplast (Section 4.5: Plastids).

In the carbon fixation reactions, carbon from CO_2 is added to an existing carbon skeleton and reduced, making sugars. NADPH made in the light reactions is used as a reducing agent; an extra input of free energy comes from ATP. Some of the sugars are siphoned off as products. The rest are recycled to a form which can accept more CO_2, continuing the process. All this takes place in the stroma of the chloroplast. The carbon fixation reactions are sometimes, misleadingly, called the dark reactions. They do not require light directly but require NADPH and ATP made by the light reactions. In the dark, these are quickly used up, so the carbon fixation reactions stop shortly after the light reactions. The light reactions and carbon fixation reactions are summarized in Table 10.1 on page 156.

10.3 The light reactions

The chemical machinery of the light reactions is embedded in the membranes of the thylakoids: flattened sacs within the chloroplast. It has several components. First, there are large complexes made up of hundreds of chlorophyll molecules bound to proteins. There are two types of complex: photosystem I (PSI) and photosystem II (PSII). Secondly, there are two electron transport complexes which receive high energy electrons passed on by the photosystems. One pumps protons

Table 10.1 What's made and what's used: a comparison of the light reactions and the carbon fixation reactions

	Light reactions		Carbon fixation reactions	
	Used	**Made**	**Used**	**Made**
ATP		√	√	
ADP	√			√
P_i	√			√
NADPH		√	√	
$NADP + H^+$	√			√
H_2O	√			
O_2		√		
CO_2			√	
Glyceraldehyde-3-phosphate				√

across the thylakoid membrane, using the energy released as electrons are passed from one carrier site to the next. This is used to set up an energy-storing proton gradient across the membrane. The other passes its electrons to $NADP^+$, reducing it to NADPH. Thirdly, there are three smaller electron-carrying molecules which are able to move around within the membrane, linking the photosystems and the electron transport complexes. Finally, the thylakoid membrane contains ATPases which span the membrane, using the energy stored in the proton gradient to make ATP.

This apparatus can work in two distinct ways. The first uses both photosystems, makes both ATP and NADPH, and splits water, releasing waste oxygen. This mode of operation is known, cumbersomely, as **non-cyclic photophosphorylation**. The second only uses PSI: only ATP is made, and water is not split. This is **cyclic photophosphorylation**. The starting point for both processes, however, is the absorption of light by chlorophyll molecules.

Chlorophyll and the photosystems

The chlorophyll molecule has two parts, a large ring structure, known as a porphyrin, with a magnesium atom at its centre, and a long hydrophobic tail (Fig. 10.1). There are two subtly different types, chlorophyll a and chlorophyll b, but they work in very similar ways. Chlorophyll absorbs red and blue light well, but green light poorly (Fig. 10.2). The green light which is not absorbed is what gives leaves their usual colour. In many plants, chlorophylls are the only really significant pigments. In these, the absorption spectrum for the leaf (showing how well light of different wavelengths is absorbed) is strikingly similar to the action spectrum for photosynthesis, which shows how effective light of different wavelengths is in driving photosynthesis (Fig. 10.3). The conclusion must be that light energy absorbed by chlorophyll molecules drives photosynthesis.

Non-cyclic photophosphorylation: Light reactions which require light absorption by both photosystems. ATP and NADPH are made. Water is used, oxygen is made.

Cyclic photophosphorylation: Light reactions which require light absorption only by PSI. Only ATP is made. Water is not split.

The porphyrin ring is the business end of chlorophyll. It is a flat ring, surrounded

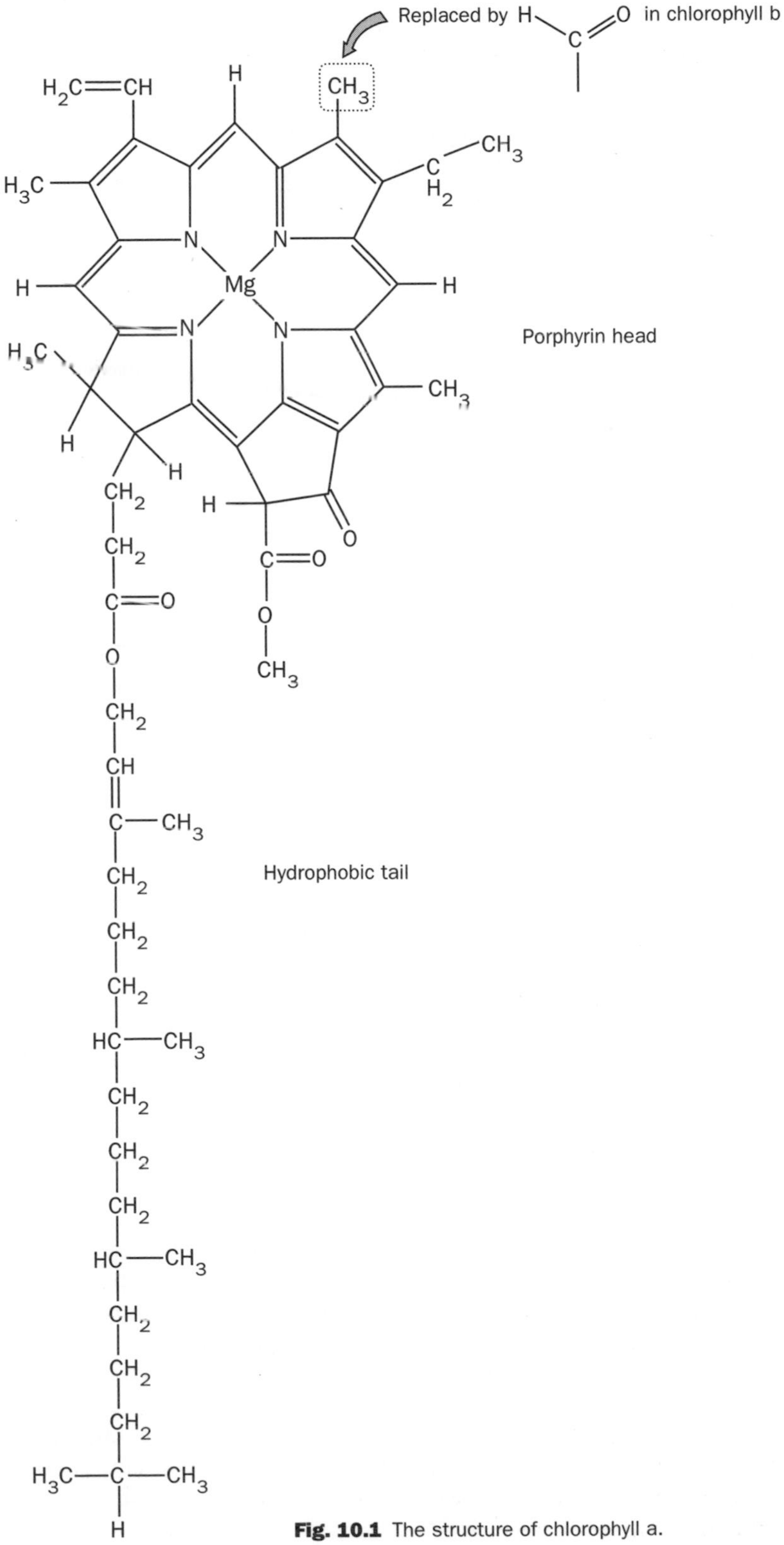

Replaced by H–C=O in chlorophyll b

Porphyrin head

Hydrophobic tail

Fig. 10.1 The structure of chlorophyll a.

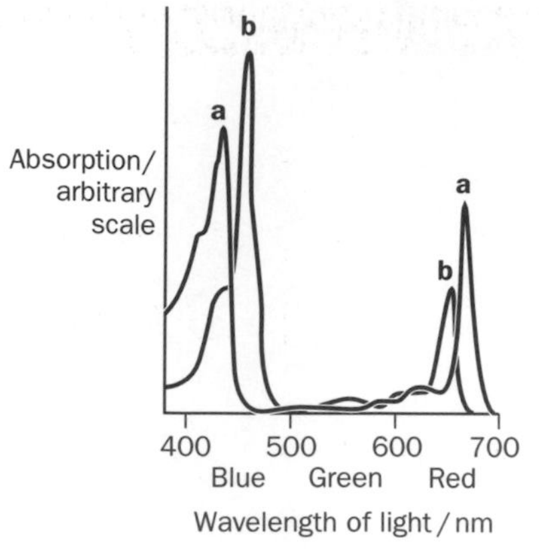

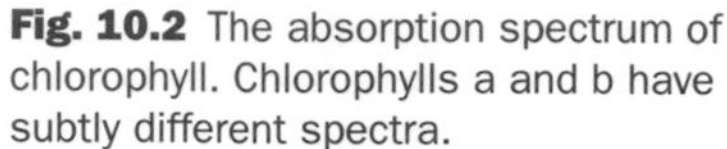

Fig. 10.2 The absorption spectrum of chlorophyll. Chlorophylls a and b have subtly different spectra.

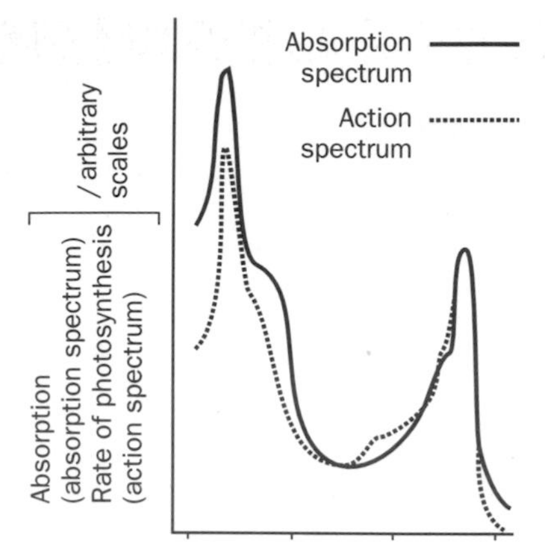

Fig. 10.3 The absorption spectrum of pigments from a chlorophyll-rich leaf compared with the action spectrum for photosynthesis.

by electrons in orbitals (Section 3.2) shared by many atoms of the ring. When light is absorbed, an electron moves to an outer orbital in which it has more energy. Chlorophyll is then said, delightfully, to have become 'excited'. An excited chlorophyll molecule must lose its energy, quickly. This may happen by the electron falling back to its original orbital and giving up its extra energy as heat, or by giving out light (fluorescence). Chlorophyll extracted from a leaf in a solvent like propanone does just this when it absorbs light. The red fluorescence is attractive, and quite useless. In the chloroplast, however, the chlorophylls are bound tightly together by proteins to form **photosystems**. The energy of one excited chlorophyll molecule may be transferred to another. This happens either when energy released as an electron falls back and is used to excite an adjacent molecule, or when the excited electron actually moves from one chlorophyll to the next. This direct transfer is highly efficient if the molecules are extremely close together: 2.5 nm is an optimum.

Energy can only be transferred to a chlorophyll molecule which requires slightly less energy for excitation than the previous one. The energy required is modified slightly by the proteins that bind chlorophyll. The photosystem is organized so that energy absorbed by chlorophyll anywhere in the system is transferred towards a single, special molecule of chlorophyll a, somewhere in the complex. This special molecule, the reaction centre, is positioned so that its excited electrons may be passed on to a quite different molecule which can carry its energy and reducing power away through the membrane.

Chlorophyll is essential; various other pigments are more or less common as well. β-carotene absorbs mainly blue light, giving it a yellow colour. It forms part of the photosystems, and some of the energy it absorbs can be passed on to chlorophyll. It also has a protective role: one particular excited state of chlorophyll is able to pass its energy to oxygen, producing superoxide (O_2^-). This ion, a type of free radical, is most dangerous since it is highly reactive, reacting with almost anything in sight within the chloroplast. β-carotene minimizes the risk by taking the energy from chlorophyll in this state. Xanthophylls are found, in one form or another, in both plants and algae. Like chlorophyll, they are bound within the thylakoid membrane.

Photosystems: Highly organized arrays of chlorophyll molecules, other pigments and proteins, in the thylakoid membrane.

Box 10.1

Phycobilins

These pigments are found in cyanobacteria and the Rhodophyta (red algae). They can absorb light and pass its energy on to chlorophyll. The wavelengths which they absorb suit the cells they are in.

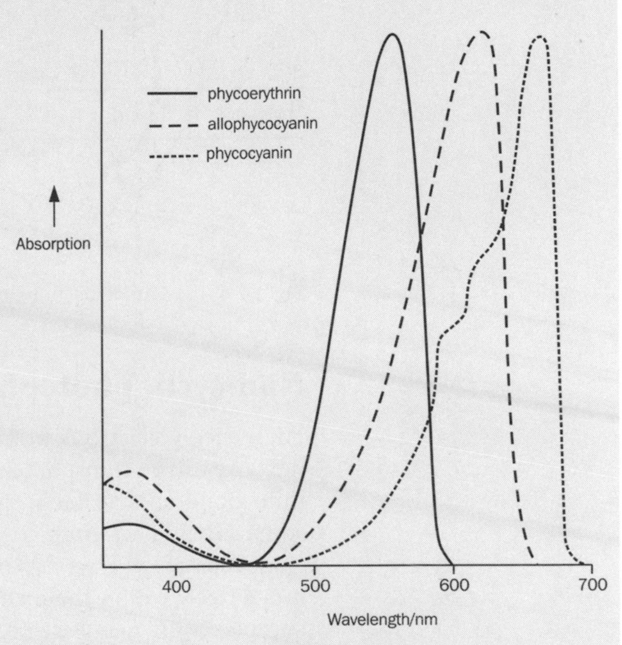

Absorption spectra of three phycobilins

What and where?

The light-absorbing part of a phycobilin molecule is a tetrapyrrole, much like the porphyrin head of chlorophyll but lacking the central magnesium atom, and not folded into a ring. It is hydrophilic, and is attached to a hydrophilic protein molecule. This prevents it being inserted into the thylakoid membrane like other photosynthetic pigments, which have at least some hydrophobic regions. Instead, they are organized into structures called phycobilisomes, which are attached to the outer surface of the thylakoid membrane.

How and why?

Three types of phycobilin have distinctive absorption spectra. Phycoerythrin is found in the Rhodophyta. These algae are almost all marine, and most live in deep water, often at greater depths than other algal groups can tolerate. Sea water absorbs light, but green is absorbed less than other wavelengths. Phycoerythrin, with its absorption peak in the green, helps adjust the photosynthetic machinery of red algae to deep water. The deeper the alga grows, the more phycoerythrin it contains.

Phycocyanin and allophycocyanin are found in cyanobacteria, which typically live in the surface plankton of the sea and fresh water, and on rocks and soil. The blue absorption peaks of these phycobilins are suited to the abundant blue light component of full sunlight.

The light-absorbing tetrapyrrole component of phycocyanin and allophycocyanin.

They absorb light in the blue-green region of the spectrum, and channel energy to chlorophyll. Phycobilins are found in certain protoctists, notably the red algae, and in cyanobacteria. They are bound to the outer surface of the thylakoid membrane, absorb green light particularly effectively, and again channel energy to chlorophyll (Box 10.1).

Whether or not these **auxiliary pigments** are present, the central idea is the same: chlorophyll molecules within photosystems channel the energy of absorbed light into a reaction centre which gives out high-energy electrons. The energy and reducing power of these electrons is used to make ATP and NADPH, as now follows.

Auxiliary pigments: Light-absorbing molecules other than chlorophyll which take part in photosynthesis.

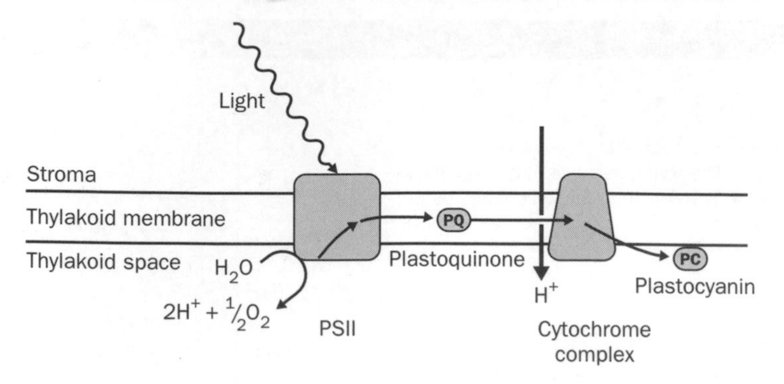

Fig. 10.4 The fate of electrons emitted by photosystem II.

Non-cyclic photophosphorylation

High-energy electrons from the photosystems are passed from one carrier to the next in electron transport chains. The similarity with the electron transport chain of mitochondria is striking: the two almost certainly shared a common ancestor in some ancient bacterium.

PSII passes its electrons to a relatively small molecule, plastoquinone, which can move freely within the membrane. This passes electrons on to a large cytochrome complex with many electron-carrying sites (Fig. 10.4). The free energy released when plastoquinone passes electrons to this complex is used to pump protons into the space inside the thylakoid sac. Energy is stored in this proton gradient. The flow of protons back out of the thylakoid space releases energy. ATPases within the membrane couple proton movement to the formation of ATP from ADP and phosphate. Once electrons have been passed through the cytochrome complex, they are passed on to a small, mobile protein, plastocyanin.

All this leaves PSII short of electrons. These are provided by the splitting of water into oxygen molecules, H^+ ions and electrons.

$$2H_2O \rightarrow O_2 + 4H^+ + 4e^-$$

This is catalysed by an elegant enzyme, the water-splitting clock, which is attached to PSII and releases electrons as they are needed.

PSI, once light is absorbed, passes electrons onto another mobile molecule, ferredoxin (Fig. 10.5). This is a small protein which carries electrons in a prosthetic

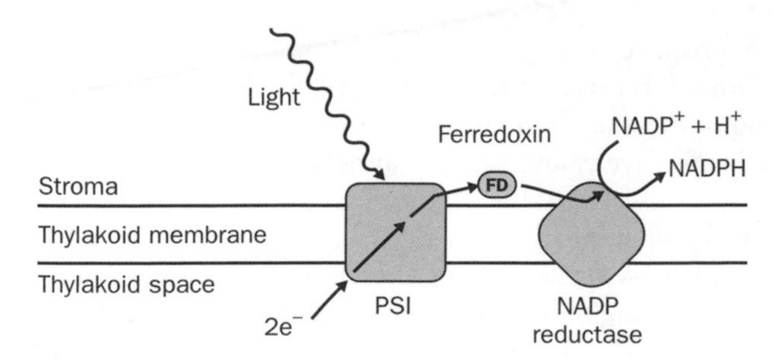

Fig. 10.5 The fate of electrons emitted by photosystem I.

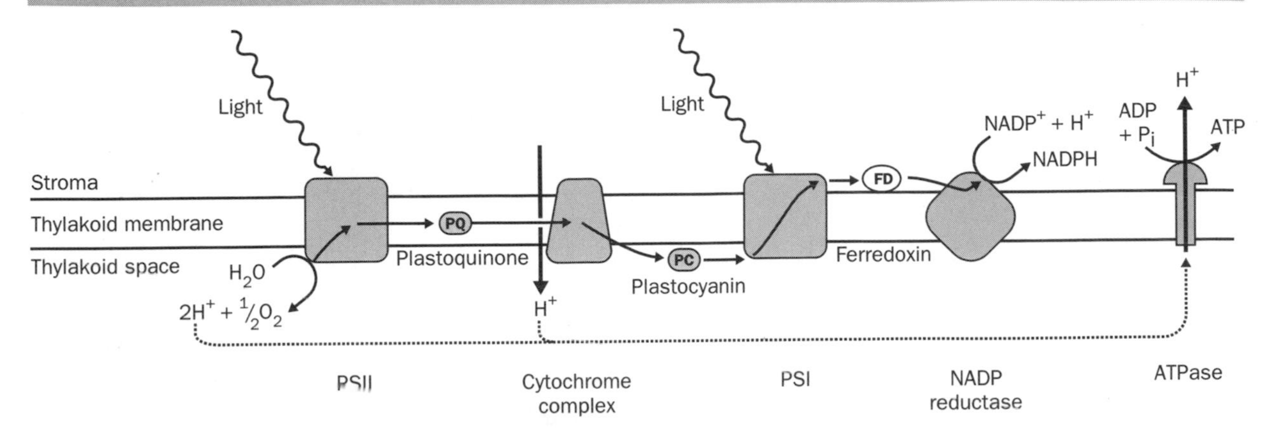

Fig. 10.6 Electron transport and proton pumping in the thylakoid membrane.

group which contains iron and sulphur. Ferredoxin passes its electrons to another electron-transporting complex, NADP reductase. Here, the electrons are used to reduce $NADP^+$ to NADPH.

Two problems remain: where do electrons from PSII go after they reach plastocyanin, and how does PSI get electrons to replace those it has lost? The answers are linked: plastocyanin passes its electrons on to PSI, solving both problems. So electrons flow through the system from water to NADPH; energy released on the way is used to make ATP. This is non-cyclic photophosphorylation: 'non-cyclic' because electrons do not pass round a closed cycle, 'photo' since light is required, 'phosphorylation' because ADP is phosphorylated to make ATP. Fig. 10.6 shows the chemical apparatus of the thylakoid membrane operating this way. The complexes are not arranged into single transport chains like this, but are mixed and scattered through the membrane. Random diffusion of the three mobile carriers links them.

Proton pumping by the cytochrome complex has already been described. However, something else assists in building up the proton gradient. H^+ made by the splitting of water is released into the thylakoid space. H^+ used in the making of NADPH is taken from the stroma, on the opposite side of the membrane. The overall result is an accumulation of protons in the thylakoid space, a sort of indirect second proton pump.

Why are two photosystems needed? One approach is to look at the amount of reducing power given to electrons excited by the photosystems. Electrons released by the splitting of water have a very low reducing power: electrons of a high reducing power are needed to reduce $NADP^+$. Light absorbed by one photosystem on its own cannot give a big enough energy input to bridge this gap (Fig. 10.7).

Cyclic photophosphorylation

The same electron transport apparatus can operate in a different way. Instead of passing electrons to NADP reductase, ferredoxin passes them to plastoquinone (Fig. 10.8). Excited electrons from PSI can then be passed through a cycle. Protons are pumped as electrons are passed from plastoquinone to the cytochrome complex,

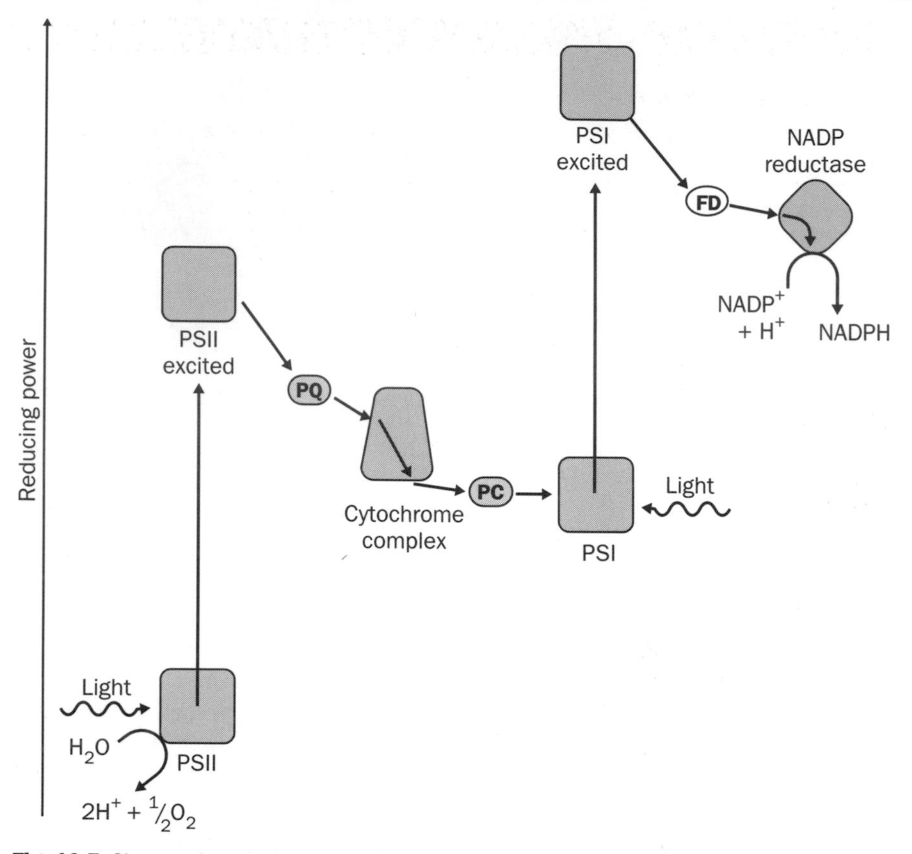

Fig. 10.7 Changes in reducing power during non-cyclic photophosphorylation, as electrons are excited in the photosystems and passed between carriers.

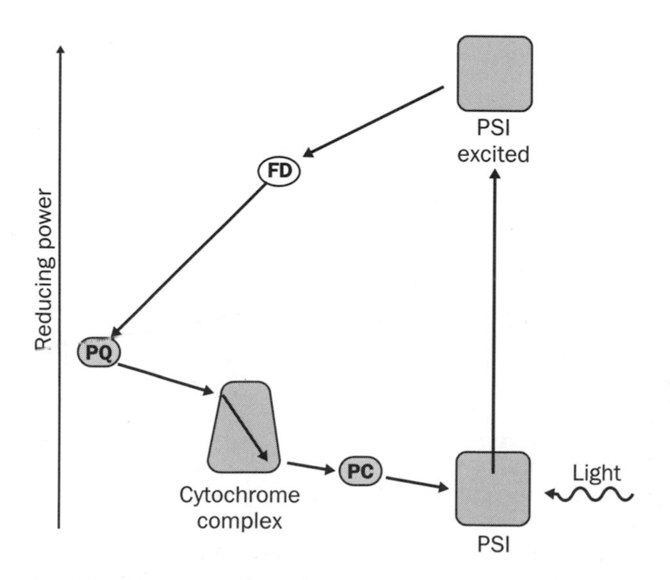

Fig. 10.8 Changes in reducing power during cyclic photophosphorylation.

Box 10.2

Old friends in new places

We are eukaryotes, and arrogant with it. We tend to see plants and animals as being 'normal', fungi as a bit different, and the prokaryotes as really rather obscure. On the contrary, the prokaryotes were here first, and our chemistry is derived from theirs. The eukaryotic plasma membrane, the inner membrane of the mitochondrion, and the thylakoid membrane all have bacterial plasma membranes as ancestors, if we accept the endosymbiotic theory (Section 4.8). Many familiar components of the chloroplast's photosynthetic apparatus turn up in unfamiliar contexts, within various types of bacteria.

Case 1: PSI, ferredoxin and NADP reductase in green sulphur bacteria

PSI acts much as it does in the thylakoid membrane, absorbing light and using the energy to excite electrons, which pass to NADP via ferredoxin. The novelty is that electrons come direct from hydrogen sulphide, rather than from water via PSII. Hydrogen sulphide is a much stronger reducing agent than water: the extra energy input from the second photosystem is not required in order for PSI to accept its electrons. These bacteria hang out in anaerobic environments where hydrogen sulphide is abundant.

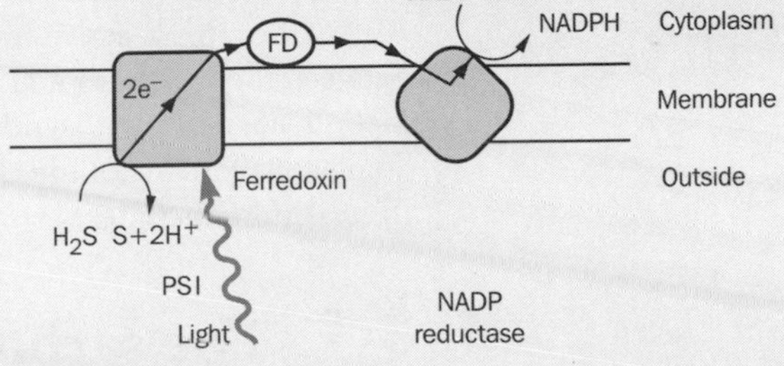

Electron transport during photosynthesis in green sulphur bacteria.

Case 2: Proton-transporting ATPases in bacteria

Most bacterial membranes contain proton-transporting ATPases like those found in thylakoids and mitochondria. They can work in two directions, either using a proton gradient and making ATP, or using ATP to pump protons, setting up a gradient. In some cases, another system will set up the gradient. This could be the respiratory chain of an aerobically respiring bacterium, or the photosynthetic apparatus of a purple non-sulphur bacterium (Case 3). The ATPase will then be used to make ATP. In other cases, ATP made within the cell through anaerobic respiration will be used to set up a proton gradient. This is then used to drive active transport of food molecules into the cell through symports. Bacterial flagella are also powered by the transmembrane proton gradient.

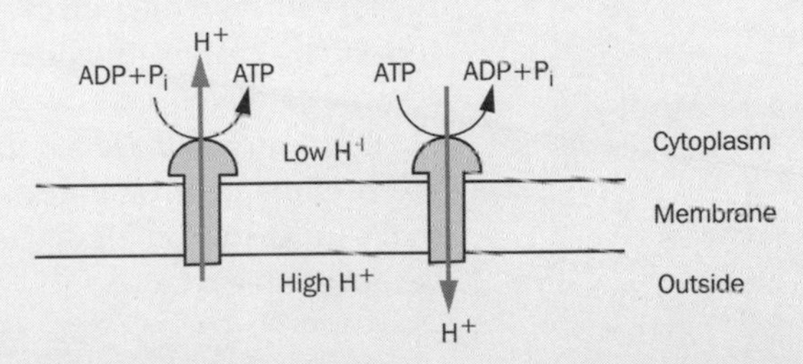

Proton-transporting ATPases can work in both directions.

Case 3: The cytochrome complex, a quinone and ATPase in purple non-sulphur bacteria

The cytochrome complex in the membranes of these bacteria is closely related to the one in the thylakoid membrane, and to the cytochrome reductase complex in the mitochondrion. The quinone is similar to both plastoquinone and ubiquinone: like them, it passes electrons on to the cytochrome complex. The familiar proton-transporting ATPase is here too, using the proton gradient set up by the cytochrome complex to make ATP.

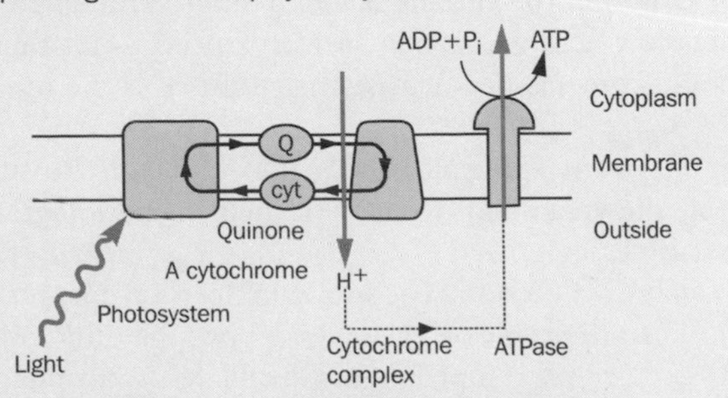

Cyclic electron transport and ATP synthesis during photosynthesis in purple non-sulphur bacteria.

Interestingly, the proton gradient also drives NADPH synthesis, despite there being only one type of reaction centre, if organic molecules like succinate are available as electron donors.

The challenge is to work out when, and in what contexts, these components first appeared. It is a challenging task, piecing together events of 3 billion years ago. We can only make intelligent guesses, based on what we see in contemporary bacteria like these.

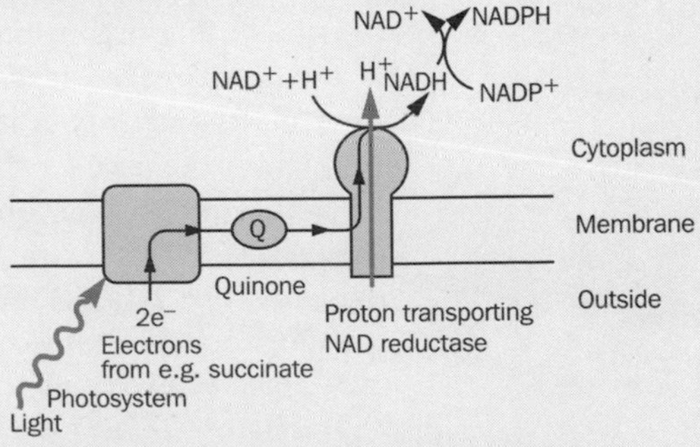

Non-cyclic electron transport and NADPH synthesis during electron transport in purple non-sulphur bacteria.

so ATP can be made. The electrons ultimately pass back to PSI. No NADPH is made, and PSII is not involved.

Isolated chloroplasts under experimental conditions can be persuaded to carry out either cyclic or non-cyclic photophosphorylation, or both together. It is harder to determine the importance of cyclic photophosphorylation in the intact plant. It seems likely that by adjusting the balance between the two modes of operation, the chloroplast can finely tune the ratio of ATP to NADPH being manufactured. The carbon fixation reactions require a slightly higher ratio than that made by non-cyclic photophosphorylation alone. Other processes within the chloroplast also require ATP, for example active transport across membranes. Whatever the precise balance between the two, both ATP and NADPH from the light reactions are required for the second stage of photosynthesis, the carbon fixation reactions. Box 10.2 shows how some components of the electron transport apparatus are also found in bacteria, in quite different contexts.

10.4 The carbon fixation reactions

These are the reactions in which carbon from carbon dioxide is added to carbon skeletons. The photosynthetic eukaryotes all have one central pathway in common, the C_3 carbon reduction cycle, which takes place in the stroma of the chloroplast. In this cycle, carbon dioxide is added to a pentose sugar phosphate, forming two molecules of a 3-carbon acid. This is reduced to a triose sugar phosphate; both NADPH and ATP are required. Much of the triose phosphate is recycled to reform the original pentose phosphate so that the cycle can continue. The rest is used to make other carbohydrates, amino acids or fats. The starting point for the cycle is the carbon-fixing step.

Rubisco

Ribulose 1,5-bisphosphate (5 carbons) is carboxylated to give two molecules of glycerate-3-phosphate (3 carbons) as shown in Fig. 10.9(a). This is catalysed by the enzyme ribulose bisphosphate carboxylase/oxygenase, known to its friends and admirers as Rubisco. Its significance on a global scale is awesome. Almost every carbon atom in every organic molecule in every cell of my body has passed through its active site at some stage (the rest will have passed through the active sites of other carbon-fixing enzymes in photosynthetic bacteria). What is true of my body is also true of all other organisms, apart from photosynthetic bacteria and the things that eat them. Moreover, it is a rather slow-acting enzyme, so plants need to fill their chloroplasts with lots of it. It should come as no surprise that Rubisco is the most abundant protein on Earth.

There is another carbon-fixing enzyme in some plant cells, phosphoenolpyruvate carboxylase. It is an extra, however, allowing plants to solve particular physiological problems (Box 10.3). Rubisco remains essential.

Unfortunately, Rubisco also catalyses a second reaction. Ribulose 1,5-bisphosphate reacts with oxygen to give just one molecule of glycerate-3-phosphate and one of glycollate-2-phosphate (Fig. 10.9(b)). This is a problem. Glycollate-2-phosphate can be recycled to produce glycerate-3-phosphate in a cycle whose reactions are spread between chloroplasts, mitochondria and peroxisomes, but ATP is used and

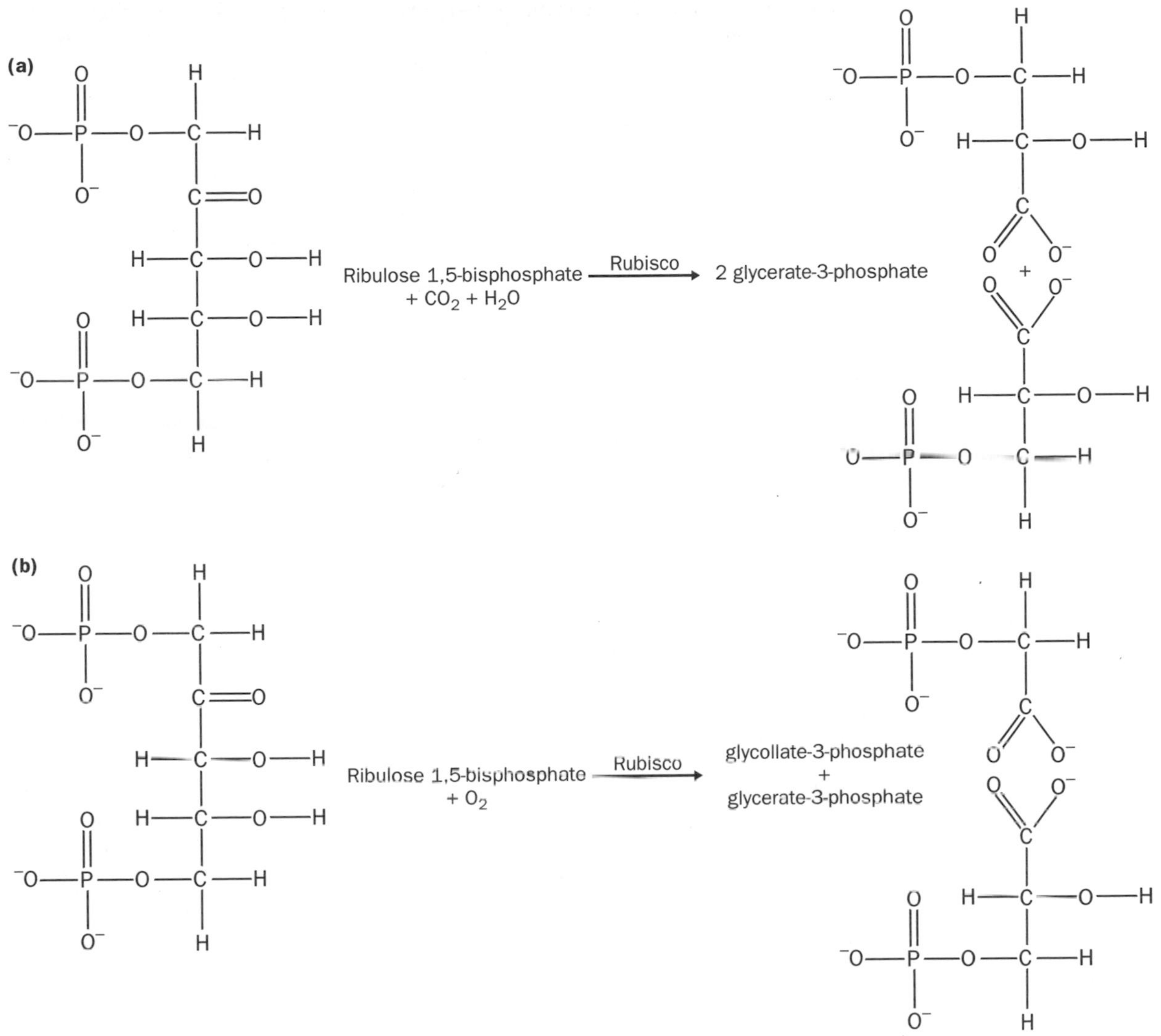

(a)

Ribulose 1,5-bisphosphate + CO_2 + H_2O $\xrightarrow{\text{Rubisco}}$ 2 glycerate-3-phosphate

(b)

Ribulose 1,5-bisphosphate + O_2 $\xrightarrow{\text{Rubisco}}$ glycollate-3-phosphate + glycerate-3-phosphate

Fig. 10.9 Reactions catalysed by Rubisco. (a) carboxylation, (b) oxygenation.

carbon dioxide is made. Since oxygen is used and carbon dioxide is made, the process looks superficially like aerobic respiration, and is called photorespiration. It goes against the flow of photosynthesis: both energy and carbon are wasted. It seems to be an unavoidable cost of using Rubisco to fix carbon.

The C_3 carbon reduction cycle

The fixing of carbon by Rubisco is only the first step on the way to making carbohydrates. The glycerate-3-phosphate which has been made is then phosphorylated, using ATP, then reduced to glyceraldehyde-3-phosphate (a triose phosphate) using NADPH (Fig. 10.10). By this stage, carbon has been reduced from carbon dioxide to sugar. Some glyceraldehyde-3-phosphate can be removed from

A matter of context

A specific enzyme and its reaction may have quite different roles in different situations.

PEP carboxylase

Rubisco is the most familiar carbon-fixing enzyme in plants. Phosphoenolpyruvate carboxylase catalyses another, quite different, carbon-fixing reaction. The 3-carbon molecule phosphoenolpyruvate (PEP) is carboxylated to oxaloacetate. Carbon dioxide itself does not react: instead, it is the hydrogencarbonate ion formed when CO_2 reacts with water. The significance of this reaction is linked both to the plant's environment and to its leaf structure.

$$\underset{\text{PEP}}{{}^{-}O{-}P(=O)(O^{-}){-}O{-}C(=CH_2){-}COO^{-}} + \underset{\text{Hydrogen carbonate}}{HCO_3^{-}} \longrightarrow \underset{\text{Oxaloacetate}}{{}^{-}OOC{-}CH_2{-}C(=O){-}COO^{-}} + \underset{\text{Phosphate}}{{}^{-}O{-}P(=O)(O^{-}){-}O{-}H}$$

The carbon-fixing reaction catalysed by PEP carboxylase.

C_4 plants

PEP carboxylase forms part of a CO_2-pumping pathway which concentrates CO_2 in a small proportion of leaf cells, the bundle sheath. There, CO_2 concentration is kept high, so Rubisco acts almost exclusively as a carboxylase. The wasteful process of photorespiration is minimized. C_4 plants are typically found in tropical habitats where photosynthesis is rapid, so CO_2 reaches low levels within the leaf. Several types of pathway occur: in the one shown here, malate (4C) and pyruvate (3C) are exchanged between the cell types and the system carries reducing power from NADPH as well as carbon.

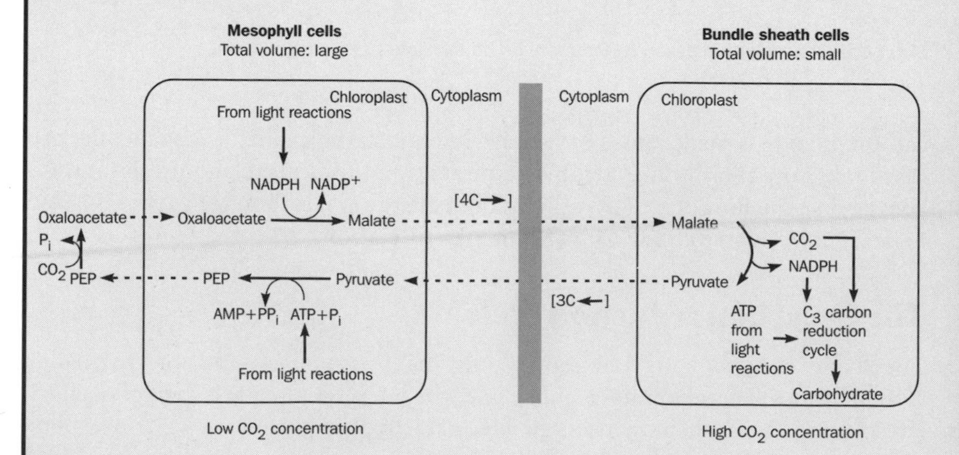

The carbon-pumping pathway of a typical C_4 plant.

Box 10.3

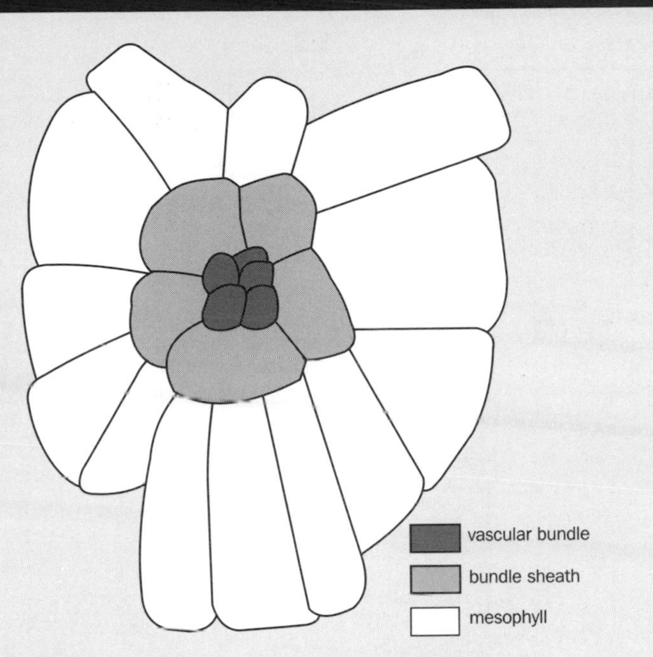

Kranz anatomy: the arrangement of cells around a typical vascular bundle in a leaf of maize, a C_4 plant.

CAM plants

Plants which exhibit crassulacean acid metabolism (CAM) generally live in arid conditions. They minimize water loss in transpiration whilst continuing to carry out gas exchange by opening their stomata (Section 17.2) only at night. PEP carboxylase fixes carbon at night. Carbon is stored as malate, which accumulates in the vacuoles. During daylight hours, the stomata close. Malate is removed from the vacuoles and decarboxylated. The carbon released is refixed by Rubisco, now that the light reactions are in operation.

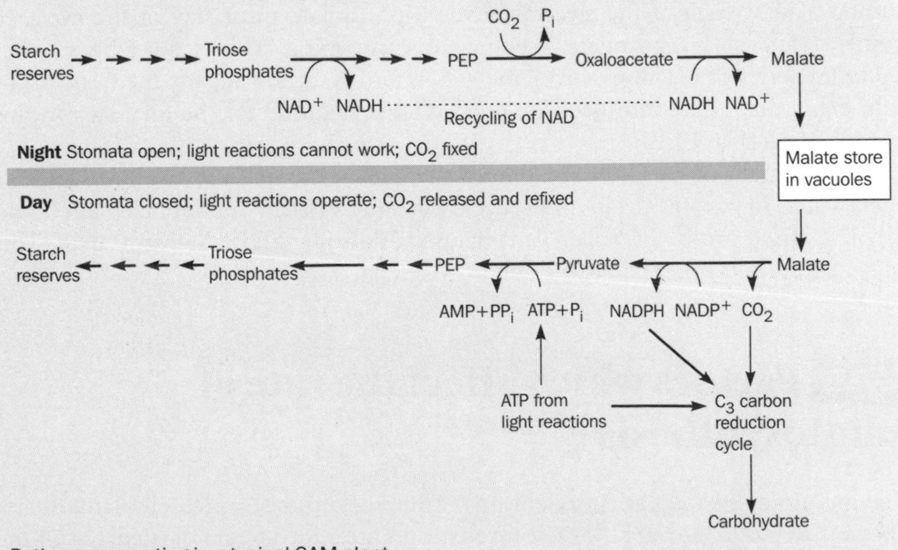

Pathways operating in a typical CAM plant.

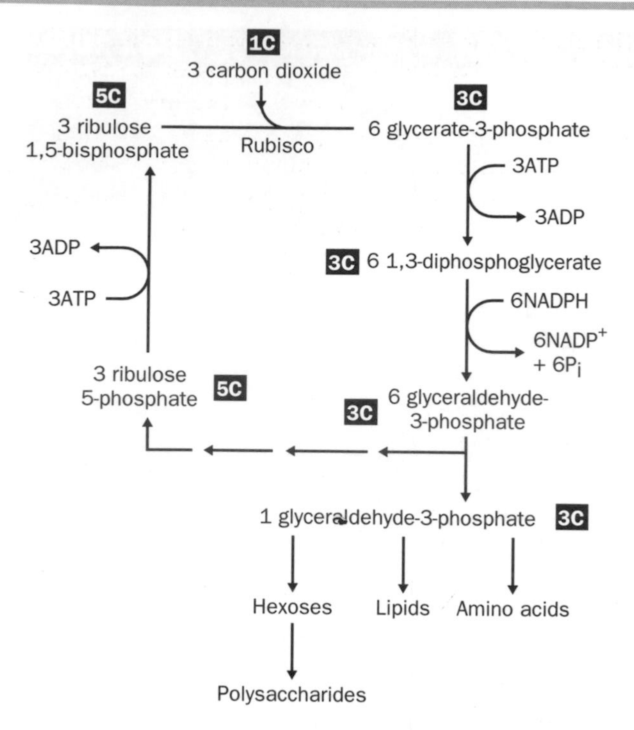

Fig. 10.10 The C_3 carbon reduction cycle. Numbers of molecules indicate the number required to produce a single molecule of glyceraldehyde-3-phosphate as product, whilst retaining enough to continue the cycle. Boxes show the number of carbon atoms in each molecule.

the cycle. It may be converted to hexoses, then on to starch, for storage within the chloroplast, or exported to the cytoplasm. In the cytoplasm it may also be converted into hexoses, then on to sucrose for longer distance transport within the plant. Alternatively, it may be used to provide reduced carbon skeletons for fats or amino acids. Much of the glyceraldehyde-3-phosphate must stay in the cycle, to remake ribulose bisphosphate, keeping the cycle going. This is done by a rather complex series of reactions, involving 4, 5, 6 and 7 carbon sugars, the basic problem being that 'three into five won't go'. ATP is required for the final step in this regeneration (Fig. 10.10).

The reactions of the C_3 carbon reduction cycle are common to all photosynthetic eukaryotes. In fact, the details of the cycle were first worked out not in a plant but in a single-celled protoctist, *Chlorella* (phylum Chlorophyta), by Melvin Calvin. This gives the cycle its other name, the Calvin cycle.

10.5 Factors which affect the rate of photosynthesis

Having discussed the basic mechanisms of photosynthesis, which all plants share, the next step is to look at how these mechanisms are added to or adjusted, to suit the particular conditions in which the plant is growing. To understand how a plant becomes adapted (over evolutionary time) or acclimatized (during the lifetime of an individual) to a particular environment, requires an understanding of the factors which affect the rate of photosynthesis as a whole. Any of these factors may limit the rate of photosynthesis. If a factor is **limiting** the rate of a process, an increase in the level of that factor will lead to an increase in the rate of the process. For example, when photosynthesis is limited by light intensity, brighter light will lead to faster photosynthesis.

Internal factors

Various things about the photosynthetic cells and the tissues they form affect the potential rate of photosynthesis. Chiefly, the amount of photosynthetic apparatus, chlorophyll for example, puts a ceiling on rate. When any one piece of apparatus is working at its maximum rate, the whole process can go no faster. Levels of important enzymes like Rubisco may be important. The three-dimensional distribution of chloroplasts may have an effect: a thin leaf with a large surface area may trap more light than a smaller, thicker leaf containing as many chloroplasts. These factors are, at least to some extent, under the plant's control. Others may not be.

Light intensity

Light is required for photosynthesis: crudely, brighter light means faster photosynthesis. This is true only up to a certain light intensity (Fig. 10.11(a)). Above this point, a further increase in light intensity has no effect on rate. Photosynthesis is no longer limited by light: it is said to be light saturated. At very high intensities, rate declines again as the photosynthetic apparatus becomes damaged; sometimes this damage cannot be reversed.

Limiting factor: If a factor limits the rate of some process, increasing the level of the factor will increase the rate of the process.

(a)

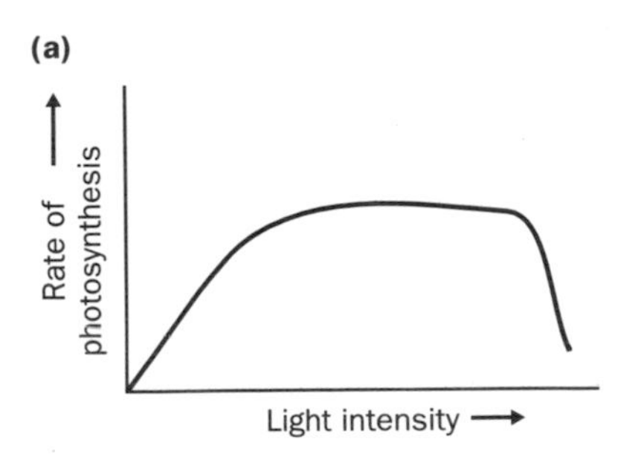

(b)

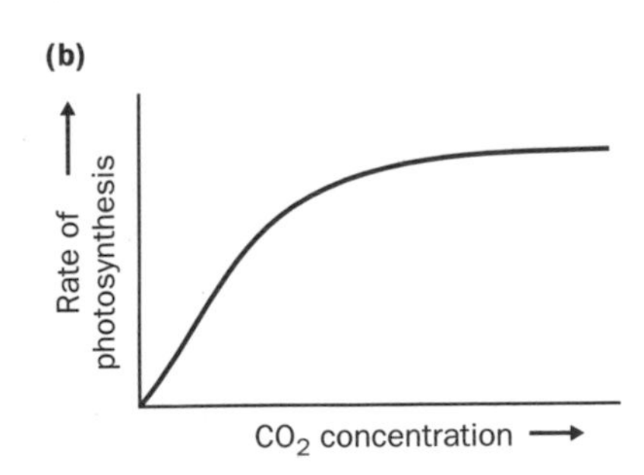

(c)

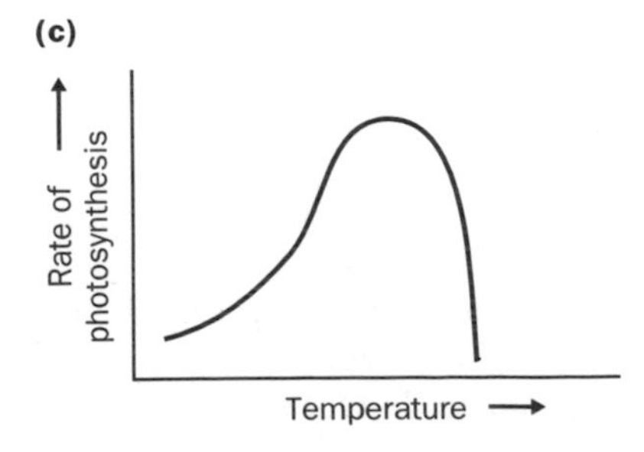

Fig. 10.11 Factors affecting the rate of photosynthesis. (a) light intensity, (b) CO_2 concentration, (c) temperature.

CO_2 concentration

Photosynthesis also requires CO_2, and as CO_2 concentration increases, photosynthetic rate also increases, tailing off as photosynthesis becomes CO_2 saturated (Fig. 10.11(b)).

Temperature

Like most processes involving enzymes, there is an optimum temperature for photosynthesis in a given cell or tissue (Fig. 10.11(c)). As temperature rises towards the optimum, rate of reaction increases, since the molecules involved move more rapidly and have a greater chance of colliding. Above this temperature, rate declines as enzymes are denatured.

Water

Water is used up in the reactions of photosynthesis, but the effect of water shortage is felt indirectly. The physiology of a plant under drought conditions is dominated by the need to conserve water. Stomata will generally close to reduce water loss long before water shortage affects photosynthesis directly. When stomata close, gas exchange is drastically reduced, and it is likely to be lack of CO_2 which first affects the rate of photosynthesis.

Multiple factors

Photosynthesis may be limited by a single factor, or by several at once. Fig. 10.12 illustrates conditions under which light intensity, CO_2 concentration, or both are limiting.

The precise shapes of the curves sketched in Fig. 10.11 vary considerably between plants growing under different conditions. These differences are best understood when linked to another type of measurement: compensation points.

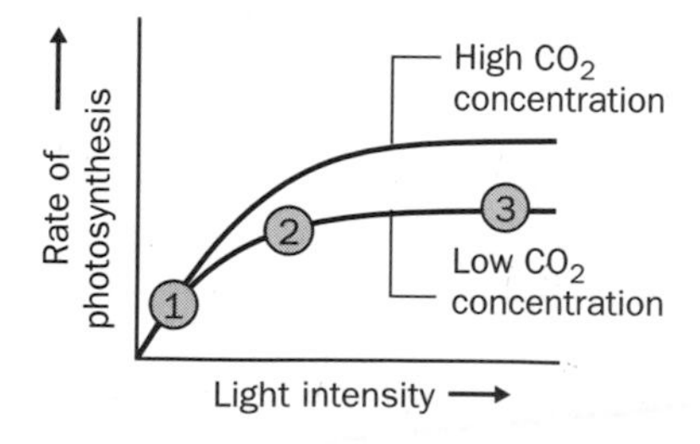

Fig. 10.12 Factors limiting the rate of photosynthesis. 1: Light intensity limits rate; 2: Both light intensity and CO_2 concentration limit rate; 3: CO_2 concentration limits rate.

10.6 Compensation points

Monitoring gas exchange is a common way of estimating the photosynthetic rate of a plant. The problem is that photosynthesis and respiration have opposite effects. Aerobic respiration, which is going on all the time in most plant cells, uses O_2 and makes CO_2. Photosynthesis, which goes on only in the light, uses CO_2 and makes O_2. In complete darkness, then, the plant will give out O_2 and make CO_2. In bright light

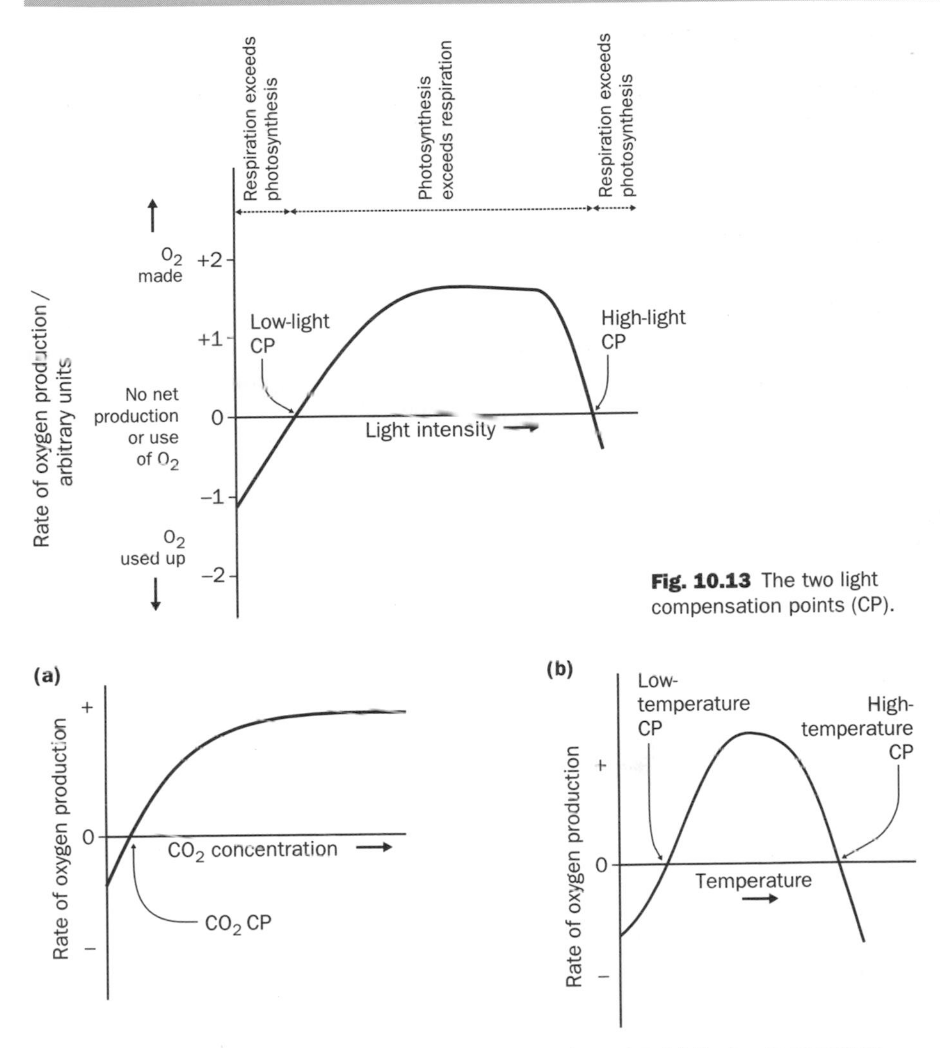

Fig. 10.13 The two light compensation points (CP).

Fig. 10.14 Compensation points. (a) the CO_2 compensation point, (b) the two temperature compensation points.

the reverse will be true. It follows that there will be some light intensity, more or less dim, at which there is just enough photosynthesis to balance respiration. This light intensity is called the low-light **compensation point** (Fig. 10.13). Below it, CO_2 is released and the plant's stores of carbohydrates are decreasing. Above it, CO_2 is taken in and carbohydrates are building up. Because the rate of photosynthesis declines in extremely bright light, there will also be a high-light compensation point where rate has fallen until it only just balances respiration.

In a similar way, one can estimate both CO_2 and temperature compensation points. The CO_2 compensation point is the concentration at which there is no net gas exchange (Fig. 10.14(a)). There are two temperature compensation points, due to the humped shape of the graph (Fig. 10.14(b)).

Photorespiration complicates the issue. This is the energy-wasting oxygenation of ribulose bisphosphate by Rubisco, and the series of damage limitation reactions which follows it. The process uses O_2 and releases CO_2, so mimics respiration even

Compensation point: The environmental conditions under which rates of photosynthesis and respiration are equal.

though ATP is used rather than made. When compensation points are estimated, one is really finding the conditions under which photosynthesis balances respiration *plus* photorespiration. Photorespiration is rapid when O_2 levels are high, CO_2 is low and temperature is high: photorespiration is a significant component of the balance at the high temperature and high-light compensation points.

10.7 Living with the environment

The photosynthetic apparatus of a cell, or of a whole plant, can vary to suit the conditions. Populations or entire species which usually experience particular conditions, shade for example, may show adaptations to photosynthesis under low light conditions. These are genetically-controlled characteristics which have been favoured by natural selection. Other populations only sometimes experience shade, or have some leaves in shade and some in full sun. Here, tactical responses may be seen: the plant, leaf or cell adjusts its characteristics to make itself more efficient in photosynthesis.

Sun or shade

Plants growing in shade tend to respond to light intensity in a different way to plants growing in full sun (Fig. 10.15). In dim light, shade plants can generally photosynthesize faster than sun plants. They also tend to have lower low-light compensation points. In brighter light, however, shade plants quickly become light-saturated, while sun plants are able to exploit this extra light. These differences can be measured in whole plants. Their causes are seen on a smaller scale, and have to do with both structure and chemistry.

Shade plants tend to have low-light compensation points which are low because they have lower respiration rates and can absorb light efficiently. Shade leaves of many species are thinner than sun leaves. There are two ways of seeing the advantage of this. First, each unit area of leaf which absorbs light has fewer cells below it, so the respiratory costs of maintaining the leaf will be lower. Secondly, in a thinner leaf, the same amount of photosynthetic apparatus can be spread out over a wider area, intercepting more light. In brighter conditions, light penetrates deeper into the leaf, making investment deeper down worthwhile.

In deep shade it is likely to be light which is limiting, rather than CO_2. This affects the ways in which plants invest their limited resources in different parts of the photo-

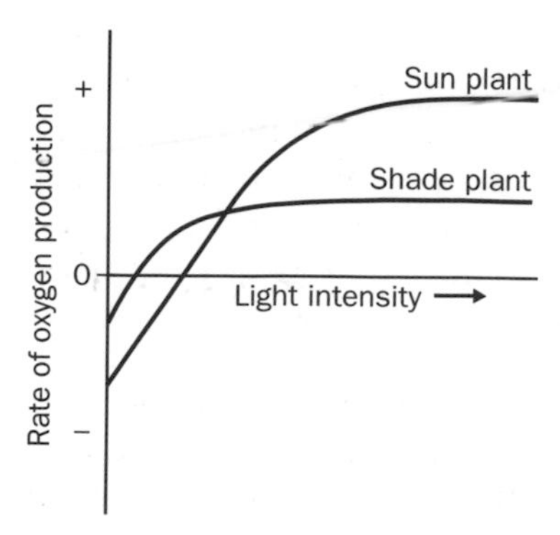

Fig. 10.15 Typical responses to light intensity in sun and shade plants.

synthetic apparatus. Shade chloroplasts tend to be packed with thylakoids, carrying the machinery of the light reactions. There is no point investing in extra carbon-fixing enzymes in the stroma if CO_2 is not limiting but ATP and NADPH from the light reactions are scarce. Even within the thylakoids there are variations. Shade leaves generally have a higher ratio of photosystems to electron-transport complexes. If the absorption of light limits rate, it makes sense to invest in the complexes which do just that, allowing more photosystems to funnel electrons into the same number of transport complexes. Most shade plants are living in shade created by other plants: trees are more numerous than caves. The light passing through a leaf canopy has been adjusted as chlorophyll absorbs particular wavelengths: little red or blue light will get through, but there will be plenty of green. Even within the red band, very long wavelength red light (far red) passes through the canopy rather better than the shorter wavelength red light. PSI is slightly better at absorbing far red than is PSII. In a shade leaf, then, any given PSI complex is likely to release excited electrons more often than a given PSII complex. One might predict that the plant could afford to invest slightly more heavily in PSII, at the expense of PSI. This proves to be correct: shade leaves tend to have slightly higher PSII/PSI ratios than do sun leaves. It may be awe-inspiring, but should not be surprising that natural selection acts on such minute details. Plants live in a competitive world, where the efficiency of photosynthesis means life or death.

Hot or cold

Plants may adapt and acclimatize to hot or cold conditions. The optimum temperature for photosynthesis varies quite widely, tending to be higher in plants from hot climates, and higher during warmer seasons of the year. This is not unique to photosynthesis: the plant will be adjusting the temperature responses of many important chemical processes. At least two mechanisms seem to be involved. Firstly, as temperature falls, the composition of membranes changes. Components which increase fluidity (Section 4.2), such as polyunsaturated fatty acids, become more abundant. This counteracts the slower movement of molecules at lower temperatures. Secondly, at least some enzymes seem to exist in more than one form, with different optimum temperatures. These may be produced as required.

Moist or dry

Water shortage is a problem to many plants, but it is most acute in deserts. The effect of drought on photosynthesis is indirect: shortage of CO_2 due to stomatal closure is one of the major factors. One chemical variation of photosynthesis which has a bearing on water conservation is Crassulacean Acid Metabolism (CAM). This is found in some plants of arid habitats, for example many tall, column-forming cacti, some fleshy leaved plants of the family Crassulaceae and, interestingly, the pineapple. The problem is one of when to open the stomata. Gas exchange is needed by day, for photosynthesis. However, this is also the warmer part of the day and water loss through transpiration will be significant. Some method which allows gas exchange to take place in the night is needed. CAM allows this. Stomata open at night, and carbon dioxide is fixed by an alternative pathway, using the enzyme phosphoenolpyruvate carboxylase. Malate is formed, and put into the vacuoles, which become quite acidic. In the day, with stomata closed, malate is brought back out of the vacuoles and decarboxylated. The carbon dioxide released is then refixed by Rubisco in the usual C_3 cycle using ATP and NADPH from the light reactions. Full details of the CAM pathway are given in Box 10.3. CAM is not particularly energy-efficient, but for some plants the saving in water makes it worthwhile. Many CAM plants revert to straightforward C_3 photosynthesis when water is abundant.

Hot, sunny and moist (or even quite dry . . .)

When conditions are ideal for photosynthesis, with high temperatures, bright light and a reasonable supply of water, conditions are also ideal for photorespiration. Rapid photosynthesis tends to reduce CO_2 concentration and raise O_2 concentration: this favours photorespiration, as does high temperature. Under these conditions, photorespiration can severely restrict the rate of photosynthesis. Many tropical and subtropical plants greatly reduce photorespiration by the C_4 carbon fixation pathway, so called because carbon dioxide is fixed as a 4-carbon molecule, rather than a 3-carbon molecule, as in the C_3 cycle. These plants belong to many flowering plant families and include many crops, such as maize, and some of the worst weeds of tropical agriculture.

C_4 plants share a particular way of dividing the leaf into two unequal compartments. The cells around the vascular bundles form one compartment, the bundle sheath. The other mesophyll cells of the leaf form the other. The mesophyll compartment is much larger than the bundle sheath. In the mesophyll, carbon dioxide is fixed by phosphoenolpyruvate carboxylase, just as in CAM plants, but this time by day. Malate is transported to the bundle sheath and decarboxylated. The CO_2 released is refixed by Rubisco. The effect is to pump CO_2 from the whole leaf into just a small proportion of cells. Carbon dioxide concentration is then high in the bundle sheath where Rubisco is found, so the enzyme acts primarily as a carboxylase. Photorespiration is far lower than in a C_3 plant, so photosynthesis is more efficient. Further details of the C_4 cycle are given in Box 10.3.

C_4 may also be at an advantage where water is limited. Partial closing of the stomata to save water will lead to a much smaller increase in photorespiration than in a C_3 plant.

10.8 Photosynthesis as a weapon

Plants rarely live in harmony with their neighbours. Plants do not need many things in life: light, carbon dioxide, water, minerals and the space in which these things are to be found. All plants, however, need these same things. For most plants, competition with neighbours for limited resources is an important and unavoidable part of life. Above ground, the shoots of neighbouring plants compete for light. Once a shoot has extended into a new bit of habitat, and started to photosynthesize there, it is able to continue outwards, conquering more habitat. The best way for its neighbours to defend their territory is to deprive invaders of light, shade them out. Leaves absorb precisely what other leaves require. One's own leaf above one's neighbour's leaf is the perfect weapon in the struggle for space. Beneath an established tree only the most shade-tolerant seedlings can become established: the list of potential competitors is drastically reduced. The leaf canopies of neighbouring trees rarely overlap to any extent. The boundary is the front line along which local battles may be won

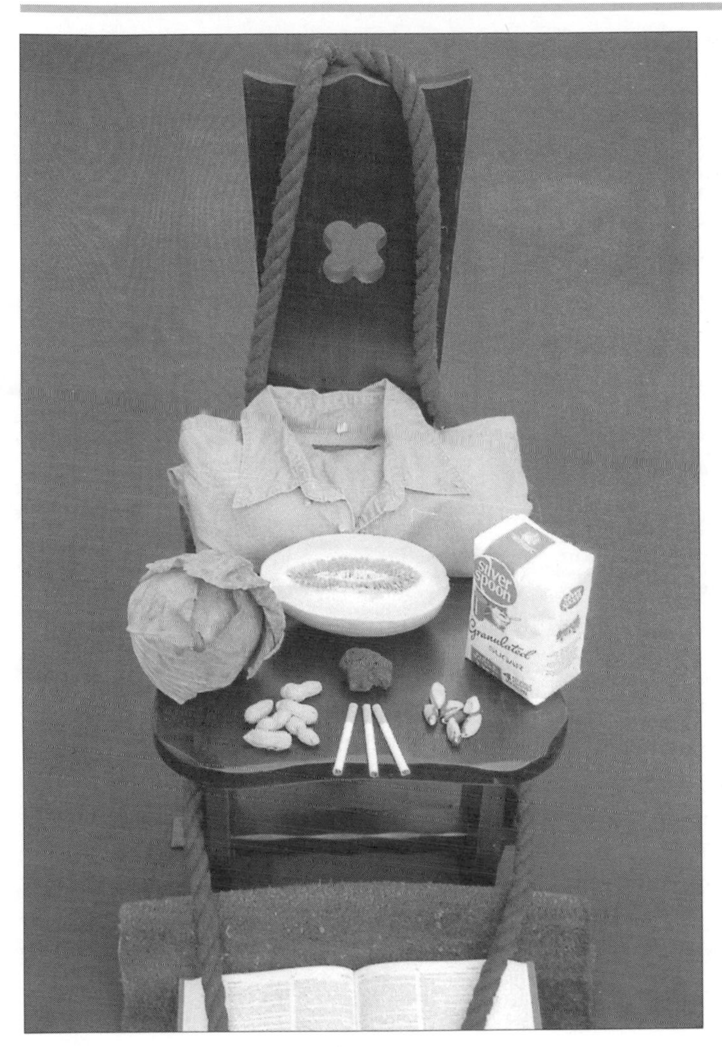

Still life with useful products of photosynthesis.

and lost in a continuing war over territory. Photosynthesis provides both the weapons of the war and the means by which the local winner exerts its dominance. This highlights the importance of adaptation and acclimatization to local conditions. Well-adapted photosynthetic apparatus does not simply provide the plant with a more abundant food supply with which it can produce more offspring. It is a necessity if the plant is to hold its own within the community. The price of inefficiency may be death, overrun by more efficient neighbours. There is more to life as an autotroph than sitting around, soaking up the sun.

Summary

- ◆ Plants make their own food by photosynthesis.
- ◆ Light is used as an energy source.

- Carbon dioxide is chemically reduced and used to build the carbon skeletons of sugars.
- Water is used; oxygen is made as a waste product.
- Photosynthesis occurs in the chloroplasts of plant cells.
- Photosynthesis has two stages: the light reactions and the carbon fixation reactions.
- In the light reactions, chlorophyll and other pigments absorb light.
- An energy store, ATP, and a reducing agent, NADPH, are made in the light reactions.
- Chlorophyll gives out high-energy electrons which can be used to reduce NADP. Their energy is also used to set up an energy-storing proton gradient, which is used to phosphorylate ADP.
- Water molecules are split, releasing electrons which replace those given out by chlorophyll. Oxygen is made at this stage.
- In the carbon fixation reaction, the enzyme Rubisco adds carbon dioxide to an existing carbon skeleton.
- The product is then converted to a sugar phosphate using ATP and NADPH.
- Rubisco also catalyses a wasteful oxygenation reaction: photorespiration.
- C4 and CAM photosynthesis also use an additional carbon-fixing enzyme. They increase the plant's efficiency in particular environmental conditions.
- The rate of photosynthesis is affected by light intensity, carbon dioxide concentration, temperature, availability of water and internal factors like the amount of chlorophyll.
- Plants and even single leaves are adapted to local environmental conditions.
- Compensation points are environmental conditions under which the rates of respiration and photosynthesis are equal.
- Photosynthesis deprives neighbouring plants of light.

Exercises

10.1. Chloroplasts can be isolated from ground up leaves, although the envelope is normally damaged. These chloroplasts do not release oxygen when light is shone on them. However, if they are given an electron acceptor such as potassium ferricyanide, and then illuminated, they do produce oxygen. Electron acceptors which can do this are called Hill reagents.

(i) Why can no oxygen be produced without a Hill reagent?
(ii) What is the natural Hill reagent?

10.2. The details of the C_3 carbon reduction cycle were worked out as follows. Green algae were given carbon dioxide made with a radioactive isotope of carbon. Samples of the algae were killed at intervals. Sugars and organic acids were separated and tested for radioactivity.

(i) Which organic molecule would you expect radioactive carbon to appear in soonest?
(ii) Would you expect any radioactive carbon to appear in ribulose bisphosphate?

10.3. It is summer. The days are warm and sunny, the nights are dark (just in case you thought we were in the Arctic!). How many times would you expect a plant to be at its low-light compensation point in each 24 hour period?

10.4. Explain the following observations:

(i) C_4 plants have a lower CO_2 compensation point than C_3 plants.

(ii) In C_4 plants, the bundle-sheath chloroplasts carry out more cyclic phosphorylation than do the mesophyll chloroplasts.

(iii) In CAM plants, photorespiration is only detectable late in the afternoon.

Further reading

Hall, D.O. and Rao, K.K. *Photosynthesis* (5th ed.) (Cambridge: Cambridge University Press, 1994). A simple but authoritative introduction: small and affordable!

Salisbury, F.B. and Ross, C.W. *Plant Physiology* (4th ed.) (Belmont CA: Wadsworth, 1992). A big undergraduate textbook.

PART
4

Genes and Life Cycles

Life Cycles, Cell Division and Variation

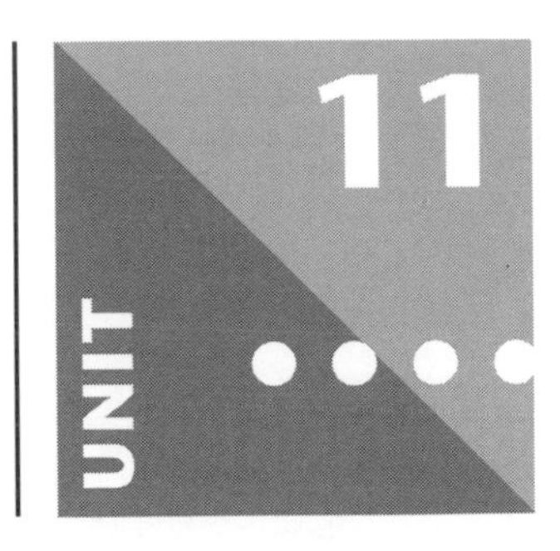

Connections

► You should already have a basic understanding of natural selection, cell structure and the workings of DNA (Units 2, 4 and 7). This unit is an important introduction to plant and animal life cycles (Units 13 and 14) as well as classical genetics (Unit 12).

Contents

11.1 One set or two?

Cells have **chromosomes**, long DNA molecules. Shorter sections of each chromosome are **genes**. Each gene codes for the amino acid sequence of one polypeptide (Section 7.2). Eukaryotic cells have several chromosomes, in the nucleus. Cells normally have either one or two copies of each. Cells with one set of chromosomes are described as **haploid**. **Diploid** cells have two sets. A chimpanzee sperm has 24 chromosomes, all different; it is haploid. Other cells in the chimp's body are diploid. Each has 48 chromosomes, two complete sets. They are not completely alike, however. One set came from the chimp's father, in a sperm. The other came from its mother, in an egg. Each copy of a particular chromosome carries the same genes, but they may have different versions (**alleles**) of these genes. Our chimpanzee may have inherited the normal version of the gene coding for the enzyme hexokinase from its mum, and an allele coding for a non-functioning version of the enzyme from its dad: same gene, different alleles.

Chromosome: A long DNA molecule.

Gene: A section of a chromosome which codes for one polypeptide, and so for some characteristic.

Haploid: Having one set of chromosomes.

Diploid: Having two sets of chromosomes.

Allele: One version of a gene.

Sexual reproduction: Reproduction involving fertilization. Offspring are variable.

Gametes: The haploid cells which join in fertilization.

Fertilization: Two haploid cells join, forming a diploid cell.

Zygote: The diploid cell formed in fertilization.

11.2 Sex and cell division

Two cells come together, bringing their own genetic information. A new cell is formed, with genes from both original cells. This is the heart of **sexual reproduction**. If it does not take place, sex has not happened. All the other exciting bits, like stuffing a beakful of dead fish down your partner's throat, bribing a bee to carry your pollen round to the neighbours, the how and why of erection in the tom cat penis, and so on, are all steps on the way to this central event. Two cells (**gametes**) join together (**fertilization**) to produce a new cell (**zygote**). One of the gametes is normally bigger: this is the female, by definition. The smaller male gamete typically swims.

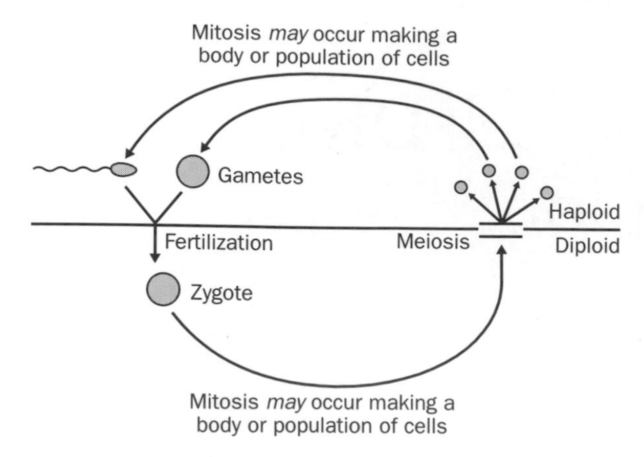

Fig. 11.1 The sexual life cycle.

Gametes are haploid. In fertilization, two gametes join and their chromosomes are pooled. This means that the zygote must be diploid.

Asexual reproduction is quite different. New individuals have just one parent. They are genetically identical to that parent. No gametes or fertilization are involved. A new body develops just like part of the old body. Clover spreads through our lawn asexually. It has horizontal stems which grow roots at intervals. If the older stems die or get eaten, we are left with several independent plants which share the same genes. Some hydrozoans such as *Hydra* (Phylum Cnidaria) reproduce asexually: a new animal grows out from the side of the parent. Eventually it separates. Female greenfly produce clones asexually all summer. Dandelions produce seeds using flowers. It looks as if sex is involved, but this is a form of asexual reproduction called apomixis. Cells of the mother plant invade the embryo sac (Section 14.3) and form a seed.

Cell division is necessary in any life cycle. Life cycles involving sexual reproduction need two sorts of cell division, **mitosis** and **meiosis**. Strictly speaking, mitosis and meiosis are divisions of the nucleus, but the entire cell generally divides straight afterwards (cytokinesis: Box 11.3). Division of the nucleus is what matters here, because the chromosomes are inside it. Mitosis is a way of producing identical clones of a cell. It is used to build bodies and in asexual reproduction. Meiosis makes haploid cells from a diploid cell. It balances the change from haploid to diploid at fertilization.

So, sexual life cycles always include fertilization and meiosis (Fig. 11.1). Mitosis may occur in either the haploid or diploid stage, or in both. In animals, mitosis only occurs in diploid cells, making a diploid body (Fig. 11.2). The haploid cells made by meiosis are gametes. In plants, however, mitosis occurs in haploid and diploid cells. This means that two types of body can be formed, one haploid and one diploid. In flowering plants, the haploid bodies are tiny, living inside the pollen grains and the ovules. In mosses, the diploid body is a spore capsule on a stalk, growing out of the leafy haploid body. In some seaweeds, the two bodies live independently, and may look very different. Chapter 13 explores these startling ideas in more detail.

Asexual reproduction: Reproduction which makes clones of a single parent.

Mitosis: A nucleus divides to make two identical clones.

Meiosis: A diploid nucleus divides to make four haploid nuclei.

Mitosis

Mitosis produces genetically identical cells. A parent cell divides to form two clones of itself. They are smaller, of course, but can grow later. If the parent cells were haploid,

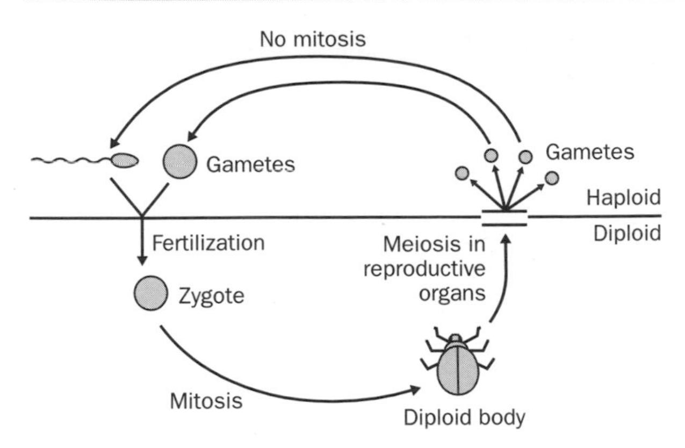

Fig. 11.2 Sexual life cycle of any animal.

the daughter cells will be haploid. A diploid parent cell produces diploid daughter cells. Mitosis is important in producing bodies, and in asexual reproduction. Even unicellular organisms use mitosis to build up a population of clones.

Just before mitosis, each chromosome is copied. The two copies, called **chromatids**, remain linked. The chromosome becomes shorter and thicker. This must involve very high levels of coiling and folding of DNA, which is poorly understood (Sections 4.5, 7.4). Then the nuclear envelope disintegrates. The chromosomes have the run of the entire cell. The centrosome (Section 4.5) replicates: one goes to each end. They organize a series of microtubules running the length of the cell: the spindle. The chromosomes attach to the spindle. Motor proteins move one copy of each chromosome to each end of the cell, where new nuclei form. Box 11.1 shows the process in more detail.

Chromatids: Identical copies of a chromosome, formed by DNA replication before mitosis or meiosis.

Homologous chromosomes (=homologues): Two chromosomes of the same type in one cell, one inherited from each parent.

Meiosis

Meiosis produces haploid daughter cells from a diploid parent cell. It halves the chromosome number. Each haploid cell gets one of the two copies of each chromosome. The mechanism of meiosis (Box 11.2) is very like mitosis. However, the two copies of each chromosome (**homologous chromosomes**) pair up and are separated

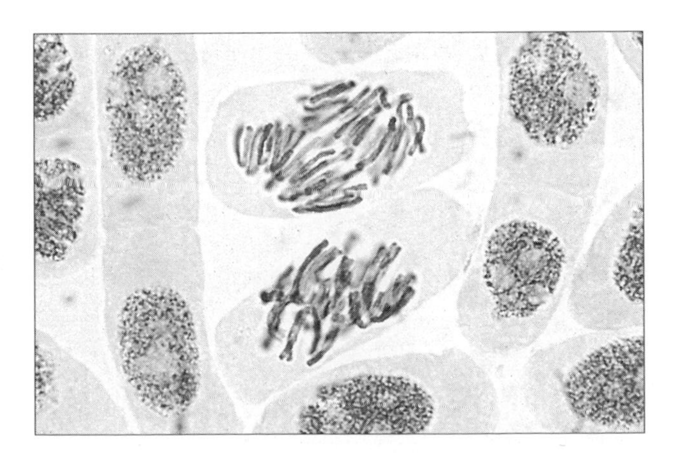

Mitosis in plant cells: which stages can you spot?

Box 11.1

Stages of mitosis

Mitosis produces two daughter cells. They are genetic clones of the parent cell. These diagrams show mitosis in a cell whose diploid chromosome number is four (2n=4).

Interphase

This is the normal life of the cell when it is not dividing; not part of mitosis at all. Late in interphase, DNA is replicated (Section 7.3).

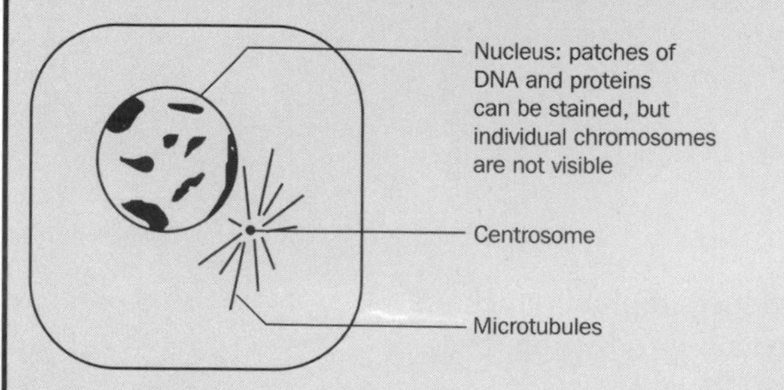

Prophase

Chromosomes 'condense': high level folding makes them short and wide. Each chromosome is made up of two identical copies: chromatids. Chromatids are linked at the centromere, a specialized DNA sequence. The centrosome replicates. At the end of prophase, the nuclear envelope breaks down. The centrosomes at the poles of the cell organize the spindle. This is made up of parallel microtubules running the length of the cell.

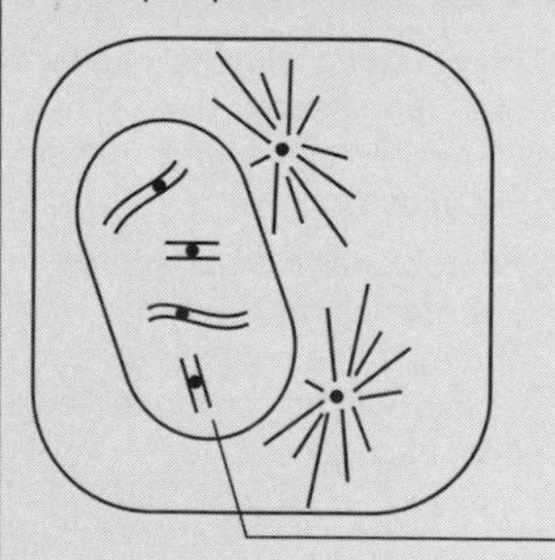

Metaphase

The centromeres become linked to the spindle fibres by protein complexes: kinetochores. The chromosomes are lined up across the cell equator.

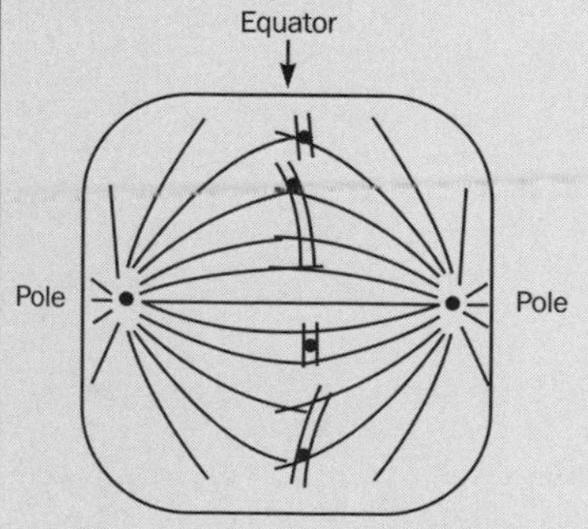

Anaphase

The centromeres separate, and chromatids move to opposite ends of the cell. This is driven by motor proteins in the kinetochores and, perhaps, shortening of the microtubules. Spindle poles move further apart as microtubules which are not linked to chromosomes lengthen.

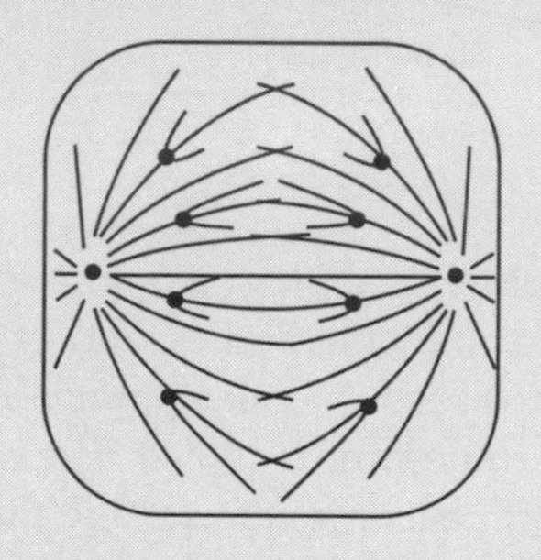

Telophase

A nuclear envelope forms around each set of chromosomes. Chromosomes expand again.

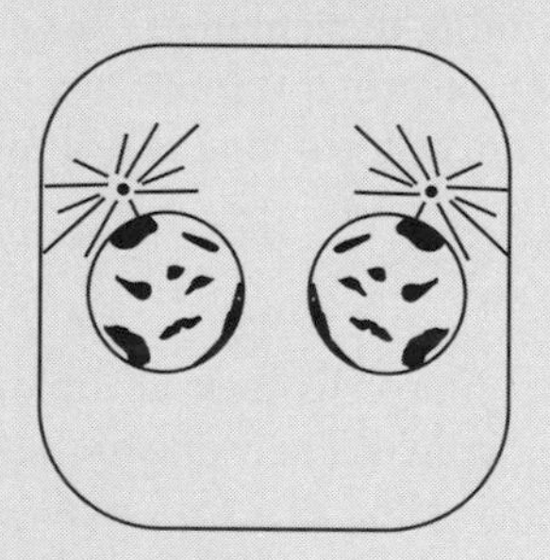

Cytokinesis

This is cell division, following mitosis or meiosis (see Box 11.3).

Box 11.2

Stages of meiosis

Meiosis produces four haploid daughter cells. Each is genetically unique. Many details of the process are much like mitosis. These diagrams show meiosis in a cell whose diploid chromosome number is four.

First meiotic division

Prophase I: Chromosomes condense. Homologous chromosomes pair. Crossing over takes place.

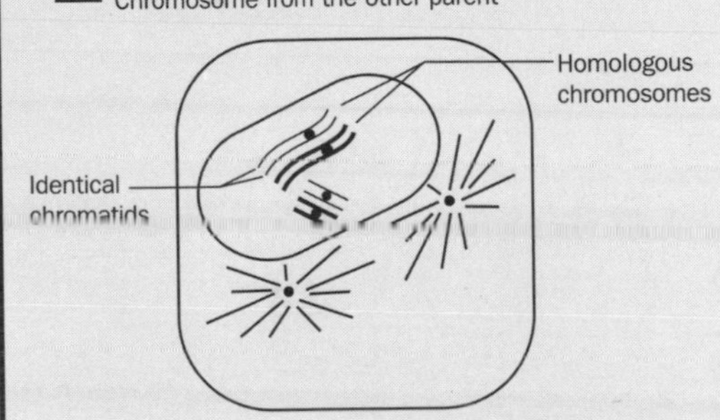

Anaphase I: Homologous chromosomes move to opposite poles of the cell.

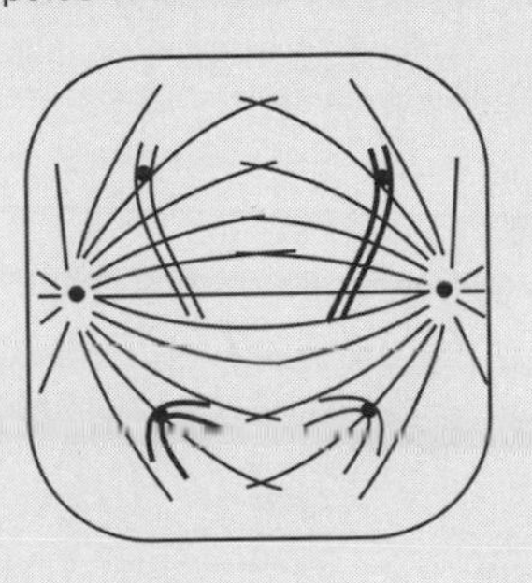

Metaphase I: Homologous chromosomes attach to spindle in pairs.

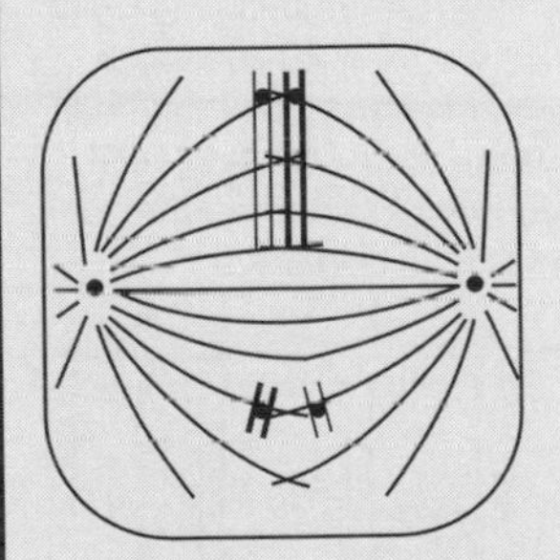

Telophase I and cytokinesis: The mechanism is just as in mitosis. Two haploid cells are formed.

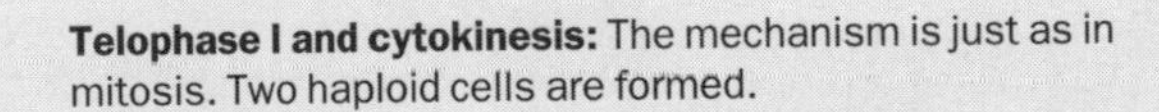

Second meiotic division

Prophase II, metaphase II, anaphase II, telophase II and cytokinesis are as in mitosis. The chromatids separate at anaphase.

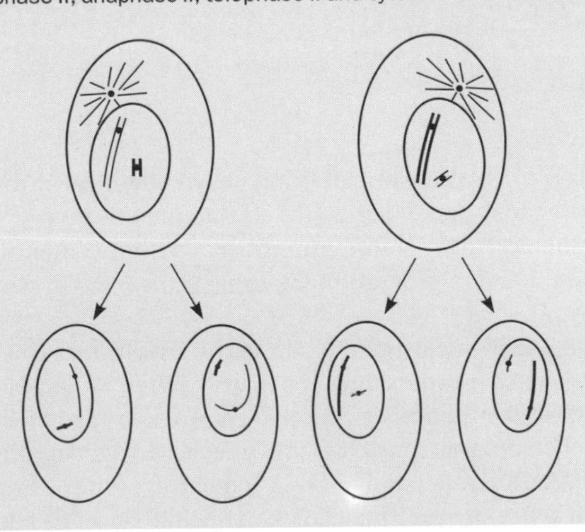

Box 11.3

Cytokinesis

Mitosis and meiosis are nuclear divisions. They are followed by cytokinesis, when the cell itself divides. Plant and animal cells divide in quite different ways.

In animal cells

The plasma membrane is pulled inwards, forming a narrow waist in the cell. A ring of actin and myosin filaments (Section 4.5: The cytoskeleton) lying under the plasma membrane contracts. The filaments slide against one another, much like in muscle. Eventually, the membranes become separate. The centrosomes seem to control where the contractile ring forms.

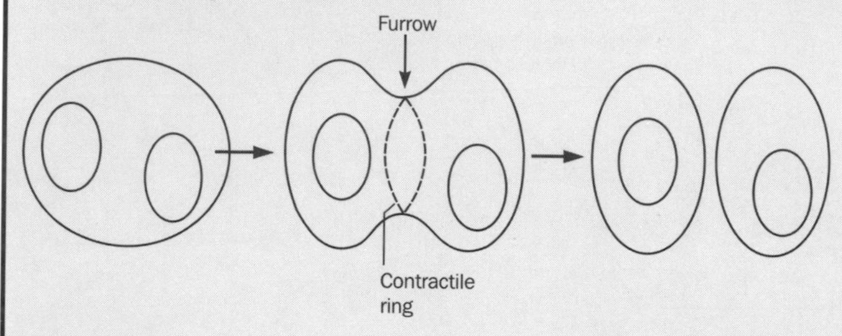

In plant cells

A new wall must be made, as well as some extra plasma membrane either side of it. A middle lamella and plasma membranes are made first. The primary and secondary walls are added later. Pectins for the middle lamella are made by the Golgi apparatus. The microtubules left over from the spindle move the Golgi vesicles to the equator of the cell. A waistband of actin filaments seems to mark the position, but does not contract. The Golgi vesicles fuse. Their membranes become a continuous sheet of plasma membrane. their contents end up between the two new cells, forming the middle lamella. Each cell can then build a primary wall on its own side.

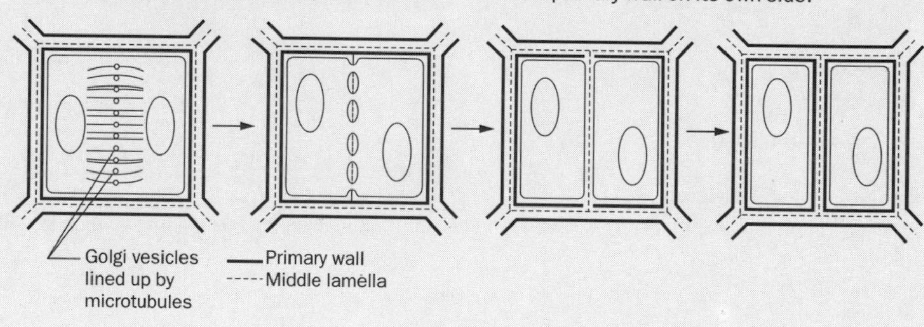

on a spindle. This is the first meiotic division. Then, identical chromatids are again separated, producing four haploid cells (the second meiotic division).

Fertilization doubles chromosome number. Without meiosis, chromosome number would double every generation. So long as meiosis happens somewhere in the cycle, this is avoided.

The four cells made by meiosis are all different. Two things cause this variation. Firstly, the homologous chromosomes may carry different alleles. Some daughter cells will get one allele, some another. Secondly, a process called **crossing over** occurs early in meiosis. The homologous chromosomes exchange lengths of DNA while they are paired up. At this stage, similar DNA sequences in the homologous chromosomes are brought very close together (Fig. 11.3). Quite how this happens is unclear,

Crossing over: Exchange of sections of DNA between homologous chromosomes.

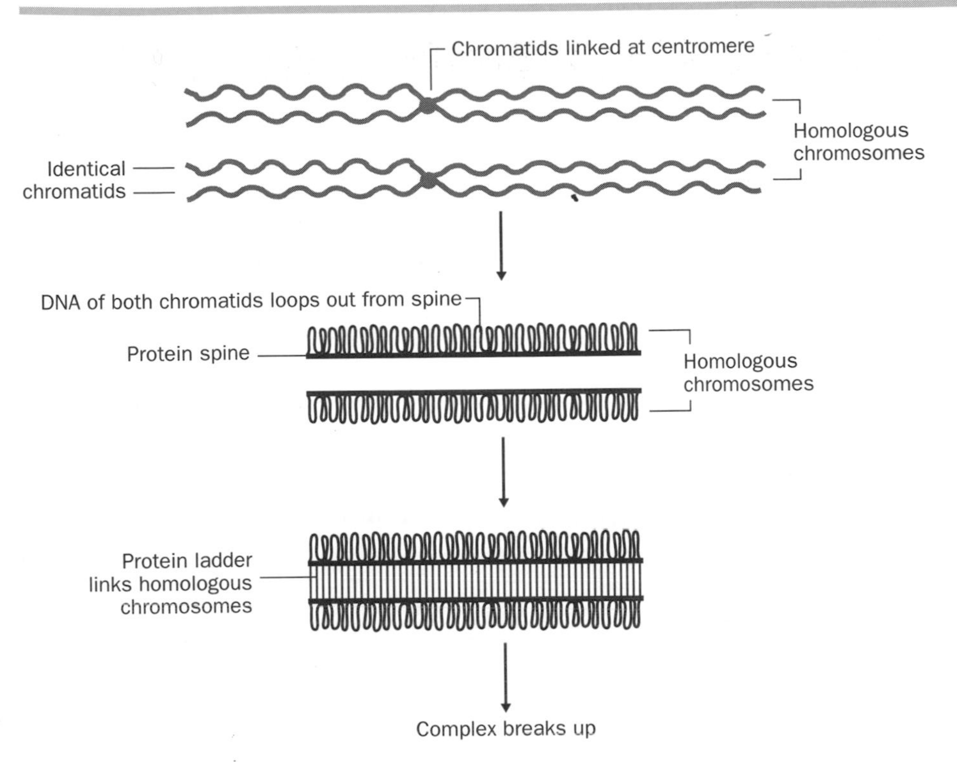

Fig. 11.3 DNA of homologous chromosomes comes very close together during the first meiotic division.

but it seems likely that specific DNA sequences in the homologues bind together by complimentary base pairing. While the homologues are paired, an enzyme complex called the recombination nodule binds to the protein ladder linking the homologues. Exactly how it works is very uncertain, but the overall effect is simple. The DNA of one chromatid on each homologue is broken at exactly the same point. The broken ends are rejoined in a new way. The chromosomes have swapped a section (Fig. 11.4). The point at which crossing over took place can be seen later as the homologues move apart. Each section of DNA stays paired with its original sister chromatid. The strands appear to cross where crossing over happened (Fig. 11.5). This is a **chiasma** (plural: chiasmata): the physical manifestation of crossing over. If several chiasmata form at intervals along a pair of chromosomes, several sections are swapped. Each chromatid becomes a unique mosaic of DNA from the two homologues. This leads to a mixing of DNA from each of the organism's parents (Section 12.3). Meiosis contributes to variation.

11.3 Variation

Organisms vary. Some of this variation is caused by differences in the environment: how much food they get, the type of soil they are growing in, how they see their parents behaving, and so on. The rest is caused by genetic differences. The phenotype is the characteristics of the organism, rather than the genotype, which is the genetic information coded in its DNA. Sometimes, one gene on its own is enough to

Chiasma: Two strands of DNA visibly crossing where crossing over has happened.

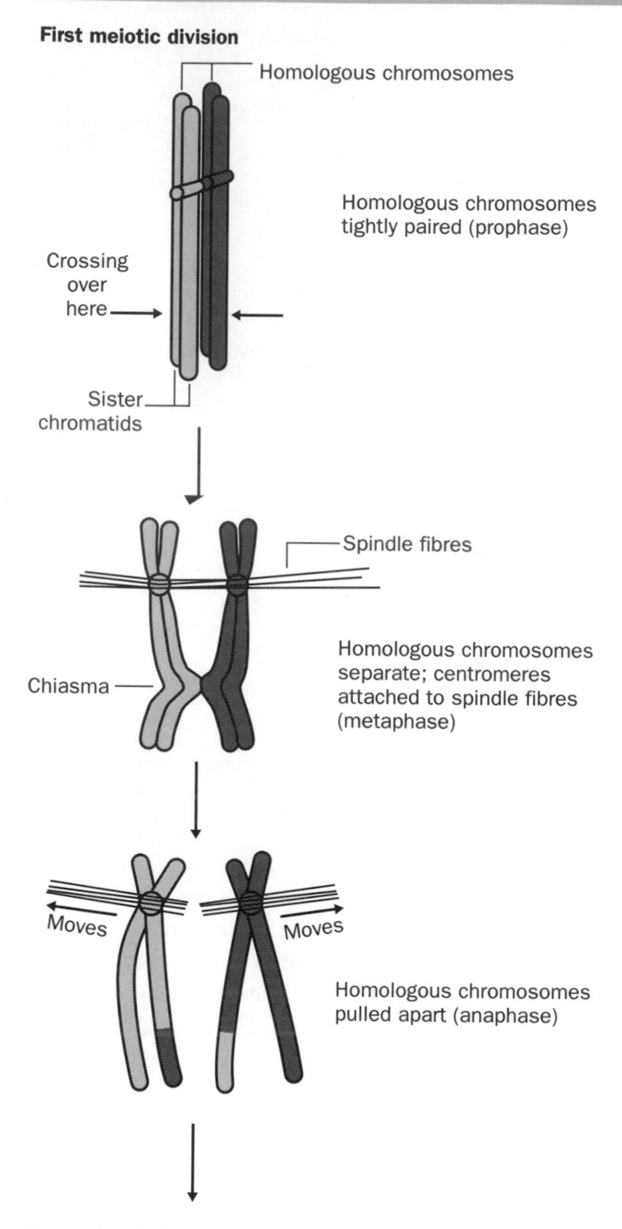

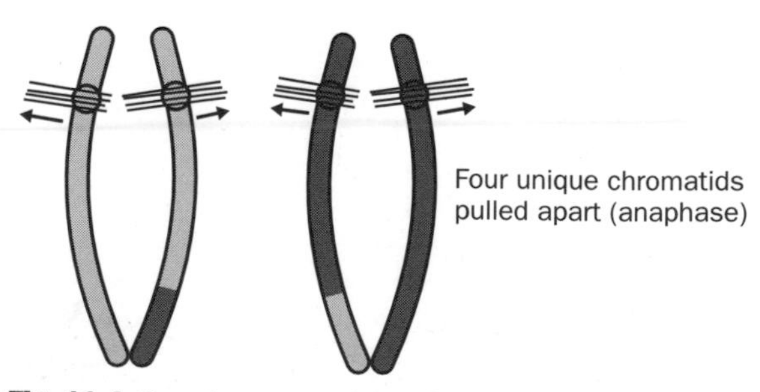

Fig. 11.4 Crossing over and its effects (see text for details).

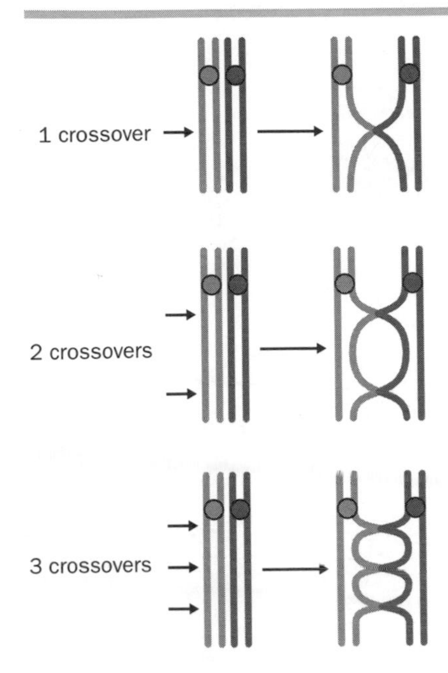

Fig. 11.5 Chiasmata are formed where crossing over took place.

cause a distinctive phenotype. More often, several or many genes together determine a phenotype. Change in any one of them affects the phenotype.

Genetic variation is possible because of **mutation**. Mutations are random changes to genes, larger-scale changes to chromosomes, or even changes in chromosome number. If mutation did not happen, new alleles would never crop up. Beneficial alleles would become common. Harmful alleles would become rare and vanish through natural selection. They would not reappear.

Once mutation has led to genetic diversity within a population, other factors such as crossing over can combine alleles in all sorts of ways, so that individuals are different. Mutation creates diversity in the gene pool: the potential for variation. Crossing over and a group of other factors release that variation into the phenotypes of the population. Now we can see the variation, and it is open to natural selection.

Think of a pack of cards. There are only so many ways in which we can arrange the cards in a stack. Now we print an extra card. It could be another copy of the three of spades, or a completely new one, like the 26 of hearts. A whole lot more sequences are possible now. However, those sequences can only be realized by shuffling the cards again and again. Printing a card introduces the potential for variation. Shuffling releases that variation.

Mutation and the release of variation are examined in turn.

Mutation

Mutations are random changes to a cell's DNA. Point mutations are on a small scale, affecting just one gene. They are changes to the base sequence at a single point. Chromosomal mutations are changes affecting larger parts of chromosomes, or the chromosome number.

Some mutations are caused by mistakes in replication. Occasionally, DNA polymerase incorporates the wrong nucleotide. Others are caused by damage to DNA. This can be caused by some types of radiation, including ultraviolet and X-rays, and

Mutation: Random changes to a cell's DNA.

Exposure to radioactive materials increases mutation rate.

by chemical mutagens such as acridines. Cancer is the result of particular combinations of mutations (Box 30.16), so any carcinogenic chemical must be a mutagen.

Point mutations come in three varieties: deletion, insertion and substitution. These are discussed in Section 7.5.

There are more classes of chromosomal mutation (Fig. 11.6). Duplication, deficiency and inversion involve sections of a chromosome. In translocation, a part of a chromosome becomes attached somewhere else. Duplication is similar, but a copy moves somewhere else; the original stays in position. Deficiency is when a section is lost completely. The situation in which a section of chromosome is cut out and reinserted the other way round is called inversion. Some of these mutations are more serious than others. Deficiency means that, in a haploid cell, some genes are

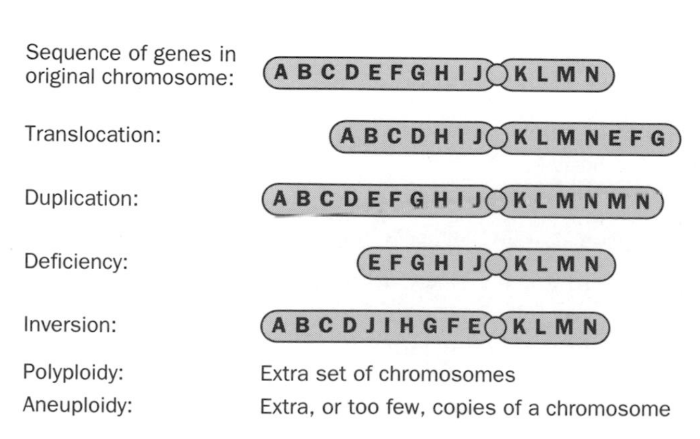

Fig. 11.6 Types of chromosomal mutation. The sequence of letters represents a sequence of genes, not bases, on the chromosome.

lost altogether. In diploid cells, there is only one copy of some genes, instead of two. Duplication increases the number of copies: this can lead to problems if too much of the gene products are made. Inversion and translocation only affect the genes at the point where DNA is broken or rejoined. They may be split or separated from their promoters. These mutations can lead to problems when chromosomes pair in meiosis. The mutated chromosome does not match its homologue for part of its length. Isolated populations may accumulate chromosomal mutations over a long time. Eventually, this may lead to a failure of meiosis when they hybridize with other populations. This can be important in the evolution of new species (Section 2.5).

Aneuploidy and **polyploidy** involve changes in chromosome number. Aneuploidy is the situation where a cell has one or several too many or too few chromosomes. Aneuploidy is usually harmful because lots of genes have the wrong number of copies. Most human aneuploidies lead to the embryo failing to develop, and an early miscarriage. Trisomy 21 (a third copy of chromosome 21) is a famous exception. It causes Down's syndrome. People with this condition have distinctive facial features and learning difficulties, but otherwise lead normal, if rather short, lives. Other human aneuploidies involve the sex chromosomes (Section 12.4).

In polyploidy, a cell has whole extra sets of chromosomes. Instead of being haploid (chromosome number n) or diploid (2n), it may be triploid (3n), tetraploid (4n), and so on. Box 11.4 shows how polyploidy can arise. Polyploidy is quite

Aneuploidy: Cells having one or a few extra or too few chromosomes.

Polyploidy: Cells having whole extra sets of chromosomes.

Box 11.4

The origin of polyploids

Polyploids have extra sets of chromosomes

Polyploids may be able to survive, especially in plants and occasionally in animals. They tend to have larger cells and so larger bodies. This can be useful in crop plants.

Some polyploids are fertile, others sterile. Ones with an odd number of chromosome sets (triploids, pentaploids and so on) are generally sterile. An odd number of chromosomes cannot make pairs at meiosis. In the resulting confusion, cells get strange mixtures of chromosomes and cannot survive. Even-ploids are often fertile, especially in plants.

Polyploid cells arise when chromosomes do not separate at cell division. Failure of mitosis in a diploid body makes a tetraploid cell. Failure of meiosis makes diploid gametes.

From polyploid cell to polyploid organism: two scenarios

In **autopolyploids,** all sets of chromosomes are similar. In **allopolyploids,** they come from different sources. A hybrid, diploid plant will have one set of chromosomes from one species, and a second set from another species. Many hybrids are sterile because the chromosomes cannot pair at meiosis. They may be able to reproduce asexually, however. A tetraploid cell appearing in a hybrid will have two sets of chromosomes from each original species. These can pair, so fertile polyploids can be formed as in scenario 2.

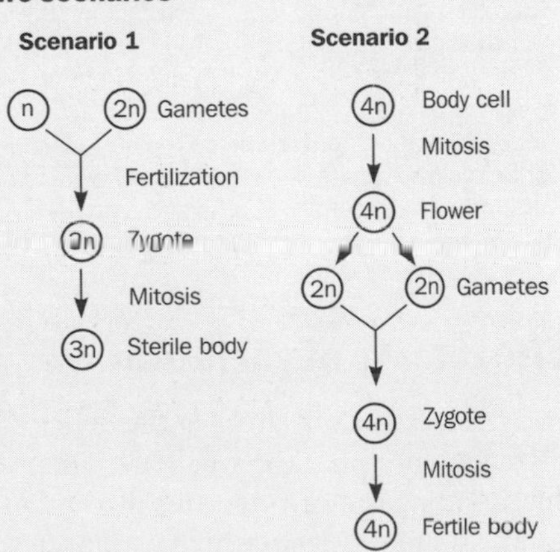

common in plants, especially in some groups like the ferns. It is much rarer in animals. Perhaps this is because the development of the tightly integrated animal body is more sensitive to changes in the amounts of gene products. Perhaps it is because self fertilization and asexual reproduction, both common in plants, make some steps in the origin of polyploids more likely. However, polyploid animals are known, amphibians in particular. The full reasons behind these patterns remain a mystery. Polyploidy can sometimes lead to speciation in a very short time. Box 11.5 shows a case history in which plant polyploidy has had dramatic evolutionary and ecological consequences.

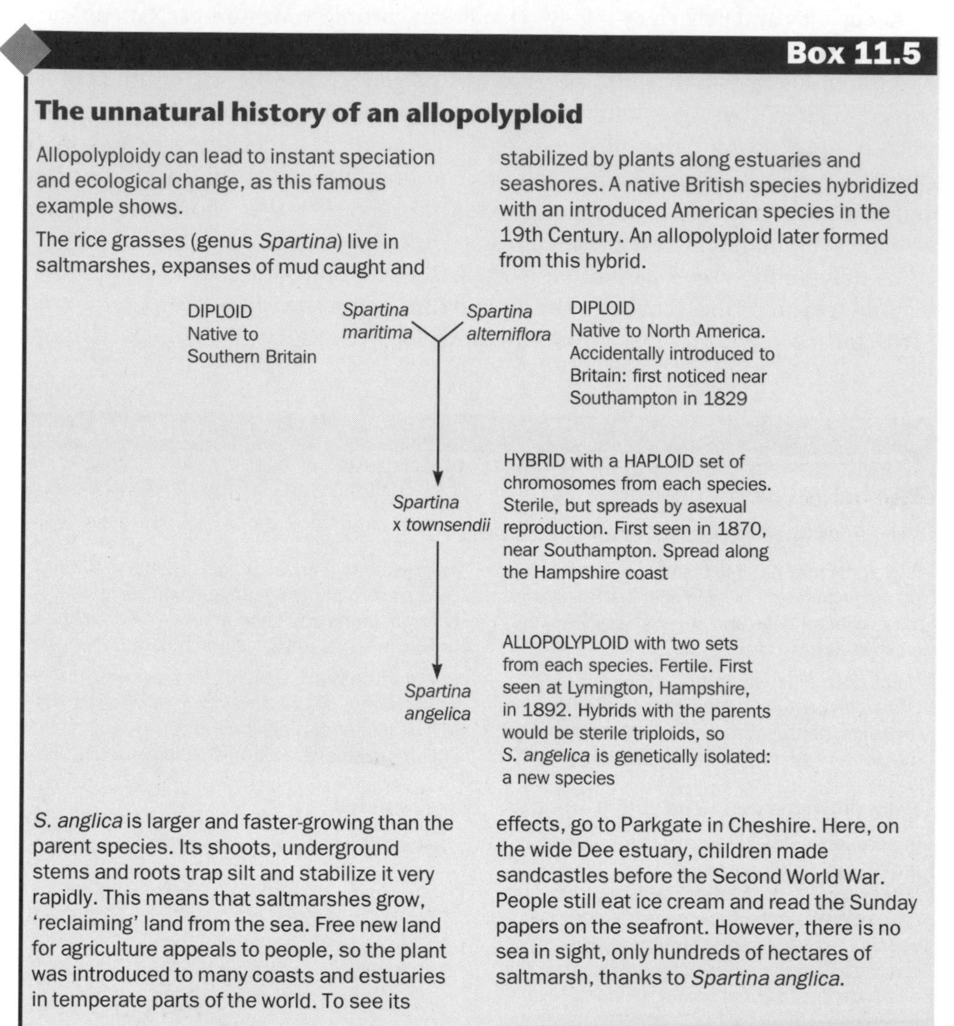

Box 11.5

The unnatural history of an allopolyploid

Allopolyploidy can lead to instant speciation and ecological change, as this famous example shows.

The rice grasses (genus *Spartina*) live in saltmarshes, expanses of mud caught and stabilized by plants along estuaries and seashores. A native British species hybridized with an introduced American species in the 19th Century. An allopolyploid later formed from this hybrid.

DIPLOID
Native to
Southern Britain
Spartina
maritima
Spartina
alterniflora
DIPLOID
Native to North America.
Accidentally introduced to
Britain: first noticed near
Southampton in 1829
Spartina
x townsendii
HYBRID with a HAPLOID set of
chromosomes from each species.
Sterile, but spreads by asexual
reproduction. First seen in 1870,
near Southampton. Spread along
the Hampshire coast
Spartina
angelica
ALLOPOLYPLOID with two sets
from each species. Fertile. First
seen at Lymington, Hampshire,
in 1892. Hybrids with the parents
would be sterile triploids, so
S. angelica is genetically isolated:
a new species

S. anglica is larger and faster-growing than the parent species. Its shoots, underground stems and roots trap silt and stabilize it very rapidly. This means that saltmarshes grow, 'reclaiming' land from the sea. Free new land for agriculture appeals to people, so the plant was introduced to many coasts and estuaries in temperate parts of the world. To see its effects, go to Parkgate in Cheshire. Here, on the wide Dee estuary, children made sandcastles before the Second World War. People still eat ice cream and read the Sunday papers on the seafront. However, there is no sea in sight, only hundreds of hectares of saltmarsh, thanks to *Spartina anglica*.

The release of variation

The business of sex shuffles the genetic variation in populations. New individuals have new combinations of alleles. The potential for variation is released. Sometimes this shuffling is more vigorous than others. Several factors affect it.

The simplest way in which variation is released is in the separation of homologues during meiosis. Which homologue goes to a particular end of the cell is normally down to chance. Each haploid cell produced by meiosis has a mixture of chromosomes from the organism's two parents. The more chromosomes there are, the more possible ways in which the parents' genes can be remixed. So, a large chromosome number releases variation faster.

Crossing over swaps genes between homologues. It sets up new combinations of alleles from the two parents. The more chiasmata there are along the length of each chromosome, the more mixing there will be. There certainly is variation from one species to another in the frequency of crossing over.

Ploidy makes a difference, too. In a species with a haploid body, every gene contributes to the phenotype. In a species with a diploid body, there are two copies of each gene. Some individuals may have two different alleles of a gene, but only one affects the phenotype. This is called a dominant allele. The other, recessive allele is masked (Section 12.2). An individual with two dominant alleles has the same phenotype as one with a dominant allele and a recessive allele. Some of the potential variation fails to reach the phenotype. So, variation is released faster in species with a haploid body. As for species with a diploid *and* a haploid body: well, you can think through the issues that raises.

Individuals have more genes in common with close relatives than with distantly-related individuals. Inbreeding, that is mating with yourself or a close relative, leads to fewer new allele combinations than outbreeding. Outbreeding releases variation more quickly.

These factors release subtly different types of variation. We do not have an easy way of combining them into a single figure which tells us how quickly variation is released. If we did, there would be all sorts of interesting questions to investigate. Is variation released at a similar rate in all species? Might an increase in chromosome number lead to selection for, say, lower chiasma frequency? Do long-lived plants and animals release more variation per generation? Do species living in changeable habitats release variation more or less quickly than others? Without a single measure of the rate of release of variation, it is very hard to find out.

11.4 Weird sex in bacteria

Bacteria do not have conventional 'boy meets girl' sexual life cycles. Bacteria reproduce asexually. The cells simply divide, each one gaining an identical copy of the chromosome. This is not mitosis. Bacteria are prokaryotes: they have no nucleus to be split up, or microtubules to form a spindle.

However, genes do sometimes pass from one bacterial cell to another. The result is new genetic combinations, so this is a sort of sex.

Pure DNA is sometimes passed from cell to cell. This is usually a plasmid, a small 'optional extra' chromosome (Box 31.18). Sometimes, DNA from the main chromosome enters another cell. Then, it may pair up with the homologous bit of the cell's own chromosome, and recombination occurs. The chromosome becomes a mosaic of DNA from the two cells.

The process of conjugation is even more like sex. Two cells come together, and DNA passes directly from one to the other. It happens in various bacteria, but is best understood in *E. coli*. Some *E. coli* cells have a plasmid, the F factor. These F^+ cells can produce pili: hollow protein tubes which link the cell to a nearby F^- cell, which lacks the plasmid. The plasmid is copied, and one copy spools through a pilus. The

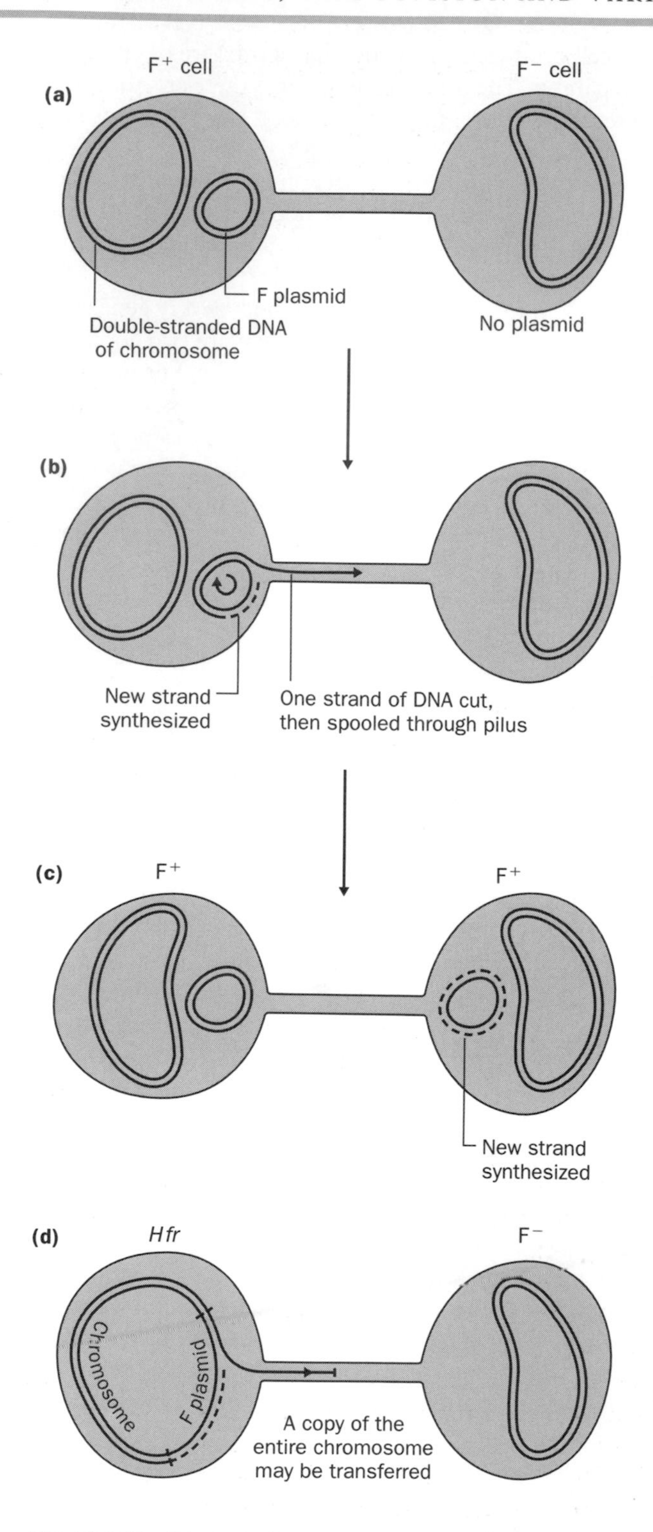

Fig. 11.7 The F factor in *E. coli*. (a) A copy of the plasmid can cross from one cell to another through a pilus. (b) In Hfr strains, the plasmid is incorporated into the chromosome, which is transferred with the plasmid.

second cell becomes F^+ (Fig. 11.7(a)). Occasionally, the plasmid becomes linked to the chromosome. Then, part of the plasmid and part of the chromosome are spooled through (Fig. 11.7(b)). The chromosome fragment pairs up with its counterpart in the other cell, and recombination takes place. This is as close to sex as prokaryotes get. Conjugation is a rare event, however. It is probably an accidental consequence of the way the plasmid behaves, not an adaptation for producing variable offspring.

11.5 Why sex?

Sex leads to variable offspring. Is this an advantage? The answer seems to be that it can be, but there is a cost. How can it be that the advantages so often outweigh the costs? After all, most of the organisms we see around us are sexual.

This is an exciting but tricky topic. A complete theory must explain the evolution of each part of a sexual system: fertilization, meiosis, recombination and the differences between the sexes. Here, we simply discuss the costs and benefits of the overall process. Follow the references in 'Further Reading' once you are hooked!

Sex is a costly business. Imagine a well-adapted organism (and if it were not well-adapted, would it be alive today?). It mates and produces variable offspring. Many of these offspring will be less well-adapted than the parent. A few may be better adapted. Over time, natural selection will favour these variants. Fewer and fewer improvements are possible: more and more offspring will prove to be poorly adapted. A mutant individual which reproduced asexually would leave more viable offspring: they would be equally well-adapted. This mutant form would increase. The population would become asexual.

Another argument reinforces this idea. Consider a population with equal numbers of males and females, in which males do not care for the young. Imagine a mutant gene which inhibits meiosis in females. Eggs develop asexually, as clones of the female. If a female has this allele, every one of her offspring will also be a female carrying this allele. In a sexual population which stays at a constant size, two fertilized eggs per female must survive, on average. In effect, one replaces the female parent, one the male. If two eggs survive from each mutant female, the new allele will double in frequency each generation, at least at first. The sexual population would be overrun by asexual mutants.

We can only explain sex if variable offspring continue to be an advantage, generation after generation. One idea depends on the species living in a patchy environment. Imagine a habitat which is a mosaic of many subtly different environments. An individual may be well-adapted to one sort of patch, but less well-adapted to all the others. Variable, sexual offspring may be better suited to life in a range of other patches. Variation allows the offspring to live in more of the habitat, competing effectively with other species. More offspring survive.

The **Red Queen hypothesis** (Section 2.2) provides another, complementary explanation. We assumed that an organism which is well-adapted now will be well-adapted in the future. This will not be true if the environment is constantly changing. Even if the physical environment stays the same, the biological environment evolves. Organisms have predators, prey, pathogens and parasites. Selection for strains of a parasite which escape from the body's defences will lead to selection for defence systems which can deal with this new strain. This in turn will put a selection pressure on the parasites, and so on. This has often been observed in the cell-surface

Red Queen hypothesis: The idea that interactions between species may drive ongoing evolution in both species.

molecules of fungal pathogens, and the receptor molecules which plants use to detect them (Section 23.5).

Populations are constantly evolving, just to keep their place in the world. Once this evolutionary treadmill starts, there is no way off. Variation remains an advantage. Asexual reproduction becomes a risky strategy. If there are sometimes short-term advantages, asexual species are doomed to early extinction. It is sobering to speculate that we have sex because our ancestors had parasites.

Summary

- Haploid cells have one set of chromosomes.
- Diploid cells have two sets of chromosomes.
- Sexual life cycles involve fertilization and two sorts of cell division: mitosis and meiosis.
- At fertilization, two haploid gametes fuse to form a diploid zygote.
- Mitosis produces two genetically identical clones of the original cell.
- Mitosis is involved in forming a body.
- Diploid and haploid cells can divide by mitosis, although animal life cycles have only diploid mitosis.
- Many plants have a haploid body and a diploid body.
- In meiosis, a diploid cell forms four different haploid cells.
- Meiosis balances the doubling of chromosome number at fertilization.
- In mitosis, identical copies of the chromosomes are pulled to opposite ends of the cell on a spindle of microtubules.
- In meiosis, homologous chromosomes pair up and are separated on a spindle, making two haploid nuclei. These divide again in a similar way to mitosis, giving four haploid cells.
- While homologous chromosomes are paired, they exchange sections of DNA: crossing over. Each chromosome becomes a mosaic of sections of the original pair.
- Variation is caused by environmental differences and genetic differences.
- Mutations are changes to DNA sequence.
- Point mutations affect a single gene.
- Chromosome mutations involve whole chromosomes or large sections of them.
- Mutations are random events. They can be caused by mistakes in replication, by radiation or by chemical mutagens.
- Mutation generates genetic variation.
- Genetic variation is 'released' into the phenotype by factors such as crossing over, random separation of chromosomes in meiosis and outbreeding.
- Polyploidy is a special type of chromosomal mutation: cells have one or more extra sets of chromosomes.

- ◆ Polyploidy has been important in plant speciation.
- ◆ Bacteria do not have conventional sex, but DNA is sometimes transferred between cells.
- ◆ The cost of sex is that many of the variable offspring will be poorly adapted. Sex is an advantage when there is benefit in having variable offspring. This is true in patchy environments, and when populations suffer an evolving parasite or predator.

Exercises

11.1. The Christmas rose, *Helleborus niger*, is a plant of woods and alpine pastures in and around the European Alps. A cell from one of its roots has 32 chromosomes.

(i) How many chromosomes would there be in a nucleus from its pollen grains?

(ii) A single gene codes for an important enzyme in *H. niger* cells. How many copies of this gene are there in a root cell which is not about to divide? Are these copies identical?

(iii) How many copies of this gene are there in a root cell which has just begun mitosis?

11.2. Compare mitosis and meiosis by filling in this table.

Feature	Mitosis	Meiosis
Number of divisions		
Number of daughter cells		
Is parent cell haploid or diploid?		
What happens to chromosome number?		
Are daughter cells identical to one another?		
Are chromosomes replicated before division?		
Do homologues pair up?		
Are chiasmata formed?		

11.3. There is considerable variation in the chromosome number of dog whelks on the shores of northern France and southern England. The genetic material is essentially the same, but some of the shorter chromosomes are sometimes linked end to end, forming fewer, longer chromosomes. What is the effect of this variation on the release of variation?

11.4. The genus *Meconopsis*, in the poppy family, includes the Himalayan blue poppies.

(i) A survey of the chromosome numbers in many *Meconopsis* species showed that diploid numbers of 28, 56 and 84 were each found in more than one species. Speculate on how these different chromosome numbers could have evolved.

(ii) Some plants of *M. simplicifolia* have a diploid number of 84. Others have just 82. How might this difference have arisen?

Further reading

Alberts, B., Bray, D., Lewis, J., Raff, M., Roberts, K. and Watson, J.D. *Molecular Biology of the Cell* (3rd ed.) (New York: Garland, 1994). An enormous but excellent textbook. It puts the reader in touch with current ideas on the mechanism of cell division, the systems which control it, and the mechanism of recombination without too much pain on the way. Molecular genetics is covered in the context of the life of the cell.

Hartl, D.L. *Essential Genetics* (Sudbury MA: Jones & Bartlett, 1996). Bright, clear and well explained. It goes well beyond the scope of this book, without an unbearable acceleration in pace.

The Gene as the Unit of Inheritance

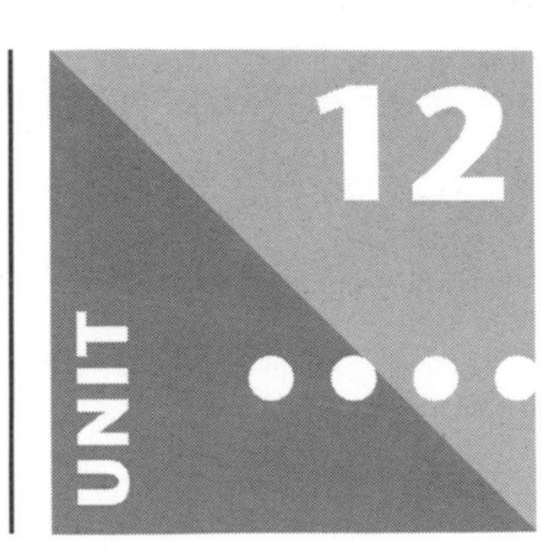

Connections

► This unit tackles one branch of the science of genetics. Another, molecular genetics, is introduced in Unit 7. It is most important that you have a clear understanding of sexual life cycles and cell division (Unit 11) before you study this unit. A basic understanding of natural selection (Unit 2) would also be helpful.

Contents

12.1 The history of the gene

Question: Why does the giraffe have a long neck? Answer: Because its parents had long necks. This answer is not as silly as it first appears. Something is being passed on from one generation to the next, parent to offspring. Something in the egg and the sperm is controlling development, doing the same thing that it did in the parents as they developed. Yet each giraffe is unique. Some of these features are unlike its parents, and not all these are caused by the environment. The sexual process is combining something from each parent to produce offspring which are unique, yet fundamentally similar. What is going on?

One possibility is that some sort of 'essence' of each parent is being combined in the zygote. The offspring's characteristics will be half-way between those of the parents. This idea is called blending inheritance. It makes some sense. After all, we think we can recognize characteristics of both parents in a child. However, blending inheritance would decrease, not increase variation. Surely all the children of two parents would be identical? The population would become more and more uniform.

The alternative is particulate inheritance. Tiny physical structures are passed on, unchanged, from generation to generation. An individual inherits particles from each parent. The combination determines its characteristics. In some cases, characteristics may seem to blend, but each particle remains the same down the generations.

Classic research by Mendel, a Czech monk, demonstrated particulate inheritance in peas (Box 12.1). Mendel was clever, lucky, or most probably both, in selecting his material. He chose inbred strains of a cultivated plant. These show very few differences between individuals. He crossed them with very different inbred strains. This minimizes the confusion of genetic variability. He selected an outbreeding species (many crop plants automatically fertilize themselves, unhelpful in genetic

Box 12.1

Inheritance is particulate

Mendel's experiments showed that something was passed unchanged from generation to generation. He called these things 'factors'; we call them genes.

For example, Mendel crossed peas from a strain which was always tall with peas from a dwarf strain. All the offspring were tall. When these were crossed with one another, some offspring were tall but others were dwarf.

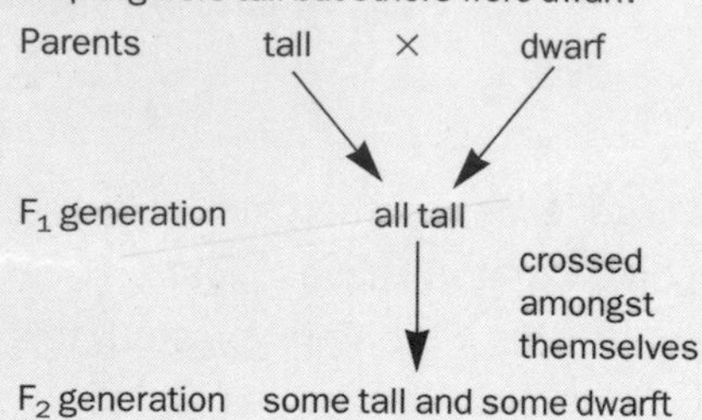

We can conclude:

1 The offspring show the phenotypes of a parent or grandparent, but not intermediates. This does not look like blending inheritance.
2 In the F_2, dwarf plants appeared even though both their parents were tall. The phenotype has 'skipped a generation'. Mendel suggested that a factor coding for dwarfness was present in the F_1 generation, but masked by another factor.
3 If one type of factor can mask another, there must be more than one copy of each in a plant. All Mendel's results could be explained by there being just two copies (Fig. 12.2).

experiments). He investigated clear-cut characteristics: these peas were either tall or dwarf, had smooth or wrinkled seeds, coloured or white petals, and so on. Characteristics showing continuous variation were ignored. Mendel maximized the chance of finding what he was looking for. This is common sense, not cheating! Mendel called the particles of inheritance 'factors'. He also established the rules of how genes are passed on (Sections 12.2, 12.3).

Mendel's experiments were carried out in the mid 19th century. Other workers took up the idea in the early 20th century. They quickly showed that the same rules applied to lots of other species. Mendel's factors became known as genes. Nobody knew what they were.

It was soon realized that chromosomes behaved just like genes. Diploid cells had two of each type. Only one copy was passed to each daughter cell at meiosis, randomly. The link was made. Genes are bits of chromosomes.

Chromosomes are made of DNA and proteins. Which carries genetic information? In the 1940s, early experiments on gene transfer between bacteria showed that purified DNA taken from one strain of a bacterium could change the characteristics of another strain. Chromosomal proteins could not. Genes must be made of DNA. This surprised everyone. Understanding the structure and workings of DNA became a priority.

Genetics then took two separate courses. Molecular genetics was the science of the gene at work (Unit 7). Classical genetics was the study of inheritance of genes. Only now are the two disciplines meeting in a really fruitful way (Section 12.7).

12.2 Inheritance of one gene

A new zygote inherits tens of thousands of interacting genes from its parents. To understand what is going on, we need to simplify the situation. We will focus first on just one gene, and ignore all the rest. We will concentrate on genes which have their effects in diploid cells. This is fair enough: it covers nearly all the genes in animals and flowering plants, as well as many genes in other plants. We will also talk about an organism having genes, rather than the cells that make it up having genes. So, if I write 'Sid the snake has two copies of the gene for ...', I really mean that each cell in Sid's body has two copies of that gene. All his cells carry the same genes as the zygote from which he grew. This sloppy way of talking keeps explanations simple, without hiding the meaning. So, down to business.

A diploid individual has two copies of each gene. They are carried on a pair of homologous chromosomes, at the same place on each. They carry information about the same type of polypeptide, so they affect the same characteristics of the organism. The place on a chromosome where a gene is found is called a **locus** (plural: loci). Genes may come in more than one version, called **alleles**. Peas have a gene which controls the colour of the unripe pods. Quite what polypeptide it makes is unclear, but the effect on the plant is obvious. The gene has two alleles. One allele codes for green pods. We could choose the symbol **G** for this form. The other allele (g) codes for yellow pods. A pea plant has two copies of this gene. There are three possible **genotypes** (combinations of alleles) that a pea plant could have: **GG, gg** and **Gg**. What will be the **phenotypes** (characteristics) of these plants? Two are obvious. A **GG** plant has green pods. A **gg** plant has yellow pods. However, we cannot guess the phenotype of a **Gg** plant. It turns out that **Gg** peas always have green pods. The **G** allele is having its effect, the **g** allele is not. **G** is said to be **dominant**; **g** is said to be **recessive**. Organisms which have only one type of allele at a particular locus are said to be **homozygotes** (here **GG** and **gg**: Fig. 12.1). Organisms with two different alleles at one locus are **heterozygotes**.

	Genotype	Phenotype
Homozygotes	GG	green pod
	gg	yellow pod
Heterozygote	Gg	green pod

Fig. 12.1 Genotype and phenotype: the 'green' allele (**G**) is dominant.

Locus: The place on a chromosome where a particular gene is found.

Allele: One version of a gene.

Genotype: The combination of alleles an organism has.

Phenotype: The characteristics of an organism.

Dominant allele: An allele which has an effect on the phenotype in homozygotes and heterozygotes.

Recessive allele: An allele which does not affect the phenotype in heterozygotes.

Homozygote: An individual which has two identical alleles at one locus.

Heterozygote: An individual which has two different alleles at one locus.

Fig. 12.2 is a standard way of showing how one gene is inherited. Consider a cross between a pea plant from a strain which always makes green pods and one from a yellow podded strain. Each plant will make gametes. They are haploid, with only one copy of each gene. The **GG** parent can only make gametes carrying the **G** allele, and so on. When the gametes combine, all the offspring must have **G** from one parent and **g** from the other. They will all be heterozygotes with green pods. This generation is called the F_1 (first filial: 'filial' means 'to do with offspring'). Now we decide to cross two, any two, of the F_1 individuals. They can make gametes carrying different alleles. The diagram shows two snapshots in the sexual cycle: the F_1 adults and the gametes they make. Between the two, meiosis has happened. The homologous chromosomes have separated. Each haploid daughter cell has just one of each pair. Half carry **G**, the other half **g**. So, half the gametes carry **G**, half **g**. The diagram shows two gametes. This means two *types* of gamete, not two individuals. The gametes from the two parents combine randomly to form the F_2 generation. There are four possible combinations, equally common. Two combinations give **Gg** offspring, so these are twice as common as either **GG** or **gg**. **GG** and **Gg** both give green-podded

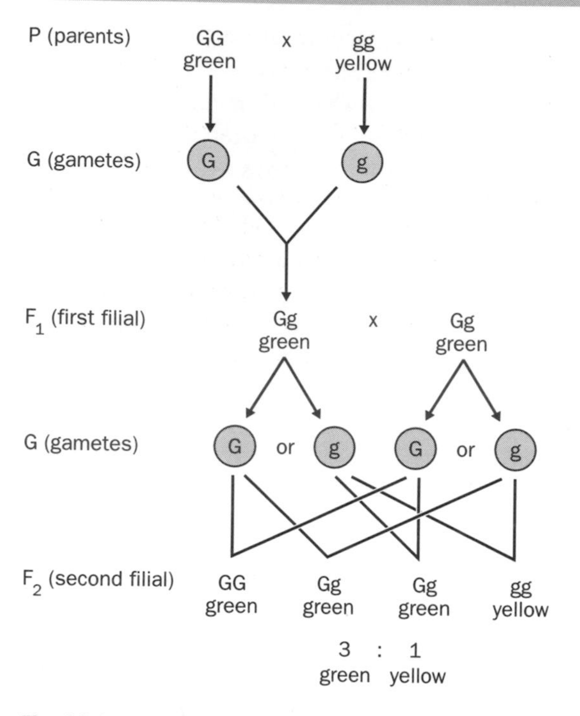

Fig. 12.2 A cross in which only one gene is studied (see text for details).

plants, so about three quarters of the F_2 generation will have green pods. The phenotypes will be in a ratio of 3:1. This is not a certainty, but an estimate based on the probabilities of each type of gamete combining. Box 12.2 shows how another type of cross gives other classic ratios.

In many situations like this, one allele is dominant to another. Heterozygotes show the dominant phenotype. With some genes, however, heterozygotes have a unique phenotype. A serious human disease, sickle cell anaemia, is a good and

Box 12.2

The backcross

An individual shows the dominant phenotype. Is it heterozygous, or homozygous for the dominant allele? The only way to find out is to cross it with an individual showing the recessive phenotype.

For example, in the fruit fly *Drosophila melanogaster*, normal wing (**V**) is dominant to vestigial wing (**v**). Vestigial wings are pathetically small: their owners cannot fly. A normal fly could be either homozygous (**VV**) or heterozygous (**Vv**). To find out, we cross it with a vestigial-winged fly (**vv**).

Of course, if you get just a few offspring which are all normal, you cannot *prove* that the parent was homozygous. There is a small chance that this would happen even with a heterozygote. The more offspring there are, the less likely this is.

If the normal parent is homozygous:

P VV × vv
G V v
F_1 Vv
normal

All offspring are normal

If the normal parent is heterozygous:

P Vv × vv
G V or v v
F_1 Vv vv
normal vestigial

Offspring include normal and vestigial flies in a 1:1 ratio

famous example. A mutation affects the structure of haemoglobin. A point mutation leads to a single amino acid substitution in the β chain (Section 19.2). When combined with the α chain, the modified protein it forms is called haemoglobin S (HbS). So there are two alleles of the gene coding for the β chain: the normal one (**s**) and the mutant one (**S**). **SS** homozygotes have only HbS, which crystallizes easily inside the red blood cells. They collapse into a sickle shape and carry oxygen very poorly. This has a devastating effect on the body, and early death is common. The condition is known as sickle cell anaemia. Heterozygotes (**Ss**) have one copy of each allele. They make some normal β chains, and some mutant ones. Some haemoglobin molecules are normal, some are HbS. These people suffer a different set of symptoms, sickle cell trait. Much of the time, red cells behave as normal, but they may collapse when oxygen levels are low. This genetic situation is called incomplete dominance. Fig. 12.3 shows how it can alter the classic 3:1 ratio.

What is dominance? Why are some alleles dominant? Why do we see incomplete dominance in some cases? Imagine a gene which codes for an enzyme. The enzyme makes a red pigment, found in petals, from a colourless substrate. Mutation could lead to an enzyme which works just like the normal one, an enzyme which works slightly differently (probably less well), a protein which does nothing, or no protein at all. If the mutant protein behaves just like the normal one, we will see no effect on the phenotype. Many genes seem to have lots of equivalent alleles like this. Natural selection does not affect them. The different versions of the enzyme are called isozymes. The concept of dominance does not apply, because the phenotypes are identical. If the enzyme does not work, or is absent, it is harder to predict what will happen. Certainly, normal homozygotes will have red petals. Homozygotes for the mutant allele will make no pigment, so will probably be white. Heterozygotes will have only half the normal number of copies of the functional gene (one instead of two). This may lead to only half as much normal enzyme being made, but that depends on how transcription and translation are controlled. If the usual amount of working enzyme is made, heterozygotes will be just like normal homozygotes. If less enzyme is made, we might see less pigment made. The heterozygotes would then have paler coloured petals: incomplete dominance. However, this depends on how the enzyme is controlled. We might just see fewer enzyme molecules, working much harder. The normal allele would be dominant again.

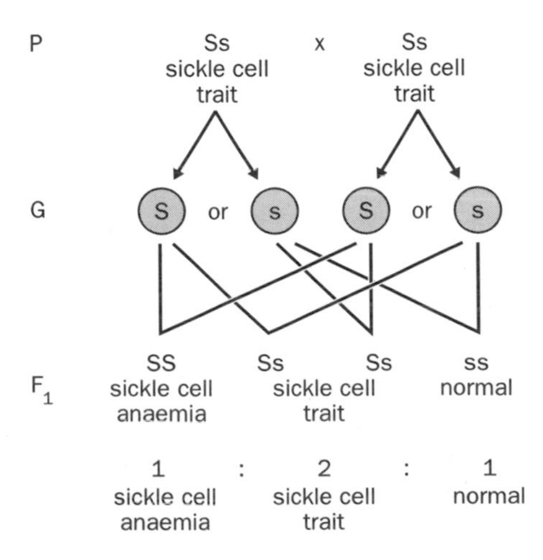

Fig. 12.3 A cross illustrating incomplete dominance.

Occasionally, a single polypeptide affects the phenotype in more than one way. For example, one gene in the pea affects starch synthesis from sugars in the seed. There are two alleles (**A** and **a**). **AA** homozygotes synthesize starch effectively. They have large starch grains. As the peas ripen, they dry to a rounded shape. **aa** homozygotes synthesize starch much less effectively. Their starch grains are small, and the ripening seeds collapse into a wrinkled shape (sweet-tasting garden peas have this phenotype). Heterozygotes produce round seeds, so **A** seems to be dominant. However, if we look at starch grains, they are intermediate in size in **Aa** peas. On this basis, the alleles show incomplete dominance. So, dominance is not a built-in feature of a gene or the protein it codes for. It depends on how the protein works, how it is controlled, and which aspect of the phenotype we choose to look at.

This section started with an extremely simple system: a single gene with two alleles, showing clear dominance. The concept of incomplete dominance allows us to explain some more real situations. Further subtleties are genes with more than two alleles (Box 12.3) and lethal alleles, where some genotypes fail to survive (Box 12.4). The next step is to look at more than one gene at a time.

Box 12.3

Multiple alleles

Sometimes genes have more than two alleles. Whether there are three, four, five or fifty, the rules are the same. A diploid organism still has just two copies of each gene. There are just more possible genotypes.

For example, a gene with three alleles controls the ABO blood group system. The gene codes for a glycoprotein in the plasma membrane of red blood cells. Two alleles, $\mathbf{I^A}$ and $\mathbf{I^B}$ code for different versions of the glycoprotein. A third allele, $\mathbf{I^o}$, does not produce the glycoprotein at all. Blood group depends on which versions of the glycoprotein

Genotype	Phenotype (blood group)
I^AI^A	A
I^AI^o	A
I^BI^B	B
I^BI^o	B
I^oI^o	O
I^AI^B	AB

are present, so $\mathbf{I^A}$ and $\mathbf{I^B}$ show incomplete dominance. $\mathbf{I^o}$ is recessive.

Some crosses can give surprising outcomes, for example:

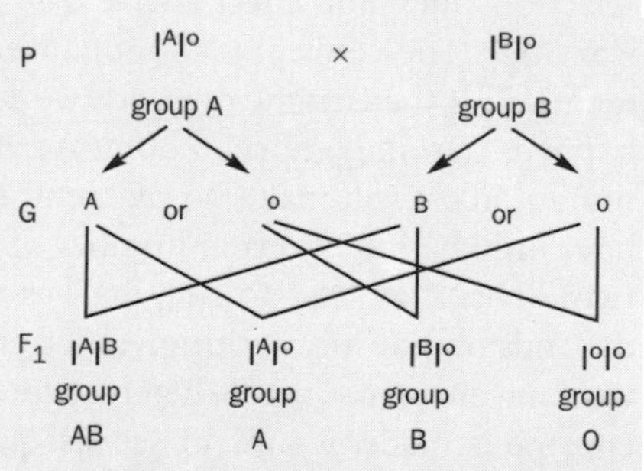

So some combinations of parents can produce offspring with any blood group. The explanation need not involve the milkman.

Box 12.4

Lethal alleles

Lethal alleles are mutations which cause the individual to die before maturity. Some are dominant, some are recessive. Their effect on genetic crosses is normally very simple: there are fewer offspring than expected, and they all have the normal (living) phenotype. Strange effects are seen in situations where the lethal allele has a second, apparently unrelated effect on the phenotype (in other words, the allele is pleiotropic). A famous, and probably very unusual example is an allele in mice which causes yellow fur. All yellow/yellow matings lead to a ratio of 2 yellow : 1 normal in the F_1. This is because the yellow allele (**Y**) has two effects. In a homozygote (**YY**) it causes death in the uterus. In a heterozygote (**Yy**) it causes yellow fur.

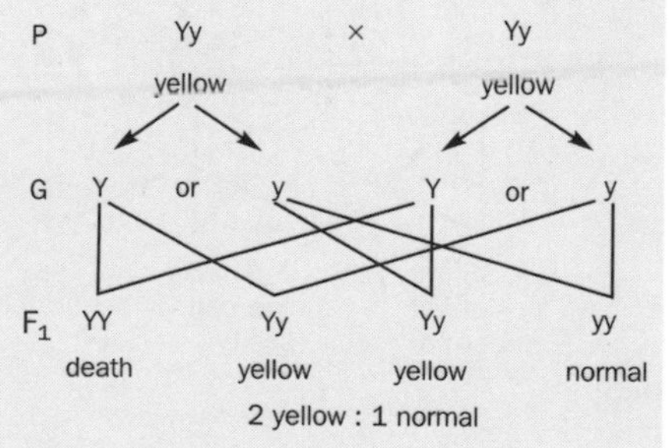

12.3 Inheritance of two genes

If two genes are on different chromosomes, they are inherited independently. For example, in the pea, yellow seed (**Y**) is dominant to green seed (**y**) whilst round seed (**R**) is dominant to wrinkled seed (**r**). Fig. 12.4 follows the same pattern of crosses as Fig. 12.2, but this time these two genes are followed together. A homozygous dominant parent (**YYRR**) is crossed with a homozygous recessive parent (**yyrr**). Each

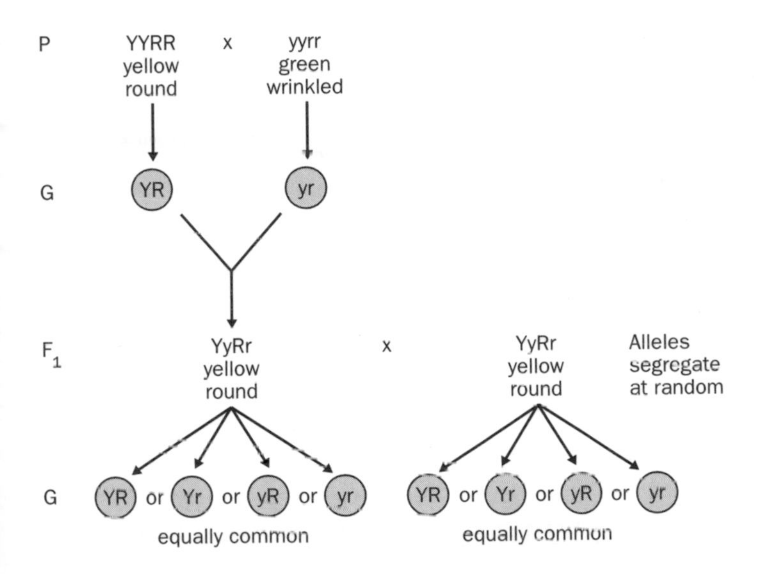

Gametes from first parent (columns); gametes from second parent (rows)

	YR	Yr	yR	yr
YR	YYRR yellow round	YYRr yellow round	YyRR yellow round	YyRr yellow round
Yr	YYRr yellow round	YYrr yellow wrinkled	YyRr yellow round	Yyrr yellow wrinkled
yR	YyRR yellow round	YyRr yellow round	yyRR green round	yyRr green round
yr	YyRr yellow round	Yyrr yellow wrinkled	yyRr green round	yyrr green wrinkled

9 yellow round : 3 yellow wrinkled : 3 green round : 1 green wrinkled

that is: 12 yellow : 4 green (= 3:1)

and: 12 round : 4 wrinkled (= 3:1)

Fig. 12.4 A cross in which two genes are studied (see text for details).

makes gametes with just one genotype: **YR** and **yr**, respectively. All the F_1 are heterozygotes (**YyRr**). Two F_1 plants are crossed. Each makes four types of gamete: **YR**, **Yr**, **yR** and **yr**. They are equally common. Even though **Y** and **R** originated in the same parent, nothing is holding them together. During meiosis, the homologues go to opposite ends of the cell (segregate) at random. If a haploid daughter cell gets the father's copy of one chromosome, it is equally likely to get the father's or the mother's copy of another chromosome. Alleles of genes carried on separate chromosomes segregate at random, so the four gamete genotypes are equally common. Given this, the gametes can combine in 16 different ways, giving four offspring phenotypes. The table used to show these combinations in Fig. 12.4 is called a Punnett square. The phenotypes are in a ratio of 9:3:3:1. Notice that we can get this by combining two 3:1 ratios (3×3:3×1:1×3:1×1). If we only look at seed colour, the ratio becomes 3:1 as expected. The same is true of seed shape. Box 12.5 shows how a different cross can give another classic ratio, 1:1:1:1.

When two genes affect the same characteristic, different ratios may appear (Box 12.6). The inheritance of genes is not affected, however, only the way they determine the phenotype. When two genes are on the same chromosome, the pattern of inheritance changes. The genes are said to be linked.

The inheritance of linked genes depends on crossing over in meiosis (Section 11.2). First, consider two linked genes (**A/a** and **B/b**) in an imaginary species with no crossing over (Fig. 12.5(a)). The parents have genotypes **AABB** and **aabb**. The fact that the genes are linked has no effect on the genotypes of the gametes or the F_1 generation. All the F_1 will be **AaBb**. The difference comes when the F_1 heterozygotes

Box 12.5

Backcross: two genes involved

Just as the classic 3:1 ratio becomes 9:3:3:1 when two genes are considered, the 1:1 backcross ratio becomes 1:1:1:1.

For example, in *Drosophila*, normal wing (**V**) is dominant to vestigial wing (**v**). Also, grey body (**E**) is dominant to ebony body (**e**). Is a normal-winged, grey-bodied fly homozygous or heterozygous? To find out, cross them with a vestigial ebony fly. There are two possibilities:

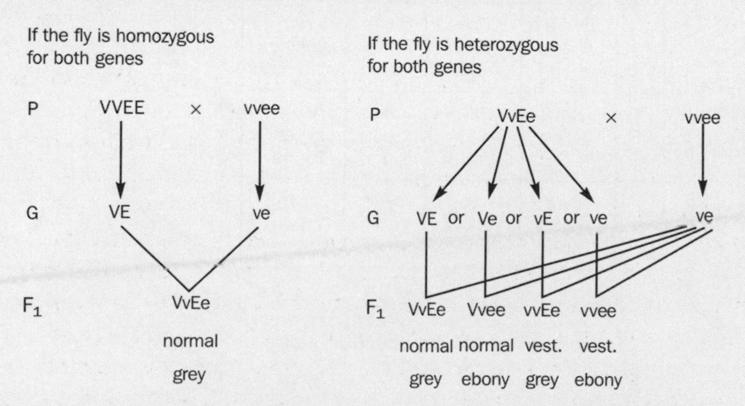

There are two other possibilities, however, because the fly could be heterozygous for just one of the genes. If the fly was **VVEe**, the F_1 phenotype ratio would be 1 normal grey : 1 normal ebony. If the fly was **VvEE**, it would be 1 normal grey : 1 vestigial grey. Try setting out the cross diagrams to show this.

Box 12.6

Epistasis

Two genes may produce a phenotype together. A common situation is when one gene controls whether or not a second gene affects the phenotype. The first gene is epistatic to the second.

For example, several genes affect fur colour in mice. Normal grey mice are referred to as agouti. Their hairs have black melanin pigment for most of their length, with a yellow band near the tip. Albino (white) mice have no melanin. Black mice have melanin throughout each hair.

One gene affects melanin production. **C** (coloured coat) allws melanin production. It is dominant to **c** (albino). Another gene affects melanin distribution. **A** (agouti) is dominant to **a** (black). So any **cc** mouse will be albino, whatever its genotype at **A/a**. **C/c** is epistatic to **A/a**.

Some interesting phenotype ratios can occur, for example in the cross **CcAa** (agouti) × **CcAa** (agouti):

Gametes (2nd parent)	Gametes (1st parent) CA	Ca	cA	ca
CA	**CCAA** agouti	**COAa** agouti	**CcAA** agouti	**CcAa** agouti
Ca	**CCAa** agouti	**CCaa** black	**CcAa** agouti	**Ccaa** black
cA	**CcAA** agouti	**CcAa** agouti	**ccAA** albino	**ccAa** albino
ca	**CcAa** agouti	**Ccaa** black	**ccAa** albino	**ccaa** albino

9 agouti : 4 albino : 3 black

make gametes. Whole chromosomes are inherited, not separate genes. If there is no crossing over, gametes will inherit alleles in their original combinations: **AB** or **ab**. If these are combined with **ab** gametes from an **aabb** parent, just two F_2 genotypes are produced: **AaBb** and **aabb**, in a 1:1 ratio. They have the same phenotypes as their parents. If the genes had not been linked, we would have expected four genotypes, equally common, as in Box 12.5. Now consider the same situation, but with crossing over in meiosis (Fig. 12.5(b)). F_1 individuals can now make four classes of gamete. Crossing over takes place at more or less random sites on the chromosome. Sometimes, there will be no cross-over between two loci. Then, the gametes will inherit the parental combinations of alleles. If there is a cross-over in this part of the chromosomes, the alleles will be recombined, giving **Ab** and **aB** gametes. Sometimes, more than one cross-over event will take place between the loci. Two, or any other even number, will cancel each other out, giving parental gametes. Any odd number will give recombinant gametes (Fig. 12.6).

The number of cross-overs between two loci depends on how far apart they are. If the loci are very close, the chance of a cross-over is small. Only a tiny proportion of the gametes will be recombinant. If they are further apart, the chance of a cross-over will be higher, and the chance of a second or third becomes significant. When two loci are very far apart, it is virtually certain that cross-overs will occur between them. There is an equal chance of an odd or even number of cross-overs, so on average half the gametes will be recombinant. The gamete and F_1 ratios will then be the same as they would be if the genes were not linked. The frequency of crossing over can be used to make maps of chromosomes (Box 12.7). Other mapping techniques which do not rely on breeding experiments are also available.

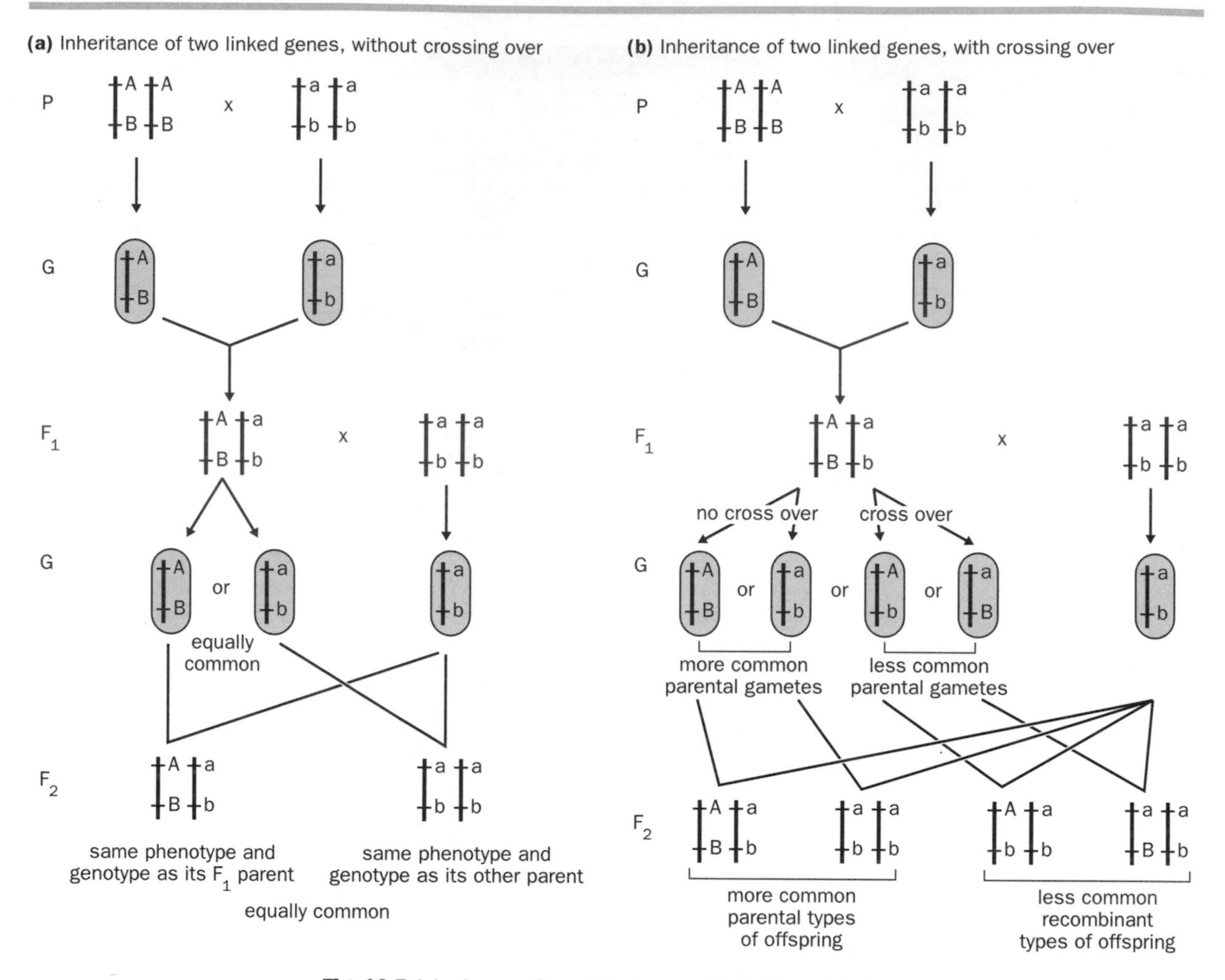

Fig. 12.5 Inheritance of two linked genes, first without (a), then with (b) crossing over.

12.4 Genes and sex

In most animal species, and some plants, individuals are either male or female. Every individual has the genes it needs to develop as either sex. Something switches on one genetic programme, and switches off the other. Occasionally, the switch is environmental, as in some annelid worms, fish and reptiles. Usually, it is a genetic switch.

In mammals, the switch is a single gene on the **Y** chromosome. One type of chromosome comes in two versions, the sex chromosomes. The longer **X** chromosome carries many genes which have nothing to do with sex. It is a typical chromosome. The **Y** chromosome is much shorter. It has few genes apart from the male-determining gene. Females have two **X** chromosomes. Males have one **X** and one **Y**. All matings are between **XX** females and **XY** males, so half of all offspring are male, and half female. The cross diagram in Fig. 12.7 should make this clear. The sex ratio stays at 1:1 from generation to generation.

Human mutants with too many or too few sex chromosomes show that the **Y**

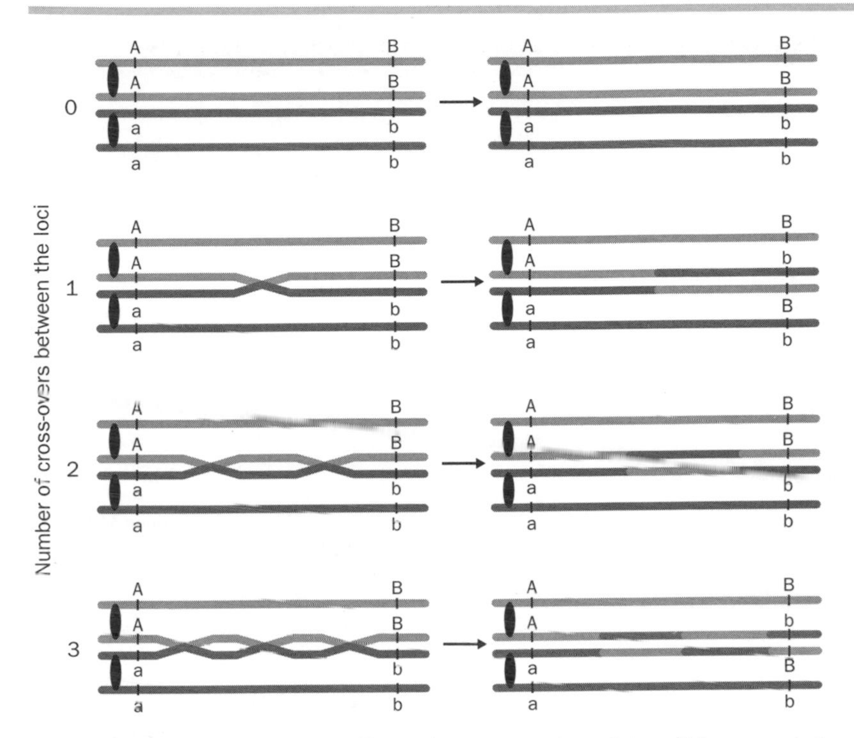

Fig. 12.6 Number of cross-overs affects whether or not two alleles will be separated.

chromosome is crucial. In Klinefelter's syndrome, patients have one **Y** chromosome and two or more **X**s (**XXY**, **XXXY**, and so on). They are sterile, but the **Y** chromosome ensures that they develop as males. Metafemales have no **Y** but three or more **X**s (**XXX**, **XXXX**, and so on). They are female, though with limited fertility. Turner's syndrome patients have just one sex chromosome, an **X** (**XO**). They are sterile

Individuals of this slipper shell, *Crepidula fornicata*, begin life as males and become female as they grow larger. Several other molluscs, crustaceans and fish change sex during life.

Box 12.7

Mapping genes

A genetic map shows where genes are on a chromosome. We need to know their order and distance apart. There are several ways of doing this. The classical way involves breeding experiments.

Take two linked genes, **A/a** and **B/b**. The further apart they are, the more likely a cross-over is to come between them at meiosis. The proportion of gametes which are recombinant is a measure of how far apart the genes are. We cannot usually see the genotypes of the gametes, but a cross between **AaBb** and **aabb** shows them in the F_1 (Fig. 12.5). Every offspring with a recombinant phenotype has come from a recombinant gamete. Every offspring with a parental phenotype comes from a parental gamete.

Cross-over value (COV) can be calculated from a cross like this. It is a measure of how far apart on the chromosome two linked genes are.

$$\text{COV} = \frac{\text{no. of recombinant offspring}}{\text{total no. of offspring}} \times 100\%$$

Cross-over value ranges from zero (*very* close together) to 50% (far enough apart that there is an equal chance of an odd or even number of cross-overs).

A map can be constructed from several cross-over values. For example:

Genes	COV
A–B	28.2
A–C	7.9
B–C	22.3

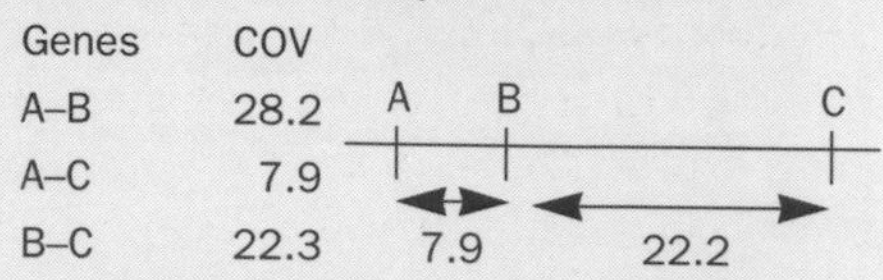

Notice that the cross-over value between A and B is less than the total for A–C and B–C. This is because the chance of two cross-overs in the longer distance is more than the chance of a double cross-over in either one of the shorter distances.

Part of a genetic map

This map shows the positions of some genes on chromosome 3 of *Drosophila melanogaster*. The names of the mutant genetic markers are shown, along with their phenotypes and the normal equivalents. These are not necessarily important genes, but ones with phenotypes that are easy to see in breeding experiments.

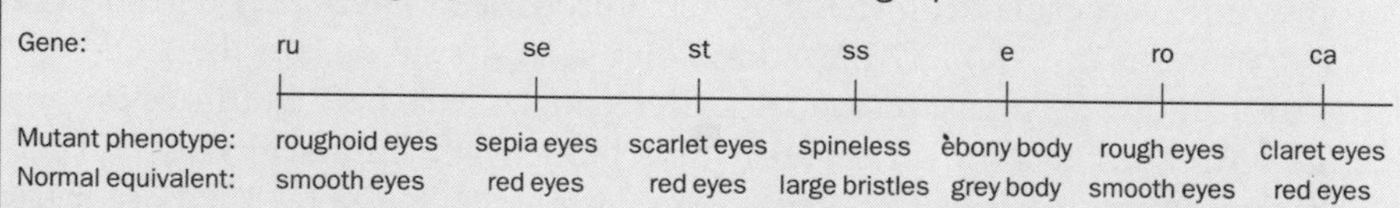

More modern techniques

Molecular biology provides new ways of mapping. Genes whose phenotypes are hard to see can be mapped. Organisms like bacteria, and humans, which are hard to map in the classical way can be investigated.

Restriction enzymes cut DNA at specific sequences (Box 31.15). Each type will break down an organism's DNA into a fixed number of fragments. Gel electrophoresis can separate them and show their length (Box 31.16). DNA probes can identify which fragment carries a particular gene (Box 31.17). If two genes are on the same fragment, they can be no further apart than the length of the fragment. By breaking an organism's DNA in different ways by different restriction enzymes, one can work out the sequence of genes.

This strategy works best for smaller genomes, or fragments of a big genome which have been cloned (Boxes 31.19, 31.20). On the finer scale, DNA sequencing can show the precise order and position of genes. It can also show up genes which nobody knew were there.

females. So, the **Y** chromosome with its sex-determining master gene acts as a switch. The gene itself is enough. Genetically engineered **XX** mouse embryos which carry the gene on another chromosome develop as males.

Sex chromosomes are also seen in other animal groups. In birds, females are **XY** and males are **XX**. Otherwise, the system works in the same way. In the fruit fly *Drosophila*, males are **XY** and females are **XX**. However, maleness is determined by the ratio of **X** chromosomes to autosomes (ordinary non-sex chromosomes). An **XO** mutant is female in mammals, but male in *Drosophila*. It has the same number of **X** chromosomes as a normal **XY** male.

A gene carried on the **X** chromosome is described as **sex linked**. This affects the way the gene is inherited. For example, the genes coding for the red and green pigment proteins of cone cells in the human retina are sex linked. Some mutations lead to only one type being produced. Sufferers are red-green colour-blind. We can simplify the situation by thinking of a single gene. The colour-blindness allele (**v**) is recessive, the normal allele (**V**) is dominant. A woman, with two **X** chromosomes, has two copies of the gene. Both copies must be the recessive allele (X^vX^v) if she is to be colour-blind. A man has only one copy. If it happens to be the recessive allele (X^vY) he will be colour-blind. This is why colour-blindness is more common in men. If the chance of inheriting the recessive allele is small, the chance of inheriting two will be far less. Fig. 12.8 shows two crosses. Firstly, a homozygous normal woman and a colour-blind man have a family. None of the children could be colour-blind. This is because the boys cannot inherit their father's **X** chromosome. (If they did, they would not be boys!) In the second cross, the parents are a heterozygous normal woman and a normal man. There is a 50% chance that a son will be colour-blind. Most colour-blind men come about this way.

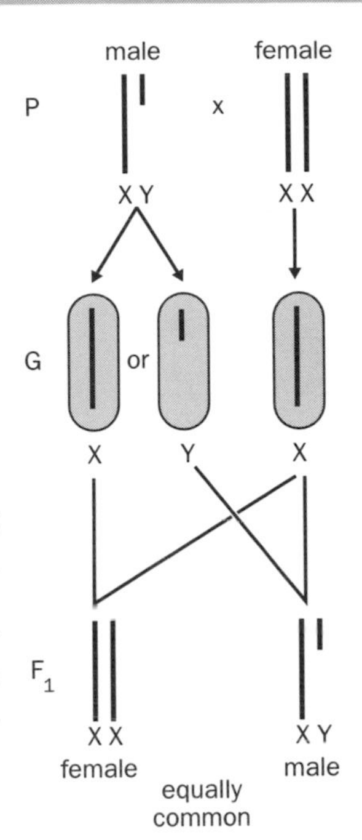

Fig. 12.7 The system of sex determination in mammals leads to a sex ratio of about 1:1.

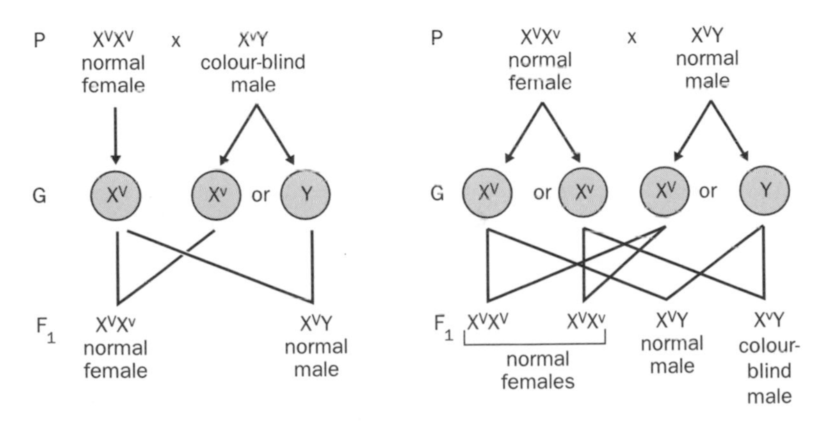

Fig. 12.8 Inheritance of a sex-linked gene. **V** (normal) is dominant to **v** (colour-blind) (see text for details).

12.5 Genes, environment and continuous variation

Many characteristics vary continuously: there is a spectrum of values, rather than two or more distinct groups. Differences in the environment can cause this, but different genes may too. When many genes each have a small effect on a character, there is likely to be continuous variation. Even a few genes combine to give many possible genotypes. If one gene has two alleles, there are three possible genotypes (**AA**, **Aa** and **aa**). If two genes each have two alleles, there are $3 \times 3 = 9$ possible combinations. For just six genes, there are over 700 possible genotypes. This, combined with differences in the environment, would give continuous variation in phenotype.

Sex-linked genes: Genes carried on the X chromosome.

Even a single gene may have a variable effect. Some alleles may show a greater effect in some individuals than others (variable expressivity). Sometimes, the effect is not seen at all in some individuals (incomplete penetrance). For example, a single dominant allele causes Huntington's chorea. This human disease involves degeneration of the nervous system. People who inherit the allele may first get the symptoms any time from childhood to late in life. Some live to a ripe old age without getting symptoms at all. The gene shows incomplete penetrance and variable expressivity. We must assume that differences in other genes and/or the environment cause this variation.

How much of continuous variation is due to genes and how much to the environment? In most species this can be investigated by seeing how well the characteristic responds to artificial selection. The less it responds, the more important the environment must be. In humans, identical twins can be compared. Each pair shares the same genes, so any differences must be due to the environment. This method is fraught with difficulties, however.

Either way, a statistic called **heritability** is estimated. First the variance of the characteristic is estimated. This is a statistical measure of the variability in that character. Heritability is the proportion of the variance which is due to genetic differences. A heritability of 1.0 means it is all due to genes; zero means it is all down to the environment. For example, white spotting on Friesian cows was found to have a heritability of 0.95. At the other extreme, multiple birth in Shropshire ewes had a heritability of just 0.04.

Heritability is easily misunderstood. It is not a measure of how important genes are in controlling a characteristic, only how much of the *variation* in that characteristic is controlled by genes. We know that genes are important in controlling multiple birth in sheep, yet this characteristic showed a low heritability in Shropshire ewes. Presumably, that breed has little variation in the genes responsible, so what differences there are must be caused by the environment. Other breeds will have quite different genotypes, and so very different rates of multiple birth. Heritability tells you nothing about differences between groups of individuals, only differences within the group.

Multiple birth in sheep is important to farmers, but few people get emotional about the causes. If we move the debate to intelligence and human races, everyone gets excited. Suppose we find a difference in IQ between two races: is this due to genes or environment? Racists, as well as those who suffer racism or fight against it, all want to know. Suppose we now find that IQ is highly heritable in both racial groups. All this tells us is that variation within each race is mainly due to genes. It tells us nothing about what causes the difference between the races. Perhaps it is caused by cultural factors: upbringing and education. Perhaps there are genetic differences. Maybe there are both. We just cannot tell by looking at heritability.

12.6 Genes in strange places

Heritability: A measure of how much of the variation in a characteristic is controlled by genes.

All genes are pieces of DNA; not all genes are parts of chromosomes. Prokaryotic cells have plasmids, small 'optional extra' chromosomes which carry one or a few genes. Eukaryotes have DNA in their mitochondria and plastids (Section 4.5).

Genes in these places are inherited in distinctive ways. Plasmids replicate independently within the cell. There may be more than one copy, but there is usually an upper limit. When the cell divides, some plasmids may go to each daughter cell. Alternatively, one daughter cell may get no plasmids at all. Plasmids can sometimes transfer from one cell to another.

Mitochondrial and plastid DNA is passed on when the organelles divide. However, the organism as a whole inherits organelle genes in a peculiar way. In animals, sperm contain less cytoplasm than eggs. It follows that most, or even all of the zygote's mitochondria will come from the female parent, via the egg. This seems to be true, at least in mammals. Only the mother's mitochondrial genes are passed on. Maternal inheritance of chloroplast genes is seen in most flowering plants, where pollen grains contain no plastids.

In each case, the simple rules of Mendelian genetics just do not apply.

12.7 From gene to genome

When genes were found to be made of DNA, the science of genetics split in two. Classical and molecular genetics went their separate ways. One dealt with the inheritance of genes, the other with how they work. Only now are the techniques of molecular biology sophisticated enough to reunite them. If classical genetics shows that a gene produces a particular phenotype, and maps to a certain region of one chromosome, there is a good chance that a molecular biologist can track it down, work out its sequence, discover what polypeptide it produces, introduce it to new contexts, knock it out to see what happens, and examine how it is controlled. This approach takes time and effort. Much of it is concentrated on genes which underlie medical conditions and on genes which may control development.

When classical genetics meets molecular biology, a whole new field of study opens up: the genome. The genome is the whole of an organism's genetic material. If we believe that living things share a common ancestor, then their genomes must all have evolved from a simple ancestral genome. Genes, like species, do not pop out of thin air, but evolve.

Prokaryotic genomes are relatively simple. There seems to have been a rapid increase in genome complexity with the evolution of the eukaryotes, and another with the evolution of the vertebrate animals. Genes must somehow be duplicated within the genome, before the two copies can evolve into distinct genes. This is very much like speciation on the genetic level, but how does it happen? How many familiar genes actually belong to families with related but undiscovered genes? Can we piece together the evolution of a gene family from an ancient ancestor? Why are some parts of the genome more prone to mutation than others? Is the mutability of a gene something that is open to natural selection? If so, when would it be an advantage? What about the non-coding DNA? We know that some holds information of a sort: binding sites for control proteins; information on splicing when a single gene can make more than one product; centromeres; telomeres. What about the rest? Is there information which controls the higher levels of DNA coiling? Are the repetitive sequences which seem to copy themselves around the genome really just junk? Even if they are purely 'selfish', surviving simply because they spread, do they have any interesting effects on the genome? And how *do* they spread? To understand genes, we need to see them in the context of their environment, the genome. There has never been a better time to study the natural history of the gene.

Summary

- Pieces of information, genes, are passed intact from generation to generation.
- Chromosomes are long DNA molecules. Genes are pieces of chromosomes coding for one polypeptide.
- A gene may come in different versions: alleles.
- Diploid organisms have two copies of each gene, one from each parent, via the gametes. These two copies need not be the same allele.
- Individuals with two identical alleles of a particular gene are called homozygotes. Individuals with two different alleles are heterozygotes.
- The combination of genes in an individual is its genotype. The characteristics of that individual are its phenotype.
- In a heterozygote, one allele (the dominant one) has its effect on the phenotype, while the other (recessive) is masked.
- In some genes, two alleles show incomplete dominance. The heterozygote has an intermediate phenotype.
- Many genes have more than two alleles.
- Crosses between individuals with different genotypes give distinctive phenotype ratios in the offspring.
- At meiosis, only one copy of each gene goes to each daughter cell.
- Genes which are carried on different chromosomes are inherited independently.
- Genes which are carried on the same chromosome are described as linked.
- When two linked genes are very close together, there is only a small chance that the copies will be recombined by crossing over in meiosis.
- The further apart linked genes are, the greater the chance of crossing over between them. This is the basis of genetic mapping.
- When one gene controls whether or not a second gene has an effect on the phenotype, the first gene is described as epistatic. Epistasis produces some unconventional phenotype ratios.
- In mammals, sex chromosomes are involved in determining the sex of an individual. The longer X chromosome is an ordinary chromosome, with many genes. The Y chromosome is short and has few genes.
- A master gene on the Y chromosome causes the male phenotype. If it is absent, the individual is female.
- Female mammals have two X chromosomes. Males have an X and a Y.
- Genes carried on the X chromosome are 'sex linked'. Males only have one copy of these genes, so recessive phenotypes are seen much more often in males.
- Where several genes combine to produce a phenotype, the effect is often continuous variation in the phenotype.

◆ Environmental differences also contribute to variation.

◆ Heritability is a measure of how much of the variation in a character is caused by genes.

◆ The genome is the whole of an organism's genetic material. At present we have more questions than answers about it.

Exercises

12.1. In the brown rat, a dominant allele (**R**) codes for resistance to the rat poison warfarin. The recessive allele (**r**) leads to rats which are susceptible.

(i) A homozygous, warfarin-resistant rat mates with a susceptible rat. What proportion of the offspring are likely to be resistant?

(ii) The offspring of cross (**i**) mate amongst themselves. What proportion of their offspring are likely to be resistant?

(iii) How would you discover whether a resistant male rat was homozygous or heterozygous?

12.2. Artificial insemination is widely used in dairy cattle. One excellent bull can become the father of thousands of cattle, over many years. The offspring of one Friesian bull suffered a high incidence of a rare, lethal, genetic disorder. Male and female calves were equally likely to be affected. However, it was only seen in calves whose mothers were fathered by that same bull. Try to explain the genetics behind this agricultural nightmare.

12.3. In the fruit fly, *Drosophila melanogaster*, grey body (**E**) is dominant to ebony body (**e**). Normal wing (**D**) is dominant to dumpy wing (**d**). A series of genetic crosses gave the following results. In each case, the father was a grey, normal fly, and the mother was an ebony, dumpy fly. What were the genotypes of the parents in each cross?

Cross	Offspring			
	Grey, normal	Grey, dumpy	Ebony, normal	Ebony, dumpy
(i)	19	22	18	21
(ii)	58	0	0	0
(iii)	36	41	0	0
(iv)	45	0	40	0

12.4. In humans, a rare mutant allele leads to deafness. Study the information in this family tree. A square represents a male, a circle represents a female, and shading indicates this type of deafness. Horizontal lines link sexual partners. A vertical line leads down from this, and branches out to their children.

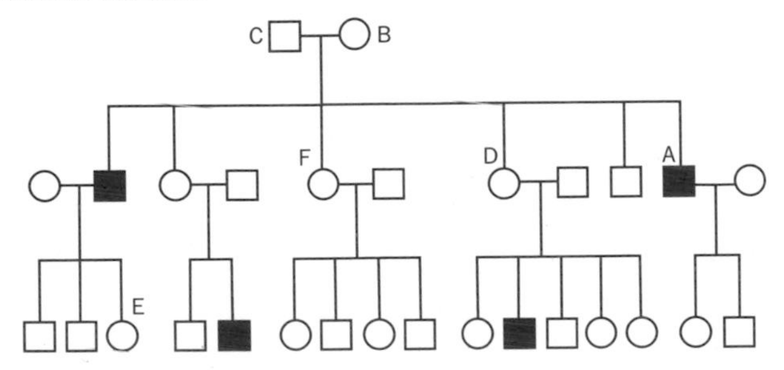

(i) Is this allele dominant or recessive? On which chromosome is it found? Give reasons for your answers.

(ii) Choose suitable symbols for the normal and mutant alleles. Then work out the genotypes of individuals A–F.

12.5. Use the following information to map the relative positions of five linked genes, A–E.

Pair of genes	Cross-over value
AD	3.6
AE	8.1
BC	11.0
BD	7.0
BE	2.1
CD	17.2
DE	5.2

Further reading

Textbooks to study:

Clark, M.S. and Wall, W.J. *Chromosomes* (London: Chapman & Hall, 1996). Clear, staight-forward genetics at the chromosome level. Topics covered include chromosome structure, mutation, mapping sex chromosomes, chromosomes in evolution

Hartl, D.L. *Essential Genetics* (Sudbury MA: Jones & Bartlett, 1996). Bright, clear and well explained. It goes well beyond the scope of this book, without an unbearable acceleration in pace.

A book to enjoy:

Jones, S. *The Language of the Genes* (London: HarperCollins, 1993). A literate, intelligent explanation of genetics for the general reader. Full of ideas and gentle wit.

Reproduction in Animal Life Cycles

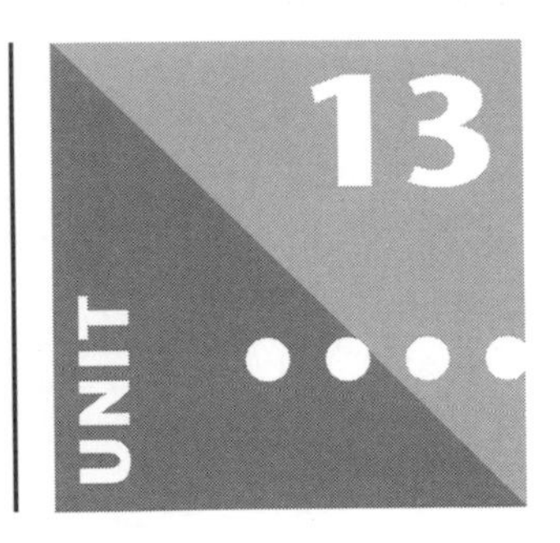

Connections

▶ Unit 11, which deals with cell division and life cycles in general, is an essential introduction to this unit. You should also have a basic knowledge of natural selection, cell structure, homeostasis and communication between cells (Units 2, 4, 15, 16 and 17).

Contents

13.1 Variations on a simple theme

Nearly all animals have the same type of life cycle. They are diplontic, that is mitosis only happens in diploid cells. Animal bodies are diploid. Meiosis makes gametes directly, with no mitosis in between. This is quite unlike the situation in plants, which have both haploid and diploid mitosis (they are haplo-diplontic). A very few animal species, including some bees and wasps, are haplo-diplontic (Box 13.1). When animals have very different bodies at different stages of the life cycle, they are usually stages in the development of a single diploid body. The various life cycles of insects, some with larvae and pupae, illustrate this (Box 13.2).

Almost all animals reproduce sexually. Some reproduce asexually as well, for example greenfly in summer. In this unit, we follow the stages of sexual reproduction in animals: gamete formation, mating, fertilization, and how the embryo is supported before it can live independently. In each case, examples of invertebrates and vertebrates are used, but mammals are always included. Then we explore the control and timing of reproduction, methods and significance of asexual reproduction, and finally the sex ratio: why do most species have equal numbers of each sex? We start at the beginning of the sexual process, with gamete formation.

13.2 Making sperm

Sperm are highly specialized, highly reduced, haploid cells. The haploid nucleus carries the genes which will become half the offspring's genome. The rest of the cell is a delivery system which attempts to get the sperm nucleus to the egg nucleus.

Sperm (= spermatozoa): The male gametes of animals.

Box 13.1

Haplodiploidy in insects

Many species of ants, bees and wasps (order Hymenoptera) have an unusual form of sex determination. This means that males are haploid whilst females are diploid. In this strange system, evolution has frequently led to true social behaviour. The insects cooperate in caring for the young, and other aspects of life. Individuals come in several distinct forms (castes) specialized for different roles. Some castes are sterile.

Females are diploid. They make haploid eggs by meiosis. Males are haploid. They make haploid sperm by mitosis. If an egg is fertilized, it develops into a female. Unfertilized eggs develop into males.

Males make gametes by mitosis, so they all have the same genes. This means that if a female mates with one male, her daughters will be unusually closely related. Relatedness is the probability that two individuals will share a particular gene as a result of being related (as opposed to by chance). This tricky concept is explained in Section 2.3. In sexual diploids, parent and offspring have a relatedness of 0.5. Full sisters have the same. However, in haplodiploid insects, full sisters have a relatedness of 0.75.

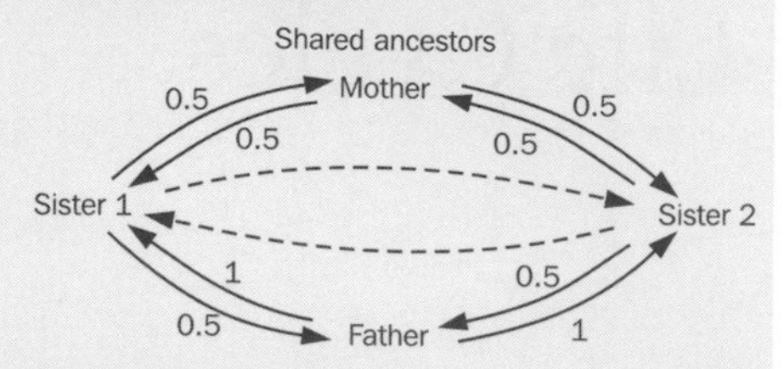

Notice that in haplodiploidy, the relatedness of father to daughter is not the same as daughter to father (can you think why?)

Sister–sister relatedness = $(0.5 \times 0.5) + (0.5 \times 1) = 0.75$

So, these sisters are more closely related to one another than they are to their own children. If sisters can live to reproduce more than once, there may be more benefit in keeping a sister alive than in replacing oneself. This is the basis of sterile castes.

In ants and bees, the common 'worker' caste consists of sterile females. They support a fertile sister, the queen and her offspring.

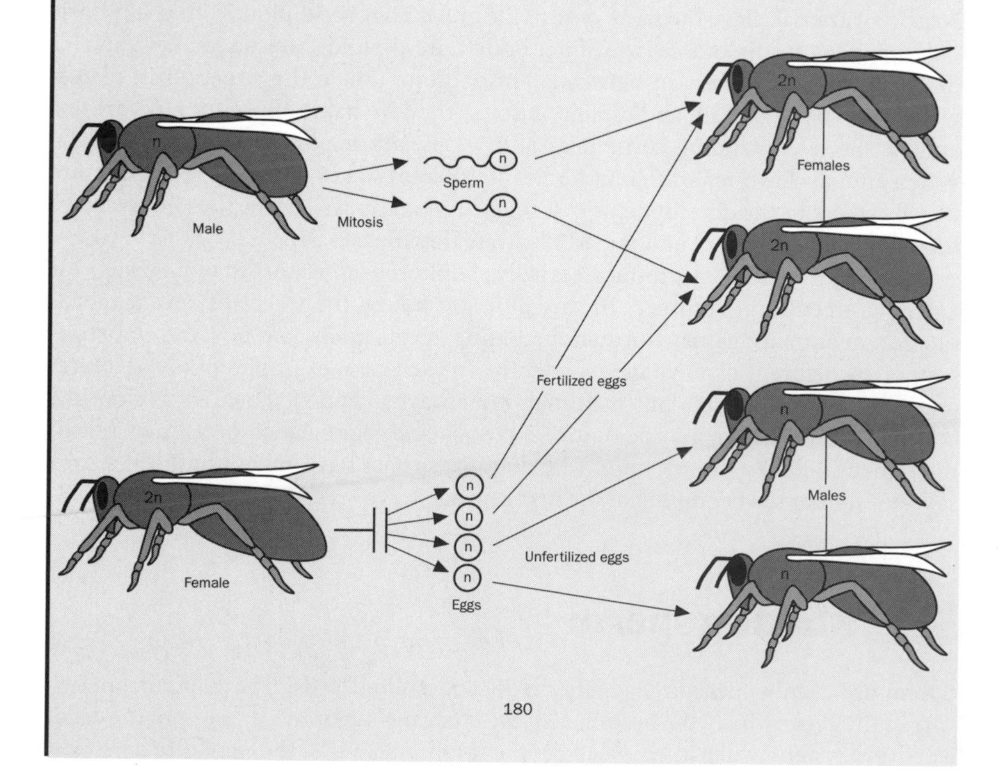

Box 13.2

Some insect life cycles

Insect life cycles are very diverse. Some of the differences are related to whether the adults and juveniles eat the same food or not.

Grasshopper (order Orthoptera)

Adults and juveniles look similar and feed on similar foods. Juveniles cannot fly.

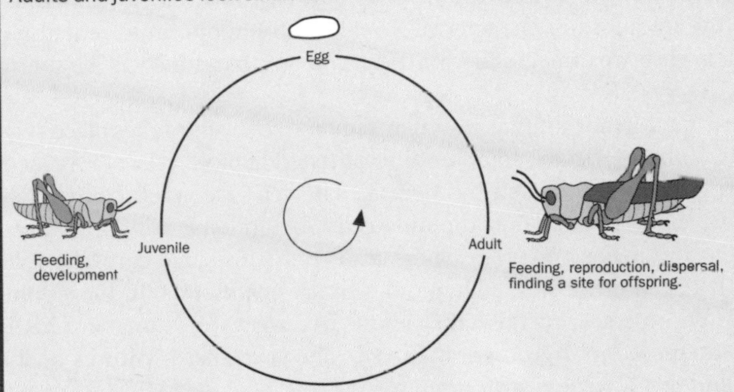

Fly (order Diptera)

Adults and juveniles feed on very different foods.

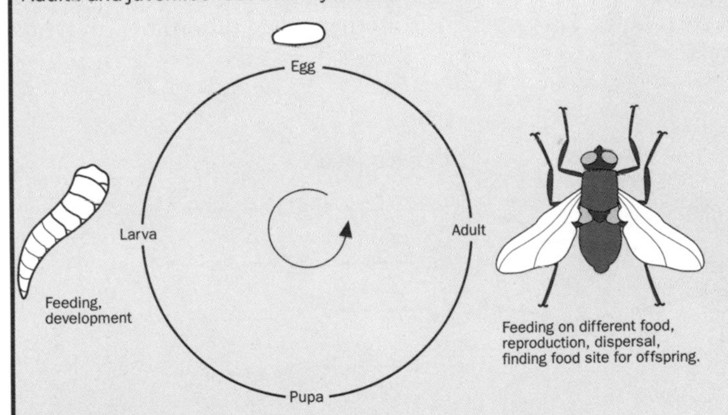

Mayfly (order Ephemeroptera)

Juveniles feed in water. Flying adults live a short time and do not feed.

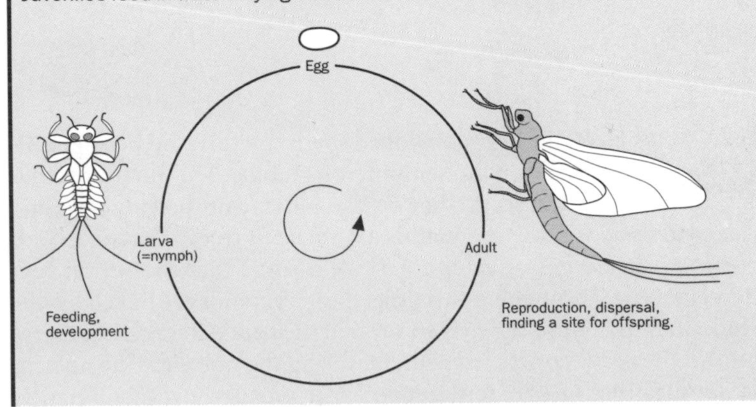

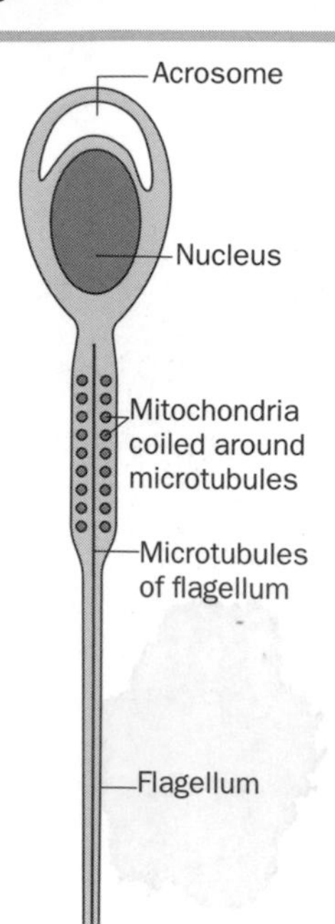

Fig. 13.1 A human sperm.

Fig. 13.1 shows a human sperm. The head contains the nucleus, which is small as nuclei go. The DNA is highly condensed. In some species, histones are absent in sperm, and other charged proteins bind DNA. Apart from chromosomes, there is little in the nucleus, keeping its volume small. In front of the nucleus is a large vesicle, the acrosome. Acrosomes contain digestive enzymes, much like lysosomes. Proteases, and enzymes which digest proteoglycans, are especially important. They digest the outer layers of the egg before fertilization. The tail of the sperm is a flagellum (Section 4.5: The cytoskeleton, Box 24.4). Mitochondria are coiled around its base, the midpiece of the sperm. There is no endoplasmic reticulum, Golgi or ribosomes. They are unnecessary burdens, and are abandoned as the sperm develops.

When sperm are made in any animal, a similar sequence of events takes place (Fig. 13.2). There are two phases. Firstly, some undifferentiated haploid cells are formed, by several rounds of mitosis followed by meiosis. The cells are given confusingly similar names at each stage. The diploid cells before meiosis are called **spermatogonia**. During meiosis they are **spermatocytes** (this stage can take four weeks in humans). The undifferentiated haploid cells are **spermatids**. In the second phase, spermatids become sperm: this takes about five weeks in humans. This is purely cell differentiation. No division is involved. The specialized features of the sperm are made, and the unwanted structures are destroyed.

Most animals make sperm in a highly organized testis. In mammals, the testis is made up of several hundred tightly coiled tubes, the seminiferous tubules (Fig. 13.3). Sperm are formed in the tubules. They open into another coiled tube, the epididymis,

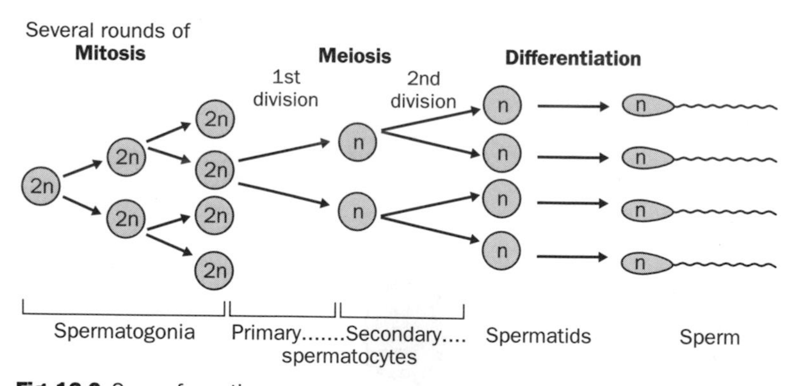

Fig. 13.2 Sperm formation.

which is outside the testis. Here, the fully-developed sperm mature and become able to swim. Fig. 13.4 shows a section through a seminiferous tubule. The outermost cells are spermatogonia, dividing by mitosis. As they divide, cells are pushed inwards, and continue to divide. After meiosis, the spermatids are shielded from their surroundings by Sertoli cells, which have long extensions. We assume that the Sertoli cells control sperm development. Throughout this time, the spermatids are linked by fine cytoplasmic threads: they did not completely separate at meiosis. After meiosis everything that happens to the cell is concerned with delivering the nucleus. The nucleus itself is ready for fertilization. ***In vitro* fertilization** experiments show that a newly-

Spermatogonia: Diploid cells in the testis which divide by meiosis, in order to make sperm.

Spermatocytes: Sperm-forming cells in the process of meiosis.

Spermatids: Haploid cells which are differentiating as sperm.

***In vitro* fertilization:** Fertilization under laboratory conditions, outside the body.

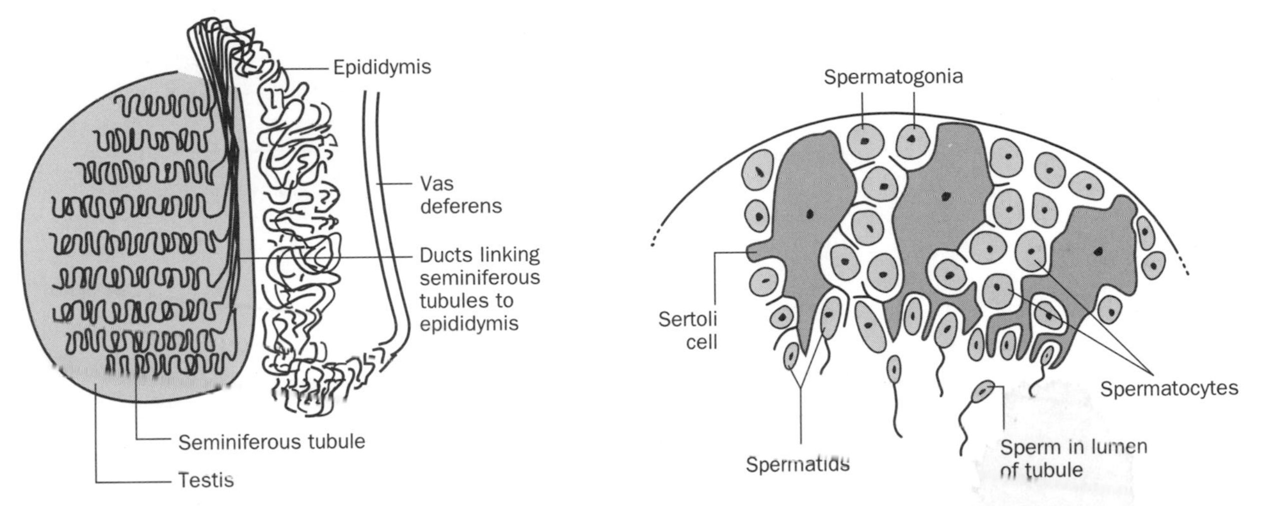

Fig. 13.3 Structure of the mammalian testis.

Fig. 13.4 Part of a seminiferous tubule, in section.

formed spermatid can lead to a viable embryo if injected right into an egg. Only the delivery system is immature.

In some simple animals, sperm are not made by a distinct organ. Hydrozoans (phylum Cnidaria) are a good example. The polyp of *Hydra*, and the medusa larvae of other hydrozoans make sperm in a region of the body wall. Unspecialized cells in the outer layer start to divide, first by mitosis, then by meiosis. The result is a mass of sperm between the two layers of the wall.

13.3 Making eggs

An **egg** is a large haploid cell. Just how big it is depends on the species, but an egg's volume is always many times greater than the sperm's. Sperm and egg make equal genetic contributions to the zygote, but the egg provides much more cytoplasm. If the sperm's cytoplasm is a stripped down delivery system, the egg's cytoplasm is a food parcel for the embryo. The cytoplasm contains clusters of proteins and lipids: the yolk. The bigger the egg, the more food it contains, so the bigger the embryo can grow before it has to fend for itself. The large, shelled eggs of birds are an extreme. Mammalian eggs are much smaller, because the embryo can get food from the mother through the placenta from an early age (Section 13.7). The primitive echidna and duck-billed platypus are exceptional egg-laying mammals: they have large eggs.

In some simple animals, egg cells develop on their own. In most animals, including the vertebrates, they are surrounded by a layer of epithelial cells: a follicle. Fig. 13.5 shows a mammalian egg developing in its follicle, and after it has been released. Even after release, the egg is surrounded by two layers. The outer layer is a covering of follicle cells. The inner layer, the zona pellucida, is a thick extracellular matrix rich in glycoproteins.

Egg-making involves several rounds of mitosis, followed by meiosis (Fig. 13.6). The cells entering meiosis are called **primary oocytes**. In most animals, meiosis halts at prophase I. This is when the cytoplasm fills up with food and expands. In most

Egg (= ovum): The female gamete of an animal.

Primary oocyte: An egg-forming cell in the first stage of meiosis.

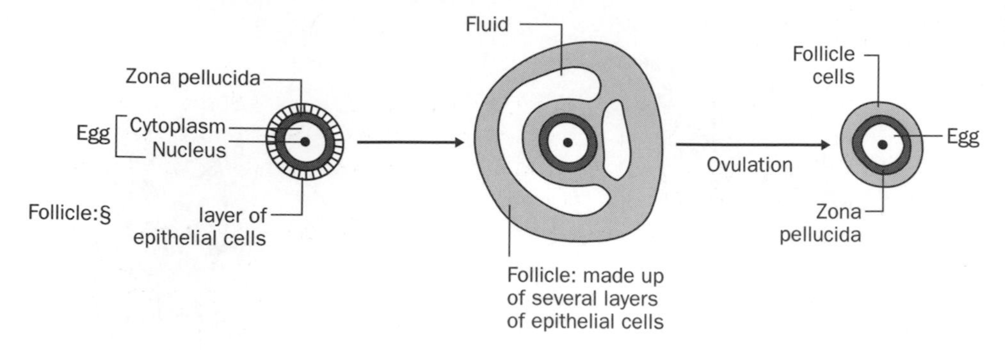

Fig. 13.5 The mammalian egg develops inside a follicle.

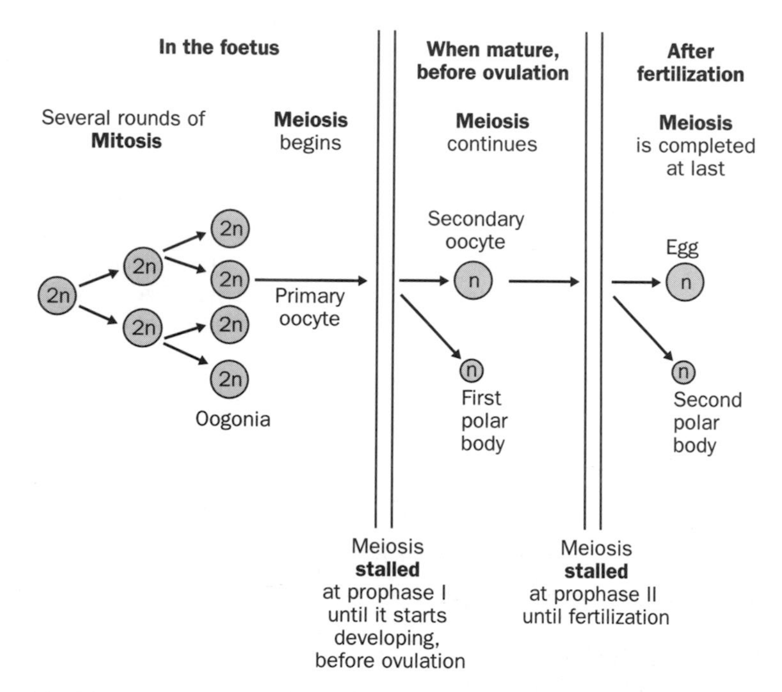

Fig. 13.6 Egg formation.

animals, including the vertebrates, proteins and lipids are imported from surrounding cells by endocytosis. In mammals, meiosis stalls at prophase I for an indefinite period. All the primary oocytes are formed in the ovary before birth. They only grow and carry on dividing just before they are released. The first meiotic division is unequal. The nucleus divides equally, but almost all the cytoplasm stays with one nucleus. The other nucleus is essentially a waste product: the first **polar body**. Only the large cell undergoes the second meiotic division. Again it is unequal, making the haploid egg cell and a second polar body. The point of this is to make big gametes, rather than many gametes. This is at the heart of being female.

Polar body: Tiny cell made during egg formation, containing an unwanted nucleus.

Vertebrates and many invertebrates have specialized ovaries which make eggs, but as with testes this is not the case in the simplest animals.

13.4 Mating

Sperm and eggs must get together. Mating is an important part of this in most, but not all animals. Fertilization may occur inside or outside the female body. Mating is essential: something has to put the sperm in there. Animals with external fertilization may or may not go in for mating.

Many marine invertebrates release their gametes into the water. There is no need for the parents to make direct contact. The sperm swim to the eggs. In some species the eggs release chemicals which attract sperm. In sea urchins, the attractant is a short peptide. Other species may rely on sperm swimming in random directions.

Frogs have external fertilization, yet a form of mating takes place. A male clings to a female's back for hours, until she releases eggs. Inmmediately, he covers them with semen. Even fish, which cannot easily hold on to one another, may have courtship leading to a pairing of male and female, before eggs are laid and fertilized. Courtship is an important prelude to mating. It gives females an opportunity to assess the quality of a male. Also, it allows time for coordination of the male and female reproductive organs. Sperm and eggs must be ready at the same time.

Internal fertilization requires direct contact between the sexes. In most birds, the openings of the reproductive systems are simply brought together while sperm are transferred. This is a brief encounter, typically a very few seconds. Far more commonly, a male has a penis, that is an organ which can be put into the female

How do porcupines mate? Very carefully.

reproductive opening. Snails, insects and mammals are typical examples. A penis may be more than a stiff tube for sperm injection. Some dragonfly penises have attachments for pushing sperm from previous matings out of the way, or scraping them out altogether.

Mating in mammals is a subject close to our own hearts, and we know a great deal about it. Here, we follow events from copulation (the sexual act itself) to the time when sperm meet eggs. Humans are specifically mentioned, but much of this refers to any mammal. Figs. 13.7 and 13.8 show the male and female reproductive systems.

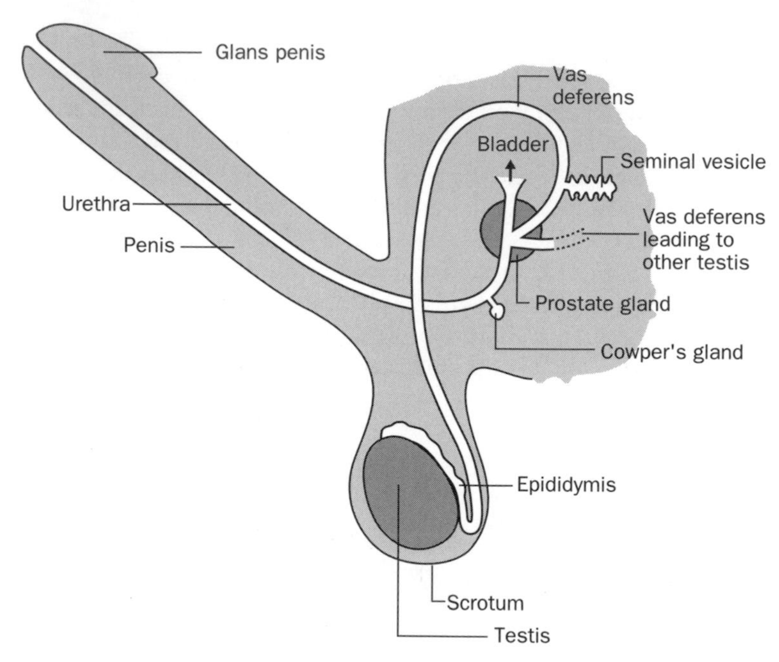

Fig. 13.7 Human male reproductive system.

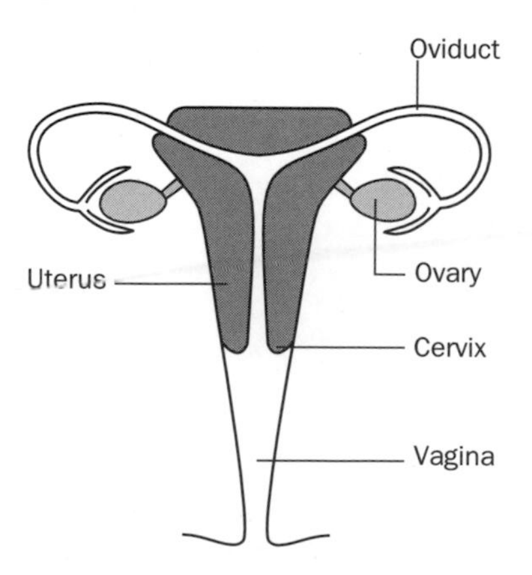

Fig. 13.8 Human female reproductive system.

Sexual excitement leads to changes in both sexes. In the male, the penis becomes erect, that is longer and stiffer. It has three blood-filled chambers running length-wise. Blood pressure rises in these chambers, making the penis erect (Box 13.3). In the female, the lining of the vagina begins to secrete more mucus than normal. Together, these changes mean that the penis can enter the vagina without buckling, and can move back and forth freely. As it moves, the glans of the penis is stimulated. This eventually leads to ejaculation of semen.

Box 13.3

Erection? Just say NO

Nitric oxide (NO) is a local signalling molecule, discovered quite recently (Section 16.5). One of its roles is in controlling erection of the penis. This relies on increasing blood flow into the penis.

The penis has three large sinuses, spaces which can fill with blood. If the arteries leading to the sinuses dilate, more blood flows into them. They swell, and the penis becomes longer and stiffer.

Erection is controlled by part of the parasympathetic nervous system (Section 17.6). When stimulated, the nerve fibres release NO. It causes smooth muscle in the walls of arteries to relax. The arteries dilate, causing erection.

NO is an effective signalling molecule because it is hydrophilic and can pass right into the target cell, where it binds to an enzyme. This releases a second messenger in the cell (Section 16.4), in this case cyclic GMP. It is short-lived, because it is easily oxidized: this makes it effective on a local scale.

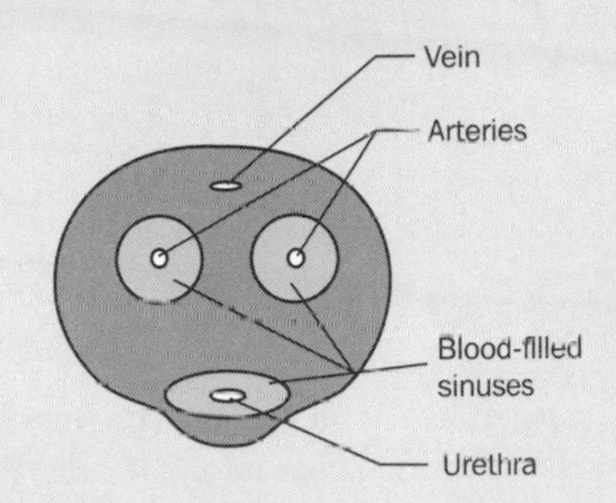

Section through the penis

Semen is a mixture of sperm and fluids secreted by several glands. The sperm are stored in the vas deferens under acidic conditions. This inhibits movement. They are mixed with other secretions in the seconds leading up to ejaculation. These secretions contain several things: alkalis to neutralize the acid, allowing sperm to swim; sugars and other nutrients; fibrinogen, the plasma protein which forms clots (Section 23.3); an enzyme which promotes clotting; and another enzyme which digests fibrin clots. The last three items ensure that semen forms sticky masses on ejaculation. These stay in place high in the vagina, and are digested slowly to release the sperm. The whole business can give intense physical pleasure to both sexes: the reasons are discussed in Box 13.4.

From the top of the vagina, sperm must get into the uterus through a narrow opening, the cervix. Then they must pass through the uterus and up one of the oviducts. Fertilization takes place in the oviduct. This journey is only possible because sperm can swim. However, they are helped through the uterus by rippling movements of its lining, the endometrium, and perhaps by rhythmic contractions of the uterus wall.

Box 13.4

Why does sex feel good?

In both men and women, sexual intercourse can lead to intense physical pleasure: an orgasm. In men, this coincides with ejaculation. There is some evidence that other species of mammal also experience orgasms. Why?

At first sight, orgasms make good evolutionary sense. If something has survival value, linking it with a pleasant feeling will encourage it. It is no surprise that food smells nice but faeces smell horrible. Eating food is necessary for survival; eating faeces could lead to disease. In the same way, mating has survival value, so orgasms might encourage it. This would be particularly relevant to males, because they produce more gametes and, in most cases, can leave more offspring if they mate with many females.

However, intercourse leads to orgasms far more often in men than women. Some surveys suggest that, over a whole population, only a minority of matings lead to female orgasm. Perhaps we should not be looking for a function for the female orgasm.

Men and women belong to the same species. They develop as variants on the same body plan, controlled by the same genome. A man's waist and hips are subtly different from a woman's, but they develop from the same structures in the embryo. In the same way, the male penis and the female clitoris develop from the same structure. The clitoris has blood sinuses allowing erection, and lots of sensory neurones, much like the penis. It is stimulation of the clitoris which leads to orgasm. Perhaps a female orgasm is not adaptive at all. It may simply be a consequence of the way the body develops (Box 2.4).

13.5 Fertilization

Once sperm have reached the egg they have to get into it. In mammals, they have to cross the covering layer of follicle cells, the zona pellucida, and the plasma membrane (Figure 13.9).

Once they reach the egg, sperm go through a phase of chemical change. This is poorly understood, but new proteins appear in the plasma membrane, and the sperm's swimming movements become more violent. They can then force their way between the follicle cells to the zona pellucida. Sperm have receptors in the plasma membrane for one of the glycoproteins which makes up the zona pellucida. When they bind, the enzymes in the acrosome are unleashed, by exocytosis. Enzymes digest the zona pellucida, exposing the plasma membrane. Quite how many sperm reach the egg is unclear, but there are hints that a single sperm cannot get as far as fertilization on its own. Lots of acrosome reactions at once might be needed.

Sperm also have receptors for a protein in the egg's plasma membrane. When they bind, the sperm and egg membranes can fuse. The sperm nucleus is released into the egg cell cytoplasm. Fertilization has now happened: the zygote is formed. At this stage the zygote has two haploid nuclei, from the egg and sperm. They never fuse. Instead, the cell goes into mitosis. First, a spindle is formed. It is organized by the centrosome from the sperm, which replicates when it enters the egg. Then both haploid nuclei disintegrate, and their chromosomes attach to the spindle. From now on, they behave as a single, diploid set of chromosomes.

What stops other sperm fertilizing the egg too? When the first sperm fuses with the egg plasma membrane, it triggers two changes. The first is electrical. The plasma membrane is polarized before fertilization. The voltage across the membrane is probably set up in the same way as in neurones (Section 17.3). On fertilization, the mem-

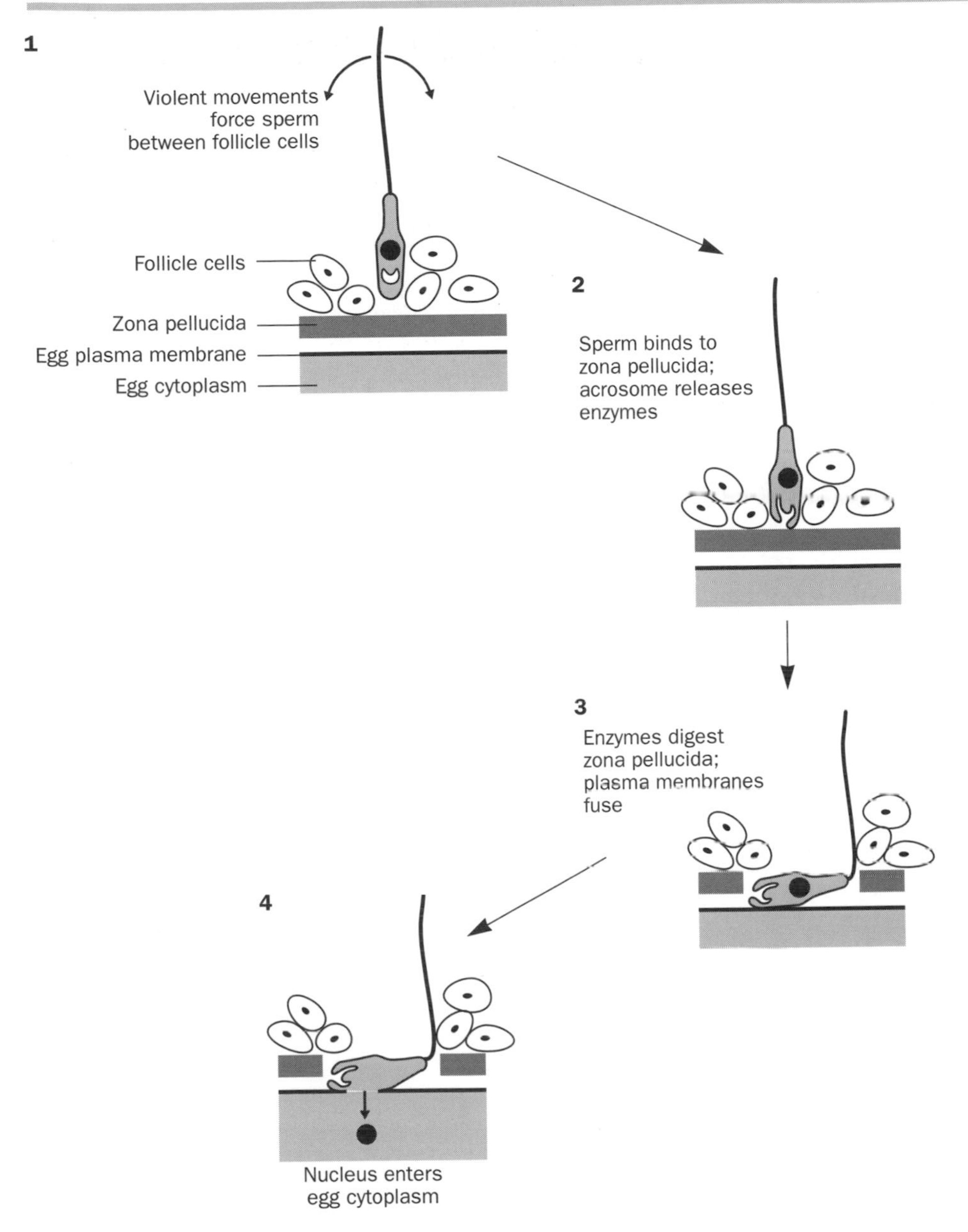
1
Violent movements force sperm between follicle cells
Follicle cells
Zona pellucida
Egg plasma membrane
Egg cytoplasm
2
Sperm binds to zona pellucida; acrosome releases enzymes
3
Enzymes digest zona pellucida; plasma membranes fuse
4
Nucleus enters egg cytoplasm

Fig. 13.9 Fertilization in mammals.

brane depolarizes. This seems to prevent other sperm binding. Depolarization is quick but short-lived. The second change is slower but longer lasting. Vesicles under the egg's plasma membrane contain enzymes which modify some of the glycoproteins of the zona pellucida. They are released at fertilization. The modified glycoproteins cannot be recognized by sperm, so from this point on no more of them should bother the egg. The act of fertilization causes the vesicles to fuse with the plasma membrane through one of the cell's usual signalling pathways (Section 16.4). Inositol triphosphate is released from the plasma membrane, which triggers calcium release into the cytoplasm. The pulse of calcium is the signal for vesicles to fuse.

13.6 The egg as a home

Eggs are large, as cells go. We can generally see them, or what they become, for some time after fertilization. All eggs contain food, the yolk, to support early growth of the embryo. There must be enough to last until the embryo can get its food another way. In most animals, apart from mammals, the embryo leaves to feed itself once the yolk runs out: the egg 'hatches'. Until that happens, the egg is the embryo's home.

Amphibians and birds have two very different types of egg (Fig. 13.10). The amphibian egg is simple, and much more like the eggs of invertebrates. Think of frog spawn. It is made up of blobs of jelly, each with a dark centre. The centre is a yolky egg cell, about one millimetre across. The jelly is a protective layer added after the egg leaves the ovary. After fertilization, the cell divides unevenly, so that most of the yolk ends up in a few big cells. The other cells go on to make the frog's body, using the reserves of the yolk cells. By the time the embryo is a tadpole ready to swim off, it fills the space occupied by the egg. The egg has become the embryo.

Birds' eggs are much more complex. Outside, they have a hard shell added after fertilization. It allows gas exchange, despite being rigid. Inside, there are two obvious compartments, the yolk and the albumen ('white'). The yolk is a huge egg cell. The albumen is a water store, contains some extra protein, and cushions the yolk. After fertilization, the yolk does not divide completely. The first mitotic division makes two diploid nuclei which are boxed off by plasma membranes to form small cells. They form an embryo perched at one end of the yolk. Then tissues belonging to the developing gut grow out around the yolk, and blood vessels link it to the body. As the embryo grows, both yolk and albumen shrink. Another layer of cells grows out from the embryo to form a pouch, the allantois. This is used for gas exchange. It is

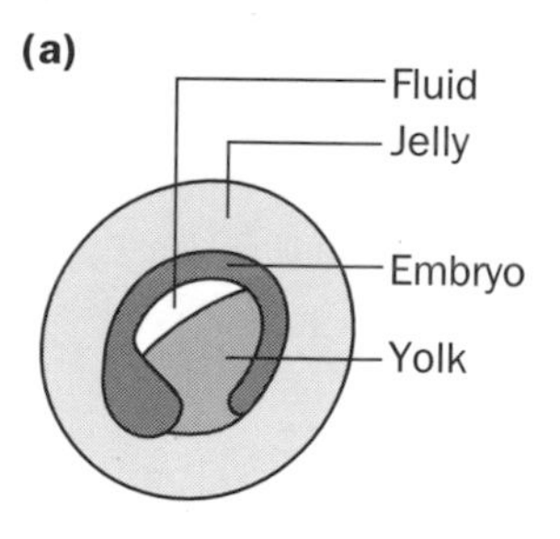

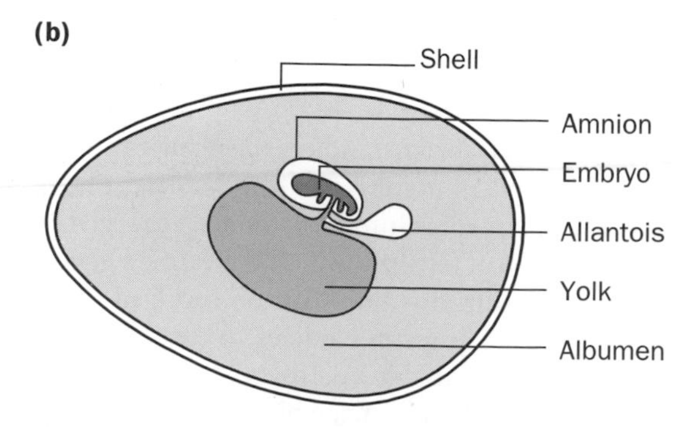

Fig. 13.10 Embryos developing within eggs. (a) amphibian, (b) bird.

also used as a dump for nitrogenous waste (Section 22.2) which is left behind on hatching. Reptiles and the primitive egg-laying mammals have very similar eggs to those of birds.

Most mammals have small eggs with little yolk. The embryo soon needs food from outside. In the eutheria, the biggest group of mammals, this food comes from the mother, via the placenta.

13.7 Life in the uterus

Eutherian mammals have small eggs. The uterus, not the egg, is the embryo's home as it feeds and grows. During pregnancy, the way in which the embryo feeds goes through three stages.

The embryo arrives in the uterus soon after fertilization, about three days in humans. It has been carried down the oviduct in a gently flowing fluid. The flow is set up by cilia on the epithelial cells lining the oviduct. Once in the uterus, it needs food. For the first day or two, it uses nutrients secreted by cells of the endometrium, the uterine lining. Then the embryo starts to feed on the endometrial cells themselves. A layer of cells on the outside of the early embryo secretes enzymes which digest the endometrium. At this stage, the endometrial cells are swollen and store plenty of nutrients. The embryo sinks into the endometrium as it digests it: it has implanted. After implantation, the embryo's tissues invade the uterus wall. Over a period of weeks, the placenta develops. By the time the placenta is mature, the embryo is recognizable as a young mammal: it is now called a foetus. The placenta is an amazing organ. It is an intimate mixture of the foetus' and the mother's tissues. The point is to bring the two blood systems very close together to exchange nutrients and waste products. The placenta is connected to the foetus by the umbilical cord, which contains arteries and a vein. In the placenta, they branch into a system of villi. Foetal blood flows through capillaries in each villus. Around the villi, maternal blood flows in big open cavities, blood sinuses (Fig. 13.11).

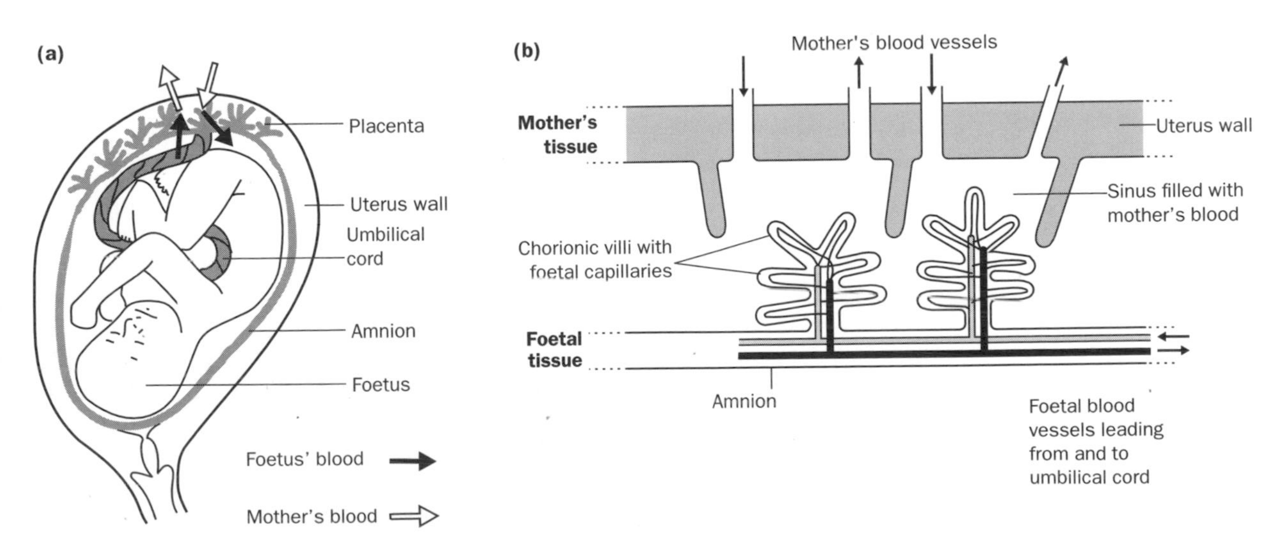

Fig. 13.11 The placenta. (a) relationship between foetus, mother and placenta, (b) blood flow in part of the placenta.

Transport across the placenta seems to be simply by diffusion. Glucose, fatty acids and mineral ions diffuse from mother to foetus. Nitrogenous waste, including urea, uric acid and creatinine (Section 22.1) diffuse out. Carbon dioxide also leaves the foetus, oxygen diffuses in.

A long period in the uterus means that mammals are born relatively large. Birth is a major upheaval. Very simply, it involves strong rhythmic contractions of highly organized muscles in the uterine wall. The cervix dilates to allow the foetus through. The head or front legs, depending on the species, act as a wedge to push the birth canal open. This sounds easy, but requires a level of coordination which certainly involves several hormones and probably includes a positive feedback system within the uterus wall. There is a lot of scope for things to go wrong. Birth is a dangerous time for mother and foetus, as farmers and 19th Century graveyards testify.

13.8 Controlling reproduction

The events of reproduction take place in cycles. In animals which reproduce more than once, the cycles take place within a single lifetime. An animal reaches maturity, then goes through cycles of making and releasing gametes. Other species reproduce, then die. Here, one stage of the life cycle is the reproductive phase. Either way, cycles may be tied in to the changing seasons. Some times of year are good for having babies. A season when food is plentiful and the weather is not too harsh might be ideal. In Western Europe, spring is that time: birds are nesting, rabbits are breeding like rabbits and, they say, a young man's fancy turns to love. Moreover, it makes sense for individuals to go round the cycle together. There is no point having gametes ready for action now, if all potential mates will have theirs ready in six months' time.

Reproduction needs to be regulated. There must be some system in animals' bodies to control what happens when, and signals from outside to set off the system. When reproduction is tied to the seasons, day length is a reliable signal. Patterns of rainfall and temperature may vary from year to year, but day length does not. Animals ranging from polychaete worms to mammals have been shown to respond to day length.

In mammals, the internal control system has several levels, involving the nervous system and hormones. As in many other situations, the hypothalamus (part of the brain) and the pituitary gland are involved (Section 16.3). The hypothalamus secretes gonadotropin releasing hormone. This stimulates the anterior pituitary to release the gonadotropic hormones: luteinizing hormone (LH) and follicle stimulating hormone (FSH). These hormones act on the ovary. Simply, they stimulate follicle development and ovulation (the release of eggs). The follicles and other ovarian tissues make two more hormones, oestrogen and progesterone, when they are stimulated. These hormones control changes in the ovary which prepare it for pregnancy. They also have an effect on the hypothalamus and pituitary, decreasing release of LH and FSH. This is an example of negative feedback (Box 15.1). LH and FSH set off a chain of events which eventually leads to less of these hormones being produced. Box 13.5 shows how the human female sexual cycle is controlled in more detail.

In humans, the female sexual cycle carries on all year, with ovulation about once a month unless the woman becomes pregnant. In mammals such as sheep, which reproduce seasonally, there is an extra level of control. Nerve signals from the eyes do not simply go to the parts of the cerebral cortex dealing with sight (Section 16.4:

Box 13.5

The female sexual cycle

Events in the ovaries and uterus are coordinated by hormones. They need to be, if the uterus is to be ready for a fertilized egg.

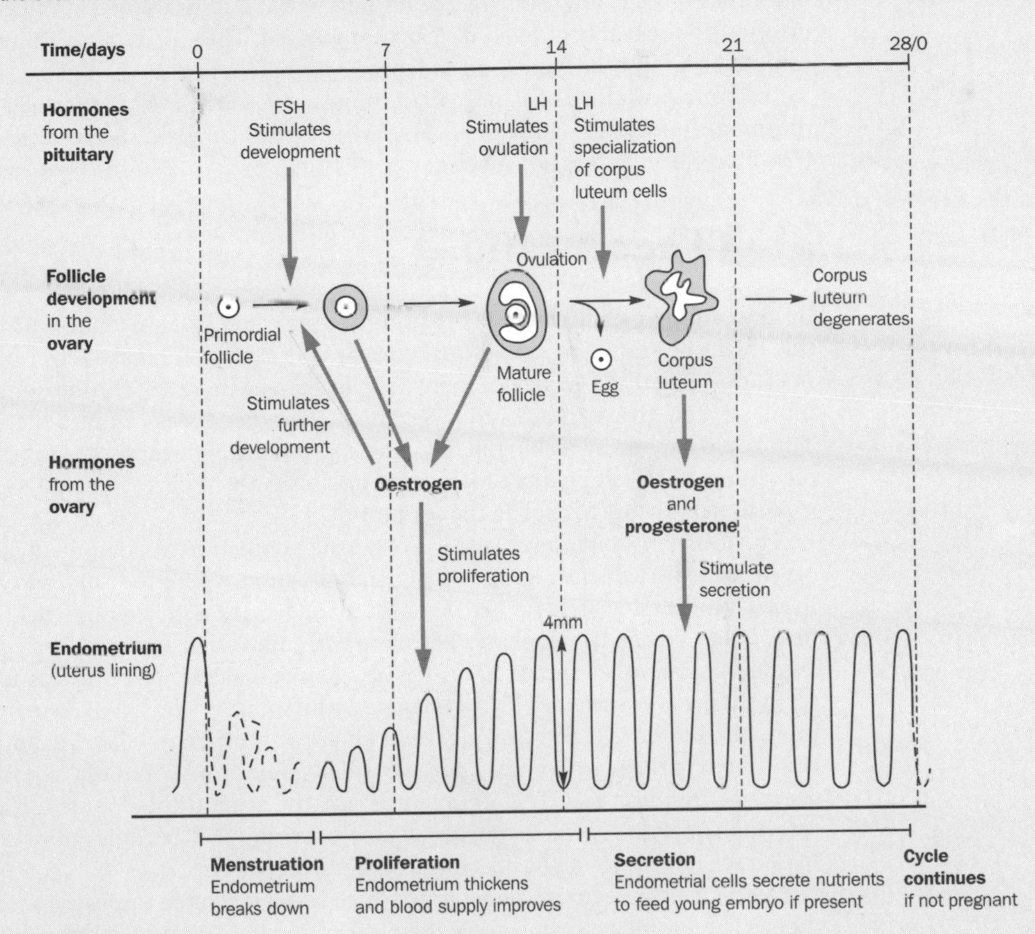

Controlling the cycle

Crudely, the anterior pituitary controls events in the ovary by FSH and LH. The follicle and corpus luteum control events in the uterus by oestrogen and progesterone. This leaves a big question: what controls the pituitary? We know that the hypothalamus secretes gonadotropin-releasing hormone, but this does not explain why there is a cycle. We need some sort of feedback loop (Box 15.1).

Oestrogen and progesterone control the cycle by rather complex feedback effects on the pituitary and hypothalamus.

1. Early in the cycle, FSH and LH are released. FSH stimulates follicle development.
2. Before ovulation, oestrogen seems to *stimulate* secretion of FSH and LH. As the follicle grows and releases more oestrogen, FSH and LH rise to a sharp peak in the middle of the cycle. This LH surge stimulates ovulation. It also triggers the development of a corpus luteum from the remains of the follicle.
3. After ovulation, oestrogen and progesterone *inhibit* FSH and LH secretion. The corpus luteum also makes another inhibitory hormone, inhibin.
4. As FSH and LH fall, the corpus luteum degenerates. Progesterone and oestrogen levels fall, so the pituitary is able to release FSH and LH again as the cycle starts.

Why does menstruation stop during pregnancy?

Menstruation in pregnancy would be a disaster, washing the embryo away with the endometrium. It is prevented because the embryo makes a hormone called chorionic gonadotropin, which behaves like LH. This maintains the corpus luteum, and so maintains oestrogen and progesterone levels. Much later in pregnancy, the placenta itself makes progesterone.

Common themes in cell signalling). They also go to a group of cells in the hypothalamus, and on to the pineal gland nearby. These parts of the brain detect day length. The pineal gland secretes melatonin, which inhibits release of LH and FSH. By varying melatonin release, the pineal gland can make the animal ovulate at particular times of year. Interestingly, the human pineal also makes melatonin, although no effects on sexuality have been clearly established. There is certainly more to melatonin than simply the regulation of reproduction. It appears to be linked, somehow, to the biological clock of mammalian brains. This is the system that tells you it is morning, lunchtime or bedtime, and which gets confused in jet-lagged people. However, the picture is not yet clear.

13.9 The sex ratio

In most animals, about half the offspring are male and half are female. At first sight this might be surprising. Males make many sperm; females make fewer eggs. Why not have a few males to service many females? Even where sex chromosomes determine sex and give a 1:1 sex ratio (Section 12.4) other genes can bias the sex ratio. This could happen by making sperm carrying, say, the X chromosome more likely to die young, or by killing more foetuses of one sex early in development. We need an evolutionary theory to explain the sex ratio.

The theory is simple and elegant. Forget the population for a moment, and focus on individuals and their genes. Imagine a situation where the sex ratio is highly biased towards females, say ten females to one male. The average male gets ten mates, the average female shares her mate with nine other females. Every newborn animal has a mother and a father, so the average male has ten times as many offspring as the average female. Now imagine a mutant gene which makes an individual more likely to have male offspring. Individuals with the gene will have more grandchildren than those without it. The gene will become more common, and males will get more common too. The argument is just the same if the population is biased towards females. A gene favouring female offspring will become more common. Biased sex ratios seem to be unstable in evolutionary terms.

Occasionally one finds a population which is clearly biased towards one sex or the other. The problem is explaining these exceptions to the rule. It is an interesting business, and can turn up fascinating insights into the reproduction and life cycles of individual animal species.

Summary

- In almost all animals, mitosis only takes place in diploid cells.
- Gametes are made by meiosis.
- Animal sperm are small haploid cells with a small nucleus, a flagellum, a vesicle containing digestive enzymes, but very little other cytoplasm.
- Sperm are made in seminiferous tubules, within the testes.
- Haploid spermatids are made by meiosis, which then develop the characteristic structure of sperm.
- Eggs are made by ovaries.

- Eggs are larger than sperm. They contain food reserves, the yolk. This feeds the embryo after fertilization.
- The large-shelled eggs of birds and reptiles provide a large amount of food in a protected environment. Mammalian eggs are relatively small, because the embryo feeds from the mother via the placenta.

◆ Most eggs develop in a layer of diploid cells: the follicle.

- Mammalian eggs leave most of the follicle behind in the ovary at ovulation.
- Fertilization may take place inside or outside the female body, according to species. Mating must take place before internal fertilization.
- A penis is a device for introducing sperm to a female body. The mammalian penis becomes erect by filling with blood.
- In mammals, fertilization takes place in the oviduct.
- A system of receptors allows the sperm to recognize the egg before it breaks down the outer covering with digestive enzymes and its plasma membrane fuses with the egg's.
- Two systems prevent further sperm fertilizing the egg.
- The embryo develops in the uterus. It feeds first on nutrients secreted by the uterus lining. Then it grows into the uterus wall, feeding on the cells themselves. A placenta is formed, in which substances are exchanged between mother's and foetus' blood.

◆ The timing of reproduction is controlled. Many species use day length as a reliable guide to the time of year.

- In mammals, the female sexual cycle involves coordinated changes in the ovary and the uterus.
- In the ovary, follicles develop until the egg is released. The follicle becomes a corpus luteum, which eventually breaks down.
- In the uterus, the endometrium becomes thicker, and its blood supply improves. After ovulation, the endometrium secretes nutrients in case an egg is fertilized. If no egg implants, the endometrium breaks down and is washed out through the vagina by blood.
- The hypothalamus controls the ovarian cycle, by FSH and LH released by the anterior pituitary. The follicle and corpus luteum control the cycle in the uterus, by oestrogen and progesterone.

◆ A feedback system ensures that the cycle continues.

◆ Most animal species have a 1:1 sex ratio. This is the result of natural selection.

Exercises

13.1. What are the advantages of internal fertilization?

13.2. A human sperm fertilizes an egg. List in order all the structures it has passed through, from the time it was made.

13.3.

(i) Why are birds' eggs bigger than fish eggs?

(ii) Why are birds' eggs bigger than mammals' eggs?

13.4. Many cases of female sterility are caused by the pituitary secreting too little of the gonadotropins, FSH and LH.

(i) Explain how this could lead to sterility.

(ii) One treatment for this condition is to give the patient the hormone chorionic gonadotropin. Explain why this works.

13.5. Most contraceptive pills contain oestrogen and progesterone. Explain how they work.

13.6. Dairy farmers make sure that their cows have calves at intervals, so they keep on producing milk. Female calves are more valuable than male calves, because they can be kept for milk production. A sex ratio biased towards females would be desirable in dairy cattle. In theory, could breeders select for this characteristic, or does the argument of Section 13.9 make this unlikely? What else would you need to know?

Further reading

Barnes, R.S.K., Calow, P. and Olive, P.J.W. *The Invertebrates: a New Synthesis* (2nd ed.) (Oxford: Blackwell Science, 1993). This book contains a good chapter on invertebrate life cycles and reproduction.

Guyton, A.C. and Hall, J.E. *Textbook of Medical Physiology* (9th ed.) (Philadelphia: Saunders, 1996). I like this hefty textbook for its thoroughness and clear, no-frills diagrams. A good general physiology book for mammalian reproduction.

Guyton, A.C. *Human Physiology and Mechanisms of Disease* (5th ed.) (Philadelphia: Saunders, 1992). Effectively a shorter, boiled-down version of Guyton and Hall.

Johnson, M.H. and Everitt, B.J. *Essential Reproduction* (3rd ed.) (Oxford: Blackwell Science, 1988). A more specific, detailed introduction to reproduction in mammals, written for medical, veterinary and biology students.

Reproduction in Plant Life Cycles

Connections

▶ You should have a clear understanding of the principles of life cycles and cell division (Unit 11) before studying this unit.

Contents

14.1 Plant reproduction

You are sitting high on a hillside on a sunny spring morning. Admire the view, then look down at the grassland around you. It is a seething mass of plants, fragrantly engaged in the business of reproduction. The grasses are flowering. Their tiny flowers are hanging out their male and female reproductive organs, releasing and catching pollen grains: microscopic haploid males. Among the grasses, there are other plants which have put resources into larger, more colourful flowers to attract animals which will carry the pollen for them. Spring-flowering plants have already made seeds: baby diploid plants in an easily dispersed case. These are shaken, blown or carried across the hillside. Some will find a suitable spot to grow up, and reproduce for themselves.

Look closely at these grass plants. They are clumps of similar units: leaves, a tiny stem and roots. These units are produced asexually. Who knows whether this plant started life as a seed or as part of another clump? Either way, asexual reproduction is far more common in plants than in animals. Now rummage around in the grass. Push the dense mat of leaves to one side and try to see the soil. Almost every grassland has mosses down in there, hidden from view. The moss is almost certainly reproducing asexually, but it is likely to have sexual organs as well. However, they are microscopic. Spores are made in a clearly visible capsule, and dispersed, but they are not seeds and not pollen. A spore is a haploid cell which will grow into a haploid body, the leafy bit of a moss. Yet the capsule is diploid. Clearly, we need to understand plant life cycles before we can make sense of the details of plant reproduction. They often feel strange and unfamiliar. This is not surprising: we are animals, not plants.

This unit examines reproduction in the life cycles of a range of plant groups, as well as some of the more plant-like protoctists. Then the reproduction of the most successful group of modern plants, the flowering plants, is looked at in more detail.

14.2 Plant life cycles

In plants, both haploid and diploid cells can divide by mitosis (Section 11.2). This means that plants can have two different bodies in the same life cycle (Fig. 14.1). The haploid body produces gametes by mitosis. It is called the **gametophyte** (gamete plant). After fertilization, the zygote divides by mitosis to form a diploid body, the **sporophyte**. This makes haploid spores by meiosis. After they have dispersed, spores divide by mitosis to form gametophytes again. The gametophyte and sporophyte usually look very different. In some groups, the gametophyte is larger and lives longer than the sporophyte. In others, the sporophyte is dominant. Depending on the species, asexual reproduction may happen in either gametophyte or sporophyte. So, as we follow a plant around its sexual life cycle, it alternates between a haploid body and a diploid body. This is known as alternation of generations. It happens in all sexual plant species, but is more obvious in some than in others.

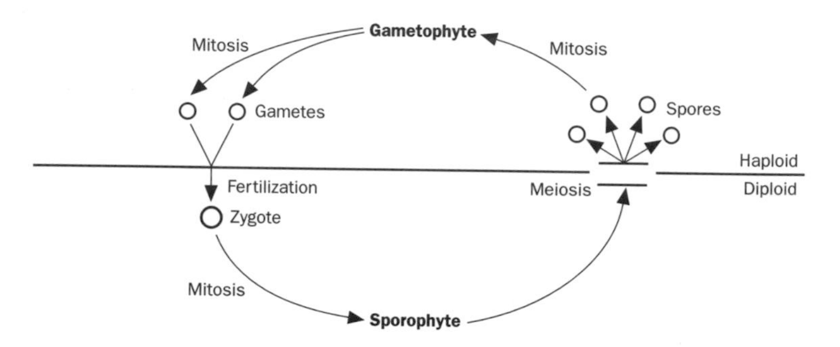

Fig. 14.1 The sexual life cycle of any plant.

Fig. 14.2 on pages 238 and 239 summarizes eight life cycles, drawn from plants and protoctists. All the protoctists included belong to phylum Chlorophyta. These are arguably the most plant-like protoctists. Since plants evolved from protoctistan ancestors, it makes sense to include these here.

Protoctists

Starting with the simplest, the unicellular green alga *Chlamydomonas* (phylum Chlorophyta) has a dominant gametophyte (Fig. 14.2(a)). The swimming cells are haploid. They divide by mitosis to form populations of genetically identical individuals. Some of these cells may act as gametes. In most species the gametes are similar in size. After fertilization, the zygote does not form a sporophyte. It divides by meiosis to form four haploid cells, which become swimming gametophytes. Before this it may form a resting zygospore for a while, but no division is involved.

The filamentous green alga *Spirogyra* (phylum Chlorophyta) also has a dominant gametophyte (Fig. 14.2(b)). Haploid cells are cylinders which divide by mitosis to form a filament of cells linked at their ends. In this Order, the Conjugales, a strange sort of mating called conjugation takes place. Two genetically different filaments line up together. The nuclei act as gametes. A tube forms between each pair of cells. The entire contents of one cell moves through the tube. The nuclei act as gametes: one

Gametophyte: A haploid plant body which makes gametes.

Sporophyte: A diploid plant body which makes spores.

fuses with the other. The zygote may become dormant, but its first division is meiosis. There is no sporophyte, as in *Chlamydomonas*.

Ulva, the sea lettuce, also belongs to the Chlorophyta, but has both a gametophyte and a sporophyte (Fig. 14.2(c)). The haploid body is a green, crinkly sheet of cells, two cells thick, attached to seashore rocks at one end. The diploid body looks just the same! The gametophyte makes gametes by mitosis. They have flagella, so can swim to one another. There is no clear distinction between male and female gametes, but gametes will only fertilize gametes from another individual. The zygote divides by mitosis to form the sporophyte. This produces spores by meiosis. The spores also have flagella. Spores divide by mitosis to form the gametophyte.

Mosses and liverworts

Moving on to true plants, the mosses and liverworts (phylum Bryophyta) have both gametophytes and sporophytes (Fig. 14.2(d)). The leafy shoots of a moss, anchored by rhizoids, are the gametophyte. This is the longest-lived stage of the life cycle. These 'leaves' and rhizoids have little tissue differentiation and are not related to the true leaves and roots of 'higher' plants. The gametophyte makes gametes by mitosis, in specialized flask-shaped structures: **archegonia** and **antheridia**. The gametes are recognizably male and female. The female gametes are larger and cannot swim, so stay in the archegonia which make them. Male gametes are smaller and have flagella. They swim from the antheridia which make them, to the archegonia. This relies on a film of water across the plant surface. After fertilization, the zygote divides by mitosis to form a sporophyte which grows straight out of the archegonium. It consists of a stalk, the seta, with a capsule on top. The capsule makes spores by meiosis. After dispersal (Box 14.1), they divide to form filaments of cells, the protonema. The differentiated gametophyte body develops from this later.

The moss sporophyte relies on the gametophyte for support, probably for water and minerals, but not for carbohydrates. The sporophyte can photosynthesize. Most moss capsules also have stomata, which are unknown in protoctists and plant gametophytes. No part of a moss or liverwort has xylem or phloem, although some moss gametophytes have cells which carry out similar functions.

Ferns

In ferns (phylum Filicinophyta) the situation is quite different (Fig. 14.2(e)). The familiar leafy plant is a sporophyte, with xylem and phloem, true leaves, stems and roots. It makes spores by meiosis on the backs of the leaves (fronds). They are made in structures called **sporangia**, which are grouped into sori (Box 14.1). Most ferns produce just one type of spore.

After dispersal, the spores divide by mitosis to form a small gametophyte: the prothallus. This is a thin, green sheet of tissue, anchored by rhizoids. It makes gametes in a similar way to mosses. After fertilization, the sporophyte grows out of the archegonium. The shoot goes up, the roots go down, and the gametophyte withers away around it.

In ferns, the sporophyte is dominant, but the gametophyte does live an independent life for a while. The next example shows two further developments: an even more reduced gametophyte, and not one but two types of spore.

Archegonia: Structures in gametophytes which make female gametes.

Antheridia: Structures in gametophytes which make male gametes.

Sporangia: Structures in sporophytes which make spores.

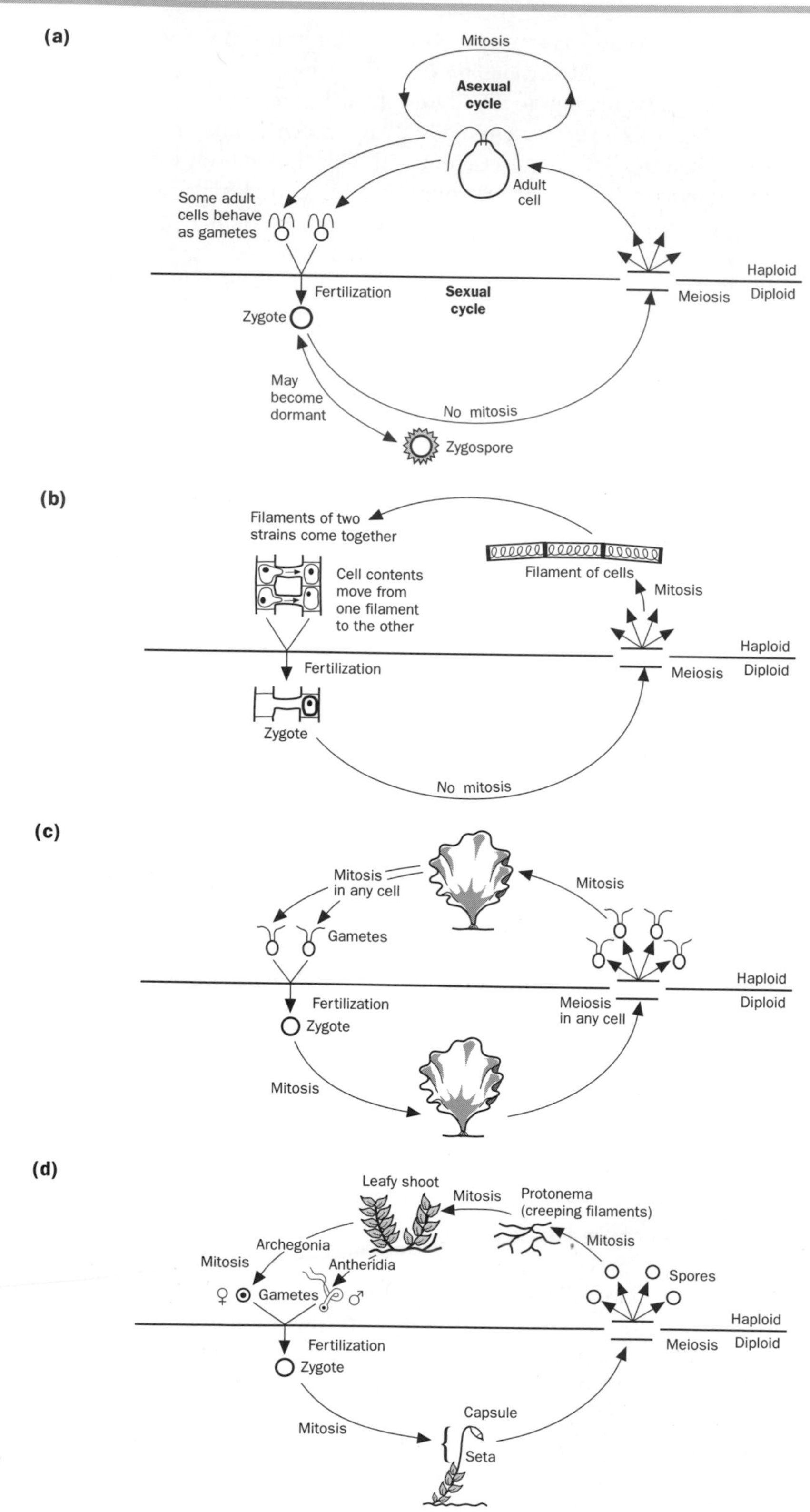

Fig. 14.2 Life cycles of plants and protoctists mentioned in the text. (a) *Chlamydomonas*, (b) *Spirogyra*, (c) *Ulva*, (d) a moss, (e) a fern, (f) *Selaginella*, a clubmoss, (g) a conifer, (h) a flowering plant.

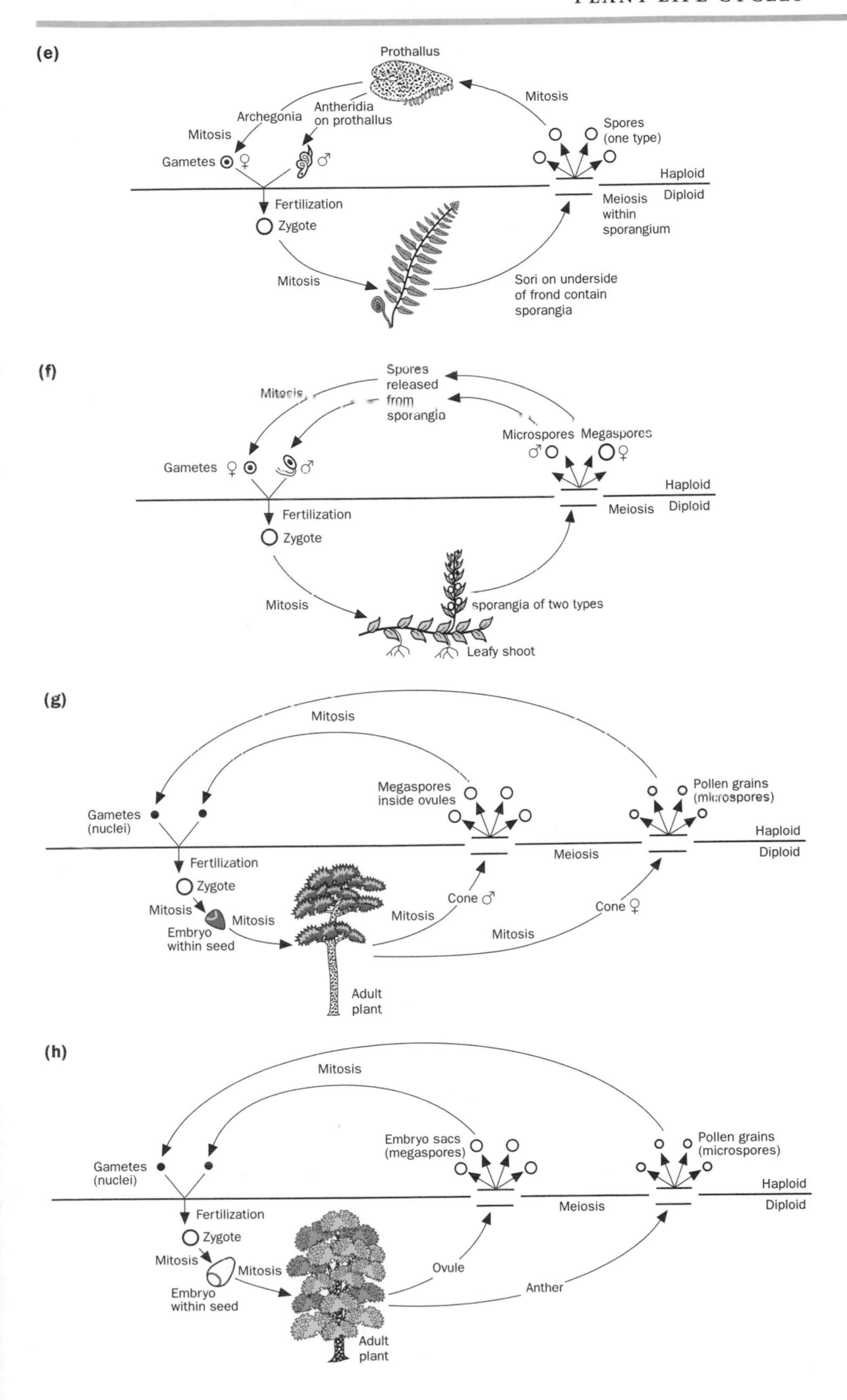
(e)
Prothallus
Mitosis
Archegonia
Antheridia on prothallus
Mitosis
Gametes ♀
♂
Spores (one type)
Haploid
Diploid
Fertilization
Zygote
Meiosis within sporangium
Mitosis
Sori on underside of frond contain sporangia
(f)
Spores released from sporangia
Mitosis
Microspores Megaspores
♂ ♀
Gametes ♀
♂
Haploid
Meiosis
Diploid
Fertilization
Zygote
Mitosis
sporangia of two types
Leafy shoot
(g)
Mitosis
Megaspores inside ovules
Pollen grains (microspores)
Gametes (nuclei)
Haploid
Meiosis
Diploid
Fertilization
Zygote
Mitosis
Embryo within seed
Mitosis
Mitosis
Cone ♂
Cone ♀
Mitosis
Adult plant
(h)
Mitosis
Embryo sacs (megaspores)
Pollen grains (microspores)
Gametes (nuclei)
Haploid
Meiosis
Diploid
Fertilization
Zygote
Mitosis
Mitosis
Embryo within seed
Ovule
Anther
Adult plant

Box 14.1

Spore dispersal in bryophytes and ferns

Fern, moss and liverwort spores are dispersed by air. Each has a different mechanism for throwing spores out to catch the wind, but they all rely on structures which move as they dry out.

Ferns

Fern sporangia are clustered under the leaves. Each one is a thin-walled bag on a stalk, full of spores. A strip of special cells, the annulus, runs part way around it. The annulus is a catapult. The cells of the annulus have a thin bendy wall outside, but thicker, stiffer walls on the other side. To start with, the annulus cells are turgid (a). As they lose water, they shrink. because the outer walls are weaker. The annulus bends (b). However, the annulus acts as a spring. It puts the water in the cells under tension. Suddenly, cavities form which fill with water vapour: the annulus springs back, hurling out the spores (c).

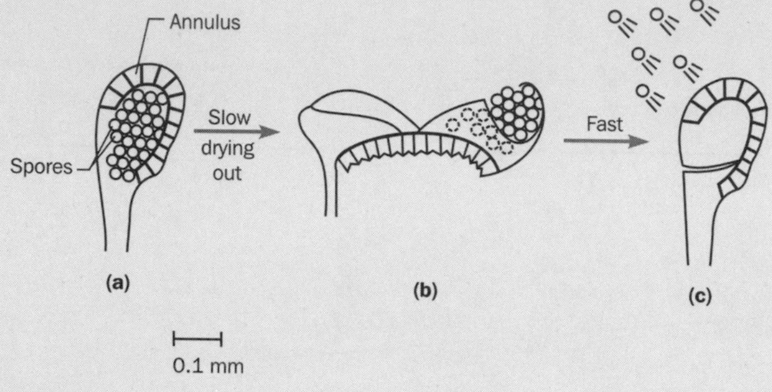

Liverworts

Liverworts have capsules which open up to reveal a mixture of spores and thread-like cells called elaters. Elaters have spiral thickenings in the wall. They make the cells twist as they dry and shrink. The thickenings act as a tightening spring. In most species, the tension becomes too great after a while. Cavities form and the elater untwists violently. The seething mass of elaters and spores throws itself apart over a few minutes.

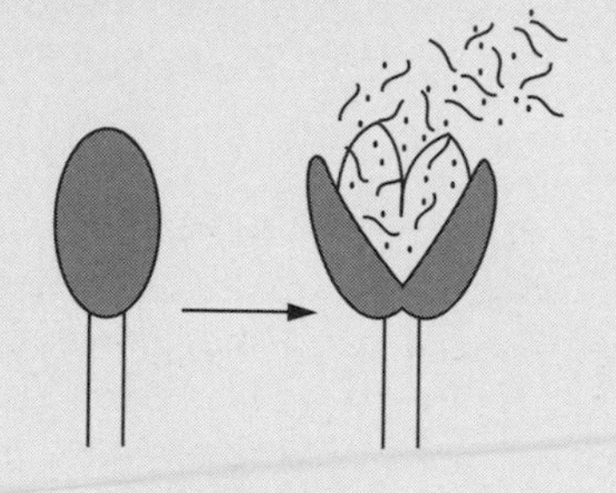

Mosses

The ripe moss capsule has two sets of teeth over a hole in its end. These teeth are made of cell walls, not whole cells. A sheet of cells covering the hole dies, and the inner and outer walls come apart as two sheets. These split up to form two rings of teeth. Walls are multi-layered structures (Section 4.6). The two surfaces of the outer teeth shrink by different amounts as they dry. This makes the outer teeth bend inwards when wet and outwards when dry. They grate between the fixed inner teeth, so move by little jerks. This flicks out any spores sticking to them.

There are exceptions!

Ferns, liverworts and mosses can be found with quite different mechanisms. For example, mosses in the genus *Sphagnum* fire spores from an air gun.

Clubmosses

The phylum Lycopodophyta includes the clubmosses. Only some clubmosses, such as the genus *Selaginella*, have two types of spore (Fig. 14.2(f)). The diploid *Selaginella* plant has more or less creeping stems, small scale-like leaves, and branches carrying sporangia. There are two types of sporangia, making two types of spore. One type makes a single large **megaspore**. The other makes many small **microspores**. Both spores divide by mitosis, but the cells simply divide without growing. The gametophyte develops inside the thick wall of the spore. The spores, containing their gametophytes, fall to the ground. Megaspores make one big female gamete by mitosis. Microspores make many small gametes which swim across the soil surface to the megaspores. Because the gametophyte lives in the spore, we can think of the megaspore as female and the microspore as male. Moving back further, we can even think of the sporangia as being male and female, because they produce the male and female spores. So sex differences move back from haploid tissues to diploid tissues.

Conifers

In conifers (phylum Coniferophyta) this process is taken even further by making seeds. The sporophyte is a tree which makes sporangia on highly modified shoots: cones. Female cones make megaspores, called ovules. These cones become the familiar woody pine and fir cones. The ovules at the base of each scale become the seeds. Male cones are smaller and not woody. They make microspores with wings to help them drift in the wind: pollen grains. Gametophytes develop inside both ovule and pollen grain. Pollination is the transport of pollen grains to the ovule. It relies on wind blowing the pollen, so huge numbers of pollen grains are released, to ensure that some get there. After pollination, the gametophyte grows out of the pollen grain, towards the female gamete, forming a pollen tube. A nucleus near the tip of the tube is the male gamete itself. After fertilization, the embryo sporophyte develops inside the ovule to form a seed. Only then does it leave the parent tree.

Flowering plants

Flowering plants show even more advanced features (Fig. 14.2(h)). Most importantly, the ovule (a megaspore) is completely surrounded by the ovary, which is part of the sporophyte. This, not pretty petals, is what defines a flower. The ovary later becomes the fruit, surrounding the seed. The female gametophyte is even more reduced than in conifers. It is a single cell with eight identical haploid nuclei. Flowering plant reproduction is discussed in more detail in Sections 14.3–14.5.

Advancing features

These life cycles have been put in order so they show more and more advanced features. The dominant sporophyte is an advanced feature. In *Chlamydomonas* and *Spirogyra* the gametophyte was dominant. Through bryophytes and ferns the sporophyte became more and more important, until in flowering plants the gametophytes were microscopic and dependent on the sporophyte. This is linked to life on land, and surviving in dry conditions. Only sporophytes have xylem, stomata and highly organized water-conserving leaves. All these features allow the plant to carry water up from soil to leaves and to minimize loss. It is true that some mosses live in very dry conditions, but they rely on surviving drying out, not on conserving water. In the same way, non-swimming male gametes are advanced. Swimming male gametes need a wet surface to reach the female gamete. Every life cycle we have seen, except the conifer and flowering plant, relies on a covering of moisture at this stage of the cycle. Reproduction only becomes independent of water when pollen grains are

Megaspore: A large female spore.

Microspore: A small male spore.

made. Pollen is carried to the ovules by wind or animals, and pollen tubes grow out, carrying the gamete nuclei with them.

Heterospory is the production of two types of spore, micro- and megaspores. It is an advanced feature. We have only seen it in three life cycles, the clubmoss, conifer and flowering plant. It is part of the trend for differences between the sexes to stretch further and further back, from the gametes to the gametophyte, then to the spore, and finally to the sporophyte.

Do not be fooled into thinking that each group evolved from the one before it in this sequence. These are modern plants. Some have more primitive features than others, some may be more like an ancient plant group than others, but all are alive today. The plants they evolved from have been dead for a very long time.

The rest of this unit examines the sexual reproduction of flowering plants in more detail.

14.3 Flower, anther and ovary

A flower is a highly-modified shoot (Fig. 14.3). Instead of leaves arranged around a stem, there are one or more ovaries containing ovules, stamens making pollen grains, and the perianth (more varied structures which give the flower its shape). A small proportion of species have stamens and ovaries on separate plants.

Heterospory: The practice of making two types of spore.

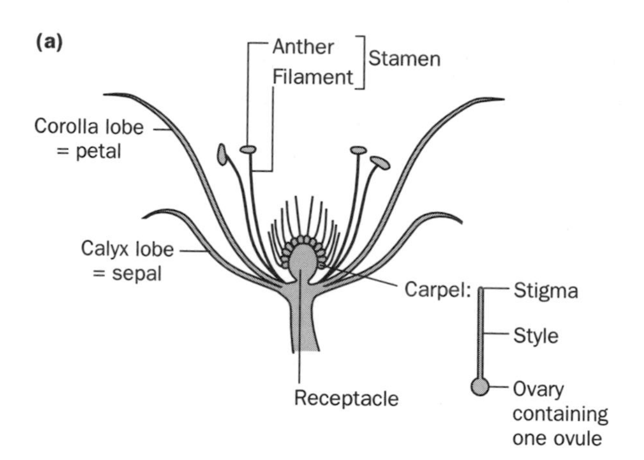

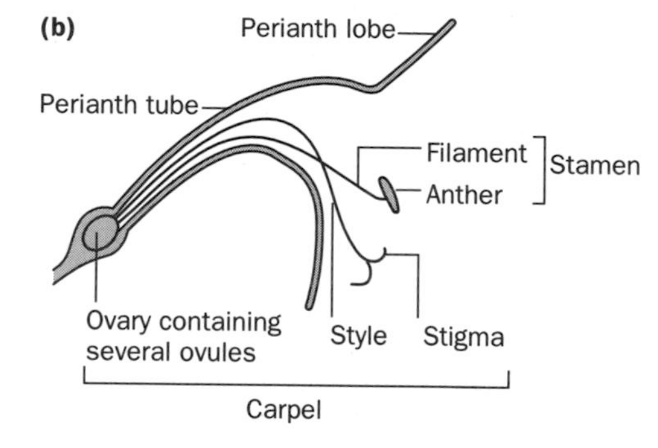

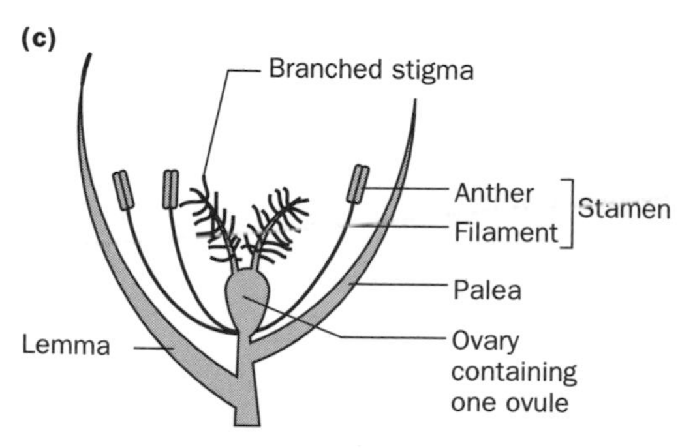

Fig. 14.3 Diagrammatic vertical sections through some flowers. (a) *Geum* species, (b) *Watsonia* species, (c) a grass.

In many flowers, the perianth can be divided into two rings or whorls (Fig. 14.3(a)). The inner whorl is the corolla. If it is divided into distinct segments, they are known as petals, but often it forms a corolla tube, sometimes divided into corolla lobes for part of its length. The outer whorl is the calyx, sometimes divided into distinct sepals. However, in other flowers there is no clear distinction between calyx and corolla (Fig. 14.3(b)). Further, more leaf-like, structures called bracts may surround the perianth. Some flowers, like the wind-pollinated grasses (Fig. 14.3(c)), are so distinctive that it is hard to draw parallels with the perianths of other flowers.

A stamen has two parts. The business end is the anther, which makes pollen. It is usually attached by a thin stalk: the filament. An anther is made up of usually four, sausage-shaped tubes: pollen sacs (Fig. 14.4(a)). The pollen sacs are filled with diploid cells, made by mitosis. These cells divide by meiosis. Each haploid cell made by meiosis becomes a pollen grain. The nucleus divides once by mitosis, so the mature pollen grain has two nuclei (Fig. 14.4(b)). These are called the generative nucleus and the tube nucleus. The generative nucleus will later divide again to form the male gamete. The mature pollen grain has a thick, deeply fissured secondary cell wall. Its patterns are distinctive and are important in binding pollen grains to one another, to animals and to the stigma. Some pollen grains have wing-like extensions, allowing them to drift in the wind.

An ovary contains ovules, each one attached to it by a short stalk, the funicle. Some ovaries contain just one ovule, some a few, some thousands. Ovaries also have a pollen-collecting device: the stigma. Sometimes this is on the end of a stalk: the style.

Most of an ovule is diploid, part of the sporophyte. This is called the nucellus, and is surrounded by two tougher layers, the integuments (Fig. 14.5). The integuments do not quite cover the ovule. A small hole is left: the micropyle. Inside the ovule, a cell divides by meiosis. One of the haploid cells goes on to become the embryo sac, which is the female gametophyte. Within the embryo sac, three rounds of mitosis

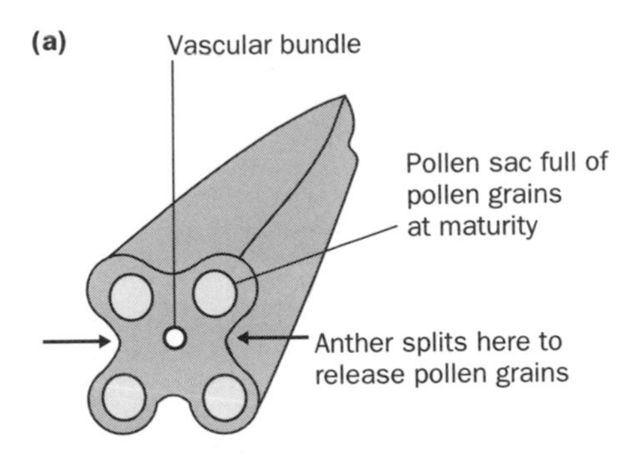

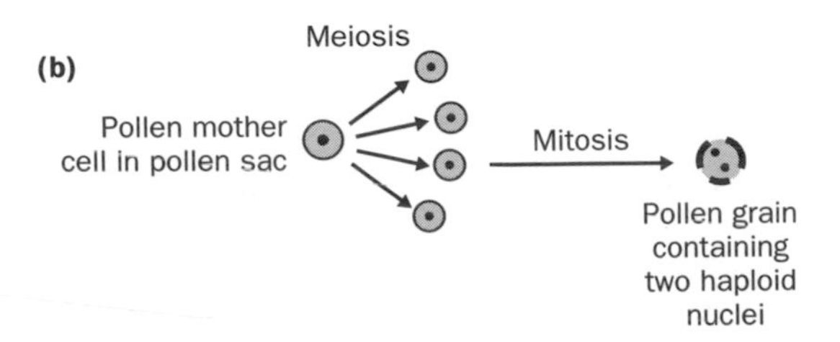

Fig. 14.4 The flowering plant anther. (a) anther structure, (b) pollen development.

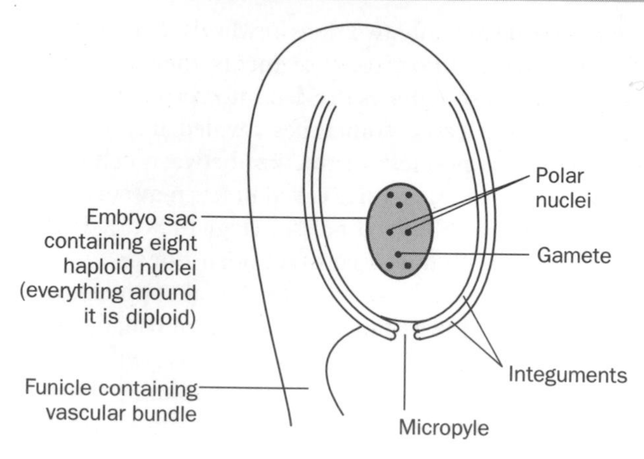

Fig. 14.5 The flowering plant ovule.

form eight identical haploid cells. They take up distinct positions within the embryo sac (Fig. 14.5). One of them is destined to be the female gamete.

14.4 Pollination

Pollination: The transfer of pollen from anther to stigma.

Fertilization: The fusion of gametes to make a zygote.

Pollination is the transfer of pollen from anther to stigma. **Fertilization** is the fusion of gametes to make a zygote. Pollination is the equivalent of mating in flowering plants, male and female getting together before fertilization. Plants cannot walk. Either the flower must pollinate itself, or some vector must carry the pollen to another flower. Most flowers are cross-pollinated at least some of the time. Many have to be, because various devices prevent self-pollination and self-fertilization (Box 14.2).

Box 14.2

Avoiding self-fertilization

Self-fertilization looks easy when flowers have male and female parts. However, it cuts out one of the big advantages of sex: outbreeding. Flowers avoid 'selfing' in several common ways.

Timing

The stigma stops receiving pollen before the anthers ripen, or vice versa. This prevents selfing within one flower, but not between different flowers on the same plant.

Chemical incompatibility

'Self' pollen is prevented from germinating on the stigma. In many species this has a genetic basis. One gene exists in loads of forms (alleles). Each diploid stigma has two copies of the gene; each haploid pollen grain has just one. A pollen grain carrying an allele which the stigma also has, cannot germinate. The drawback is that some non-self pollen will be rejected, if it happens to share an allele with the stigma.

Separation of the sexes

If different plants have only male or only female flowers, selfing is impossible. There are intermediate situations, for example where some plants are female but others are hermaphrodite (male and female). In other cases, separate male and female flowers are made by a single plant.

Some factors prevent selfing, others just make it happen less often. Many plants have a balance between cross- and self-fertilization. The violets are an extreme example. Their spring flowers are cross-pollinated by insects, but their summer flowers never open, and are selfed while in bud.

Plants use all sorts of pollen vectors. Some, like grasses, use wind. A very few waterweeds are pollinated by water, releasing pollen grains or even whole male flowers into the surrounding water. Most flowering plants use animals to carry pollen. This is because they move actively and have sense organs, so can home in on flowers. If an animal carries pollen to other flowers efficiently, the plant can make less pollen, or can pollinate more flowers.

Animal-pollinated flowers cannot rely on the goodwill of their vectors. An animal will only pollinate a flower if it benefits from doing so. Flowers give food to animals, and also signal to them. The food rewards and the signals depend on the pollinator. The most common rewards are nectar and pollen itself.

Nectar is a sugary solution made by nectaries: small structures at the base of the corolla. Butterflies and moths can use nectar as an energy supply. Many flowers which they pollinate, like the butterfly bush (*Buddleia davidii*), have long, thin corolla tubes which fill with nectar. The insect's long proboscis reaches the nectar, but pollen brushes onto its body at the same time. When it visits another flower, pollen may brush off onto the stigma. Hummingbirds in America, and sunbirds in Africa are also nectar feeders. They are larger animals, needing more energy, and bird-pollinated flowers tend to have corolla tubes which hold a lot of nectar.

Pollen contains more protein then nectar. As a food, it can support growth, not just animals' energy needs. Beetles which pollinate *Magnolia* flowers feed on pollen; the flowers do not make nectar. Plants which use pollen as a reward rely on animals feeding inefficiently. Pollen which is eaten cannot pollinate another flower. Some must get stuck on the outside of the animal's body.

Bees are specialists at flower feeding. They have evolved on the same time scale as the flowering plants, and very many flowers are specialized for bee pollination. Bees use both pollen and nectar. They use nectar as an energy supply, and carry pollen back to the nest to feed their larvae.

Bird pollination is important in many parts of the world, though not in Europe.

Flowers signal to their pollinators by colour, shape and smell: 'there is food here – visit me!' This is good for the plant because it encourages visits. It is particularly important when the plant uses a specific, scarce pollinator. Random searching by the pollinator might not be enough to find some flowers. It is also good for the animal. Searching for food uses energy. Signals make flowers more obvious, so less energy is wasted.

The signal must suit the pollinator. For example, bees can detect part of the visible spectrum, and ultraviolet light, but not red light. Bee-pollinated flowers are often blue, violet, yellow or pink, but very rarely red. Birds, however, can detect red light. Very many bird-pollinated flowers have red or orange flowers. They are obvious to birds, but not to bees. Bright colours are not much use in the dark, however. Flowers which are pollinated by night-flying moths or bats may have a distinctive smell, but only a pale or drab colour. For example, the sweetly-scented flowers of honeysuckle are visited by moths. It is possible to identify sets of flower features which reflect particular pollinators, but pollination strategies can be classified in another way too (Box 14.3).

Not all relationships between flowers and animals benefit both species. Some-

Box 14.3

Pollination strategies

Flowers are often classified according to exactly what pollinates them. However, looking at the efficiency of pollen transfer shows three fundamental strategies for cross-pollination.

Pollination by wind or water

Wind and water are not alive. They do not seek out flowers to visit. This has two consequences. First, flowers do not produce attractants. The perianth tends to be small and green. Nectar and scents are not made. Second, pollen must be made in huge amounts, relying on some grains landing on a stigma by chance. The grains tend not to stick together, so are dispersed as individuals. Typically, wind-pollinated flowers have long feathery stigmas and stamens hanging out of the flower. Resources are put into pollen quantity, not attractants. Wind-pollination is more efficient when the species lives at a high density. A scattered population of isolated individuals might not be pollinated efficiently by wind.

Unspecialized animal pollination

Many pollinating species visit a wide range of flowers. Similarly, many flowers are visited by more than one animal species. Free petals and stamens, with an open bowl shape, as in buttercups, allow many insects in.

Even though animals can search for flowers, unspecialized pollinators may not find the same species each time. Pollen still needs to be spread widely, with separate pollen grains and many pollinator visits. However, investment in attractants may allow less investment elsewhere, such as in pollen quantity.

Specialized animal pollination

Other flowers have a close relationship with a single pollinating species. Complex three-dimensional shapes may exclude most species on the grounds of body size or tongue length. Specific attractants may attract only one group. The more specific the relationship, the fewer pollinator visits are needed: a high proportion will reach a stigma of the same species. If pollen grains stick together, an animal can carry more away after a single visit. Investment in attractants tends to be high, but there is less need for lots of pollen.

The orchids include some extremely specialized examples. Many have highly specific pollinator relationships, and complex flower structures. All the pollen from each anther sticks together in masses, the pollinia. One insect may carry all the pollen from a flower in one visit.

Specific relationships are ideal in plants which live in scattered populations. A specialized pollinator seeks out the plants over a wide area.

times, animals feed on pollen or nectar without transferring pollen. These 'robbers' are using the flower as a resource, like any other flower visitor. There is nothing unselfish about visits by the 'right' pollinator. In other cases, only the plant benefits. Some orchids attract insects with flowers that look superficially like a female insect. English names like bee orchid and fly orchid reflect this. The flower is making a signal, but not a truthful one: 'I am a female of your species – mate with me'. Males of one or more species are attracted, and pollen sticks to them while astride the flower.

Pollination is useless unless followed by fertilization.

14.5 Fertilization

Pollination brings pollen grains to the stigma. They are now millimetres, or even centimetres from the ovules. The male gametophyte covers the distance by growth. A pollen tube grows out from the pollen grain. It is just an extension of the cell, with a thin wall around it, rather like a root hair (Section 18.5). The pollen tube grows towards regions of lower oxygen concentration, so buries itself in the stigma. It grows down through the style. It may follow a channel in hollow styles, or force its way between cells, along the line of the middle lamella. Pollen tubes secrete digestive enzymes to make this possible. Typically, the pollen tube grows up to the micropyle of the ovule (Fig. 14.6).

At some stage, the generative nucleus in the pollen grain divides by mitosis. This may happen during pollen development or while the pollen tube grows. Either way, these two nuclei follow behind the growing tip of the tube. The tube nucleus trails behind, or may even disintegrate.

The act of fertilization is hard to study, and poorly understood. However, one amazing fact is clear. In flowering plants, fertilization happens twice. One of the male nuclei from the pollen tube fuses with the female nucleus of the embryo sac,

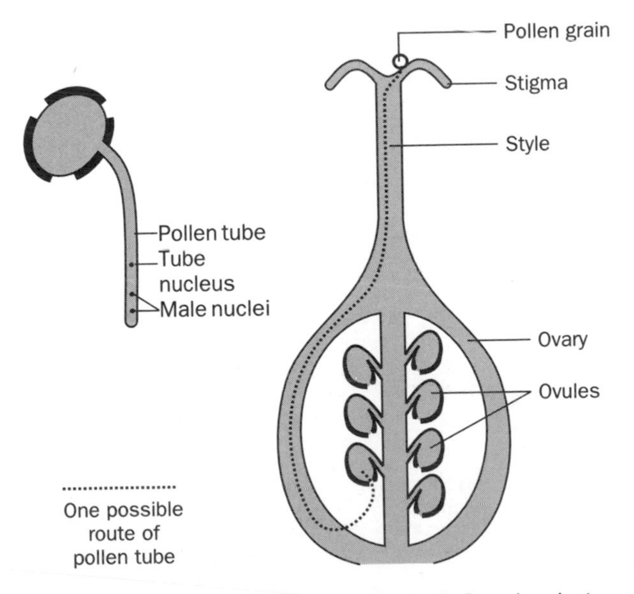

Fig. 14.6 Pollen tube growth and fertilization in flowering plants.

forming a diploid zygote. This goes on to form the embryo: the baby plant. The other male nucleus fuses with both polar nuclei of the embryo sac, forming a triploid nucleus. The triploid nucleus goes on to form the endosperm tissue around the embryo.

14.6 Seed and fruit

A **seed** is a plant in a box. The plant is there in miniature, ready to grow. Seeds have three functions: firstly, they allow dispersal; secondly, they can lie dormant in soil for days, months, years or even centuries, depending on the species; thirdly, they carry some food reserves to support the plant until it begins to photosynthesize and take up minerals itself.

The ovule develops into a seed after fertilization (Fig. 14.7). The embryo, the baby plant itself, develops from the zygote. It has distinct organs: the plumule, a small shoot; the radicle, a root; and one or two cotyledons. The cotyledons, or seed leaves, have various functions. In some seeds they act as a food store, in others they are the seedling's first photosynthetic organs: in yet others, they do both jobs.

Seed: A tiny sporophyte, usually with food stores, in a tough outer coat.

The triploid endosperm nucleus divides by mitosis to form the endosperm tissue. Sometimes the nuclei become boxed into separate cells by plasma membranes and walls. Sometimes the endosperm is a soup of nuclei and cytoplasm. In seeds where the cotyledons are food stores, the endosperm usually disappears as the seed

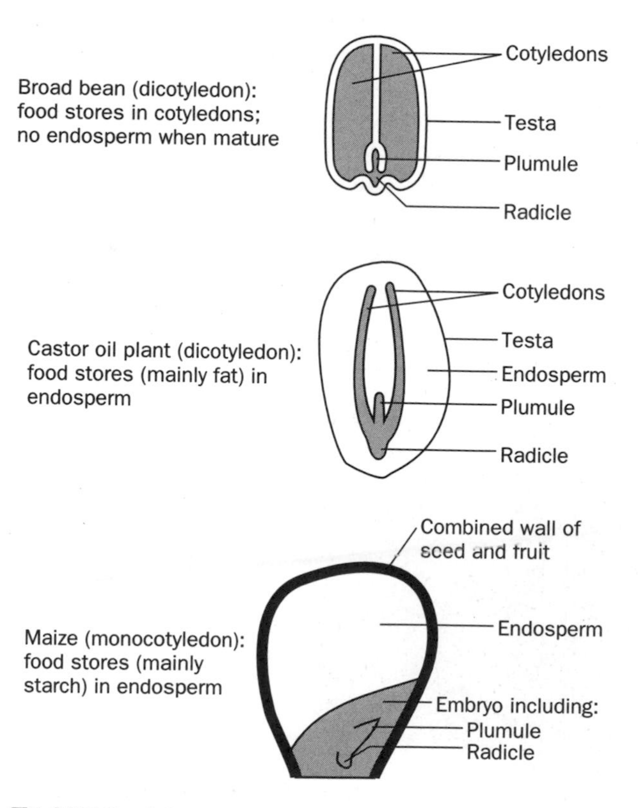

Fig. 14.7 Seed structure: three examples.

develops. The embryo uses it as food, along with the rest of the ovule. Otherwise, the endosperm becomes a food store in the seed. The embryo will only use it when the seed germinates. A few seeds have almost no food reserves. The orchids are an extreme example. They rely on a relationship with a fungus for food during germination. The fungus benefits only later, once the plant is established.

Imagine eating a sweetcorn kernel, then a bean seed (broad, kidney or baked, whichever you fancy). The sweetcorn kernel is unripe when you eat it. It has a rather tough outer coat. Bite through it, and lots of endosperm oozes out. Most of the food value of corn, and of cereals like wheat, is in the endosperm. Now feel around in your mouth. There should be a soft little spear, a few millimetres long. This is the embryo, which was embedded in the endosperm. Now pick up the bean and gently bite it. If you angle it right, the whole bean inside the seed coat should fall into two halves. These are the cotyledons, which contain the food reserves. The radicle and plumule are attached, but relatively small. The endosperm is long gone.

The seed coat, or testa, surrounds the seed. It develops from the integuments. It is generally waterproof and gives the seed some protection. In some species, an extra outer layer called the aril develops from a third integument. Arils tend to be coloured or juicy. Either way, they attract animals which disperse the seeds. The juicy bit of a lychee is an aril. Violet seeds have much smaller, fatty arils which attract ants.

Fruits are formed from ovaries. They are not all colourful, juicy things! Hazelnut shells and pea pods are fruits too. Some, like mangos, have just one seed. Kiwi fruits, and the dry capsules of poppies, contain lots of them. Some one-seeded fruits develop with the seed coat to form a single structure around the embryo and endosperm. This is seen in grasses, so the outer covering of a wheat grain is an intricate mixture of fruit and testa. Occasionally, fruit-like structures develop from another part of the flower. In apples and pears, the ovary is buried inside the flower base: the receptacle. The receptacle swells to form a fleshy layer around the fruit. The juicy bit is the receptacle, the core is the fruit and the pips are the seeds. In the same way, a strawberry is a big fleshy receptacle, covered in small, dry, one-seeded fruits.

Fruit: A structure which surrounds one or more seeds, formed from an ovary.

The fruit of the coco de mer, which grows in one valley on one island in the Seychelles. It contains just one seed, the world's largest.

Seed dispersal

Seeds are dispersed in the hope of finding a brighter future somewhere else, away from one another and the parent plant. There are many dispersal mechanisms. Sometimes the seed itself has adaptations for dispersal; sometimes the fruit has them.

Dispersal by animals I

A juicy fruit with a coloured skin advertises itself as food. Most fruits like this have tough, indigestible seeds which pass through the animal's gut and germinate elsewhere. Many edible fruits have toxic seeds (plums and apples, for example) which makes it even less likely that the seed itself will be used as food.

Dispersal by animals II

Seeds themselves are full of food. In some species, seed-eating animals carry away and hoard seeds, but only return to eat some of them: perhaps they lose track of where the others are, or find they do not need them. Jays help disperse acorns like this. There is a cost to the oak tree in terms of seeds eaten, but an edible fruit also has a cost in terms of energy and raw materials.

Dispersal by animals III

Some fruits have hooks which cling to animals' coats. They may hang on for a long time, slowly shedding seeds. I still remember, aged eight, putting a burdock fruit in my sister's hair. Scissors were needed in the end, but she carried the seeds at least two kilometres.

Seeds are normally dispersed before they germinate. Dispersal allows seeds to colonize new areas, and avoids competition with the female parent: two sides of the same coin. Box 14.4 shows some of the many ways seeds are dispersed. Another consequence of dispersal is gene flow.

Moving genes

Gene flow is the movement of genes through the area where a population lives. In vertebrate animals, genes move in just one way: animals move around. Plants cannot do this. However, gametes, seeds, pollen and other spores can move. When this happens, genes flow and are mixed. Populations remain as interbreeding units, not genetically isolated individuals. Even scattered plant populations, where individuals may be hundreds of metres apart, can be held together by long-distance pollen flow or seed dispersal. The individuals may not compete for resources, communicate, or get together for mating, but they do interact genetically. Plants do it – at a safe distance.

Summary

- In plants and many plant-like protoctists, mitosis can take place in haploid or diploid cells.

Box 14.4

Dispersal by wind I

Wings (e.g. sycamore) and parachutes (e.g. dandelion) on fruits maximize the time they spend in the air when they fall off the plant. That way, the wind blows them further. Small seeds will also blow further if the fruit is a dry capsule which only lets them shake out in windy weather.

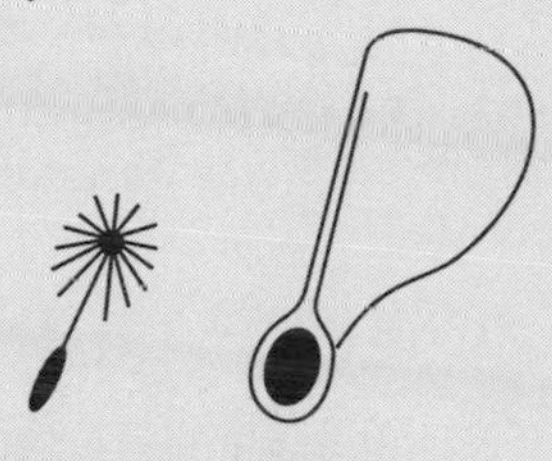

Dispersal by wind II

Tumbleweeds are plants with fruits or entire flowering shoots which break off and roll along, scattering seeds. This works well in open, flat areas like deserts and steppe grassland, where these species are found.

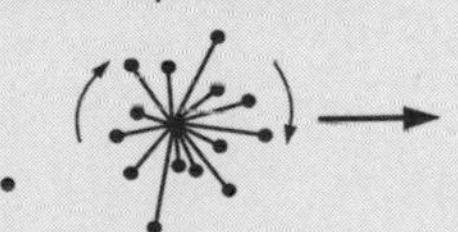

Dispersal by water

Rivers and the sea can disperse floating seeds. However, the fact that seeds can travel right across oceans need not mean effective dispersal: few seeds can live for long in salt water.

Self dispersal

Exploding fruits and catapults fire seeds away. The squirting cucumber fruit (*Ecballium elaterium*) develops a high pressure inside. When ripe, it breaks off. Seeds and pulp squirt out, sometimes for several metres. The cranesbills (*Geranium* species) fire seeds from a sling.

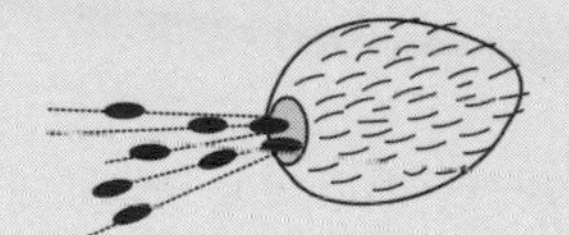

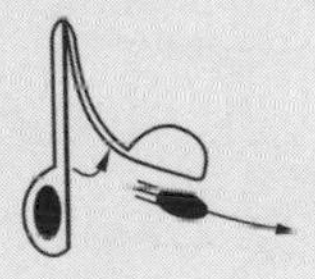

- This means a plant can have two different bodies in a single life cycle: alternation of generations.
- The haploid body makes gametes by mitosis. It is called the gametophyte.
- The diploid body makes spores by meiosis. It is called the sporophyte.
- In *Chlamydomonas* and *Spirogyra* (phylum Chlorophyta), mitosis takes place only in the gametophyte. The zygote divides by meiosis.
- In *Ulva* (phylum Chlorophyta), mitosis takes place in both generations. The gametophyte and sporophyte live independently, and look very similar.
- Mosses and liverworts (phylum Bryophyta) have a longer lived, usually leafy gametophyte. The sporophyte is short-lived and grows out of the gametophyte.
- In ferns (phylum Filicinophyta), the sporophyte is dominant. The gametophyte is small and short-lived.
- Fern, moss and liverwort spores are dispersed by various systems involving structures which move as they dry out, sometimes violently.
- In clubmosses (phylum Lycopodophyta) the gametophyte is so small that it stays inside the spore wall.
- Some clubmosses have two sorts of spore, a large megaspore containing the female gametophyte, and a microspore containing the male. This is also the case in conifers and flowering plants.

- In conifers (phylum Coniferophyta) and flowering plants (phylum Angiospermophyta), the male gamete is a nucleus, not a swimming cell.
- In conifers and flowering plants, the microspore is the pollen grain. The tiny gametophyte is inside and grows a pollen tube towards the female gamete.
- The female gametophyte is the embryo sac, inside the ovule.
- Pollen is transported from male to female structures: pollination. Conifers and some flowering plants are pollinated by wind. Others are pollinated by animals including various insects, birds and bats.
- In flowering plants, ovules are enclosed in ovaries. In conifers, they are at the bases of cone scales.
- The flowering plant embryo sac has eight haploid nuclei. One, the female gamete, fuses with the male gamete to form the zygote. Two more fuse with another nucleus in the pollen tube to form the triploid endosperm.
- The seed is made up of the embryo (baby plant), endosperm and the testa, a seed coat made of maternal cells.
- Some seeds store food in the endosperm. Others store it in the cotyledons, part of the embryo.
- Flowering plant seeds develop inside a fruit, formed from the ovary.
- Seeds are dispersed from the parent plant by wind, animals or explosion.
- Pollination and seed dispersal both lead to movement of genes within a population.

Exercises

14.1. Here are a number of statements about plant life cycles and reproduction. Which of the following plants and protoctists does each refer to? (Statements may refer to several, one or none of the plants):

flowering plant, conifer, *Selaginella*, fern, moss, *Ulva*, *Spirogyra*, *Chlamydomonas*.

(i) Mitosis takes place in diploid cells.
(ii) Mitosis takes place in haploid cells.
(iii) Fertilization occurs in the sexual life cycle.
(iv) Pollination occurs in the sexual life cycle.
(v) The male gamete can swim.
(vi) The female gamete is inside an ovule.
(vii) Ovules are enclosed in ovaries.
(viii) The gametophyte is dominant.
(ix) There are two sizes of spore.

14.2. Which of the following flowers would you expect to produce nectar?

(i) a cactus pollinated by hummingbirds;
(ii) an orchid pollinated by males of a small wasp which try to mate with the flowers;
(iii) a pondweed pollinated by water;
(iv) honeysuckle, a climbing plant pollinated by moths;
(v) a wind-pollinated grass.

14.3. Suggest why bird pollination is much more common in tropical and sub-tropical areas of the world than in cold-temperate areas such as northern Europe and northern Asia.

14.4. What do the following structures within the flower develop into?

(i) ovary;
(ii) ovule;
(iii) fertilized female gamete;
(iv) endosperm nucleus;
(v) integuments.

Further reading

Ingrouille, M. *Diversity and Evolution of Land Plants* (London: Chapman & Hall, 1992). The major plant phyla are covered; life cycles are included.

Proctor, M.C.P., Lack and Yeo, P.F. *The Natural History of Pollination* (London: Collins, 1992). Written for naturalists as well as biologists, this is probably the best introduction to an exciting but specialized field.

South, G.R. and Whittick, A. *Introduction to Phycology* (Oxford: Blackwell Science, 1987). A book on the algae, organized according to areas of biology rather than taxonomic groups. Life cycles are included.

PART 5

The Working Organism

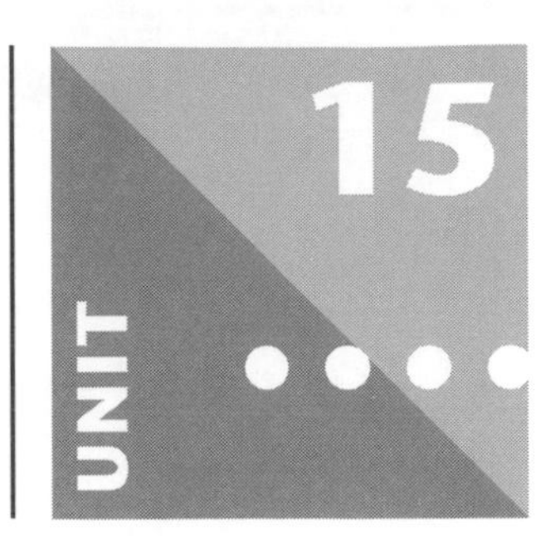

Principles of Homeostasis

Connections

▶ This unit introduces the idea that organisms control their internal conditions. Units 16 and 17 go on to explore how cells communicate in order to do this. This unit is an important introduction to animal physiology, especially Units 19, 20 and 22. Similar principles also apply to the regulation of population size (Unit 26), though for very different reasons.

Contents

15.1 Some things change, some stay the same

Chemical reactions are central to life. The rates of chemical reactions change according to environmental conditions, and most organisms live in changeable environments. This is a great problem: cells are at the mercy of environmental change. The solution is to resist these changes, to keep conditions inside the cell constant, whatever may be going on outside.

The rate of a reaction is affected by the concentrations of the molecules involved in the reaction, as well as by temperature. Add more of the reacting molecules, and the reaction speeds up. Allow the products to build up, and it slows down. The higher the temperature, the higher the rate of reaction. Hence, many key molecules are kept at constant levels within cells and bodies. Many animals also control their body temperature. The business of keeping these variables at a constant level is called **homeostasis**. All kinds of homeostasis, in nature and in engineering, use the concept of negative feedback (Box 15.1). As a result of homeostasis, cells and organisms are islands of predictability in a restless, changing world. It helps keep them different from their environment, different and alive.

15.2 Homeostasis within the cell

Homeostasis: The ways in which variables are kept at a constant level within a cell or body.

Single-celled organisms can do little to change their environment. They can only attempt to keep things constant inside the plasma membrane. Many freshwater pro-

Box 15.1

Negative feedback

Many control systems in nature and in engineering keep some variable at or near a fixed value: the set point. The controlled variable could be anything: furnace temperature, ATP level in a cell, the altitude at which an aircraft is flying, the amount of cholesterol in the blood. If the level increases, a chain of events is set in motion which decreases the level again. If the level falls too far, changes happen which bring it back to the set point. This is negative feedback: feedback because the chain of events feeds back to the variable which caused it; negative because an increase causes a decrease, and vice versa.

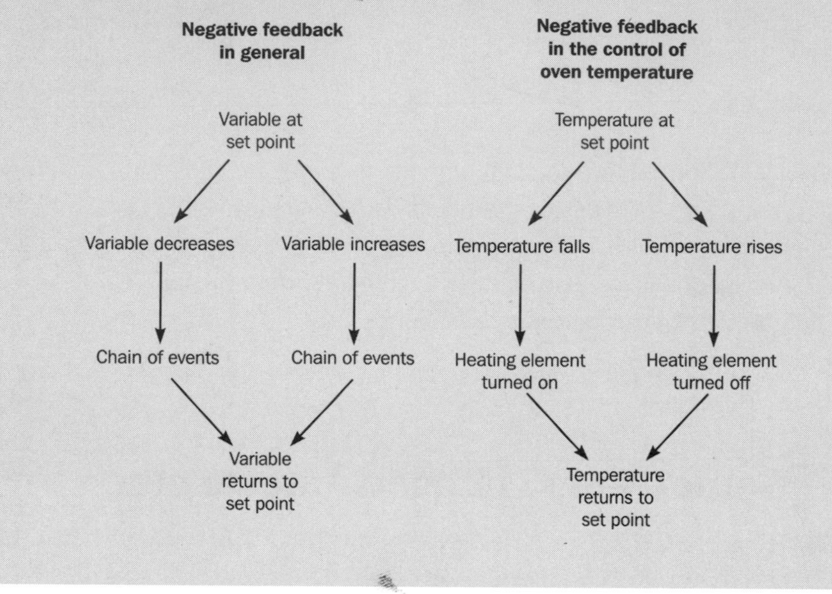

tozoa, such as *Amoeba*, clearly show this in the way they regulate the solute potential of their cytosol. Water constantly enters by osmosis, diluting the cytosol. It is sent out again by a device called a contractile vacuole. This is a large membrane sac within the cytoplasm. Water is actively pumped into it all the time, so it swells up. Periodically it contracts, spewing its water back out of the cell. If the cell is put into a more concentrated solution, water will enter less quickly. The cell responds by filling the vacuole more slowly, so it contracts less frequently. When the cell is put into a more dilute solution, the vacuole pumps water out of the cell faster. Either way, the salt concentration of the vacuole is kept constant: a homeostatic mechanism (Fig. 15.1).

Many important molecules are not passed freely from cell to cell within an organism. Individual cells need to control levels of these molecules themselves, whether or not they are part of a larger body. A good example is ATP. Cells make a great effort to keep levels of this short-term, energy-storing molecule constant. ATP is used as an energy supply for many cell processes (Sections 8.3, 8.4), which will not always work at the same rate. Supply must match demand if the amount of ATP in the cell is to stay constant. In most cells, the majority of ATP is made by respiration. If ATP is being used more quickly, the reactions of respiration must speed up. Some key respiratory enzymes, notably phosphofructokinase (PFK, Box 9.2), are inhibited by ATP. When ATP gets low, PFK is more active, so more ATP gets made. If ATP is

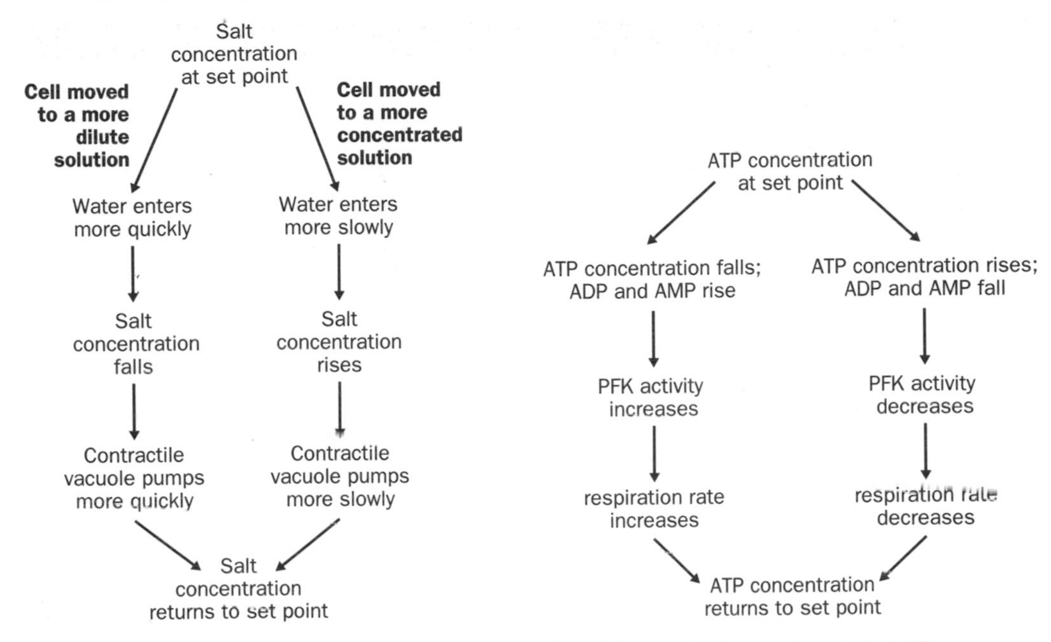

Fig. 15.1 Negative feedback control of salt concentration within a protozoan.

Fig. 15.2 Negative feedback control of ATP concentration within the cell.

abundant, PFK is less active, so less ATP is made. There is only a certain number of adenine nucleotides in the cell. Those that are not in the form of ATP are either ADP or AMP. Hence, high ADP or AMP levels indicate low ATP. Both ADP and AMP activate PFK. This is a neat homeostatic mechanism (Fig. 15.2). It is also absolutely typical of the way in which important molecules are kept at a constant level by feedback control of the pathways that make them.

15.3 Homeostasis within bodies

In all but the simplest animal bodies, cells conspire to hide from the environment. Specialized tissues like epithelia keep the outside world at bay. Within the body, cells are bathed by an internal ocean. In some animals such as earthworms, this is the blood. In vertebrates, it is the tissue fluids derived from the blood. Other specialized cells make sure the composition of the blood remains very, very predictable. This is homeostasis at the whole body level, sometimes involving cells whose only role in life is homeostasis.

Three examples show how diverse homeostatic mechanisms in mammals can be. Just one cell type, the liver cell, is involved in regulating the amount of cholesterol circulating in the blood (Box 15.2). Two organs are involved in controlling blood glucose level. It is detected by the pancreas but removed or added by the liver. The pancreas communicates with the liver by hormones: chemical signals released into the blood (Box 15.3). All sorts of mechanisms are involved in keeping body temperature constant. They are orchestrated by the brain, by way of nerves and hormones (Box 15.4).

Box 15.2

Regulation of blood cholesterol

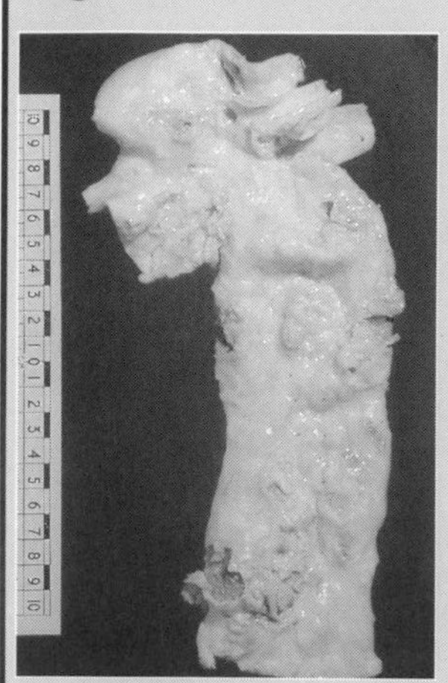

When cholesterol is deposited on the wall of an artery, the artery becomes narrower, restricting blood flow. This is very serious in the coronary artery leading to the heart muscle: it can lead to angina and heart attacks.

Here, a big artery has been split lengthwise and opened out. Warty deposits of cholesterol can be seen on the inner surface.

Cholesterol in the body

Most cholesterol in the human body is made by liver cells. Some also enters from food. All cells need it as a component of membranes (Section 4.2). The liver uses a lot to make bile salts (Section 21.4: Guts). Significant amounts are deposited in the skin, helping to make it waterproof, and some is used to make steroid hormones.

Cholesterol in the blood

Cholesterol is non-polar, and so is insoluble in water. It is carried in the blood as part of complexes called lipoproteins, mostly low density lipoprotein (LDL). A protein and phospholipid coating separates a ball of cholesterol from the aqueous environment.

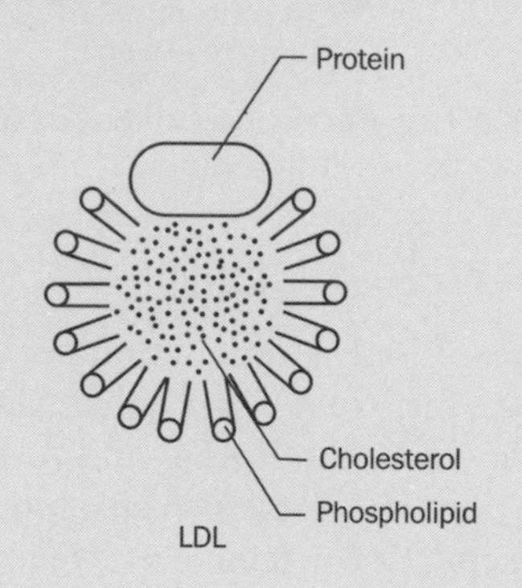

Why regulate it?

Cholesterol is needed by all cells, so cannot be allowed to get too low in the blood. However, at higher concentrations it tends to be deposited in the linings of artery walls. This narrows the artery, and could make heart attack or stroke more likely. Patients with an inherited inability to regulate blood cholesterol would not survive beyond childhood without treatment.

Regulation

Liver cells regulate cholesterol levels in their own cytoplasm. Cholesterol inhibits one of the enzymes involved in its own synthesis. This is a typical example of negative feedback within the cell. However, by doing this, liver cells are also regulating the amount in the blood. They export cholesterol to the blood, so when levels rise in the cell, levels in the blood rise as well. Just like other body cells, liver cells also take in LDL, containing cholesterol, by receptor-mediated endocytosis (Section 6.6). This means that when blood cholesterol rises, cytoplasmic cholesterol increases too. In terms of cholesterol concentration, the blood can be seen as an enormous extension of the liver-cell cytoplasm. By regulating levels inside the cell, they also regulate the amount in the blood. This is an unusually simple example of whole body homeostasis. One cell type both detects the change and brings about the response.

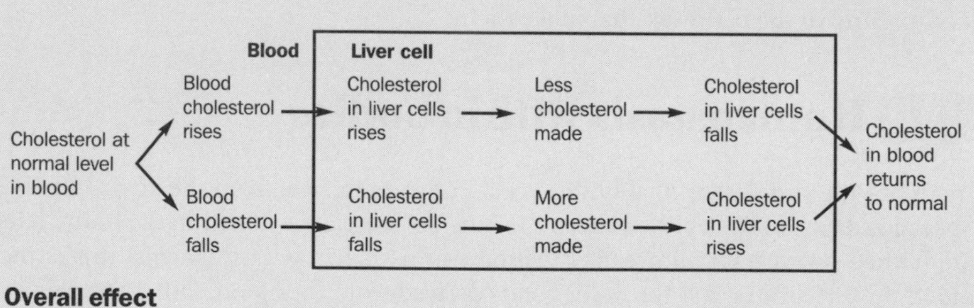

15.4 Why not plants?

Individual plant cells control levels of molecules within themselves, just like any other cell. However, it is hard to find any example of homeostasis at the whole plant level. In general, plant cells are more independent than animal cells. Some have specialized functions: xylem vessels for water transport, guard cells for controlling gas exchange, and so on, but there are only a few types, forming only a few types of

Box 15.3

Regulation of blood glucose

Why?

In humans, blood glucose concentration is normally about 900 mg l^{-1}. It rises considerably after a meal, but returns to this set point within a couple of hours. This is important! It must not fall too low because some organs, like the brain, rely on glucose as their major energy supply. It must not get too high, because cells could become dehydrated by osmosis, and valuable glucose could start to leak out in the urine.

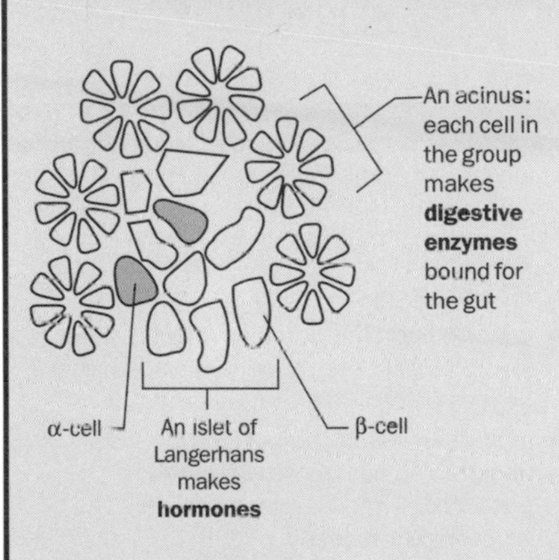

A small piece of pancreas

Detection

Two cell types in the pancreas detect glucose level. They communicate with the rest of the body by releasing two polypeptide hormones. β-cells secrete insulin when blood glucose rises above the set point. α-cells release glucagon when glucose gets too low. (Note that most cells in the pancreas have a quite different function: making digestive enzymes.)

The response

The most important short-term effects of insulin and glucagon are seen in the liver. Insulin stimulates liver cells to take up glucose and convert it to glycogen (Section 3.4). It does this by activating enzymes which make glycogen, and inhibiting an enzyme which breaks it down. Muscle cells respond in a similar way. Glucagon has the opposite effect.

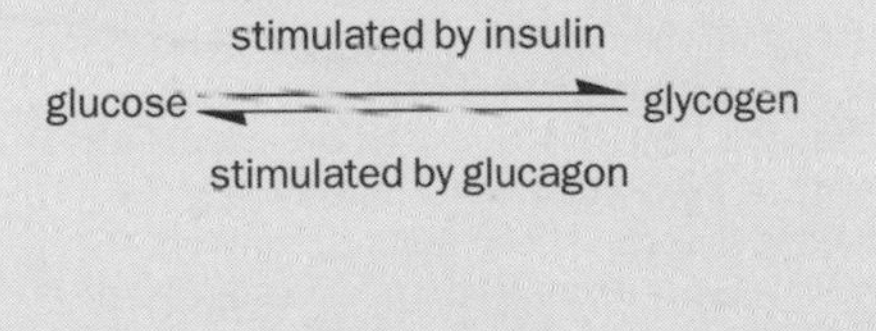

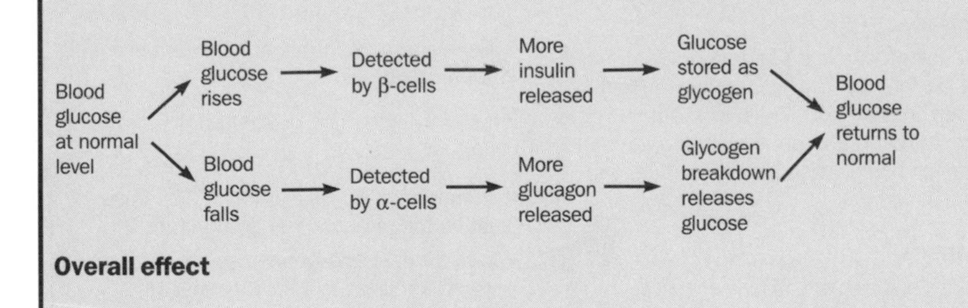

Overall effect

organ. Plant organs form bodies in a different way, too. The typical animal body is highly integrated. A badger has two kidneys, one liver, a set number of muscles in precise positions, which work together beautifully, but would certainly not work if jumbled up and reassembled in a new way. A flowering plant body is really only made up of only four organs: root, stem, flower and leaf. In most plants there is no fixed number of each organ, and new ones can be added in all sorts of places, according to a few simple rules of growth. Larger plants can meet quite a range of environmental conditions, all at once, as a result of this modular growth pattern. Some shoots of a pine tree may be shaded by a neighbour, others may be in full sun. A creeping buttercup plant, spreading through grassland by horizontal stems, may

Mammals and birds regulate their body temperatures homeostatically. No invertebrates do this, but some large insects like this bumble bee raise their body temperature before flight by buzzing their wings.

Box 15.4

Control of body temperature

Why?

Mammals and birds keep their bodies at a more or less constant temperature. This means that important chemical reactions do not speed up or slow down just because the weather has changed. Tadpoles kept in a tank in the 'fridge just can't swim fast, while others living in a heated tank can speed around. Squirrels, however, can escape up a tree just as fast on a cold day as in a heat wave. The set point for human body temperature is about 37°C.

Detection

Nerve endings which are sensitive to either hot or cold are found in the hypothalamus (part of the brain) and the skin. The hypothalamus puts together information from these receptors and controls the body's response, by nerves and hormones.

Responses

If the body is too cold, the following responses occur:

- erection of hairs, trapping a layer of air: insulation (by nerves);
- constriction of arterioles leading to outer layers of skin: less heat delivered to skin, so less can be lost (by nerves);
- increased contraction of individual cells in skeletal muscle: more waste heat produced: when extreme, becomes shivering (by nerves);
- increased respiration rate of most cells, making more heat (by the hormone thyroxine, Table 16.1);
- behavioural responses, for example finding a warmer place.

If the body is too hot, the following responses occur:

- flattening of hairs, dilation of arterioles, no shivering, reduced respiration rate;
- sweating: evaporation of sweat carries heat away, cooling the body (by nerves).

meet wetter and drier soils, different nutrient levels, some places with many competitors and others where the plant can use resources unchallenged. A typical animal's body faces one set of conditions at a time, as an integrated unit. Blood circulating through the body ensures that all the cells are experiencing similar conditions, and are working together. There is no place for this in the modular, extended plant body. Parts of the plant which have found 'good' conditions will flourish. Other parts will respond to poorer conditions in a different way, perhaps channelling all their resources into escaping from the problem, splurging them in a final burst of reproduction, or simply failing to thrive. If one shoot dies, the rest of the plant need not be threatened. The plant can be thought of as a population of loosely associated units, a metapopulation. Homeostasis is simply not an issue here.

Summary

- Organisms keep internal conditions constant even in a changing environment.
- Homeostasis is the ways in which variables are kept at a constant level within a body or cell.
- Homeostasis typically involves negative feedback: a change brings about events which reverse the original change.
- Homeostasis goes on within individual cells.
- Homeostasis goes on at the whole body level in animals but not plants.

Exercises

15.1. Which of these variables would you *not* expect to be regulated by homeostasis in the human body?

(i) amino acid concentration in the blood;
(ii) amount of fat stored in fat cells;
(iii) concentration of the hormone adrenaline in the blood;
(iv) number of red blood cells per litre of blood;
(v) rate of heat production by the body.

15.2. Predict the effect of killing all the β cells in the pancreas.

15.3. One of the body's responses to infection is to raise the set point for temperature regulation, by up to 2°C. The patient is said to 'have a high temperature'. Explain the following observations.

(i) Early on in the course of an infection, the patient complains of feeling cold, and may shiver. However, her skin feels normal, or even warmer than usual.
(ii) Late in the course of infection, there comes a point when the patient begins to sweat, while her skin develops a pink flush and feels hot.

Further reading

Guyton, A.C. and Hall, J.E. *Textbook of Medical Physiology* (9th ed.) (Philadelphia: Saunders, 1996). I like this hefty textbook of human physiology for its thoroughness and clear, no-frills diagrams.

Guyton, A.C. *Human Physiology and Mechanisms of Disease* (5th ed.) (Philadelphia: Saunders, 1992). Effectively a shorter, boiled-down version.

Schmidt-Nielsen, K. *Animal Physiology: Adaptation and Environment* (4th ed.) (Cambridge: Cambridge University Press, 1990). One of the all-time great textbooks! Exciting, authoritative and easy to read, it puts animal physiology in the context of the environment.

NB: The basics of homeostasis are covered in any textbook of physiology.

Cell Communication: Chemicals

Contents

Connections

► Before studying this unit you should have a basic understanding of how ions and molecules move across membranes (Unit 6). Many examples of communication between cells have to do with homeostasis, so you should understand the principles of this (Unit 15). This unit introduces the roles of chemical signalling molecules like hormones. Signalling molecules have important roles in animal reproduction, defence and development (Units 13, 23 and 25). Communication involving the nervous system is covered in Unit 17.

16.1 Cell communication

Bodies are colonies of genetically similar cells. There is no point in this if the cells simply hang out together in a formless blob. To get the benefits of communal living they must cooperate. They must divide, grow, differentiate and work in a way which helps the body develop and operate effectively, and so improves the chances of the cells' genes surviving to another generation. In order to cooperate, cells must communicate. When communication breaks down, cooperation ceases. This is seen in cancer: some tumours come close to being that same formless blob.

Cells frequently communicate through chemicals released by one cell and detected by others. In animals these chemicals can be classified according to the distance over which they operate. Hormones are carried throughout the body in the blood. Paracrines operate very locally, typically over a few millimetres. Neurotransmitters carry information across synapses, the gaps between neurones. They diffuse across distances measured in nanometres, and are discussed in Unit 17. A number of molecules seem to be involved in communication within plants. These plant growth substances are notoriously difficult to investigate, and whilst some of their major effects are well known, no clear understanding of how they interact has emerged. Since most of their effects are on growth and development, they are discussed in Unit 25. In animals, specialized cells carry information between other cells relatively quickly. These are neurones, the cells of the nervous system, which deserve a unit of their own (Unit 17).

16.2 Hormones

Hormones are molecules which are released into an animal's bloodstream by an organ or cell type. They are chemical signals which can be detected by other specific cell types. When a hormone is detected, some change in the working of the cell comes about. Different cell types may respond to the same hormone in different ways. Most hormones belong to one of three groups of chemicals (Fig. 16.1).

Firstly, there are steroids such as testosterone, aldosterone and the group of molecules known collectively as oestrogens. All these are hydrophobic and can cross the plasma membrane to enter target cells. Secondly, there are various amines derived from the amino acid tyrosine. Some of these, including thyroxine, are hydrophobic. Others, like adrenaline, are hydrophilic. Finally there are proteins such as insulin, and shorter peptides like antidiuretic hormone (ADH). All these are hydrophilic and can only be detected by receptor molecules at the cell surface.

The cells which make and release hormones are called **endocrine cells**. The organs they are found in are endocrine organs. Table 16.1 summarizes the major endocrine organs of the human body, and some of the hormones they make. A number of these hormones have clear roles in homeostasis, for example insulin and glucagon in the regulation of blood glucose level (Box 15.3). Others intervene in some system which is already regulated, under particular circumstances. For example, adrenaline causes constriction of blood vessels (apart from those in muscles), increased heart rate, increased metabolic rate, and other changes which together prepare the body for intense activity in an emergency. Yet other hormones are signals to another endocrine organ, stimulating release of a second hormone. The significance of this is best seen in terms of the control of the endocrine system as a whole.

Hormones: Chemical signals made in one cell type, carried in the blood, and detected by another cell type.

Endocrine cells: Cells which make hormones.

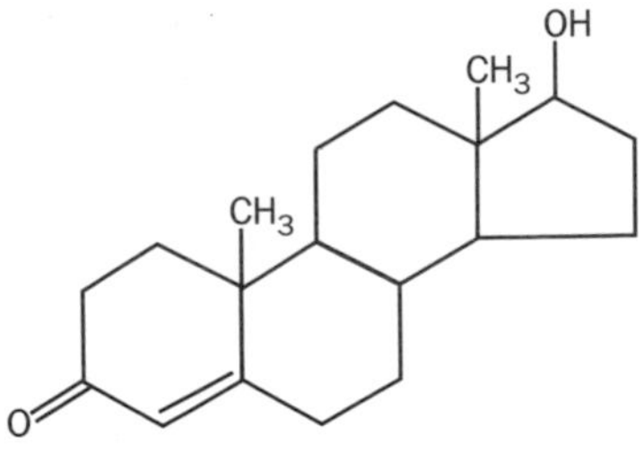

Testosterone, a steroid

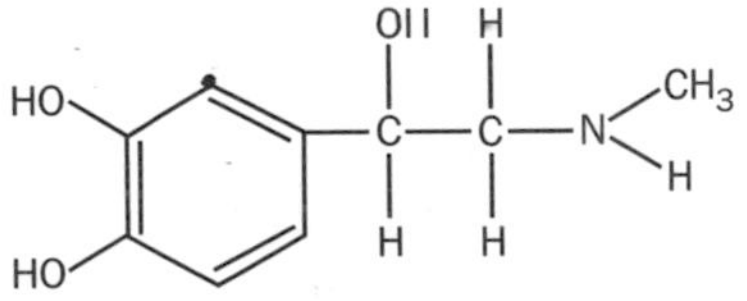

Adrenaline, an amine

HOOC—Cys—Tyr—Ile—Gln—Asn—Cys—Pro—Leu—Gly—NH_2

ADH, a short peptide

Fig. 16.1 Hormone structure: three examples.

Table 16.1 Some major mammalian endocrine organs and their products

Endocrine organ	Hormone	Major role of hormone
Hypothalamus	various releasing hormones	controls secretion of hormones by the anterior pituitary
Hypothalamus (manufacture); posterior pituitary (release)	antidiuretic hormone (ADH)	osmoregulation: stimulates water reabsorption in kidney
	oxytocin	stimulates contraction of uterus during labour
Anterior pituitary	growth hormone	stimulates liver to make somatomedin
	adrenocorticotrophin	stimulates adrenal cortex
	thyrotropin	stimulates thyroid
	follicle stimulating hormone (FSH)	stimulates ovarian follicle development or testis development
	luteinizing hormone (LH)	stimulates development of corpus luteum from ovarian follicle. Also stimulates progesterone or testosterone production
Thyroid	thyroxine and tri-iodothyronine	increase metabolic rate
Pancreas	insulin	decreases blood glucose concentration
	glucagon	increases blood glucose concentration
Liver	somatomedin	stimulates growth
Adrenal cortex	glucocorticoids, e.g. cortisone	control metabolic rate and blood glucose
	aldosterone	stimulates sodium reabsorption in kidney
Adrenal medulla	adrenaline and noradrenaline	preparation of body for activity in emergency
Kidney	renin	part of a blood pressure control system
Ovary	oestrogens	stimulate growth of uterus lining development of female sexual characteristics
	progesterone	stimulates maintenance of uterus lining development of female sexual characteristics
Testis	testosterone	development of male sexual characteristics

16.3 Endocrine control: the hypothalamus and the pituitary

Some endocrine organs are relatively independent. The β-cells in the pancreas, which secrete insulin, seem to detect glucose concentration (the major factor which controls insulin release) themselves. Insulin release leads to a drop in blood glucose, which in turn means that less insulin is released.

Many other endocrine organs are under the indirect control of the brain. The hypothalamus, part of the base of the brain, is crucially important. The pituitary gland, immediately below it, is essentially the slave of the hypothalamus. The pituitary has two parts, the posterior and anterior pituitary (Fig. 16.2). The hypo-

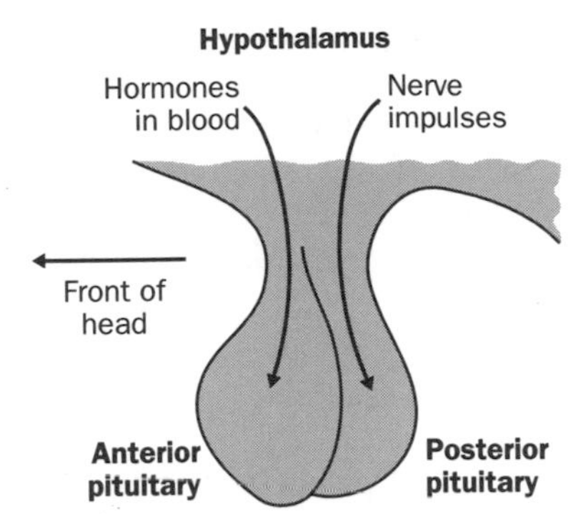
Hypothalamus
Hormones in blood
Nerve impulses
Front of head
Anterior pituitary
Posterior pituitary

Fig. 16.2 The relationship between the hypothalamus and the pituitary.

thalamus may detect some things directly, such as the concentration of solutes in intercellular fluid, but it is certainly connected to other parts of the brain which handle inputs from sense organs. It makes two important hormones, ADH and oxytocin, which are transported inside nerve cells running down to the posterior pituitary. From here they are released into the blood. Nerve impulses from the hypothalamus control their release. Other hormones are released directly from the hypothalamus into the blood vessel that leads to the anterior pituitary. These trigger or inhibit the release of other specific hormones from the anterior pituitary. Many of these stimulate yet another endocrine organ to release a hormone which, at last, has a physiological effect. The advantage of a cascade system of this sort is amplification. Only a tiny amount of hormone is released by the hypothalamus, but it can still be detected by the anterior pituitary since it has travelled there in the blood directly, before being diluted in the general circulation. The pituitary releases much more of its hormones, which stimulate even greater production of the final hormone. Tiny amounts of a substance produced by a few cells in the brain can result in massive secretion by an entire organ. Box 16.1 shows an example of this.

Box 16.1

The control of thyroxine release

The hypothalamus, pituitary and thyroid cooperate in controlling the release of this hormone.

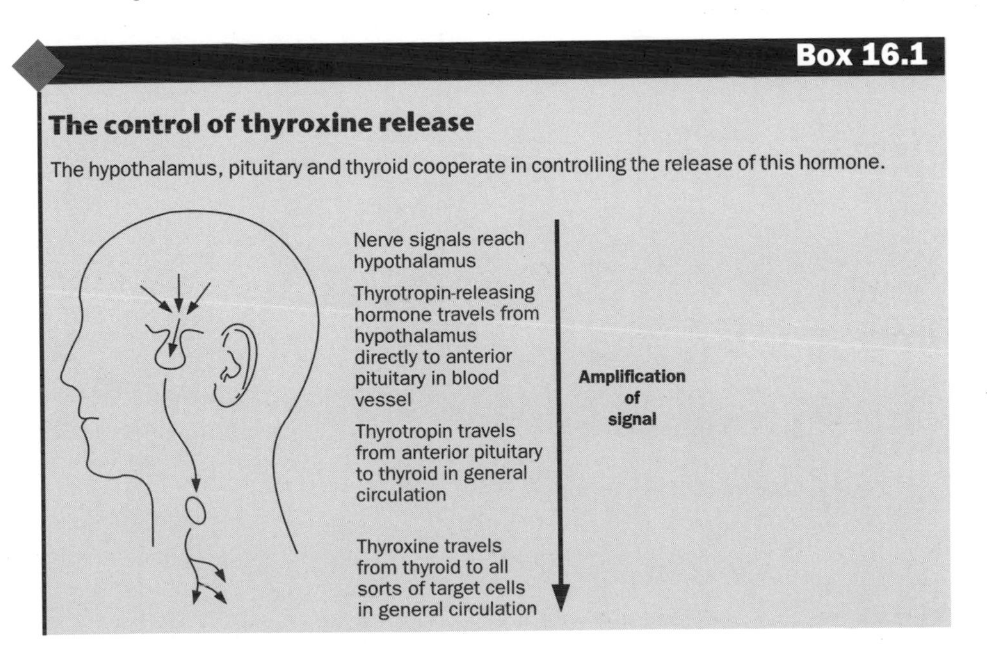
Nerve signals reach hypothalamus
Thyrotropin-releasing hormone travels from hypothalamus directly to anterior pituitary in blood vessel
Thyrotropin travels from anterior pituitary to thyroid in general circulation
Thyroxine travels from thyroid to all sorts of target cells in general circulation
Amplification of signal

What goes on at the molecular level when hormones are released? Hormones are packaged in clathrin-coated vesicles (Section 6.6) which bud off the Golgi apparatus. They accumulate beneath the plasma membrane until they receive the signal to fuse with the membrane and release their contents. When the cell detects the appropriate releasing hormone, or environmental stimulus, a second messenger (Section 16.4), typically Ca^{2+}, stimulates the vesicle to fuse with the plasma membrane.

16.4 How cells respond to hormones

Cells use specific receptor proteins to detect hormones. Only the hormone will bind to the receptor. Only the cells that respond to the hormone have the receptor. Steroid hormones, which can slip through the plasma membrane, bind to receptors inside the cell. Hydrophilic hormones are bound by receptors embedded in the plasma membrane. The big question is, how does the simple act of a receptor protein binding a hormone lead to such a range of metabolic, physiological and developmental effects?

With hydrophobic hormones, like steroids and thyroxine, the mechanism is relatively simple (Fig. 16.3). The receptors are gene-control proteins found in the nucleus. One region of the molecule binds the hormone; another binds a specific DNA sequence. Hormone binding allows the protein to bind DNA, switching on one or more genes. The proteins coded for by these genes do not bring about the response directly. Instead, they bind to and activate a larger set of genes. It is the proteins coded for by this second wave of genes which bring about the effect. Since two rounds of transcription and translation are required before the final response is seen,

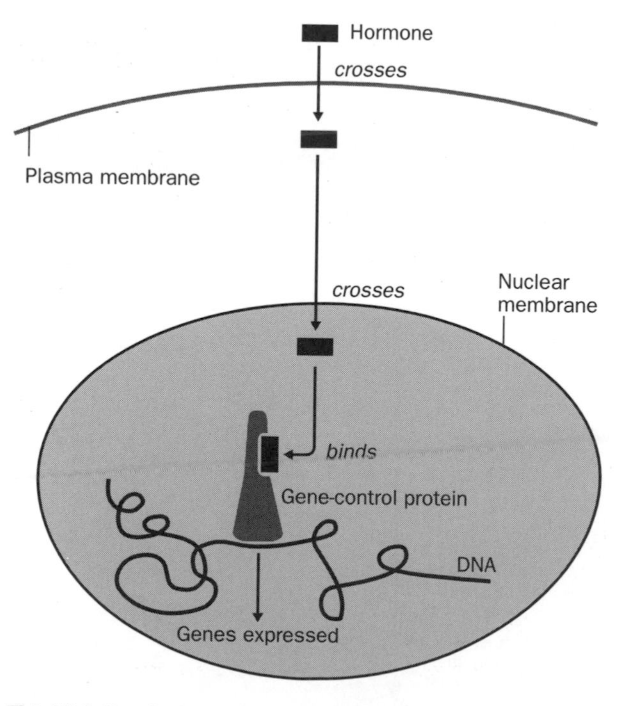

Fig. 16.3 How hydrophobic hormones act in the target cell.

these hormones do not act quickly. An hour or more passes from hormone binding until the first effects are seen.

Hydrophilic hormones like insulin are detected by cell-surface receptor proteins. These fall into two very distinct classes. G protein-linked receptors set in motion a chain of events leading to the release of a small signalling molecule, a second messenger, within the cell. Enzyme-linked receptors activate a cascade of protein kinases, each activating the next until the final kinase activates important control proteins.

G proteins and cell signalling

When G protein-linked receptors bind a hormone or other signalling molecule, they activate another protein inside the cell, called a G protein. These proteins can bind either guanosine diphosphate (GDP) or triphosphate (GTP). The active form is the one with GTP attached. The activation of the G protein sets off a chain of events leading to the appearance of unusually high levels of a second messenger in the cell. The most widely used second messengers are cyclic AMP (cAMP; Fig. 16.4) and Ca^{2+}. These in turn trigger a whole range of chemical changes which bring about the response.

If cAMP is the second messenger, the G protein activates adenylyl cyclase (Figs. 16.4 and 16.5), which makes cAMP from ATP. The cAMP level in the cytosol suddenly rises. In animal cells, at least, this is detected by a protein kinase called A-kinase. Protein kinases are enzymes whose role it is to activate other enzymes by phosphorylating certain amino acid residues (Box 16.2). A-kinase activates enzymes which bring about the final response. For example, when muscle cells detect adrenaline, cAMP is released as a second messenger. A-kinase, in these cells, activates two of the enzymes involved in glycogen breakdown, releasing glucose in preparation

Box 16.2

Protein kinases

Many proteins are controlled by the addition of a phosphoryl group to a particular amino acid side chain. In some cases this activates the protein. In others, it inhibits its activity. Protein kinases are enzymes which add phosphoryl groups to proteins. Protein phosphatases reverse the effect by removing phosphoryl groups.

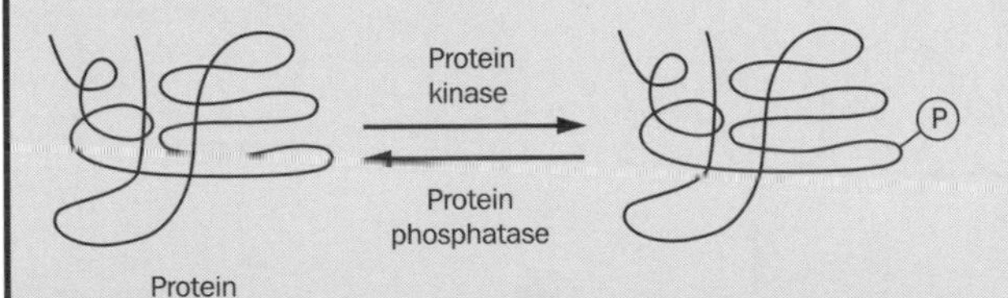

The proteins that are phosphorylated may be enzymes, including other protein kinases as part of a signalling cascade, or DNA-binding proteins. Most protein kinases phosphorylate serine and/or threonine residues. A few phosphorylate tyrosine residues. Many of these tyrosine kinases are receptor proteins.

Some protein kinases are highly specific, others act on a wide range of proteins. They are all related: the part of the active site which catalyses phosphorylation is extremely similar in all protein kinases. This implies that the DNA sequences of the genes coding for them are also similar: they belong to a gene family (Section 12.7). The parts which recognize the amino acid to be phosphorylated are more variable. The parts which determine whether or not the kinase is active, possibly as a result of phosphorylation by another kinase, are much more diverse.

The first protein kinase to be discovered was phosphorylase kinase. It controls glycogen breakdown in animal cells by phosphorylating a key enzyme.

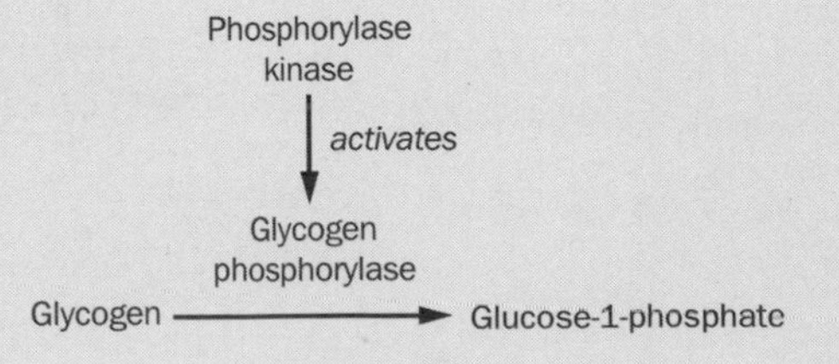

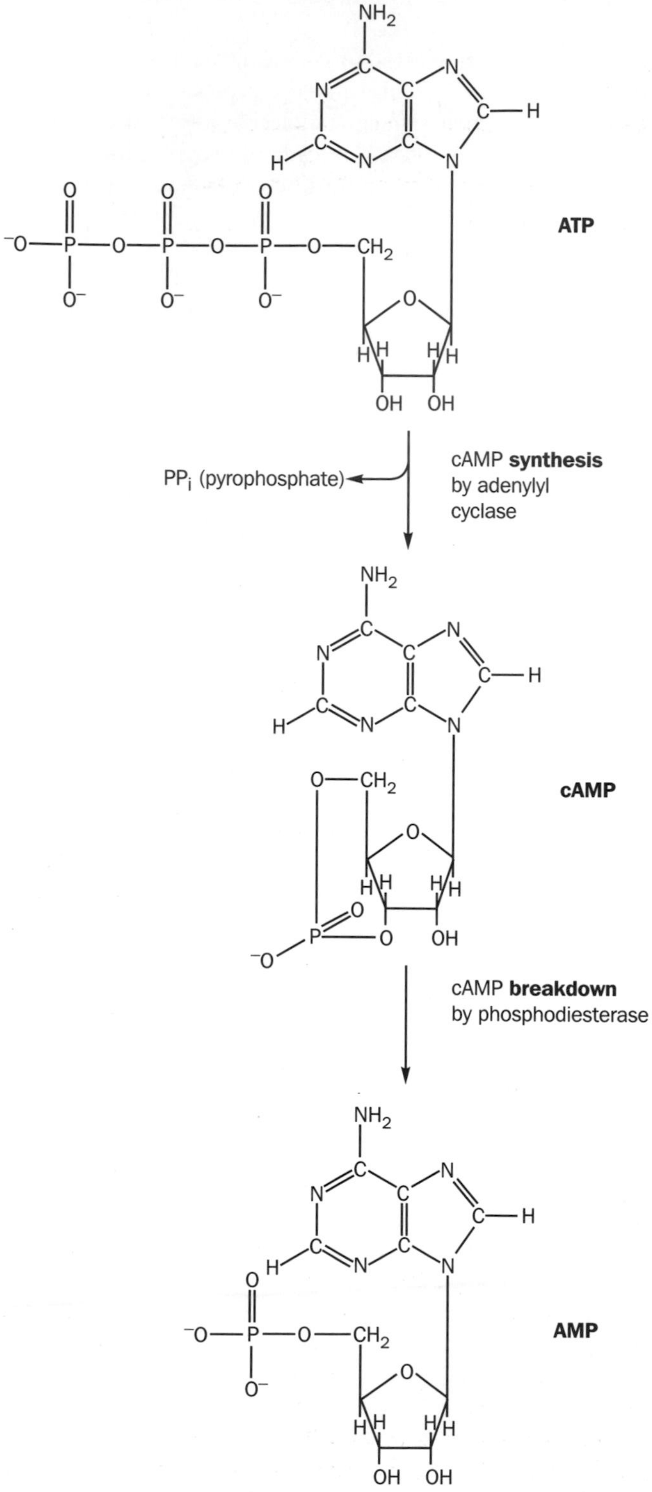

ATP

PP_i (pyrophosphate)

cAMP **synthesis** by adenylyl cyclase

cAMP

cAMP **breakdown** by phosphodiesterase

AMP

Fig. 16.4 Cyclic AMP (cAMP), its formation and breakdown.

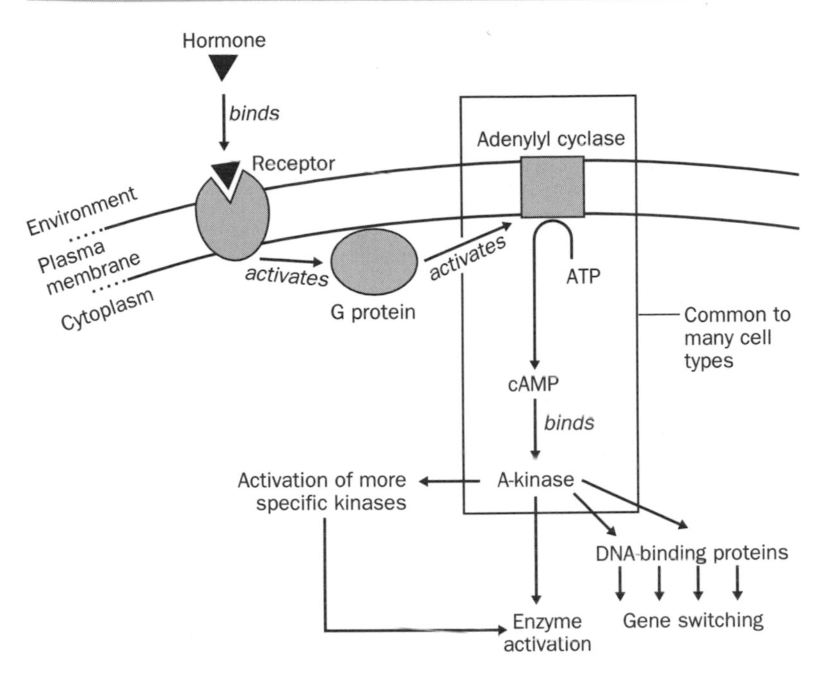

Fig. 16.5 Cyclic AMP as a second messenger.

for sudden muscle activity. In some cells, cAMP is the second messenger in situations where gene switching brings about the response. In these cases, A-kinase activates DNA-binding proteins.

Most of this signalling pathway is common to all animal cell types. Each can only respond to a few hormones. It is the presence of different receptors and different proteins activated by A-kinase, which makes a cell type respond to hormones in a unique way.

Cells usually need to respond to hormones in a reversible way. When the hormone goes, the response should stop. This can only happen if the signalling pathway is switched off. Several steps in the cAMP pathway are quickly reversed. G proteins rapidly lose one phosphoryl group from the bound GTP, which reverts to GDP, inactivating the protein. cAMP is quickly converted to simple AMP by an enzyme. The effects of A-kinase are constantly being reversed by protein phosphatases which remove the phosphoryl groups it has added. Only the continued presence of the hormone maintains the response.

The signalling pathway within the cell which uses Ca^{2+} as a second messenger is quite similar, but an extra stage is involved. Whereas cAMP was manufactured, Ca^{2+} is released into the cytosol from the endoplasmic reticulum where it is stored, as well as entering through channels in the plasma membrane. An extra messenger is needed to carry the signal from the G protein at the inner face of the plasma membrane to all parts of the calcium-sequestering ER. This messenger is usually inositol trisphosphate (IP_3; Fig. 16.6). The G protein activates a phospholipase which breaks IP_3 off a membrane phospholipid: phosphatidylinositol bisphosphate (Fig. 16.7). IP_3 can diffuse around the cytosol and activates calcium channels in the ER membrane. The result is a rapid influx of Ca^{2+} into the cytosol. This is a one-off event, of course: once calcium has diffused out and reached equilibrium, no more can follow.

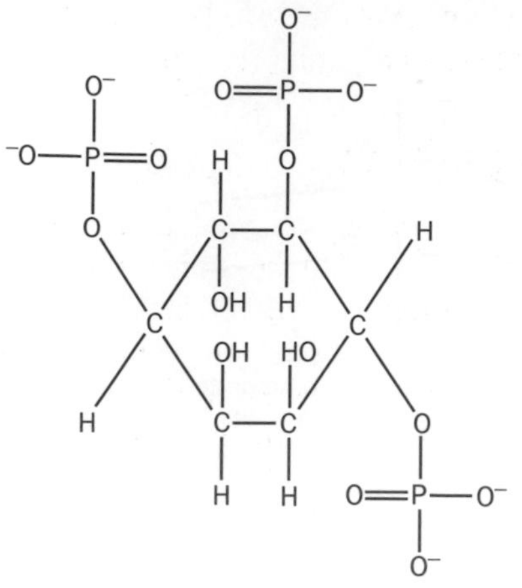

Fig. 16.6 Inositol trisphosphate (IP_3).

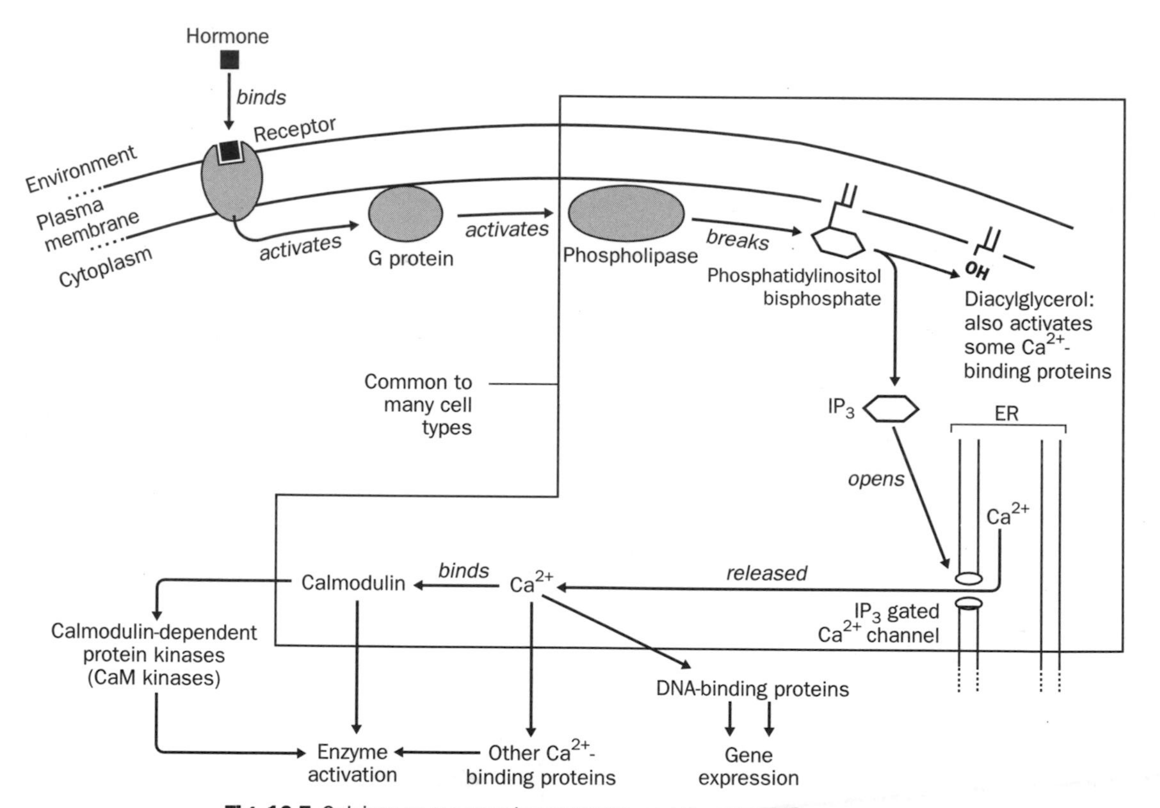

Fig. 16.7 Calcium as a second messenger, and the role of IP_3.

However, the ion channels close, active pumps continually put calcium back into the ER, and the system is quickly primed ready for the next activation. In this way, detection of a hormone results in repeated bursts of Ca^{2+} in the cytosol. How is calcium detected? A protein called calmodulin binds Ca^{2+} and then activates protein kinases by binding to them. Some of these calmodulin-dependent protein kinases (CaM kinases) are highly specific; others are found in many cell types. ADH receptors work via this pathway. Non-hormonal signalling systems may also use this mechanism.

For example, Ca^{2+} is the second messenger when platelets in the blood recognize thrombin and clump together as part of the clotting mechanism (Section 23.3).

Enzyme-linked receptors in cell signalling

These receptors are mostly either tyrosine kinases themselves, or have one intimately attached to them. Many sorts of receptor seem to activate the same enzyme cascade (Fig. 16.8), in which a series of serine-threonine kinases activates the next in the cascade until the last activates gene-control proteins or more specific protein kinases. Receptors are linked to the cascade by Ras proteins, which are active when bound to GTP and inactive when bound to GDP. Receptors which work like this include those for insulin and for various obscure but vital hormones which stimulate growth in particular cell types.

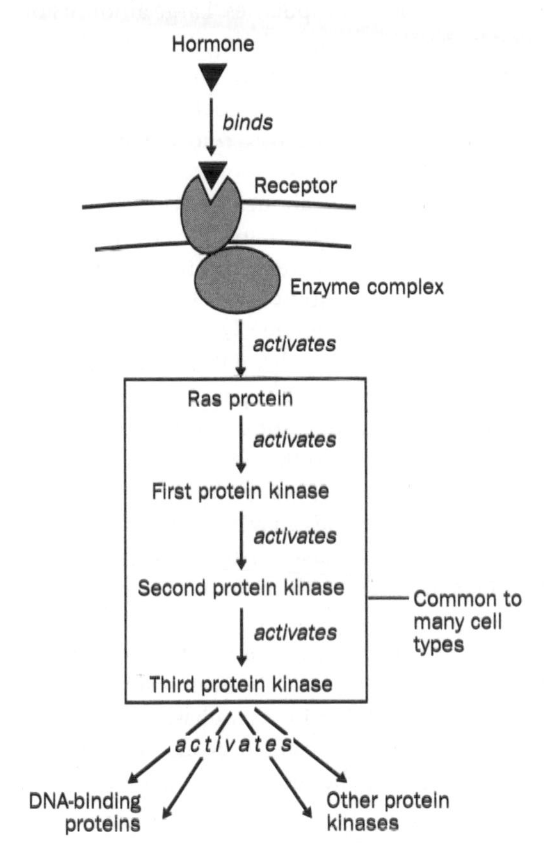

Fig. 16.8 Enzyme cascade activated by an enzyme-linked receptor.

Common themes in cell signalling

There are strong similarities between the ways in which enzyme-linked and G protein-linked receptors bring about an effect. Many types of receptor found in various types of cell each activate a common signalling pathway. The 'meaning' of the signal within the cell depends upon the call type, upon which receptors are linked to it and what the effects of the signal happen to be. In each case, GTP-binding proteins make the link between diverse receptors and the shared signalling pathways. G proteins and Ras proteins may be quite distinct in structure, but their roles within signalling pathways are very similar. This is a striking example of parallel evolution at the molecular level. Each of these signalling pathways has a number of steps. This allows great amplification of a signal within the cell, just as the system of releasing hormones allows signal amplification within the body as a whole.

16.5 Paracrines

Paracrines are local signalling molecules which operate on the scale of micrometres to millimetres. Histamine (Fig. 16.9(a)) is made by mast cells, which lurk outside capillaries in most parts of the mammalian body and look much like white blood cells. It is released in the event of infection or injury and causes local inflammation. In particular, it causes dilatation of blood vessels and leakiness of capillaries. Large amounts of plasma accumulate between the cells, and fibrin clots form, isolating the affected area.

Many sorts of cells make prostaglandins. This family of paracrines is manufactured from certain membrane phospholipids and has many members (Fig. 16.9(b)). Their effects are diverse and still poorly understood. Some have a role in pain. Others seem to act through stimulating or inhibiting smooth-muscle contraction, particularly smooth muscle in blood-vessel walls. This brings about vasoconstriction or vasodilatation (Section 19.5). Three examples illustrate their specific effects in one organ: the uterus. Just before menstruation begins, blood vessels in the uterine wall become highly constricted. The failure of the blood supply leads to localized cell death, initiating the breakdown of the uterine lining. High levels of certain prostaglandins seem to trigger this. Other prostaglandins act directly on the smooth muscle of the uterine wall during labour, stimulating the complex pattern of rhythmic contraction which leads to childbirth. There is also a suggestion that the high levels of prostaglandins in semen may stimulate contractions in the non-pregnant uterus, helping to carry sperm towards the Fallopian tubes by a sort of peristalsis.

To look for a general role for prostaglandins in the body may be to miss the point. They are local signalling molecules and can therefore have distinctive local effects. The same signal can have different meanings in different places only if the signalling molecule does not travel too far.

Paracrines: Local signalling molecules.

Neurotransmitters: Molecules which carry information across synapses.

Synapses: Tiny gaps between adjacent neurones.

Neurotransmitters carry information across **synapses**, the gaps between adjacent neurones. They are, then, operating on the scale of nanometres. Neurotransmitters will be discussed in the context of the nervous system, but it should be emphasized that there is no clear distinction between chemical cell communication and nervous communication. Information transfer in nerves is still, essentially, a chemical phenomenon.

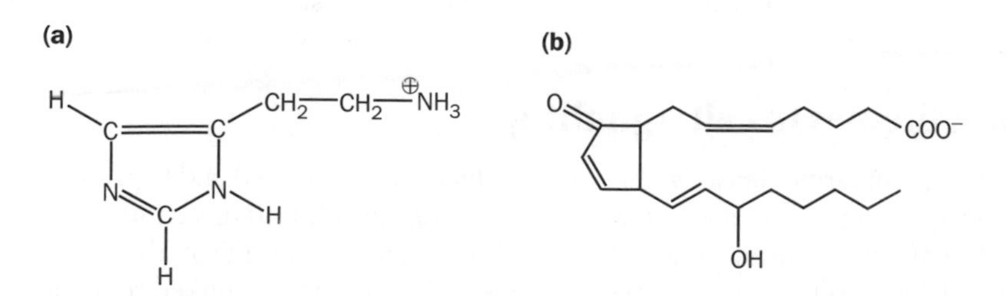

Fig. 16.9 Paracrines. (a) the structure of histamine, (b) the carbon skeleton and functional groups of prostaglandin A_2.

Nitric oxide (NO) has only recently been shown to act as a local signalling molecule. It is a neurotransmitter in some distinct regions of the brain, but can also act as a paracrine. When the endothelium lining small arteries (Box 19.2) detects rapid blood flow, it releases NO. It is a signal for smooth muscle cells in the artery wall to relax, so the artery dilates. NO is being implicated in more and more processes: Box 13.3 shows one example.

16.6 Speed at a price

Within cells, signalling molecules move by diffusion. The same is true when cells communicate with their neighbours. Small molecules diffuse effectively over micrometre distances, but not over millimetres or more. Transport in the blood is the only effective way to spread signalling molecules throughout an entire body. In humans the fastest acting hormones, such as adrenaline, take seconds to have an effect, which lasts for seconds. The slowest, like thyroxine, take hours or days to have an effect which may last for days or weeks. When quicker responses are needed, the nervous system becomes involved. Neurones can carry signals at up to 30 m s^{-1}. The entire process of responding to some change in the environment takes only a fraction of a second.

Communication by neurones is also highly specific. Hormones carry information from one cell type to another. Paracrines carry information to all the cells of one type in a local area. Neurones carry information between two individual cells, which may be metres apart. This rapid, specific communication comes at a price. Each receptor cell, or group of co-operating receptor cells, has to have its own neurone to carry information away. Each group of muscle cells needs its own neurone to carry incoming signals. It takes materials and energy to build and maintain a neurone. Hormones, on the other hand, are carried everywhere by a transport system which already exists: the blood. In situations where speed and extreme specificity are not critical, hormones are a slower but cheaper option.

Summary

- Cells communicate by chemicals and nerve impulses.
- Hormones are chemical signals carried in an animal's blood. They are made in one organ or cell type, and are detected by another.
- Steroid hormones enter the cell and bind to gene-control proteins.
- Most other hormones are detected by receptors in the plasma membrane.
- G protein-linked receptors are linked to second messengers, such as Ca^{2+} or cyclic AMP. They appear in the cytoplasm when the receptor detects a signal.
- Enzyme-linked receptors activate a cascade of enzymes.
- Paracrines are local signalling molecules. They do not travel throughout the body.

Exercises

16.1. There are no G proteins linked to the receptors for aldosterone or thyroxine. Why not?

16.2. Thyroxine controls metabolic rate only on a long time scale. Its level is controlled by homeostasis, preventing wild fluctuations. Predict the effect of high thyroxine level on the anterior pituitary.

16.3. Dwarfism is a condition in which the body grows far more slowly than normal. Most children with this condition also fail to become sexually mature, and have a low metabolic rate. In one third of cases, however, only growth is affected. What is the pituitary failing to do in each case?

16.4. For each hormone there is a system which removes it from the blood. It may be broken down by cells, excreted in the urine by the kidneys, or excreted in the bile by the liver. Why is this necessary?

Further reading

Alberts, B., Bray, D., Lewis, J., Raff, M., Roberts, K. and Watson, J.D. *Molecular Biology of the Cell* (3rd ed.) (New York: Garland, 1994). An enormous but excellent textbook which puts the reader in touch with current ideas without too much pain on the way. A long chapter on cell signalling is especially useful for its treatment of signal transduction within the cell (G proteins, enzyme cascades, second messengers and so on).

Guyton, A.C. and Hall, J.E. *Textbook of Medical Physiology* (9th ed.) (Philadelphia: Saunders, 1996). I like this hefty textbook of human physiology for its thoroughness and clear, no-frills diagrams.

Guyton, A.C. *Human Physiology and Mechanisms of Disease* (5th ed.) (Philadelphia: Saunders, 1992). Effectively a shorter, boiled-down version.

Schmidt-Nielsen, K. *Animal Physiology: Adaptation and Environment* (4th ed.) (Cambridge: Cambridge University Press, 1990). One of the all-time great textbooks! Exciting, authoritative and easy to read, it puts animal physiology in the context of the environment.

Withers, P.C. *Comparative Animal Physiology* (Fort Worth: Saunders, 1992). More detailed than Schmidt-Nielsen, with wider-ranging examples: the inevitable cost is that it is a less easy read.

Cell Communication: Nerves

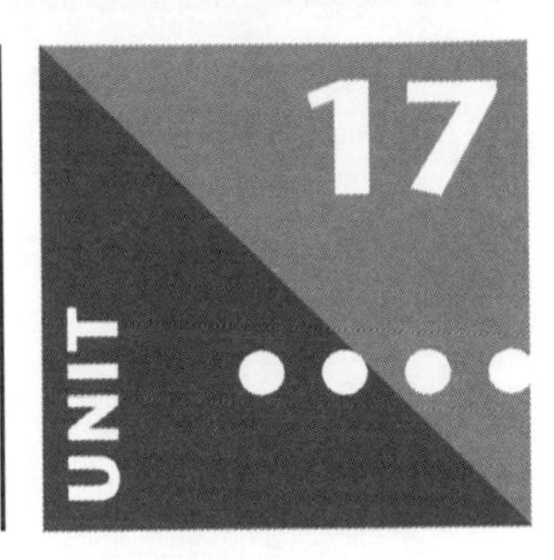

Connections

▶ Before studying this unit you should have a basic understanding of how ions and molecules move across membranes (Unit 6). Many examples of communication between cells have to do with homeostasis, so you should understand the principles of this (Unit 15). The nervous system has close links with chemical communication systems (Unit 16) and you should already have some knowledge of this topic. The nervous system underlies much of animal behaviour and the control of muscle contraction (Units 29 and 24).

Contents

17.1 Communication and more

I am sitting on a park bench eating chips. (Chips are French fries, but British and more greasy.) I open the paper parcel carefully. In order to do this, several muscles in my arms are controlled very precisely by my brain. Some relax, others contract. Some contract more than others, and the degree of contraction changes rapidly. This sort of control requires signals that are fast, directed at a precise target, and short-lived: nerve impulses. A chip slides out of the soggy mass and begins to fall. I see it and lunge at it; so do the waiting pigeons. Only nerve impulses can carry information from eye to brain fast enough to make reactions like this possible.

Neurones are the cells of the nervous system. They have an obvious role in communicating between sense organs and the brain, as well as between the brain and muscles. However, neurones can interact with other neurones. This means that the nervous system does much more than transfer information: it can also process information. For example, when light reflected by a chip enters the eye, an array of light-sensitive cells sends very simple information to the brain. Each cell simply tells the brain how much light it is detecting. It is for the brain to sort out this mass of incoming data, identify patterns in it and draw a useful conclusion: 'I see a chip before me'.

Even the simplest nervous systems can coordinate the actions of cells within a body. The most advanced nervous systems seem to be capable of processes which have no immediate input or output, but are confined to interactions between neurones. Now imagine sitting on a park bench yourself on a sunny evening, eating chips.

You've pictured it? No information from sense organs entered the brain to help with this; no muscles moved as a result. Yet something clearly went on in your brain as you imagined the scene.

17.2 Neurones

Neurones are the cells of the nervous system. They are specialized for carrying information quickly. Each has a cell body containing the nucleus, and a number of fine extensions. Most have one long extension, an axon, and several shorter, branched extensions, dendrites. The plasma membrane covers all these extensions. Fig. 17.1 shows a range of mammalian neurones.

Axons carry signals from one place to another. They can be very long: some axons in a giraffe or sperm whale are several metres long. Dendrites and the ends of axons make connections with other neurones. These connections are called synapses. Information passes across synapses from one neurone to another. Differences in the ways in which input signals bring about output signals in synapses are the basis of information processing in the nervous system. Some axons in vertebrates are coated

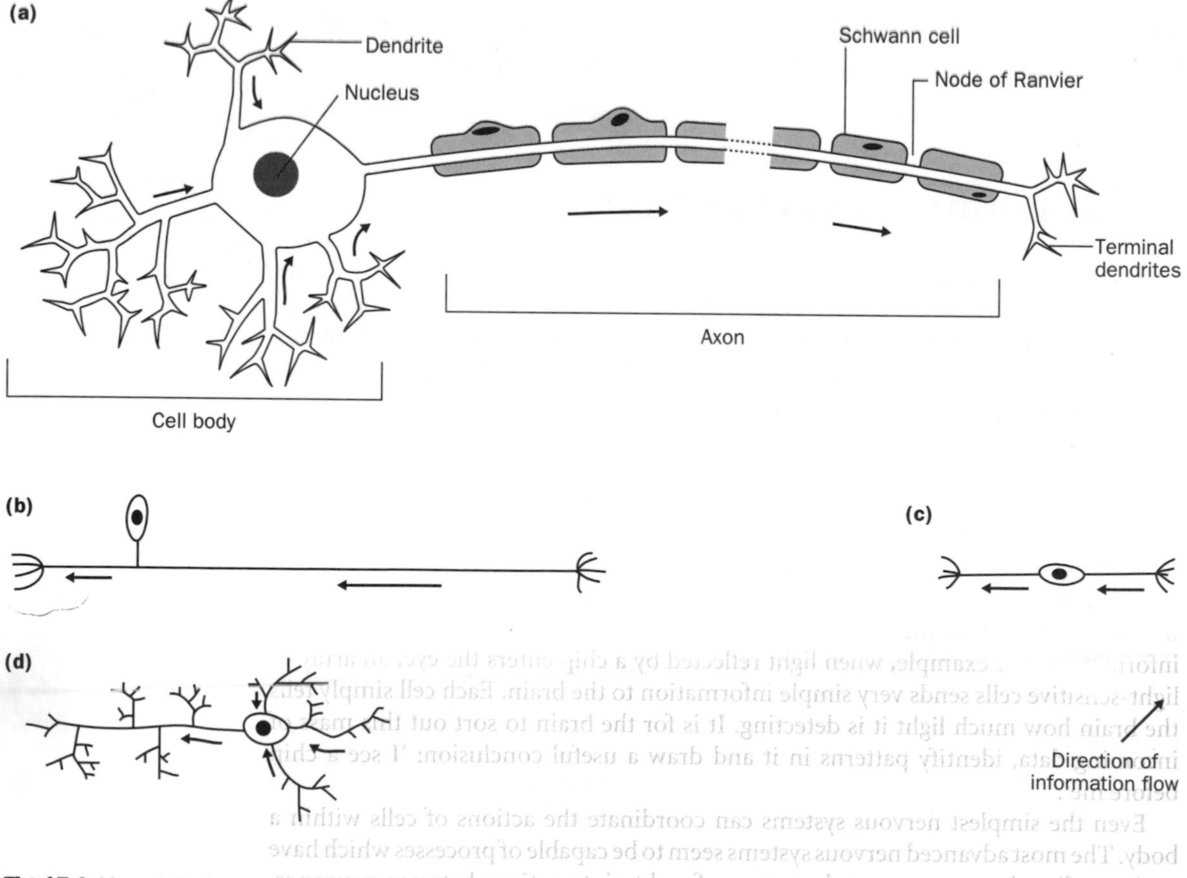

Fig. 17.1 Mammalian neurones. (a) a motor neurone in detail, (b) a sensory neurone, (c) a bipolar neurone from, for example, the retina, (d) one of many types of neurones in the brain.

in a fatty layer: a myelin sheath. This is formed by Schwann cells which grow coiled around the axon like Swiss rolls. Once mature, they contain very little cytoplasm, so the myelin sheath is essentially many layers of membrane. Myelin allows faster conduction (Section 17.3: How quick is quick?).

Nerves are bundles of axons. Cell bodies are found only in clusters called ganglia (singular: ganglion) or within the **central nervous system** (CNS) of vertebrates. Many entire neurones are found within the CNS. Others hang out of the CNS into the body, their axons forming parts of nerves. Most contain a mixture of cells which bring information from receptor cells to the CNS (sensory neurones) and motor neurones which carry signals out to effector organs, notably muscles. Any one axon could carry information in either direction, but in practice they are one-way streets. It is the cells to which the neurone is connected that determine which way information normally flows.

17.3 Nerve impulses

Neurones carry information faster than hormones. The destination of the information is also much more specific. The cost of this is in the energy and materials needed to build and maintain neurones. This is justified if rapid cell communication has survival benefits, for example in detecting and escaping from predators, or in sizing up the qualities of a potential mate before it wanders off or is snapped up by a rival. It may not be justified if a response on the time-scale of seconds or minutes is adequate, or if there are many target cells, for example when every liver cell needs, individually, to be told to increase the rate of glycogen synthesis.

Information travels along neurones as a stream of pulses. Each pulse consists of a cycle of changes in the axon membrane. These changes involve the opening and closing of ion channels and the movement of sodium and potassium ions. The upshot is that the electrical gradient across the membrane breaks down for a few milliseconds before being restored. This depolarization triggers depolarization of the next piece of membrane, and so the impulse moves down the axon.

Nerve impulses, or action potentials, are more or less identical. There is not one sort of impulse meaning 'bright light has been detected' and another meaning 'this muscle cell should contract'. The meaning is determined by what the neurone is attached to and how frequently impulses are passing along it. A sensory neurone leading from a group of rod cells in the retina of the eye to the brain carries information on the intensity of light falling on one small region of the retina. The more intense the light, the more frequent the impulses.

The first problem to tackle is the nature of the nerve impulse and the way it is passed down the axon. The important changes in the axon membrane concern its permeability to sodium and potassium ions. As in any other animal cell, a Na^+/K^+ ATPase constantly pumps Na^+ out of the cell in exchange for K^+. This sets up concentration gradients: both ions would diffuse back again if the membrane allowed them to. However, the membrane need not allow both ions to move to the same extent. If only Na^+ could cross freely, it would tend to re-enter the cell, even though K^+ could not leave. This would result in the cytosol becoming more positively charged. Calculations suggest that once equilibrium was reached, there would be a voltage across the membrane (a membrane potential) of about 60 mV (Fig. 17.2). On the other hand, if the membrane was only permeable to K^+, the cytosol would be

Central nervous system: The brain and spinal cord.

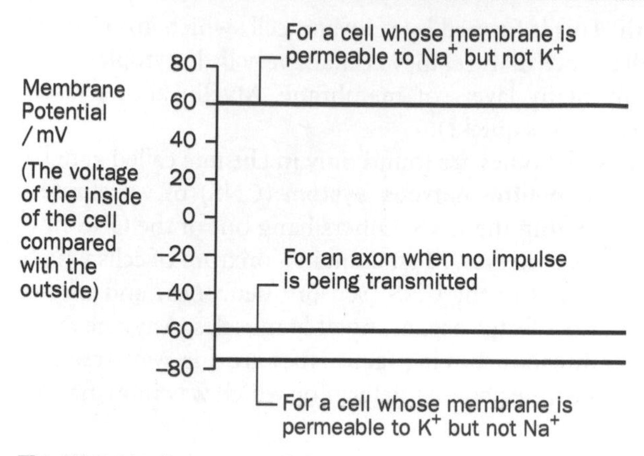

Fig. 17.2 Membrane potentials.

more negative, with a membrane potential of about −75 mV. Real axons, when not carrying a nerve impulse, have membrane potentials of about −60 mV. This is called the resting potential. This value is close to the second theoretical value, suggesting that the resting axon membrane is much more permeable to potassium than sodium (Fig. 17.3). This is because the membrane contains some K^+ channels which are permanently open.

Now consider what would happen if the membrane contained sodium channels which were normally shut, but could be opened for very short periods of time, about 1 ms each time. Sodium ions would diffuse in, making the cytosol less and less negative, or even slightly positive. The membrane would have been depolarized (Fig. 17.4). Once the channels closed again, the Na^+/K^+ ATPase would re-establish the resting potential. All this really happens in each part of the axon membrane as a nerve impulse passes, but that is not the whole story.

The membrane contains some closeable potassium channels as well as sodium channels (Fig. 17.5). Both are controlled by voltage. The sodium channels are shut when the membrane is at resting potential, but even if the membrane is depolarized only a little, the channels open, allowing sodium ions to diffuse in. This brings about further depolarization, and so on. The result is a positive feedback system in which a tiny initial change in membrane potential triggers a cataclysmic depolarization. The sodium channel seems to have a voltage-controlled gate at the outer face of the membrane. There is a second gate at the cytoplasmic end of the channel, which was initially open but slams shut as the membrane depolarizes and remains closed for

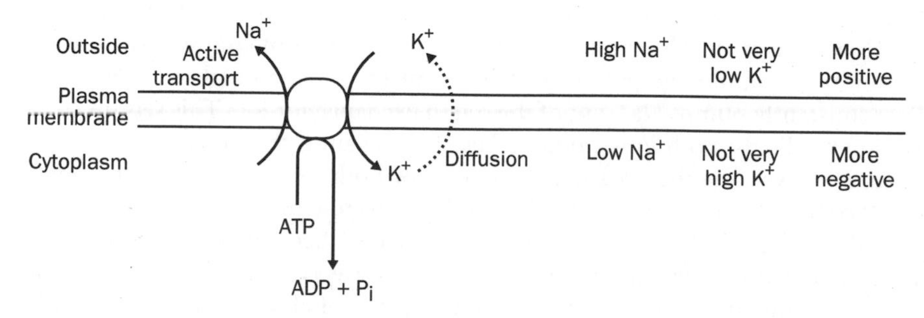

Fig. 17.3 Movement of sodium and potassium ions across the resting axon membrane.

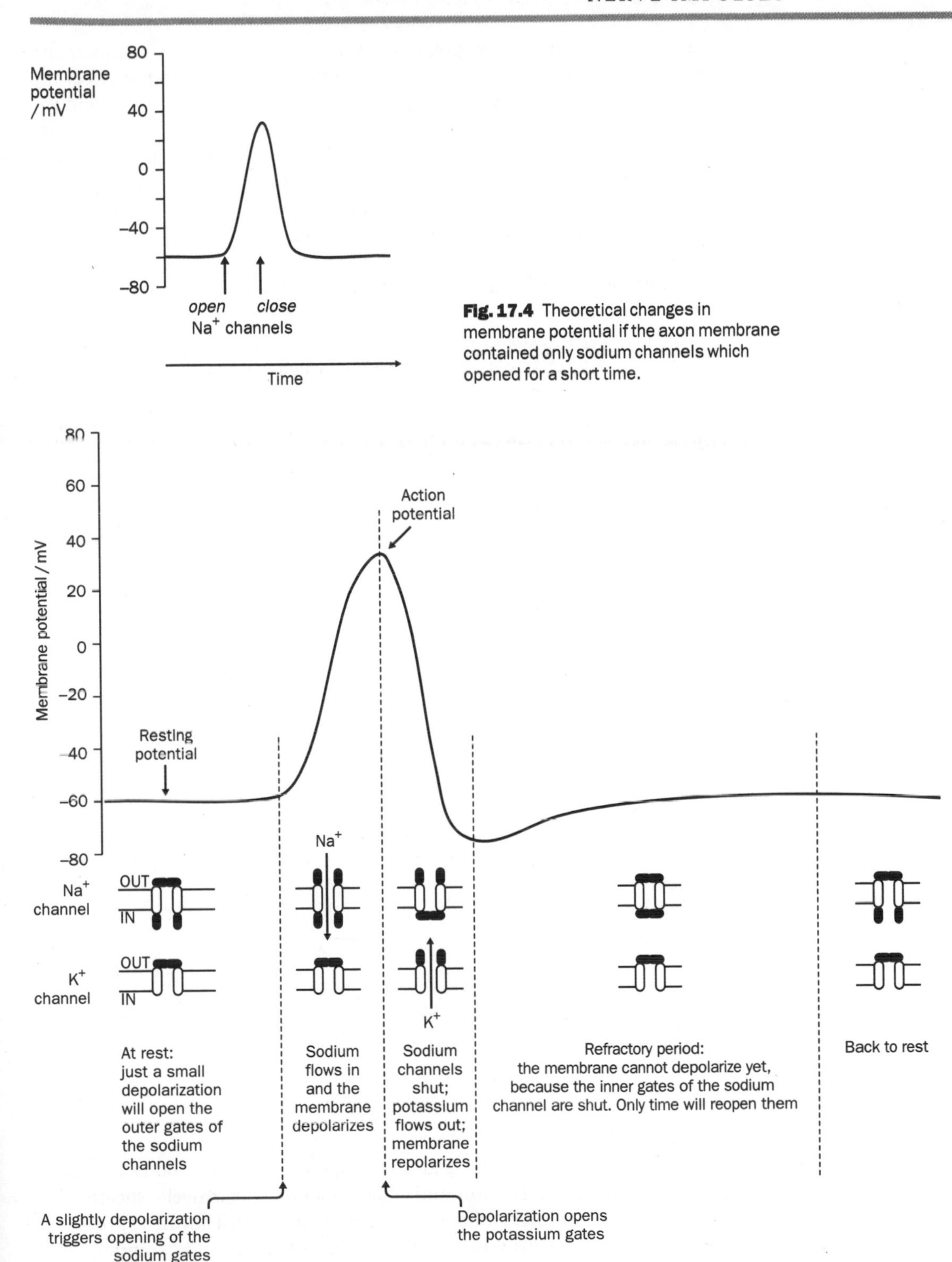

Fig. 17.4 Theoretical changes in membrane potential if the axon membrane contained only sodium channels which opened for a short time.

Fig. 17.5 Opening and closing of ion channels in the axon membrane as an action potential passes.

several milliseconds after resting potential is restored. Potassium channels are also opened by depolarization, but there is a slight delay, so they do not affect the situation until the membrane has fully depolarized. Once open, they allow rapid K^+ diffusion, which helps in speedy repolarization. This increased K^+ permeability means that the membrane becomes hyperpolarized for a short time, with the cytosol even more negative than usual. As K^+ channels close, the membrane returns to its resting potential. Even then, the membrane of this section of axon cannot carry another impulse for a while, until the second gates of the Na^+ channels have reopened. This hiatus lasts several milliseconds and is called the refractory period.

All this is quite useless unless an action potential at one point in the axon can start to depolarize the next bit of membrane, so allowing the impulse to move down the axon. This is very easy to explain given that positive charges (such as sodium and potassium ions) are attracted towards negatively charged places.

Imagine a length of axon carrying an impulse (Fig. 17.6). At this moment, the action potential is halfway along it. At that point in the axon, the cytosol is slightly more positive than the extracellular fluid. Everywhere else, the cytosol is more negative. Na^+ and K^+ ions in the cytosol will be attracted away from the point with the action potential. Even a slight movement of ions will mean that the cytoplasm on either side of the action potential will become slightly less negative: the membrane will have been slightly depolarized. Ahead of the action potential, this will trigger a full depolarization, so the action potential is propagated on down the axon. Behind the action potential, the membrane is still in its refractory period, so cannot depolarize. Hence, the action potential moves steadily, in one direction. It is not a material thing, but a state of the membrane, as real and as insubstantial as a Mexican wave.

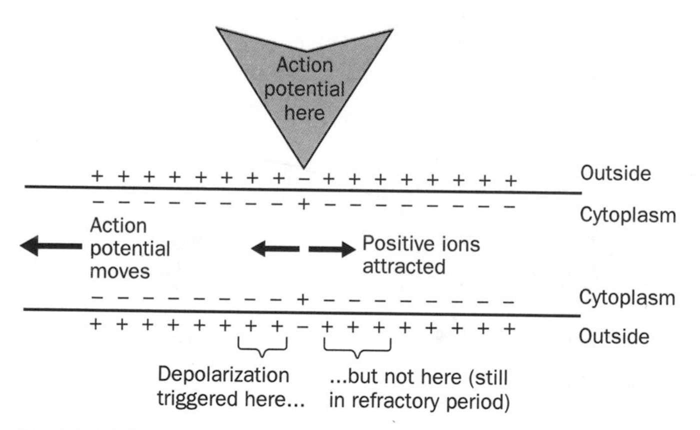

Fig. 17.6 Why the action potential moves along the axon.

How quick is quick?

Three things affect the rate at which an action potential moves along an axon: temperature, the thickness of the axon, and whether or not it has a myelin sheath. The higher the temperature, the higher the speed of conduction. Most chemical processes respond to temperature in this way, and the action potential is no exception. The thicker the axon, the faster the impulse travels. This is clearly seen in many invertebrates including insects and, most famously, the squid, which have a few giant axons up to 1 mm in diameter, as much as 100 times thicker than their other axons. These

Much of our knowledge of nerve impulses came from studies on the giant axons of squids.

carry information fast: impulses travel at around 30 m s^{-1}. They are frequently involved in bringing about escape responses, where milliseconds count. Vertebrates do not have giant axons, although some vertebrate axons are thicker than others. A myelin sheath greatly speeds up conduction. Many neurones outside the CNS of vertebrates have thin, myelinated axons which carry signals as quickly as invertebrate giant axons but with a lower cost in materials and space. Each section of axon has only one Schwann cell Swiss-rolled around it, and there is a short gap, the node of Ranvier, between one Schwann cell and the next. The sections of axon membrane between the nodes have scarcely any sodium channels. These membranes cannot depolarize so the action potential jumps from node to node almost instantaneously.

One neurone meets another at a synapse. The next problem to tackle is how information crosses synapses.

17.4 Synapses and information processing

Each neurone has synapses with other neurones. Some have thousands. In vertebrates, most are in the brain and spinal cord. They are not just awkward gaps which must somehow be bridged, but have an important role in information processing. Impulses stream, unstoppable, down the axon of a neurone. Their frequency, and so their meaning, does not change as they go. Individual neurones have to process incoming information and 'decide' at what frequency impulses should be sent off down their own axons. Information enters the neurone through its synapses. The signals need not be identical on the two sides of the synapses. Incoming information is summed and passed to the axon hillock, a region which acts as the gateway to the axon and fires off action potentials on the basis of this information.

There are two quite different types of synapse, electrical and chemical. Electrical synapses are quick, simple and relatively uncommon. The two neurones meet at a gap junction (Section 4.7). The closeness of the membranes, and the pore complexes which allow free diffusion of ions mean that action potentials pass across unchanged. Electrical synapses are found in small numbers in both vertebrates and invertebrates. Their general significance is not well understood although their role in the 'tail flick' escape response of fish, where speed is everything, is well documented.

Chemical synapses pass information more slowly and allow the possibility of some change in the signal. Also, unlike many electrical synapses, information can only cross a chemical synapse in one direction. The neurone carrying information into the synapse is referred to as the presynaptic neurone; the one carrying it onwards is the postsynaptic neurone.

Neurotransmitters

The plasma membranes of the pre- and postsynaptic neurones are not physically linked, as they are in electrical synapses. Instead, there is a gap of 20–30 nm between them: the synaptic cleft. Information passes across the cleft in the form of a diffusing molecule: a neurotransmitter (Fig. 17.7). Neurotransmitters are stored in vesicles close to the presynaptic membrane. They fuse with the membrane when action potentials arrive, releasing the neurotransmitter into the cleft. It is bound by receptors in the postsynaptic membrane, which leads to a response. There are not many possible responses: the postsynaptic membrane can either depolarize, or not. Some neurotransmitters usually depolarize (excite) the postsynaptic membrane. These include acetylcholine, found in many parts of the CNS as well as where motor neurones meet skeletal muscle; noradrenaline, used widely in the CNS as well as in the sympathetic nervous system; and glutamate, probably the most common excitatory neurotransmitter in the brain. Others usually inhibit depolarization. These include γ-aminobutyric acid (GABA) and dopamine, both restricted to the CNS. It is not the transmitter itself that is inhibitory, but the combination of transmitter and receptor. So, less commonly, noradrenaline may be inhibitory and dopamine excitatory.

What triggers the release of a neurotransmitter? The arrival of an action potential triggers a flow of Ca^{2+} into the end of the presynaptic neurone, due to voltage-gated calcium channels in the plasma membrane. This acts as a second messenger, leading ultimately to vesicle fusion.

Neurotransmitters work by binding to and opening ion channels in the postsynaptic membrane. Excitatory transmitters open sodium channels, leading to

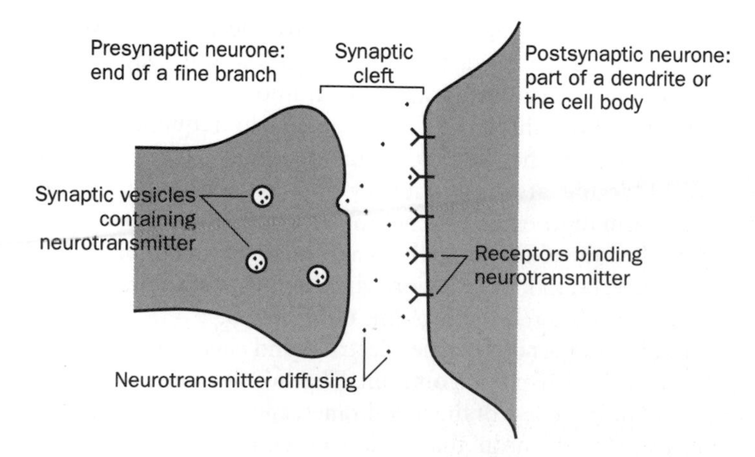

Fig. 17.7 Operation of a generalized chemical synapse.

depolarization. Inhibitory transmitters open other channels which allow chloride ions to diffuse in, making the cytosol even more negative than before.

Neurotransmitters need to be removed from the synaptic cleft. Otherwise, the postsynaptic membrane would keep on responding indefinitely. Some are destroyed by enzymes in the synaptic cleft, for example acetylcholine by acetylcholinesterase. Others, like noradrenaline, are actively reabsorbed into the presynaptic terminal for reuse.

Drugs and synapses

All sorts of interesting drugs act on molecules in synapses. The arrow poison curare causes paralysis by blocking acetylcholine receptors. The tranquillizer Valium® binds to GABA receptors, making them bind GABA more easily. As a result, Valium® tends to open chloride channels. Another poison, strychnine, makes neurones more excitable by inhibiting the effect of glycine in some inhibitory synapses: the excitatory inputs to the postsynaptic neurone then dominate, and muscles go into spasm. Nerve gases and the more alarming types of flea spray for cats prevent coherent muscle control by inhibiting acetylcholinesterase.

Effects of synapses on signals

The significance of the inhibitory synapse is that it reverses the signal: 'signal' becomes 'no signal', 'weak signal' becomes 'strong signal'. It acts much like a 'NOT' gate in electronics. A simple example shows their importance. I sit in the dark,

The arrow poison curare flies into action.

holding a sack of potatoes above my head. I stay like this, feeling foolish, until my colleague turns the light on, a prearranged signal for me to drop the bag, to rapturous applause from my warped audience. Intense light causes frequent action potentials in sensory neurones leading from the eye, yet relaxing the muscles in my arms requires the opposite: impulses arriving down the motor neurones must become much less frequent. What goes on in between the two, in the CNS, is certainly highly complex, but there must be inhibitory synapses in there somewhere. Somehow, 'signal' has become 'no signal'.

The second way in which synapses can affect signals lies in their sensitivity. There is not a simple one to one link between incoming and outgoing impulses. The rate at which neurotransmitter arrives at the postsynaptic membrane determines the frequency of outgoing impulses. The more sensitive the synapse, the less transmitter is needed to produce a given frequency of impulses. Synapses can, then, either amplify or muffle a signal.

The postsynaptic membrane does not depolarize catastrophically then quickly repolarize, like the axon membrane. A different cocktail of ion channels means that it can depolarize a little or a lot, and returns to the resting potential more slowly. The membrane potentials set up by each of the synapses in the postsynaptic region of the neurone are added together to give an overall picture of the degree of stimulation. This overall potential informs the axon hillock how frequently to fire off action potentials down the axon; how this happens is another exciting, but complex story.

The next problem to tackle is that of the inputs to the nervous system: receptors.

17.5 Receptor cells

Receptor cells trigger nerve impulses when they receive a particular environmental stimulus. In mammals alone, there are quite a few types of receptor cell. The retina of the mammalian eye (Box 17.1) contains light receptors called rod cells. These respond to light intensity. Mammals with colour vision also have cone cells mixed in with the rods. These come in three versions, each responding most strongly to a different colour of light: red, green and blue. This is what determines the primary colours of light: it is nothing magical about the light itself. Other colours are perceived on the basis of how much each cell type is stimulated. Each cone is linked to its own neurone, whilst any one of a group of rods can fire a single neurone. The result is that cones are less effective for vision in dim light, but allow finer resolution.

The cochlea of the ear (Box 17.2) contains receptor cells which respond to vibrations – sounds – at particular frequencies. A quiet, high-pitched sound will be detected by one small group of receptor cells. They will fire off action potentials at a low frequency. If the noise gets louder, the same cells will fire more frequently. If the pitch changes, different cells will fire.

The skin contains a wonderful range of receptors. Together, they provide the complex 'sense of touch'. Specialized receptor cells or simple nerve endings detect light pressure, heavy pressure, warmth, coldness (yes, they are different receptors!), pain and hair movement.

Muscles have receptors called spindles wound around their cells. Similar things are found in tendons. They respond to stretching and are central to the perception of where your body is. The state of contraction of every skeletal muscle is used to build up a constantly changing picture of its posture and movements. Before writing this

Box 17.1

Receptor cells in context I: the eye

The human eye is typical of mammalian eyes. Other vertebrate eyes are broadly similar.

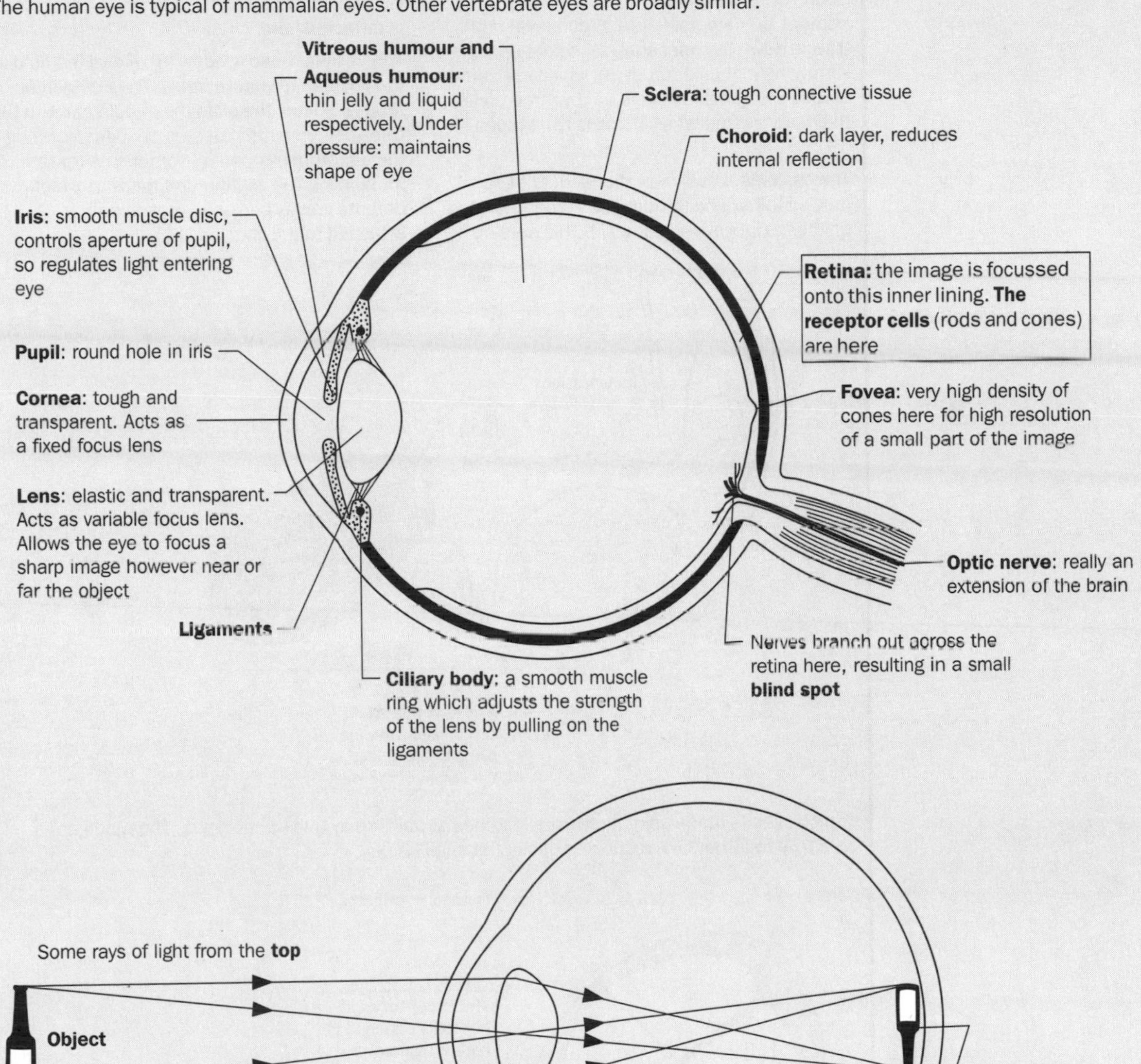
Vitreous humour and
Aqueous humour: thin jelly and liquid respectively. Under pressure: maintains shape of eye
Sclera: tough connective tissue
Choroid: dark layer, reduces internal reflection
Iris: smooth muscle disc, controls aperture of pupil, so regulates light entering eye
Retina: the image is focussed onto this inner lining. **The receptor cells** (rods and cones) are here
Pupil: round hole in iris
Cornea: tough and transparent. Acts as a fixed focus lens
Fovea: very high density of cones here for high resolution of a small part of the image
Lens: elastic and transparent. Acts as variable focus lens. Allows the eye to focus a sharp image however near or far the object
Optic nerve: really an extension of the brain
Ligaments
Nerves branch out across the retina here, resulting in a small **blind spot**
Ciliary body: a smooth muscle ring which adjusts the strength of the lens by pulling on the ligaments
Some rays of light from the **top**
Object
and from the **bottom**
Light is entering the eye from all over the object, of course. These rays are just to show what happens.
Image: on retina: It's upside down, but all the information is there for the brain to work on

How an image is made

Box 17.2

Receptor cells in context II: the ear

Sound enters the pinna (1) (a fixed flap in humans but directable in e.g. donkeys) and travels down the ear canal (2). It sets up vibrations in the ear drum (3), which are carried across the air-filled middle ear by three bones, the ossicles (malleus (4), incus (5), stapes (6)).

The ossicles amplify vibrations, but not so much if the sound is loud. The stapes vibrates on the flexible oval window (7). This sets up vibrations in the fluid-filled cochlea (9). Sound is detected here.

The flexible round window (8) allows the fluid in the cochlea to vibrate freely. The Eustachian tube (10) links the air in the middle ear with the throat. It prevents pressure changes bursting the ear drum by opening sometimes (ears 'popping'). The vestibular apparatus which detects gravity and head movements is attached to the cochlea (11).

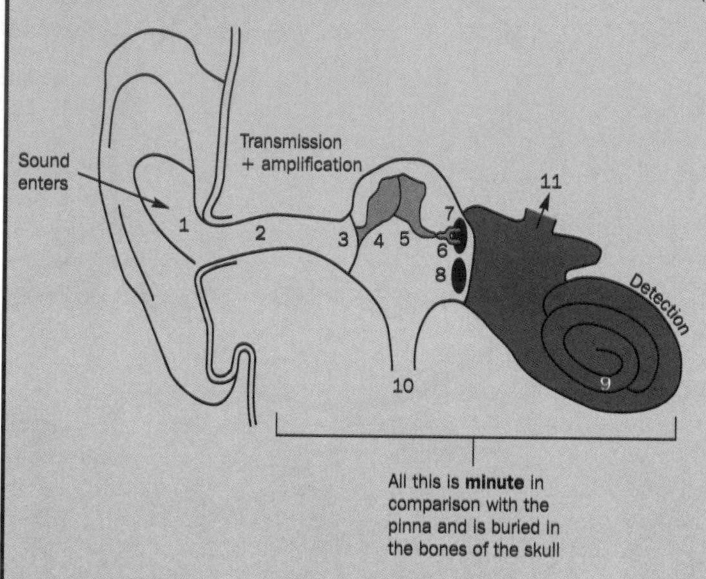
Sound enters
Transmission
+ amplification
1
2
3
4
5
6
7
8
9
10
11
Detection
All this is **minute** in comparison with the pinna and is buried in the bones of the skull

The cochlea is made up of three parallel tubes separated by thin membranes. The tubes are coiled up together, like a Chelsea bun or flat snail shell.

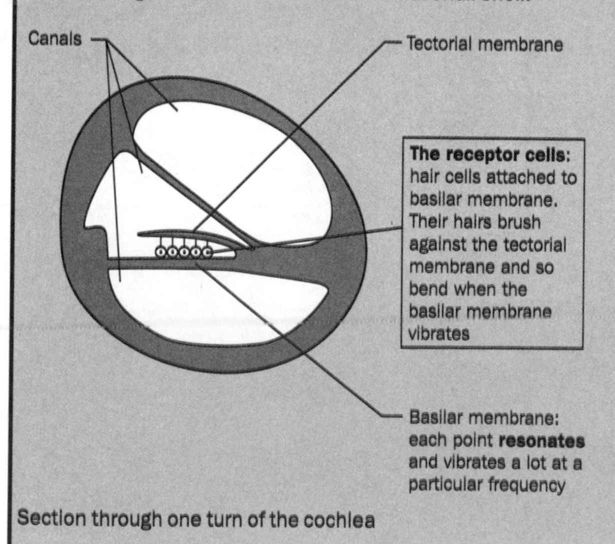
Canals
Tectorial membrane
The receptor cells: hair cells attached to basilar membrane. Their hairs brush against the tectorial membrane and so bend when the basilar membrane vibrates
Basilar membrane: each point **resonates** and vibrates a lot at a particular frequency

Section through one turn of the cochlea

sentence I closed my eyes and touched my big toe with the end of my pen. It worked first time. This would not have been possible without muscle spindles.

Gravity is detected by receptor cells in the vestibular apparatus of the inner ear. The principle is simple: the receptor cells have hairs which are all embedded in a calcium carbonate crystal. The crystal can flop around on its hairs, within a chamber. Gravity pulls it downwards, and receptor cells detect the hairs bending. Which cells fire depends on which way is down. Movement of the head is detected by very similar cells, without the crystal, in another part of the vestibular apparatus. Their hairs project into loops of tube, filled with fluid. As the head moves, fluid moves in the tubes, deflecting the hairs. The direction of movement can also be worked out, because the tubes are arranged in three different planes.

Various receptor cells monitor the state of blood and tissue fluids. Cells exist which detect levels of oxygen, carbon dioxide, salts, glucose, amino acids and fatty acids, as well as temperature and pressure. They are central to homeostasis (Unit 15).

Other cells detect chemicals in the environment. In land mammals these chemical senses are divided into smell (airborne molecules) and taste (dissolved molecules). There are probably about four types of taste receptor on the human tongue, corresponding to the primary tastes: sour, sweet, salty and bitter. The sense of smell has been a great mystery for years. No one can identify a clear set of primary smells. At last, molecular biology techniques are being applied to the problem. There appears to be a huge number of different types of receptor proteins, hundreds or even thousands. Each receptor cell probably has just one sort of receptor protein, so responds to one type of odour molecule. The genetics underlying this diversity are still not clear. Nor is the matter of how the brain knows which neurones are connected to receptor cells with the same sort of receptor protein. Is it 'hard wired' (fixed according to set rules during development), or does the brain learn once the nose starts work, and certain groups of neurones always fire together?

How a receptor cell works

The rod cell of the retina is a model system for investigating how receptor cells work at the molecular level. How does light lead to a neurone firing? A rod has four distinct regions (Fig. 17.8). The outer segment contains flattened membrane bags with the light-sensitive pigment rhodopsin embedded in the membrane. Light is detected here. Next is the inner segment, full of mitochondria providing ATP, and other organelles. The two parts are linked by a narrow region of the cell containing an arrangement of microtubules very like those in a flagellum. This strange similarity deserves attention, especially since in both cases ATP and other metabolites need to be transported very efficiently through a narrow channel. Third is a region containing the nucleus. Finally, nearest the inside of the eye, is the synaptic region. Neurotransmitters are released across synapses, causing the attached neurones to fire.

Somehow, a signal needs to be passed from the outer segment, which detects light, to the synaptic region. This signal is a polarization of the membrane. In the dark, the membrane is depolarized because sodium channels are open. Light, even a single photon, can lead to the channels closing. This causes a short-lived polarization of the membrane, which in turn causes neurotransmitter release at the other end of the cell. Note that this is the reverse of what happens in neurones.

Light absorption is linked to hyperpolarization by a signalling pathway within the cell (Fig. 17.9). This involves a G protein and a second messenger, cyclic guanosine

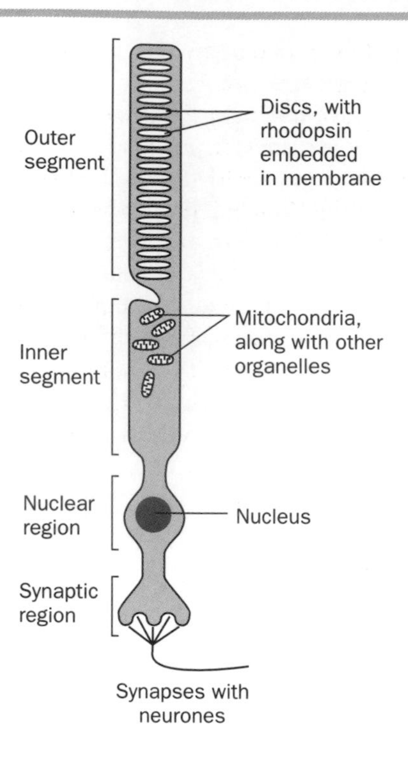

Fig. 17.8 A rod cell from the retina.

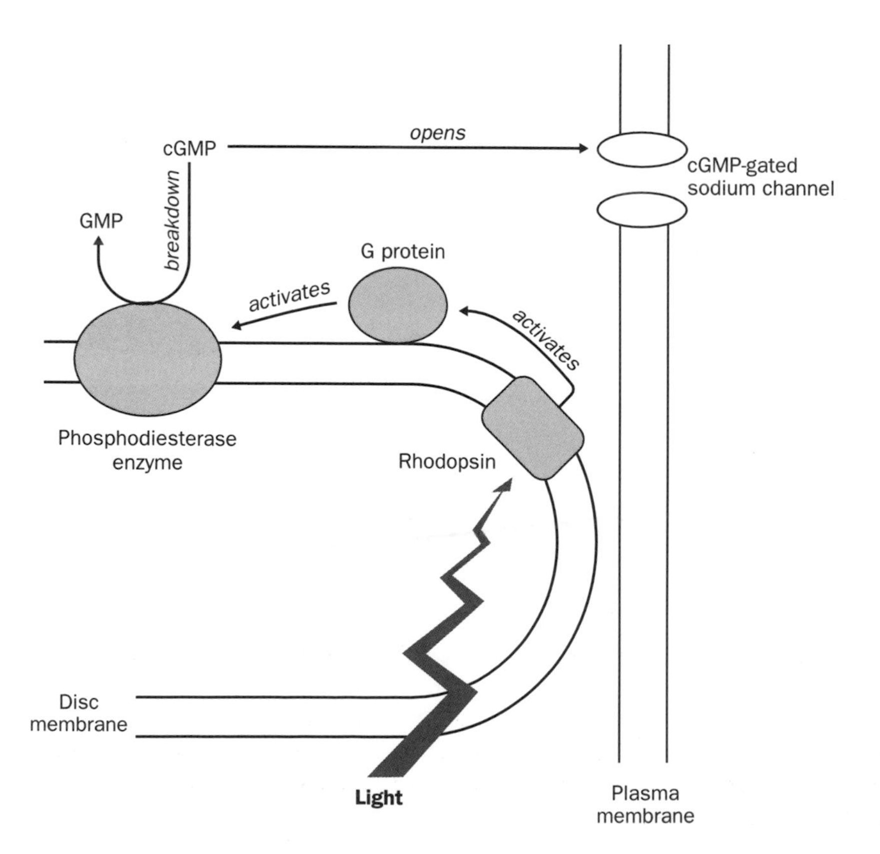

Fig. 17.9 How the rod cell plasma membrane polarizes when rhodopsin absorbs light.

monophosphate (cGMP, much like cAMP). Rhodopsin is made up of a protein covalently linked to a small molecule: retinal. In the dark, retinal is in the 11-*cis* form with a kinky chain (Fig. 17.10). When it absorbs a photon of light it changes to the straighter all-*trans* form. This forces a change in the protein, activating it. This in turn activates a G protein which in turn activates an enzyme which destroys cGMP. Sodium channels in the membrane are opened by cGMP. In the dark, there is more cGMP in the cytosol, and channels are open. In the light, cGMP falls so channels close, hyperpolarizing the membrane (Fig. 17.11).

The workings of individual neurones and some receptor cells are quite well understood. The structure of the nervous system as a whole is also well known. While this is discussed in the next section, it is worth remembering the gulf between these large scale and molecular views of the nervous system. To integrate them is the task of neuroscience.

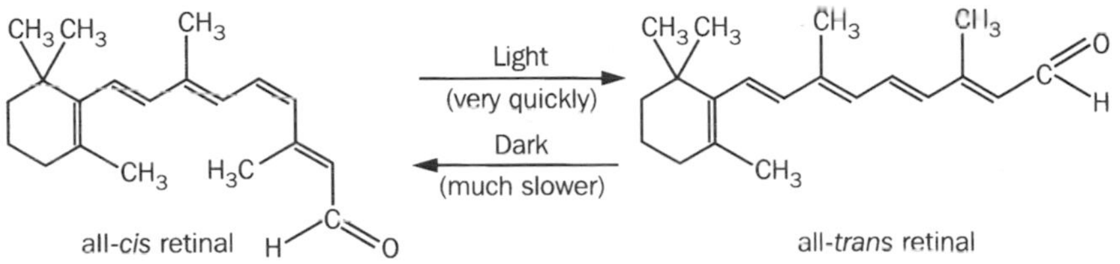

CH_3 CH_3 CH_3 CH_3 H_3C all-*cis* retinal

Light (very quickly)

Dark (much slower)

CH_3 CH_3 CH_3 CH_3 CH_3 all-*trans* retinal

Fig. 17.10 The two forms of retinal

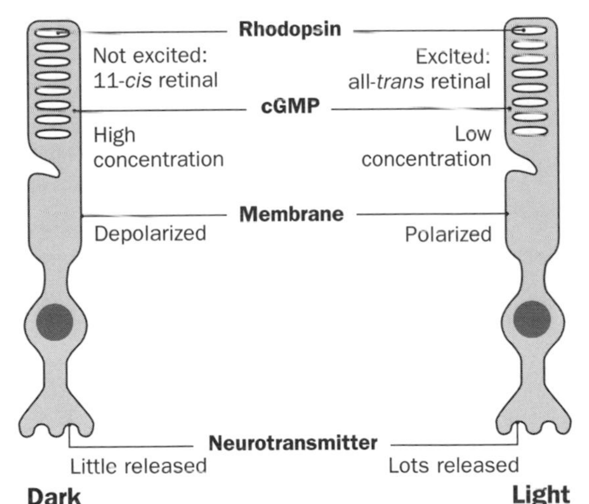

Fig. 17.11 Rod cells in dark and light conditions: summary of events.

17.6 The organization of the nervous system

The nervous system can be divided into the central nervous system (CNS) and the peripheral nervous system. The CNS is the brain and spinal cord. The peripheral nervous system has three main elements: sensory neurones, motor neurones and the autonomic nervous system.

Sensory neurones carry signals from receptor cells to the CNS. They may be linked directly to receptor cells, or there may be other local neurones in between, carrying out low-level information processing within the receptor organ. Motor neurones carry signals out from the CNS to skeletal muscles, the muscles that move bones. They are largely involved in controlling voluntary actions.

The autonomic nervous system affects the action of internal organs, including the cardiac muscle of the heart, and the smooth muscle of blood-vessel walls, sphincters and the gut wall. The autonomic nervous system is rather independent of the rest of the nervous system. Only a few distinct parts of the brain are involved with it. Heart rate, the breathing cycle, control of body temperature by controlling blood flow to the skin and hair erection, are all largely under autonomic control. They can be influenced by voluntary parts of the nervous system (hold your breath – now breathe out long and hard) but the autonomic system dominates.

The autonomic nervous system has two distinct parts, the sympathetic and

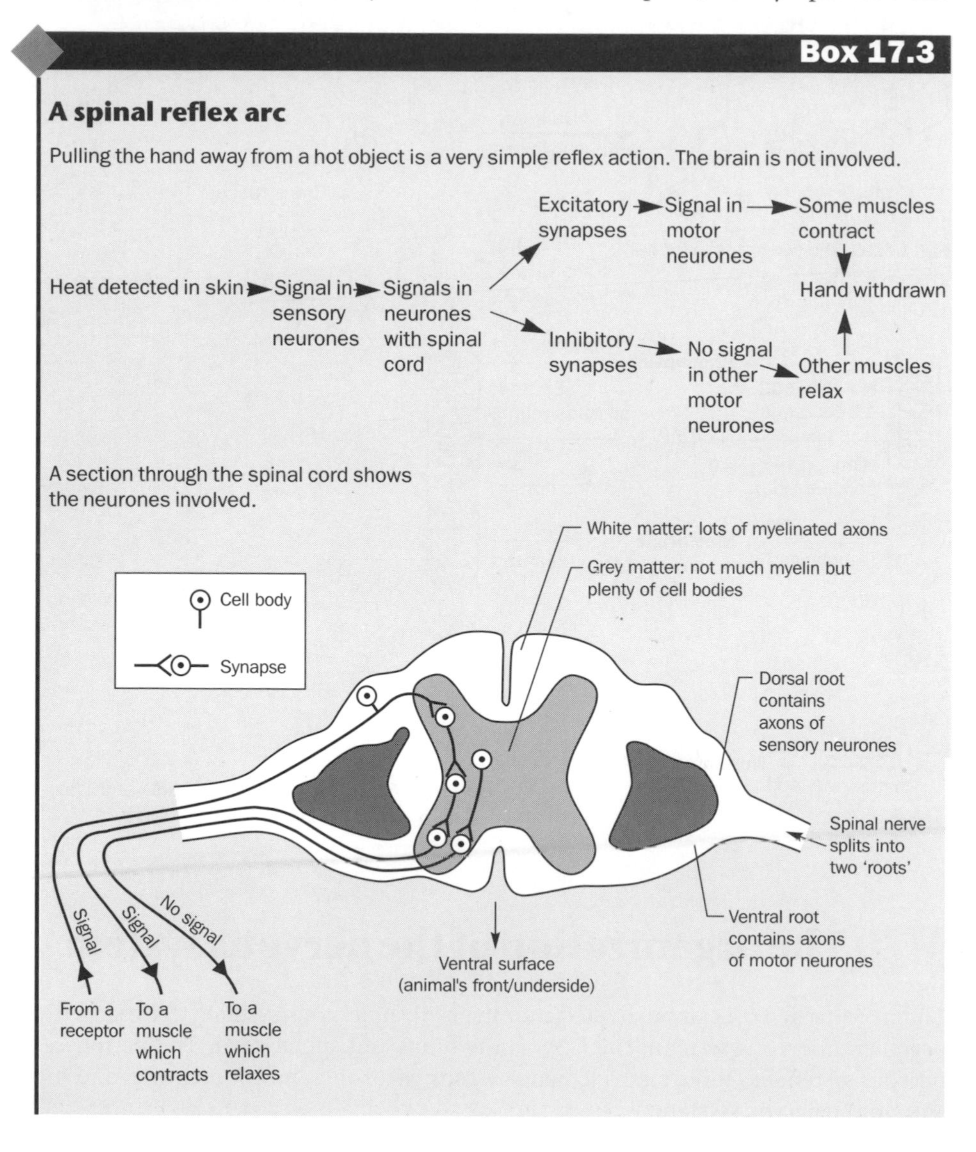

Box 17.3

A spinal reflex arc

Pulling the hand away from a hot object is a very simple reflex action. The brain is not involved.

A section through the spinal cord shows the neurones involved.

parasympathetic systems. In most situations they affect the same process in opposite ways: they are antagonistic. For example, sympathetic neurones increase heart rate, parasympathetic neurones decrease it. Autonomic neurones do not form tight neuromuscular junctions. Instead, they release neurotransmitters more diffusely into the tissue. Parasympathetic neurones release acetylcholine, whilst sympathetic neurones usually release noradrenaline.

The spinal cord runs through a large hole in each vertebra. The spine as a whole forms a protective tube around the cord. The spinal cord has two roles. Firstly, it acts as a trunk route for information flowing between brain and peripheral nervous system. Secondly, some information processing goes on here. Some simple reflexes (the chain of events leading from the detection of some stimulus to the body's response) do not involve the brain at all. These spinal reflexes include recoiling from hot things and walking movements. Box 17.3 shows the machinery of a very simple spinal reflex.

17.7 The brain

The brain is a centre for information processing of nightmarish complexity. It is essentially a hollow dome made up of neurones. Some parts are thicker than others, and it is deeply folded. Fig. 17.12 shows some visibly distinct regions of the brain.

Three brutally simple, though technically tricky methods have given information on what particular parts of the brain are doing. Firstly, if part of the brain is damaged, which functions are affected? Secondly, electrodes placed in one part of the undamaged brain can detect increased electrical activity when that region is

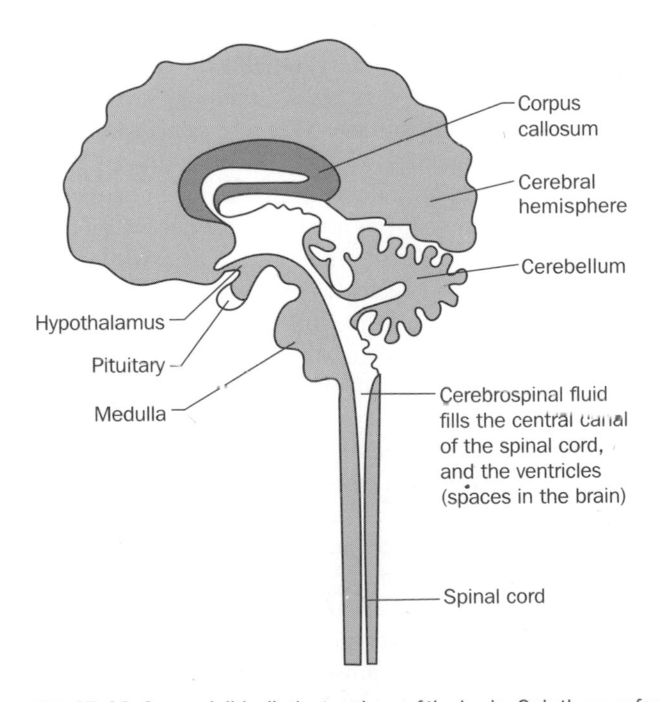

Fig. 17.12 Some visibly distinct regions of the brain. Only those referred to in the text are labelled.

particularly active. Thirdly, various brain-scanning techniques find active areas by detecting increased blood flow, without having to break into the skull.

The first distinction is between the cerebral cortex and the other 'lower' parts of the brain. Some brain functions seem to be located in specific parts of the lower brain. For example, the cerebellum controls many reflex actions which need to be learned, including the control of the body's posture, running, and the mechanics of speech. The autonomic nervous system is largely controlled by the hypothalamus and medulla.

The cerebral cortex has two halves: the cerebral hemispheres. These communicate via a belt of neurones below them: the corpus callosum. Groups of neurones within the cortex have extraordinarily precise functions, but they are not visibly distinct. For example, there is a large area in each hemisphere involved in motor control. It is sub-divided into areas controlling particular parts of the body. The left motor area controls the right side of the body, and vice versa. Similarly, there is a large area at the back of each hemisphere devoted to vision. Particular groups of neurones within this region detect different elements of visual images. Information starts off in centres detecting simple features like horizontal lines. It moves on to centres detecting more useful, high-level features like shape and movement. Damage to one centre will affect perception in a highly specific way. Cases are known in which patients can see perfect still images yet cannot perceive movement. There is even a region of the cortex whose main role seems to be recognizing faces.

Knowing what each part of the brain does is not the same as knowing how it works. How does the brain learn? What is memory? What is thought? How does consciousness arise? These are the fundamental questions, and are still largely mysteries.

Towards an understanding of memory

Some parts of the brain have a particular role in memory, notably the hippocampus and limbic system, deep in the cortex. However, they are not memory stores in the way that a computer has sites for information storage. Memory seems to be built into the way the brain works, the ways in which individual neurones interact. When the brain learns, the way it works changes in a subtle way. This seems to happen by changing the importance of particular synapses. If a group of neurones interact in a way that brings about a useful outcome, the set of synapses involved will gain in influence. If the outcome is worthless, they will decrease in importance. The hippocampus and limbic system seem to have a role in judging the value of outcomes and either punishing or rewarding the neural circuits which brought them about. Pushing this view to the extreme, one could argue that many of the details of brain function are not in the brain's developmental instructions. Instead, the brain is programmed to develop as it begins to operate. Inbuilt standards are used to judge the success or failure of its first attempts at perception and control. Neural network computers are electronic circuits designed to work like interacting neurones. They can learn to operate in a startlingly brain-like fashion, in just this way.

There are also some interesting indications of what may happen at the molecular level when neurones learn. Box 17.4 shows two examples. Both are poorly understood, both are certainly important in some specific types of learning, but their place in the overall picture is unclear. Both are mechanisms for the potentiation of synapses, that is they make a synapse fire more often, or for longer.

Box 17.4

Molecular memory in synapses

Some synapses 'remember' an incoming nerve impulse by continuing to fire after it has gone, or firing more easily next time.

Example 1

A synapse can be potentiated by prolonging neurotransmitter release. Many presynaptic terminals contain a calmodulin-dependent protein kinase, CaM-kinase II. It seems to be part of the signalling pathway linking Ca^{2+} influx to neurotransmitter release. This amazing enzyme phosphorylates itself as well as other proteins further down the pathway. Hence, once an action potential arrives and Ca^{2+} flows in, the signalling pathway is switched on and stays on until other enzymes remove the phosphoryl group from the kinase.

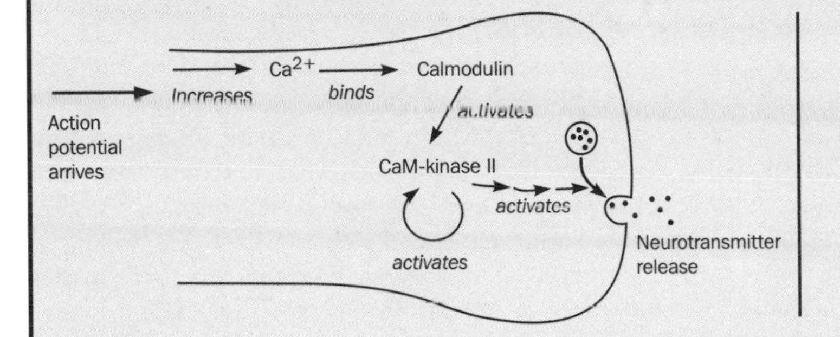

Example 2

This system prolongs neurotransmitter release in the long term. It operates in synapses using glutamate as the transmitter, and is certainly important in the hippocampus. Glutamate binds to two types of receptor in the postsynaptic membrane. One is a typical transmitter-gated sodium channel which brings about depolarization. The second, the NMDA receptor, is more unusual. It is a calcium channel, which only opens if it binds glutamate when the membrane is already depolarized – in other words, when the synapse is firing rapidly. Calcium floods into the postsynaptic neurone. It acts as a second messenger. By some unknown pathway, which must involve a signal passing back across the synapse, it increases glutamate release in the longer term. So, a rapid burst of synaptic firing potentiates the synapse: it 'remembers' its activity.

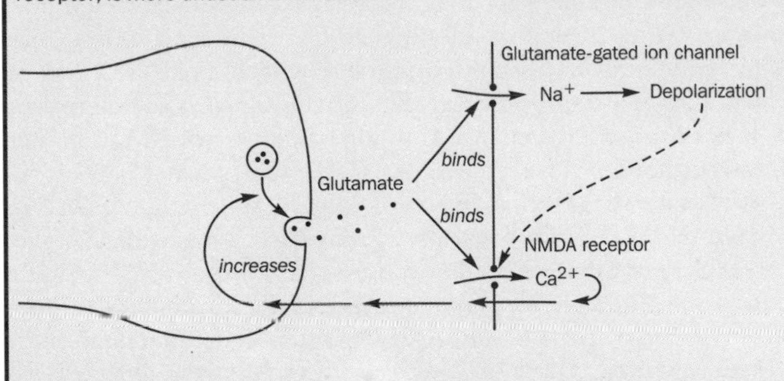

No one pretends that these molecular titbits explain learning and memory. They are just the first glimpses of the sorts of molecular mechanism which are involved. The harder questions – what is a thought? What is consciousness? – remain scarcely touched, in the area where biologists, psychologists, philosophers and even mathematicians meet.

17.8 Together forever

At first sight, the nervous system and endocrine system seem unrelated. One is slow, the other quick; one is chemical, the other electrical. A closer look makes it clear that they not only depend upon one another, but have many chemical similarities.

The hypothalamus–pituitary system clearly links the two. Here, neurones manufacture, transport and control the release of hormones. Neurotransmitters are another link. In their release, movement between cells and binding to receptors, they are just like very local hormones or paracrines. In the adrenal medulla, neurotransmitters actually become hormones. The sympathetic neurones which end in the medulla convert most of their noradrenaline to adrenaline before releasing both into the blood. They act on the same types of target cells, but throughout the body instead of locally, and for longer because they are removed from the blood more slowly.

The cell-signalling pathways involved in responding to hormones are shared by most types of cell, including neurones. The role of calcium as a second messenger in neurotransmitter release is well known, and it is also involved in the NMDA receptor memory mechanism (Box 17.4).

The discovery of a quite different type of information transfer in neurones emphasizes the links with the endocrine system even more. Neuropeptides are short peptides made in the cell bodies of some neurones, by chopping up larger proteins. They travel down axons in vesicles which are actively transported along microtubules. This is quite slow, tens of millimetres per day. The neuropeptides are released at synapses, just like neurotransmitters. Instead of binding to ion channels, they bind to G protein or enzyme-linked receptors. They bring about slower changes than neurotransmitters by gene switching or enzyme activation. They are chemical signals which happen to travel long distances inside rather than outside cells.

The enkephalins are neuropeptides which reduce the perception of pain. Pain is an early (or not so early) warning system for problems in the body, and several types of neurone are concerned with it. Enkephalins are released onto some types of pain neurone at synapses, and are strongly inhibitory. Interestingly, opiate drugs like morphine work by binding to enkephalin receptors. ADH and oxytocin are made and released by neurones in just the same way, but into the blood. They are neuropeptides which have become hormones, just as adrenaline is a neurotransmitter which has become a hormone.

It would be surprising if these two types of long-distance communication between cells were unrelated. Natural selection does not bring about totally novel solutions to each new problem that cells and organisms face. The only way that these problems can be solved is through mutations which alter the workings of existing molecular systems. When cells first began to live together and cooperate within animal bodies, there would have been a need for both rapid, highly-directed cell communication and slower, more widespread signalling. The same genetic raw material was there to be adjusted through natural selection in order to solve both problems. The simplest multicellular animals have both nervous and chemical communication between cells. Presumably they evolved together; certainly they depend upon one another.

Summary

- Neurones are the cells of the nervous system. Information travels down their long axons.
- The axon membrane is electrically excitable.
- A nerve impulse is a wave of depolarization (an action potential) which travels along the axon. All action potentials are alike.
- Information is coded as the frequency of action potentials.
- Synapses are connections between neurones. They are essential in information processing and learning.
- Signals are carried across synapses by diffusing neurotransmitters.
- Receptor cells set up nerve impulses when they detect a stimulus.
- The nervous system is made up of several components with more or less distinct functions.
- Chemical and nervous signalling work together in controlling animals' bodies.

Exercises

17.1. Cyanide inhibits ATP production in the cell, by interfering with oxidative phosphorylation. Predict the effect of cyanide on an axon which is not carrying impulses.

17.2.
(i) Explain why it is hard to see the colour of a bus at night, in dim light.
(ii) Explain why it is hard to read a newspaper in dim light.

17.3. An excitatory synapse is stimulated by very rapid action potentials in the presynaptic neurone, over a period of several seconds. At first, this leads to rapid impulses in the postsynaptic neurone, but these become slower as time goes on. This reduction in response is called fatigue of synaptic transmission. Use your knowledge of synapses to suggest a cause of fatigue.

Further reading

Alberts, B., Bray, D., Lewis, J., Raff, M., Roberts, K. and Watson, J.D. *Molecular Biology of the Cell* (3rd ed.) (New York: Garland, 1994). An enormous but excellent textbook which puts the reader in touch with current ideas without too much pain on the way. The molecular biology of neurones and synapses is dealt with in a clear, direct way.

Guyton, A.C. & Hall, J.E. *Textbook of Medical Physiology* (9th ed.) (Philadelphia: Saunders, 1996). I like this hefty textbook of human physiology for its thoroughness and clear, no-frills diagrams.

Guyton, A.C. *Human Physiology and Mechanisms of Disease* (5th ed.) (Philadelphia: Saunders, 1992). Effectively a shorter, boiled-down version.

Schmidt-Nielsen, K. *Animal Physiology: Adaptation and Environment* (4th ed.) (Cambridge: Cambridge University Press, 1990). One of the all-time great textbooks! Exciting, authoritative and easy to read, it puts animal physiology in the context of the environment.

Withers, P.C. *Comparative Animal Physiology* (Fort Worth: Saunders, 1992). More detailed than Schmidt-Nielsen, with wider ranging examples: the inevitable cost is that it is a less easy read.

Plant, Soil and Atmosphere

Contents

Connections

► Much of this unit deals with movement of materials into, out of and within the plant. A thorough understanding of the movement of molecules is essential (Unit 6). A basic knowledge of biological molecules, cell structure, respiration and photosynthesis is also important (Units 3, 4, 9 and 10). Other aspects of plant physiology are covered in Units 23–25.

18.1 Life at the edge

Plants live at the edge of things, where earth meets sky. Light comes from above; water and minerals are found below. Plant structure is intimately related to this asymmetrical environment. A typical plant has roots in soil, anchoring the plant and extracting water and mineral ions. The roots are linked to stems near the soil surface. Stems stick up into the air. They hold up leaves which absorb light for photosynthesis. All parts of the plant need a gas supply. Every part needs oxygen for respiration, and photosynthetic parts need carbon dioxide. Leaves take gases from the air; most roots rely on air in the soil for their oxygen. Photosynthesizing leaves make carbohydrates which are transported to the roots and other regions which need more energy than they can get from the sun through photosynthesis.

Not all plants fit this simple pattern. Some live under water, others live on trees and have roots which hang in the air. This unit deals mainly with plants rooted in soil. The emphasis is on vascular plants, those with specialized transport tissues (xylem and phloem). These include flowering plants, conifers, ferns, horsetails and clubmosses. The very different lives of mosses, which have no xylem or phloem, are touched on in Section 18.8.

18.2 Plants, water and gas exchange

Most leaves have air spaces inside them (Fig. 18.1). This allows photosynthesizing cells to take in CO_2 and excrete O_2. The air is in contact with the atmosphere. More CO_2 can diffuse in, and O_2 can diffuse out. However, this creates a water problem. A

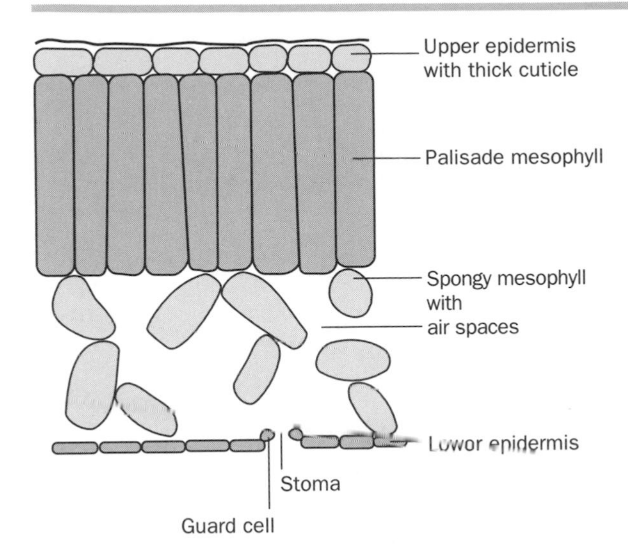

Fig. 18.1 Part of a typical leaf, in vertical section.

very large area of moist cell wall is in contact with air inside the leaf. The plant could lose a lot of water by evaporation.

The solution is a compromise. The leaf is covered in a continuous layer of cells: the epidermis. This sometimes has a cuticle of waxy substances (Section 3.7) to reduce evaporation even further. The epidermis has pores, called stomata, which allow limited gas exchange and limited water loss. They can be opened and closed to optimize the balance (Section 18.3). Stems, leaf veins and roots contain xylem. This tissue is specialized for water transport (Section 18.4). Water is absorbed by the roots (Section 18.5) which have a large surface area in contact with the soil. It enters the xylem and flows upwards to replace water lost by the leaves.

The whole process of water loss, movement and uptake is called **transpiration**. Transpiration affects most plants most of the time, but it is a special problem in very dry places, such as deserts (Box 18.1). Each part of the process is dealt with in turn.

18.3 Stomata

Stomata are holes in the leaf epidermis. Most leaves have more in the lower surface, because the air spaces are in the lower part of the leaf. Water diffuses out of stomata. This is because the air inside the leaf is always saturated with water vapour, but the atmosphere is usually not. In a photosynthesizing leaf, CO_2 diffuses in through the stomata, and O_2 diffuses out. Stomata can be opened and closed to optimize the balance between the cost of losing water (which may be scarce) and the benefits of photosynthesis. How do they do it?

Stomata have a pair of special epidermal cells around them: **guard cells**. When the cells are turgid, full of high-pressure water, the stoma opens. When guard cells lose water again, the stoma closes. The structure of the guard-cell walls makes this possible. In grasses, the guard cells are sausages with bulging ends (Fig. 18.2(a)). When water enters the cell, the pressure rises. The bulbous ends swell, because the walls can stretch slightly. They push against one another. The middle section cannot

Transpiration: The uptake, transport and loss of water by plants.

Stomata (singular, stoma): Holes in the leaf epidermis surrounded by pairs of **guard cells**.

Box 18.1

Xerophytes

Xerophytes are plants which live in places where water is scarce. Deserts (regions with little rainfall), sharply-drained soils and rock faces are all home to xerophytes. Halophytes (Box 18.5), which live in salty habitats, often show similar adaptations to xerophytes. This is because the plant has to absorb water despite the very negative solute potential in the soil. Whatever is absorbed must be used efficiently.

Four terms, used to classify xerophytes more than 70 years ago, still hold good today.

Avoid

Plants may get at water which others cannot reach. They avoid drought altogether. The long roots of the mesquite (*Prosopis glandulosa*), an American desert shrub, reach down to the water table many metres below the surface.

Resist

Many xerophytes store water and/or minimize losses. They make good use of whatever water they can get. Their cells never experience drought. The wide, shallow root systems of many desert shrubs absorb rainfall over a wide area. Succulent tissues, such as the stems of cacti, are water stores. All sorts of adaptations limit water losses, including small leaf area, hairy leaves which limit movement of humid air from the stomata, low stomatal density, stomata sunk in pits which trap humid air, leaves which drop off in dry seasons and CAM (Section 10.7: Moist or dry) which allows stomata to close during the day.

Escape

Desert annuals have short life cycles. Dormant seeds lie in the desert soil until rain falls. They germinate, grow, flower and set seed within a few weeks. Their seeds return to the soil for, perhaps, many years. By growing in the rare windows of time when the desert is not dry, these plants escape drought.

Resist

Some desert plants can tolerate drying. Their cells stop metabolizing when dry, but they are not killed. When they absorb water again, metabolism restarts. Some mosses do this (Section 18.8). The creosote bush (*Larrea divaricata*) of American deserts is another example.

swell, because cellulose microfibrils are lined up to run around the cell, preventing stretching. If this is unclear, imagine a long balloon, the sort you can twist into rude shapes at parties. Partly blow it up, then twist loops of string round the middle bit. Now blow it up a bit more. Only the ends get fatter, making a sort of dumbbell. If two guard cells do this next to one another, a gap opens between them: the stoma.

In most other plants, the guard cells are bean shaped (Fig. 18.2(b)). Microfibrils run along, rather than around the cell. They appear to radiate from a point next to the stoma itself. When the guard cell becomes turgid, it lengthens but does not get fatter. The arrangement of microfibrils means that the outer wall lengthens more than the inner wall: the cell bends. When two guard cells do this together, the stoma between them opens.

Several things stimulate opening or closing of stomata. Typical stomata respond as follows:

- **Light:** The brighter the light, the faster stomata open, and the wider they become. This makes sense. Brighter light means faster photosynthesis, so more gas exchange is needed. Stomata of CAM plants (Section 10.7: Moist or dry) behave quite differently.
- **CO_2:** Increasing CO_2 concentration within the leaf leads to stomata narrowing, at least above a threshold level. Higher CO_2 concentration means that less gas exchange is needed to maintain photosynthesis, at least for a while.

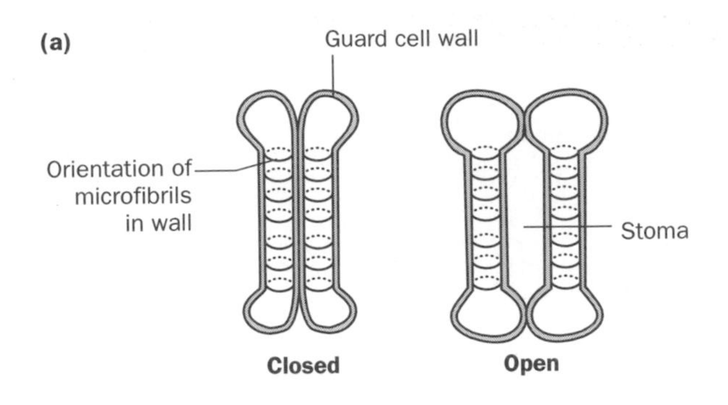

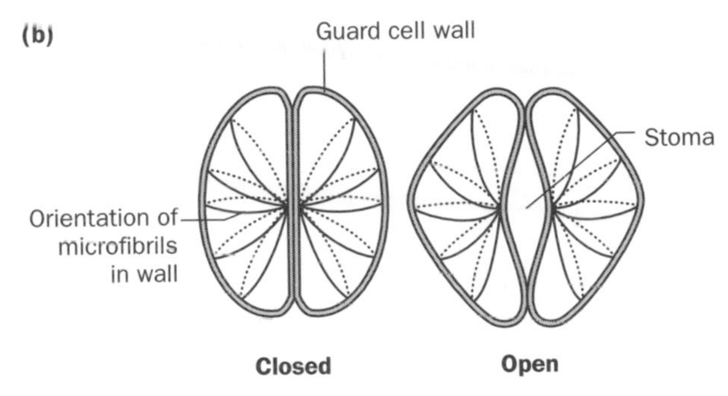

Fig. 18.2 Guard cells and stomata in (a) grasses and (b) most other flowering plants.

- **Temperature:** Higher temperatures tend to make stomata open wider. This reflects stimulation of photosynthesis and photorespiration (Sections 10.4, 10.6), leading to a need for more gas exchange. The cost is more water loss, although this can help cool the plant.
- **Water stress:** When water levels in the leaf (or even the root) fall below a threshold, stomata quickly close. This reduces water loss, but photosynthesis must also decline. High rates of evaporation from the guard cell itself can also lead to closure. This can happen in dry air and high winds.
- **A biological clock:** Plants grown in continuous light show cycles of opening and closing. This suggests that some internal mechanism 'anticipates' the need to open and close at dawn and dusk. The clock mechanism is in the guard cell itself, but its chemistry is still unclear.

The next step is to determine what makes guard cells become turgid, and how these environmental factors control it. The following things take place when stomata open (Fig. 18.3). An active proton pump (Section 8.10) in the guard cell plasma membrane pumps H^+ ions out of the guard cell. This sets up a proton gradient. The inside of the cell becomes alkaline and negatively charged. The outside becomes acidic and positively charged. Potassium ions enter the cell through a passive carrier, down the electrical gradient. Water follows K^+ into the cell by osmosis. The guard cell becomes turgid.

This simple story leaves a problem. The H^+ concentration gradient would remain. After a while, no more H^+ could be pumped out. The charge would be balanced by the K^+ which had moved into the cell and the whole process would stop.

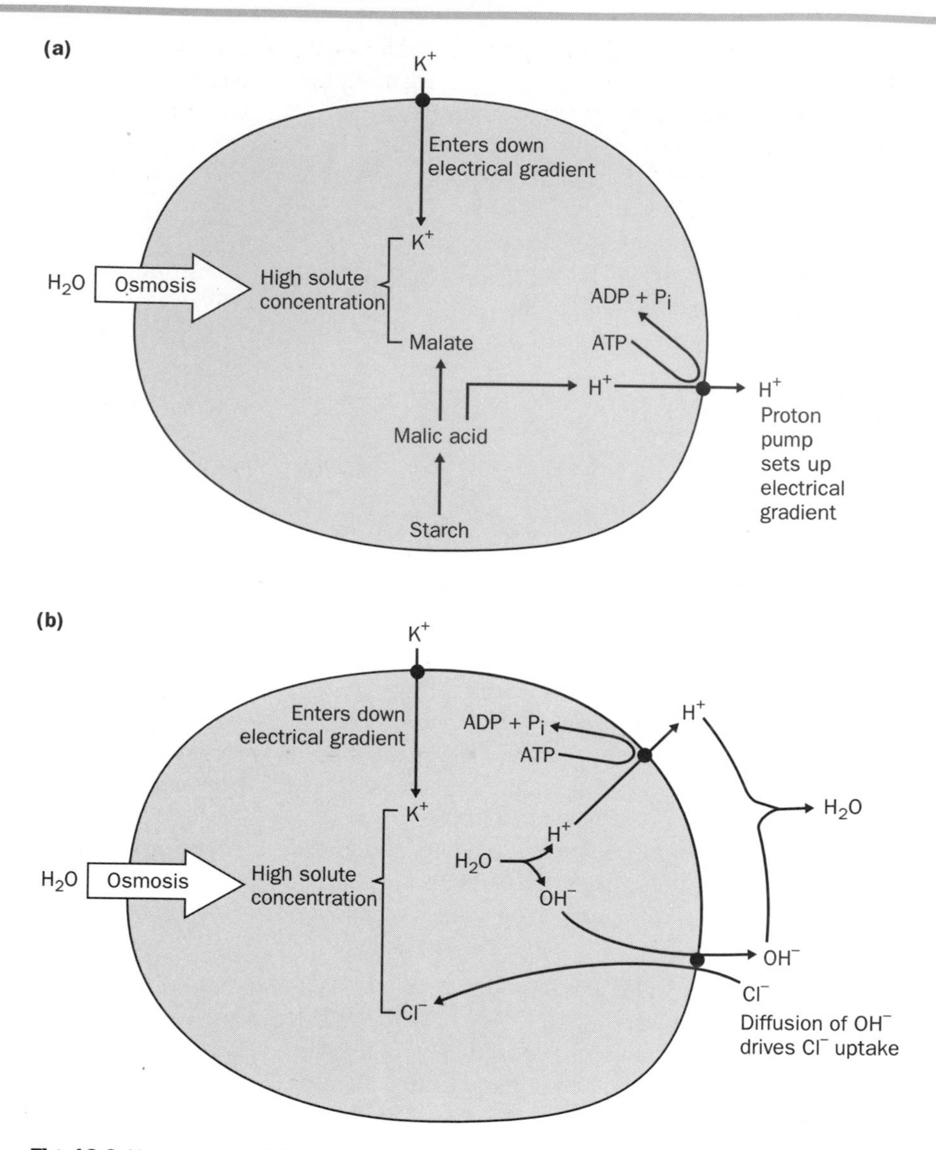

Fig. 18.3 How guard cells become turgid, in order to open stomata. (a) the typical situation, (b) an alternative, seen in onion, for example.

There are two possible solutions. Firstly, more H^+ ions could be released in the cell. Most guard cells do just this. Starch is broken down to sugars, then on to organic acids such as malic acid. Starch is insoluble. Malic acid is soluble, and dissociates to malate and H^+ (Section 3.9). This supply of extra H^+ allows pumping to continue. The malate stays in the cytosol and adds to the osmotic effect of K^+.

The second solution does not involve H^+ directly. Water dissociates to form H^+ and OH^- ions. H has been pumped out of the cytoplasm. If OH^- could also leave, both sides of the membrane would return to neutral pH. In some guard cells, OH^- is exchanged for Cl^-. The result of the entire process is that K^+ and Cl^- accumulate in the cytosol.

We know less about exactly how environmental factors affect this process. Light is detected in two ways. The first is photosynthesis. The second is by cryptochrome, a blue-light absorbing pigment which has no role in photosynthesis. Only the light reactions of photosynthesis take place in guard cell chloroplasts. They probably provide the ATP needed to drive proton pumps. The blue-light receptor may stimulate oxidative phosphorylation (Section 9.2), which makes ATP. However, the precise control system is not understood.

Water stress in the leaf or root leads to production of a signalling molecule, abscisic acid (ABA; Section 25.7). Loss of cell turgor seems to be the stimulus. ABA makes stomata close, in an unknown way. Guard cells lose water to the atmosphere more easily than other epidermal cells. Their outer walls have patches without a cuticle, in at least some species. This makes them particularly sensitive to dry air. Water loss by evaporation will cause pressure in the cell to fall, directly. Stomata will close.

Temperature probably acts indirectly, by affecting rates of photosynthesis and respiration. We have no clear idea how CO_2 has its effect.

18.4 Transport in the xylem

Xylem is a plant tissue specialized for water transport. Two cell types carry water: tracheids and vessels. There are also fibres present (Section 24.3), which have a strengthening role, and other less specialized cells. Tracheids and vessels have thick secondary walls impregnated with lignin (Fig. 18.4). This makes them waterproof. Once mature, the cell dies, leaving the wall as a hollow tube. The secondary wall is thickened with spirals or rings. This resists collapse.

Xylem: A plant tissue specialized for water transport.

Tracheids have pointed ends. Their walls are covered in pits. These are holes in

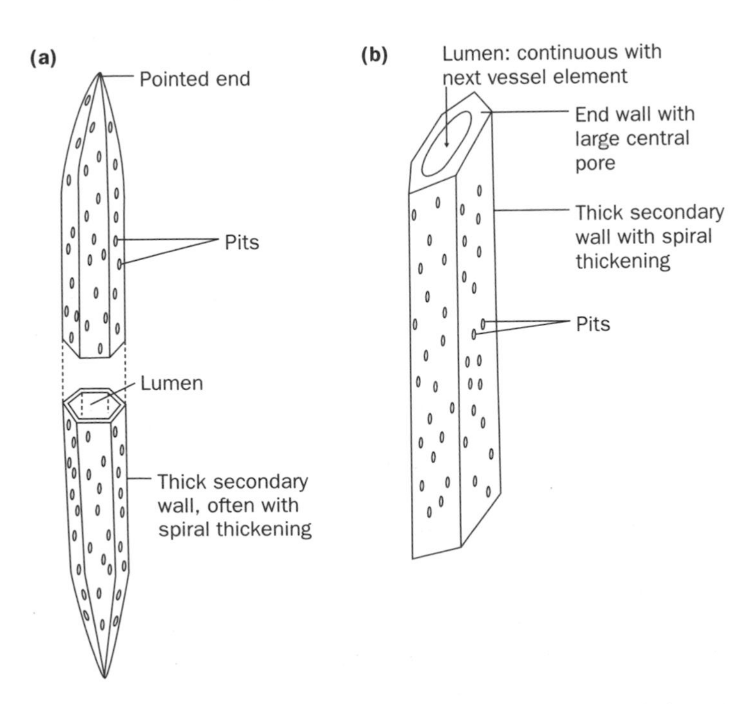

Fig. 18.4 Water conducting cells in xylem. (a) tracheid, (b) vessel element.

the secondary wall (Fig. 18.5). The primary wall is still present in the pits, but is not lignified. Water can pass from one tracheid to another through pits. Some pits have a more interesting structure. These bordered pits have secondary walls which extend out over the pit. A central disc of primary wall, the torus, is thick but the rest is porous. Bordered pits may act as valves (Fig. 18.5(b)). If pressure on one side is higher, the torus is pushed aside and prevents flow. Water can only flow through if pressure is equal on both sides. Bordered pits and simple pits are found together. The role of bordered pits in the living plant is not at all clear.

Vessels are only found in flowering plants (with one or two freaky exceptions). Cells called vessel elements are cylinders with flat end walls. Each end wall has a large hole. They are joined end to end, like sections of drain pipe, to form a vessel. Water can flow freely along the vessel, from cell to cell. One vessel may be centimetres or even metres long. Vessels also have simple pits, bordered pits, and spiral or ring-shaped thickenings.

Water flows through the hollow lumen of vessels and tracheids. Mineral ions are dissolved in the water, so they are transported too. The xylem provides a continuous route from the inside of each root, through the stem, to the veins of each leaf, as well as to flowers and fruit.

What drives the flow? Is water sucked from the top, or pumped from the bottom? Both can happen, but suction is much more important. Evaporation in the leaf lowers water potential (Box 6.3) so water moves out from the xylem to other leaf tissues. The water potential in the leaf's xylem is now lower than in the soil or root tissues. Water flows up through the xylem, from higher to lower water potential. The water column in the xylem is under tension, stretched. Why does it not snap? Water molecules are polar (Section 3.2: Bonds) and are held together by the attraction between positive and negative charges. In the same way, they are attracted to charges in the xylem walls. The whole cell wall–water system sticks together (cohesion) and resists the tension. This is the cohesion–tension model. It is supported by direct measurements of water potential in different parts of plants.

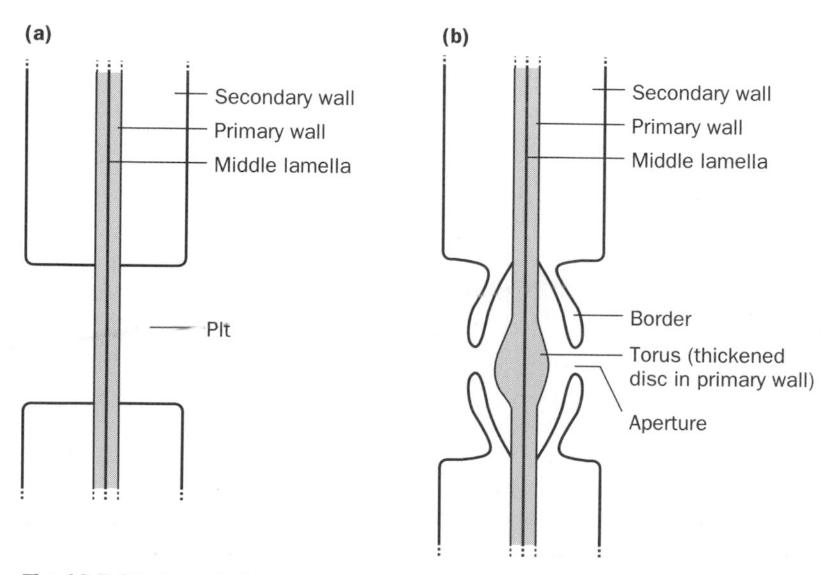

Fig. 18.5 Pits in tracheids and vessels. (a) simple pit, (b) bordered pit.

Sometimes the water column in a vessel breaks. This happens when tension becomes too great, usually when water is evaporating from the leaves very rapidly. Tiny amounts of air are sucked in through pits, from air spaces between cells. This acts as a centre of evaporation, and a bubble of water vapour forms, breaking the column. The tiny clicking sound this makes can be detected and used as a measure of water stress. The bubble of vapour will condense and vanish when tension falls, but it stops flow while it lasts. Transport can continue only because water can zigzag between cells, through pits.

Plants lift water high above the ground. This needs energy, but where does it come from? Vessels and tracheids are dead, passive pipes which supply no energy. Transpiration is driven by evaporation. The more energy a molecule has, the faster it moves. Temperature is simply a measure of the average energy of the molecules. When a liquid evaporates, only the fastest molecules can escape the forces pulling them towards other molecules in the liquid. The slower molecules are left behind: temperature falls. Fewer molecules can then evaporate, so the process would slow down and stop. In a plant, the leaves are warmed by radiation from the sun. This replaces the lost energy and allows more evaporation. So, transpiration is solar powered.

Pressure from below is sometimes significant. Mineral ions are actively pumped into the xylem in roots (Section 18.6). Water follows by osmosis. This creates a small positive pressure in the xylem. It is not enough to explain most water movement. However, when evaporation is very slight and there is a lot of water in the soil, it may drive a little movement. It is more likely to be important in reducing the tension in the water column, in order to reabsorb vapour cavities.

Trees are machines which carry water high into the air, powered by solar energy.

18.5 Water uptake in the root

Water is absorbed by younger, thinner roots (Fig. 18.6). The thicker, older roots simply carry water away from their younger branches. The root epidermis is in contact with soil water. Some epidermal cells have long outgrowths, root hairs, which increase surface area for absorption. The xylem, which carries water away, is found near the centre of the root. Water must pass from the soil to the xylem, through the outer part of the root. What drives this, and what path does it take?

There are two mechanisms for water uptake. When plants are transpiring, the tension in the water column is enough to set up a water-potential gradient across the root, from soil to xylem. Water is simply dragged in, just as it is pulled up the xylem. This happens in most leaves, most of the time. The second mechanism is important when there is little transpiration. Mineral ions are actively absorbed by cells in the root (Section 18.6). Water enters by osmosis. This could happen in the root hairs, drawing in water straight from the soil. However, it would also work if cells inside the root absorbed minerals from the solution in the cell walls. Water would follow by osmosis. This would set up a water-potential gradient through the interconnecting cell walls, back to the soil. Water would still be drawn in.

Water (and mineral ions) can follow two paths through the root, the **symplast** and the **apoplast**. The symplast pathway is the network of cell contents, interconnected by plasmodesmata. Water can move from the cytosol of one cell to the cytosol of the next, without crossing a membrane. The apoplast pathway is the network of porous, interconnecting cell walls. It is in contact with the soil water, and

Symplast: The network of cell contents in a plant tissue, linked by plasmodesmata.

Apoplast: The network of cell walls in a plant tissue.

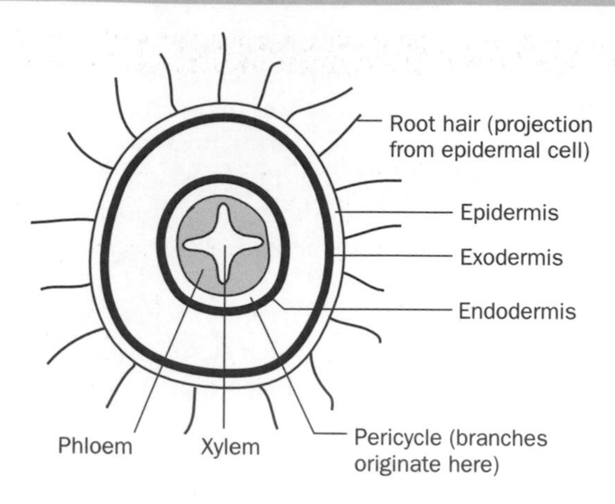

Fig. 18.6 Young root of a dicotyledonous plant, in section. The exodermis and endodermis prevent diffusion through the apoplast.

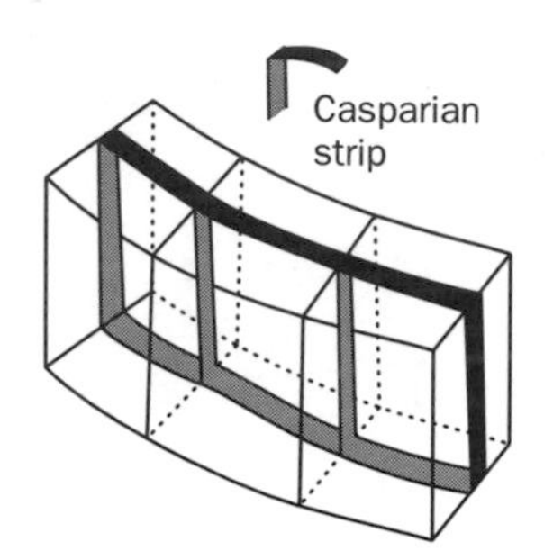

Fig. 18.7 Three cells of the endodermis, showing the Casparian strip.

with the xylem contents. Water moves along both pathways, but a special feature of the root forces it to enter the symplast at one point, at least.

The endodermis is a hollow cylinder of cells, one cell thick (Fig. 18.7). The xylem is found inside the cylinder. The cell walls of the endodermis are impregnated with suberin. This makes them waterproof. Suberin is concentrated in a thick band around each cell: the Casparian strip. At this point, water and anything dissolved in it must cross the plasma membrane to enter the symplast. This gives the plant some control over what enters. Viruses can be kept out, for example. Otherwise, they could be carried straight into the xylem and up into the shoots. A similar layer is sometimes found just under the epidermis. This is called the exodermis. Many, probably most flowering plants have one.

Overall, then, water must cross at least two plasma membranes in the root, once to enter the symplast and once to leave it inside the ring of endodermis.

18.6 Mineral uptake and transport

Plants need a wide range of mineral ions, including potassium, nitrate and phosphate (Box 18.2). They normally get them from the soil. Ions are absorbed in the roots and transported to the shoots in the xylem, dissolved in water as it moves through the plant.

Just like water, minerals can cross the root in the symplast or the apoplast. Like water, they must cross membranes at least twice. First they enter the symplast, then they leave it again to reach the dead xylem cells. Most ions are transported actively on both occasions.

Most plant cells take in mineral ions actively. There are a few exceptions, such as calcium ions which are pumped *out* of the cytosol as part of a chemical signalling system (Section 16.4). The cells in the outer part of the root, including the endodermis, accumulate ions in the normal way. Active transport is driven indirectly, by a proton pump (Section 8.5). Energy is used to pump H^+ across the plasma membrane, out of the cell. A proton gradient is set up which acts as an energy store. Symports allow H^+ to diffuse back in, along with another specific type of ion, nitrate for

Box 18.2

Plant mineral requirements

Plants need 17 (perhaps more) chemical elements, some in bulk, some in tiny amounts. Carbon, hydrogen and oxygen are acquired by photosynthesis, from carbon dioxide and water. The others are absorbed from the soil as mineral ions. Table 18.1 summarizes their functions in the plant and the chemical form(s) which can be absorbed. Detecting and identifying mineral deficiencies has as much to do with the craft of horticulture as it does with science, but one simple distinction applies to most plants. Some minerals cannot be recycled from older leaves and moved to growing tips, so deficiency symptoms (often yellowing or mottling) appear in new growth ('new' in table). Others can be mobilized, so the symptoms appear in the old leaves first ('old' in table). The more that is needed, the higher the element is on the list.

Table 18.1 Summary of plant mineral requirements

Element	Usable forms	Some uses in plants	Deficiency
nitrogen	nitrate (NO_3^-), ammonium (NH_4^+)	amino acids, proteins, nucleic acids	old
potassium	K^+	contributes to solute potential of cells	old
calcium	Ca^{2+}	pectates in cell walls, signalling within cells	new
magnesium	Mg^{2+}	part of chlorophyll molecule, cofactor for some enzymes	old
phosphorus	phosphates ($H_2PO_4^-$, HPO_4^{2-})	nucleic acids, phospholipids, sugar phosphates, ATP	old
sulphur	sulphate (SO_4^{2-})	amino acids, proteins, coenzyme A	new
chlorine	chloride (Cl^-)	contributes to solute potential of cells, cofactor for some enzymes	probably never deficient in nature
iron	mainly Fe^{2+}, also Fe^{3+}	part of some electron transport proteins and enzymes	new
boron	boric acid (H_3BO_3)	unclear	new

Manganese, zinc, copper, nickel and molybdenum are needed in smaller amounts, mainly as enzyme cofactors.

example. Diffusion of H^+ into the cell drives uptake of other ions, even against a concentration gradient (Fig. 18.8).

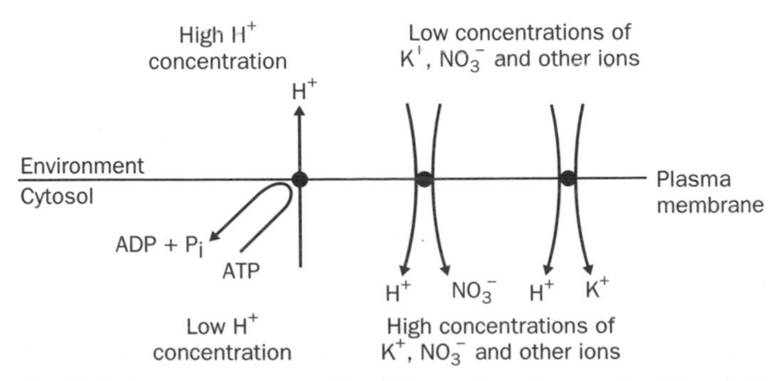

Fig. 18.8 A proton pump, working with a series of symports, drives active uptake of mineral ions into plant cells.

Box 18.3

Nitrogen fixation

Plant growth is often limited by the amount of nitrogen available in the soil. The air around plants is almost 80% nitrogen as N_2, yet they cannot use it. N_2 is extremely unreactive. Some prokaryotes, however, can reduce N_2 to ammonia (NH_3) which becomes ammonium ions (NH_4^+) in solution. Nitrogen fixers include some true bacteria, photosynthetic cyanobacteria and filament-forming actinomycetes. Some are free-living in water or soil. Others live in association with plants. This is a mutualism: both plant and microbe benefit.

Root nodules

Root nodules are swellings on plant roots which contain nitrogen fixers. Most members of the family Fabaceae (including peas, beans, lupins and clover) have root nodules containing *Rhizobium* bacteria. Some other leaves including species of *Alnus* (alders) and *Hippophae* (sea buckthorn) have nodules too, mostly containing actinomycetes.

Rhizobium lives right inside the plant cells of the nodules. The bacteria are in groups in the cytosol, within vesicles. The plant gains NH_4^+ exported by the bacteria. The bacteria gain carbohydrates from the plant as an energy supply.

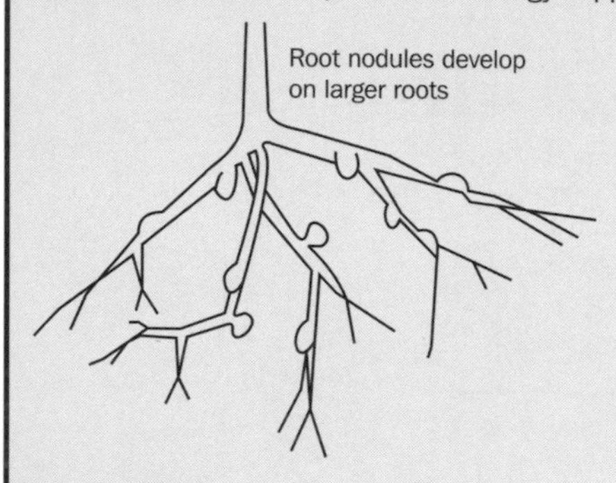

The reaction

The overall fixation reaction is as follows:

$$N_2 + 8H^+ + 16ATP \xrightarrow{\text{8 electrons}} 2NH_3 + H_2 + 16ADP + 16P_i$$

Nitrogen is being reduced (Section 8.2). A reducing agent is needed to supply the electrons. This is usually NADH or NADPH. Note that hydrolysis of ATP provides a large energy input. Also note that H_2 (a reduced molecule and energy store) is produced. Some bacteria can recycle it. Others release it, wasting energy.

The enzyme(s)

Nitrogenase catalyses the reaction, but it turns out to be two distinct proteins. One has an iron-containing prosthetic group (Fe protein). The other has a prosthetic group containing iron and molybdenum (Fe–Mo protein). Fe–Mo protein is the site where N_2 binds and is reduced. Fe protein, along with ferredoxin (Section 10.3) transports electrons to it.

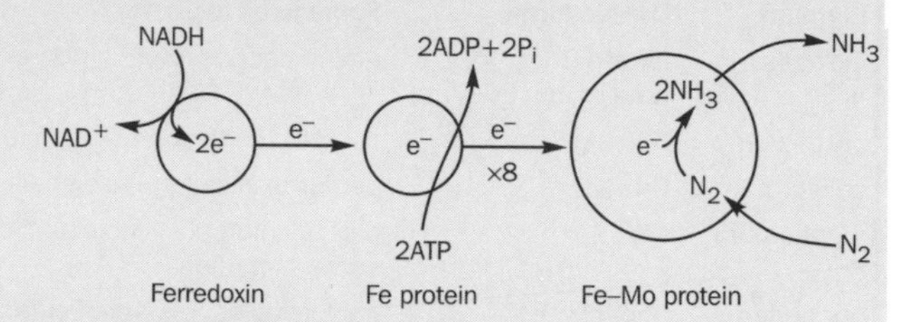

Every time Fe protein passes an electron to Fe–Mo protein, two molecules of ATP are hydrolysed. This happens eight times for every N_2 molecule.

This energy-expensive pathway is worthwhile if nitrogen limits growth. Plants with nodules will grow better and so compete better for space.

Keeping out oxygen

Nitrogenase is damaged by oxygen. Root nodules provide a barrier to oxygen diffusion, protecting it. However, the bacteria need some oxygen for respiration. Nodule cells contain an oxygen-binding protein, leghaemoglobin, in the cytosol. This buffers oxygen level, keeping it low, but not *too* low.

This is just a taste! Other exciting aspects of nitrogen fixation include its genetics, communication between plant and microbe before the association is formed, the weird way bacteria get into plant cells, the evolution of the relationship, and the prospects for transferring genes for nitrogen fixation from microbes to plants.

Ions must be actively pumped out of cells to reach the vessels and tracheids. Cells of the pericycle (Fig. 18.6) and perhaps living cells in the xylem do just this, but the precise mechanism is not clear.

Some plants get minerals in special ways. Some have a mutualistic relationship with nitrogen-fixing bacteria. This allows them to use nitrogen from the air (Box 18.3). Others form mycorrhizae. These are mutualistic relationships between roots and fungi which allow efficient absorption from the soil (Box 18.4). The carnivorous plants digest animals as a source of minerals. They live in nutrient-poor bogs and

Box 18.4

Mycorrhizae

A mycorrhiza is a mutualistic association between plant roots and fungi. Most plants have them. A wide-spreading network of hyphae penetrates the soil between young roots and absorbs minerals. Hyphae are much thinner than roots, so a given volume of hyphae can spread much more widely than the same volume of roots. This makes mineral uptake more efficient. Minerals are passed from fungus to plant. The fungus benefits by receiving carbohydrates from the plant.

There are several sorts of mycorrhiza, but two are very common.

Ectomycorrhizae

Hyphae form a dense mantle around the root. They extend between the cells of the cortex, but not into them. This internal network is the Hartig net. It allows materials to pass between cells and hyphae. This type is seen in many trees, including beech, oak, pine and birch.

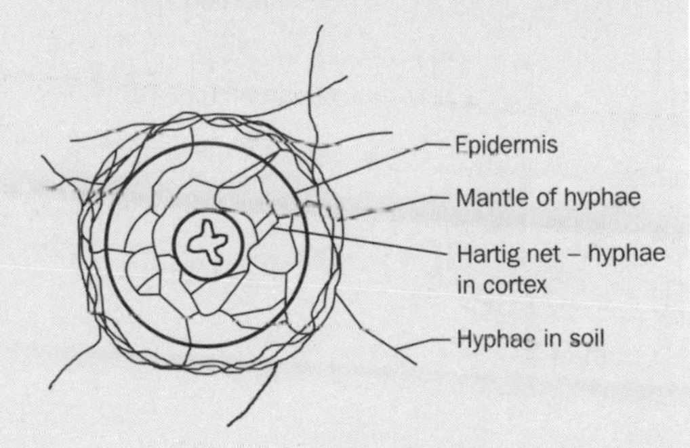

Vesicular–arbuscular mycorrhizae

Most non-woody plants have these. There is no mantle, but hyphae do extend through the cortex. They cross cell walls and form branched networks (arbuscules). The plasma membrane is folded in around the arbuscule, forming an open vesicle.

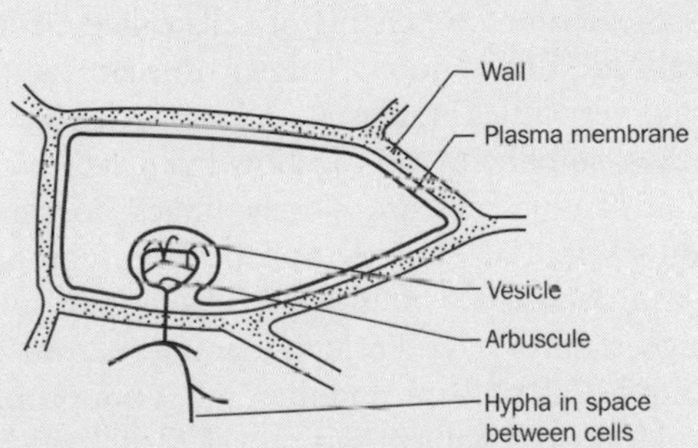

pools, trapping insects with sticky leaves, slippery-sided buckets, or traps which slam shut.

18.7 Transport in the phloem

Energy needs to be transported around the plant. It is carried in the form of a sugar: sucrose. Sucrose is carried from places where it is made – sources – to where it is being used – sinks. The green leaves and young stems of the apple tree outside my window are sources. The roots, woody stems, growing tips and developing fruits are sinks. However, back in the early spring, there were no green leaves. Sugar was carried up from stores in the roots to feed the buds as they started to grow.

Sucrose is transported in solution, in the **phloem**. The solution is called phloem sap, and may also contain amino acids and some minerals. Phloem is found in roots, stems and leaf veins. It is made up of several cell types. There are sieve tubes which carry sucrose, companion cells which are closely linked to them, strengthening fibres, and other less specialized cell types.

Phloem: A plant tissue specialized for transporting sucrose and amino acids around the plant.

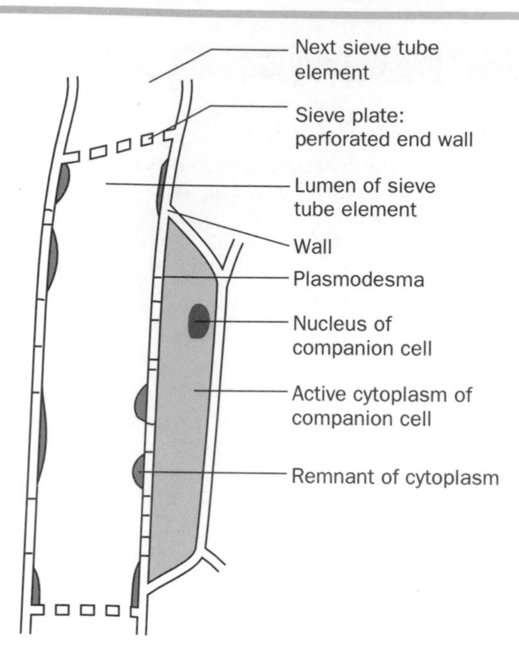

Fig. 18.9 A sieve tube element and companion cell, from phloem.

Sieve tubes are chains of cells called sieve tube elements (Fig. 18.9). Their end walls are full of pores. This gives them the name sieve plates. Sieve tube elements have very little cytoplasm, and no nuclei. The plasma membrane passes through the pores, so phloem sap can flow from cell to cell, much like water in a xylem vessel. Companion cells are closely linked to sieve tubes. Their cytoplasm has many mitochondria. The cell wall between companion cells and sieve tubes is rich in plasmodesmata. Companion cells tend to surround the sieve tubes in sources. This suggests they have a role in loading sucrose into the sieve tubes. However, they are found in all phloem, and their job is not properly understood.

How does phloem sap move? Most evidence suggests that it simply flows down a pressure gradient from source to sink. Pressure measurements, and movement of small particles with the flow, like viruses, support this. Aphids feed by sticking a hollow, pointed mouthpart, the stylet, into a sieve tube. Phloem sap flows into their mouths under pressure: they do not need to suck.

What sets up the pressure gradient? Sucrose is loaded into sieve tubes at sources. This is active: energy is needed. Water follows sucrose by osmosis, so pressure in the sieve tube rises. At sinks, sucrose is removed. Unloading seems to be active too, at least in part. Water follows the sugar out, so pressure in the sieve tube falls. The overall effect is a pressure gradient from source to sink. If new sinks develop, or sources fail, the direction of transport can be reversed.

This simple model of phloem transport is very different to some of the weird and wonderful ideas in the older literature. Those suggestions were developed to overcome two big objections to the pressure flow model. Firstly, some calculations suggested that the pressure gradient was too small to overcome friction at sieve plates. Secondly, some experiments hinted that material might flow along a sieve tube in both directions at once. However, newer calculations suggest that the gradient is sufficient, and most data suggesting transport in both directions can be interpreted in other ways. Once the objections disappeared, the complicated explanations vanished too.

18.8 Another view from the edge

Most mosses live on surfaces. Their shoots are in air, while cells called rhizoids anchor them to soil, stone or bark. Mosses have swimming male gametes (Section 14.2), so they rely on water for reproduction. Some species do live in very wet places. For example, the vegetation of bogs is dominated by mosses of the genus *Sphagnum*. Others, however, may suffer severe drought. My mother's house has a slate roof where a *Grimmia* species thrives. Some summers, the plants bake for weeks on end, but carry on growing when the weather breaks.

Mosses do not have an efficient transport system. They have no xylem or phloem. A few genera do have specialized water-conducting cells, but these are the exception. Most mosses rely on rain falling on the shoots rather than water in the soil. When it rains, the leaves absorb water. The small leaves are often close together, so water clings to whole shoots. Dense cushions of stems act as sponges. *Sphagnum* even has specialized water-storing cells. These dead cells fill up with water and hang onto it very strongly. When it does not rain for a long time, mosses simply dry up. Many species are amazingly tolerant of desiccation. While dry, they exist in suspended animation. When they absorb water, photosynthesis begins again. This is all very different from most flowering plants, which take up water from the soil, shed rain water from the waxy cuticles of leaves, and try to maintain the turgor of cells at almost any cost. Mosses remind us that we must not assume that we can learn everything about plants by studying flowering plants. Life really is diverse.

Box 18.5

Trouble at the root

Extreme soil conditions can cause problems for plants. Dry soil (Box 18.1) causes one set of difficulties; waterlogged or salty soils cause quite different ones.

Waterlogging

Soil contains pockets of air. Plant roots use them as a source of oxygen for respiration. When soil becomes waterlogged, the air is lost. How can the root survive? Plants which are not adapted to waterlogging tend to start respiring anaerobically. They produce ethanol which eventually poisons them. The plant dies. There are two types of adaptation which avoid this. Firstly, some roots have air channels in the cortex, which allow diffusion of gases from shoot to root. Stems and petioles (leaf stalks) of water plants have similar adaptations. In some, an amazing system leads to a mass flow of gas down through one part of the plant and back up through another. Secondly, some wetland plants have roots which respire anaerobically, but produce waste products which are less toxic, such as malate.

Salty soils

Salinity can be a problem on coasts, and in some deserts where rivers and streams carry in mineral ions which are left in the soil when the water evaporates. Sodium and chloride ions are most significant. Salt causes two related problems. First, a high salt concentration in the cytosol changes patterns of hydrogen bonding, and so affects protein shape and function. Salt has to be kept out of the cytosol. Secondly, the soil solution has a much more negative solute potential than the cytosol. This makes it harder for the plant to take in water. The tension in the xylem water column must be that much greater in order to drag water into the root.

Halophytes are plants which can tolerate salt in the soil. A few halophytes, including some mangroves, keep salt out of the roots. Their xylem water column has to be under extreme tension. Many other halophytes absorb salt but put it into the vacuoles of cells. This would tend to draw water out of the cytosol by osmosis. The cell counteracts this by putting high concentrations of other solutes in the cytosol. These solutes, which do not disrupt protein structure, include betaine and the amino acid proline.

$$H_3C\text{—}N^+\text{—}C(CH_3)_2\text{—}COOH$$

(structure: N^+ bearing CH_3 above and CH_3 below, bonded to H_3C and CH_2—COOH)

Betaine

Summary

- Leaves carry out gas exchange, but they lose water at the same time.
- Water evaporates into air spaces inside the leaf. It diffuses out through stomata, holes in the leaf epidermis.
- The rest of the epidermis is covered in a waxy cuticle to minimize water loss.
- Stomata can be opened and closed so that there is enough gas exchange to support photosynthesis, without losing too much water.
- Each stoma is surrounded by a pair of guard cells. When they become turgid, the stoma opens.
- Stomata open and close in response to environmental stimuli including light intensity (brighter leads to opening), carbon dioxide concentration (higher leads to closing), temperature (higher leads to opening) and water stress (causes closure).
- Guard cells become turgid as follows: protons are pumped out of the cell; potassium ions enter down the electrical gradient; starch is broken down to organic acids, which dissociate to form more protons, and organic ions; overall, the cell accumulates potassium and organic ions; water enters the cell by osmosis, making it turgid.
- Water and mineral ions move up the plant in the xylem.
- Xylem vessels are chains of dead cells which transport water like pipes. Tracheids are similar, but water moves from cell to cell through small pits.
- Water in the xylem is at low pressure in the leaf where it evaporates, but at higher pressure in the root where it is absorbed. Water in the xylem is under tension, but the column does not break because water molecules are attracted to one another and to the walls of the xylem.
- Water and mineral ions are absorbed in the young roots. Root hairs give a large surface area.
- Mineral ions are absorbed actively. Water is normally sucked in, driven by evaporation in the leaf. It also follows mineral ions into the root, by osmosis.
- Water and minerals cross the root either in the interconnecting cell walls (apoplast) or the interconnecting cytoplasm (symplast). A waterproof layer in the cell walls of the endodermis forces material to enter the symplast.
- Many plant roots are associated with fungi which help absorb minerals: mycorrhizae.
- Some plants have nitrogen-fixing bacteria in root nodules. They can reduce nitrogen gas to usable ammonia.
- Sucrose is transported around the plant in the sieve tubes of the phloem.
- Sucrose is actively loaded into the phloem, and probably unloaded actively too. Water follows by osmosis. This sets up a pressure gradient. Sucrose solution moves along by mass flow.
- Unlike flowering plants, most mosses cannot transport water, and rely on absorbing rain directly. Many can tolerate drying out.

- Plants which live in dry habitats have adaptations which avoid their cells drying out.
- In waterlogged soils, plant roots can be short of oxygen. Many wetland plants have air channels in the roots, or make non-toxic waste products in anaerobic respiration.
- Plants of salty soils either keep salt out of the root cells, or confine it to the vacuoles.

Exercises

18.1. A tree trunk is carrying water and minerals from the roots to the leaves, and sugar solution in the opposite direction. A section of the trunk is heated to 100°C. Transport of sugars stops, but the upward movement of water continues. Explain this observation.

18.2. The lignin in the walls of vessels and tracheids makes them very strong. Why is this important?

18.3. In the 1960s, Scholander and colleagues made direct measurements of water pressure inside the xylem of stems, in order to test the cohesion tension theory.

(i) At 7 am they used a rifle to shoot twigs from different heights on a Douglas fir tree.

Height/m	27	79
Water pressure/kPa	−500	−1000

Explain how these data support the theory.

(ii) At 1 pm, results were as follows:

Height/m	27	79
Water pressure/kPa	−1700	−2300

What has caused the changes?

18.4. Nectar contains sucrose, glucose and fructose. The sugars are there to give pollinating animals an incentive to visit the flower. Explain why these three sugars in particular are present.

18.5. In these examples, what are the sources and sinks?

(i) The leaves and stems of a potato plant are photosynthesizing. Starch is being stored in the tubers.

(ii) It is early spring. An ash tree is flowering, before its leaves have appeared. Energy stores in the roots are being mobilized.

Further reading

Ridge, I. *Plant Physiology* (London: Hodder & Stoughton, 1992). A lighter textbook aimed at new undergraduates, written for self-study.

Rudall, P. *Anatomy of Flowering Plants* (2nd ed.) (Cambridge: Cambridge University Press, 1992). A concise summary of flowering-plant structure and how it develops.

Salisbury, F.B. and Ross, C.W. *Plant Physiology* (4th ed.) (Belmont CA: Wadsworth, 1992). A big undergraduate textbook. Any text in this field has to navigate a minefield of conflicting ideas and evidence in so many topics. This book does as well as you could hope in drawing out conclusions without avoiding the arguments.

Transport in Animals

Contents

Connections

► You should have a basic knowledge of biological molecules, cell structure and the movement of molecules (Units 3, 4 and 6) before studying this unit. Some knowledge of hormones and the nervous system (Units 16 and 17) is also required.

19.1 Why have blood?

Animal bodies need to transport molecules from place to place. A sheep is wandering slowly about a field, grazing. Its cells are respiring, releasing energy from food molecules to allow them to go about their business. Oxygen is used, carbon dioxide is made. The sheep is transporting oxygen from its lungs to cells in every part of its body. Carbon dioxide is brought back. The gases are carried by the blood as it flows around the body. In the sheep, like other vertebrates, oxygen is carried in the red blood cells, while most of the carbon dioxide is transported in the liquid part of the blood, the plasma.

The sheep's gut is absorbing sugars and amino acids, along with other organic molecules and minerals from the food it has digested. These need to be carried to the liver, where some will be stored and some modified. Others will continue on to the rest of the body where they are needed. Some of these chemicals are dissolved in the plasma, just as if it were water. Others are bound to blood proteins. Insoluble cholesterol can only be transported when bound to a hydrophilic protein in the plasma. Blood can hold far more oxygen than can water, because it is bound to a protein, haemoglobin, in the red cells.

A dog has appeared in the corner of the field. The sheep spots it, and watches warily. Now the dog starts to run towards it. An emergency is approaching for the sheep. The adrenal glands release a hormone, adrenaline, into the bloodstream. Within seconds, adrenaline has been carried all around the body. The cells which can detect it are preparing the body for some hard exercise. The sheep turns and runs. Now its leg muscles are working hard. They need lots more oxygen and glucose than a moment ago. This transport system has to be regulated, matching the rate of delivery and collection to the body's changing requirements. The muscles are generating waste heat now, just as the liver was doing all along. The blood carries it away before the organs overheat. The body surface is losing heat, even under all that wool, and heat delivered by the blood from internal organs keeps it up to temperature.

So blood is a chemical transport system, carrying dissolved gases, food molecules, waste products, minerals and hormones, as well as heat, around the body. Only the simplest animals have no blood. They must rely on diffusion, which means they must be small, or at least very thin (Section 6.1).

Blood does more than simply carry things from place to place. It is important in defence (Unit 23). It has a complex clotting mechanism which seals up wounds. The various types of white blood cell are involved in recognizing and combating foreign molecules and cells. Blood also transmits forces. Earthworms can crawl because when muscle rings contract, the blood within is squeezed out lengthwise (Section 24.7). When a ram's (or a man's) penis becomes erect, it is all due to blood pressure. The force is generated by the beating heart and transmitted by the blood. This unit, however, is about the transport functions of blood. And if you are wondering what happened in our story, the dog's owner appeared and called it off before the farmer could take aim. We all like a happy ending.

19.2 Oxygen transport

Oxygen is carried around the body in the blood. This is true for all animals with a blood system, apart from the insects. Insects bring air directly to the tissues in the tracheal system (Section 20.4). Some invertebrates carry dissolved oxygen in the blood, just as if the blood were water. Most animals have an oxygen-binding protein in their blood. The most common is haemoglobin, found in the red blood cells (Fig. 19.1) of all vertebrates except a few fish, as well as in some invertebrates.

Haemoglobin is a protein. It is made up of four polypeptides, two α-chains and two β-chains. Each has a prosthetic group: an iron-containing haem group. Haemoglobin can bind up to four oxygen molecules. However, when it binds one oxygen, its shape changes. This makes it much easier for three more to bind, so once one has bound three more will quickly join it. In effect, haemoglobin changes between two stable states, deoxyhaemoglobin (with no oxygen bound) and oxyhaemoglobin (with four oxygens).

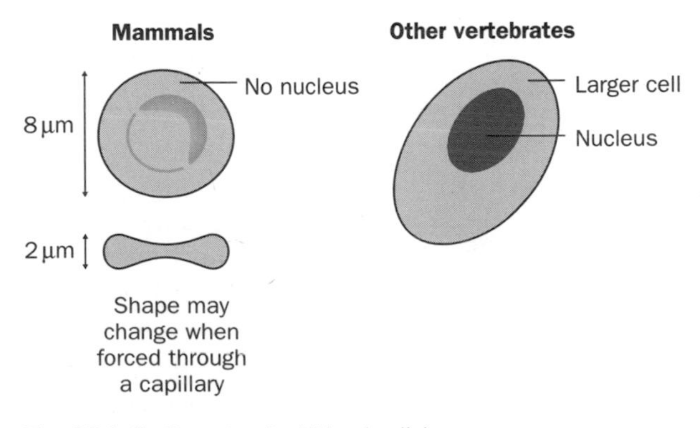

Fig. 19.1 Erythrocytes (red blood cells).

Haemoglobin is important in three ways. Firstly, it allows blood to hold far more oxygen. Blood saturated with oxygen contains all the oxygen the same volume of water could hold, plus all the oxygen the haemoglobin can bind. This makes blood a more effective oxygen carrier. Secondly, oxygen can diffuse more quickly through a haemoglobin solution than through pure water. This is largely because there is more oxygen in the solution. The free oxygen diffuses just as it would in water. The haemoglobin, with its bound oxygen, also diffuses. The total rate of diffusion is higher. This is not very significant in blood, where flow is more important than diffusion. It is important in other tissues containing haemoglobin. Mammalian muscles contain myoglobin, a related oxygen-binding protein. This helps oxygen diffuse through the muscle from the capillaries. It also provides a 'reserve tank' of oxygen, because it binds oxygen more strongly than haemoglobin, and only releases significant amounts when oxygen levels in the muscle drop below normal, for example at the start of exercise. Thirdly, haemoglobin loads and unloads oxygen in an interesting way. This is best understood by looking at how much oxygen haemoglobin holds at different oxygen concentrations.

Partial pressure is an indirect measure of the concentration of a gas. Atmospheric pressure at sea level is about 100 kilopascals (kPa). Air is a mixture of gases. Each gas contributes to the total pressure. If 21% of the total volume is oxygen, 21% of the pressure is due to oxygen. The partial pressure of oxygen is 21% $\times$ 100 kPa = 21 kPa. If the total pressure was lower, the partial pressure would be lower, too. We sometimes talk about the partial pressure of a gas in a solution, even though dissolved gases do not exert any pressure. Take a gas sample and shake it up with water, until no more will dissolve: the system has reached equilibrium. If the oxygen in the gas has a partial pressure of, say, 15 kPa at equilibrium, we say that the dissolved oxygen also has a partial pressure of 15 kPa.

Fig. 19.2 shows a graph of oxygen saturation of haemoglobin against partial pressure of oxygen. This is called an **oxygen dissociation curve**. It is sigmoidal (S-shaped). The middle section of the graph is steeper than the parts on either side. This is important. Over the middle range, a small change in oxygen partial pressure will lead to quite a large change in oxygen saturation. This is caused by the all-or-nothing way in which haemoglobin binds its four oxygens.

In humans, haemoglobin is loaded with oxygen in the alveoli of the lungs (Section 20.6). This has a partial pressure of about 15 kPa. The curve shows that haemoglobin will be almost fully saturated. A typical partial pressure elsewhere in the body might be 4 kPa. Then, haemoglobin will be about 50% saturated. So, as oxygenated blood arrives in a tissue, it unloads about half its oxygen. This is on the steepest part of the curve. If the tissue starts respiring faster, oxygen partial pressure will fall. The steepness of the curve means that a disproportionate amount of extra oxygen will be unloaded. In summary, blood has a reserve oxygen capacity which can be used efficiently when local oxygen demand rises.

Not all oxygen dissociation curves are in the same position on the axes. Fig. 19.3 shows that the rat blood curve lies to the right of the human curve. At any given partial pressure of oxygen, rat blood is less saturated. This means that rat blood must bind oxygen less strongly: it has a lower oxygen affinity. This is important, because rats have a higher metabolic rate than humans (Section 9.6). Each gram of rat tissue uses more oxygen per minute. Blood in the capillaries is able to unload 50% of its oxygen at a relatively high partial pressure of oxygen. This means that the diffusion

Partial pressure: The component of total pressure exerted by one gas in a mixture.

Oxygen dissociation curve: A graph of oxygen saturation of haemoglobin against partial pressure of oxygen.

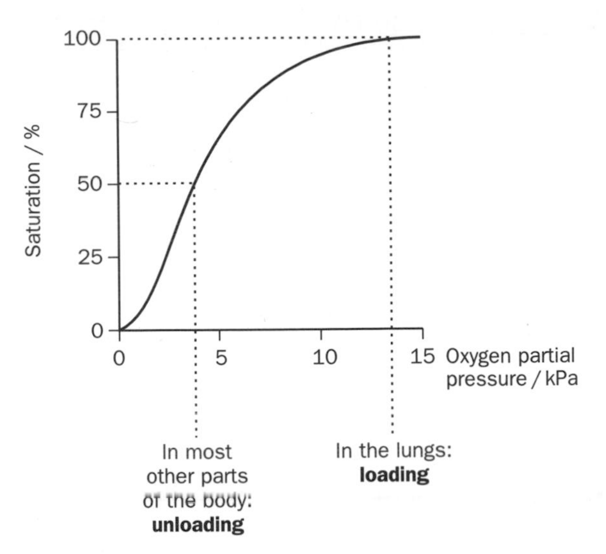

Fig. 19.2 Oxygen dissociation curve for adult human blood.

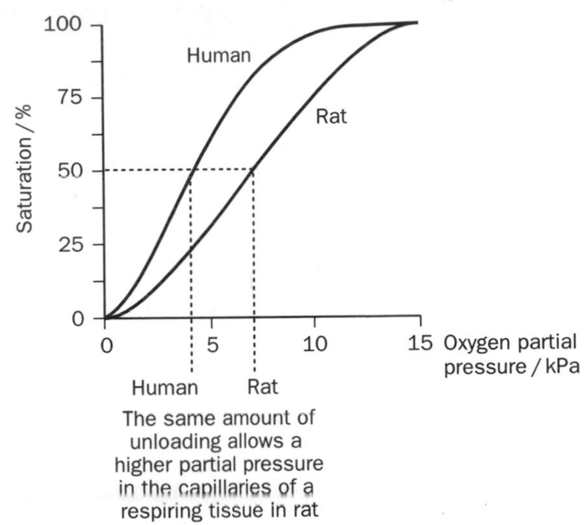

Fig. 19.3 Rat blood has a lower oxygen affinity than human blood.

gradient from the blood to the cells which are using oxygen is relatively steep. Oxygen diffuses at a faster rate. This allows it to carry on its high rate of respiration.

Carbon dioxide concentration affects the oxygen dissociation curve. Increasing CO_2 decreases the oxygen affinity of blood. The curve shifts to the right (Fig. 19.4). This is called the Bohr effect. It is important, because an increase in CO_2 is a sign that the tissue is respiring faster. The curve shifts to the right, so more oxygen is unloaded. How does this happen? There are two linked reasons. Firstly, CO_2 solutions are acidic. In acidic solutions, carboxyl groups in the haemoglobin molecule are less likely to dissociate (Box 3.5). Secondly, CO_2 itself can bind to the haemoglobin molecule. Both these events subtly change the shape of the molecule, affecting its oxygen affinity. Box 19.1 shows some other factors which shift the oxygen dissociation curve in interesting and sensible ways.

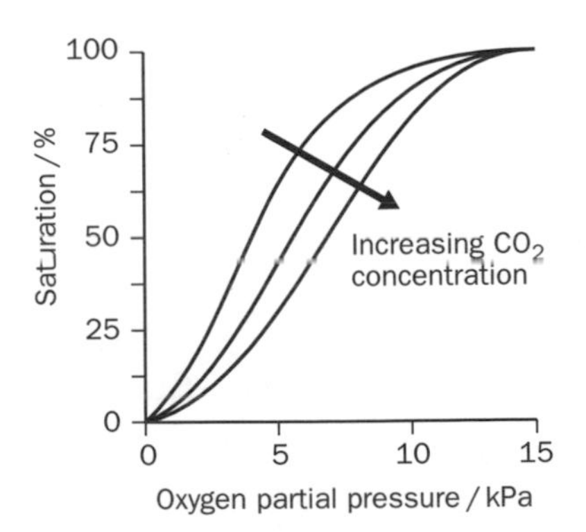

Fig. 19.4 The Bohr effect: increasing carbon dioxide concentration leads to a lower oxygen affinity.

Box 19.1

Warm, pregnant and high: oxygen dissociation curves in action

Comparing oxygen dissociation curves gives insights into several aspects of an animal's life.

Pregnancy

Foetal blood takes oxygen from the mother's blood in the placenta. It has a higher oxygen affinity. Foetal haemoglobin is different to the adult form. It is made up of two α and two γ polypeptide chains, instead of two α and two β chains. Foetal blood also has a higher haemoglobin concentration. This helps it to carry enough oxygen, although it does not affect the curve.

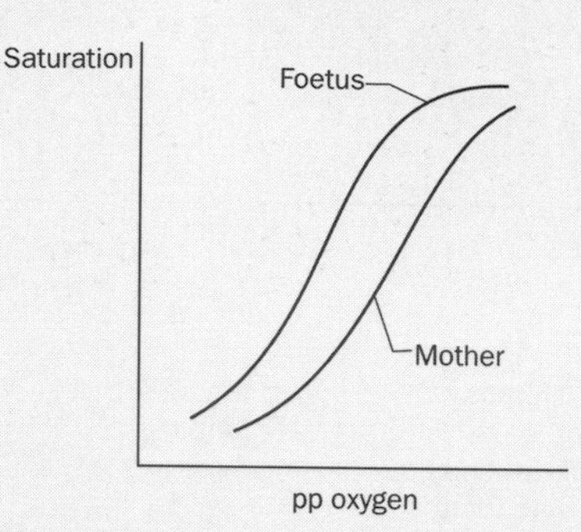

Temperature

Higher temperatures weaken the bonds between oxygen and haemoglobin. Oxygen affinity falls, shifting the curve to the right. This is important in cold-blooded animals. When warm, metabolism speeds up, so they use more oxygen. Haemoglobin unloads oxygen more efficiently, matching this rise.

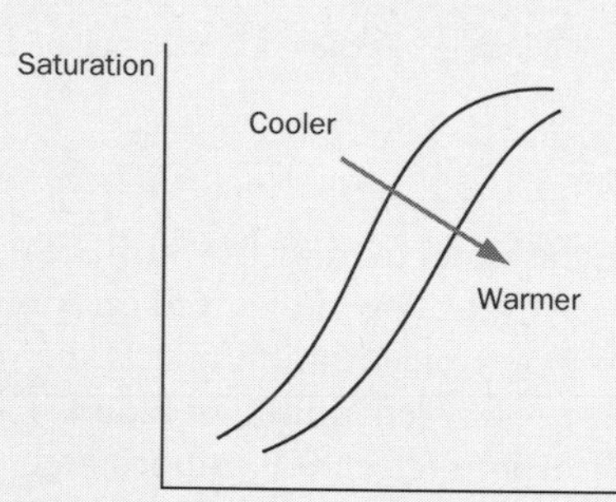

Altitude

Some high-altitude mammals, like the llama, have curves well to the left of most mammals. This could be explained as a response to the low partial pressure of oxygen at high altitudes. Llama blood binds oxygen at low partial pressures more strongly than other mammals. When humans acclimatize to high altitudes, the curve shifts to the *right*. It could be argued that this means that oxygen will be unloaded more efficiently. However, we cannot have it both ways. There is a trade-off between loading and unloading. When the curve shifts left, oxygen is loaded more efficiently, but unloaded less efficiently. When it shifts right, the opposite is true. These changes make no sense unless considered along with the other changes that take place at high altitude.

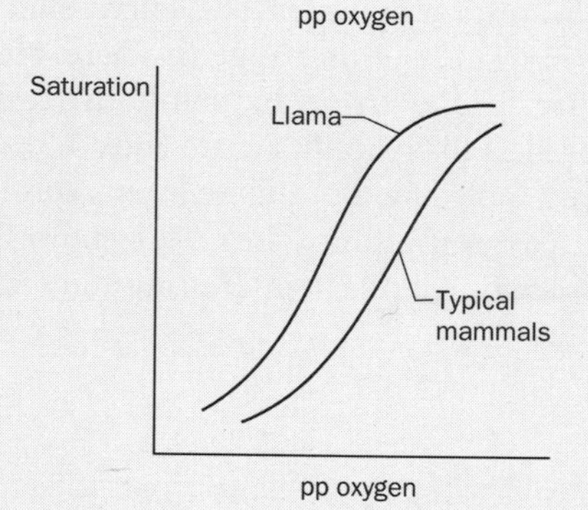

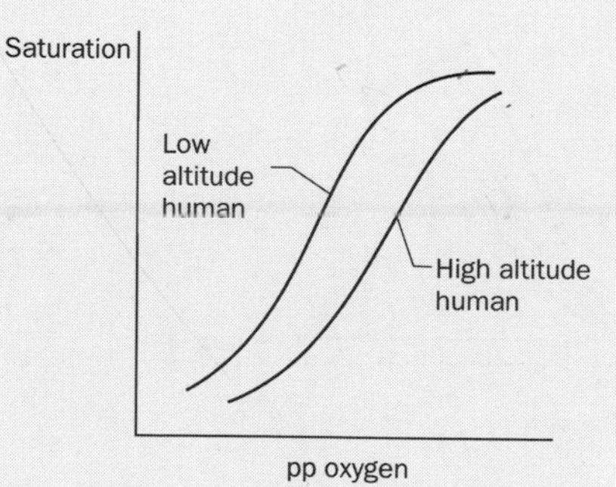

Llamas and the humans who look after them are adapted to life at high altitude, yet their oxygen dissociation curves are shifted in opposite directions (see Box 19.1).

19.3 Carbon dioxide transport

When carbon dioxide dissolves in water, it reacts with the water. Carbonic acid is formed, which dissociates to form hydrogencarbonate ions. These can dissociate further, to form carbonate ions.

CO_2	+	H_2O	$\rightleftharpoons$	H_2CO_3	$\rightleftharpoons$	HCO_3^-	+	H^+	$\rightleftharpoons$	$CO_3^{2-} + 2H^+$
carbon dioxide		water		carbonic acid		hydrogen-carbonate				carbonate

At blood pH, very little carbonate is formed. Most carbon dioxide is actually transported as hydrogencarbonate. Most of this is in the plasma. A small amount of CO_2, perhaps 20% of the total, is carried weakly attached to all sorts of blood proteins.

The reaction between CO_2 and water is rather slow. Mammalian blood contains an enzyme which speeds it up. The enzyme is called carbonic anhydrase, and it is found in the red blood cells. When the blood picks up CO_2 in the tissues, it diffuses into the red cells, where it reacts to form carbonic acid. This quickly dissociates to form hydrogencarbonate. Hydrogencarbonate is now at high concentration in the

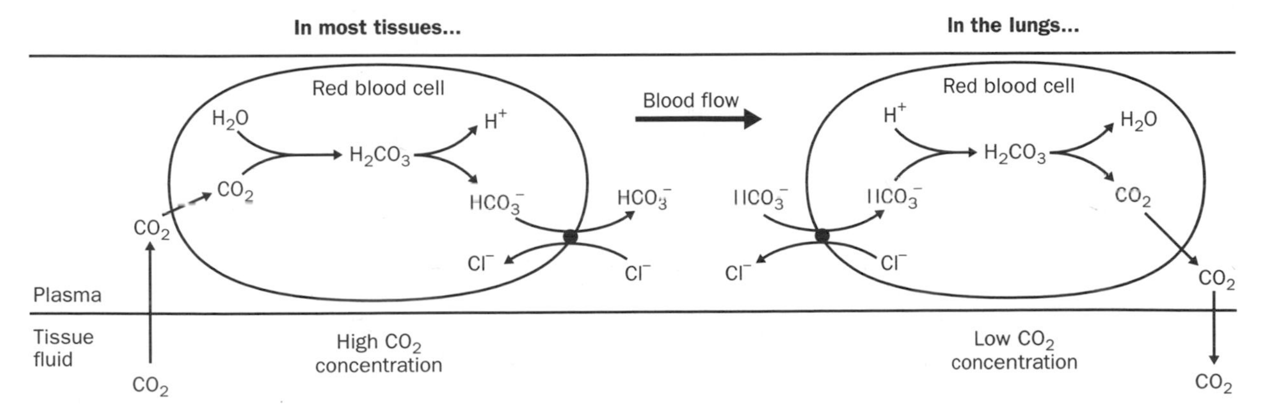

Fig. 19.5 Carbon dioxide transport in the blood.

cells, so it diffuses out into the much larger volume of plasma. It is exchanged for Cl^-, to avoid the red cells becoming charged. In the lungs, the process is reversed. Fig. 19.5 summarizes the whole sequence.

This is what happens, but it turns out that the whole system could work without carbonic anhydrase. If the enzyme is completely inhibited by a drug, CO_2 transport is hardly affected. The blood can carry enough CO_2 without accelerating the reaction. The enzyme seems to be necessary only when the blood has to carry very large amounts of CO_2. More significantly, the Bohr effect (Section 19.2) depends on the enzyme. CO_2 forms an acidic solution only because carbonic acid dissociates to release H^+. If too little carbonic acid is formed, the Bohr effect will be less marked.

19.4 Transport of solutes

Most solutes other than gases are simply carried dissolved in the plasma, as they would be in water. Hydrophilic molecules which will dissolve freely and will not react with other things in blood are carried in this way. Glucose, amino acids, urea, peptide hormones like oxytocin, and most mineral ions are good examples. Some solutes, however, are escorted by plasma proteins.

Hydrophobic molecules like oestrogen and cholesterol are bound to proteins simply to make them more soluble. The proteins can release these steroids quickly. Thyroxine, a hydrophobic hormone, is bound strongly to a plasma protein. The hormone is released slowly, over a period of days. A short pulse of thyroxine release can have a long, drawn-out effect.

Iron is strongly bound to a plasma protein, forming a complex called transferrin. Otherwise, iron ions would form insoluble compounds with some other ions present in plasma. The body's entire iron stock would then become unavailable. The protein, apotransferrin, is made in the liver and secreted into the small intestine via the bile duct. It binds iron ions and some iron compounds in the gut. The whole complex is taken into the epithelial cells by receptor-mediated endocytosis (Section 6.6) and sent on into the blood. A similar protein complex, ferritin, protects iron in the cytoplasm.

19.5 How blood circulates

Blood is useless for transport unless it moves. An animal needs something to pump the blood and somewhere for it to flow: a circulatory system. There is a surprising amount of variation between the circulatory systems of different animal groups.

In many invertebrates, blood can flow anywhere in the body, within and between the organs. Every cell is bathed in blood, and gets what it needs directly from the blood. This is called an **open circulatory system**. Arthropods and most molluscs are good examples. Other circulatory systems, including those of the vertebrates and annelids, are **closed**. Blood circulates within a network of tubes: blood vessels. The cells of the body are not surrounded by blood but by tissue fluid, which is derived from blood. It is basically whatever is squeezed out of the blood in capillaries, the finest blood vessels.

Open circulatory system: Blood bathes all body cells.

Closed circulatory system: Blood stays inside blood vessels.

Open circulatory systems

Even in an open system, a pumping organ is needed to set up a flow. In insects, the 'heart' is a string of muscular chambers running lengthwise in the body (Fig. 19.6). Each chamber has one-way valves: ostia. These let blood in from the surrounding tissues, but not out. When the chambers contract, blood is pushed forward, from chamber to chamber. At the front end of this chain of pumps is a blood vessel which carries blood forward through the thorax and head. Open-ended branches carry blood out to different areas. Once it leaves the vessels, blood moves within the tissues, in an overall backward direction. Local barriers channel the blood, so that all organs get the flow they need. Try it when you have a bath! Sit facing the taps and slosh the water forward on your left. Put in a floating object (rubber duck?) and see where it goes. Even though your body is an incomplete barrier, the duck will tend to circle you, clockwise. By putting barriers to flow on your right, you should be able to direct the duck along the edge of the bath, under your knees, and so on. The message is clear. Even though an open circulatory system allows blood to go anywhere, flow need not be random. There is scope for the body to control how much goes to each organ. We know too little about invertebrate circulations to be able to say how effective this control is, in general.

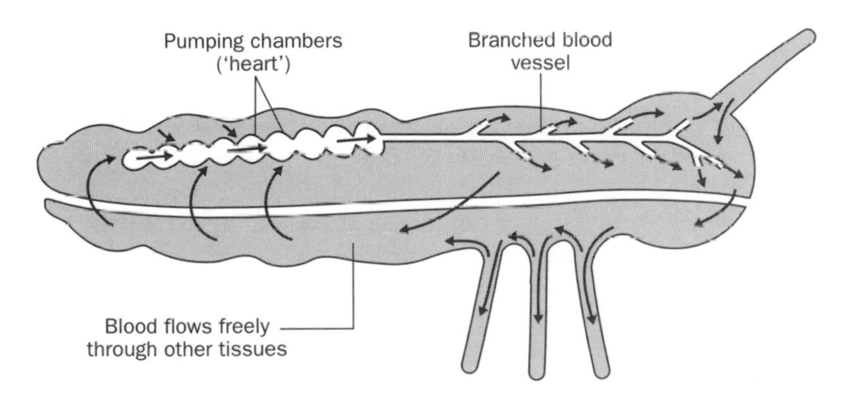

Fig. 19.6 Circulation in an insect.

Closed circulatory systems

Three examples show the range of closed circulatory systems. In annelid worms, the body is clearly divided into segments. The blood system in each segment is similar, with no central pump. In vertebrates, segmentation is less obvious. A central heart drives the entire system. In fish, blood passes through the heart once in each complete circuit. In mammals, it goes through twice.

Annelids

Blood circulation in annelids happens at two levels. (Fig. 19.7 shows the circulation in the ragworm, *Nereis*.) Firstly, blood circulates around the organs of one segment. Particularly important are the skin, where gas exchange occurs; the gut, where food is absorbed; and the nerve cord. Secondly, blood needs to flow between segments, keeping the different parts of the body in contact. The circulation in each segment is linked to two vessels which run the length of the worm.

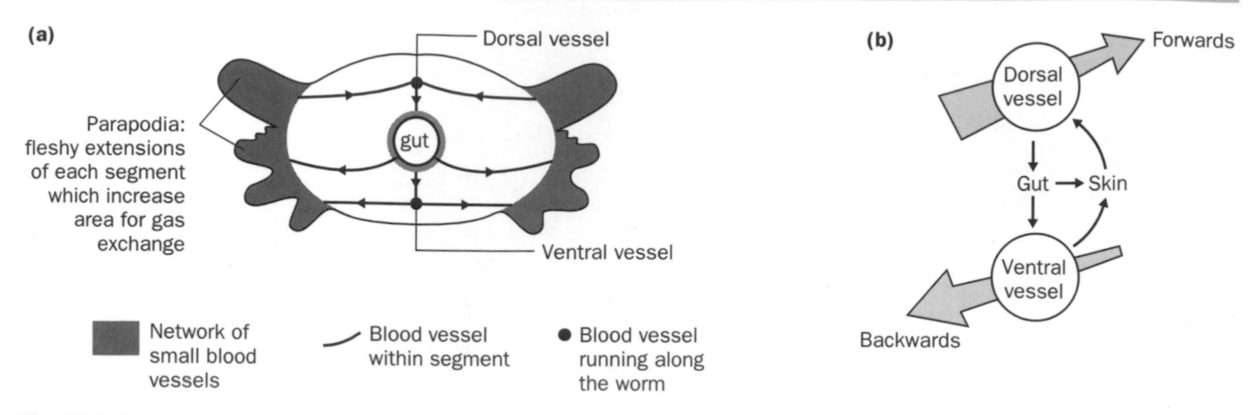

Fig. 19.7 Circulation in the ragworm. (a) within one segment; (b) how this fits in with circulation between segments.

The dorsal vessel, running along the worm's back, carries blood towards the head. It has a muscular wall. Waves of contraction pass forward along it. Blood is squeezed forward, like hand-milking a cow. This sort of movement is called peristalsis. Food is moved through the gut in the same way (Section 21.4). The ventral vessel carries blood back again. So, the segments of the worm are partly, but not fully, independent.

Fish

In fish, a single heart pumps blood throughout the body (Fig. 19.8(a)). It has two chambers: an atrium and a ventricle. Blood flows into the atrium from a vein which brings deoxygenated blood back from all parts of the body. The atrium pumps blood into the ventricle, which in turn pumps it out into an artery leading to the gills. The artery divides again and again. Each branch ends up in a capillary bed, where oxygen is gained and carbon dioxide is lost (Section 20.5). Blood, now oxygenated, flows on into a vessel which branches out to supply other parts of the body. Deoxygenated blood returns to the heart. Overall, blood flows through two capillary beds in each complete circuit. One is in the gills, the second in some other organ. This is called a **single circulation**.

Single circulation: Blood passes through the heart once per circuit of the body.

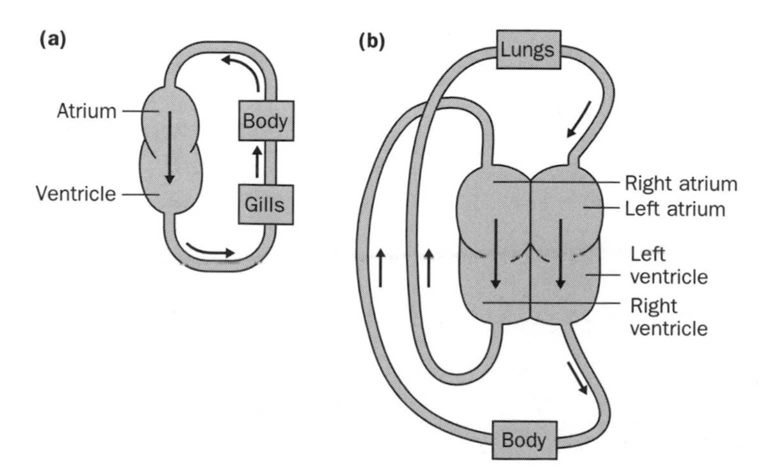

Fig. 19.8 Single circulation of a fish (a) compared with double circulation of a mammal (b).

Mammals

Mammals have a **double circulation** (Fig. 19.8(b)). The heart is not one but two pumps, the left and right sides. Blood cannot flow from one to the other. Each has an atrium and a ventricle. The right side pumps deoxygenated blood to the lungs. Oxygenated blood returns to the left side. This pumps it to the rest of the body. Deoxygenated blood comes back to the right side. Blood passes through the heart twice in each circuit. It only passes through one capillary bed between visits to the heart. This means that the pressures in the lung circulation and the general circulation can be controlled independently. Also, the beating of the heart has to overcome the friction on only one, not two sets of capillaries.

Double circulation: Blood passes through the heart twice per circuit of the body.

Arteries: Blood vessels which carry blood away from the heart.

Fig. 19.9 shows the main blood vessels of a mammal in more detail. **Arteries** carry

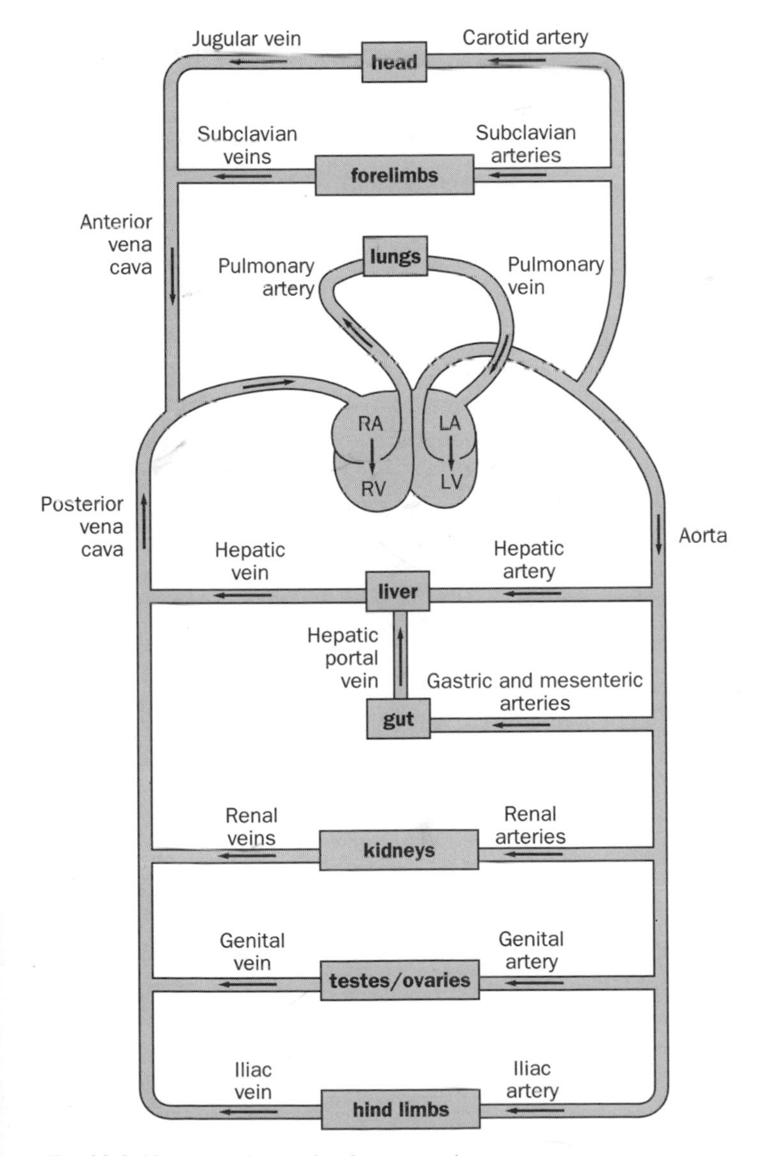

Fig. 19.9 Major blood vessels of a mammal.

blood from the heart, **veins** carry blood to the heart. Box 19.2 compares arteries, veins and capillaries. Four major vessels are connected to the heart. The pulmonary artery goes from the right ventricle to the lungs. Blood returns to the left atrium in the pulmonary vein. The aorta carries blood out from the left ventricle. Smaller arteries branch off it, going to specific organs. Blood returns through veins, which join two trunk veins. These are the posterior (inferior) vena cava, coming up from below, and the anterior (superior) vena cava, coming down from the head and arms. They lead back to the right atrium. The biggest exception to this pattern is blood flow from the gut wall. This contains variable amounts of substances absorbed from food. These are carried straight to the liver in the hepatic portal vein. The liver absorbs substances which are in excess, before blood flows on to the vena cava.

Veins: Blood vessels which carry blood to the heart.

Box 19.2

Not just dumb pipes . . .

Blood flows from the heart in arteries, and back in veins. In between, it passes through capillaries, where materials enter and leave. Each type of blood vessel has distinctive features. They all do more than simply carry blood.

In general

All vessels except the capillaries have the same general structure. The wall has three layers. The tunica intima consists of the endothelium, a single layer of cells in contact with the blood, plus a thin layer of connective tissue beneath it. It forms a barrier. The tunica media gives strength, the power of contraction, and some degree of stretchiness. It contains smooth muscle, elastin fibres and tougher collagen fibres. The tunica adventitia is a sheath of connective tissue which holds the blood vessel in position. Nerves and small blood vessels run through it, leading to the smooth muscle of the media.

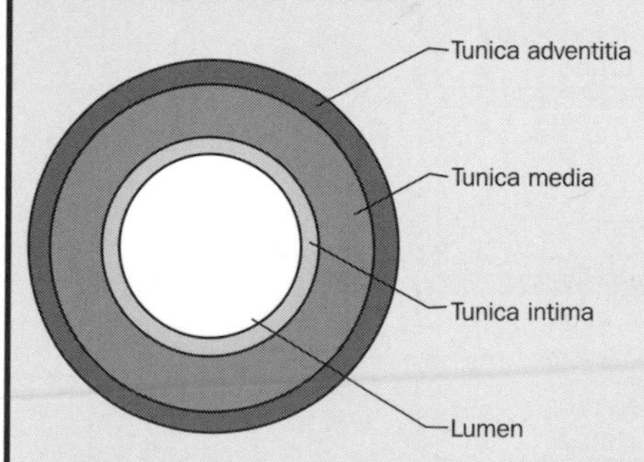

Arteries and arterioles

- Pressure and flow rate are higher than in the veins. As a result, the lumen is narrower than in the equivalent vein.
- No valves are needed.
- Tunica media is thick.
- The largest arteries have lots of elastin in the media. The wall is stretchy. This evens out the pulses of pressure and flow generated by the heart.
- Smaller arteries and arterioles have more muscle in the media.
- The muscles of the arteriole wall relax or contract (vasodilatation and vasoconstriction) to control the flow of blood to a particular area.

Veins and venules

- Pressure and flow rate are lower, so the lumen is relatively large.
- Semilunar valves prevent backflow, just like where the great arteries leave the heart.
- Tunica media is thin with some muscle and collagen.
- Vessels may flatten, so blood volume can vary.
- The veins are an adjustable blood reservoir. Some can contract to reduce blood volume (venoconstriction).

Capillaries

- Very thin (typical lumen diameter 5 μm).
- Short (typical length 0.5 mm).
- Endothelium only.
- Tissue fluid is formed here (Section 19.7).

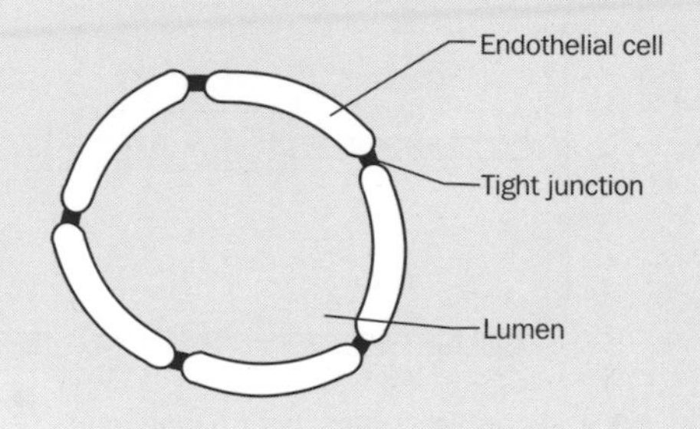

Before birth, mammals exchange gases in the placenta (Section 13.7). The lungs are useless, since there is no air in the uterus. There needs to be radical differences between the circulatory systems before and after birth. Box 19.3 explains them.

Box 19.3

Circulation before birth

The foetus has a very distinct pattern of blood flow. It must switch to the adult pattern at birth. For the foetus, the placenta does the job of both lungs and gut. It absorbs oxygen, excretes carbon dioxide and absorbs nutrients. The lungs and gut need hardly any blood flow, yet need to be 'plumbed in' ready for birth.

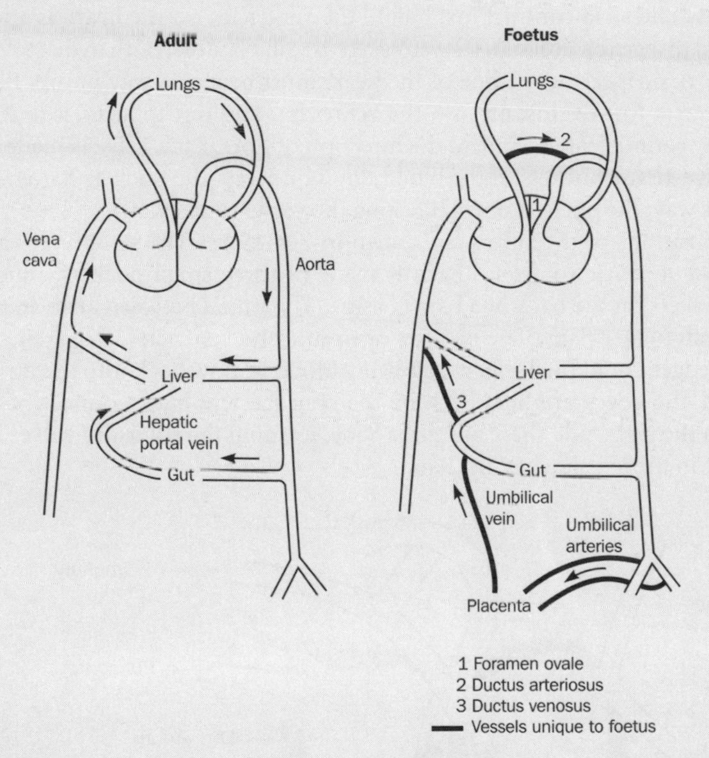

The blood system of the foetus has several special features. The umbilical arteries and vein run through the umbilical cord to and from the placenta. They are connected to the aorta and the vena cava. This means that oxygenated and deoxygenated blood are mixed in the vena cava. The mixture returns to the heart. There is no point in it all going straight to the lungs, so two features bypass them. First, the foramen ovale (1 on diagram) links the two atria. It is a hole with a valve, letting blood flow from right to left atrium, avoiding the lungs. Second, the ductus arteriosus (2 on diagram) carries blood straight from the pulmonary artery to the aorta.

Food-rich blood is arriving from the placenta, not the gut. Some of this needs to go straight to the liver, but there is no point in blood from the gut going there too. The ductus arteriosus (3 on diagram) solves all these problems.

At birth, the special foetal vessels must close. Within hours, they constrict strongly, and over a period of months they become blocked by fibrous tissue. The foramen ovale stops working, because pressure becomes higher in the left atrium than the right atrium: the valve stays closed. Eventually, tissue may grow over it, blocking it permanently.

19.6 The mammalian heart

The closed, double circulation of mammals raises two big questions. Firstly, the centralized heart must be efficient and well controlled. How does it work? Secondly, the blood is trapped in blood vessels. How do substances get from the blood to the cells of the body, and vice versa? These questions are tackled in turn.

The beating heart

Each chamber of the mammalian heart has a muscular wall. Each goes through a cycle with two phases: diastole (pronounced die-ass-tole-ee) and systole (sis-tole-ee). In diastole, the muscles are relaxed. The chamber fills with blood. In systole, the muscles contract, forcing blood out. The two atria go through the cycle together. So do the ventricles. The atria contract just before the ventricles, so the heart as a whole goes through a three-step cycle: atrial systole; ventricular systole; both in diastole.

To work properly, each side of the heart must be a one-way pump. Blood flows into the atria from veins, on into the ventricles, and out to arteries, but never the other way round. Valves control the direction of flow. Each valve is made up of thin connective-tissue pockets. If the blood flows one way, the pockets flatten. If it flows the other way, the pockets open, blocking the vessel (Fig. 19.10).

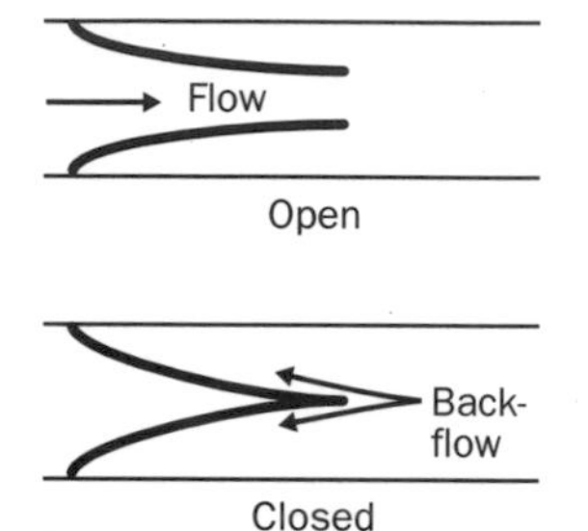

Fig. 19.10 Semilunar valves prevent backflow of blood in veins, and where the great arteries leave the heart.

The positions of the valves are shown in Fig. 19.11. The semilunar valves allow flow from ventricle to artery. Each is made of three small pockets, much like the patch pockets on the back of a pair of jeans. The valves between atria and ventricles have much bigger flaps. The pockets open into the ventricles, but their size means that the edges could buckle back, allowing blood to flow back into the atria. To prevent this, the edges are held back by the chordae tendinae, connective-tissue guy ropes. In the right side there are three flaps, forming the tricuspid valve. The bicuspid valve in the left side has just two.

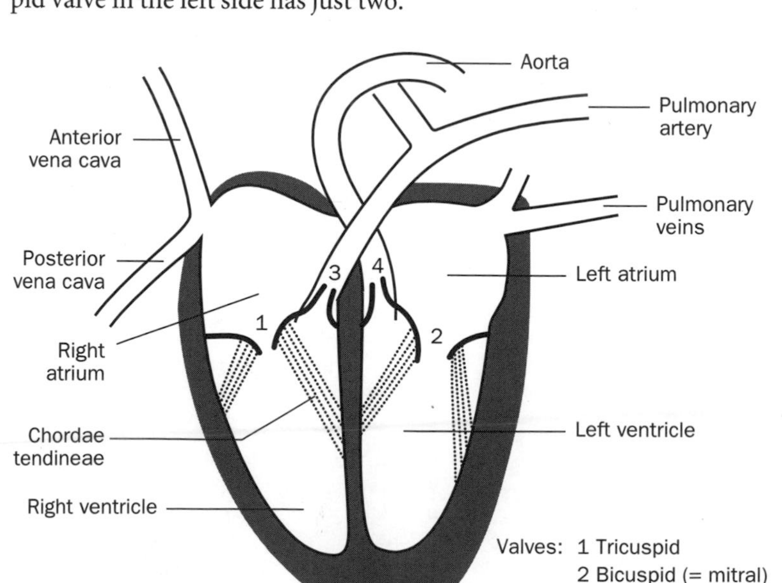

Fig. 19.11 The mammalian heart.

The heart beats by itself. If all the nerves to the heart are cut, it carries on beating. The nerves simply adjust the rate. The cardiac muscle of the heart must, then, be able to set up a signal to contract, and to carry that signal throughout the heart, as well as the simple action of contracting. Cardiac muscle cells come in two types. Work cells contract. The others are specialized for setting up and carrying electrical signals.

Work cells are very like skeletal muscle cells in the way they contract (Section 24.5). Actin and myosin fibres run lengthwise through the cell. When the myosin fibres 'walk' along the actin fibres, the cell contracts. This uses energy from the hydrolysis of ATP. A pulse of calcium ions entering the cytosol is the signal for contraction. At rest, the cardiac muscle cell has a voltage (resting potential) across its plasma membrane, just like the axon of a nerve cell (Section 17.3). Depolarization triggers a flow of calcium into the cytosol. Box 19.4 examines these membrane potentials more closely.

Work cells can propagate impulses. A wave of depolarization spreads along the plasma membrane, just like the sequence in a neurone, and can pass from cell to cell. However, this is relatively slow, about 1 m s^{-1}. Specialized conduction cells carry impulses much more quickly.

Impulses are set up by a group of cells in the right atrium wall. This is called the sino-atrial node (SAN). It measures about 20×4 mm in adult humans. The membranes of these cells behave in a subtly different way to work cells. They depolarize automatically, about once per second (Box 19.4). This sets off an impulse which spreads through the work cells of the atria, making them contract (Fig. 19.12). The impulse cannot pass straight to the ventricles. This is because a ring of non-muscular tissue separates the ventricle walls from the atria. Impulses can only cross at one point. A second group of cells is here, the atrio-ventricular node (AVN). The AVN is connected to a bundle of big, fast-conducting cells, the bundle of His. The bundle crosses into the ventricle walls, in the part which divides the two ventricles. The bundle branches out into each ventricle wall. It ends in a diffuse network of conducting cells, the Purkinje fibres, near the inner surface of the ventricle walls.

The AVN conducts impulses slowly. There is a delay of about 0.1 seconds before the signal crosses into the ventricles in the bundle of His. It spreads rapidly through the ventricle walls in the Purkinje fibres. This means that all parts of the ventricle wall contract more or less simultaneously, but after the atria.

All cardiac muscle cells can set off impulses spontaneously. However, the SAN does this faster than other parts, so its impulses spread across the heart before other cells ever get to fire spontaneously. The SAN is the heart's pacemaker.

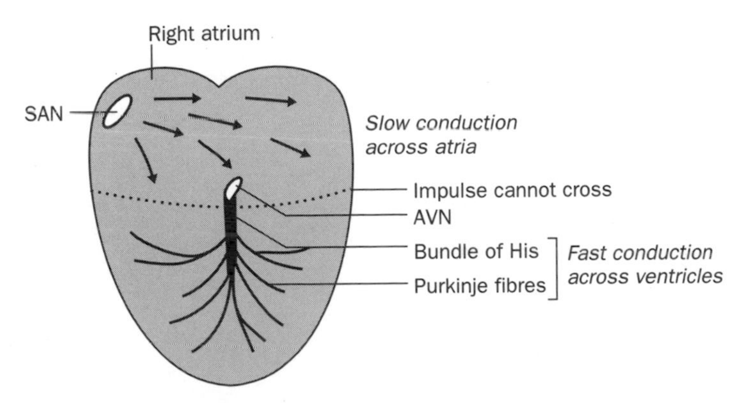

Fig. 19.12 The heart's electrical system (see text for details).

Box 19.4

Membrane potentials in cardiac muscle

Remember how movement of ions across the axon membrane is the basis of nerve impulses? (Section 17.3). Something very similar happens when cardiac muscle carries the signal to contract. Just a slight modification makes the cell fire automatically, as seen in the SAN.

Axon membrane

1. At rest. Na^+ pumped out, K^+ pumped in. K^+ diffuses out, Na^+ cannot diffuse in. Cytosol negatively charged.
2. Depolarization. A slight depolarization opens Na^+ channels. Na^+ floods in.
3. Repolarization. Na^+ channels shut, so Na^+ concentration falls in cytosol. More K^+ channels open, so cytosol becomes even more negative for a while.

Small currents ahead of the action potential make it move on down the axon.

Confused? Stop right now, read Section 17.3, then try again.

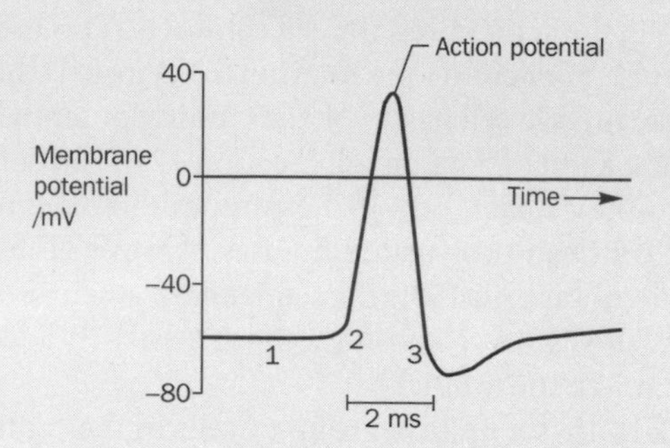

Cardiac work cell membrane

Ca^{2+} is important here, as well as Na^+ and K^+.

1. At rest. Na^+ and K^+ are exchanged. K^+ diffuses out, just as in the axon. An antiport allows in some Na^+, in exchange for Ca^{2+}.
2. Depolarization. A slight depolarization opens Na^+ channels. Na^+ floods in. As it does so, Ca^{2+} channels open, so Ca^{2+} flows in too.
3. Repolarization. The Na^+ channels close quickly, but Ca^{2+} channels stay open for a longer time. Repolarization is *much* slower than in the axon (note the timescales).

The burst of Ca^{2+} in the cytosol, lasting for about 300 ms, is the trigger for contraction. This is much like the situation in skeletal muscle (Box 24.5).

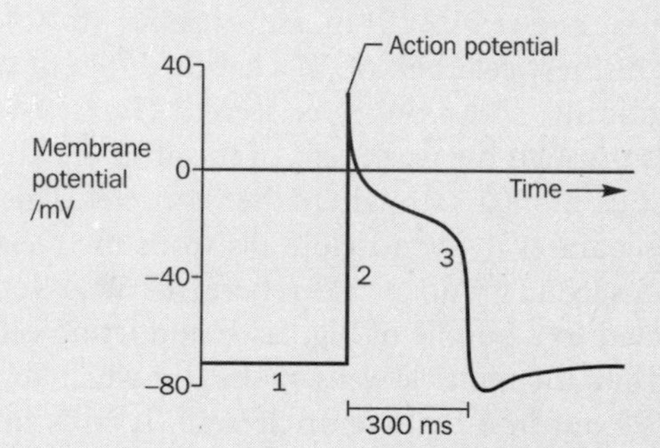

Sino-atrial cell membrane

1. Pacemaker potential. There is no stable resting potential: it slowly decays. The membrane slowly becomes less permeable to K^+, but a slow inward flow of Ca^{2+} and Na^+ continues.
2. Action potential. Once membrane potential has fallen to a threshold value, calcium channels open. Ca^{2+} flows in, giving the action potential. They open much more slowly than the sodium channels of work cells and axons.
3. Repolarization. This is slow, as in work cells.

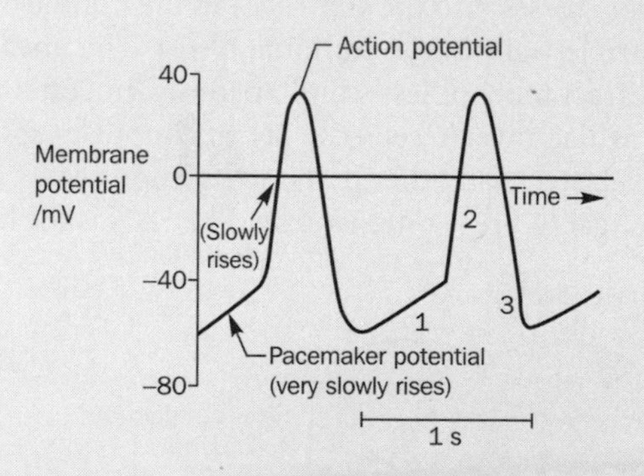

Nerve signals and the SAN

Sympathetic and parasympathetic neurones adjust the rate of SAN firing. They affect the steepness of the pacemaker potential, by means of neurotransmitters.

Parasympathetic neurones release acetylcholine. This binds to a G protein-linked receptor, leading to K^+ channels opening.

The outflow of K^+ increases, so the pacemaker potential falls more slowly. It takes longer to reach the threshold, so the SAN fires more slowly, and the heart beats more slowly.

Sympathetic neurones release noradrenaline. This leads to an increased flow of Na^+ and Ca^{2+} into the cell. Pacemaker potential decays faster, so the SAN fires more often. Adrenaline has a similar effect.

Controlling heart output and blood flow

The heart does not push blood out at a constant rate. For example, my heart needs to pump a greater volume of blood each minute when I run for the bus than it does now, as I sit at the kitchen table, writing. The body as a whole is using oxygen and glucose more quickly when I run, so the blood must deliver it more rapidly.

Two things determine heart output: heart rate (how many beats per minute) and stroke volume (how much blood is pumped per beat). Both are controlled.

The autonomic nervous system (Section 17.6) controls heart rate in a big way. Parasympathetic neurones from the vagus nerve, and sympathetic neurones, both reach the SAN. Parasympathetic neurones release acetylcholine, which slows the heart down. Sympathetic neurones release noradrenaline which increases heart rate. Adrenaline in the bloodstream also increases rate. The two systems work against one another: an accelerator and a brake. The balance determines heart rate. If the nerves are cut, the heart chugs along at about 60 beats per minute, whether or not this is appropriate. The only control remaining is slow and slight, by hormones such as adrenaline.

Stroke volume is controlled in a neat way. The more cardiac muscle is stretched, the more strongly it contracts. The mechanism is not fully understood, but seems to involve stretched work cells being more sensitive to calcium, the intracellular signal to contract. So, the more the ventricle fills during diastole, the more strongly it contracts in systole. The amount of blood entering each side of the heart is mainly determined by the pressure in the vein leading to that side. Sympathetic neurones reaching the heart muscle also stimulate stronger contractions.

Exercise increases heart output, by increasing heart rate and stroke volume. This increases the flow of blood to both lungs and muscles, picking up and delivering more oxygen. When exercise begins, there is a change in the pattern of signals reaching the heart in the autonomic nervous system. Sympathetic activity increases; parasympathetic activity decreases. The heart beats faster. Two factors increase stroke volume. Firstly, the sympathetic neurones terminating in heart muscle also stimulate stronger contraction. Secondly, there is an increase in vein pressure filling the heart. When walking or running, the movement of the body squeezes the veins in the limbs rhythmically. This pumps blood along, increasing pressure in the veins. Most of us can increase stroke volume by as much as four times when we exercise vigorously. Really fit athletes can increase it by up to six times. We all increase heart rate to as much as 200 beats per minute with exercise. Training has little effect on this.

Away from the heart, blood is directed where it is needed. If more flows to one organ, less must flow to another. The muscular walls of small arteries can contract, causing a narrowing (vasoconstriction). Relaxation causes a widening: vasodilatation. Some small veins work in the same way, notably in the skin and gut wall. In exercise, more blood goes to the skeletal muscles and less to the gut. After a big meal, more goes to the gut and less elsewhere. When the body is cold, less goes to the skin, to reduce heat loss, and so on.

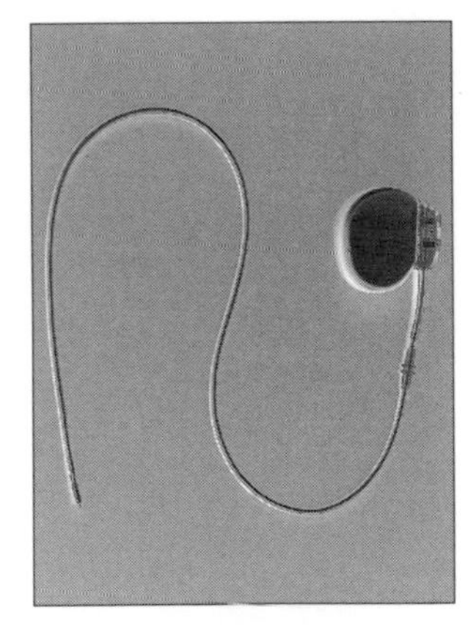

Electrical impulses from an artificial pacemaker implanted under the skin and connected to the heart can be used to control the heart. It overrides or replaces the signals from the SAN, when the node itself or some other part of the heart's electrical systems fails.

19.7 From capillary to cell and back

In a closed circulation, molecules have to get from the blood to all the body's cells. Others must pass from cell to blood. Blood itself cannot leave the capillaries, but there is an efficient exchange system. The single layer of cells forming the capillary wall is called the endothelium. Very simply, the endothelium acts as a molecular sieve. Molecules up to the size of small proteins can get through. Blood cells and most proteins cannot. As blood enters the arterial end of the capillary, some plasma is squeezed out, forming tissue fluid which surrounds the cells. Some of this re-enters the capillary at the venous end. The rest returns to the bloodstream by another route, the lymphatic system. Each stage will be discussed in turn.

Formation of tissue fluid

Pressure is high in the arterial end of the capillary. The tissue fluid has a much lower pressure, in fact slightly below atmospheric pressure. (Wiggle your thumb: see how the loose skin is sucked into the little dent that forms at its base.) The pressure difference tends to drive water and solutes out of the capillary. Most solutes are at equal concentrations in plasma and tissue fluid. However, proteins are much more concentrated in the plasma, because they cannot get out. This means that the endothelium acts as a selectively permeable membrane. Water tends to *enter* the capillary by osmosis, along the gradient of protein concentration. Other solutes follow the water by diffusion. So, pressure and osmosis are working in opposite directions. The movement of water and solutes depends on the balance between them, just as in a plant cell (Box 6.3). At the arterial end, pressure is higher, so plasma leaves. At the venous end, pressure is usually lower, and tissue fluid re-enters (Fig. 19.13).

How does the endothelium act as a sieve? It seems that there are plenty of tiny pores which let smaller molecules through, plus fewer large pores which allow small proteins to pass. The trouble is, no one is quite sure what the pores are. The small pores are probably in the gaps between adjacent endothelial cells. The cells are tied together by tight junctions (Section 4.7), but smaller molecules may be able to slip round the ends of the protein strands which bond the membranes. The gap also contains an extracellular matrix (Section 4.6), basically a mesh of proteoglycans. The size of the pores in this mesh may be important. What are the large pores? Perhaps a few of the gaps are bigger than normal. Perhaps there are occasional tunnels running right through an endothelial cell, from blood to tissue fluid. Perhaps there is

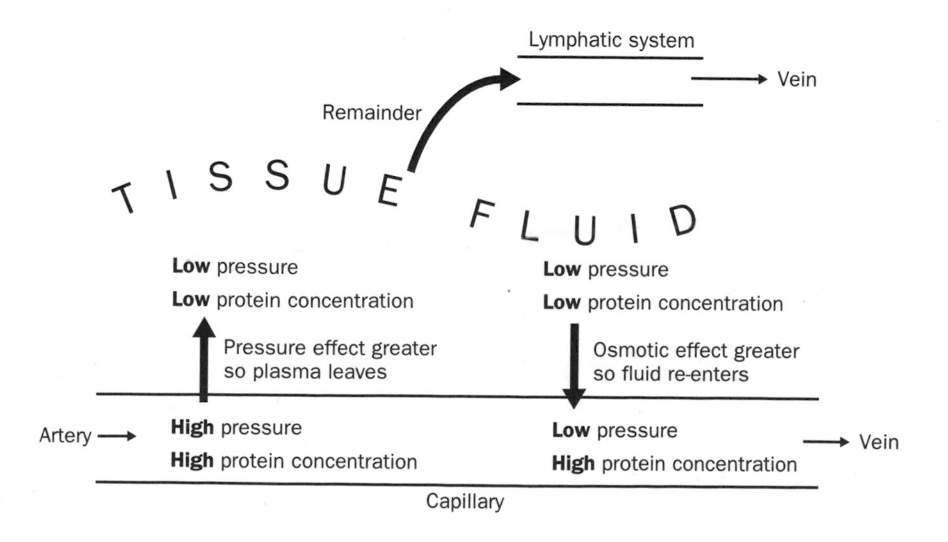

Fig. 19.13 Tissue fluid: how it is made and where it goes.

some system involving endocytosis then exocytosis (Section 6.6). It is most unclear.

The blood capillaries supplying the brain are much more fussy. Pores are less significant. Specific substances enter and leave the endothelial cells by active transport and diffusion through carrier proteins. This keeps the fluid around brain cells very constant.

Tissue fluid to lymph

Once in the tissue fluid, molecules move by diffusion. There is little scope for mass flow, because the spaces between cells are full of the proteins and GAGs of the extracellular matrix. Tissue fluid is essentially a thin gel.

More fluid leaves a capillary than re-enters it. The balance flows into a series of vessels, the lymphatic system, which eventually return it to the blood system (Fig. 19.14). Parts of the lymphatic system are dense masses of small vessels: the lymph nodes. Many are found in the neck, groin and around the gut. White blood cells hang out in the lymph nodes. They are checking the entire body for foreign molecules simply by sitting in the lymph returning from the tissues. The lymph nodes are just one example of how a circulatory system allows a body to work as an integrated unit, not just a population of smaller units doing everything for themselves.

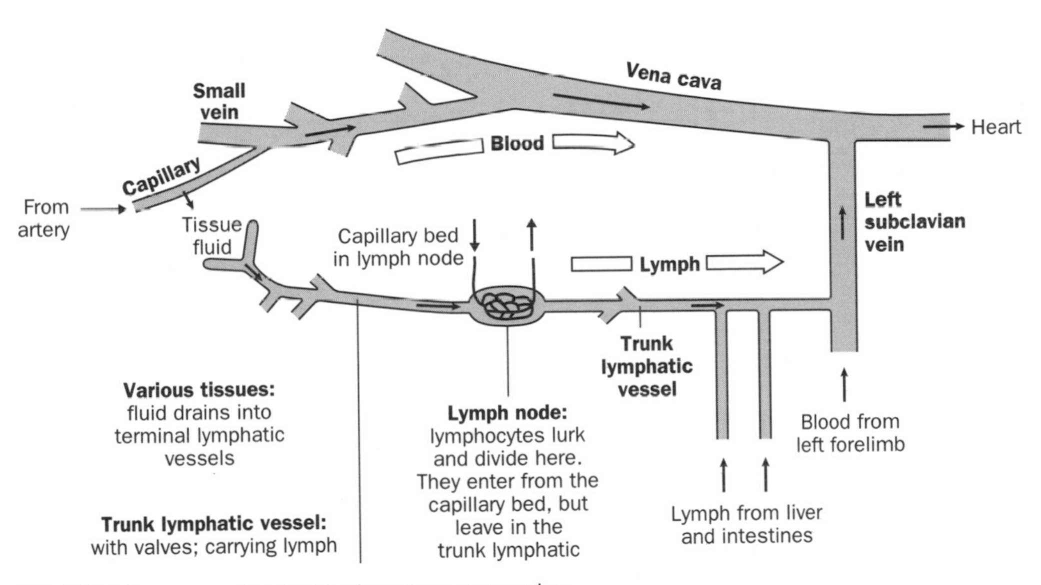

Fig. 19.14 The mammalian lymphatic system: an overview.

19.8 Transport and uniformity

I live in Devon. Once, this was a remote part of England. Devon people had their own customs, their own dialect, even set their clocks differently to Londoners – after all, the sun rises and sets later here. Things have changed, and transport was the cause. The Great Western Railway linked Devon to London in the 19th century. Anyone who could afford the ticket could get to or from London in half a day. Road development continued the process in the 20th century. The result is that Devon is less distinct, culturally, socially and economically. It is less a unit on its own and more part of an integrated whole.

Transport does the same for animal bodies. Important molecules are at similar levels throughout the bloodstream, whether in an organ which makes them or uses them. Heat is carried around. Warm-blooded animals are more or less the same temperature all over. Hormones (along with nerve impulses) help coordinate the action of different organs. The concept of organs doing jobs on behalf of the whole organism depends on transport.

In plants, transport is confined to a few key substances moving in particular directions (Unit 18). This is inextricably linked to the modular construction of plants. They are made up of relatively independent working units, a European Union rather than a United States. Most animal bodies are about unity and mutual dependence. Transport is at the heart of what it is to be an animal.

Summary

- A transport system is essential if an animal body is to work as an integrated whole.
- Blood transports dissolved gases, food molecules, minerals ions, waste products, hormones and heat around animals' bodies.
- Vertebrate blood consists of a liquid (plasma), red cells, white cells and platelets.
- Oxygen is carried bound to a protein, haemoglobin, in red cells.
- Haemoglobin allows blood to carry more oxygen.
- Haemoglobin has special properties which allow it to unload more oxygen when it is needed.
- Carbon dioxide is transported in the plasma as hydrogencarbonate ions.
- Plasma proteins bind to hydrophobic molecules like steroids, making them soluble.
- Some invertebrates have an open circulation: blood can flow in and around the organs.
- Other animals have a closed circulation: blood is trapped in blood vessels.
- Mammals have a double circulation. Blood flows from heart (right side)→lungs→ heart (left side)→other organs→heart (right side).
- Arteries carry blood from the heart under pressure. They have thick walls.
- Within the organs, blood flows through tiny capillaries.
- Plasma leaves capillaries, forming tissue fluid, which surrounds the cells.
- Some tissue fluid returns to the blood, which goes back to the heart in thinner walled veins. The rest of the tissue fluid returns via the lymphatic system.
- The heart pumps blood around the body.
- The heart has muscular walls, and valves to control blood flow.
- Each side of the heart has two chambers: an atrium and a ventricle.
- The atria contract first, pushing blood into the ventricles. Then the ventricles pump blood out into the arteries.

- The heart beats by itself.
- The sino-atrial node is a part of the heart which sets up electrical impulses about once a second. These impulses are carried through the heart wall by specialized muscle cells, and stimulate contraction.
- Nerve impulses and hormones adjust both heart rate and stroke volume, to suit the body's needs.

Exercises

19.1. Flatworms, sponges, corals and hydrozoans have no blood. What do they have in common which allows this?

19.2. Some very active fish have oxygen dissociation curves which are to the right of fish which lead slow-moving lives. Explain the significance of this.

19.3. Carbon monoxide is a gas produced when fuels are burnt, especially at low oxygen concentrations. It is very harmful to mammals. It binds to haemoglobin, in the same place as oxygen, and in a similar way, but much more strongly.

(i) Carbon monoxide poisoning can be fatal. Why?
(ii) One way of treating carbon monoxide poisoning involves the patient breathing pure, or nearly pure oxygen. Explain how this might help.
(iii) Carbon dioxide, at a higher level than in the atmosphere, may be added to the patient's oxygen supply. Explain how this could help. (Hint: think more widely about the issues, and read Section 20.6.)

19.4. How would the following things affect heart output?

(i) an injection of adrenaline into the bloodstream;
(ii) electrical stimulation of the vagus nerve;
(iii) severe bleeding from a major vein?

19.5. Sometimes, some part of the heart other than the SAN starts to act as a pacemaker: an ectopic pacemaker.

(i) Would you expect an ectopic pacemaker to set off impulses faster or slower than the SAN?
(ii) Why might an ectopic pacemaker be harmful?

Further reading

Guyton, A.C. and Hall J.E. *Textbook of Medical Physiology* (9th ed.) (Philadelphia: Saunders, 1996). I like this hefty textbook of human physiology for its thoroughness and clear, no-frills diagrams.

Guyton, A.C. *Human Physiology and Mechanisms of Disease* (5th ed.) (Philadelphia: Saunders, 1992). Effectively a shorter, boiled-down version.

Schmidt-Nielsen, K. *Animal Physiology: Adaptation and Environment* (4th ed.) (Cambridge: Cambridge University Press, 1990). One of the all-time great textbooks! Exciting, authoritative and easy to read, it puts animal physiology in the context of the environment.

Withers, P.C. *Comparative Animal Physiology*. (Fort Worth: Saunders, 1992). More detailed than Schmidt-Nielsen, with wider ranging examples: the inevitable cost is that it is a less easy read.

Gas Exchange in Animals

Contents

Connections

▶ Before studying this unit you should have a basic knowledge of cell structure (Unit 4) and how substances move (Unit 6). Animals need to exchange gases because they respire (Unit 9), and most have a transport system (Unit 19) coupled to the gas-exchange system.

20.1 Oxygen in, carbon dioxide out

Cells respire. Most cells respire aerobically, most of the time. They break down food molecules, releasing energy. As they do this, they use oxygen and make carbon dioxide. Cells need to take in oxygen from the environment and give out carbon dioxide. This is **gas exchange**. When animal cells live together in bodies, the body as a unit deals with its environment, so bodies also carry out gas exchange. Parts of plants which do not photosynthesize, like roots, have exactly the same problem. In photosynthetic organs, like leaves, things are a little more complex (Section 10.6). So, cells do it, bodies do it, even germinating beans do it: but where, and how?

20.2 Dimensions and diffusion

Cells use up oxygen, so there is normally less inside the cell than outside. The reverse is true for carbon dioxide. Oxygen diffuses in, carbon dioxide diffuses out. In very small animals, the same is true for the body as a whole. Oxygen diffuses in through the body surface. It then diffuses through the body itself, towards the middle. Some is used as it moves inwards, so the further in you go, the less oxygen arrives. There would come a time when a cell was so far from the surface that it received too little oxygen to survive. So, there is a maximum thickness for organisms which rely on diffusion. As a rule of thumb, the maximum distance from cell to surface is about 1 mm. This is why flatworms are thin and flat. Some types may be tens of millimetres long, but because they are so thin, no cell is far from the surface.

Gas exchange: Taking in gases used by the body and getting rid of waste gases to the environment.

In other animals, blood systems carry oxygen from the surface to the cells, and carbon dioxide back again (Unit 19). This takes away the limit on thickness. Now the total area available for gas exchange becomes important. Gases are exchanged over a surface, but made and used throughout the animal's volume. The ratio of gas-exchange area to volume is critical. Too low, and the cells do not get enough oxygen to survive.

Earthworms use their entire body surface for gas exchange. They are more or less cylindrical. As a cylinder gets fatter, its volume increases faster than its surface area. This means that there is a limit to how fat an earthworm can be. Worms the size of tree trunks exist only in films.

The critical area/volume ratio will depend on how quickly cells use oxygen. An animal which spends its life skulking on a muddy river bed will use much less oxygen for each unit of volume than one which dashes about chasing prey. The second animal would require a higher area/volume ratio.

Many animals, including the vertebrates, do not exchange gases over the whole body surface. They use special organs which combine extremely large surfaces with compactness.

20.3 Gas exchange organs

When you scale up an animal, volume increases faster than surface area. A shape which works when small is less effective when large. There are several ways that a really big animal could have sufficient surface area. One would be to have a body like a huge pancake. Another would be to have a deeply-folded body: a fleshy hairbrush, perhaps. Either way, the shapes would not suit most ways of life, and would be very vulnerable. Coating the surface with a protective layer, like skin or a shell, would prevent gas exchange. A much better solution is to fold one or more small regions of the body very deeply indeed, and to exchange gases very efficiently in these organs, but nowhere else.

Mammals, birds and reptiles have lungs. Fish and crustaceans, which get their oxygen from water, have gills. Insects have an internal network of air-filled tubes: the tracheal system. In each case, the problems are similar. The first is how to achieve a surface which exchanges gases effectively. The second problem is how to ventilate the surfaces, that is constantly to supply them with fresh, oxygen-rich water or air. Thirdly, because ventilation requires energy, there needs to be some control system, ensuring that ventilation provides enough oxygen, but does not waste energy by overventilating when the body is less active.

20.4 Gas exchange in insects

Insects have a network of air-filled pipes in their bodies: **tracheae** (singular: trachea). Fine branches, **tracheoles**, go into the tissues, so no cell is far from one. Muscle cells even have tracheoles inside them. Small holes in the body wall, spiracles, allow air to enter or leave in bulk, or for gases to diffuse in and out. Spiracles can be opened or shut. Fig. 20.1 shows the tracheal system of a typical insect living in air rather than water. Gas exchange takes place in the walls of the tracheoles. Insects have blood, but it is not important in gas transport, since the tracheal system spreads throughout the body.

Tracheae and tracheoles: Air-filled pipes in insects' bodies.

How do gases move between the tracheal system and the atmosphere? In small or

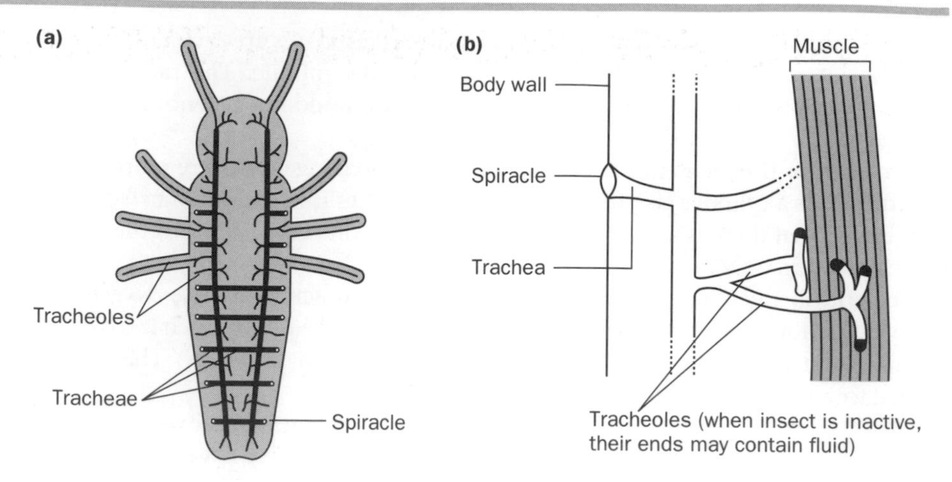

Fig. 20.1 Tracheal system of an insect (a) with a small part in detail (b).

inactive insects, diffusion through the spiracles may be enough. Large or active insects, such as locusts in flight, need to set up bulk air movements. Locusts have internal air bags attached to the tracheae. As the locust flies, its body bends back and forth, squeezing the bags like bellows. At the same time, spiracles open and close rhythmically. Fig. 20.2 shows a simple model of how this could set up a flow of air.

Gas exchange in insects is strikingly similar to gas exchange in most leaves (Section 18.2). Exchange takes place between internal air spaces and the cells. Adjustable pores link air spaces with the atmosphere: stomata in leaves, spiracles in insects. In each case, water loss is a serious problem, since internal air is in close contact with the body fluids and becomes saturated with water vapour. When stomata or spiracles close, partly or completely, they are balancing the need for gas exchange against the dangers of water loss.

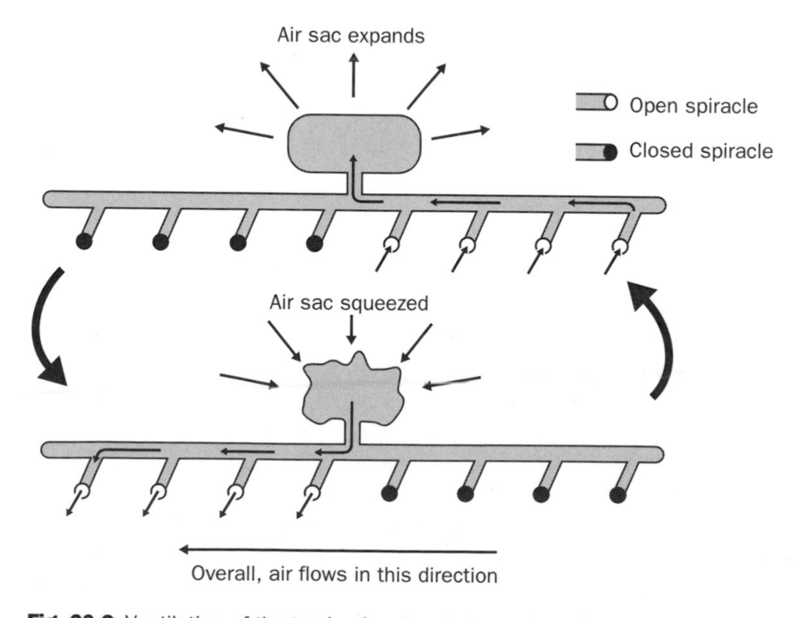

Fig. 20.2 Ventilation of the tracheal system in some large insects.

Box 20.1

Aquatic insects

Insects evolved in air. Their tracheal systems carry air from the atmosphere right into the tissues. Insects which live under water have a problem: there is no atmosphere! They solve it in two main ways. Some carry air down with them, and keep coming up for more. Others exchange gases between the air in the tracheal system and the water outside.

Gills

Dragonfly, damselfly and mayfly nymphs have gills. These are thin plates of tissue with tracheae inside. Gas exchange with the water takes place here.

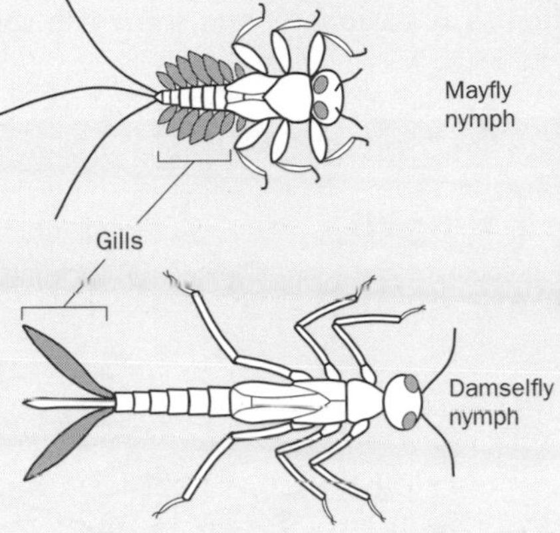

External air bubble

The great diving beetle traps an air bubble between its wings and its back. The spiracles open into this space. The bubble acts as an oxygen store. However, as oxygen is used, more diffuses into the bubble from the surrounding water. Nitrogen slowly dissolves, so the bubble gradually shrinks and must be replaced.

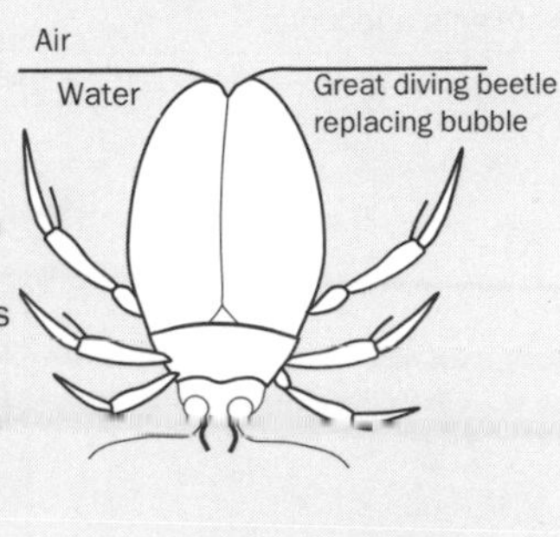

Air stored in tracheal system

Mosquito larvae exchange gases with the atmosphere directly. Diffusion takes place through a tube at the tip of the abdomen, which breaks through the surface film. The insect can hang there for some time.

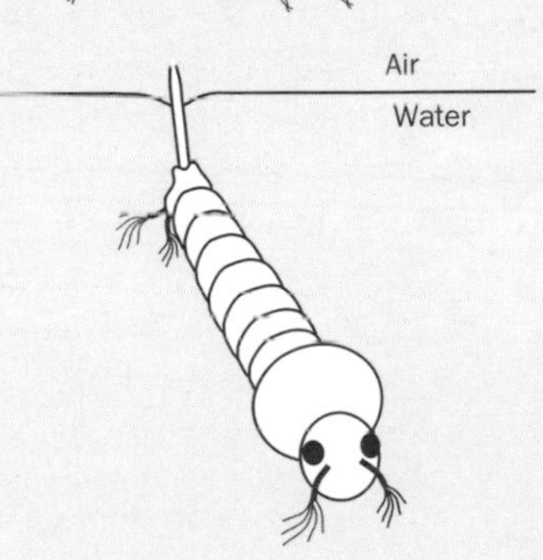

Plastron

Some aquatic bugs have a carpet of unwettable hairs, a plastron, over part of their surface. This traps a flat bubble which does not shrink, so it acts as a kind of gill for long periods under water.

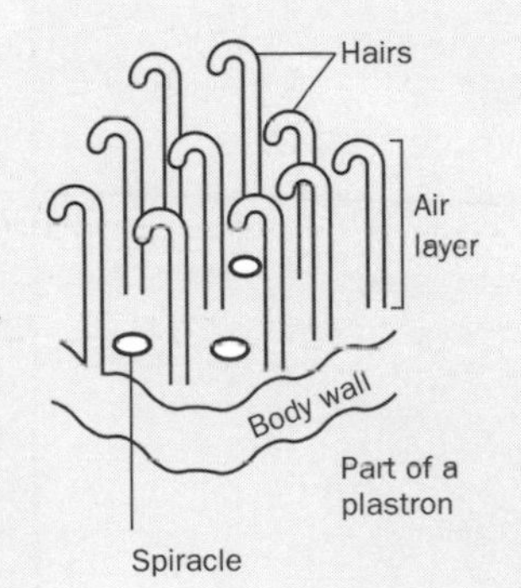

The tracheal system strongly suggests that insects evolved in air, not water. Aquatic insects still use the system, but use various cunning dodges to get air in and out of it (Box 20.1).

20.5 Gas exchange in bony fish

Fish are well adapted for exchanging gases with the water they live in. Gas exchange takes place in the **gills**. These are stacks of thin layers of tissue. They are stretched across what is, essentially, a hole in each side of the fish's pharynx (throat). Water comes in through the mouth, into the pharynx, over the gills and back out at the side. The gills of bony fish are covered by a flap on each side: the operculum.

Gills: Organs for gas exchange between animals' bodies and water.

The gills are fragile, complex, beautiful structures (Fig. 20.3). There are several gill arches on each side. They are tough and support the rest of the structure. Each gill arch has two rows of gill plates sticking out, like teeth on a comb. Each plate has a series of gill lamellae sticking up and hanging down from its surfaces. The lamellae are thin sheets of tissue with a capillary bed in each. Water is forced through the tiny gaps between the lamellae: gas exchange takes place here. Box 20.2 shows how water and blood flow in opposite directions to make gas exchange most efficient.

Forcing water over the gills needs energy. There are two different ways of doing this. The simplest is for the fish to open its mouth and swim forwards. The faster the fish swims, and the wider open the mouth, the faster water flows across the gills. Energy is provided indirectly, by the muscles used in swimming. This is called ram **ventilation**. It is no use if the fish is stationary, and some fish spend most of their

Ventilation: The ways in which water or air is brought to and from gills and lungs.

Box 20.2

Efficient gas exchange in fish gills

Blood and water flow in opposite directions along each gill lamella. This countercurrent system leads to very efficient gas exchange.

Blood
Water
Blood
Water
Blood
Water
Vein
Artery

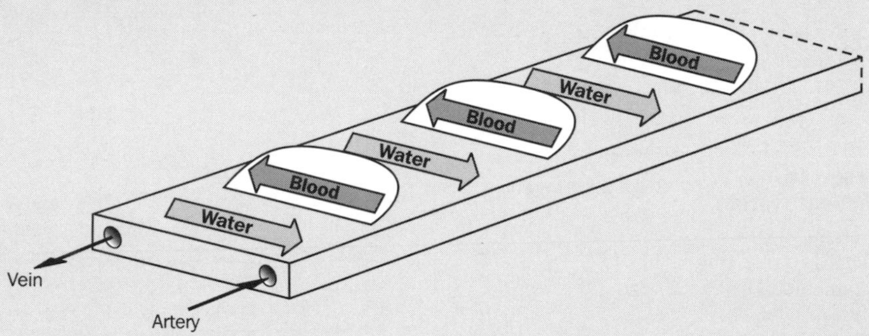

Water is forced between the gill lamellae, towards the inside of the V formed by each gill. Blood flows across each lamella, from artery to vein, in the opposite direction.

Deoxygenated blood entering the lamella meets water which has already lost most of its oxygen. A little oxygen diffuses into the blood. As it flows on, it meets water which has progressively more oxygen. It continues to diffuse in. By the end of the lamella, blood which is almost saturated with oxygen meets fully saturated water. Oxygen has diffused into the blood throughout the length of the gill.

If blood and water flowed in parallel, one blood mass would follow the same bit of water along the lamella. They would both end up with the same oxygen concentration, only partially saturated.

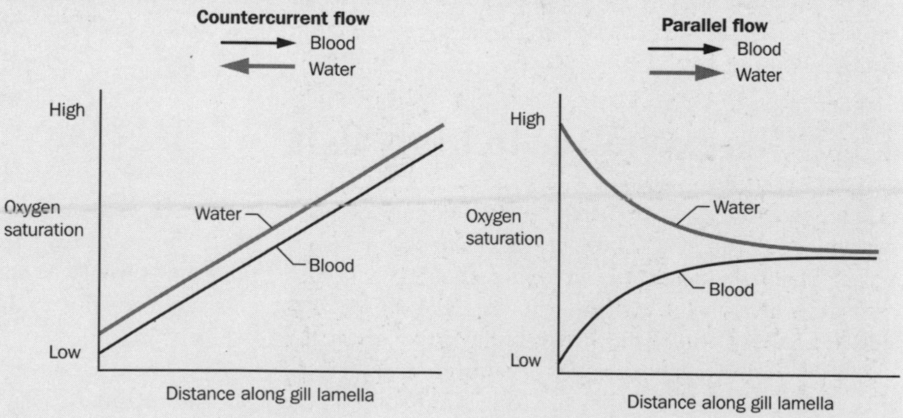

Countercurrent flow
Blood
Water
High
Oxygen saturation
Low
Water
Blood
Distance along gill lamella

Parallel flow
Blood
Water
High
Oxygen saturation
Low
Water
Blood
Distance along gill lamella

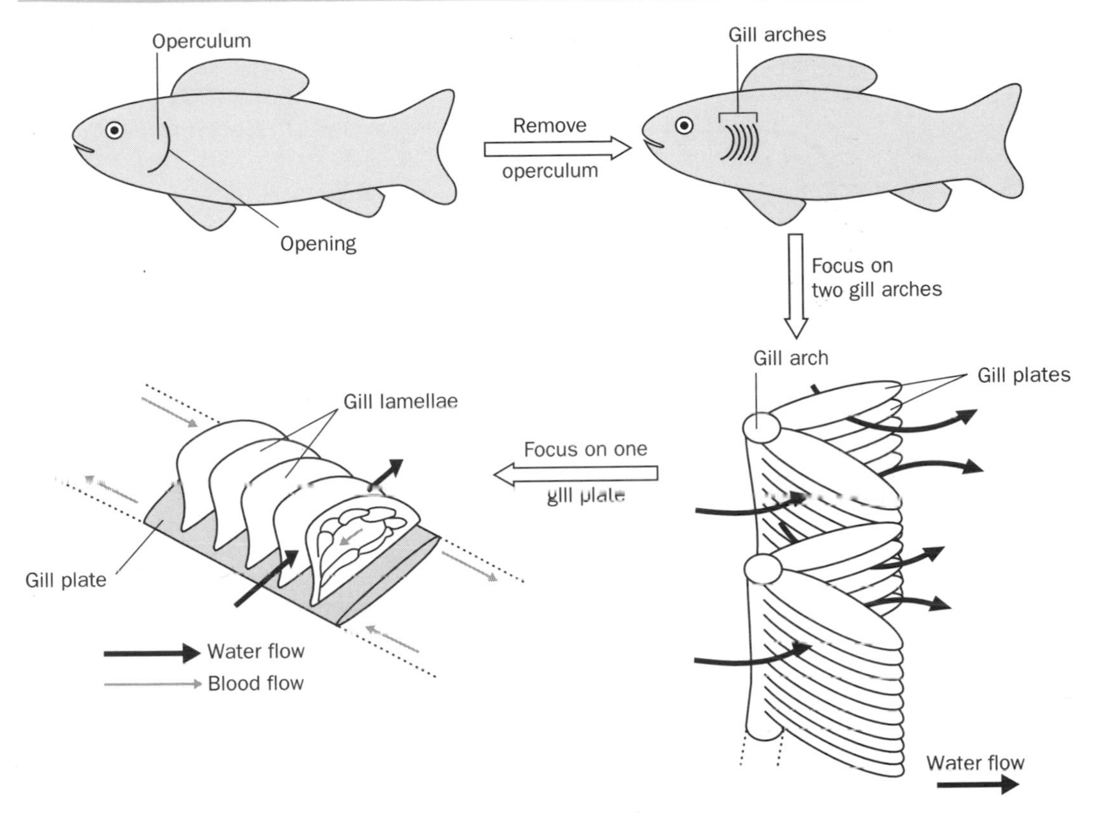

Fig. 20.3 The gills of bony fish.

time hiding, or lying in wait. Then, a second method is used. Water is pumped into the mouth and through the gills. The mouth cavity and operculum act as pumps. The mouth itself and the operculum can be opened or shut. They act as valves. A neat cycle takes place (Fig. 20.4). First, the floor of the mouth is lowered, so the volume of the mouth cavity increases. The mouth is open, so water flows in. Then the mouth shuts. The mouth floor rises, increasing pressure. At the same time, the

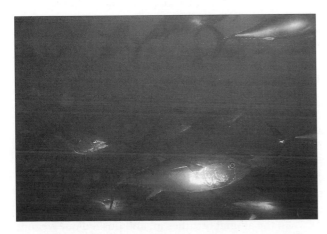

Tunas rely on ram ventilation. They die in aquaria unless given room to swim.

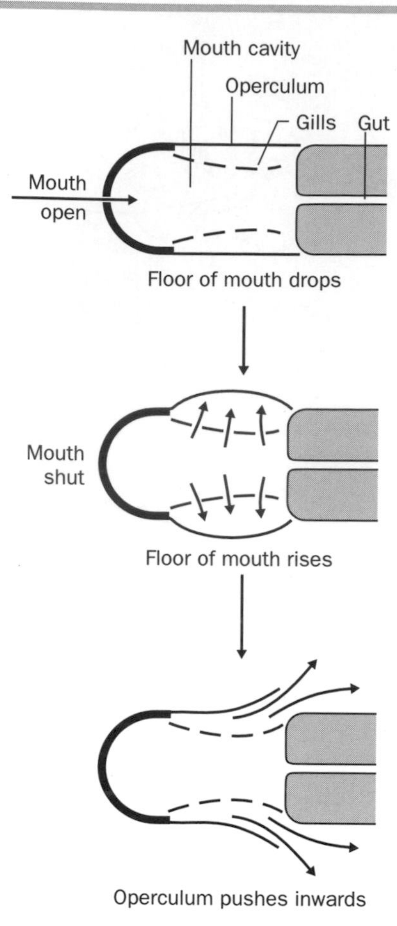

Fig. 20.4 Ventilation of the gills in bony fish.

operculum moves outwards to decrease pressure in the opercular cavity. Water moves across the gills, from high pressure in the mouth to low pressure in the opercular cavity. Finally, the operculum opens and pushes inwards, forcing waste water out. The cycle can then start again. This is called opercular ventilation.

Some sluggish fish only ever use opercular ventilation. Some fast swimmers use only ram ventilation, and die if they are prevented from swimming. Others switch between the two, according to speed.

20.6 Gas exchange in mammals

In mammals, gas exchange takes place in the lungs. These are essentially infoldings of the body surface. The folding is quite extreme: in adult humans, the total surface area of the lungs is around 100 m^2. A tube, the trachea, leads from the throat towards the lungs. It divides to form two bronchi, one leading to each lung. This branches many times into smaller and smaller bronchioles. The smallest bronchioles each end in a tiny sac: an alveolus (Fig. 20.5). This has a thin wall with blood capillaries on the other side. Gas exchange takes place in the alveoli.

Air is pulled into the lungs when the animal breathes in, and pushed out when it exhales. If breaths are shallow, most of the air in the lungs stays there from one

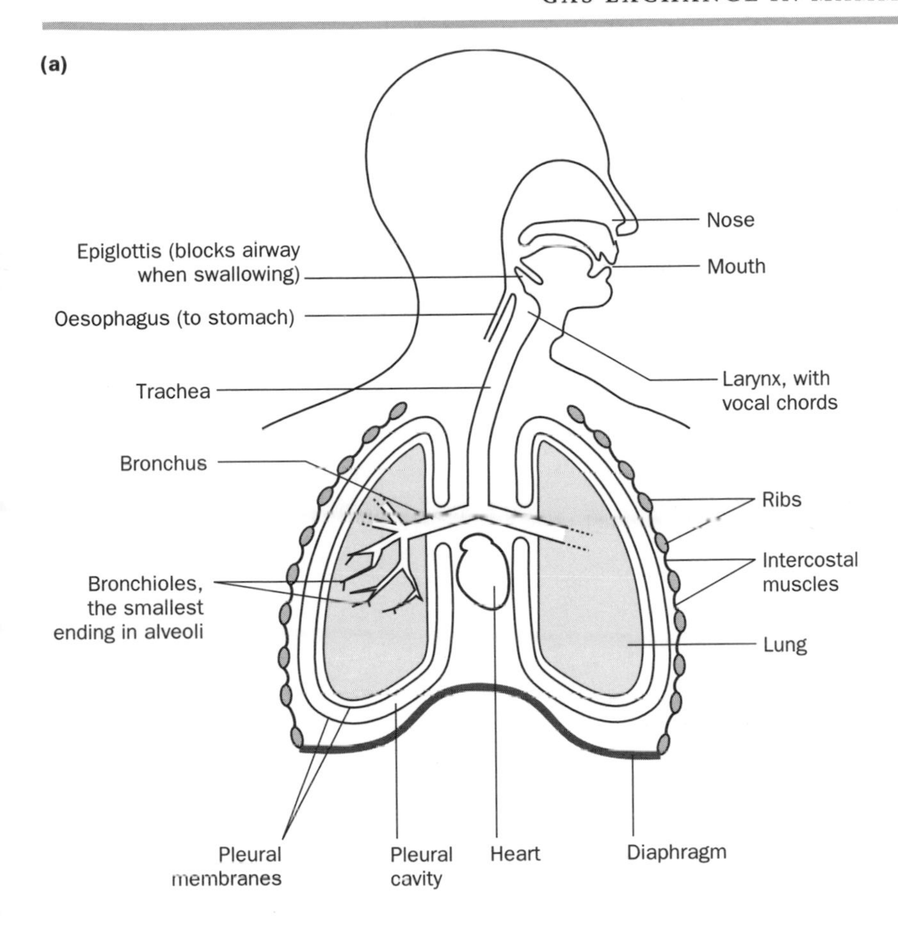

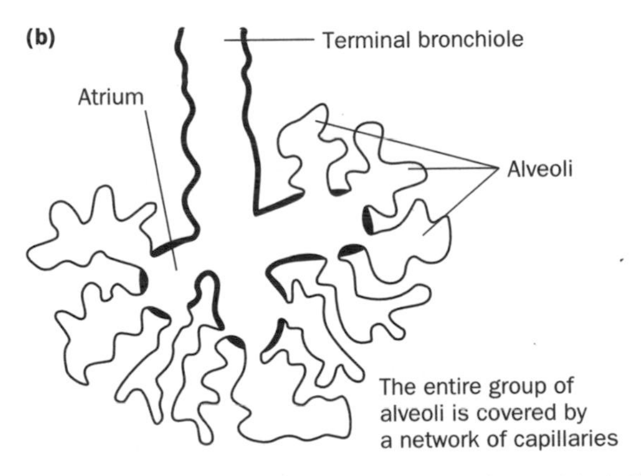

Fig. 20.5 Lungs of a mammal. (a) lungs and associated structures in human; (b) alveoli.

breath to the next. Even if breaths are very deep, some air always stays in the lungs (Fig. 20.6). 'Fresh' air is mixed with 'stale' air each time. Because of this, the composition of air in the alveoli stays almost constant. It contains less oxygen and more carbon dioxide than the air which is inhaled (Table 20.1). Exhaled air contains less oxygen than fresh air, but more than alveolar air. This is because some inhaled air

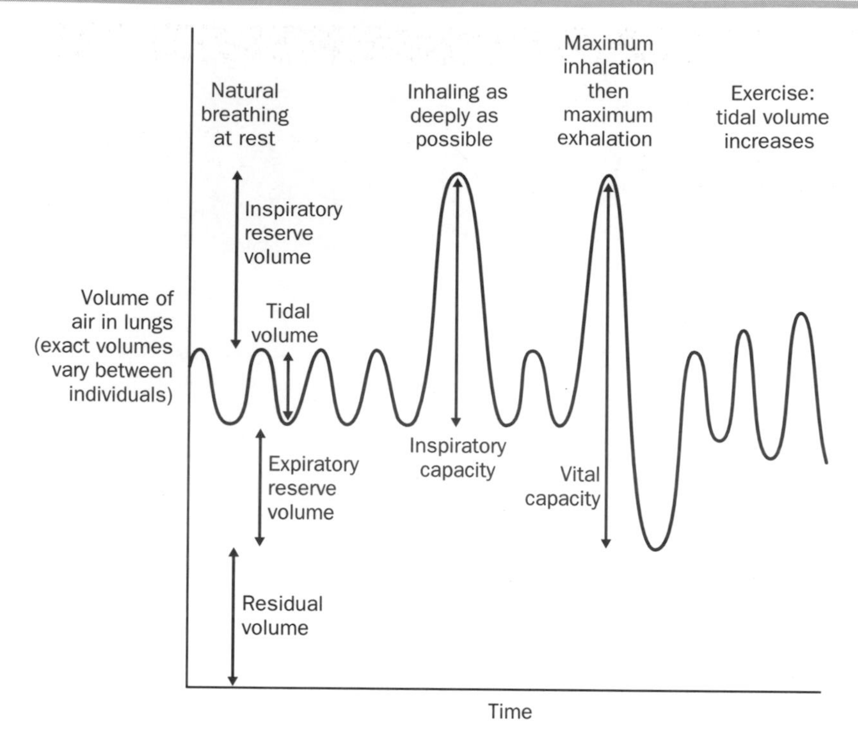

Fig. 20.6 Components of lung volume.

Table 20.1 Approximate composition of air (excluding water vapour) in the atmosphere, air in the alveoli, and exhaled air

Gas	Atmospheric air	Air in the alveoli	Exhaled air
oxygen	21%	15%	17%
carbon dioxide	0.04%	6%	4%
nitrogen	79%	79%	79%

will immediately be exhaled, without mixing fully. This is a built-in limitation to the efficiency of mammals' lungs. Box 20.3 shows how birds have a quite different type of lung which does not have this problem.

The lungs are ventilated by alternately stretching and squeezing them. Muscles are not attached to the lungs directly. Instead, the outer surfaces of the lungs are stuck firmly to the walls of the chest. The chest cavity is then made to expand and contract. The lungs expand and contract with them.

The chest cavity is surrounded by the ribs to the sides and the diaphragm below. The diaphragm is a muscle sheet separating the chest from the abdomen. The chest wall is covered by a thin membrane; the outer surface of the lungs is covered by another. These are the pleural membranes. They are stuck together by a film of liquid. This holds the lungs out against the walls and prevents them collapsing. It is almost impossible to collapse a lung without puncturing the pleural membranes. I convinced myself of their power by trying to pull two wet microscope slides apart

Box 20.3

Efficient ventilation in bird lungs

Gas flows through birds' lungs, rather than in and out. 'Fresh' and 'stale' air do not mix. Efficient ventilation means that birds' lungs are smaller than equivalent mammalian lungs. However, a complicated system of internal air sacs is needed.

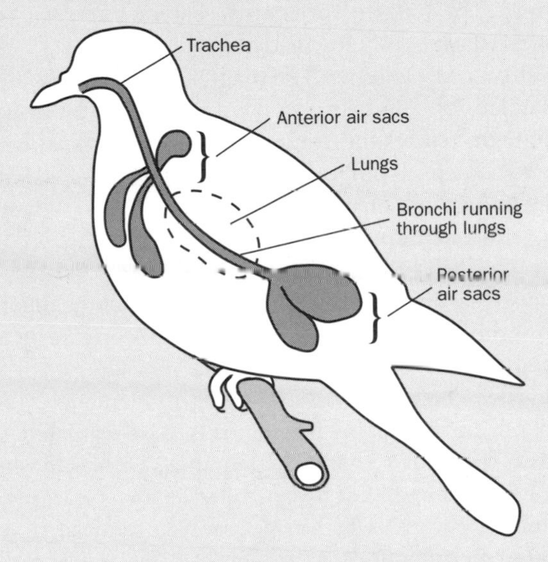

The air sacs contract and expand, drawing air into the posterior sacs, then through the lungs to the anterior sacs. From there, it is exhaled. It takes two cycles for air to flow all round this one-way system.

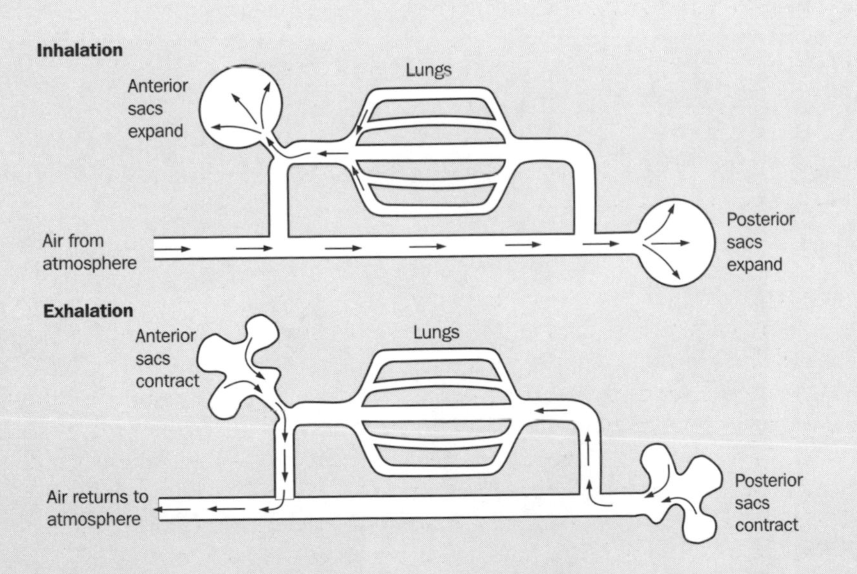

Gas exchange takes place in double-ended tubes in the lung. Different capillaries bring blood to different parts of each tube. At every stage, air is meeting a fresh supply of deoxygenated blood. This means that oxygen is removed very efficiently, even though flow is not countercurrent.

without sliding them. I failed, and do not recommend you do the same! The force needed is so great that broken glass is a real danger.

A similar problem affects individual alveoli. They are tiny, thin-walled bubbles which are open at one end. Their surface tension might make them collapse, but it does not. This is because the alveolar wall has a phospholipid coating which acts as a surfactant, lowering surface tension. Damage the cells which make surfactant, and the lungs collapse. The lungs of premature babies often fail to expand properly, because too little surfactant is being made.

Two things affect chest volume. The diaphragm is slightly domed when relaxed. As it contracts, it flattens. This increases chest volume, so air is sucked into the lungs. When the diaphragm relaxes the volume drops again, so air is forced out. The rib cage has a role, too. Each rib hinges on the spine and the sternum. The rib cage as a whole can swing up and out, then back down and in. This is driven by the intercostal muscles: belts of muscle linking adjacent ribs. Box 20.4 shows how. So, inhalation is caused by a raising of the rib cage and contraction of the diaphragm (Fig. 20.7). Exhalation is caused by the rib cage falling, relaxation of the diaphragm and a downward pull from some of the abdominal muscles. To convince yourself, feel your rib cage as you breathe.

This system needs controlling. Firstly, the movements must be coordinated. Secondly, the depth and rate of breathing needs to be regulated, so that oxygen supply is matched to the body's changing demands. Unlike the heart, the breathing apparatus does not run by itself. It is controlled by a respiratory centre in part of the brain, the medulla. Neurones which trigger inhalation lead from this centre to the diaphragm and intercostal muscles. Another set of neurones triggers exhalation. The two sets fire alternately. The respiratory centre detects carbon dioxide concentration in the fluid within the brain: this reflects carbon dioxide concentration in the blood.

Air moving in and out of the lungs is used to generate sound in the syrynx of birds and the larynx of mammals.

Box 20.4

Moving the ribs

Each pair of ribs has two bands of muscle running between them: the intercostal muscles. When the external intercostal muscles contract, the ribs swing up. When the internal intercostal muscles contract, the ribs swing down. Why? The answer lies in the angle at which the muscles meet the ribs.

The ribs are hinged where they meet the spine and sternum. The rib cage as a whole swings up and down in relation to the spine. When an intercostal muscle contracts, it puts the same force, F, on both ribs. One is pulled up, the other down. However, it is the turning force, or torque, which is important, since we are looking at rotation about the joint with the spine.

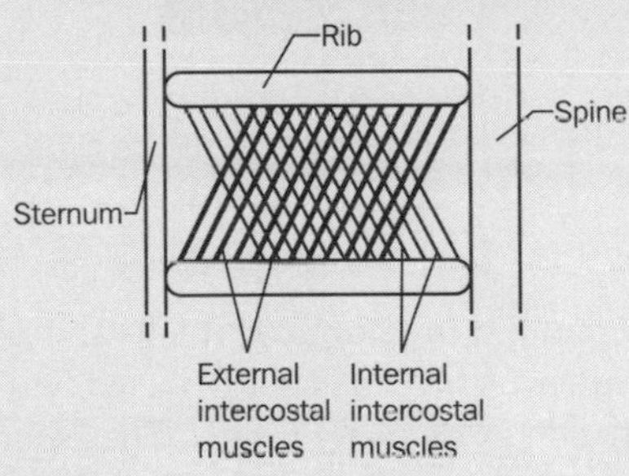

Torque = force × distance from the pivot at which force is applied.

What happens when an external intercostal muscle contracts?

Upper rib experiences: torque = $F \times D_U$ (anticlockwise)

Lower rib experiences: torque = $F \times D_L$ (clockwise)

Distance D_L is longer than D_U, so the lower rib experiences a greater torque than the upper one. Each rib has ribs above and below it, so is attached to two sets of intercostal muscles.

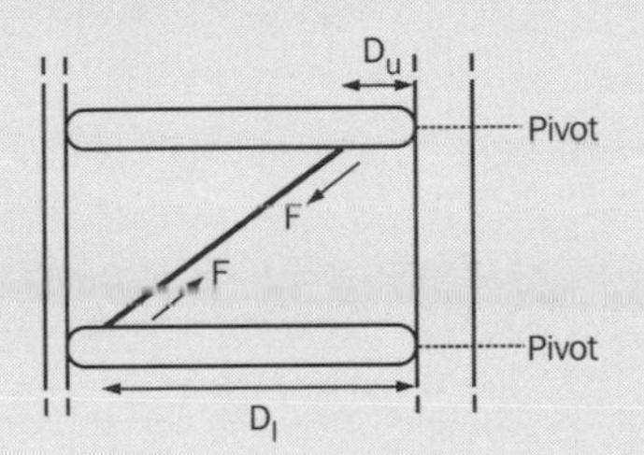

The clockwise torque applied by the muscles above it is greater than the anticlockwise torque applied from below. Each rib turns clockwise: the rib cage swings up. This happens during inhalation.

Gravity can be enough to pull the rib cage back down in exhalation. However, forced exhalation is possible. Breathe a deep, theatrical sigh. Go on, do it! The internal intercostal muscles contracted, pulling the rib cage down. Some of the abdominal wall muscles also contracted, pulling down on the lower ribs.

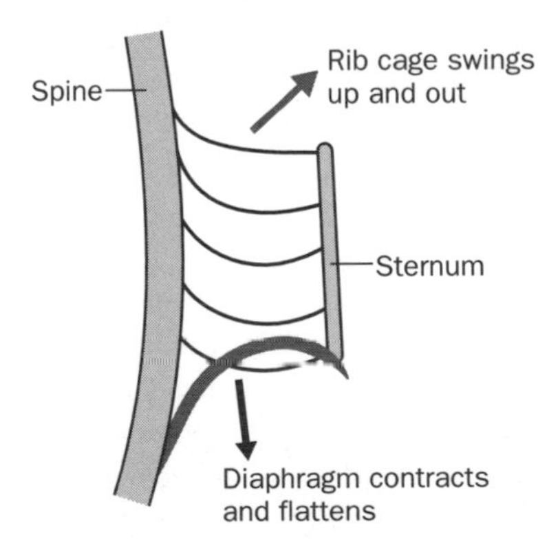

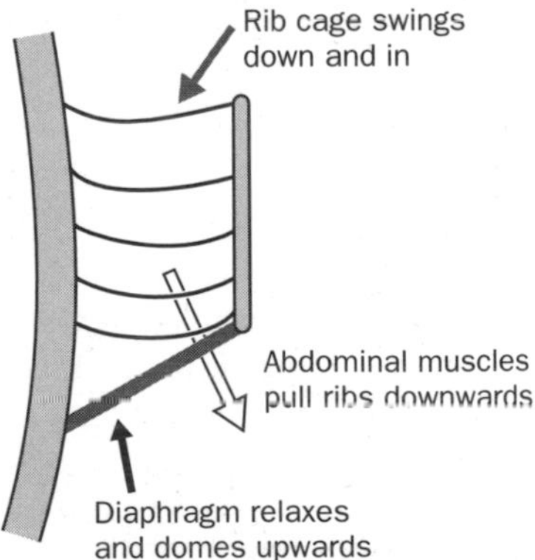

Fig. 20.7 Ventilating the mammalian lung.

Air at high altitudes contains the same percentage of oxygen as it does near the ground. The pilot wears an oxygen mask because air pressure is lower at high altitude: there are fewer molecules of any gas in each litre of air.

If carbon dioxide rises, it sets off deeper and more rapid breathing. If it falls abnormally low, breaths become shallower and less frequent. Oxygen is not detected here, but by chemoreceptor cells in parts of the aorta and carotid artery walls. They send information to the respiratory centre via neurones, but the effect is relatively small. However, when the body starts to exercise, it does not wait for carbon dioxide to fall before increasing breathing rate. Signals from the parts of the cortex dealing with voluntary movements reach the respiratory centre and set off the appropriate changes.

20.7 Drowning

Drowning happens when water gets into the lungs and is not coughed out again. Water, even when saturated with dissolved oxygen, holds far less oxygen than air does. This means that a mammal could get less oxygen from a water-filled lung than from one containing air. In principle, this could explain why mammals die when they drown. In fish, the situation is reversed. Out of water, a fish's gills are surrounded by more oxygen than normal, yet still the fish dies. This is because the thin gill filaments and lamellae are no longer supported by water. They collapse and stick together, so the surface area for gas exchange falls drastically. The fish suffers a fatal fall in oxygen and rise in carbon dioxide: asphyxia. The fish has drowned in air.

Pathologists ruin our simple story of how mammals drown, by gruesome observations from real life. Victims of drowning rarely show the symptoms of asphyxia: something else has killed them first. The mechanism differs between fresh and sea water, but either way the problem is in the heart.

When fresh water enters the lungs, the cells of the alveolar wall, and the blood beyond, meet very dilute solutions for the first time in their lives. Water enters the cells by osmosis and they burst. Potassium, normally abundant inside cells but at low levels outside, floods into the plasma from burst blood cells. High K^+ levels in the blood cause a heart attack.

This does not happen with sea water: it is too concentrated a solution. However, calcium and magnesium ions from the water diffuse into the blood. They affect the ability of heart muscle to beat, and the heart simply stops.

When victims of freshwater drowning are rescued and saved, they sometimes suffer a related problem a few hours later. Firstly, damaged alveolar cells stop making surfactant, and alveoli collapse. Secondly, irritation of lung tissue leads to histamine release, and the leakiness of blood capillaries that goes with it (Sections 16.5, 23.4). Elsewhere, the build-up of fluid around cells would cause swelling. Here, it fills alveoli and bronchioles. Fluid and collapse together reduce the surface area for gas exchange. This time, the result really is asphyxia. It is called secondary drowning, and can be fatal.

The story of drowning illustrates an important principle in science. We observe an effect – death by drowning – and we can see an obvious, simple explanation for it: asphyxia. The fact that it *could* cause the effect does not mean that it *does* cause the effect. Only observation of the real world can show if our idea is correct. As usual, scientists are in the business of asking questions of Nature. We should not be too surprised if some of the answers are not the ones we expect.

Summary

- Animals use oxygen and make carbon dioxide in respiration. They take in oxygen from the environment, and get rid of carbon dioxide: gas exchange.
- Only the smallest and thinnest animals, such as flatworms, rely on diffusion of oxygen through the body from the body surface.
- Larger bodies have a lower surface area/volume ratio. Specific organs with a large surface area are needed for gas exchange.
- The tracheal system of insects brings air right to the tissues.
- The gills of bony fish exchange gases between water and blood.
- Water is forced over the gills by swimming with the mouth open, or by pumping movements of the mouth and operculum.
- Gas exchange in the gills is very efficient because water and blood flow in opposite directions.
- Gas exchange in mammals takes place in the alveoli of the lungs, between the air and blood.
- The lungs expand and contract, drawing in fresh air to mix with what is already there, and forcing out some of the mixture.
- When breathing in, the ribs swing up and out, and the diaphragm contracts and flattens.
- Intercostal muscles move the ribs.
- The respiratory centre in the medulla controls breathing movements.
- The respiratory centre also controls breathing rate. This depends mainly on carbon dioxide levels in the blood, much less on oxygen levels.

◆ Drowning causes death because changed ion concentrations in the blood affect the heart, not because of asphyxia.

Exercises

20.1. An oxygen molecule moves from the atmosphere to the blood inside a human lung capillary. List, in order, all the structures it passes through on the way.

20.2. Chronic pulmonary emphysema is one of the diseases caused by long-term smoking. It happens like this. Substances in smoke irritate the bronchi and bronchioles. This upsets their protective systems, and they become infected. This in turn leads to swelling and mucus production, which partially block the airways. It becomes harder for air to leave the alveoli, which become overstretched. This leads to a slow destruction of the walls of the alveoli: eventually, as much as three quarters may be lost.
What effect will this condition have on:

(i) breathing; **(ii)** gas exchange; **(iii)** the heart?

20.3. Road accident victims sometimes suffer from a collapsed lung. The lung shrivels to a fraction of its normal volume, and does not inflate when the patient breathes in. The space which it normally occupies is filled with air. Suggest what could have caused this.

20.4. A person is sitting down. She records her breathing rate over one minute. Then she hyperventilates for a while: that is, she breathes more deeply and more rapidly than she needs to. Then she goes back to natural breathing, and records her breathing rate for one minute. The rate was slower on this second occasion. Why?

20.5. The Atlantic eel, *Anguilla vulgaris*, normally lives in water, but can survive for hours or even days out of water, if kept moist and cool. In water, it gets most of its oxygen through the gills. In air, more than half its oxygen uptake is through the skin. Explain the difference.

Further reading

Guyton, A.C. and Hall, J.E. *Textbook of Medical Physiology* (9th ed.) (Philadelphia: Saunders, 1996). I like this hefty textbook of human physiology for its thoroughness and clear, no-frills diagrams.

Guyton, A.C. *Human Physiology and Mechanisms of Disease* (5th ed.) (Philadelphia: Saunders, 1992). Effectively a shorter, boiled-down version.

Schmidt-Nielsen, K. *Animal Physiology: Adaptation and Environment* (4th ed.) (Cambridge: Cambridge University Press, 1990). One of the all-time great textbooks! Exciting, authoritative and easy to read, it puts animal physiology in the context of the environment.

Withers, P.C. *Comparative Animal Physiology* (Fort Worth: Saunders, 1992). More detailed than Schmidt-Nielsen, with wider ranging examples: the inevitable cost is that it is a less easy read.

Heterotrophic Nutrition

Connections

▶ This unit is about how many cells get the molecules they need. It is important to start with a basic knowledge of biological molecules, cell structure, enzymes and how molecules move (Units 3–6). Some knowledge of cell communication (Units 16 and 17) is also useful.

Contents

21.1 Cell eat cell

Heterotrophic nutrition is the business of eating ready-made food. We do this, along with all other animals from hydroid to hyena. Fungi do it too, along with most bacteria and a very few plants which do not photosynthesize. In **autotrophic nutrition**, organisms make their own food using simple chemicals from the environment and an energy supply, usually light (Unit 10). Photosynthetic plants and protoctists, photosynthetic bacteria and chemosynthetic bacteria all fall into this category.

All cells have much the same things in them: proteins, lipids, carbohydrates, nucleic acids and so on. Autotrophs have built up these molecules from scratch. Their cells are perfect chemical larders for heterotrophs. Everything they need is in there, in more or less the right proportions. They need to break up the molecules just enough to allow them into their cells: this is **digestion**. Digestion also allows some chemical fine tuning. For example, cells do not need any old proteins, they need their own proteins. By breaking down hazelnut proteins as far as single amino acids, a squirrel can absorb them and convert them to squirrel proteins.

Heterotrophs rely on autotrophs. They all eat food originally made by autotrophs, even though it may have been through animal bodies since then. For heterotrophs are vulnerable to other heterotrophs. They are made of cells, so are good food for other cells. It's a tough world.

Heterotrophic nutrition: The use of complex organic molecules as food.

Autotrophic nutrition: Making food by building up complex molecules from simple substances.

Digestion: Physical and chemical breakdown of food.

21.2 The value of food

A list of the valuable molecules in food is a list of the molecules which make up cells. Proteins are needed as a source of amino acids for making other proteins. If the diet is rich in protein, amino acids may be deaminated (Section 22.2). The organic acids which remain can be broken down in respiration. Most amino acids can be interconverted: transamination. The amino group is taken off one amino acid and added to another organic acid, making a different amino acid. Heterotrophs can adjust the levels of different amino acids to meet their needs. Some amino acids cannot be made by transamination, but have to be eaten. In humans, there are ten of these essential amino acids. Other species may have different requirements.

Carbohydrates and lipids are required as an energy supply and a source of carbon skeletons for building other molecules. Lions take in far more fat than carbohydrate; cows receive far more carbohydrate, mainly cellulose and starch; aphids feeding on phloem sap are getting large amounts of sucrose. The effect is the same. Any one can be used in respiration. The heterotroph can build its own carbohydrates and lipids, using carbon skeletons from any of these sources. There are a few exceptions to this

Box 21.1

Vitamins

Not all animals need the same vitamins. These examples of vitamins needed by humans show that many are needed for making prosthetic groups or coenzymes, whilst others are reducing agents.

Retinol (vitamin A)

Converted to retinal, the light-absorbing pigment in the rod cells of the retina. Deficiency leads to poor vision in dim light. However, it seems to have other effects which we do not understand, perhaps as a reducing agent. Deficiency leads to poor and abnormal growth of epithelia. This brings about diverse problems including being more open to infection.

Retinol can also be made in the body from carotenoids in plants: these are known as provitamins.

Thiamine (vitamin B_1)

Used to make the prosthetic groups of several enzymes involved in respiration, especially pyruvate dehydrogenase. Deficiency means that the pathways for aerobic respiration of carbohydrates are interrupted, but respiration using fat is not affected. Some tissues are badly affected by this. Nerve cells in the brain, which rely on carbohydrates for energy, may be damaged.

Riboflavin (vitamin B_2)

Converted to the coenzyme FAD. Deficiency must lead to less efficient aerobic respiration. This in turn can affect any cell in the body, so it is no surprise that animal experiments show up a wide range of symptoms.

Ascorbic acid (vitamin C)

A reducing agent. It is particularly important to an enzyme involved in adding a hydroxyl group to proline residues in collagen. It keeps the enzyme in a reduced, active form. Insufficiently hydroxylated collagen cannot form proper fibres. Collagen is important in connective tissues and in the basal lamina beneath an epithelium. Deficiency symptoms first show where collagen molecules are replaced most often, for example in blood vessel walls. Blood vessels become fragile, and capillaries may bleed. Wounds heal slowly.

More generally, ascorbic acid is an important antioxidant in blood plasma, keeping various proteins in a reduced state.

Cholecalciferol and related molecules (vitamin D)

Converted to a steroid hormone involved in regulating the body's calcium level. The hormone acts as a signal to take up and retain calcium. Deficiency of vitamin D leads to calcium deficiency. Vitamin D can also be made in the skin from other steroids if exposed to sunlight.

Vitamin K

This is a cofactor for an enzyme system which modifies an amino acid during the synthesis of prothrombin and some other clotting factors (Section 23.3). Vitamin K deficiency leads to blood failing to clot, so an injury can lead to uncontrolled bleeding. It is rare, except in newborn babies, because it is made by bacteria in the gut.

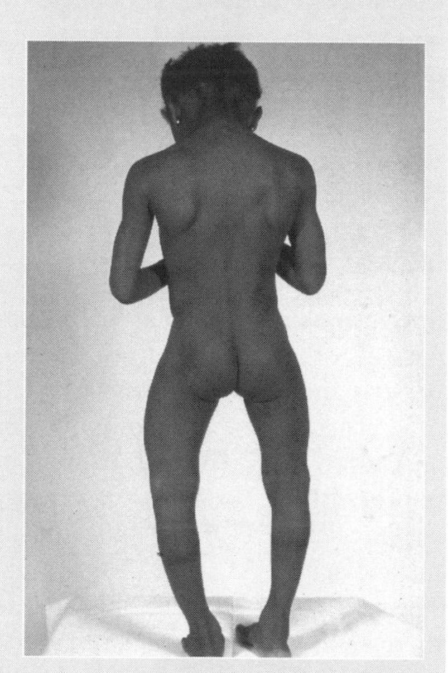

This person suffers from rickets in an extreme form. A lack of vitamin D has led to calcium deficiency. Calcium has been reabsorbed from bones, which become weaker and have bent.

rule. Some animals, including humans, need particular fatty acids which they cannot make. These have to be present in the diet.

If a heterotroph eats cells, it will gain nucleic acids, and hence nucleotides. These are valuable, but not essential, because animals can make nucleotides themselves. Heterotrophs get their mineral ions from food. They are just the same as the ones in the environment, but have been concentrated in the cells which are being eaten.

Vitamins are a strange group of molecules. The grouping is based on nutrition, not chemistry. Vitamins are organic molecules needed in tiny amounts by heterotrophs. Not every animal needs all the vitamins: some can be made in the body. Box 21.1 shows some of the vitamins needed by humans, and their chemical roles.

21.3 First catch your meal

Heterotrophs feed in many ways. The simplest is to live within the food itself. Fungi and bacteria which feed on dead matter do this. Fungal hyphae grow through their food, digesting it as they go. Gut parasites, and parasites which invade blood vessels, have a constant food supply washing over them. Some parasitic animals even grow through their food, like a fungus. The rhizocephalans are a group of crustaceans, related to barnacles. They parasitize crabs. Almost the whole of their bodies are fine branching tubes which grow through the crab's tissues.

Many animals feed by filtering. They need a filter, and a way of making water (or air) with food suspended in it pass through the filter. On the big scale, the baleen whales feed on planktonic animals. They have lines of tough plates hanging down from the upper jaw. As the whale swims along, water enters the mouth. Food is sieved out by the plates. Some fish, including herrings and basking sharks, sieve food out of the water which passes through their gills. Web-spinning spiders are filter-feeders. Some caddis fly larvae, which spin underwater nets in streams, are doing the same thing.

A very large group of filter-feeders uses cilia (Fig. 21.1(a,b)). Some polychaete worms have a fan of tentacles covered in cilia. Some bivalve molluscs suck in water through one tube, pass it across a filter of cilia, and push it back out through another

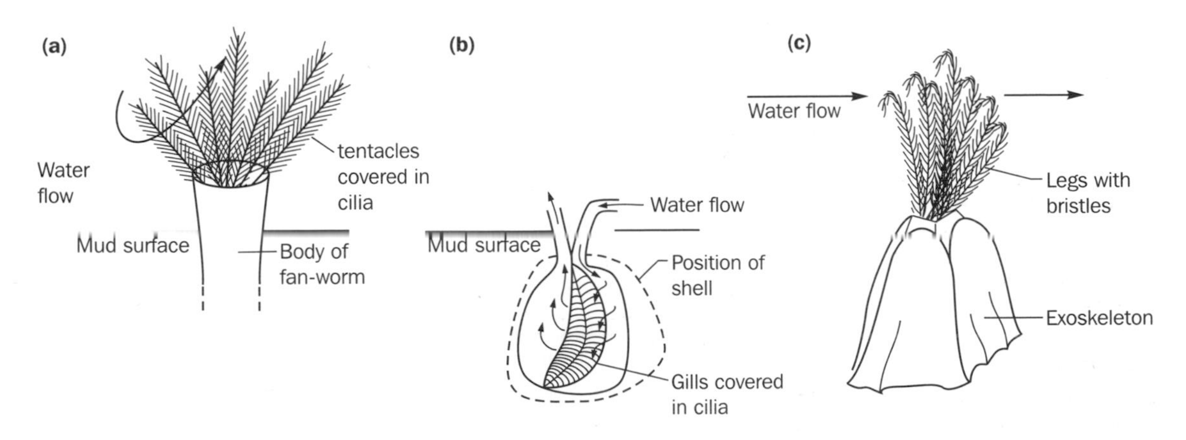

Fig. 21.1 Mechanisms of filter feeding. (a) a polychaete worm; (b) a bivalve mollusc; (c) a barnacle.

tube. Crustaceans tend to have filters made of fine chitin bristles, for example on the legs of barnacles (Fig. 21.1(c)).

Filter-feeders usually feed on prey which are smaller than themselves (spiders are an exception). Feeding on larger prey means catching and eating them one at a time. Predators, grazers and seed eaters have rather similar ways of doing this. In mammals, the mouth and teeth are most important, sometimes aided by the limbs (watch a squirrel eat a hazel nut). In fish and reptiles, the mouth is everything. Arthropods have complex, moveable, external mouthparts. They vary enormously, depending on what the species eats. In predatory insects they can be fearsome, claw-like weapons. Birds have no teeth. Their beaks handle food, and are extremely diverse (Box 21.2).

Box 21.2

Beak diversity

Birds' beaks are used to pick up food, and sometimes to tear at or crush it. Beaks are adapted to the food they work on.

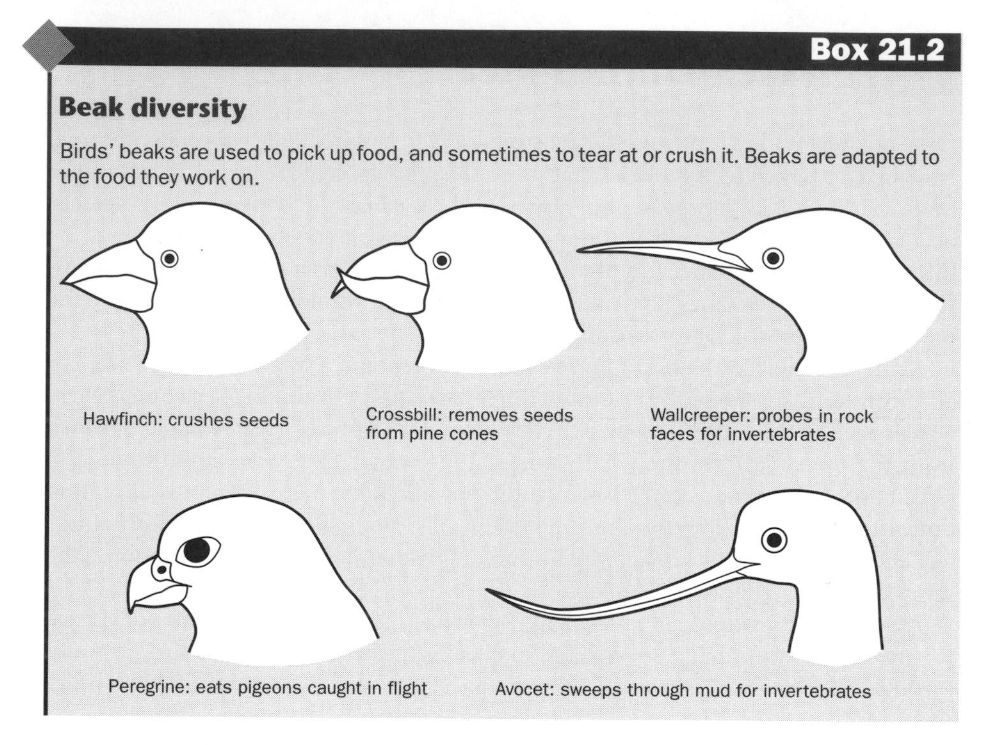

Animals that feed on liquids need tubular mouthparts. This may just need to suck, as in the proboscis of a moth which drinks nectar. It may need to pierce too, as in the stylet of an aphid, or a mosquito feeding on blood.

Once a heterotroph has caught its food, it must digest it.

21.4 Digestion

Digestion is the process of breaking down food. The aim is to make large and insoluble food particles into small and soluble ones. Digestion can be chemical or mechanical.

In chemical digestion, large insoluble molecules are broken into smaller, usually soluble subunits. Proteins are digested to amino acids, polysaccharides to sugars, nucleic acids to nucleotides, and so on. These are all hydrolysis reactions (Section 3.4: The glycosidic linkage). They are catalysed by digestive enzymes. Box 21.3 shows the main classes of digestive enzyme, and what they do.

Box 21.3

Digestive enzymes

The major groups of digestive enzymes catalyse hydrolysis reactions.

Enzymes digesting carbohydrates

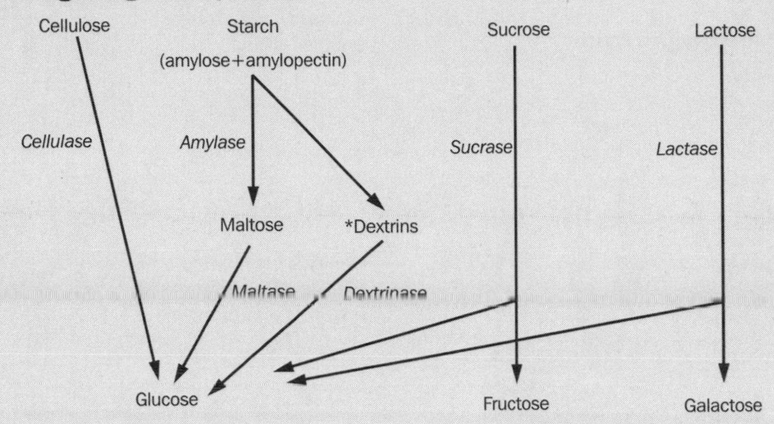

*Dextrins are short glucose polymers which include one of the branch points of amylopectin.

Enzymes digesting proteins

proteins $\xrightarrow{\textit{proteases}}$ shorter polypeptides $\xrightarrow{\textit{peptidases}}$ amino acids

Proteases break specific peptide linkages, so break a long chain into shorter fragments.
Peptidases break any peptide linkage, so the entire polypeptide is taken apart.

Enzymes digesting fats

triglycerides $\xrightarrow{\textit{lipase}}$ monoacylglycerols + fatty acids

Enzymes digesting nucleic acids

nucleic acids $\xrightarrow{\textit{nucleases}}$ nucleotides

Different nucleases digest DNA and RNA.

Most chemical digestion goes on outside cells. Fungal hyphae secrete digestive enzymes into the food around them. They let the enzymes do their work, and absorb the products. Animals with guts secrete enzymes into the gut, either from cells in the gut lining or from specialized organs like the pancreas.

Digestion can take place inside cells, because lysosomes contain digestive enzymes. This is only really important in feeding when cells take up food by phagocytosis, as in some protoctists. The cell engulfs a food particle in a fold of plasma membrane. Once inside the cell, this fuses with a lysosome and digestion begins (Section 6.6).

Mechanical digestion is the physical breakdown of food into smaller bits. When it happens, it is a prelude to chemical digestion, not an alternative.

Teeth are important in mechanical digestion. In fish and reptiles they are little more than keratin-rich scales which help catch and hang on to prey. Mammals have several types of hard, bone-like teeth specialized for cutting, ripping and grinding (Box 21.4). Food is broken down to a paste mixed with saliva in the mouth cavity of

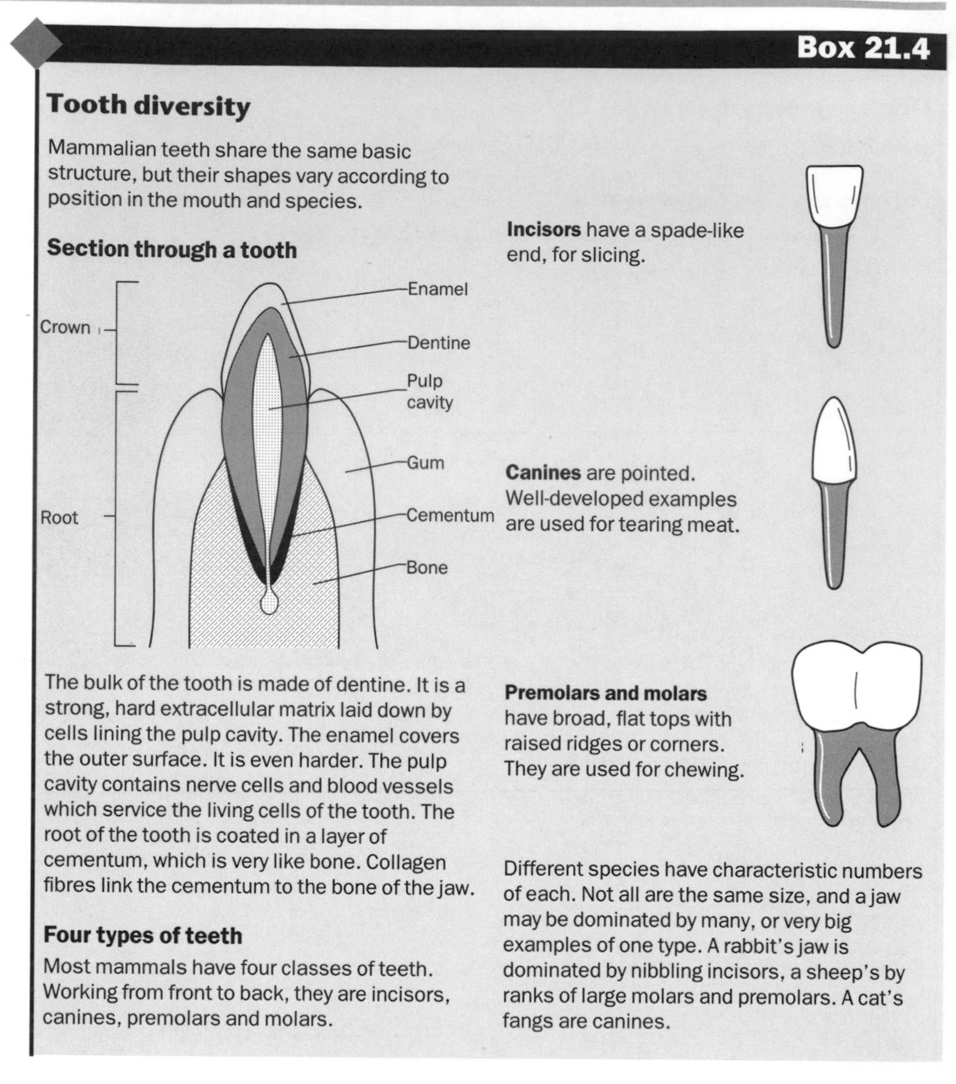

a mammal. Chemical digestion can then take place effectively in the stomach. Birds have no teeth, but food is ground in the gizzard, a muscular bag. Grit mixed with food may help this.

Box 21.4
Tooth diversity
Mammalian teeth share the same basic structure, but their shapes vary according to position in the mouth and species.
Section through a tooth
Crown
Root
Enamel
Dentine
Pulp cavity
Gum
Cementum
Bone
Incisors have a spade-like end, for slicing.
Canines are pointed. Well-developed examples are used for tearing meat.
Premolars and molars have broad, flat tops with raised ridges or corners. They are used for chewing.
The bulk of the tooth is made of dentine. It is a strong, hard extracellular matrix laid down by cells lining the pulp cavity. The enamel covers the outer surface. It is even harder. The pulp cavity contains nerve cells and blood vessels which service the living cells of the tooth. The root of the tooth is coated in a layer of cementum, which is very like bone. Collagen fibres link the cementum to the bone of the jaw.
Four types of teeth
Most mammals have four classes of teeth. Working from front to back, they are incisors, canines, premolars and molars.
Different species have characteristic numbers of each. Not all are the same size, and a jaw may be dominated by many, or very big examples of one type. A rabbit's jaw is dominated by nibbling incisors, a sheep's by ranks of large molars and premolars. A cat's fangs are canines.

All but the simplest animals digest food in specialized organ systems: guts. The structure and working of guts is examined first, before looking at the mechanisms of secretion and absorption in more detail.

Guts

Guts are where animals digest and absorb food. The gut is a hollow tube running from mouth to anus in all but the simplest animals. The inside of the gut, like the space inside any biological tube, is called the lumen. The lumen is not part of the body. It is part of the outside world, very intimate and closely controlled, but still not part of the body. Digestion takes place in the lumen of the gut. This is extracellular digestion, just as in animals which have no gut. Digestive enzymes are secreted into the lumen. Food enters the body only when it is absorbed across the gut wall.

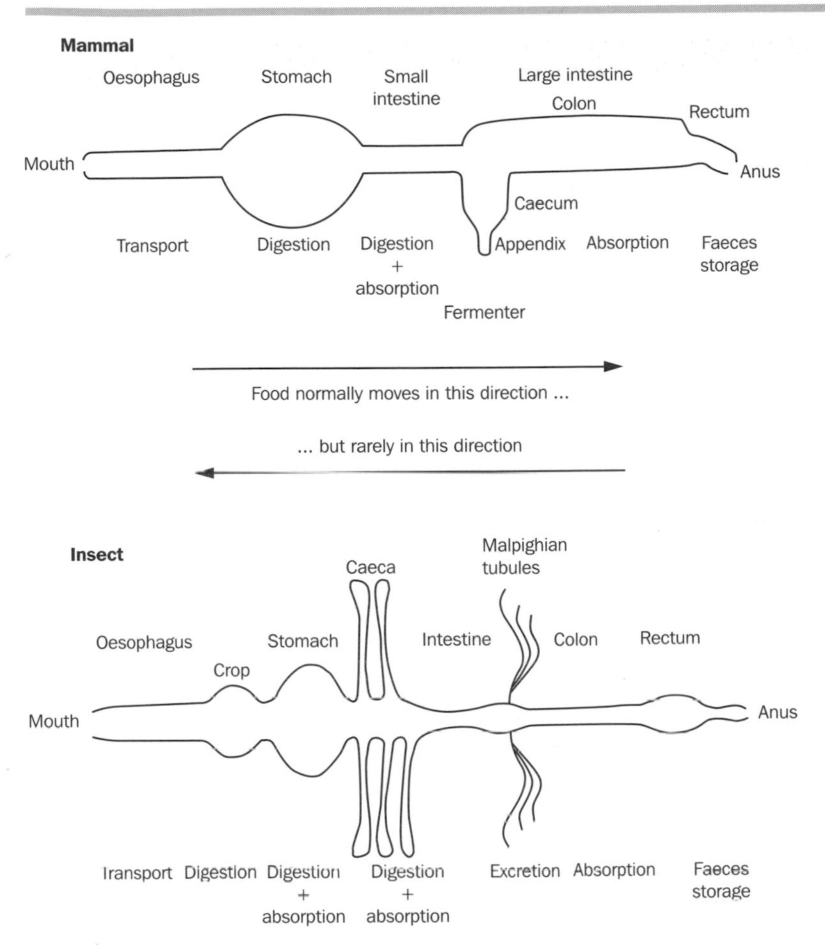

Fig. 21.2 Comparison of the mammalian and insect gut.

Fig. 21.2 shows the regions of a mammalian gut and an insect gut. These vary in size and structure between species, but are always there. Fig. 21.3 shows the human gut. It is typical of mammals, other than specialized herbivores.

Digestion begins in the mouth (the hole in the face) and the mouth cavity (the space within). Teeth and tongue turn solid food into a soft pulp, mixed with saliva. Saliva contains digestive enzymes, especially amylase, as well as lysozyme which digests bacterial cell walls. These enzymes seem to be unimportant in digesting food: they are denatured as soon as they hit the acidic stomach contents. They protect the teeth against decay, breaking down food particles left there after meals. When the enzymes are absent, bacteria on the teeth get far more food, and their acid secretions lead to a higher rate of tooth decay.

The oesophagus is just a muscular tube in mammals. It has bands of longitudinal and circular muscle in its walls. Pulses of relaxation and contraction pass along the pipe, carrying food pellets down to the stomach. This process is called **peristalsis** (Fig. 21.4). Birds have a blind-ended side branch of the oesophagus, the crop. This is simply a food store.

Peristalsis: The movement of material through a tubular organ, by waves of muscular contraction.

The stomach is a muscular bag which stores food, letting it out gradually into the intestines. It also churns food by muscular contractions, and makes some digestive

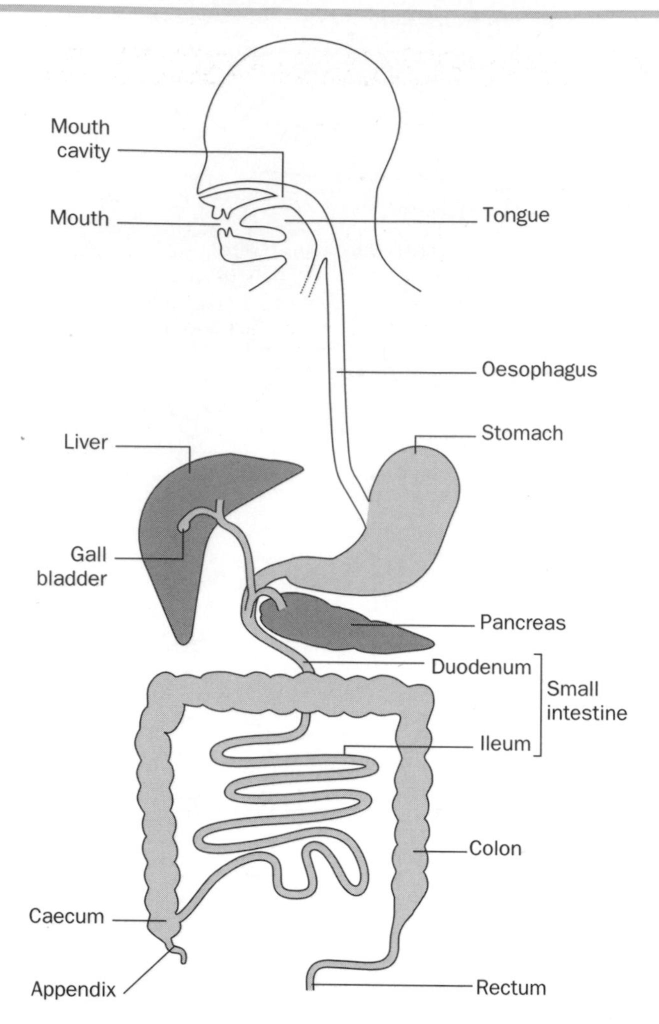

Fig. 21.3 The human gut and associated organs.

enzymes. The most important is a protease: pepsin. The optimum pH of pepsin is highly acidic. This is because the stomach secretes hydrochloric acid, which kills microbes entering with the food. A thick mucus layer protects the stomach wall against acid. Not much absorption goes on in the stomach. In humans, glucose and alcohol are two important small molecules which can cross the stomach wall. Box 21.5 compares the structure of the stomach wall with other parts of the gut.

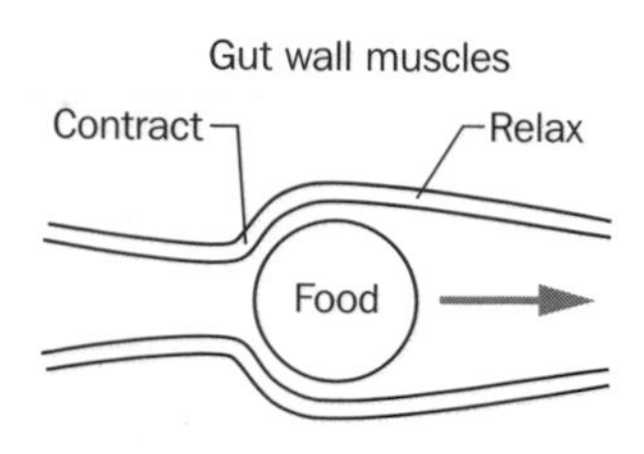

Fig. 21.4 Peristalsis.

The intestines are the main site of chemical digestion and absorption. Several secretions meet the gut contents in the first part of the small intestine: the duodenum. A range of enzymes enter from the pancreas, through a duct. The secretions of the pancreas are alkaline, neutralizing the acid material leaving the stomach. Various molecules secreted by the liver also enter the duodenum. Together they are called bile, and enter through the bile duct. Bile is alkaline. It also contains bile salts, such as sodium glycocholate (a steroid derivative), and lecithin. These act rather like detergents, binding to hydrophobic lipids and making them more soluble. They break up larger drops of fat into many tiny drops. This increases surface area so that digestion

Box 21.5

The gut wall

General structure

The wall of each part of the gut shows variations of a common structure.

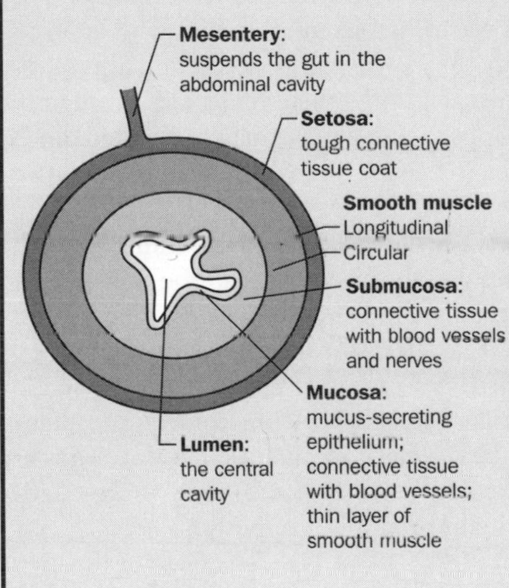

Oesophagus

The upper part has skeletal muscle, under voluntary control, as well as smooth muscle. Skeletal muscle is involved in the voluntary act of swallowing, while smooth muscle is involved in involuntary peristalsis. There are mucus-secreting glands in the submucosa.

Stomach

The mucosa is folded into deep, narrow pits: gastric glands. Different cell types secrete acid, mucus and enzymes. There is a third, oblique muscle layer.

Small intestine

The mucosa is folded into villi. The duodenum, but not the ileum, has Brunner's glands which secrete mucus. They are in the submucosa; a duct links them to the lumen.

Large intestine

There are no villi, but the mucosa has narrow, mucus-secreting pits. The muscle layer is not complete: there are three bands which run lengthwise.

by enzymes is quicker. It makes the lipids easier to absorb, too. Bile also contains waste products such as bilirubin, the remains of a haem group from haemoglobin breakdown. These have no role in digestion. They are simply being excreted via the gut. The epithelial cells lining the small intestine make several enzymes including peptidases and enzymes which digest disaccharides. The enzymes are anchored to the plasma membrane and are not released into the food. The lining of the small intestine also secretes mucus and a watery fluid.

The small intestine has a very large surface area. This is partly due to its being long and thin. The wall is also covered with finger-like folds: villi. The products of digestion are mostly absorbed through this huge surface, as the mixture of food and secretions passes through by peristalsis. Other patterns of muscle contraction chop the gut contents, or even give it the milk-shake treatment.

The large intestine is wider and has no villi. Most digestion and absorption has already taken place. Typically, the large intestine absorbs water. Some of this has entered as food and drink, but much of it comes from secretions further up the gut, particularly the stomach. The caecum and blind-ended appendix branch off the beginning of the large intestine. They are small in most mammals and have no special function. In some herbivores, however, they are large and have a role in cellulose digestion (see below).

Faeces leave the gut through the anus. They consist of indigestible material from food, excretory products from bile, dead cells from the lining of the gut, but mostly huge numbers of bacteria which live and feed in the large intestine.

Herbivore adaptations

Herbivores have two special problems. Firstly, most plant material is quite tough: the more times it is chewed, the better. Secondly, most of its energy is locked up in cellulose. No animals can make the enzyme cellulase themselves. It is only produced by some fungi, bacteria and protoctists. Herbivorous mammals solve these problems in several ways.

Ruminants, such as cows, have three chambers before the true stomach. The biggest, the rumen, contains bacteria and ciliates which produce cellulase. They benefit from the relationship by having a warm, stable environment with an excellent supply of ground-up food. They satisfy their own energy needs, and also release various organic acids and ketones. The animal absorbs these molecules through the rumen wall, and uses them as an energy supply. After a while, partly digested food is returned to the mouth and chewed again ('chewing the cud'). This is swallowed again, and enters the true stomach and the rest of the gut.

Other herbivores, like rabbits and horses, use the caecum and appendix as a home for ciliates and bacteria. Acids and ketones are absorbed through the caecum, and probably the large intestine wall. Rabbits get a second chance to digest food by passing it through the entire gut twice. When it has been through once, the animal eats the faeces, which are pale and soft. They sit in a modified part of the stomach for hours, while microbes which came through from the caecum continue to digest. Second time around, the faeces are drier, darker and contain little of value. They are left alone. Horses do not do this. As a result, horse faeces contain more coarse, partially digested plant material than ruminant or rabbit droppings.

Secretion

Secretion takes place throughout the gut. Enzymes are secreted from the salivary glands down to the small intestine. Acid is secreted in the stomach, alkali in the glands which empty into the duodenum. Mucus is secreted throughout, from mouth to anus. There are two big questions. How are substances secreted, and how is secretion controlled?

Enzymes are secreted in the same way as any other protein (Section 4.5). They are synthesized on the rough endoplasmic reticulum, transported to the Golgi apparatus, and sorted into secretory vesicles. These vesicles sit in the cytoplasm and wait for the signal to fuse with the plasma membrane, releasing the enzymes. The signal could be a nerve signal or a hormone, but either way the second messenger is calcium (Section 16.4).

Mucus is secreted by specialized cells in the epithelium lining the gut. Its composition varies slightly, but is basically a solution of mineral ions in water, thickened by proteoglycans. Proteoglycans are secreted by exocytosis. It is not clear exactly how mineral ions and water are secreted.

Hydrochloric acid is secreted by another specialized epithelial cell type in the gastric glands, little infoldings of the stomach lining. Both H^+ and Cl^- are actively secreted by carrier proteins in the plasma membrane. The precise mechanism is not yet clear.

There is no point in secreting acid, enzymes or mucus unless there is, or soon will be, food in that part of the gut. Secretion is controlled in three ways. It can be triggered by direct stimulation of the secretory cells when food is present, by chemical signals from other cells, and by nerve impulses.

Nerves control release of saliva. Various stimuli such as the taste and smell of food, the presence of small objects in the mouth, even thinking about eating, all result in nerve impulses which trigger salivation.

Hormones and nerve impulses together control acid secretion in the stomach (Fig. 21.5). Acetylcholine, a neurotransmitter, stimulates secretion. This is released by neurones originating in the brain. It is also released by local neurones which carry information along the length of the gut wall, about where food is. A hormone, gastrin, also stimulates acid secretion. Gastrin is produced by cells in the lower part of the stomach, when food is present. It travels in the bloodstream, even though the target cells are also in the stomach wall. Another molecule, histamine, has to be present if acid-secreting cells are to respond to gastrin and acetylcholine. Histamine is a paracrine, a local signalling molecule (Section 16.5). It is constantly produced by cells in the stomach lining, so strictly speaking it does not control acid secretion. It is exploited by drugs like ranitidine. This is useful to people with stomach ulcers, where acid damages patches of stomach wall which have lost their mucus protection. Ranitidine binds to histamine receptors in the acid-secreting cells. This prevents them detecting histamine, so acid production falls.

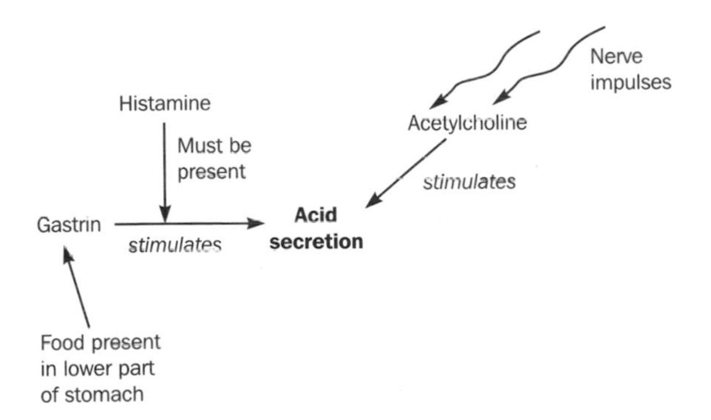

Fig. 21.5 Control of stomach acid secretion.

21.5 Absorption and afterwards

Molecules from digested food enter the body in three ways: diffusion, active transport and endocytosis (Unit 6). Small molecules which are at relatively low concentrations in the blood may simply diffuse from the lumen of the gut through the epithelium. Drugs like alcohol and aspirin enter the body this way. There is certainly no specific mechanism for absorbing them. Some molecules, such as glucose, are actively transported into the epithelial cells lining the gut, by specific carrier proteins. Endocytosis may bring in molecules which cannot cross the membrane in other ways.

Once in the epithelial cells, most absorbed molecules move on into the blood. Blood from the gut wall does not go straight back into the general circulation, but travels to the liver in the hepatic portal vein. Blood in the hepatic portal vein is very variable. What it contains depends on what has been eaten, and when. The liver removes molecules which are in excess before the blood returns to the heart.

Fats are not put directly into the bloodstream. They are digested in the gut, forming fatty acids and monoacylglycerols. These enter the epithelial cells of the small intestine. The fats are put back together in the smooth endoplasmic reticulum. They are bound with small amounts of proteins and phospholipids into more hydrophilic globules. The globules are constructed in the Golgi apparatus. The cells secrete them into the small lymph vessel within each villus, by exocytosis. They enter the blood along with the rest of the lymph (Section 19.7).

21.6 Controlling feeding

Animals are driven by a powerful motivation to feed. However, it is possible to eat too much. Appetite sometimes needs to be checked, to prevent this.

Appetite is controlled on two time scales. Having a gut full of food is a very good short-term reason not to feed. On a longer time scale, changes in body mass must be limited and reversed, to prevent the animal becoming seriously over- or underweight.

In mammals, part of the brain is dedicated to controlling appetite. The appetite centres are in the hypothalamus (Section 17.7). The hunger centre controls feelings of hunger; the satiety centre produces the feeling of being well fed. Neuropeptides (Section 17.8) have been discovered which either stimulate or inhibit feeding. They probably have some role in how the centres work.

How do the appetite centres know what to do? Information must be reaching them from the body. Stretch receptors in the gut wall detect the presence of food inside. The fuller the gut, the more they are stretched. This information reaches the brain as nerve impulses, and decreases appetite on a short time scale. How body weight is controlled in the long term has been a mystery until recently. It turns out that fat cells are signalling to the brain, using a hormone.

Fat cells are found in adipose tissue, which forms a bouncy insulating layer under the skin and around internal organs. They are large cells. A huge lipid droplet in the cytoplasm makes up most of their volume. Fat cells take up lipids (mostly triglycerides; Section 3.7) from the blood and store them in the droplet. These lipids may have been absorbed by the gut, or made in the liver from excess carbohydrate. Either way, the fat cells are putting them into long-term storage, to be used in times of hunger. Fat cells make a hormone called leptin, a protein. Leptin is released over a period of hours only if the lipid globule is growing. This will happen only if the body is well fed. Leptin is a satiety signal, detected in the hypothalamus. High leptin levels supress appetite. The details of the control system are still being worked out as I write (Spring 1997) but there is already enormous interest in the possibility of weight-control drugs which operate on the leptin system. We must be careful, however. We have discovered one way in which the brain finds out about the body's nutritional status. This does not mean that it is the only way.

21.7 Success and specialization

Digestive systems are specialized, according to the animal's diet. Generalist feeders, such as humans and brown bears, have a mixed diet and a relatively unspecialized digestive system. Herbivores show enormous development of molar and premolar teeth for grinding tough plant material. Carnivores often have large canine teeth for butchering their prey. Many herbivores have specialized chambers for cellulose-digesting microbes, and so on. The digestive system suits the animal's lifestyle. Occasionally, however, we find an animal which appears to be very poorly adapted to its way of life. The giant panda, a large woodland mammal of western China, is one of these. It has a typical, generalist mammalian digestive system, without cellulose-digesting microbes, yet it is essentially a herbivore. Pandas eat bamboo leaves and stems, with just occasional invertebrate animals. It has clear adaptations to its lifestyle, such as a paw/wrist anatomy which lets it grasp bamboo stems, but its gut is unspecialized. Leaves contain lots of cellulose and a little starch. Most of what the panda eats cannot be digested. It has to devote most of its waking hours to eating bamboo, just to meet its energy requirements. Yet the panda is a successful species. It survives today: it has not become extinct along with most of the species that have lived. Its endangered status owes more to human destruction of its habitat than to anything about the panda itself.

However, most species become extinct in the end. Some will have spawned new species which pass on their genes, others will not. A crude, but useful generalization is that more specialized species arise from less specialized ancestors. Species which are already highly specialized in one direction do not tend to give rise to specialists in other areas.

One could argue that the panda is a poor candidate for a parent species, because it is rare and leads a bizarre life. We can make intelligent guesses about its evolutionary future, but we do not know. It would be unwise to write off a species as an evolutionary dead end while it is still with us. While there's life, there's hope.

Summary

- Heterotrophs eat food ready-made. They rely on biological molecules produced by autotrophs.
- Most food is made of cells, so the important classes of molecules found in cells are found in food.
- Different heterotrophs feed by living in the food itself, by filtering food suspended in water, by eating large pieces of food, or by taking in a liquid food.
- Digestion is the process of breaking food down.
- Mechanical digestion involves breaking food into smaller pieces, for example by teeth.
- Chemical digestion involves breaking up food molecules, making them smaller and/or more soluble.
- Chemical digestion involves hydrolysis reactions, catalysed by digestive enzymes.
- Guts are specialized organ systems for digestion and absorption. The lumen of the gut is not part of the body, but an enclosed part of the environment.
- In the mammalian gut, secretions involve mucus (throughout), hydrochloric acid (in the stomach) and digestive enzymes (in the mouth, stomach and small intestine).
- Secretion is controlled by nerves and hormones, so that it only occurs when needed.
- Molecules are absorbed from the gut by diffusion, active transport and phagocytosis.

- In mammals, most absorption goes on in the small intestine. The villi covering its wall increase its surface area.
- The teeth of herbivores and carnivores suit their diets.
- Herbivorous mammals rely on microbes in the gut to digest cellulose. They may live in the rumen (cows) or caecum and appendix (rabbit).
- Appetite is controlled, to avoid over- or undereating.

Exercises

21.1. Complete this simple table summarizing the digestion of some molecules in food.

Substance	Digested to . . .	Enzyme(s) needed	Value to the body
Proteins			
Fats (triglycerides)			
Cholesterol			
Starch			
Lactose			
Sucrose			
Cellulose			

21.2. Blood leaving the intestine wall is unusual, in that it goes to the liver in the hepatic portal vein, rather than straight back to the heart. Why is this an advantage?

21.3. Some foods are rich in carbohydrates but contain hardly any proteins, nucleic acids or other molecules containing the element nitrogen. Aphids feed only on phloem sap, and some moth larvae feed only on wood, for example. How could animals overcome this problem (other than by eating something else!)?

21.4. Long before the discovery of leptin, physiologists showed that mammals had some sort of appetite-controlling system which involved the brain. (a) Low blood sugar levels lead to hunger, and this has been linked to activity in feeding centres within the hypothalamus. (b) Low levels of amino acids have the same effect. (c) So do low levels of fatty acids and other products of fat breakdown. To what extent does the leptin system explain these observations?

Further reading

Guyton, A.C. and Hall J.E. *Textbook of Medical Physiology* (9th ed.) (Philadelphia: Saunders, 1996). I like this hefty textbook of human physiology for its thoroughness and clear, no-frills diagrams.

Guyton, A.C. *Human Physiology and Mechanisms of Disease* (5th ed.) (Philadelphia: Saunders, 1992). Effectively a shorter, boiled-down version.

Schmidt-Nielsen, K. *Animal Physiology: Adaptation and Environment* (4th ed.) (Cambridge: Cambridge University Press, 1990). One of the all-time great textbooks! Exciting, authoritative and easy to read, it puts animal physiology in the context of the environment.

Withers, P.C. *Comparative Animal Physiology*. (Fort Worth: Saunders, 1992). More detailed than Schmidt-Nielsen, with wider-ranging examples: the inevitable cost is that it is a less easy read.

Excretion and Osmoregulation in Animals

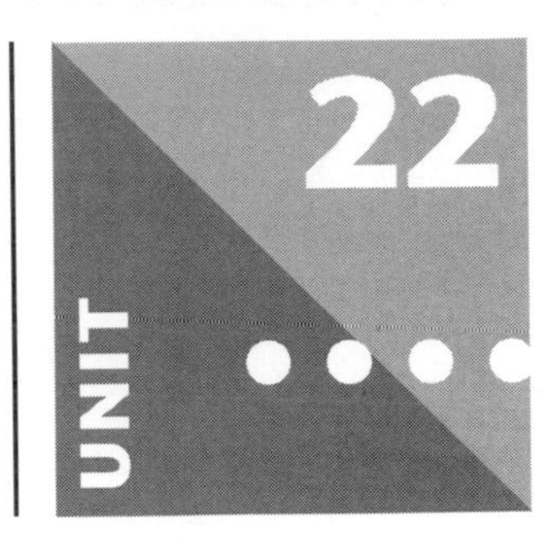

Connections

▶ Osmoregulation is one example of homeostasis (Unit 15), and is partly controlled by hormones (Unit 16). Osmoregulation involves movement of ions and molecules between cells, body fluids and the environment: Units 4, 6 and 19 provide essential background information.

Contents

22.1 The need for excretion

Cells make waste products as a result of their activities. Many of these waste products are harmful: cells must get rid of them. When cells live together as organisms, they must cooperate to expel waste from the body as a whole. The removal of waste products from a cell or organism is called **excretion**.

A number of waste products are involved:

- Carbon dioxide is a waste product of respiration. It diffuses readily out of cells through the membrane. Larger animals excrete CO_2 through specialized structures such as lungs and gills (Section 20.3).
- Nitrogen-containing waste products are of several sorts. Ammonium ions (NH_4^+) are made when amino acids are broken down. Ammonium ions and uric acid are made when nucleic acids are broken down. These, and other molecules such as urea, may be interconverted before excretion. Creatinine is a minor excretory product in many vertebrates. It seems to be formed during metabolism of phosphocreatine, an energy-storing molecule found in muscle (Section 8.4). Specialized organs such as kidneys and gills may again be involved in nitrogenous excretion.
- Various animal groups have specific excretory problems and solutions. For example, many minor waste products are excreted by the mammalian liver as bile (Section 21.4). One such molecule is bilirubin, the mangled wreck of a haem group from haemoglobin broken down when red blood cells die.

It is best not to think of the production of faeces as excretion. A better term is **egestion**. The inside of the gut is not part of the body, but a rather cosy and carefully

Excretion: The removal of waste products from cells and organisms.
Egestion: The removal of material from the gut. Egestion is not a form of excretion.

controlled corner of the outside world. Egestion is the removal of material which never entered the body. Mammalian faeces certainly contain waste products such as bilirubin, but these have already been excreted, by the liver.

22.2 Toxic waste: the nitrogen problem

Most nitrogenous waste starts out as ammonia. Some of this comes from breakdown of amino acids, when an amino group is removed, leaving a keto acid. Some comes from the breakdown of the pyrimidine bases of nucleic acids. The rest begins as uric acid, from breakdown of purine bases. Ammonia is highly soluble in water, and once dissolved in water exists mainly as ammonium ions (NH_4^+). Ammonium ions are, unfortunately, very toxic: the reason for this is unclear, but a clue comes from human disease. In some types of liver disease, ammonium ions accumulate in the blood. When two ammonium ions react with α-ketoglutarate (a Krebs cycle intermediate) glutamine is formed. Glutamine, when made in a brain cell, damages the cell. The result is an unpleasant state called hepatic coma.

Ammonium ions may be converted into the less toxic, rather less soluble molecule, urea. In mammals, this happens in liver cells. The overall reaction is

$$2NH_4^+ + CO_2 \rightarrow \underset{\text{urea}}{H_2N{-}\overset{\displaystyle O}{\overset{\|}{C}}{-}NH_2} + H_2O + 2H^+$$

In cells, this reaction takes place in several stages, as part of the urea cycle (or ornithine cycle). Carbon dioxide and ammonium ions are added to the existing skeleton of an amino acid, ornithine. Water is lost, and the enlarged molecule is broken in a hydrolysis reaction, one fragment being urea, the other ornithine, which is then available to continue the cycle. An energy input from ATP is required, rather indirectly. The key features of the cycle are summarized in Fig. 22.1.

Some animals convert ammonium ions not to urea but to uric acid, which then joins the pool formed by purine breakdown. Uric acid is only very slightly soluble and occurs in solution as urate ions. It is not toxic under normal conditions.

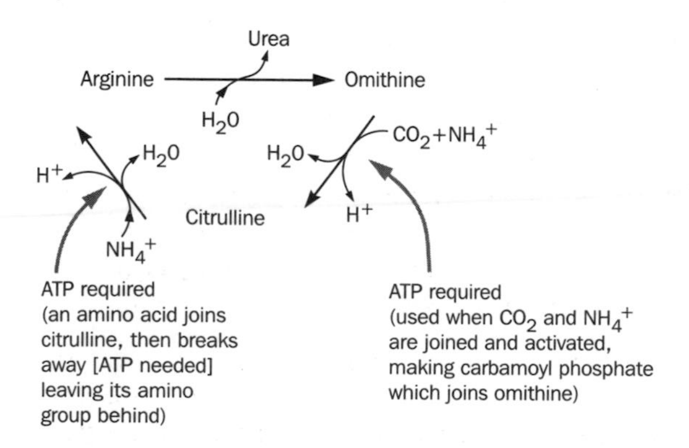

Fig. 22.1 The urea cycle.

Striking a balance

There are costs and benefits associated with excretion of any of these substances. The ammonium ion is toxic, but to excrete it carries no cost in energy terms. Urea is less toxic, but its synthesis requires energy input. The uric acid molecule contains nitrogen attached to a considerable carbon skeleton, so to excrete it is also to lose stored energy.

An organism which lives in water, or which can afford the luxury of copious urine production, may be capable of effective excretion of ammonium ions. This is seen in most aquatic invertebrates, and in most bony fish, which excrete NH_4^+ through their gills. Where water is more scarce, nitrogenous waste may have to be flushed away in less abundant, more concentrated urine. Ammonium ions might rise to toxic levels in such a situation. The energy cost of urea or uric acid excretion may then be a necessity. Mammals excrete primarily urea, reptiles and birds uric acid: these are groups of essentially terrestrial animals.

Why should reptiles and birds excrete uric acid? This is a troublesome question. The closed egg of both groups suggests one explanation. Nitrogenous waste made by the embryo cannot escape the egg. Urea could build up to osmotically dangerous levels in solution, but uric acid is far less soluble, so precipitates as harmless crystals in a stucture called the allantois (Section 13.6).

Water availability and urine production are central to these arguements. Nitrogenous excretion cannot be looked at in isolation from an animal's water balance, to which we now turn.

22.3 Water, salts and osmoregulation

We can crudely model a cell as a bag of salt solution. The bag is permeable to water, and to some extent, perhaps, permeable to salts. Depending on the environment, movement of water and salts in or out of cells by diffusion, or movement of water by osmosis, may be a problem. If cells as individuals are gaining or losing water, so will the animal body which the cells make up.

When water or salts move in or out of the body fluids, both within and between cells, an animal can respond in one of two ways:

- It can do nothing, and accept that body fluids will always be **isotonic** to the external solution. Such organisms are called **osmoconformers**, and include starfish. This is all right for marine animals which never experience much fluctuation in salinity. In estuaries, however, the cells of an osmoconformer would have to tolerate vast changes in salt concentration over the tidal cycle. On land, an osmoconformer would just dry up: the prospects for a stranded starfish are bleak.
- It can maintain a constant salt concentration within the body, either by removing water or salt which enters, or by making the body covering impermeable to water and salt which might leave. Such organisms are called **osmoregulators** and include the vertebrates as well as many invertebrates.

Box 22.1 shows four possible osmotic situations and some examples of how animals survive under them. The organs of excretion tend also to be the organs of osmoregulation. The mammalian kidney is one example of an organ which carries out this balancing act.

Isotonic: Having the same solute concentration as the surroundings.

Osmoconformers: Animals which cannot control the salt concentration of their body fluids.

Osmoregulators: Animals which can control the salt concentration of their body fluids.

Box 22.1

Osmosis and osmoregulation

Environments challenge animals in different ways, requiring different solutions.

Situation

Example of solution

1. The organism is in a moist environment, and is isotonic to its environment (for example many marine invertebrates, sharks and rays).

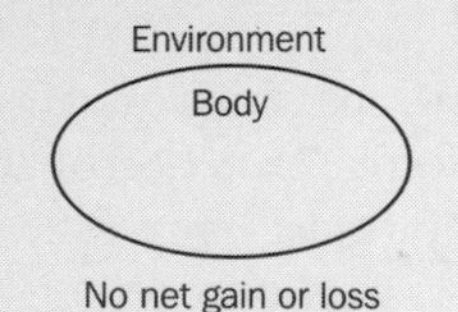

Sharks have no water problem: even though salt concentrations in cells are lower than outside, high urea levels prevent osmosis. Salt tends to diffuse in, which is then excreted by kidney and intestine.

2. The organism is in a moist environment and is **hypertonic** to its environment (for example freshwater animals, some marine animals, land animals living in moist habitats).

Environment
Body
H_2O
Salts
Water tends to enter by osmosis; salts tend to leave by diffusion

In freshwater bony fish, water gained by osmosis is removed as dilute urine. Salts are taken up and ammonium ions are excreted through the gills.

3. The organism is in a moist environment and is **hypotonic** to its environment (for example marine bony fish, some shrimps).

Environment
Body
Salts
H_2O
Water tends to leave by osmosis; salts diffuse in

Marine bony fish drink sea water to balance osmosis. They produce little urine and excrete salts and ammonium through the gills.

4. The organism is in a dry environment (for example land animals in general, desert animals as an extreme).

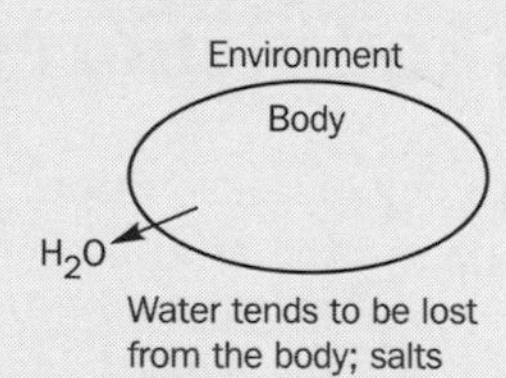

Birds produce rather small volumes of highly concentrated urine, as a slurry of uric acid crystals.

22.4 The mammalian kidney

The kidney carries out its functions of excretion and osmoregulation by making urine from materials delivered in the blood. Excretion is brought about by transfer of waste products, such as urea, from blood to urine. Osmoregulation is brought about by varying the volume and salt concentration of urine. Each kidney requires a blood supply, through the renal artery. Blood returns from the kidney in the renal vein. A third tube, the ureter, carries urine away from the kidney for temporary storage in the bladder (Fig. 22.2).

The bulk of the kidney is a tangled mass of tubes. Some are blood capillaries, some are the different regions of the **nephrons**, in which urine is formed, and some are collecting ducts through which urine drains. Kidney tissue has little mechanical strength, so each kidney is protected by a thick covering of adipose tissue. The body of the kidney is made up of two regions, the inner medulla and the outer cortex (Fig. 22.3).

Hypertonic: Having a higher solute concentration than the surroundings.

Hypotonic: Having a lower solute concentration than the surroundings.

Nephron: The working unit of the kidney, which makes urine.

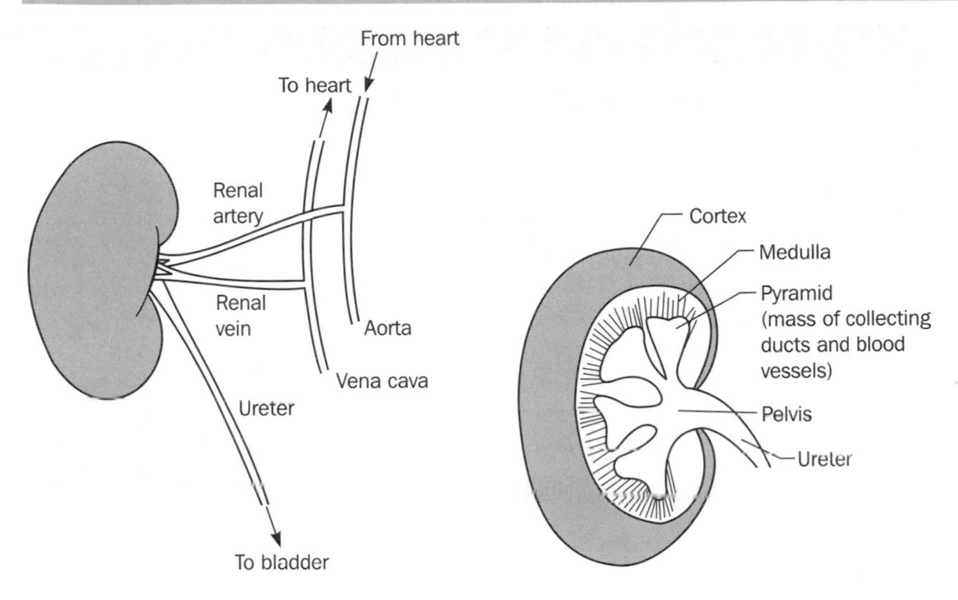

Fig. 22.2 The kidney and its connections

Fig. 22.3 The kidney in vertical section.

The Nephron

Each nephron begins in the cortex with a globular bunch of capillaries called a glomerulus, surrounded by a bag, the Bowman's capsule (Fig. 22.4). Blood is filtered under pressure through the layers of cells around each capillary of the glomerulus. The filtrate, which is much like plasma, collects in the Bowman's capsule. A tubule (literally meaning 'little tube') leads from each Bowman's capsule, winds through

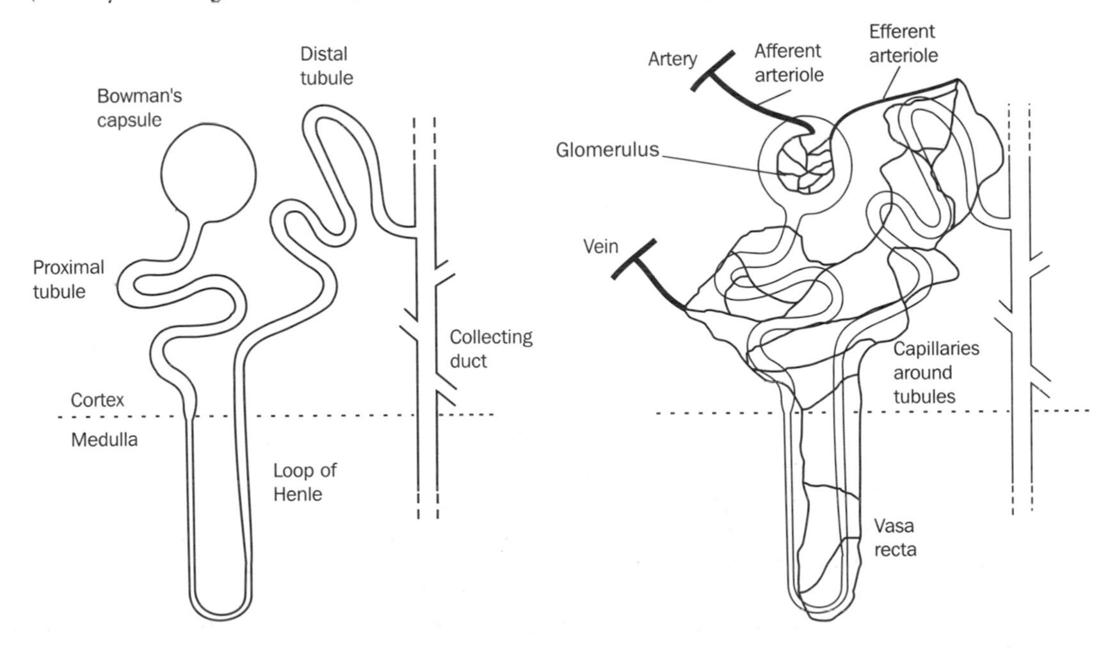

Fig. 22.4 The nephron, with and without its blood supply.

the cortex before dipping into the medulla and returning to wind through the cortex again. These regions are called the proximal tubule, the loop of Henle and the distal tubule. Nephrons drain into larger tubes, collecting ducts, which plunge straight down through the medulla and meet at the pelvis of the kidney, where urine enters the ureter.

Simply, the proximal tubule reabsorbs useful substances which should not be excreted. Salts are also reabsorbed from the distal tubule and part of the loop of Henle. The primary function of the loop of Henle is to set up an osmotic gradient down through the medulla. This gradient allows water to be reabsorbed into the body by osmosis from the collecting ducts. Blood capillaries run close to the proximal and distal tubules in order to carry away reabsorbed materials. Capillary loops, the vasa recta, dip into the medulla in just the same way as the loops of Henle.

Each kidney contains many intertwined nephrons, typically about one million in humans. Each nephron acts as an individual in terms of filtration and reabsorption, but contributes to an osmotic gradient which is built up and maintained by the nephron population as a whole.

Filtration in the glomerulus

Three layers lie between the blood in the glomerulus and the filtrate in the Bowman's capsule (Fig. 22.5). Nearest the blood is the endothelium, one cell thick, which forms the wall of any capillary. On the other side is a layer of epithelial cells, each of which has many long, narrow extensions which spread octopus-like over the surface of the capillary. Between the two is a basal lamina.

The basal lamina is not made of cells, but is a gel formed by molecules secreted by the cells on either side. It includes collagen fibres and other proteins, but most importantly it contains peptidoglycans. These are tinsel-like molecules made up of a central polypeptide strand with polysaccharide chains bristling off in all directions. The gels they form act as molecular sieves. Both endothelium and epithelium have gaps between the cells. It is primarily the basal lamina that controls what leaves the blood: if peptidoglycans are absent from the membrane, the glomerulus lets through larger molecules than normal. In the healthy kidney, only molecules up to the size of

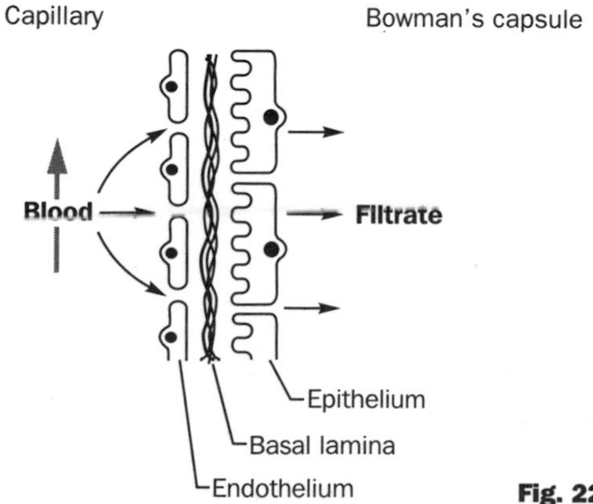

Fig. 22.5 Filtration in the glomerulus.

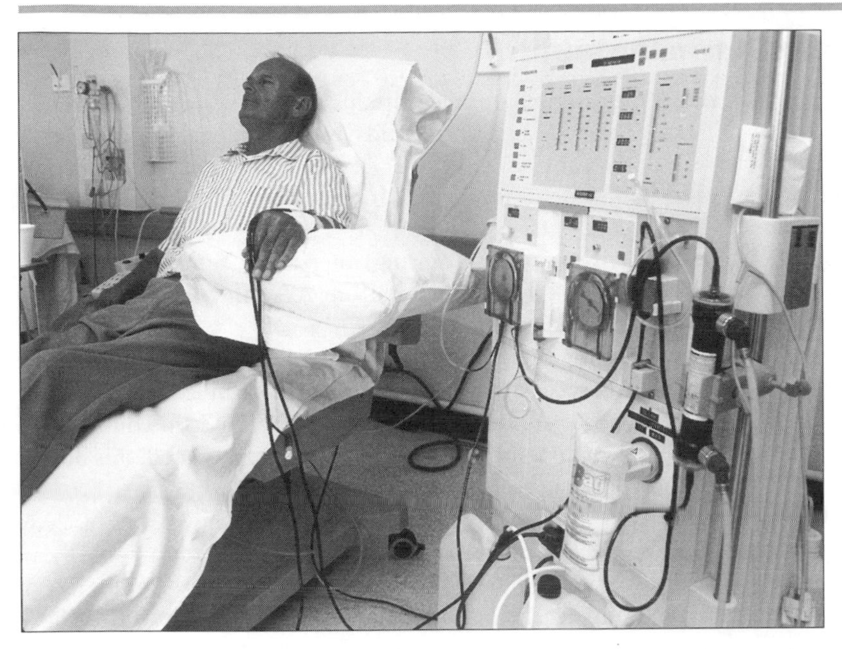

The dialysis machine is used to treat patients suffering kidney failure. Blood passes through a tube, surrounded by a solution of mineral salts, glucose and other components of plasma. The concentrations are controlled so that excess urea and excess salts diffuse out of the tube, but glucose does not.

the smallest proteins may pass through. The filtrate is essentially plasma without its proteins (Table 22.1).

As in any capillary, the blood is under pressure, so fluid is forced out into the Bowman's capsule. Filtration under pressure is called ultrafiltration. The blood vessels leading both to and from the glomerulus are arterioles, and so are capable of constriction. This provides a neat way of controlling filtration rate. If the afferent arteriole dilates and the efferent arteriole constricts, pressure in the glomerulus increases and filtration rate rises. If the afferent arteriole constricts and the efferent arteriole dilates, the major point of resistance to blood flow is now before blood enters the glomerulus, so pressure and hence filtration rate are lower.

Table 22.1 Typical composition of human blood plasma, glomerular filtrate and urine

	Concentration/g 100 cm^{-3})		
Substance	**Plasma**	**Filtrate**	**Urine**
water	90–93	97–99	96–97
proteins	7–9	0	0
glucose	0.1	0.098	0
urea	0.03	0.03	2.0
uric acid	0.002	0.002	0.05
ammonium	0.0001	0.0001	0.05
calcium	0.01	0.01	0.015
potassium	0.02	0.02	0.15
sodium	0.3	0.31	0.35
chloride	0.35	0.37	0.6
phosphate	0.003	0.003	0.12

Reabsorption in the tubules

The filtrate which enters the proximal tubule contains substances which should not be lost in the urine, such as glucose and amino acids. Normally these are completely reabsorbed into the blood (Table 22.1). Other substances are reabsorbed only partially, such as water and mineral ions: if this did not happen, the body would rapidly be drained of water and salts.

Reabsorption may be active (energy-requiring) or passive. Surprisingly, active reabsorption of many molecules requires energy from ATP only indirectly, and is driven by the active reabsorption of sodium ions.

Reabsorption is controlled by the membranes of the epithelial cells around the tubule. Each cell has one face in contact with the fluid in the tubule. This surface is folded into microvilli to increase surface area. The opposite face is in contact with fluid in the spaces between cells. So far as small molecules are concerned, this fluid is continuous with the blood plasma.

The mechanism for active reabsorption of sodium ions is shown in Fig. 22.6. Sodium is pumped out of the epithelial cells in exchange for potassium ions, by an active sodium/potassium antiport (Section 6.4) in the membrane of the outer face of the cell. This membrane is impermeable to sodium, which cannot then diffuse back in. Potassium, however, is able to diffuse out again. There has been a net outflow of positively charged ions, so the cytoplasm becomes negatively charged. It also has a relatively low concentration of sodium ions. The membrane of the inner face of the cells is permeable to sodium, which is able to diffuse down the gradient into the cells. The net effect is a pumping of sodium ions out of the tubule and into the blood. K^+ simply cycles in and out of the cells. Diffusion of sodium back into the tubule through gaps between cells is minimized by tight junctions between the cells (Section 4.7).

The net flow of positively charged ions out of the tubule results in a charge gradient. Negatively charged ions such as Cl^- follow by diffusion. Overall, salts have been reabsorbed.

The sodium gradient which has been set up between cytoplasm and filtrate is used to drive reabsorption of other molecules such as glucose, amino acids and K^+ (Fig. 22.7). A glucose/Na^+ symport (Section 6.4) in the membrane of the microvilli couples glucose transport to the inward diffusion of sodium ions. ATP is not used directly: the Na^+ gradient is used as a source of energy. This has been, in turn, set up by the Na^+/K^+ antiport which does require ATP. Other specific symports carry in other molecules, amino acids for example.

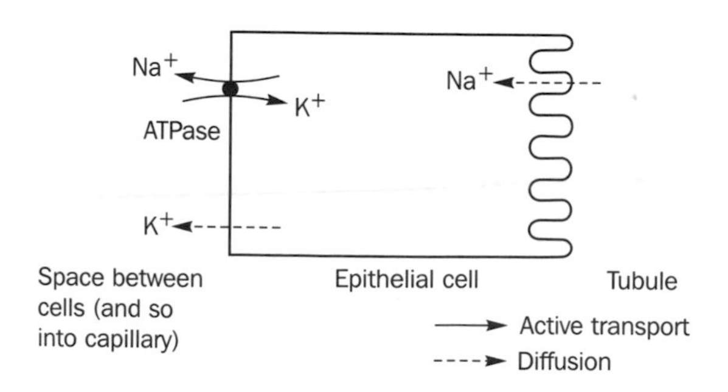

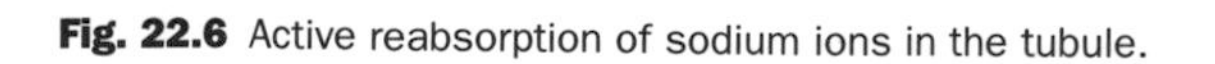

Fig. 22.6 Active reabsorption of sodium ions in the tubule.

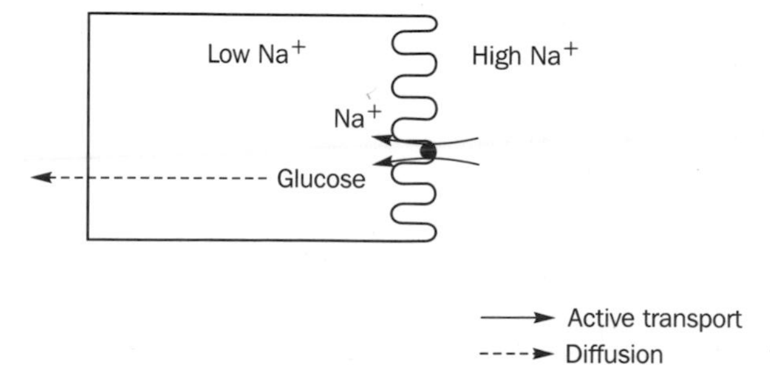

Fig. 22.7 The sodium gradient from tubule to epithelial cell is used to drive reabsorption of glucose and other molecules.

Finally, a great deal of water is reabsorbed in the tubules. The active transport of salts out of the tubule sets up an osmotic gradient: water leaves the tubule. In part, this is across the membranes of the epithelial cells. However, the 'tight' junctions between adjacent cells (Section 4.7) are loose enough to allow water to pass: this is especially true of the proximal tubule, where much water re-enters the bloodstream.

Not all molecules re-enter the blood from the tubules. Nitrogenous waste molecules are poorly reabsorbed. Creatinine is hardly reabsorbed at all, in fact it is actively secreted. Some urea is reabsorbed by diffusion, but not to the same extent as water, so it ends up at a higher concentration in urine than in blood.

Active secretion by the tubule

Some excretory systems, such as the Malpighian tubules of insects (see Fig. 21.2), do not rely on ultrafiltration, but on active secretion of molecules into the urine. This also happens to a small extent in the kidney tubule. Creatinine and urate ions are secreted actively to augment their filtration in the glomerulus. H^+ ions are also secreted as part of a mechanism to regulate blood pH. Active secretion is again driven by the sodium gradient, this time using antiports. As sodium ions move from filtrate to cytoplasm, the molecule to be secreted is carried out across the membrane into the filtrate.

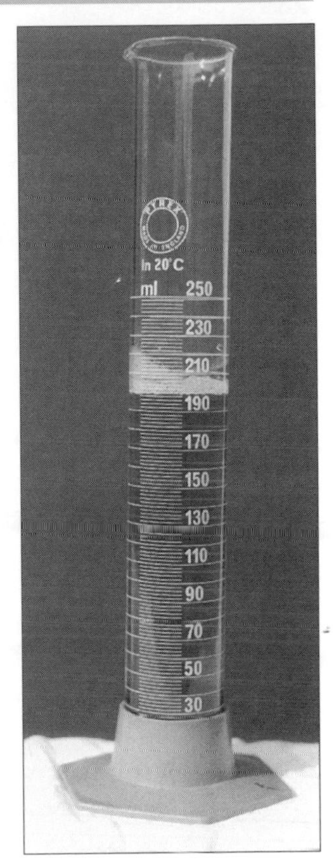

It took me three hours to produce this much urine, yet the kidneys filter this much liquid out of the blood in under three minutes. Most of the filtrate is reabsorbed (see text).

22.5 Osmoregulation by the kidney

At heart, the role of the kidney in osmoregulation is simple. If the blood has too high a salt concentration, a concentrated urine is required: the kidney reabsorbs less salt but more water from the filtrate. If blood concentration falls, a dilute urine is required: more salt and less water are reabsorbed. If blood pressure falls, less urine must be produced overall; if it rises, more urine has to be made. Two distinct but cooperating control systems are involved, one for blood salt concentration, one for blood pressure. First, the mechanism by which blood concentration is adjusted must be discussed.

How dilute urine is made

If urine is to be dilute, plenty of salts but little water must be reabsorbed from the filtrate. A mechanism for active reabsorption of salts has already been described. Water will tend to follow salts by osmosis, as in the proximal tubule. However, when dilute urine is to be made, the walls of the ascending limb of the loop of Henle, the distal tubule and the collecting duct are relatively impermeable to water: salts are reabsorbed but once the filtrate has left the proximal tubule, little water is reabsorbed (Fig. 22.8). The urine is then dilute.

How concentrated urine is made

To make concentrated urine requires large-scale reabsorption of both salt and water. The principle is really simple. Salts are reabsorbed from the filtrate by active

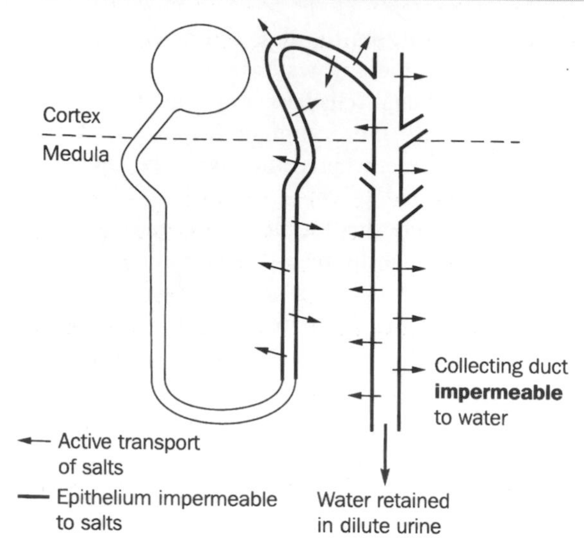

Fig. 22.8 Production of dilute urine.

transport as before. This sets up a salt gradient down through the medulla, low concentration in the outer medulla, very high concentration deep in the inner medulla. Water then leaves the collecting duct by osmosis: the collecting duct becomes more permeable to water when urine is to be concentrated. As the urine moves down the collecting duct, it becomes more concentrated as water leaves, but it constantly encounters an environment with even higher salt concentration, so water continues to leave until highly concentrated urine collects in the pelvis of the kidney (Fig. 22.9).

All this raises some big questions. How is the salt gradient built up? Why does blood flow in the capillaries of the medulla not dissipate the salt gradient? What controls collecting-duct permeability? This last question is particularly important, since

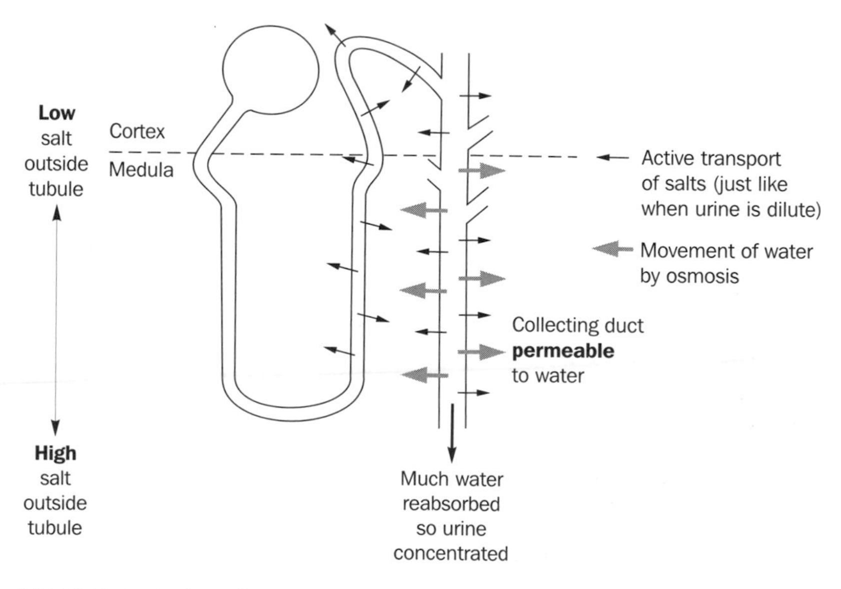

Fig. 22.9 Production of concentrated urine.

it holds the key to understanding the regulation of urine concentration, one of the major functions of the kidney.

Building up the salt gradient

The medulla is full of loops of Henle, mixed with collecting ducts and capillary loops. Filtrate flows down through the medulla in the descending limbs of the loops of Henle, and back up towards the cortex through the ascending limb. The descending and ascending limbs behave quite differently. The wall of the descending limb is permeable to salts. As filtrate moves down, it encounters increasing salt concentrations outside the tubule. Salts enter by diffusion and are carried deeper into the medulla by the flow (Fig. 22.10). The wall of the ascending limb does not allow passive diffusion of salts or water, but actively reabsorbs salts from the tubule. On average, salts that travelled down the descending limb are thrown back out of the loop of Henle at a deeper level than the point at which they diffused in. The effect is to drag salts deeper into the medulla, building up the concentration gradient.

A system of this sort is called a countercurrent multiplier: 'countercurrent' because material is transferred between fluids flowing in opposite directions, 'multiplier' because it is capable of building up a gradient.

Only loops of Henle which dip deeply into the medulla can concentrate salt. Some nephrons, with their glomeruli in the deeper part of the cortex, do just this. They are called juxtamedullary nephrons. Others have their glomeruli higher up in the cortex and have loops of Henle which only just reach the medulla: these are the cortical nephrons. Cortical nephrons cannot concentrate salt and so cannot assist in the formation of concentrated urine. Animals which need to conserve water tend to have mostly juxtamedullary nephrons: this is seen in desert rodents, for example the hopping mouse and kangaroo rat. The beaver, with an abundant supply of drinking water in its environment, can afford very dilute urine, so has only cortical nephrons. About one quarter of human nephrons are juxtamedullary.

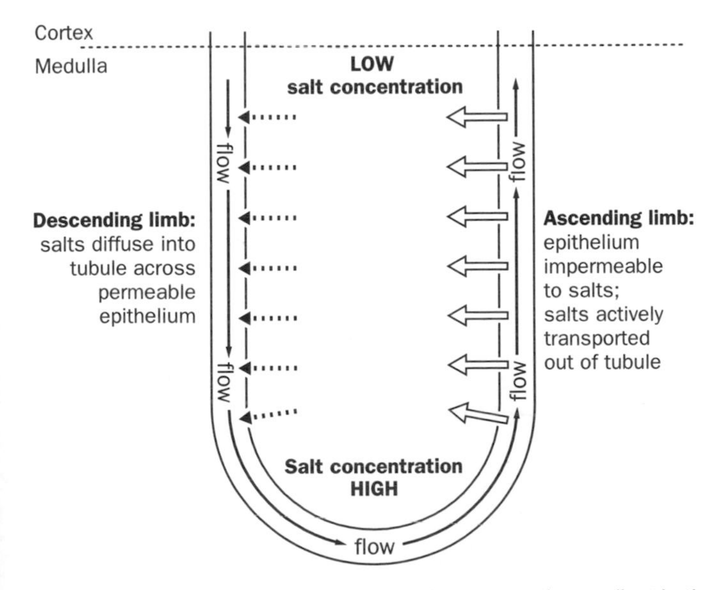

Fig. 22.10 The loop of Henle sets up a salt concentration gradient in the medulla.

Blood flow in the medulla: not a big problem

The cells of the medulla have needs of their own. As with other cells, their needs are met by a blood supply. Capillaries are permeable to salts and water: it might seem that salts would diffuse into the bloodstream and be carried away from the medulla, dissipating the gradient. Two things minimize this: firstly, only small volumes of blood pass through the medulla, at a slow rate. Secondly, the capillaries are loops, the vasae rectae, which dip into the medulla before returning to the cortex. Salts diffuse into the blood as it travels into the medulla, just as in the loop of Henle. However, all parts of the vasa recta are permeable to salts, so they will diffuse back out as the blood flows back up (Fig. 22.11). On average, salts are carried neither further down nor further up. The gradient is neither enhanced nor dissipated. This is another countercurrent system, but not a multiplier: it cannot generate a gradient.

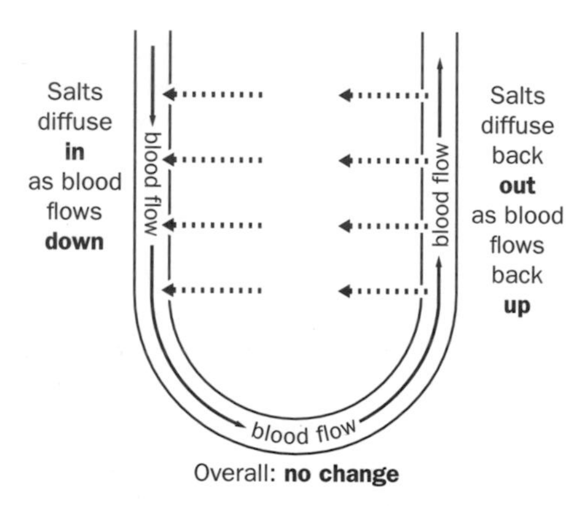

Fig. 22.11 The vasa recta and salt movement.

22.6 Controlling reabsorption

The control of reabsorption of salts and water by the kidney is dominated by two hormones. The first is antidiuretic hormone (ADH), sometimes also called vasopressin. ADH increases the permeability of collecting ducts to water, making urine more concentrated. The second hormone is aldosterone. This increases active reabsorption of sodium ions from the distal tubule and collecting ducts, with chloride following passively. The result of this is to reduce the volume of urine if ADH levels are high enough to allow water to follow the salts by osmosis. If ADH levels are very low, water cannot follow the salt, so the effect of aldosterone is to dilute the urine. Between them, these two hormones regulate urine concentration and hence blood concentration, and play a major role in regulating urine volume.

Antidiuretic hormone

ADH is a short peptide consisting of nine amino acid residues (Fig. 22.12(a)). It is produced by the cell bodies of nerve cells in a small region of the hypothalamus, on the undersurface of the brain. The hormone is carried slowly down the axons of these neurones into the posterior lobe of the pituitary, where it accumulates in the ends of the neurones.

(a)

HOOC —Cys—Tyr—Phe—Gln—Asn—Cys—Pro—Arg—Gly—NH_2

(b)

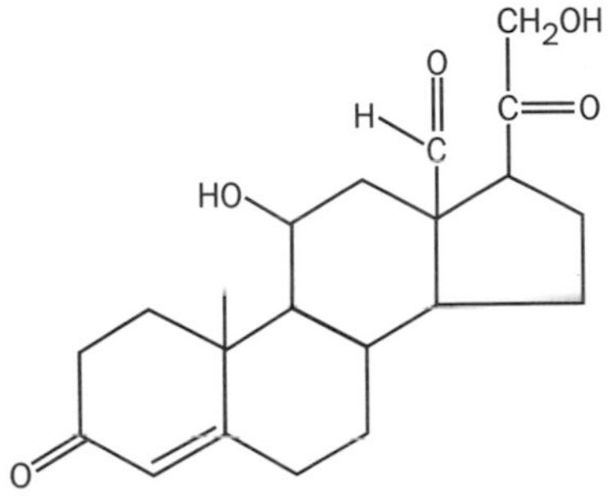

CH_2OH C=O H C O HO O

Fig. 22.12 Hormones involved in osmoregulation. (a) ADH; (b) aldosterone.

Osmoreceptor cells in the brain detect the concentration of the fluid around them, which closely reflects blood concentration. These cells may be in the hypothalamus itself, but there is a possibility that they are found in another nearby part of the brain. Whatever the location of the osmoreceptors, the hypothalamus responds to high blood concentration by sending signals down the neurones through which ADH moved to the pituitary. These signals, on reaching the ends of the neurones, trigger the release of ADH into the blood.

The target cells for ADH are the epithelial cells lining the collecting duct. Receptor molecules in the membrane bind ADH. Binding results in an increase in the level of cyclic AMP, acting as a second messenger in the cytoplasm (Section 16.4). These cells contain vesicles in whose membranes are proteins which act as pores for water. When ADH binds, high cAMP levels make the vesicles bind to the cell membrane which adjoins the inside of the collecting duct. The protein pores become part of this membrane, allowing water to move freely out of the collecting duct by osmosis. Increased water reabsorption tends to decrease blood concentration: this is then a negative feedback system (Box 15.1) which maintains blood concentration around a certain level (Fig. 22.13). The ADH system is very rapid, acting within seconds.

Aldosterone

Aldosterone is a steroid secreted by the adrenal cortex (Fig. 22.12(b)). Two main factors trigger its release. The first is a high concentration of salts, especially K^+, in the blood: this is detected by the adrenal cortex itself. The second factor is low blood pressure, but this is detected in the kidney itself, and control is via the hormone angiotensin, which also affects a wide range of other processes influencing blood pressure.

Target cells for aldosterone are the epithelial cells of the distal tubule and collecting duct. The hormone enters the cell: being a steroid, it is hydrophobic and so able to pass through membranes. It binds to a receptor protein within the cell, and brings about its effect by switching specific genes on or off. As a result, specific proteins are made by the cell: these include the Na^+/K^+ active antiport, and a passive Na^+ channel protein. The effect of these proteins is to increase sodium reabsorption; chloride will follow. Water too may follow, by osmosis, so volume and pressure of fluid in the body will tend to increase. This is another negative feedback system (Fig. 22.14). It is a much slower system than the ADH system, since the response to the hormone involves both translation and transcription. The first effects are seen nearly an hour after aldosterone is released. Box 22.2 examines another effect of aldosterone.

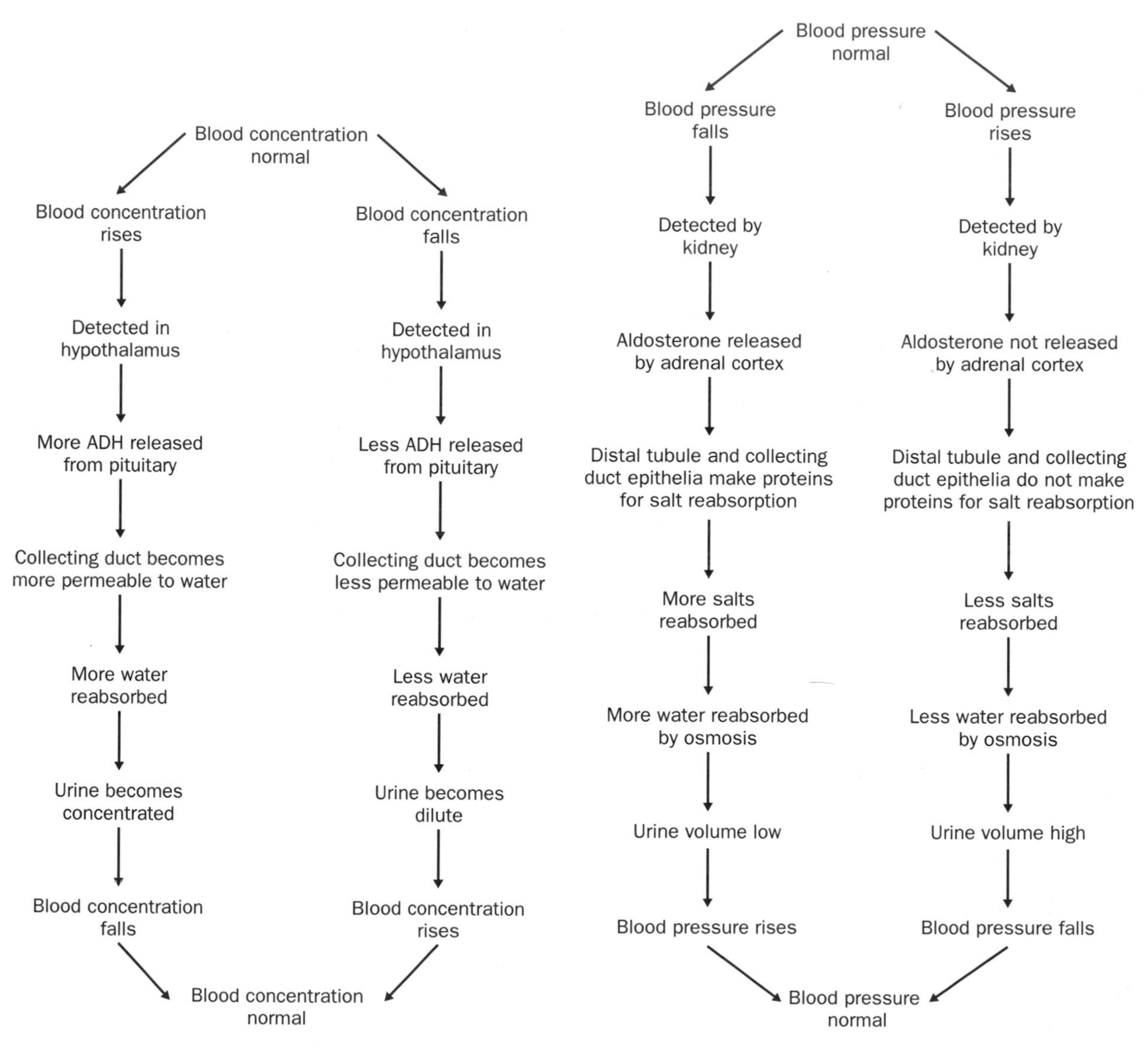

Fig. 22.13 The negative feedback loop involving ADH.

Fig. 22.14 The negative feedback loop involving aldosterone (simplified).

Box 22.2

Aldosterone again

Aldosterone stimulates reabsorption of sodium ions in the tubule, but this is not all it does. A second effect of aldosterone is to increase the secretion of potassium into the tubule by the epithelia of the distal tubule and the collecting duct. At first the active secretion of K^+ may seem odd, since the proximal tubule does the opposite: actively reabsorbing potassium. Cells and organisms generally contain K^+ at much higher levels than do their environments. This accumulation is specific to potassium: the K^+/Na^+ ratio is much higher within most cells than in most environments. Animal cells have effective means of gaining and retaining potassium, notably the ubiquitous K^+/Na^+ antiport. Organisms, too, effectively minimize its escape, for example by active reabsorption in the proximal tubule. Animals have, however, a continuous, if variable, intake of potassium in the diet. If all this was retained, potassium would build up in the body to a dangerous level. In humans, this can lead to cardiac arrest. Active secretion acts as a safety valve for potassium, operated by aldosterone.

Occasionally, the twin roles of aldosterone come into conflict. Prolonged, intense activity under hot conditions results in a great deal of sweating, and hence loss of water and salts. High aldosterone levels lead to retension of water and sodium by the kidney, the cost being excessive potassium loss in the urine. It is important, then, that athletes in endurance events should maintain potassium intake: some proprietary drinks contain K^+ for this reason.

22.7 On the integration of body systems

Excretion and osmoregulation are, in concept, quite different processes. In practice, they are inextricably linked functions of the same organs in the bodies of most animals. Similarly, control systems which can be made, on paper, to look independent, such as the ADH and aldosterone systems, do interact. A low ADH level, needed to increase blood concentration, may lead to a fall in blood pressure, and so to an increased secretion of aldosterone. Neither is aldosterone secretion isolated from other systems: its release is triggered by angiotensin, which has effects on many body processes. It is easy to dissect out particular feedback systems and to describe them in words. A description of the combined operation of many interacting systems is much more difficult. Words become inappropriate: the mathematics of control theory begins to provide a more suitable language.

It is no surprise that body systems and their controls are interdependent. Natural selection acts on the whole phenotype. A mutation which increases the efficiency of nitrogenous excretion whilst wreaking havoc with osmoregulation is unlikely to increase under natural selection. A fit individual is one that leaves many viable offspring. An effective, efficient, well-regulated body may increase fitness, but an efficient excretory system will only increase fitness if it is part of such a body. Bodies are more than collections of systems: they are integrated wholes. This integration has been acheived not through design, but through natural selection.

Summary

- Excretion is the removal of waste products from cells and organisms.
- Animals excrete carbon dioxide and nitrogen-containing waste: ammonia, urea, or uric acid.
- Ammonia is toxic. Many animals convert it to urea or uric acid, which can safely be excreted in concentrated urine.

- Osmoregulation is the regulation of the salt concentration in body fluids. Most animals do this.
- In mammals, the kidneys carry out excretion and osmoregulation.
- The kidney contains many nephrons, which make urine.
- Molecules up to the size of small proteins are filtered out of blood capillaries in the glomerulus, under pressure.
- Molecules and ions which should not be lost in the urine are then reabsorbed in the proximal and distal tubules.
- Salts and water are partially reabsorbed.
- The kidney carries out osmoregulation by varying the amount of water it reabsorbs. The more it reabsorbs, the less is lost in the urine.
- Water leaves the collecting duct by osmosis, due to a salt gradient in the medulla, set up by the loop of Henle.
- The permeability of the collecting-duct wall can be varied. If it is impermeable to water, less water is reabsorbed, making lots of dilute urine. If it is highly permeable, less urine is made.
- Antidiuretic hormone (ADH) is released by the posterior pituitary in response to high salt concentration in the blood.
- Antidiuretic hormone is a signal for the epithelial cells of the collecting duct to become more permeable, retaining water in the body.
- The hormone aldosterone increases salt reabsorption.

Exercises

22.1 The frog is an amphibian, which lives partly on land and partly in fresh water. When in the water: (a) it produces a dilute urine, and (b) it actively absorbs salt through the skin. Explain these observations.

22.2. Insects actively excrete waste products through the Malpighian tubules. Mammals filter out all sorts of solutes from blood in an unspecific way, actively reabsorbing only those which the body needs to retain. What are the costs and benefits of the mammalian system?

22.3. Complete this table to show which components of blood are filtered out, and which are wholly or partly reabsorbed by the mammalian nephron.

Component of blood	Filtered out?	Reabsorbed?
Amino acids		
Creatinine		
Glucose		
Platelets		
Proteins		
Red blood cells		
Salts		
Small, soluble toxic molecules		
Urea		
Uric acid		
Water		
White blood cells		

22.4. Sea water has a higher salt concentration than the most concentrated urine a person can produce. A round-the-world yachtsman runs out of drinking water. He is tempted to drink sea water. Would this be helpful, harmful, or simply pointless? Explain.

22.5. What symptoms will be caused by the following problems?

(i) In a patient with head injuries, the posterior pituitary stops secreting ADH.

(ii) Epithelial cells lining the proximal and distal tubules are damaged by a toxin.

(iii) The adrenal cortex secretes too much aldosterone. (Think about the effects on the urine and the blood.)

Further reading

Guyton, A.C. and Hall, J.E. *Textbook of Medical Physiology* (9th ed.) (Philadelphia: Saunders, 1996). I like this hefty textbook of human physiology for its thoroughness and clear, no-frills diagrams.

Guyton, A.C. *Human Physiology and Mechanisms of Disease* (5th ed.) (Philadelphia: Saunders, 1992). Effectively a shorter, boiled-down version.

Schmidt-Nielsen, K. *Animal Physiology: Adaptation and Environment* (4th ed.) (Cambridge: Cambridge University Press, 1990). One of the all-time great textbooks! Exciting, authoritative and easy to read, it puts animal physiology in the context of the environment.

Withers, P.C. *Comparative Animal Physiology* (Fort Worth: Saunders, 1992). More detailed than Schmidt-Nielsen, with wider ranging examples: the inevitable cost is that it is a less easy read.

Defence

Contents

Connections

► Before studying this unit, you should have a basic understanding of biological molecules (Unit 3), cell structure (Unit 4), genetics (Units 7 and 12) and cell communication (Units 16 and 17). Much of the defence system of mammals relies on blood (Unit 19). When defences fail, disease may occur. Unit 30 deals with disease and its causes.

23.1 Threats

Organisms make good food. They are made up of cells, and all cells are made of the same broad types of molecule. An organism wanting a well-balanced supply of molecules for building its own cells would do well to feed on other organisms. Animals, fungi and many bacteria do just that. This is why organisms need to defend themselves.

Plants are threatened by herbivores wanting to eat them, aphids trying to feed on their phloem sap, viruses which parasitize their cells, and fungi aiming to grow through their tissues, digesting them as they go. Animals are at risk from predators, as well as from viruses, bacteria, protoctists, fungi and animals parasitizing the body, many causing disease as they do so. Even bacteria are threatened by viruses, hungry eukaryotes and the chemical weapons of competing microbes.

A well-defended castle will have several lines of defence. 'Keep Out' signs deter casual visitors. A moat, thick walls and a good lock make it hard to force an entry. If invaders do break in, they are likely to be met by resistance: swords, boiling oil, a fire-breathing dragon perhaps. Some defenders will seal off the area, preventing the invaders running through the entire building. Others will try to repair the breached defences. What if some spies slip through the net? The well-prepared castle owner will have an internal security service, patrolling the castle and checking the credentials of everyone they meet. If anyone is suspect, they become dragon fodder.

Organisms are no less well defended. Our imaginary Mediaeval baron might be motivated by wealth, power, or damsels in distress. For organisms, food and survival itself are at stake. All these levels of defence are seen in organisms, and will be considered in turn. The role of behaviour in defence is dealt with in Unit 29.

23.2 Prevention is better than cure

Organisms make themselves difficult to eat. Hedgehogs are covered in strong, sharp spines derived from the skin. It is not easy to eat a hedgehog, although badgers seem to manage. Many grasses cover their leaves with microscopic spikes of silica. This appears to deter grazing by slugs and snails, a major threat to grasses in some habitats. Given the choice, the molluscs eat species with a less well-developed silica shield.

Poisons and nasty tastes are also used. Alkaloids are a diverse group of toxic organic molecules made by a wide range of plants, such as yew and deadly nightshade. Each species makes just one or a few types. Some clover plants have a neat way of making cyanide when bitten. One compartment of the cell contains a cyanogenic glycoside: a cyanide group bonded to a sugar. Another compartment contains an enzyme which releases the toxic cyanide. The two meet only when the cell is damaged. If they were not kept separate, the cyanide would kill the cell. Some animals use a similar strategy. Most species of ladybird produce alkaloids which smell and taste unpleasant; some may be toxic. When threatened, many ladybirds release small amounts of blood at the joints. Some predators are deterred by the alkaloids in the blood.

Poisons and foul tastes benefit the individual making them, and its relatives, on the principle of 'once bitten, twice shy'. This works best when the attacker can detect the poison before doing too much damage to its victim. It makes sense to advertize a bad taste: the castle's 'Keep Out' signs. The black and red patterns of many ladybirds seem to be warning signs. Some predators are put off just by seeing them. Presumably, they have learnt to associate the colour with the taste. The benefit will be greatest if other unpleasant species have similar warnings: there is more chance of a predator learning the meaning. This is called **Mullerian mimicry**. It could explain why several toxic ladybird species have very similar patterns. In the same way, many stinging species of bee and wasp have yellow and black stripes. Systems like this may

Mullerian mimicry: A group of harmful species look similar.

These guys will find it tough if they get hungry. The xerophytic plants all around them have small leaves, often strengthened by lignified cells, protected by spines or full of foul tasting chemicals.

break down through 'cheating'. A mutant which has the warning colours, but not the toxin, will have the benefit of protection, without the cost of making it. It will become common, unless toxicity itself gives the others extra benefit. Harmless species may cheat by having warning colours like a well-defended species. This is called **Batesian mimicry**. Some beetles and bugs mimic the ladybird pattern; the stingless hoverflies mimic bees and wasps.

If an attacker is determined, physical barriers are needed to keep it out. The epidermis of a leaf or green stem forms a continuous barrier. It is not easy for a germinating fungal spore to break through this smooth surface. However, the stomata (Section 18.3) are a point of weakness. Bark is a tough, protective layer of dead cells around a woody stem. It forms a barrier to fungi and to animals which might eat the sugar-rich phloem inside.

Cell walls are far less selective than membranes. Viruses and poisons could be carried through the plant in the continuous network of cell walls, before entering any cell. A barrier in the roots prevents this nightmare. The endodermis is a ring of cells around the central vascular tissue (Section 18.5). Its walls are impregnated with a waterproof strip of suberin. Nothing can cross from the outer part of the root to the inner part without crossing a membrane.

An epithelium is a layer of cells covering a body surface in an animal. It forms a barrier to chemicals and microbes. Epithelial cells are held close together by tight junctions (Section 4.7) which seal the gaps between them. Mammalian skin is a modified epithelium. The cornified layer of dying, drying cells lies above an actively dividing layer. As the dying cells lose water and shrivel, the keratin proteins, which strengthen them, remain (Section 4.5: The cytoskeleton). Cholesterol is deposited in the skin, waterproofing it. So the skin is a tough, flexible, impermeable body covering which is constantly being regenerated. It even becomes thicker where physical wear is greatest: look at the hands of anyone who works with a spade or pickaxe.

On less exposed surfaces, epithelia do not have a cornified layer. Various types of epithelium line the gut from mouth to anus, the tubes leading to the lungs, the urethra, bladder and ureters, the vagina, and so on. All give a first line of defence against invaders. An interesting observation shows the importance of epithelia. Infections of the abdominal cavity (the space filled by the internal organs) are more common in women than men. This is partly due to the fact that there is a route from the outside world to the abdominal cavity in women. A foreign cell could pass through the vagina, uterus, oviduct and into the abdomen without crossing an epithelium.

These defences are useful, but not perfect. What happens when they are breached?

23.3 Seal the gap

If the surface of an organism is damaged, invaders can get in. Cuts and holes must be plugged to prevent this. It is particularly important in animals, because blood would also leak out. Both plants and animals seal up injuries, but the process is especially quick in animals.

Animals

Batesian mimicry: Harmless species look like a harmful one.

Vertebrate blood and that of some invertebrates can clot to plug leaks in blood vessels. We know most about clotting in mammals, humans in particular. Everything that follows refers to humans, but the process is similar in other mammals and throughout the vertebrates.

Everything needed for clotting is in the blood. All that remains is something to set the process off. It has all the benefits and dangers of a loaded gun: ready for action at any moment, but devastating if the trigger is pulled at the wrong time. If clots were formed when there was no injury, healthy blood vessels could be blocked.

Platelets and several plasma proteins are involved in clotting. Platelets are cell fragments, made when bone marrow cells called megakaryocytes break up. They are discs, 2–4 μm across, containing most organelles except nuclei. They have a role in detecting injury, and can clump together. Fibrinogen is a soluble, globular protein which can be converted to fibrin, an insoluble, fibrous form. This is at the heart of a clot. Other plasma proteins help activate the system.

Clot formation has several stages (Fig. 23.1). Almost immediately, the wall of the injured vessel contracts. This reduces blood flow. Less blood is lost, but this also means that the clot is less likely to be washed away as it forms. Molecules released from damaged cells trigger muscle contraction. A clean cut, with fewer damaged cells, tends to bleed more than a crushed tissue.

Next, platelets begin to accumulate at the wound. They bind to collagen fibres in the damaged blood-vessel wall, and to damaged cells of the blood-vessel lining: the endothelium. They swell and become sticky. Several signalling molecules are released. One triggers constriction in small blood vessels. Others activate nearby platelets, making them stick together to form a plug. The platelet plug is enough to seal small breaks in capillaries.

Within a minute or so, the fibrin clot is developing. Soluble fibrinogen is converted to fibrin fibres, which later become cross-linked. Platelets and blood cells are trapped in the mesh, blocking it. As it develops, plasma becomes trapped too. In a very few minutes the injury is completely sealed. What converts fibrinogen to fibrin? Ultimately, the answer is molecules released by damaged cells and activated platelets. However, this is by way of a cascade of proteins called clotting factors, each one activating the next. This allows amplification. A small release of signalling molecules leads to a massive conversion of fibrinogen. Box 23.1 shows the mechanism in more detail.

Platelets are attached to the fibrin fibres. Within an hour, the platelets contract, pulling the fibres together. The clot contracts, squeezing out most of the fluid. The broken blood-vessel walls are pulled closer together.

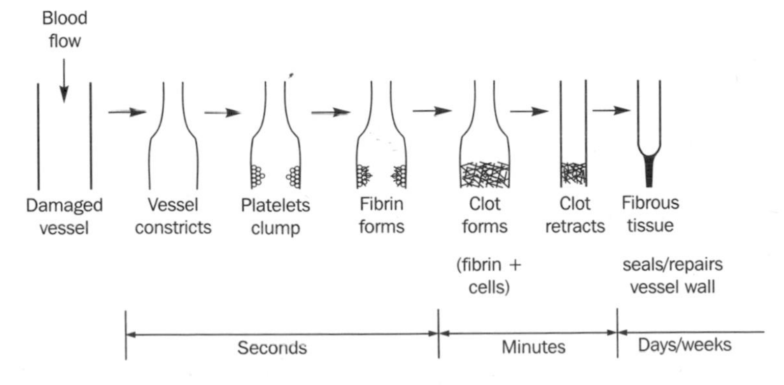

Fig. 23.1 Clot formation in a damaged blood vessel.

Finally, over days and weeks, fibrous tissue grows into the wound. This makes a permanent seal.

Box 23.1

What triggers a fibrin clot?

Soluble fibrinogen becomes insoluble fibrin at the site of an injury. Fibrin threads are the basis of the clot. They trap blood cells, making a barrier. What converts fibrinogen to fibrin? What stops this happening all the time?

Plasma contains an enzyme, thrombin. It converts fibrinogen to fibrin. Thrombin itself is activated by a cascade of proteins, each one activating the next. They are called clotting factors: most have a Roman number, not a name. Most work by cutting a larger, inactive protein into pieces, one of which is active. They are proteases, not protein kinases.

The whole cascade is set off by phospholipids from damaged cells, phospholipids from activated platelets, and contact with collagen in the walls of damaged blood vessels.

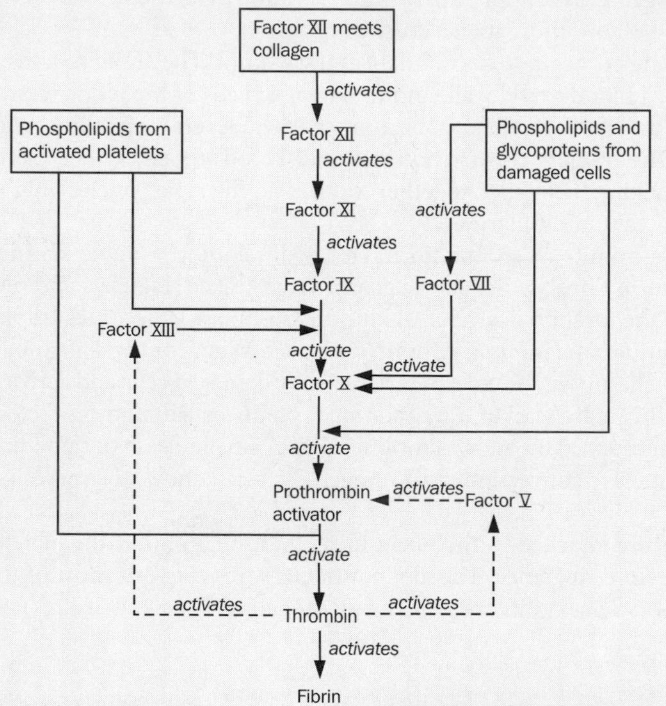

Haemophilia

Haemophilia is an inherited disease in which the blood does not clot. It is caused by deficiency of factor VIII (usually) or factor IX (more rarely). Notice that thrombin itself activates two earlier steps in the cascade: once clotting starts, it keeps on going. However, thrombin is adsorbed onto fibrin, keeping it in the area of injury. This prevents the clotting process spreading away from the wound.

What happens to old clots?

Old clots are broken down over a few days. This is especially important where the clot completely blocks a small blood vessel. Plasma contains a protein called plasminogen. It gets trapped in the clot, and is later activated to form plasmin. Plasmin is an enzyme which digests many proteins, including fibrin, fibrinogen and clotting factors. It is activated by tissue plasminogen activator (tPA), a protein released by injured cells. tPA is released very slowly, so the clot does not begin to break down for a day or so.

Plants

Most plants seal up injuries in long-lived organs like stems and roots by growth. A callus of undifferentiated cells grows across the wound. This is the swelling you see at the end of a cutting just before roots start to grow. (Never seen this? Try it!) Some plants have ducts filled with resin or latex which dries in contact with air, sealing injuries quickly. Globules of resin can be seen on pine branches; rubber is made from the latex of the rubber tree. Resins and latex often have a second role. They may contain toxic or foul-tasting chemicals, which deter herbivores as well as sealing up the holes they make. The poisonous milky latex of spurges (*Euphorbia* species) is a good example.

23.4 Confine the enemy

If invaders get into an organism they must be kept in one area. Once they start to spread out from the site of infection, the whole body is threatened.

Mammals

Mammals limit the spread of bacteria and toxins by inflammation. It happens whenever cells are damaged: by infection, burning, harmful chemicals or physical injury. A range of signalling molecules is released from damaged cells, notably histamine and some prostaglandins (Section 16.5). These bring about several changes. Firstly, blood vessels in the area dilate, increasing blood flow. Secondly, the capillaries become more leaky, allowing far more plasma proteins than normal into the tissue fluid, including fibrinogen. Thirdly, the fibrinogen leads to clotting in the fluid around damaged cells. The clots minimize movement of tissue fluid. This makes it very hard for toxins, bacteria or viruses to spread. The extra fluid also makes the area swell up. This is seen whenever a nettle stings, or the cat scratches you. The chemical signals also mobilize the immune system, but this is dealt with in Section 23.5.

Plants

Plants react to invasion in a rather similar way. When a fungal spore germinates on a leaf and starts to attack cells, the cells appear to over-react. A zone of cells about 1 mm across, centred on the site of infection, rapidly dies. This is called the hypersensitive reaction. The first event seems to be a burst of oxidation reactions in the plant cells. Hydrogen peroxide, superoxide and hydroxyl radicals are produced. They damage the plant cells. More importantly, they probably damage the fungus. They also cause cross-linking of proteins in the plant cell walls. This makes the wall less digestible, slowing the fungus's growth. Notice that this is very like the mechanism which prevents more than one sperm entering a mammalian ovum (Section 13.5). The overall effect is a disc of dead plant cells and dead fungus. Go outside now and look for these tiny brown or black spots on leaves. I bet you will find some in the first minute.

23.5 Identify and destroy

Mammals

Mammals have a powerful system for identifying foreign molecules and cells, then making them harmless. It is the **immune system**. White blood cells and various plasma proteins are involved.

White blood cells come in several types (Fig. 23.2). We can crudely divide most of them into two groups, lymphocytes and phagocytes, according to what they do. Lymphocytes are mainly concerned with identifying molecules or cells as foreign. Phagocytes are concerned with destroying invaders.

Phagocytes take up foreign cells or particles by phagocytosis (Section 6.6). They are then digested, if possible. Most phagocytes in the bloodstream are called neutrophils. They circulate, and are 'called in' to injured tissues as inflammation develops. The other phagocytes are called macrophages. They live in all sorts of tissues. Macrophages start life as a class of white blood cell, the monocytes. Once they enter a tissue, they tend to stay put and wait for infection.

An **antigen** is any foreign molecule that the immune system can detect and respond to. Most antigens are proteins, polysaccharides or glycoproteins. They could be molecules on the surface of a virus or bacterium, for example. Smaller molecules can usually only be detected if they first bind to one of the body's own molecules.

Immune system: A system which recognizes foreign molecules and cells, and makes them harmless.

Antigen: Any foreign molecule that the immune system can detect and respond to.

Antibodies: Proteins which bind to antigens, made by B lymphocytes and released to the bloodstream.

T-cell antigen receptors: Proteins which bind to antigens, made by T lymphocytes and attached to their plasma membranes.

Lymphocytes make proteins which bind to antigens. Two types of lymphocyte work in slightly different ways. B lymphocytes (B-cells) make antibodies (Fig. 23.3). These proteins are released into the bloodstream. **Antibodies** can bind to antigens throughout the body. T lymphocytes (T-cells) make rather similar molecules which stay attached to their plasma membranes. These proteins are called **T-cell antigen receptors** (TCRs). They are not released, so the T-cell has physically to meet the antigen.

There are three questions to be tackled. First, how can antibodies and TCRs recognize such a wide range of antigens? Second, how does this lead to the foreign material being made harmless? Third, why does the immune system ignore the

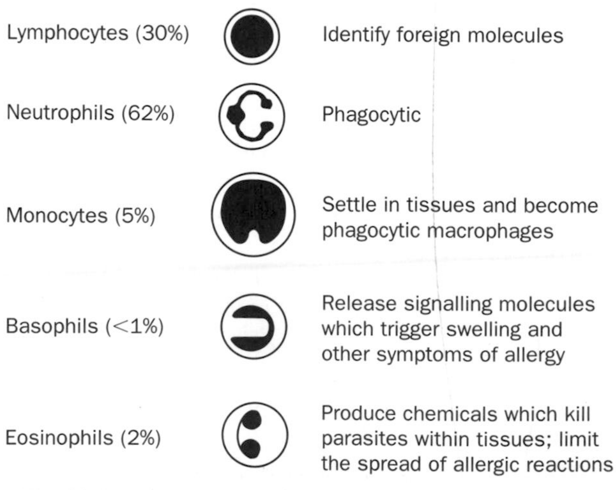
Lymphocytes (30%)
Identify foreign molecules
Neutrophils (62%)
Phagocytic
Monocytes (5%)
Settle in tissues and become phagocytic macrophages
Basophils (<1%)
Release signalling molecules which trigger swelling and other symptoms of allergy
Eosinophils (2%)
Produce chemicals which kill parasites within tissues; limit the spread of allergic reactions

Fig. 23.2 Types of white blood cell.

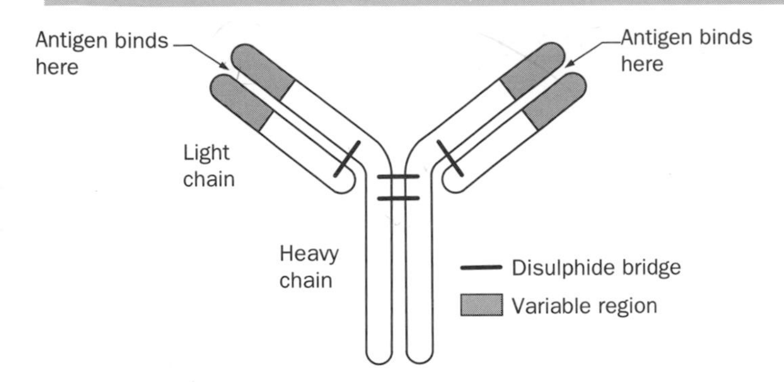

Fig. 23.3 Immunoglobulin G (IgG): the commonest type of antibody.

body's own molecules? In other words, how does it know that an antigen is foreign? These questions will be tackled in turn.

Recognizing the antigen

Antibodies are not all the same. They are all versions of the variable protein immunoglobulin. The most common type is made up of four polypeptide chains (Fig. 23.3). One end of each polypeptide chain has a variable region. This is the part which binds to antigens. A different amino acid sequence leads to a different shape and pattern of charges. This in turn means that a different antigen will, by chance, fit the pattern and bind. Each B-cell only makes one type of antibody, which binds one type of antigen. If the B-cell divides, the daughter cells make the same type of antibody. Each type of antibody was first made in the foetus. Little bits of gene coding for the variable region were randomly mixed to give a unique sequence (Box 23.2).

Box 23.2

Random generation of antibody diversity

Each clone of B-cells makes a unique type of antibody. This is the result of a genetic lottery as B-cells develop in the foetus.

In an antibody, both the heavy and the light chains have a variable region at one end (Fig. 23.4). Each B-cell has genes which, when transcribed and translated, make only one sort of antibody. These genes were assembled during development.

Each heavy chain is made up of four regions. The variable section is made up of the V region, the D region and the J region. The rest of the chain is the C (constant) region. The gene is assembled from four segments, one coding for each region. There is an awesome menu of segments available: hundreds of V segments, at least 12 D segments and four J segments. In most cells, all the segments of one type are grouped together. This does not matter, since the genes will not be transcribed. The segments are separated by short introns (non-coding sequences). As B-cells develop, recombination takes place in a highly ordered way. DNA is broken and rejoined so that one randomly selected segment of each type comes to lie together. These make up the active gene.

It is all a bit like the boy-racers* of my youth. They all drove Capris, more or less the same cars, but each had to look unique. There are only so many things you can do to a Capri. You can select one of several sorts of go-faster stripes, two names for the sunstrip (Clive and Carole? Sid and Nancy?), a colour for your furry dice, and so on. These give thousands of combinations. As the car is driven, knocks and scrapes accumulate at random: motoring mutations. Individuality within a closely defined framework is the key to understanding antibody structure, as well as boy-racers' cars.

*Non-Brits may not know that boy-racers are young men who enjoy driving fast and noisily; the Capri is a defunct model of car, which still went fast when old and cheap.

(Box 23.2 continues over page)

Box 23.2 continued

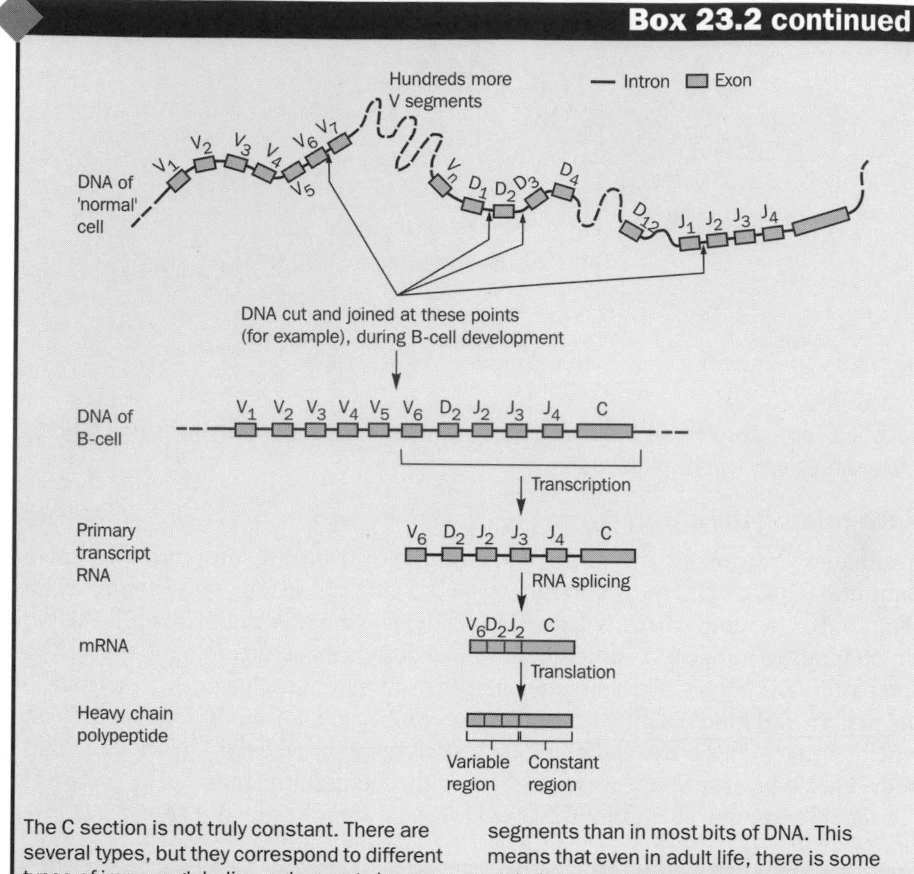

The C section is not truly constant. There are several types, but they correspond to different types of immunoglobulin, and are not chosen randomly. For example, antibodies released to the plasma have a different C segment to those anchored in the B-cell membrane.

Light chains are generated in a similar way, from different pools of V and J segments, and different C segments. There are two types of light chain, κ (kappa) and λ (lambda). Nobody really knows why.

Surprisingly, mutations happen much more frequently in the variable region gene segments than in most bits of DNA. This means that even in adult life, there is some variation in the antibodies that one clone of B-cells makes. When the body mounts an immune response, a few clones will divide enormously, and some of these cells will make subtly different antibodies. Some variants will bind to the antigen even more strongly. The B-cells that made them will be stimulated even more, and will divide more quickly than the rest. This means that better antibodies will be made as time goes on: a sort of natural selection.

Each B-cell which survives to birth will make a different antibody: there are millions of possible types. The body does not 'know' what each will bind to, and natural selection has not produced specific antibodies for particular antigens. The system relies on there being so many types of antibody that at least one will bind to any given antigen. What is more, it works. The immune system is equally able to recognize newly evolved antigens and ancient ones.

Most B-cells spend most of their time in the lymph nodes (Section 19.7) and do not release antibodies. Instead, they keep their antibodies in the plasma membrane. The antigen-binding sites stick out into the surrounding fluid. When an antigen appears in the body, a few B-cells will detect it, because it binds to their own anti-

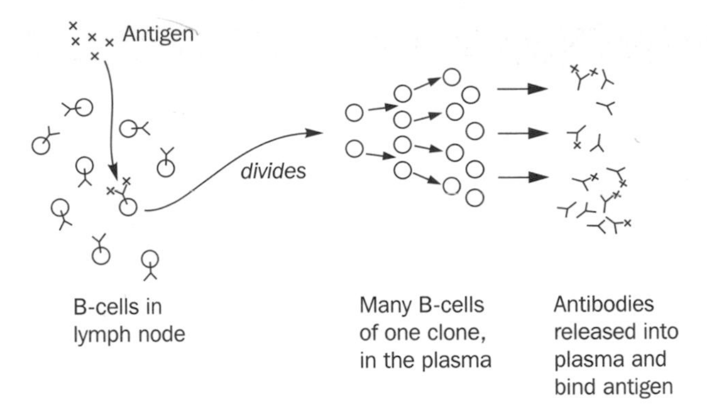

Fig. 23.4 The B-cell response to antigens.

bodies (Fig. 23.4). These B-cells will start to divide. A huge population of clones is made, all producing the same antibodies. These clones move into the bloodstream, and release antibodies into the plasma.

I have just caught a cold. I am feeling ill; the viruses are spreading almost unchecked. Already, B-cells will have detected antigens in the viral coat. Several B-cell clones are even now dividing quickly. In a day or two they will be making enough antibodies to bring the infection under control. If I meet the same type of cold virus again in a few months time, my body will be better prepared. The plasma cells produced this time will all have died, but memory cells derived from the same clones will survive. These can respond very quickly, and I will probably not suffer cold symptoms. This is the basis of vaccines (Box 23.3). If I get a cold again, I can be fairly sure that it is caused by a different strain of the virus, with slightly different antigens in its coat.

T-cell antigen receptors are also diverse, for the same reason. However, they cannot detect antigens by themselves. Other cells must present the antigen to them. A group of proteins called major histocompatibility complex (MHC) bind antigens. They are found in other types of cell. TCRs can only bind antigens which are already bound to MHC (Fig. 23.5). Macrophages are particularly important in presenting antigens, but almost any cell can do it. This is the only way the immune system can detect antigens produced *inside* infected cells. As well as binding antigens in a different way, T-cells also respond differently to antigens, as we shall see.

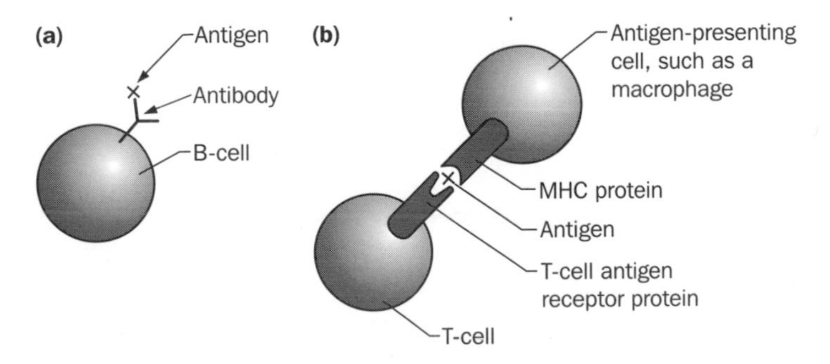

Fig. 23.5 How lymphocytes detect antigens. (a) B-cell; (b) T-cell.

Box 23.3

Vaccines

Memory cells ensure that the body responds quickly when it meets an antigen for the second time. This is why you do not normally get measles twice. Vaccines rely on this. A vaccine contains antigens from a disease-causing organism. It may be dead bacteria, viral coat proteins, a genetically engineered bacterium which does not produce toxins, or an incompetent strain of virus which cannot multiply in the body. The body does not suffer symptoms, but still mounts an immune response. The memory cells remain, ready for when the real pathogen appears.

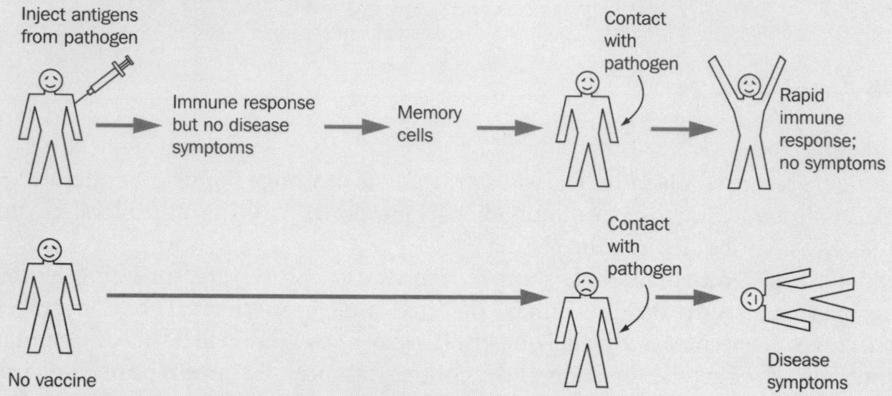

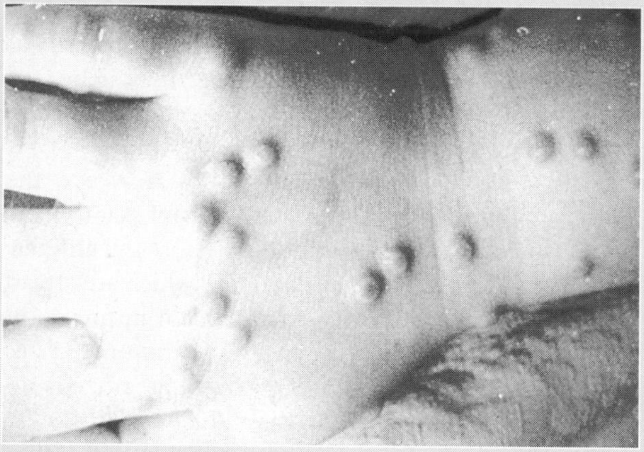

Inject antigens from pathogen → Immune response but no disease symptoms → Memory cells → Contact with pathogen → Rapid immune response; no symptoms

No vaccine → Contact with pathogen → Disease symptoms

Smallpox, a viral disease which was once a major cause of death in many parts of the world, has been eliminated by vaccination.

The theory is simple. Producing a vaccine is rarely so easy. Big problems include pathogens which have many strains with different antigens; cases where pathogen genes mutate rapidly; and pathogens which somehow hide from the immune system.

Rendering the antigen harmless

Antibodies bind to antigens. How does this help? Sometimes, the act of binding itself is enough to limit the damage. Some toxins made by bacteria may be inactivated when an antibody binds, because a binding site is covered up. Other toxins make an insoluble complex with the antibody, which cannot spread through the body. Invading cells may also be clumped together, because each antibody can bind antigens at two different sites. These effects are quite minor. More importantly, a set of plasma

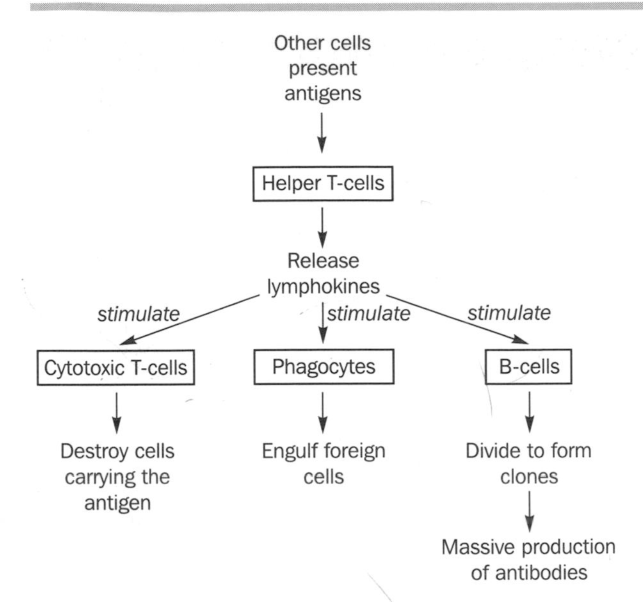

Fig. 23.6 Helper T-cells coordinate the immune response.

proteins is activated by antibodies bound to antigens. These are called the complement proteins. Each activates the next, in a cascade. Some are signals to phagocytes: 'engulf this'. Others break open the plasma membranes of invading cells.

What happens when TCRs bind to antigens? There are two distinct types of T-cell, helper T cells and cytotoxic T-cells. Helper T-cells coordinate the entire immune response (Fig. 23.6). When a helper cell detects an antigen, it releases a group of proteins called **lymphokines**. These are signals which activate the other parts of the immune system. Some stimulate the cytotoxic T-cells.

Cytotoxic T-cells are killers. They bind to antigens on the surface of a cell, and secrete proteins which destroy the cell. Some of the proteins form large pores in the plasma membrane which let toxic proteins in. Then the T-cell moves on to another victim.

Other lymphokines stimulate phagocytes. Yet others stimulate appropriate B-cells to divide.

Sorting friends from enemies

Why does the immune system ignore the body's own molecules? The body is full of proteins and polysaccharides which ought to act as antigens. Antibodies and TCRs are generated at random, so there is no way that the body could avoid making ones which bind to 'self' molecules. This is a disaster waiting to happen. If the immune system recognizes the body's cells as foreign, the organism will destroy itself.

The problem is avoided in a simple way. Before birth, the foetus's immune system works differently. If an antibody binds to an antigen, the B-cell which made the antibody is destroyed. If a TCR binds to an antigen, the T-cell is destroyed. This weeds out all the lymphocyte clones which could recognize self antigens. The system only works because the mother's immune system and the placenta shield the foetus from infection. The baby's immune system only starts to operate normally after birth.

Individuals may have subtly different versions of cell-surface glycoproteins, coded for by different alleles (Section 12.2). This means that the immune system

Lymphokines: Signalling molecules made by white blood cells.

Box 23.4

Blood groups

The body does not produce antibodies which bind its own proteins. However, transplants and blood transfusions introduce other people's proteins. If the protein is different, coded for by a different allele, antibodies may recognize it.

Human blood is classified into blood groups, according to which versions of particular proteins it has. Most important is the ABO blood group system. Here, the protein which acts as an antigen is a plasma membrane glycoprotein in the red blood cells. The gene coding for it has three alleles (Box 12.3). Two alleles code for distinct versions of the protein. I^A codes for antigen A; I^B codes for antigen B. The third (I^o) does not produce the protein at all. People with only antigen A produce antibodies to antigen B, and vice versa. Those with both antigens do not make either antibody. People with neither antigen produce both antibodies.

Fred (group A) is transfused with Freda's (group B) blood. Fred has antibodies to the B antigens. They stick to the antigens, binding red cells together. This can block the smaller blood vessels: bad news for Fred. Giving Fred's blood to Freda is equally harmful (why?). Fred, like anyone else, can be transfused with group O blood, because it contains neither antigen. Who can be transfused with AB blood? What groups can be given to an AB patient?

Other proteins may cause similar reactions, notably the Rhesus D factor. 'Rhesus positive' (Rh+) people have a form of the protein which acts as an antigen. Rhesus negative (Rh−) people have a form of the protein which is only weakly antigenic. Rh− blood should not cause problems in Rh+ patients. However, a Rh− patient can produce antibodies against Rh+ blood.

Blood group	Possible genotypes		Antigens	Antibodies
A	I^AI^A	I^AI^o	A only	anti-B
B	I^BI^B	I^BI^o	B only	anti-A
AB	I^AI^B		A and B	neither
O	I^oI^o		neither	both

may recognize cells from another member of the same species as foreign. This is why transplanted organs are sometimes 'rejected'. It is also the basis of blood groups (Box 23.4).

There is no doubt that the immune system can detect non-self molecules. It is usually assumed that this is what triggers the immune response. However, this has been challenged recently. Could it be that the immune system is first alerted by tissue damage, a sure sign of danger? Only then does the mechanism for detecting non-self molecules target the immune response on the right things. This would explain several observations which are hard to explain in other ways. For example, the immune system does not normally attack the embryo. An embryo is made up of non-self cells, but it does not cause messy cell death, as infection does. Any cells which die as the placenta is formed (Section 13.7) die tidily, by programmed cell death (Section 25.5). Also, most vaccines are effective only if they include irritant chemicals as well as antigens. Injecting an antigen alone has little effect. As yet there is little evidence to support or reject this idea, but it is attracting great interest.

The immune system of mammals is awesome in its effectiveness and complexity. Other vertebrates have immune systems organized on broadly similar lines. No invertebrates appear to make antibodies. Most rely on phagocytes to destroy invaders, and are able to tell self and foreign cells apart. How they do this is less clear.

Plants

The beauty of the mammalian immune system is that it can detect *anything* except self. New antigens will probably have an antibody waiting for them, just by chance. Flowering plants can also detect foreign cells, but in a less specific way.

The plasma membranes of plant cells have various types of receptor which can bind to cell-surface molecules of invaders. Most recognize short sections of polysaccharides, chitin or proteins in fungal cell walls. Others bind molecules in bacterial walls. Somehow, the receptor binding to a foreign molecule triggers one or more responses. These include the hypersensitive reaction (Section 23.4); secretion of lignin and easily cross-linked wall proteins; and various chemical weapons. Some chemical weapons are phytoalexins. The phytoalexins are a ragbag of molecules, all harmful to fungi or bacteria. They are produced in response to infection. This distinguishes them from the alkaloids, which are present all the time.

What goes on between detecting a foreign molecule and responding to it, is less clear. Research on cell signalling in plants lags behind work on animal cells (Section 16.4). There are hints from a range of studies on these receptors that most of the usual things are involved, in one plant or another: G proteins, IP_3, calcium, protein kinases, changes in membrane potential, though not cyclic AMP.

These specific receptors have a weakness. A mutation in the invader could lead to differences in a cell-surface molecule. The old receptor in the plant might no longer recognize it. Only a new receptor would do, probably a mutant form of the old one. Because of this, there will be some strains of plant and pathogen in which infection will not be checked: the invader wins. In other combinations, the plant will recognize the pathogen and stop the infection early. The only difference may be a single gene in either the plant or the pathogen. It is clear that natural selection goes on and on in this situation (Section 2.2). A change in the pathogen favours a change in the plant, which favours another change in the pathogen. Neither species can win the battle in the long term.

Summary

- Organisms must defend themselves because other organisms use them as food.
- Toxic and foul-tasting chemicals discourage attackers.
- Warning patterns and colours advertize chemical defences.
- Physical barriers such as the skin and leaf epidermis keep invaders out.
- Vertebrate blood clots to prevent microbes getting in and blood leaking out.
- Platelets clump together to block tiny injuries to capillaries.
- Fibrinogen, a soluble plasma protein, is converted to insoluble fibrin fibres at the heart of a clot.
- The fibrin mesh traps blood cells to form a barrier.
- Chemicals released by damaged cells and activated platelets, as well as collagen in blood vessel walls, stimulate clotting through a cascade of proteins.
- Clots are digested over a period of days.
- Plants seal injuries by tissue growth, and sometimes by resin or latex.
- Inflammation of mammalian tissues involves fibrin formation around the cells, restricting the spread of toxins and microbes.
- Plant leaves respond to infection by killing a disc of tissue around the site of infection: the hypersensitive reaction.
- The immune system of mammals responds to foreign molecules.

- Phagocytes engulf foreign cells or particles.
- Lymphocytes identify and respond to antigens, foreign proteins or glycoproteins.
- B lymphocytes make antibodies: proteins which bind specific antigens.
- T lymphocytes have similar proteins in their plasma membranes.
- Each B lymphocyte makes one type of antibody, generated randomly before birth.
- On infection, the appropriate lymphocytes multiply to make clones, all making the same antibody.
- After infection, memory cells belonging to those clones remain. They can respond quickly if that antigen ever reappears.
- T lymphocytes respond to antigens from the inside of other cells, presented to them by a plasma membrane protein.
- Some T lymphocytes kill other cells. Others coordinate the immune response.
- Plants can detect molecules in fungal and bacterial walls by specific receptors. This can trigger the hypersensitive reaction, release of digestive enzymes, or production of poisons.

Exercises

23.1. The human immunodeficiency virus, HIV, infects and kills helper T-cells. It has a high mutation rate. How does this information help to explain the observation that most peoples' immune systems cannot effectively tackle HIV?

23.2. O Rh− blood is often used for emergency transfusions. Why?

23.3. Thrombocytopenia is a condition in which the blood contains far fewer platelets than normal. Predict the symptoms.

23.4. Does the hypersensitive reaction to infection in plants have any parallels in mammals?

23.5. Alkaloids are present all the time; phytoalexins are made in response to infection or damage. What are the advantages of each?

Further reading

Alberts, B., Bray, D., Lewis, J., Raff, M., Roberts, K. and Watson, J.D. *Molecular Biology of the Cell* (3rd ed.) (New York: Garland, 1994). An enormous but excellent textbook which puts the reader in touch with current ideas without too much pain on the way. Molecular aspects of immunology are dealt with very well.

Guyton, A.C. and Hall, J.E. *Textbook of Medical Physiology* (9th ed.) (Philadelphia: Saunders, 1996). I like this hefty textbook of human physiology for its thoroughness and clear, no-frills diagrams.

Guyton, A.C. *Human Physiology and Mechanisms of Disease* (5th ed.) (Philadelphia: Saunders, 1992). Effectively a shorter, boiled-down version.

Jones, D.G. *Plant Pathology*. (Milton Keynes: Open University Press, 1987). A general introduction, straightforward and concise. Both the threats to plants and their defences are covered.

Roitt, I., Brostoff, J. and Male, D. *Immunology* (4th ed.) (London: Mosby, 1996). Detailed and up to date, yet clear and easy to learn from.

Support and Movement

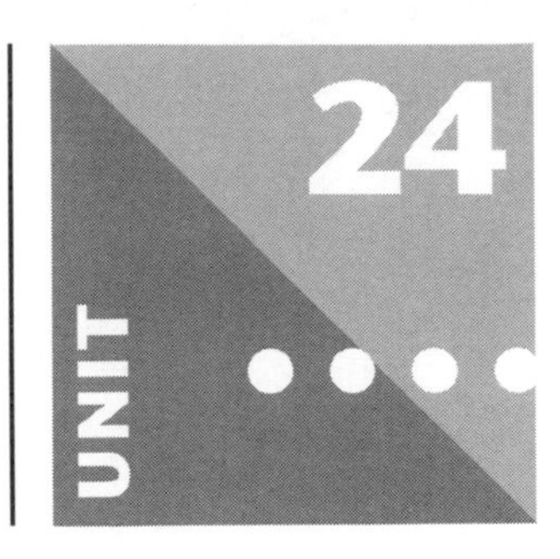

Connections

▶ This unit assumes some knowledge of cell structure (Unit 4), as well as a little *very* basic Physics. Otherwise, it should make sense on its own.

Contents

24.1 Forces and life

Forces are pulls and pushes. Forces make objects speed up or slow down. Sometimes they can change the direction of a moving object. Sometimes they distort an object's shape. Organisms live in a world of forces.

Gravity is impossible to escape if you live on land or in the air. Objects with mass are attracted towards one another. Big masses like the Earth, attract other objects strongly. Every living thing experiences this as the downward pull of gravity. An elephant is pulled firmly down against the ground surface. If it jumps, it quickly thumps down again. Tree leaves are held above the ground by strong frameworks of trunks and branches: cut a branch and the leaves fall. A sheep jumping out of a tree plummets. Even in water, the force of gravity is there. However, upward flotation forces more or less balance it, so organisms can float at or below the surface.

The force of gravity tends to distort bodies. An animal cell has the strength of jelly. A cunning parent can make a jelly for a children's party which stays in shape, quivering, when put on a plate. However, try to make a jelly the size of a sheep or the shape of a pine tree, and you have problems. The result is a squishy heap. If an organism is to be reasonably big, or have a shape other than a blob, it needs strengthening. Plants and fungi do this by supporting each cell with a wall (Section 4.6). Animals do it by supporting the whole body with a **skeleton**, either inside or outside.

Some animals move passively, carried along by ocean currents. Others, like corals, do not have to move around at all. Most animals need to move around by themselves, however. They must generate forces of their own. Several mechanisms are known, including muscle contraction and beating of flagella. All involve protein molecules moving against one another within cells.

Muscles need something to pull against if they are to move the body. If I want to catapult pieces of rubber across the room with an elastic band, I have to fix one end of the band to something which will not move. In the same way, muscles need something to pull against. That something is usually a skeleton.

Force: A push or a pull.

Skeleton: A system of structures which supports a body.

So, support and movement are linked by two ideas: forces and skeletons.

24.2 Support in animals

Almost all animals, big or small, have some sort of stiffening structure which helps support their weight and keep their shape: a skeleton. The internal skeletons (endoskeletons) of vertebrates are the most familiar. These are made up of cartilage in the cartilaginous fish (class Chondrichthyes). In the other vertebrates, bone is more important, although some cartilage is still present. Bone combines hardness with strength when compressed or stretched, and limited flexibility. Box 24.1 shows how the structure of bones helps them to do their jobs.

Box 24.1

Bone

Most vertebrates have an endoskeleton made of bone. It is hard and resists stretching and compression. It is most vulnerable to twisting forces, but even then there are big safety margins. Even when bodies are pushed to the limit in sport, it is rare to break a bone without crashing into something.

Bone materials

Bone is a composite material. It is dominated by extracellular matrix. Tough collagen fibres run through a hard, crystalline ground substance. This is a mineral containing calcium, phosphate and hydroxide ions. Collagen resists stretching; the mineral resists compression.

Bone cells

Bone contains two types of cell: osteoblasts and osteoclasts. Osteoblasts produce the bone material, and live in tiny holes within it. They are not isolated from one another because a web of fine extensions of the cell spreads out through the surrounding bone in tiny canals. These processes meet similar bits of other cells. Osteoclasts break down bone. This is important during growth. Long bones, such as the femur, need to be hollowed out. The ends of bones must be reshaped as they grow.

Two sorts of bone

Compact bone is a dense mass of bone material, with osteoblasts living in it. Cancellous bone is an open sponge of bony threads. Compact bone is strongest. Cancellous bone is very light, and bone marrow can occupy the spaces between the threads.

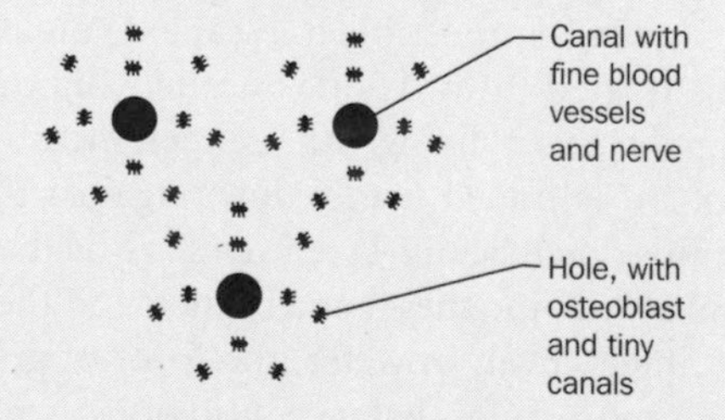

Section of a small piece of compact bone

A long bone in longitudinal section

The hollow shaft makes the bone lighter and cheaper. A hollow tube is one of the strongest structures that can be made from a limited amount of material. The cavity and cancellous bone also make a place for bone marrow.

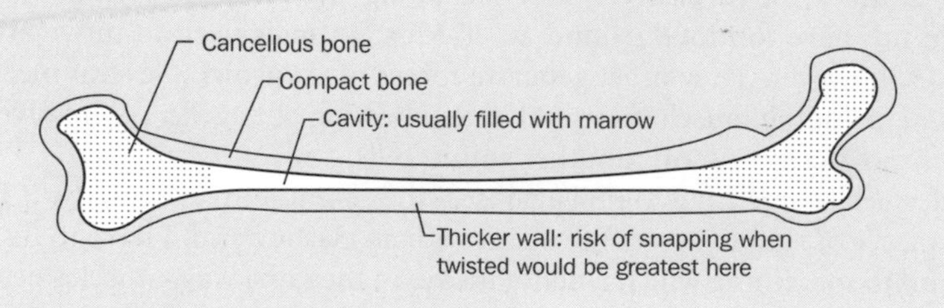

The arthropods have an external skeleton (exoskeleton). This is a more or less hard shell based on chitin (Section 3.4: Polysaccharides). It is made up of several plates which can move in relation to one another. This allows the body some movement. Exoskeletons provide excellent protection, but become inefficient in large animals (Box 24.2). Exoskeletons which protect and support are also seen in more primitive invertebrates. Within the phylum Cnidaria, some colonial hydrozoans have a chitin exoskeleton, whilst some corals have a hard exoskeleton made of calcium carbonate.

In some animals, such as nematodes and annelids, the body can be thought of as a closed bag of fluid within a muscular wall. If the wall muscles contract, the fluid will be under some pressure. This can then act as a crude skeleton. A hydrostatic skeleton may help maintain body shape, in the same way that high-pressure air gives a balloon its rounded shape. It may also transmit forces from muscle contraction when the animal moves (Section 24.7)

Box 24.2

Big animals don't have exoskeletons

A tube is stronger than a rod, if it is made of the same amount of material. If the tube is skeletal material, why not live inside the tube? This gives strength and protection.

Unfortunately, as a tube gets wider it becomes prone to buckling, and puncturing is an increasing problem. A large animal's exoskeleton would have to be disproportionately thick and heavy to overcome this.

All the largest land animals are vertebrates with endoskeletons. They still have tubular elements in the skeleton, but hang the body outside them. They must protect themselves in other ways. However, the skull is a bony box. In mechanical terms it is an exoskeleton protecting the brain.

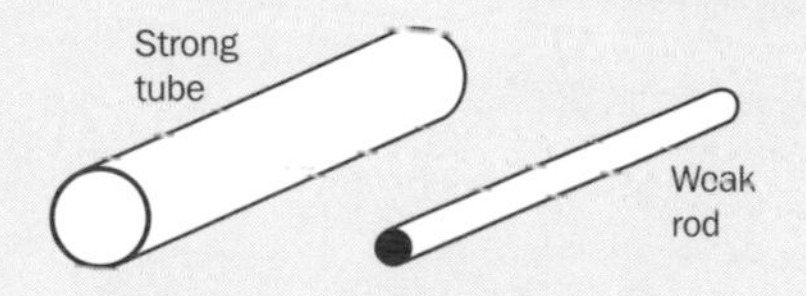

24.3 Support in plants

Plants do not have skeletons – or do they? Each cell has a wall around it. The composite structure of the primary wall, with tough cellulose fibres running through a softer matrix of other polysaccharides and proteins, gives the cell some strength. Some cells have a stronger secondary wall as well, with layers of parallel cellulose fibres, sometimes impregnated with waterproof lignin. Each cell, then, has an exoskeleton which plays a part in supporting the plant. However, some cells are specialized for support. Sometimes they are scattered amongst other cell types. Sometimes they form tough bundles (sclerenchyma tissue) and are, to all intents and purposes, part of an internal skeleton.

Specialized strengthening cells are relatively long, ranging from less than 1 mm to 0.5 m (Fig. 24.1). They have thick secondary walls, often lignified. Classifying them is a nightmare! They are all called either fibres or sclereids, but both categories are variable, and overlap. Green stems, leaves and other organs have bundles of these fibrous cells running through blocks of weaker cells. This gives a **composite structure** which can resist stretching, compression and twisting forces. Other fibrous cells are found in the phloem, among the weaker sieve tubes which transport sucrose (Section 18.7).

Composite structure: A structure made up of fibres runing through a matrix.

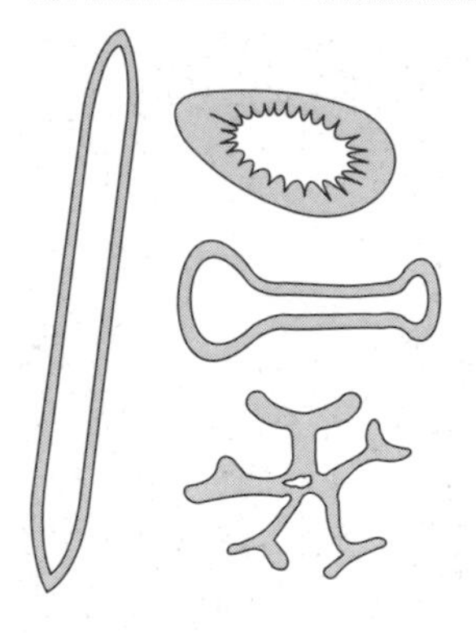

Fig. 24.1 Sclereids and fibres.

The xylem also has fibres. These are similar to the water-carrying tracheids, but have fewer, smaller pits (Section 18.4). Xylem fibres and tracheids both give strength to the plant. Xylem vessels are highly specialized for water transport, and are less effective in support.

Woody stems such as tree branches have a great core of xylem, made in layers, each new one outside the last. Phloem and other tissues are in a thin layer round the outside. The stem as a whole is not one of those composite materials in which strength and cheapness are balanced. Woody stems have very few cells with thin walls and large vacuoles. The cells of the xylem have thick, lignified walls, so these stems are expensive to construct but are enormously strong. Several things affect the strength of wood. The relative numbers and lengths of the different cells make a difference. The fibres and tracheids have thicker walls in some woods than in others. The rays, horizontal bands of weaker cells, have to be produced as a result of the way woody stems grow. They introduce weaknesses, and different woods have different distributions of rays. Fig. 24.2 shows the layout of fibrous tissues in a range of stems and other organs.

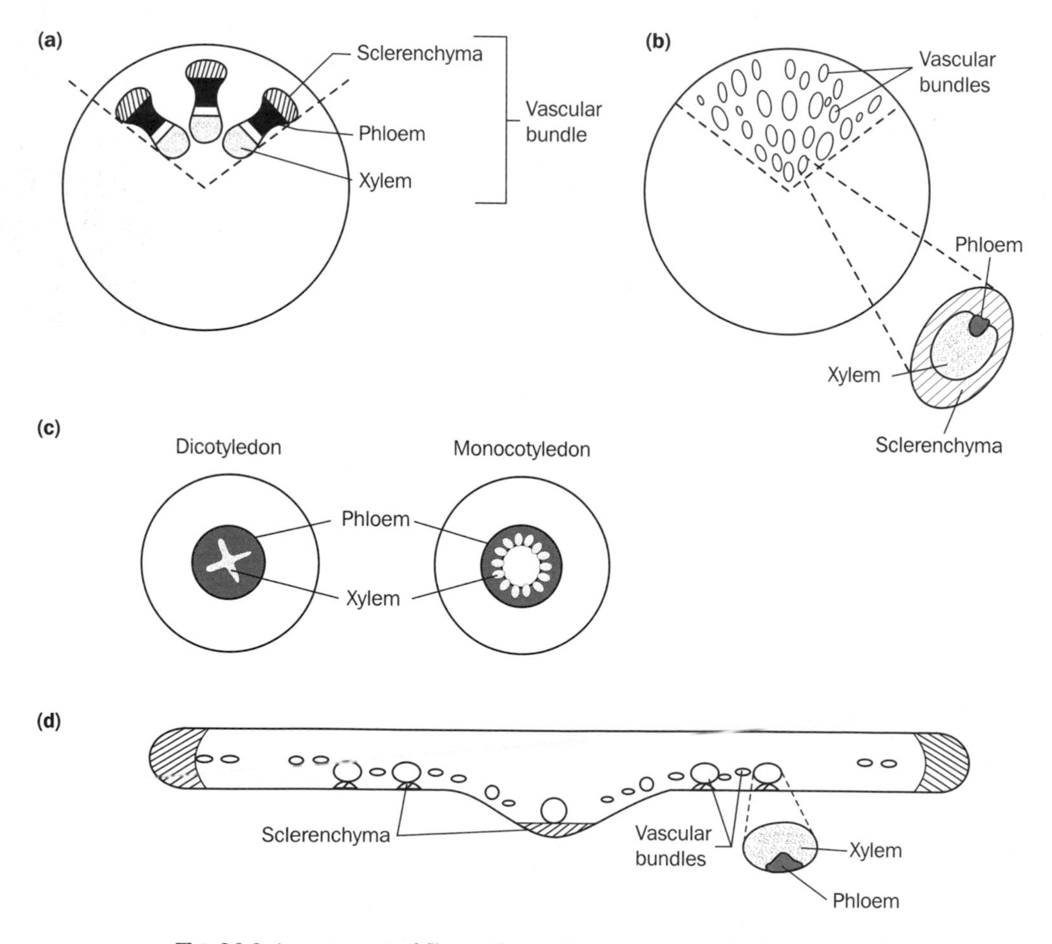

Fig. 24.2 Arrangement of fibrous tissues in some non-woody plant organs. (a) a dicotyledonous stem; (b) a monocotyledonous stem; (c) dicotyledonous and monocotyledonous roots; (d) a grass leaf.

24.4 Buoyancy

Organisms which live in water have the chance of floating. Some do, some do not. Some, like the Portuguese man o' war with its bag of gas, are less dense than sea water, and float on top. This is called positive buoyancy. This suits the man o' war, whose stinging tentacles hang down for many metres and catch food as the animal drifts along. Other animals need to move up and down in the water. Some have the same density as sea water: they have neutral buoyancy. They can stay at a particular depth without using energy. Energy is only needed to change depth or to swim about. Most cells are more dense than water, even sea water. Special measures are needed to reduce the animal's overall density. Keeping out heavy ions, such as sulphate, helps in some planktonic algae. Large stores of lipids, which have a low density, make many fish and algae more buoyant. Gas floats are especially effective. Some jellyfish have gas bags. The cuttlefish bones which budgies and beachcombers love, are the gas-holding organs of a squid-like animal. The coiled shell of *Nautilus* (another cephalopod) has air-filled chambers, and many bony fish have an internal air bag, the swim bladder (Box 24.3).

With buoyancy, size matters. Small things do not behave quite like big things. Friction in water or air becomes incredibly important to smaller organisms. If a small organism is flattened, or has spines, hairs or wings which increase its surface area, friction becomes even more important. This means that tiny organisms which are more dense than water may experience a big resistance to falling. Their lives are spent sinking very, very slowly.

A gas bag gives the Portuguese man-o'-war positive buoyancy as it drifts on the ocean surface, catching planktonic organisms with its tentacles.

24.5 Generating forces

An organism that moves by itself needs to be able to generate forces. This almost always involves many individual cells or organelles, each generating forces. Just a few molecular systems drive most sorts of movement. Prokaryotes which can move use flagella. These are long, flexible protein filaments attached to a rotating protein motor in the plasma membrane. The motor is driven by energy stored in a proton gradient (Section 8.5) across the membrane. As protons flow into the cell, the motor turns. Several rotating flagella wind together into a twisting propeller, driving the cell forward (Fig. 24.3(a)). Sometimes, the motors reverse direction, just for a moment. Then, the flagella fly apart, pointing the cell in a new direction at random

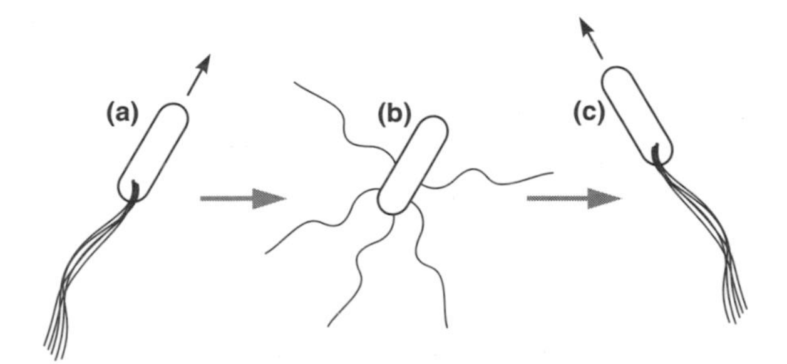

Fig. 24.3 Bacteria use their flagella to swim, and randomly change direction.

Box 24.3

The swim bladder of bony fish

Many bony fish have an internal gas bag which gives neutral buoyancy at one particular depth. This is useful for species which need to hang about at one level without working hard. It is not ideal for fish which lurk on the sea bed, or which dash up and down chasing prey.

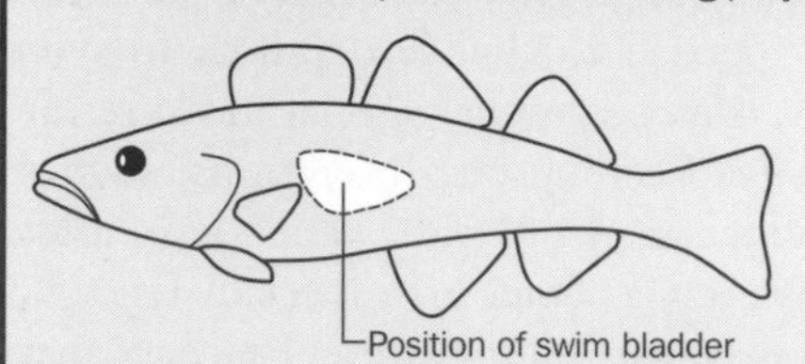

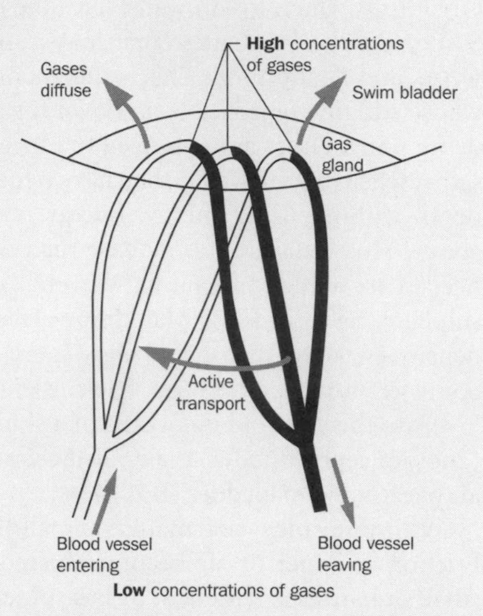

Fish secrete oxygen and other gases into the swim bladder at high pressure. It is quite tricky to understand how the gases get there, and why the blood does not carry them away, as it does in lungs. The answer lies in a cunning capillary bed near the gas-secreting gland. Capillaries carrying blood to the gland run beside others carrying blood away. Essentially, oxygen and other gases are actively transported from blood leaving the gland to blood entering it. This constant gas recycling means that very high levels of dissolved gases build up in the capillaries of the gas gland. Gas enters the swim bladder, even when it already contains gas under pressure. This system is a countercurrent multiplier, like the loops of Henle in the kidney (Section 22.5).

The mechanism of active transport is indirect. For oxygen, it relies on the fact that haemoglobin binds less O_2 when the pH is low (Section 19.2). Cells of the gas gland produce lactic acid which dissociates into lactate plus H^+, and secretes them into the blood. This makes the blood leaving the gland more acidic, so its haemoglobin binds less O_2. This means that the blood contains more free, dissolved oxygen. Even though blood leaving the gland contains less oxygen in total than blood entering, the free dissolved oxygen is higher, so it diffuses across. Energy is supplied by lactic acid secretion.

Other gases are secreted in a less specific way. Lactic acid, like any other solute, makes any gas less soluble.

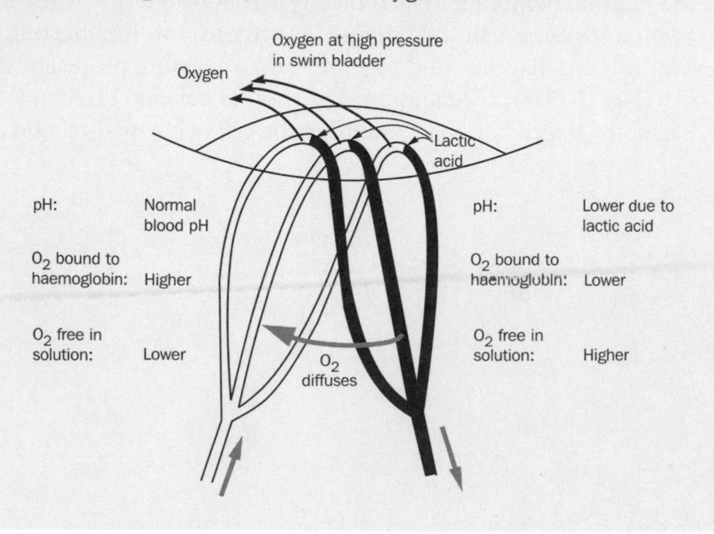

(Fig. 24.3(b,c)). The motors then continue as before, pushing the cell on its new course. The only way to steer is to change direction more often when conditions are getting worse, and less when they are improving: crude, but it works.

In eukaryotes, some movements are also driven by flagella and cilia. These are quite different to prokaryotic flagella. They are long thin extensions of the cell, with plasma membrane around a regular array of paired microtubules (Section 4.5: The cytoskeleton).

Cilia and flagella have the same structure, but beat differently. Flagella are longer (about 40 μm in a human sperm) and waves of beating move down them from the base (Fig. 24.4(a)). Cells only have one or a few flagella. Mammalian sperm have one, male gametes of mosses have two, some single-celled algae swim with two or four, and male gametes of ferns have a tuft of several. Cilia are shorter (typically 10 μm) and cells have more of them. Waves still pass down them, but the wavelength is longer than the cilium, so they row back and forth (Fig. 24.4(b)). Some protozoa swim using belts of cilia beating together. Flatworms crawl using a carpet of cilia underneath. On the epithelial cells lining a mammal's bronchus, cilia clear out the mucus which traps dirt.

Dyneins are proteins which walk along microtubules using energy from the hydrolysis of ATP. In cilia and flagella, dyneins are attached to each pair of microtubules. When they walk along the adjacent pair, the microtubules slide with them. Microtubule sliding is organized so that each part of the flagellum twists first one way then the other. This is shown in more detail in Box 24.4.

Actin filaments (Section 4.5: The cytoskeleton) can also help generate forces in two ways. Animal cells have a network of actin filaments just beneath the plasma membrane. When cells crawl, extending actin filaments push the plasma membrane forward at the leading edge of the cell. Secondly, myosins are proteins which walk along actin filaments. This uses energy from the hydrolysis of ATP. Some types of movement within cells are driven by myosins. Their most spectacular role is in muscle contraction, where myosins are linked to form fibres.

Muscles are made up of long, cylindrical cells with several nuclei around the outside. Muscle cells are made up of many myofibrils. These are smaller fibres running lengthwise. Each myofibril contains actin filaments and myosin filaments. The filaments run lengthwise, and overlap. When the myosins walk, the filaments slide together (Fig. 24.5). When filaments slide together, the myofibril contracts. If all the myofibrils contract, the muscle cell contracts. If many cells contract together, the

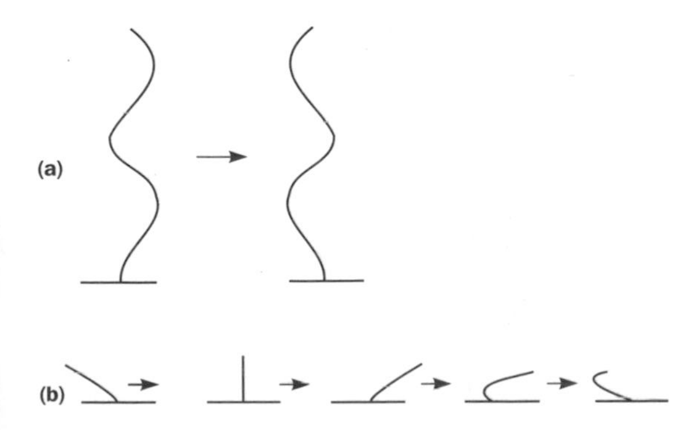

Fig. 24.4 Beating of (a) a flagellum and (b) a cilium.

Box 24.4

How eukaryotic flagella bend

In eukaryotic cilia and flagella, waves of bending move along, making the flagellum wriggle. This is driven by microtubules and dyneins which walk along them.

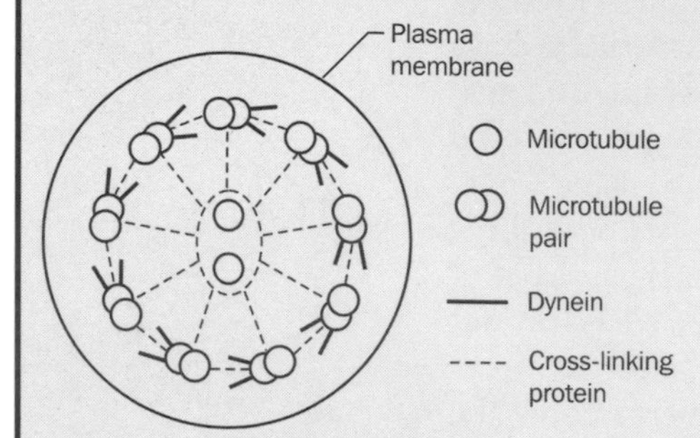

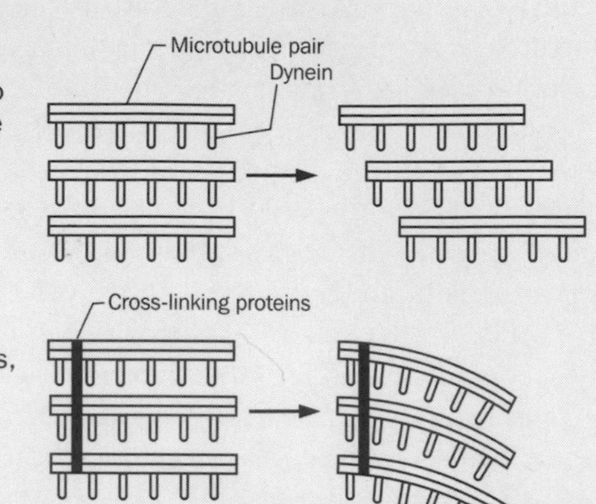

Transverse section of flagellum

Dyneins are attached to one side of each microtubule pair. They use energy from the hydrolysis of ATP to walk along the adjacent pair.

This process needs ATP all along the flagellum, all the time it is beating. ATP is made in the mitochondria, so somehow it must move down the flagellum from its base. Is diffusion enough, or is there some sort of streaming of cytoplasm within the flagellum as it beats? If there is, is it driven by actin and myosin, or could forces set up by distortion of the flagellum as it bends set up a flow? We just do not know.

muscle as a whole contracts. Force is generated at the molecular level, but its effects are seen in cell, organ and body. The problem is to make the filaments slide only when the body needs the muscle to contract. Box 24.5 shows how nerve impulses trigger contraction throughout the muscle.

Box 24.5

How nerve signals make muscle cells contract

An action potential in a motor neurone leads to a skeletal muscle cell contracting. This is no simple business. As well as simply crossing from neurone to muscle cell, the signal has to be carried throughout the huge muscle cell. There is a complex chain of events.

1. The neuromuscular junction

This is where the end of a neurone meets the outside of a muscle cell. The two plasma membranes are very close together, as in a synapse. Acetylcholine is secreted, and binds to receptors on the muscle cell.

2. Muscle action potential

Binding of acetylcholine leads to opening of sodium channels, and depolarization of the muscle cell membrane. An action potential spreads across the cell, much like in an axon.

3. Transverse tubules (T-tubules)

Tunnels of plasma membrane, the T-tubules, cross the muscle cell from side to side. No part of the cell is far from the plasma membrane in these giant cells.

4. Calcium is the second messenger

A muscle action potential leads to a pulse of Ca^{2+} in the cytoplasm. Calcium is stored in the endoplasmic reticulum, which forms a net wrapped around each myofibril. The T-tubules are closely linked to the ER. Some signal (exactly what is unclear) passes from the T-tubule to the ER when there is an action potential. It triggers calcium release.

5. Control of the sliding filaments

A complex of two proteins, troponin and tropomyosin, is bound to actin filaments. The complex can bind in two different positions, depending whether troponin has bound calcium ions. If calcium is absent, tropomyosin blocks the myosin binding sites in the actin filaments. Myosin cannot walk along it, so the muscle cell does not contract. When troponin binds calcium, tropomyosin shifts, exposing the myosin binding sites. The cell can then contract.

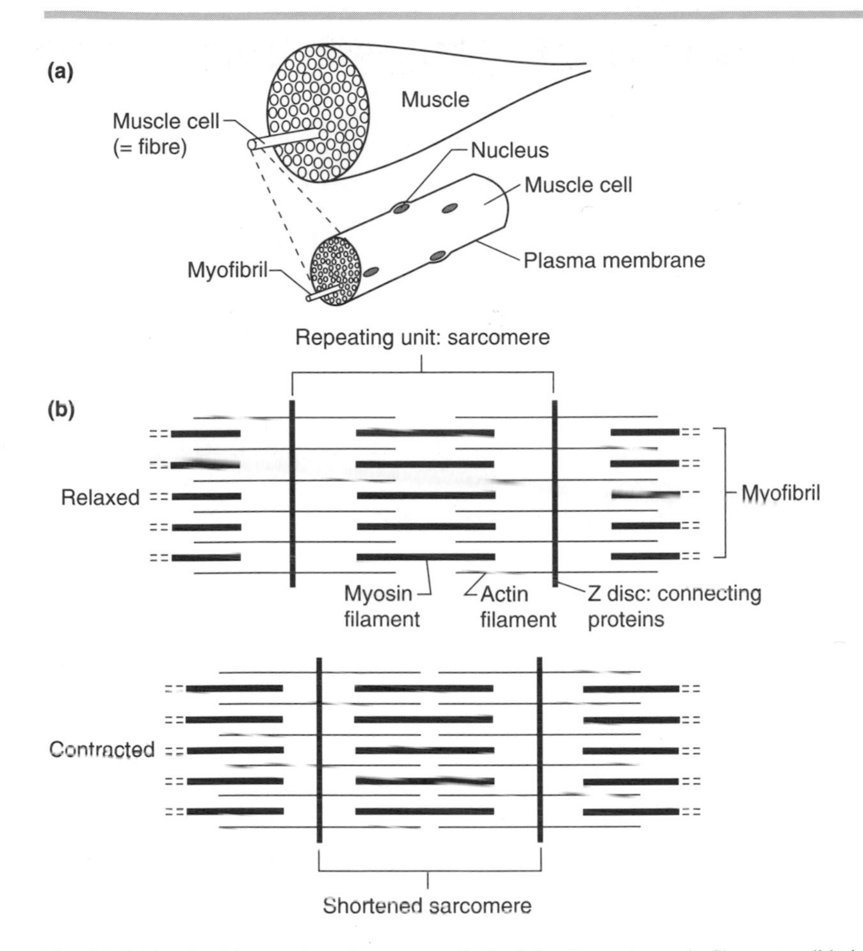

Fig. 24.5 Muscle. (a) muscle cells and myofibrils; (b) actin and myosin filaments slide between one another during contraction.

Not all muscle cells are the same. There are two extreme types, fast fibres and slow fibres, with a range of intermediates. Fast fibres respond to nerve impulses by rapid but short-lived contraction. They rely on anaerobic respiration for most of their energy needs when contracting. Fast fibres are suited to rapid, powerful movements such as jumping or sprinting. Slow fibres respond less quickly, but their response lasts longer. Aerobic respiration is much more important to them. They are suited to prolonged contraction, for example holding the body in a particular posture, or long-distance running. Muscles contain a mixture of cell types, but some have far more of one type than others, according to their needs.

In plants, most movements are caused by growth in a particular direction. There are some reversible movements, including petals folding when a flower closes at night, 'sleep' movements of some leaves, and even sudden movements such as when a Venus fly trap leaf snaps shut on a fly. All these movements are driven by changes in turgor, the pressure within cells. The arrangement of cellulose microfibrils in the wall controls how the cell expands when it becomes turgid. This can lead to changes in shape, as in the guard cells around stomata (Section 18.3).

The plant on the right is wilting. Cells throughout the plant are losing water, becoming less turgid, and so doing less to support the plant.

24.6 Levers and skeletons

Skeletons are systems of levers. Levers are machines which make forces bigger or smaller. They all involve some sort of a bar which can swivel about a pivot or hinge (Fig. 24.6(a)). Try pushing a door open at the handle end, a long way from the hinge. It was probably quite easy. Only a small force was needed, but you had to push it a long way. Now try it near the hinge end: much harder! A big force was needed, but you only pushed a short distance to open the door completely. Two things stayed the same, wherever you pushed. First, the same amount of work was done. Second, the force multiplied by the distance from the pivot stayed constant. This principle can be used in levers. If you apply a force far from the pivot, but use another part of the lever nearer the pivot to do the work, the force increases. If you put in a force near the pivot, but use a distant part of it to do work, the opposite is true. The force you get out is smaller, but it moves through a bigger distance (Fig. 24.6(b)).

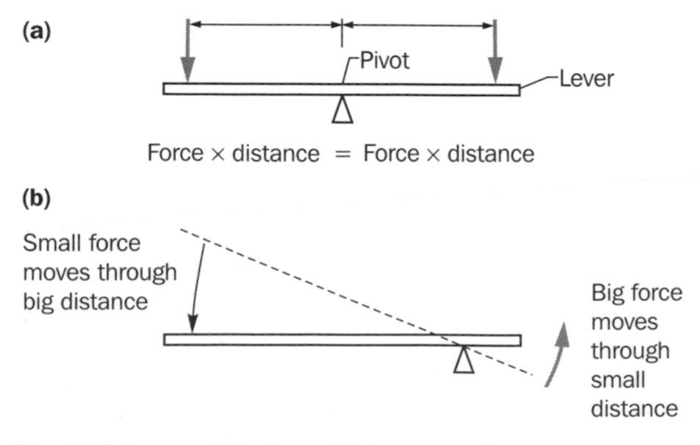

Fig. 24.6 Levers. (a) a lever; (b) levers can be used to change forces and the distances through which they move (see text for details).

The vertebrate skeleton

In vertebrate skeletons, bones are the levers and joints are the pivots. Muscles generate the forces. They are linked to bones by tendons, tough connective-tissue cables. Joints are not simply points where two bones rub together. This would wear the bones away, and friction would make the system inefficient. Also, the big forces involved might pull the bones apart. These problems are overcome in real joints. A layer of cartilage across the ends of the bones reduces wear and friction. Bones are held together by ligaments, another group of connective-tissue cables. Joints may become dislocated if ligaments are damaged. Some joints, like the knee, have a cushioning bag of synovial fluid, a liquid extracellular matrix (Section 4.3). Fig. 24.7 shows this type of joint.

Individual parts of the skeleton can be analysed as separate lever systems. Fig. 24.8 shows the system which moves the human forearm (other mammals are much the same) and Fig. 24.10 shows the mechanism which makes birds' wings flap. In each case, the muscles are attached very close to the pivot. This is because muscles can generate huge forces, but only shorten a little when they contract. The levers let the hand or wing tip move long distances. The cost is that they cannot apply such big forces.

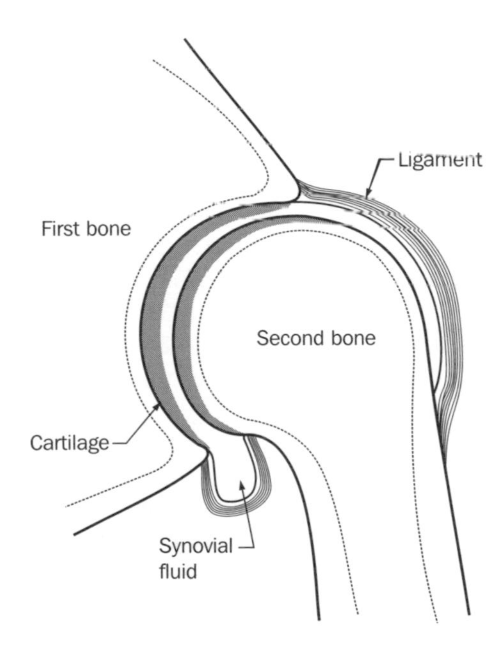

Fig. 24.7 A synovial joint.

Walking and running

Muscles and joints make legs move. Given this, how can an animal use its legs effectively? Most four-legged animals walk in a similar way. Walking is relatively slow. A walking animal needs to be able to stop instantly without falling over. This is important when stalking prey, or when being stalked by a predator. A four-legged animal moves just one leg at a time (Fig. 24.9). This leaves three legs on the ground, forming a triangle. If the animal keeps its centre of gravity above this triangle, it will be stable. It can stop at any time, even in mid-step, without falling over. Try it! Running is faster, but the animal cannot stop at once. There are several modes of running, called

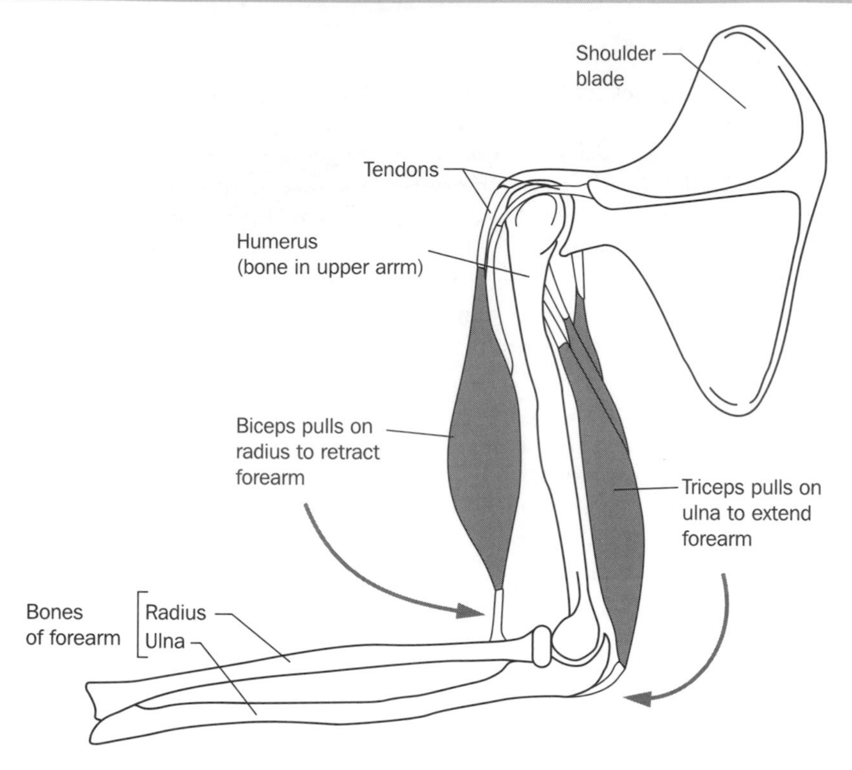

Fig. 24.8 Moving the human forearm.

trotting, cantering and galloping in horses. Two or even four legs leave the ground at some points. The best way to understand this is to watch animals moving, then to try it yourself in the privacy of your own home.

With more than four legs, balance becomes less of an issue, but coordination becomes a more complex problem.

Swimming and flying

Animals which move through water or air have two problems. Firstly, they have to drive themselves forward, just like animals which walk. Secondly, they have to generate upward forces to stop them sinking or falling, unless they are no more dense than the water or air (Section 24.4).

Birds in flapping flight push down and back against the air with their wings fully extended. This pushes the bird forwards and upwards, solving both problems. However, the wings must then go back to the starting point. If they just reversed the downstroke, the bird would be pushed backwards and downwards again: the bird would get nowhere. Instead, the bird partly folds its wings as it brings them up, so that less area is exposed to the air (Fig. 24.10). The backward/downward forces are still there, but are smaller. Over the whole wingbeat cycle, there is an overall upward/forward force on the bird. The wings are giving both lift and thrust.

A herring's tail pushes diagonally back against the water (Fig. 24.11(a,b)). The result is a diagonally forward force on the fish. This has a forward component, but also a component which pushes the fish to one side. However, when the tail swishes

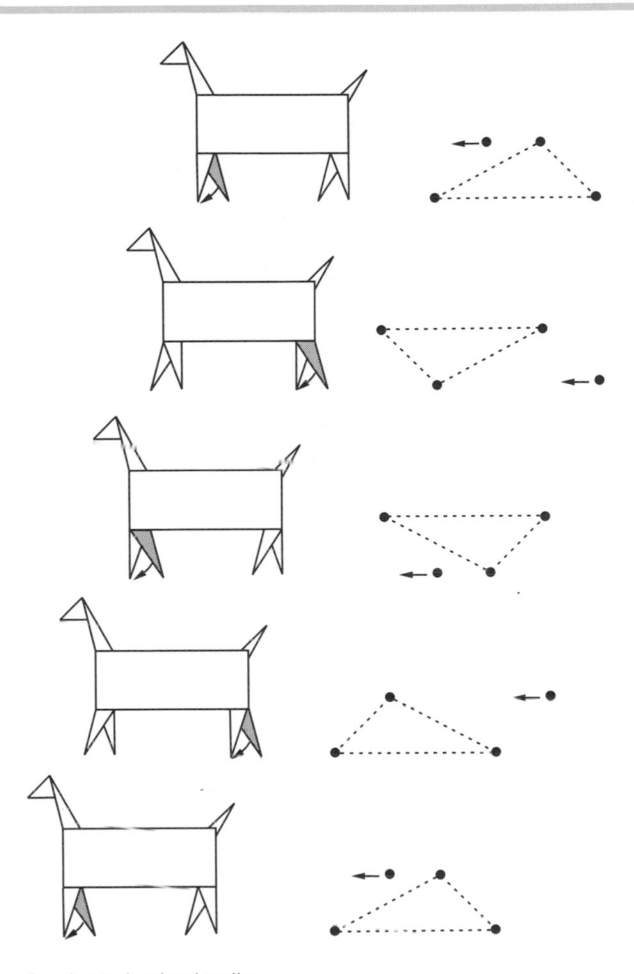

Fig. 24.9 How a four-legged animal walks.

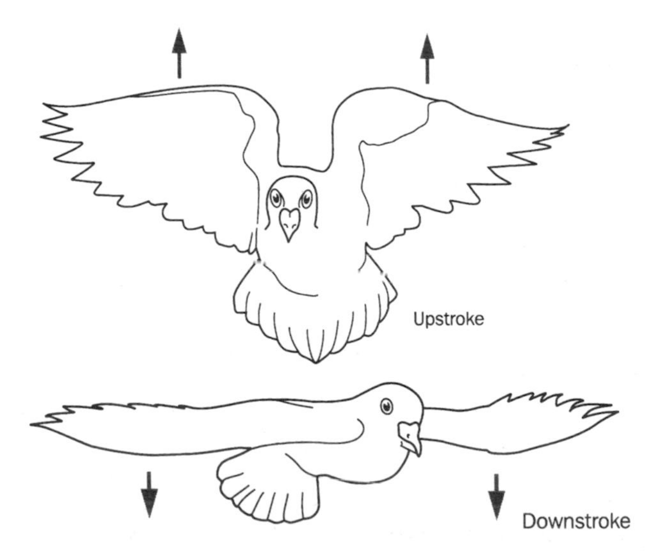

Fig. 24.10 Wing movements during flapping flight in birds.

Wing movements of a pigeon in flight.

back again, the forward component is still there, but the sideways component is in the other direction. The fish moves forward, wobbling slightly as it goes. An eel wriggles through the water in a similar way (Fig. 24.11(c)). At any one time, some points on the fish are pushing back and left on the water, while others are pushing back and right. The overall force on the eel is a push forwards.

As well as driving forward motion, flapping flight gives lift. However, a flying animal can generate some lift even when gliding. This is because wings meet the air moving past at an angle: they are aerofoils. Air moving over the wing has to travel further than air moving under, so it travels faster. The faster a fluid moves, the lower its pressure. The pressure is higher below the wing, so the bird is pushed up and back

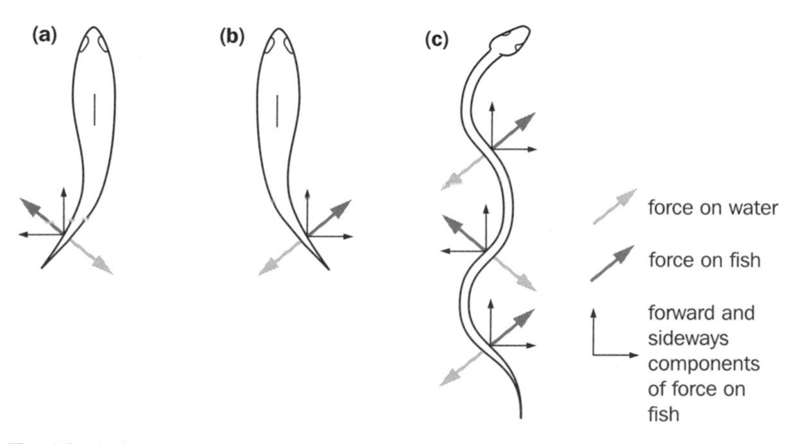

Fig. 24.11 Swimming in fish. (a) and (b) a herring; (c) an eel.

at right angles to the plane of the wing (Fig. 24.12(a,b)). Some wing shapes increase this effect (Fig. 24.12(c)). The bird gains lift without using energy; the cost is the backward drag. Gliding birds either lose height, or slow down. Many species flap for a while, gaining speed or height, then glide as they fall or slow down. Watch them and see! Cartilaginous fish with no swim bladder use some of their fins in the same way to give them lift in the water.

An aerofoil is simply an angled structure with air moving across it. Angled correctly, there is no reason why a flapping bird should not get some lift on the upstroke as well as on the downstroke, although flight like this is bound to generate less thrust. Hummingbirds hover like this. Some insects generate lift on the upstroke even when flying forward.

A few birds soar, that is they find places where warm air is rising, and use this to give them lift. They tend to have broad wings, held out flat to maximise the area over which they feel the force. Vultures are excellent examples, but in Britain buzzards and rooks can be seen soaring effectively.

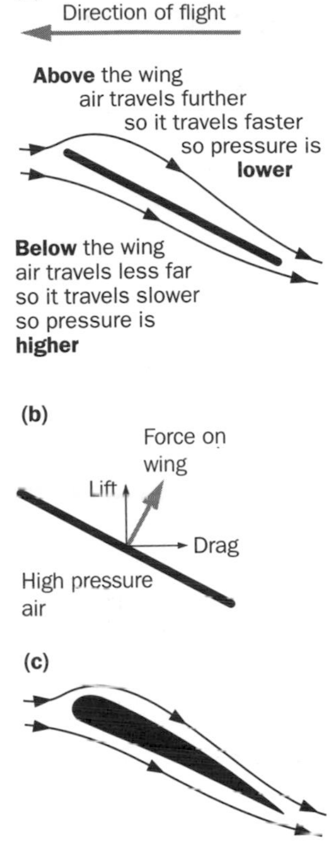

Fig. 24.12 Wings as aerofoils (see text for details).

24.7 Moving with a hydrostatic skeleton

Some invertebrates have a muscular outer wall with a bag of liquid and the internal organs inside. Liquids are extremely hard to compress. Squeeze them in one place and they try to flow somewhere else. This is the basis of movement in some animals, including earthworms and slugs.

An earthworm's body is divided into fluid-filled segments. The volume of fluid in each segment stays constant, but its shape can change (Fig. 24.13). There are two sets of muscles in the wall. Circular muscles run around each segment, and longitudinal muscles run lengthwise. When circular muscles contract and longitudinal muscles relax, a segment will be long and thin. When circular muscles relax and longitudinal muscles contract, the segment will get shorter and fatter. Fat segments cannot move along the worm's burrow for two reasons. Firstly, they are stuck like a cork in a bottle neck. Secondly, tiny chitin bristles stick out into the soil when the segment is fat. When segments are thin, they can slide through the soil, either pushing forward or being pulled in behind.

In a moving worm, groups of segments are either fat or thin. Waves of changing from one state to the other pass back down the worm. The front of the worm pushes forwards when thin for a while, then anchors when fat while the next group of

Circular muscles contract
Longitudinal muscles relax
Segments **long** and **thin**

Fig. 24.13 The volume of an earthworm segment stays constant, whether it is fat or thin.

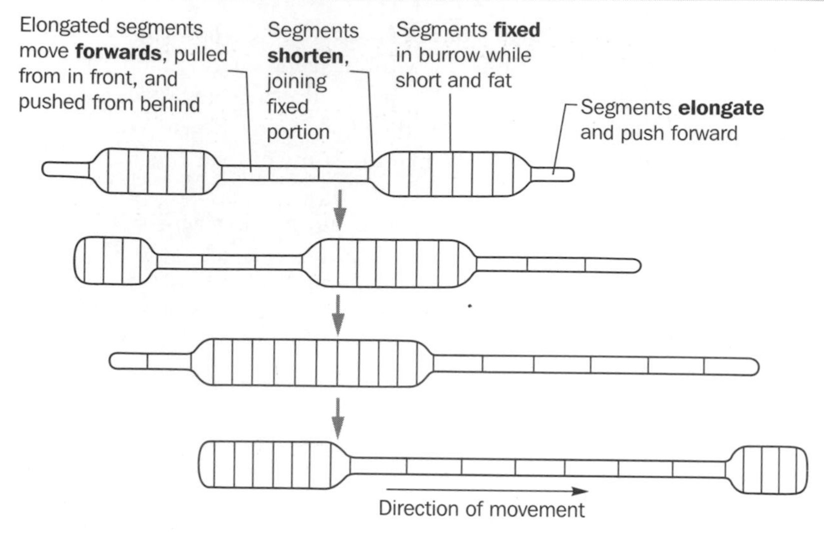

Fig. 24.14 How earthworms move.

segments catch up. Fig. 24.14 shows this beautiful mechanism better than words can say.

Rather similar things go on in the foot of a mollusc, where waves of muscular contraction pass along the foot. Things are much more complex, however, because the muscles are all at oblique angles, and the moving segments probably never leave the ground. Movement relies on the slimy mucus layer changing its thickness according to the forces acting on it.

24.8 Jet propulsion

Cephalopods, a group of marine molluscs including squids and octopuses, generate force in an exciting way. They chase their prey backwards, shooting out a jet of water which pushes them along, like a jet engine in air. The body contains a water bag: the mantle cavity (Fig. 24.15). This has a muscular wall which can contract, increasing pressure in the cavity. Water squirts out through a funnel, the siphon. Other muscles contract to increase cavity volume, slowly refilling it. Jet propulsion is also seen in some bivalve molluscs, including clams. They force water out by clapping the two halves of the shell together.

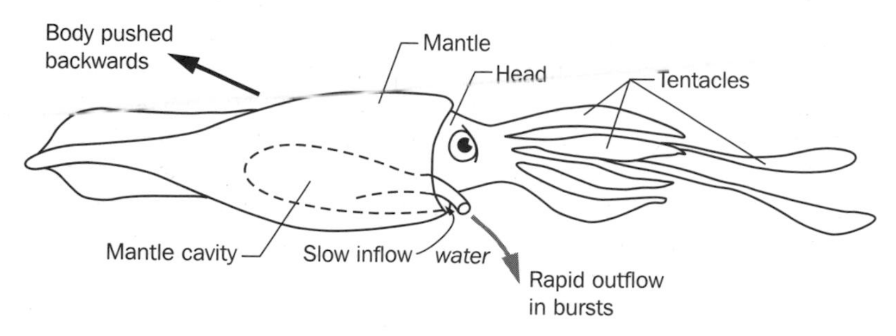

Fig. 24.15 Jet propulsion in squids.

24.9 The mystery of the biological wheel

Why don't organisms have wheels? Some organisms may roll, but a wheel is more than a hoop. The hoop rolls along as a whole, while the wheel turns in relation to whatever it is attached to. Some sort of axle and bearings are required. The question becomes a wider one: do organisms have truly rotating structures? The answer, with one big exception, is firmly 'no'.

At first sight, there appears to be a strong ecological limitation on the value of wheels. Wheeled vehicles almost always roll on surfaces which people have modified, made smoother, flatter, and freed from obstacles. A bus in a forest is useless without a road: even a mountain bike needs a track. Most habitats seem equally unsuitable for wheels, but there are exceptions: beaches, sandy deserts, steppe, parts of the sea bed. If we extend the concept of the wheel to include paddle wheels and propellers, both truly rotating structures, aquatic habitats become available. Rolling organisms, generally rolling passively at the mercy of the wind or currents, are sometimes found in such habitats, but wheels are not. This suggests that something more basic is involved.

Various structural difficulties arise. How could biological bearings be constructed? How could information, food and oxygen pass across the moving surfaces at bearings? How could lengthwise contractions of muscles be converted into rotation?

Tissues and biochemicals with the properties needed in bearings already exist. In vertebrate joints, cartilage covers the surfaces of bones, perhaps reducing friction and certainly protecting the bones. The gel-forming polysaccharides and proteoglycans of mucus lubricate many moving surfaces, for example the foot of the slug as it moves along the ground. A bony wheel rotating on a bony axle, protected by cartilage and lubricated by mucus, does not sound too far fetched. Similar structures made up of plant cells could also be imagined.

Communication across the bearing is a much greater problem. Blood supply to a vertebrate wheel would be interrupted at the bearings. So would communication by nerves. The continuity of both symplast and apoplast (Section 18.5) would be broken at the bearings of a plant wheel. The implication is that only organisms which do not rely on internal communication, and whose cells gain the molecules they need straight from the environment, could have rotating multicellular structures.

Propulsion is another difficulty. Muscles contract lengthways: this would have to be translated into rotation. One engineering solution to this is the camshaft. Several sets of bearings are required, however, adding to the problems already discussed. Eukaryotic flagella and cilia also beat from side to side, rather than rotating.

A deeper issue underlies the problem of generating rotation. The movement of an organism mirrors the movements of the cells which generate forces. Eukaryotic cells generate forces exclusively by systems in which one protein moves along a filament of a second protein. These generate linear movement. Only in prokaryotic flagella do we see true rotation: here we have a sub-cellular wheel. It is impossible to tell from our evolutionary distance why eukaryotes have not inherited this structure. Perhaps its absence from the first eukaryote was adaptive, but perhaps it was a matter of chance. Its fantastically successful descendants may have been denied rotating structures by nothing more than an ancient coincidence.

Summary

- Skeletons support the weight of a body.
- Vertebrates have an endoskeleton of bone or cartilage.
- Arthropods have an exoskeleton of chitin.
- Annelids have an internal 'water bag' which has some of the roles of a skeleton.
- Several plant tissues contain strong fibrous cells. Together, they form a sort of skeleton.
- Some aquatic animals have neutral buoyancy. They can stay at a particular depth in the water without using energy.
- Gas floats are a common way of increasing an animal's buoyancy.
- Prokaryotic flagella rotate. Force is generated by a protein 'motor' in the plasma membrane.
- Eukaryotic flagella and cilia beat. Force is generated all along the flagellum by an orderly system of microtubules and dynein.
- Muscle cells contain actin filaments and myosin filaments running in parallel.
- Muscle cells contract when the myosin filaments 'walk' along the actin filaments.
- Plant movements are driven by growth or turgor changes.
- Skeletons are systems of levers. They change the forces and distances involved in muscle contraction, to do useful tasks.
- Walking animals move their legs in a way which keeps them stable at all times.
- Running is faster, but the animal cannot stop instantly.
- Flying animals generate both lift and thrust with their wings.
- Some invertebrates use a hydrostatic skeleton to transmit force from the muscles. They rely on the fact that liquids are hard to compress.
- Cephalopods swim by jet propulsion. Muscle contraction causes water to squirt out of a body cavity.
- Animals do not have wheels. This may be because there are few natural flat surfaces to move on, biological bearings would be needed, and animal cells cannot generate rotation directly.

Exercises

24.1. **(i)** What are the advantages to an animal of being able to move around?
(ii) Think of some animals which cannot move around. How do they manage without these advantages?

24.2. List all the proteins mentioned in this unit which have a role in movement.

24.3. Fibrous proteins are important in the movement of animal bodies, but turgor is far more important in plant movements. Why?

24.4. Some trees such as birches have softer, lighter wood than others, such as beech. Very simply, the lighter wood has less lignified cell wall in a given volume of wood than the heavier wood. What are the advantages and disadvantages of each?

24.5. Human muscles contain a mixture of fast and slow fibres. The relative numbers vary between individuals. Training increases the diameter of fibres, but not the number of either type. What does this tell you about the potential of an individual in different sporting events?

24.6. Watch a four-legged animal walk and run. It could be a cat, a horse, whatever is convenient. Try to draw diagrams like the one in Fig. 24.9 to show what happens in running. How about studying a six-legged animal such as a beetle?

24.7. Spiders move their legs in an unusual way. Like other arthropods, the organs are surrounded by blood within a rigid, jointed exoskeleton. They extend them by raising blood pressure: a hydraulic system. The design of the joints means that this straightens the limb. Muscles are needed to flex the limb, that is to bend each joint again.

(i) Predict the effect on the legs of massive blood loss.
(ii) Why is this system unsuitable for vertebrates?

Further reading

Guyton, A.C. and Hall J.E. *Textbook of Medical Physiology* (9th ed.) (Philadelphia: Saunders, 1996). I like this hefty textbook of human physiology for its thoroughness and clear, no-frills diagrams.

Guyton, A.C. *Human Physiology and Mechanisms of Disease* (5th ed.) (Philadelphia: Saunders, 1992). Effectively a shorter, boiled-down version.

McNeill, Alexander R. *Exploring Biomechanics*. (New York: Scientific American Library, 1992). A wide-ranging introduction, which includes basic principles from physics with minimal mathematics. Beautifully illustrated, readable, simple yet authoritative.

Rudall, P. *Anatomy of Flowering Plants* (2nd ed.) (Cambridge: Cambridge University Press, 1992). A concise summary of flowering plant structure and how it develops.

Schmidt-Nielsen, K. *Animal Physiology: Adaptation and Environment* (4th ed.) (Cambridge: Cambridge University Press, 1990). One of the all-time great textbooks! Exciting, authoritative and easy to read, it puts animal physiology in the context of the environment.

Withers, P.C. *Comparative Animal Physiology*. (Fort Worth: Saunders, 1992). More detailed than Schmidt-Nielsen, with wider-ranging examples: the inevitable cost is that it is a less easy read.

Development and Growth

Contents

Connections

► The questions of development and growth touch on almost every area of biology. A basic understanding of genetics (Units 7 and 12) is essential if you are to make sense of how genes control development. Cell division (Unit 11) and reproduction (Units 13 and 14) are also inextricably linked to growth and development.

25.1 A marvellous problem

An elephant starts life as a single cell, the zygote. Nearly two years later, the baby elephant is born. Some amazing changes have taken place in that time. It has become much larger: **growth** has taken place. A single cell has divided by mitosis, and this has been repeated many times. A zygote which looked more or less spherical has **developed** obvious polarity and patterns of symmetry. Head (anterior) and tail (posterior) ends are distinct, back (dorsal) and front (ventral) surfaces are clear, and there is left-right (bilateral) symmetry. Many cell types have **differentiated**. They are organized into tissues and organs, fitting together in a tightly organized whole. This is development. After birth, there are fewer developmental changes, but growth continues for some years. Existing structures get bigger.

Development is a marvellous process, and a marvellous problem for biologists. How does genetic information in each cell interact with the cytoplasm, with other cells in the embryo and with the embryo's environment to form a body? We cannot yet tell a complete, satisfying story of how development is controlled in any animal. However, a few species are being studied intensively, and are giving insights into different aspects of development, which are almost certainly relevant to other animals.

Development and growth in plants are related to one another in a different way. In animals, growth takes place during and after the development of an integrated body. Plants have a modular body plan. They are made up of a few basic organs, repeated many times. A normal elephant has one liver and two kidneys, but how many leaves does a beech tree have? It depends on the size of the tree. Plants grow by adding on more organs. Growth of the existing organs is usually less important.

Growth: An increase in body size.

Development: The process by which a body takes on its mature form.

Differentiation: The process by which a cell takes on a specialized, mature form.

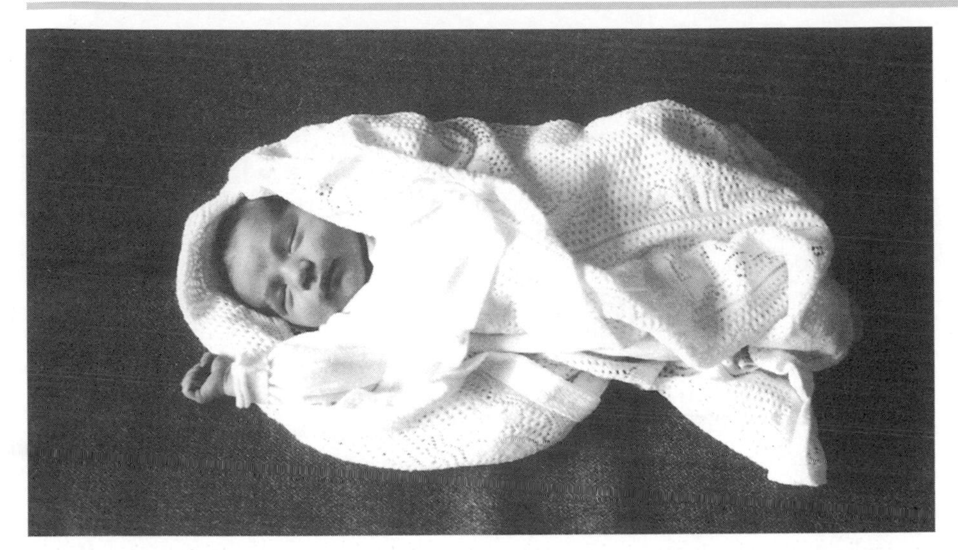

He has a lot of growth ahead of him, yet his development is all but complete.

This unit deals with general principles of development illustrated by examples, not detailed descriptions of how particular embryos develop. Animals and plants are dealt with separately, because they raise such different questions. Some of the answers, however, may turn out to be quite similar.

25.2 Division and cell movements in early animal embryos

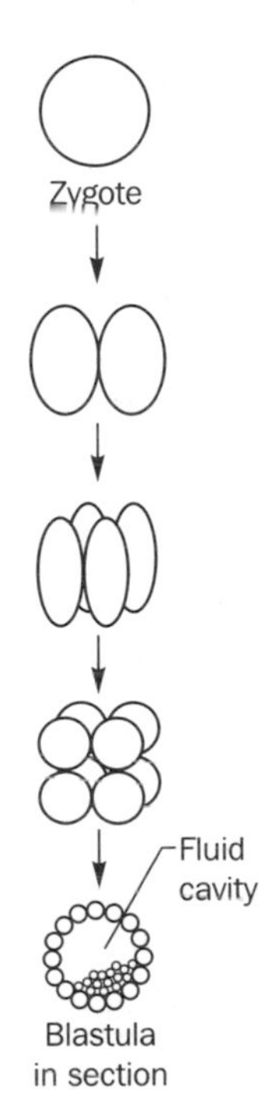

Fig. 25.1 The zygote divides to form the blastula.

Most animal embryos begin development in similar ways. Cells divide again and again, but do not grow after dividing. The zygote is a large cell, containing food stores from the egg, as well as a nucleus from each gamete (Section 13.5). It divides to produce two daughter cells, then four, eight, sixteen and so on, with cells becoming smaller with each division. This results in a **blastula**, a hollow ball of cells (Fig. 25.1). The cytoplasm of the original egg cell has been carved up, but not mixed. If one end of the egg cell had distinctive features in the cytoplasm, these features will end up in the cells at one end of the blastula.

Blastula: An early embryo which is a hollow sphere of cells.

In all but the simplest animals, the body wall has three layers (it is triploblastic). This comes about by a mass movement of cells at the blastula stage. The process is called gastrulation. It is seen at its simplest in sea urchin embryos (Fig. 25.2). Part of the blastula wall pushes inwards at one point on the surface. As it pushes, it takes on

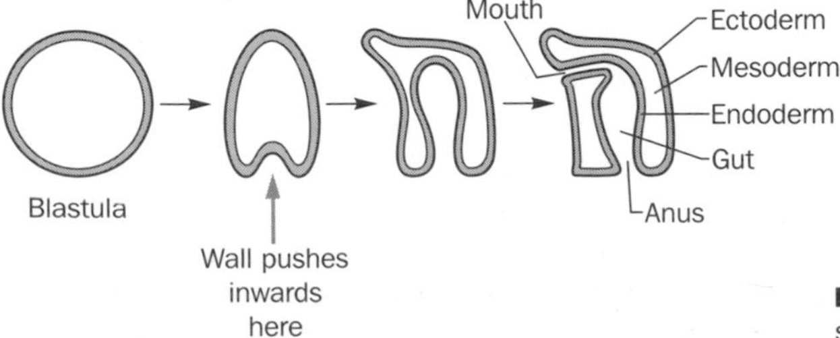

Fig. 25.2 Gastrulation in the sea urchin.

the form of a tube. Blow up a balloon just a little, and slowly push your finger into it. The balloon's skin makes a tubular infolding, just as in gastrulation. Eventually, the infolding reaches the other end of the embryo, and the cell layers fuse. The result is an open tube running right through the embryo: a gut. The embryo is now called a **gastrula**. By this stage, the body wall has three layers: the original **ectoderm**, the infolded **endoderm**, and the **mesoderm** sandwiched in between. Gastrulation is more complex in insects and vertebrates, but the basic idea is the same.

Gastrulation is just one example of cell movements in development. The neural tube of vertebrates, which becomes the spinal cord and brain, develops in a rather similar way. A long strip of ectoderm pushes inwards to form a furrow. This gets deeper, then folds in on itself, so that a hollow tube is pinched off, inside the embryo. Single cells may also move, from one part of the embryo to another, forming new structures which will become organs. Cell movements in the embryo are mostly by crawling (Section 4.5: The cytoskeleton). A single cell may crawl through the embryo. A sheet of cells, each trying to crawl over the others in particular directions, will lead to the sheet distorting, as in gastrulation.

With all this movement going on, it is important that other cells stay attached to their neighbours, keeping structures intact. The cadherins are important here. They are a family of plasma membrane glycoproteins. Similar cadherins bind together, so one cell type sticks together.

25.3 Positional information

Different parts of embryos develop in different ways. Even as early as gastrulation, it is obvious that opposite ends of the embryo are doing very different things. Cells must be dividing, moving and differentiating according to where they are in the embryo. How do they know? There are two types of answer.

The first is seen in the early embryos of molluscs and amphibians, for example. The cytoplasm of the egg itself is not uniform. It contains mRNA molecules which will be translated to produce gene-control proteins. mRNAs and other molecules may be distributed unevenly within the egg, sometimes in a gradient from one end to the other. As the egg divides into a ball of cells, each cell contains different concentrations of these molecules, according to where in the embryo it is. This is called mosaic development. The blastula is a mosaic of similar looking, but chemically distinct cells. When cells move, they carry this information with them. If a four cell mollusc embryo is broken apart, each cell can develop into an embryo. However, these embryos are strange and incomplete. They lack **positional information** which was trapped in the cytoplasm of the other three original cells.

The second answer is seen in early mammalian embryos, for example. If a mouse embryo is split up at the four-cell stage, each cell can go on to make a complete, normal embryo. Cells are not carrying positional information inside them. Instead, they establish where they are by communicating with one another. They may set up chemical gradients within the embryo. If their position changes, they can respond to their new environment. This is called regulative development.

Regulative and mosaic development are not exclusive. During development of an embryo, it is quite possible for cells to get some positional information from gradients within the egg, and the rest from communication with other cells.

Gastrula: An early embryo with a newly formed gut.

Ectoderm, mesoderm, endoderm: The outer, middle and inner layers of an animal's body wall.

Positional information: Chemical signals which tell cells where they are in an embryo.

Wherever it comes from, positional information is used to control genes. Genes control cell division. Sets of genes control the different paths along which a cell can differentiate. When a cell becomes, say, an epithelial cell, one set of genes is turned on. The sets of genes which would allow it to differentiate as a nerve cell or a liver cell are turned off. However, these are not the only genes involved in development. The polarity and body plan of the entire animal are controlled by genes.

25.4 A hierarchy of genes

We know more about how genes control the body plan of the fruit fly *Drosophila melanogaster* than any other organism. However, it is important to understand two odd features of insect development. Firstly, there is no conventional blastula. Instead, the zygote nucleus divides by many rounds of mitosis, without cell division. The result is many nuclei in one mass of cytoplasm: a syncytium. The nuclei move to the edges, and are only then boxed off by membranes. This produces a hollow ball of cells surrounding a yolky mass, instead of a fluid-filled space. Secondly, flies have two different bodies, the larva (maggot) and the imago (adult). The body goes through two phases of development. About 10 000 cells go to form the larva. Another 1000 or so cells remain in 16 clusters, strategically placed around the larva. These are the imaginal discs. They do nothing in the larva, but go on to form nearly all the adult body. During metamorphosis, when the fly is a pupa, the larval body is demolished and the materials are used for developing the imago, from the imaginal discs. Different discs give rise to different structures: a leg, a wing or an antenna, for example. Only the gut remains.

A hierarchy of genes controls the body plan of the larva and adult, on finer and finer scales. The first set of genes determines the polarity of the embryo: which is the dorsal surface, which is the anterior end, and so on. These genes are active in the mother, and help make the egg polar. Two important genes (there are others) are *bicoid* and *dorsal*. They make the proteins Bicoid and Dorsal. (These funny names were first given to mutations, which led to the discovery of the genes.)

The cells around the egg secrete *bicoid* mRNA into the anterior end of the egg as it develops. The mRNA gradient (Fig. 25.3) leads to development of a head and thorax at this end. There is no gradient of *dorsal* mRNA. It is spread evenly through the cell. However, as cells form in the syncytium, Dorsal protein becomes distributed in a peculiar way. On the ventral surface, there is more in the nuclei where it can affect

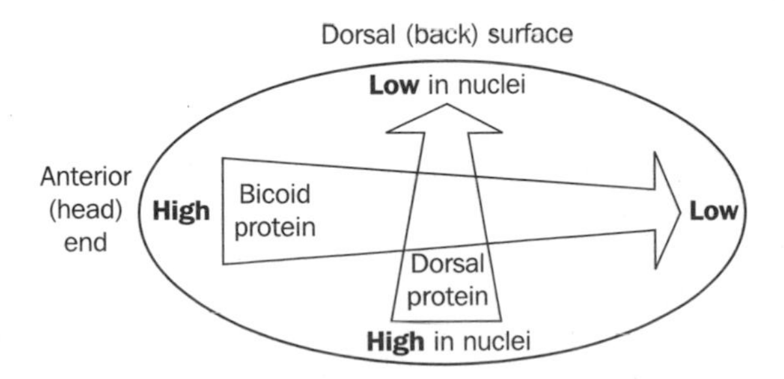

Fig. 25.3 Protein gradients in the *Drosophila* egg (see text for details).

genes. On the dorsal surface, more is in the cytoplasm where it cannot. The Bicoid gradient, then, is set up by the mother. The Dorsal gradient is set up by the embryo itself. Something else must give the original positional information.

These, and a few other genes, map out the polarity of the embryo. They make gene-control proteins, which activate a sequence of other genes. Each sets up more local gradients, and activates genes which bring about gradients which are even more localized. Eventually, the detailed segmentation of the embryo is marked out (Fig. 25.4). Detailed positional information is used to control cell differentiation, again by controlling genes.

Yet another set of genes uses this positional information to control imaginal disc development. Each disc needs to develop according to where in the larva it is. Different genes control development of different organs: they are called homeotic genes. The correct genes must be expressed in each imaginal disc. Mutations in homeotic genes lead to some weird bodies: flies with perfectly formed legs on their heads, in place of antennae, for example.

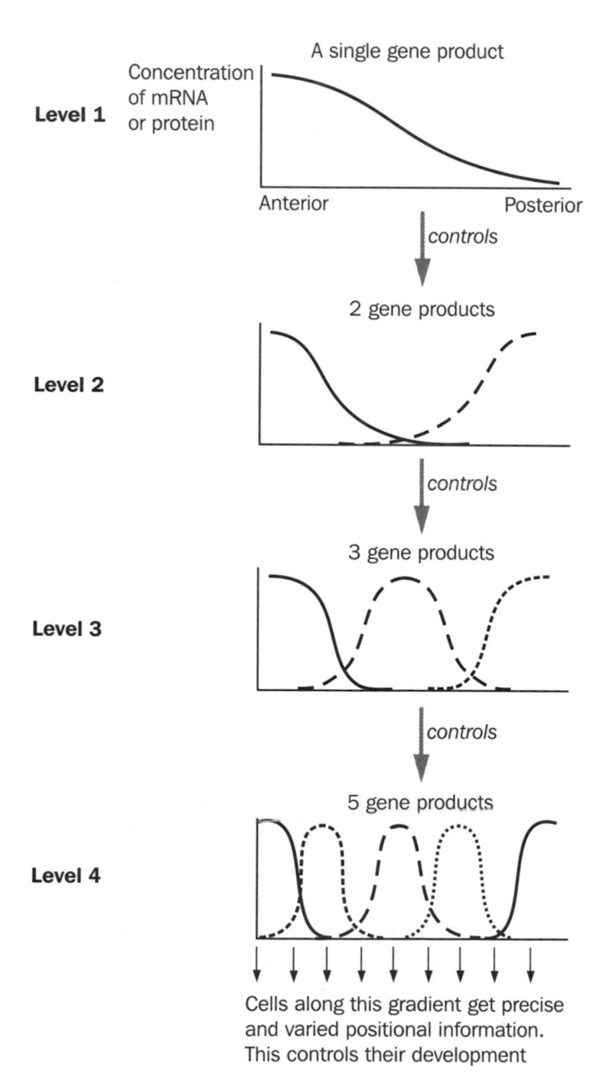

Fig. 25.4 Simple concentration gradients can give detailed positional information (see text for details).

25.5 Controlling tissue growth

The basic body plan, organs and tissue arrangement is laid out quite early in animal development. Within each tissue, growth must be carefully controlled. The number of cells must increase as the animal grows, but not too much. Some animals stop growing when mature. These include the mammals, but not the fish. Then, tissues must keep a constant number of cells.

Cell division increases cell number. Cell death decreases it (Fig. 25.5). Both processes can be controlled, in order to give slow tissue growth, faster tissue growth, no growth or even shrinkage.

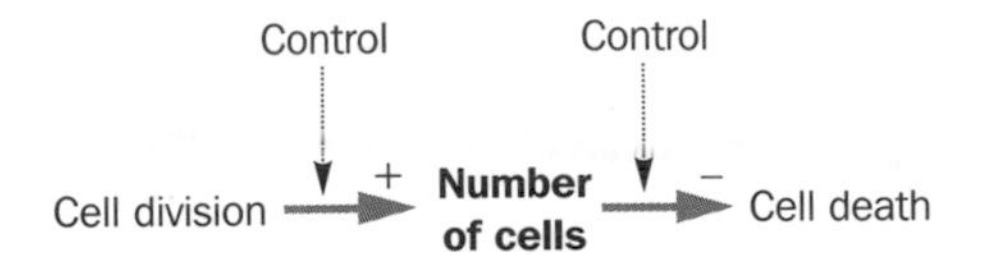

Fig. 25.5 Factors affecting number of cells in a tissue.

New body cells are made by mitotic division (Section 11.2). Some differentiated cell types can divide to make more cells like themselves. Hepatocytes, the cells which form the bulk of the liver, are an example. Others cannot divide. Instead, they are made by division of less differentiated cells called stem cells. Some types of stem cell can make just one type of differentiated cell. Epidermal stem cells in the lining of the gut only make epidermal cells. Others make more than one type. Haemopoietic stem cells in the bone marrow make cells which divide further and differentiate into any type of blood cell. To find stem cells which can give rise to any type of differentiated cell, we must go back to the early embryo.

Cell death is not solely an accidental process. Certainly, cells do die as a result of infection or physical damage. They die messily, swelling and bursting to release a cocktail of substances which leads to inflammation (Section 23.4). However, cells can also die quickly and tidily, without any sign of injury: they seem to be committing suicide. The nucleus breaks up and macrophages are called in to engulf and digest them. There is no inflammation. This programmed cell death (PCD) has a role in regulating the number of cells in some tissues. It can also occur at an early stage in viral infections, and is important in development. For example, a duct which would develop into a uterus and vagina is removed by PCD in male mammalian embryos.

The cell division cycle and PCD are each controlled by genetic programmes which are only partly understood. If a cell is to know when to divide or die, these programmes must be controlled by signals from other cells. These signals may be local signalling molecules released by neighbours, or hormones produced elsewhere in the body. In general, cells are prevented from dying by signals from their neighbours. PCD happens if the signals stop. On the other hand, cells are actively stimulated to divide by local or hormonal signals.

When a tissue stops growing, cells must divide and die at exactly the same rate. This is almost certainly controlled by negative feedback systems (Box 15.1). We know too little about these control systems: Box 25.1 shows an example which is reasonably well understood.

We now turn to development and growth in plants.

Box 25.1

Regulating erythrocyte numbers

Mammalian erythrocytes (red blood cells) are formed at the end of a chain of cell types, starting with the haemopoietic stem cells in the bone marrow. A negative feedback loop controls erythrocyte numbers in the bloodstream.

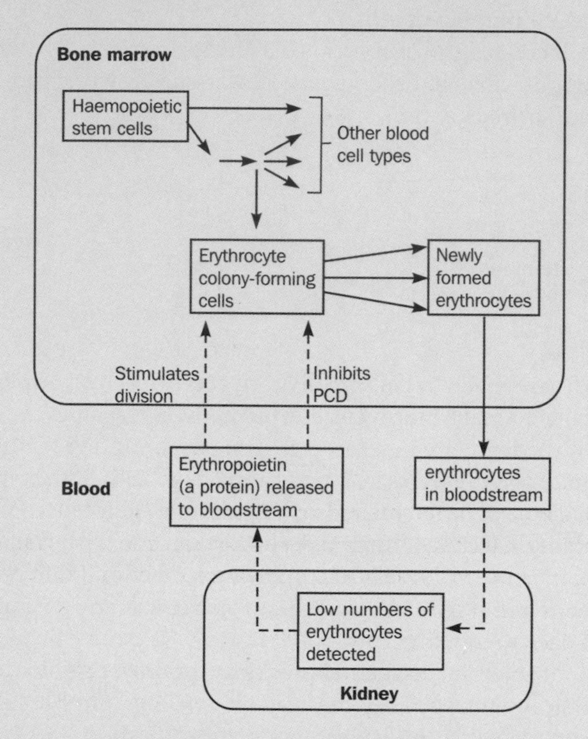

Notice that the same signalling molecule, erythropoietin, inhibits PCD as well as stimulating division. This is a negative feedback loop. If erythrocyte numbers fall, more erythropoietin is produced, stimulating erythrocyte production. If erythrocyte numbers rise, less erythropoietin is produced so erythrocyte production falls. Other factors influence the process too.

25.6 Modules and meristems

Plant bodies are made up of a few types of module, repeated again and again (Fig. 25.6). In flowering plants, leaves are attached to stems. Each stem is made up of nodes, where leaves are attached, with internodes in between. At each node, there is an axillary bud just above the leaf. The bud is a shoot in miniature, with scale-like leaves. It can grow to form a branch, or to replace the upper part of the original stem if it gets damaged. Flowers are sometimes carried on the stem, just above modified leaves. Below ground are the roots.

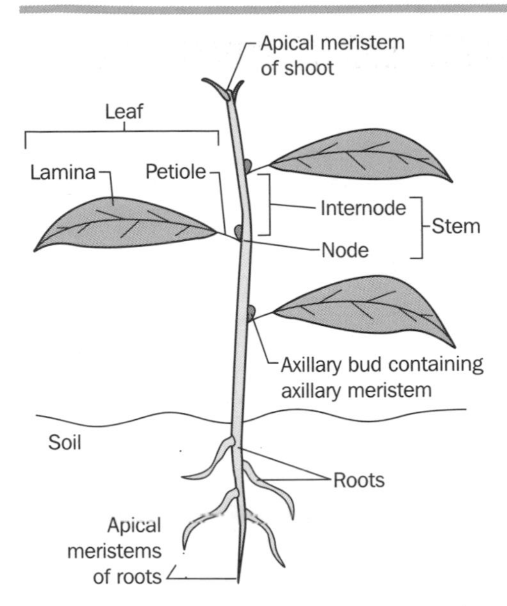

Fig. 25.6 Modular construction of a plant body.

Apical meristems

Stems and roots lengthen at the tips. Their growing points are called **apical meristems**.

Each root tip has an apical meristem very near its tip (Fig. 25.7). Cell division goes on here. As cells expand, they push one another further from the meristem, back towards the rest of the root. They are arranged in orderly lines. As they divide and expand, the root tip is pushed forward through the soil.

Cells behind the root tip divide in different ways. If the new wall is formed across the cell, at right angles to the axis of the root, the root will lengthen. If the new wall forms lengthways, in line with the root axis, the root will thicken. Once cells have expanded, they differentiate to form the tissues of the root.

Other cells are pushed ahead of the meristem. They expand to form a disordered mass of large, thin-walled cells: the root cap. These cells lubricate the root as it thrusts forward. They are scraped away as it pushes past hard particles in the soil, but the meristem stays intact.

The meristem of the shoot apex is right at the tip of the stem. It is a shallow cone of small, dividing cells (Fig. 25.8). They are organized into lines and layers. As the cells divide and expand, the meristem is pushed upwards. The cells below differentiate to form the tissues of the stem. Just behind the meristem tip, bulges form on the surface: primordia. Each primordium develops into a leaf and an axillary bud. Each bud, being a shoot, has its own apical meristem. In flowering shoots, the primordia develop into bracts (modified leaves) and flowers. The developing flower has a meristem of its own. The primordia on its surface become flower parts: petals, stamens and so on.

Flower production by meristems is one of the few areas where we know something about the genes which control plant development. A few important genes use positional information to control the development of a plant organ, presumably by controlling other genes. These genes are behaving very like the homeotic genes in *Drosophila*. We have little idea what sort of positional information they may be responding to.

Apical meristems: The growing points at the tips of roots and shoots where cell division occurs.

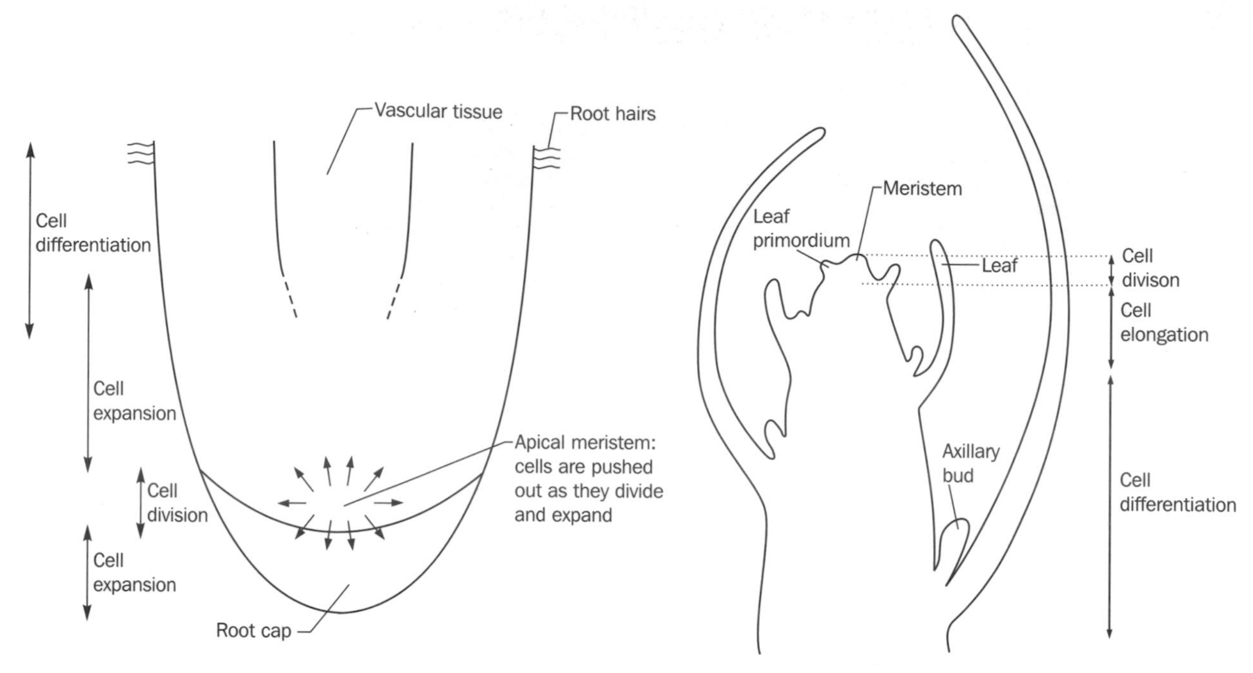

Fig. 25.7 The root tip meristem.

Fig. 25.8 The shoot tip meristem.

Fig. 25.9 shows the role of a gene called *floricaula* (*flo*) in flowering shoots of snapdragons. It is only expressed in the primordia which will become bracts and flowers. We do not know what turns it on. In normal plants, it is not expressed in the apical meristem itself, so the shoot keeps on growing and producing more flowers. Mutations are known in which the apical meristem develops into a flower and stops growing. They reveal that another gene, *centroradialis* (*cen*), is active in the apex. *cen* inhibits expression of *flo*. Simple genetics can lead to the important difference between a flowering shoot which carries a succession of flowers for weeks, and one which has a fixed number of flowers over a short period. Box 25.2 shows how three genes control the arrangement of flower parts in a floral meristem.

In 'double' flowers, primordia are developing into petals instead of stamens and other flower parts. This is the result of mutation in genes affecting flower development.

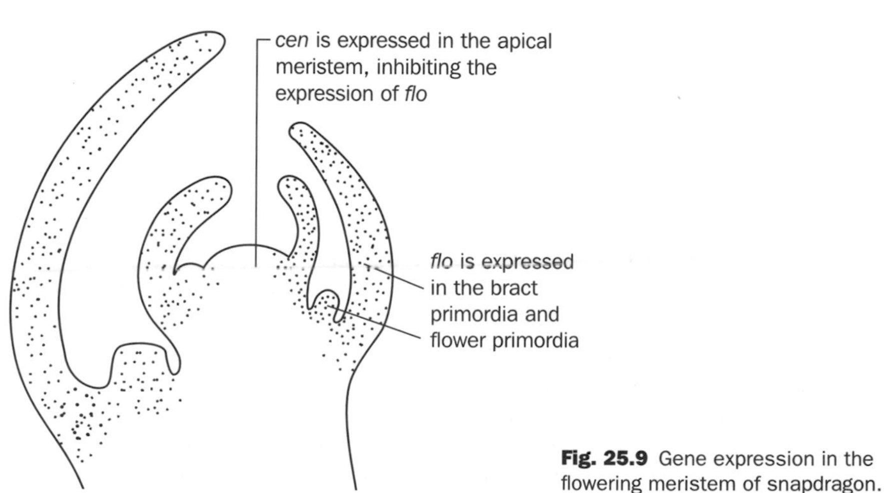

Fig. 25.9 Gene expression in the flowering meristem of snapdragon.

Box 25.2

Genes and the origin of flower parts

Arabidopsis thaliana is a small, dingy weed. It is a model system for flowering plant genetics because it can go through its life cycle in a matter of weeks, in a test tube: very convenient!

Analysis of mutants in which flowers had organs in the wrong places, led to the discovery of a set of genes which are expressed in the meristem. Some unknown positional information determines exactly where in the meristem each gene is expressed. The combination of genes which is expressed in a region of the floral meristem determines what organ will develop from those cells. In function they are very like the homeotic genes of *Drosophila*. The genes are called *apetala2*, *apetala3* and *agamous*.

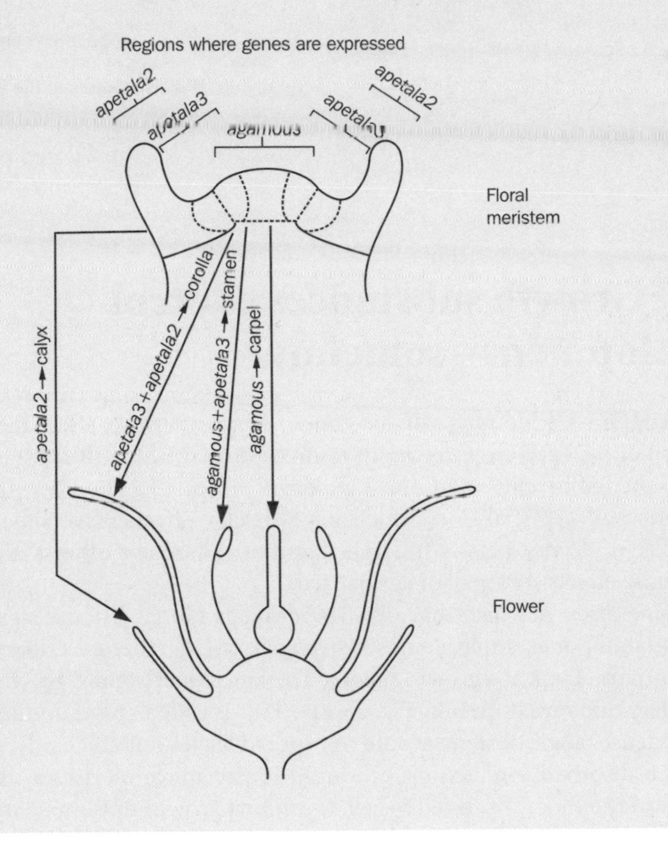

Secondary growth

Not all growth goes on at apical meristems. Woody stems such as tree trunks get wider each year. Even non-woody stems get thicker as the season goes on. Look at the base of a sunflower stem. This is the same bit of stem which was developing when the seedling was a few centimetres high, but much fatter. This is called secondary growth. Box 25.3 outlines how a woody stem thickens.

We know much less about how plant growth and development are coordinated. Chemical signals are certainly involved, but this field is deeply confusing. Light has a fundamental role in regulating development. These areas are discussed in turn.

Box 25.3

How a woody stem grows

The woody stems of dicotyledonous trees such as oaks, maples and beeches, get thicker year by year. This simple model explains the overall structure of the woody stem, though not the details of wood structure.

Cells in the cambium layer divide. If they end up inside the ring, they expand and differentiate as xylem cells. This pushes the cambium outwards. If newly-formed cells end up outside the ring, they develop as phloem. A separate layer of dividing cells forms the bark.

The central core of xylem gets bigger yearly. In seasonal climates, only a thin layer of dense, small-celled xylem is laid down in the cold or dry season. This leads to the annual growth rings of woody stems in the temperate zones and the seasonal tropics.

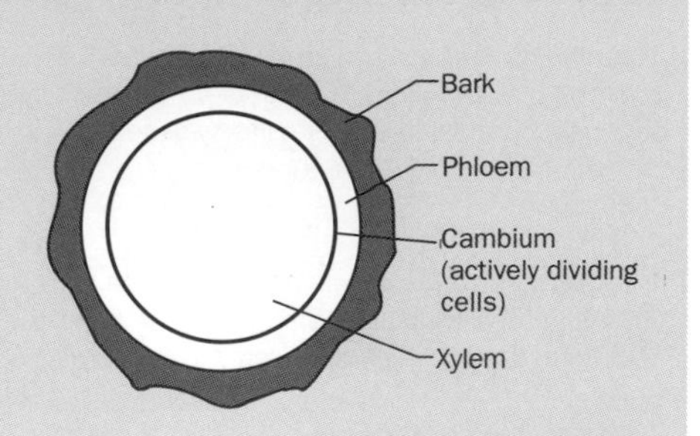

25.7 Growth substances control development – somehow

Plants contain a wide range of substances which seem to affect their growth and development. There are three main families of plant growth substance (PGS): the auxins, the gibberelins and the cytokinins (Fig. 25.10). They include natural molecules and artificial ones which have similar effects. Two others, ethene and abscisic acid, do not belong to larger families. There are others, such as jasmonic acid, whose effects are small or specialized.

In some cases, it is clear that a PGS is acting as a long-distance signal, controlling development. For example, auxins help to control whether an axillary bud grows or stays dormant. Look at a growing shoot. The apical meristem is growing, but most of the axillary buds are dormant (Fig. 25.11). This is called apical dominance. It makes sense: a dense tangle of stems would not spread leaves out efficiently, so less sunlight would be absorbed. An auxin is produced in the apical meristem, and transported back down the shoot, from cell to cell. It inhibits growth of the axillary meristems. If the apical meristem is removed, the auxin supply is cut off. One or more axillary buds will start to grow, forming new shoots. As they grow, they produce auxins and so suppress the remaining axillary buds. This is what happens when you cut a hedge to produce dense, bushy growth.

Many other roles of plant growth substances are less clear cut. We know very, very many things that can happen if a particular PGS is added to a plant in a certain way at a given concentration. However, it all adds up to a very confusing picture. Growth substances almost certainly can act as signals controlling development, probably working in combination. They bind to receptors: some lead to gene switching. When we know more about the receptors, and what genes are there to be controlled, a clearer picture should emerge. Just now, I would rather not say more.

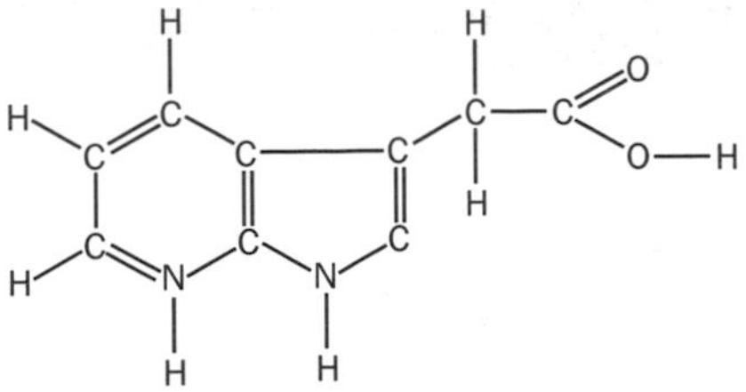

Indole-3-acetic acid (IAA): an important natural auxin

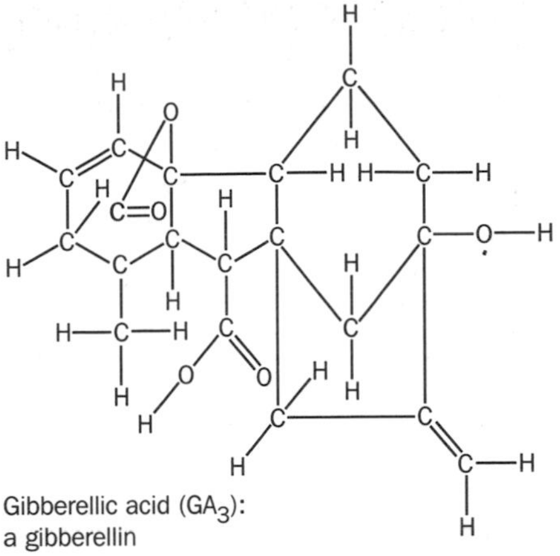

Gibberellic acid (GA_3): a gibberellin

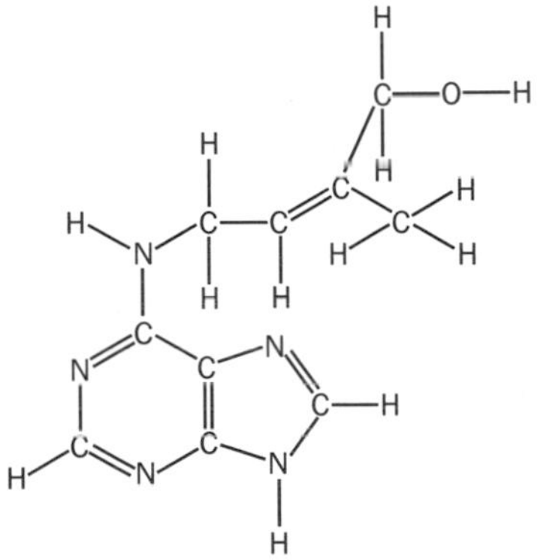

Zeatin: the most active natural cytokinin

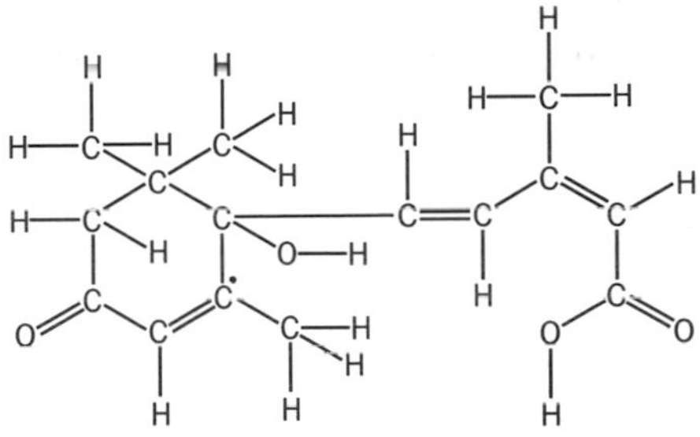

Abscisic acid

Ethene

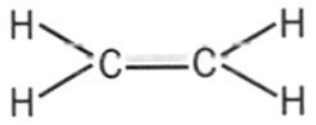

Fig. 25.10 Some plant growth substances.

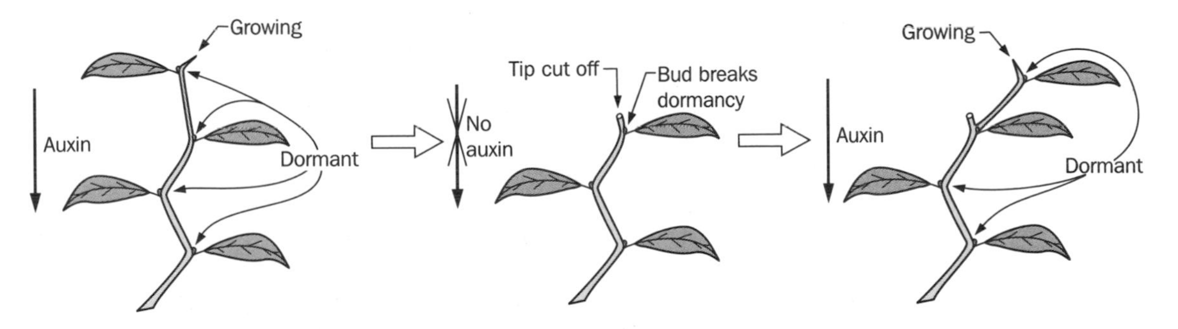

Fig. 25.11 Apical dominance.

25.8 Light and plant development

Plants need light: it is their source of energy. They trap it using canopies of leaves, arranged in a way which leads to efficient absorption. Plants develop amongst objects which block out light, neighbouring plants in particular. It is no surprise that light influences plant development in a big way.

If light falls on a shoot from just one side, the shoot grows towards it: phototropism. **Tropisms** are growth responses to the directions of stimuli. Phototropism is important, but there are others. For example, roots respond to gravity, sometimes by growing straight down, sometimes sideways: geotropism.

Leaves take up their positions in the canopy so they receive as much light as possible. Light affects the growth of the petiole (leaf stalk) as it manoeuvres the leaf into position.

Shoots which grow in the dark become etiolated. They have thin, elongated internodes, tiny leaves and have no chlorophyll. What resources they have are being channelled into growing out of this hostile environment.

Light is also the signal that a germinating seedling has broken out of the soil. In grass seedlings, the young leaves are rolled inside a sheath, the coleoptile, which spears up through the soil (Fig. 25.12(a)). Once in the light, the coleoptile stops growing. The leaves burst out and unroll, again triggered by light. Many seedlings of dicotyledonous plants push up through the soil using the hypocotyl bent into a hook (Fig. 25.12(b)). The cotyledons and shoot apex are pulled up behind it. Once in the light, the hypocotyl unhooks.

Yet other responses involve day length. This is very clearly seen in the control of flowering. Many plants can tell their flowering times have come by day length. Long-day plants only produce flowers when days are longer than a certain critical day length (CDL). Short-day plants flower when day length is below a CDL (Table 25.1). Most chrysanthemums are short-day plants. As I write, in September, dozens of chrysanthemum plants of one variety are coming into flower together on my wife's nursery. The flowers were initiated at the same time, several weeks ago, as the late summer days became shorter than the CDL. Seed germination may also be affected by day length in some species.

How do plants detect light, and how does it trigger these developmental changes? Two types of light-absorbing molecules are involved, phytochromes and cryptochrome, but not chlorophyll or other photosynthetic pigments.

Phytochromes are most important. They consist of a protein with a light-absorbing prosthetic group, very like a phycobilin in structure (Box 10.1). They are found in the cytosol. There are several types (no one is quite sure why) but all work in almost the same way.

Tropism: Growth response to the direction of a stimulus.

Table 25.1 Effect of day length on flowering in a classic short-day plant (SDP; cocklebur, CDL=15.7 hours) and a long-day plant (LDP; henbane, CDL=12 hours). The last two rows show that it is *night* length which is important. (Day/night cycles which do not add up to 24 hours are possible in a growth room with lights.)

Day length /hours	Night length /hours	Flowering in cocklebur	Flowering in henbane
10	14	yes	no
14	10	yes	yes
17	7	no	yes
14	14	yes	no
10	7	no	yes

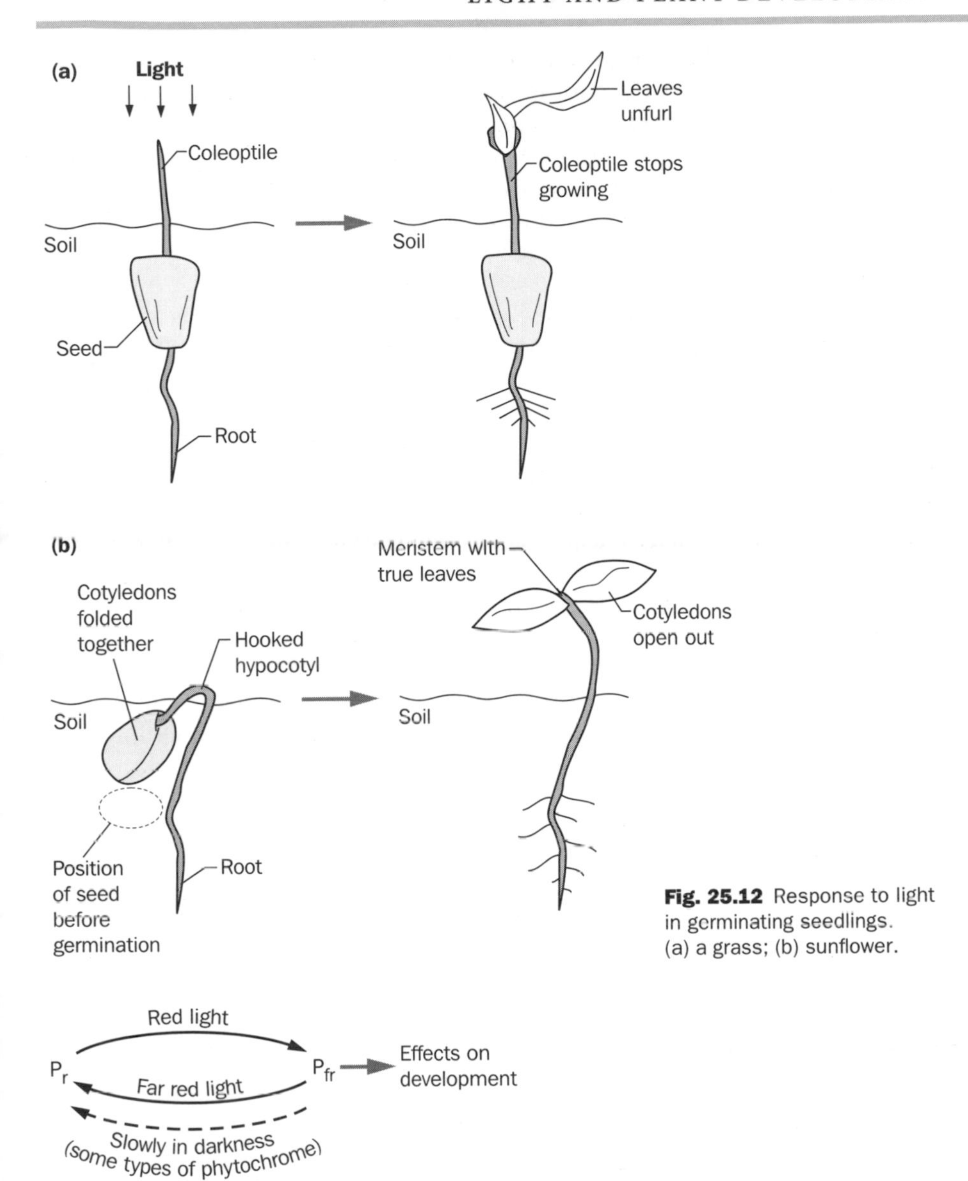

Fig. 25.12 Response to light in germinating seedlings. (a) a grass; (b) sunflower.

Fig. 25.13 The two forms of phytochrome.

A phytochrome can switch between two forms, P_r and P_{fr}. P_r absorbs red light, with peak absorption at about 660nm. P_{fr} absorbs far red light, where the visible spectrum meets the infra-red, around 730 nm. When P_r absorbs light, it becomes P_{fr}. When P_{fr} absorbs light, it becomes P_r (Fig. 25.13). In some, but not all phytochromes, P_{fr} reverts to P_r in darkness, but this takes an hour or so. Sunlight contains more red than far red light, so most phytochrome ends up as P_r in the day. At night, underground, or even in deep shade under a tree canopy which absorbs red light effectively, P_r builds up. This provides an excellent indicator of light and dark: high P_{fr} means it is light, low P_{fr} means it is dark.

Phytochrome is involved in many areas of development, including hypocotyl unhooking, germination in some seeds, etiolation and control of flowering in very many species. Whenever a plant responds to 'day length', it turns out to be night length which matters. A brief flash of red light in the middle of a long night can reverse its effect. It seems that the period of time when P_{fr} is low is being measured against some sort of biological clock. How it works is not at all clear.

Cryptochrome absorbs blue/violet light as well as near ultraviolet (UVA) radiation. It has not been positively identified, but is probably a protein in the plasma membrane. Blue light responses include phototropism, and uncurling of leaves in some grass seedlings.

We know much less about how these pigments bring about their effects, and what signals move from where light is detected to the cells which respond. The research literature is full of detailed observations which add up to almost total confusion. The most studied system is phototropism in oat coleoptiles. It has been investigated for over 100 years. The leading hypothesis runs as follows: light is detected at the coleoptile tip by cryptochrome. This leads (somehow!) to a redistribution of auxins: there is a higher concentration on the shaded side. Auxins move down the coleoptile and stimulate cell elongation. The shaded side elongates more, so the coleoptile bends towards the light (Fig. 25.14). After 70 years of testing this idea, there are still lingering doubts. It is even less clear whether we can extend the model to other grass coleoptiles, let alone stems.

Another well-studied example is the flowering stimulus. There is excellent evidence, built up over 60 years, that once the phytochrome system has detected a suitable day length, a flowering hormone, or florigen, spreads through the plant. Many people have tried to identify florigen. Nobody has succeeded.

Study of the genetic basis of plant development may one day lead to models which integrate light-detecting pigments, plant growth substances, receptors, gene-control proteins and events at the meristems. Meanwhile, the marvellous problem of plant development remains as tough as ever.

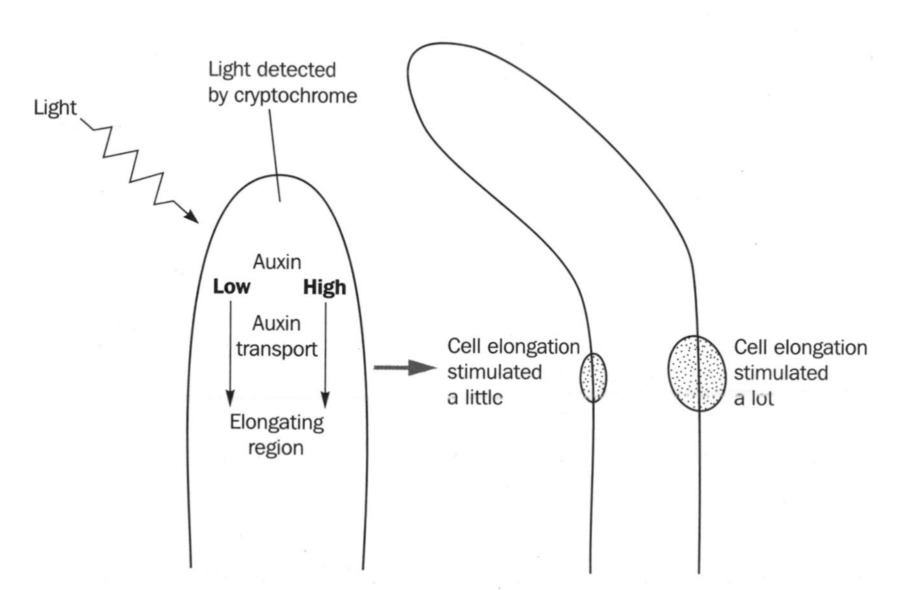

Fig. 25.14 A model of phototropism in oat coleoptiles (see text for details).

Summary

- Animals and plants start life as single cells. They divide and differentiate to form highly organized multicellular bodies: development.
- As bodies and organs develop, they get larger: growth.
- Growth involves cell division and expansion.
- Most animals have an integrated body plan, a fixed number of organs in fixed positions. Growth continues after the body plan has developed.
- Plants are made up of a few types of organs, or modules, repeated many times. Plants grow by adding on extra modules.
- Animal zygotes divide many times to form a hollow ball of cells, the blastula.
- Cell movements in one part of the blastula lead to a hollow tube passing through it, to form a gut. The embryo is then called a gastrula.
- ◆ Cells differentiate according to where in the embryo they are. They respond to positional information.
- Some positional information comes from mRNA molecules found in local regions of the egg cell. As the zygote divides, the mRNAs are trapped in certain cells.
- Other positional information comes from cell communication.
- A few genes, active in the mother, set up mRNA and protein gradients in the fruit-fly egg. These large-scale gradients control other genes which set up more local gradients, and so on. Eventually, these gradients control the genes which affect how the cell should differentiate.
- The homeotic genes in the fruit fly are switches which control the development of adult body parts.
- ◆ Some cell types make more of themselves by division. Others are formed by division of less differentiated stem cells.
- The number of cells in a tissue is tightly regulated, by controlling cell division and cell death.
- Plants grow at apical meristems. They are regions of active cell division at root and shoot tips.
- ◆ New leaves, flowers and buds are formed as bulges called primordia on the sides of shoot meristems.
- Genes which control development of floral meristems and flower parts, have been identified.
- Plant growth substances can affect growth and development in many ways. Their role in signalling is not clearly understood.
- ◆ Plant growth is deeply affected by light.
- Flowering and germination are controlled by day length in some species.
- Shoots grow towards light: phototropism.
- ◆ Seedlings undergo a set of changes as they emerge from the soil.
- ◆ Shoots grown in the dark have long internodes, small leaves and no chlorophyll: etiolation. It is an escape response.

- ◆ Light is detected by the pigments phytochrome (red and far red light) and cryptochrome (blue light).
- ◆ Day length is measured by a combination of phytochrome and a biological clock.

Exercises

25.1. **(i)** The cells of very early sea urchin embryos can be separated, and develop as normal larvae. What term describes this sort of development?
(ii) Later in development, cells taken from sea urchin embryos form incomplete embryos, or do not develop further. Explain this observation, in general terms.

25.2. Which will be the main form of phytochrome in each of the following situations:

(i) in red light;
(ii) in far red light;
(iii) after 10 hours of darkness;
(iv) after 10 hours of darkness followed by one minute of far red light;
(v) after 10 hours of darkness followed by one minute of red light?

25.3. A short-day plant has a critical day length of 14 hours. Which of these light/dark cycles will initiate flowering?

(i) 18 hours light/6 hours dark;
(ii) 16 hours light/8 hours dark;
(iii) 12 hours light/12 hours dark;
(iv) 8 hours light/16 hours dark;
(v) 16 hours light/12 hours dark;
(vi) 8 hours light/8 hours dark.

Further reading

Alberts, B., Bray, D., Lewis, J., Raff, M., Roberts, K. and Watson, J.D. *Molecular Biology of the Cell* (3rd ed.) (New York: Garland, 1994). An enormous but excellent textbook which puts the reader in touch with current ideas without too much pain on the way. One chapter introduces the molecular basis of development.

Browder, L.W., Erickson, C.A. and Jeffery, W.R. *Developmental Biology* (3rd ed.) (Philadelphia: Saunders, 1991). A big undergraduate text on animal development.

Goodwin, B. (Ed.) *Development*. (London: Hodder & Stoughton, 1991). A textbook aimed at new undergraduates, written for self-study.

Ridge, I. *Plant Physiology* (London: Hodder & Stoughton, 1992). A lighter textbook aimed at new undergraduates, written for self-study. Refer to this for plant growth substances, but do not expect easy answers from anyone!

Rudall, P. *Anatomy of Flowering Plants* (2nd ed.) (Cambridge: Cambridge University Press, 1992). A concise summary of flowering plant structure and how it develops.

Salisbury, F.B. and Ross, C.W. *Plant Physiology* (4th ed.) (Belmont CA: Wadsworth, 1992). A big undergraduate textbook. Any text in this field has to navigate a minefield of conflicting ideas and evidence in so many topics. This book does as well as you could hope in drawing out conclusions without avoiding the arguments. Plant growth substances are covered in some detail.

PART 6

Organisms in Context

Population Ecology

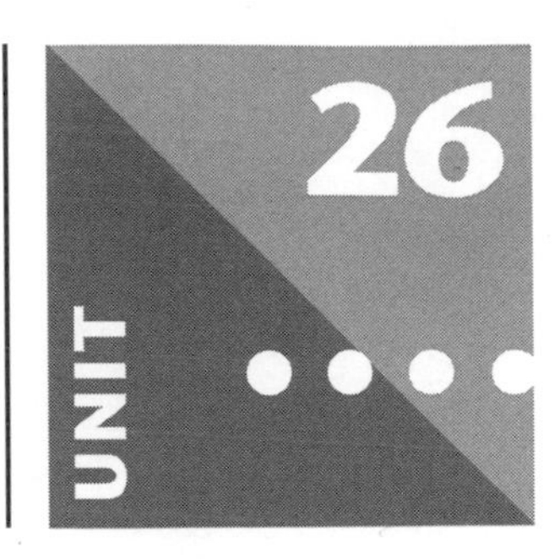

Connections

▶ This unit should be easy to follow on its own, although a basic understanding of homeostasis is useful (Unit 15). Communities are made up of populations, so much of community ecology (Unit 27) is rooted in population ecology. Ecology can also be studied on the larger scale of the ecosystem (Unit 28).

Contents

26.1 Populations

You are walking along a clifftop. Either side of the path is grassland, kept short by rabbits and other grazing animals. Like most other people, you are keeping to the path because the grassland is interspersed by prickly gorse bushes. In some places they are packed together densely, in others they are widely spaced. Some bushes are young, low and bushy; others come up to your chest and are collapsing outwards, exposing their woody stems in the centre. These gorse bushes are individuals of the same species, living in the same area: a **population**. They are not on the best of terms with one another. Each individual needs the same things in life: light, water, various minerals, and the space in which these things are to be found. As members of the same species, they take these resources in similar ways. By taking resources, individuals are not only getting themselves the raw materials for growth, but are also stopping the neighbours from getting them. This is competition. The gorse bushes are not only competing with one another. They are growing amongst other plants which are also competing with them. There are hostile animals around, too. The rabbits may be put off by the spiny leaves, but a closer look reveals froghoppers (maybe you call them spittle bugs) all over the bushes. They too form a population.

Where does one population end and another begin? Sometimes it is easy to decide. The minnows in my local river, the Dart, are completely isolated from those in the Teign, the next big river system. They are distinct populations. It is rarely so easy to decide. Badgers living in one wood may be several kilometres from badgers in the next such wood, but do they compete with one another for resources? How far and how frequently do they move? Do they interbreed regularly? For most populations we can only guess at the answers. Normally we end up defining the limits of the population on practical grounds. The population is sometimes taken to be just those

Population: Every member of a species in a habitat.

individuals living in the chosen study area. Alternatively, we assume that the sample studied represents what is going on in the wider population, however large it might be.

How large is the population? How densely packed are the individuals? What is the age structure of the population – are they mostly youngsters, oldies or a more equal mix of all ages? For how long does an individual live, and how much variation is there in this? When is the risk of dying greatest: when young or old, or is there an equal risk of dying at any age? When in their lives do individuals reproduce? Many populations do not change wildly in size, even though individuals are constantly dying and being born. What regulates population size? Is it factors which depend on the **density** of individuals, so that fewer are born or more die when the population is more dense? Competition, the risk of being eaten by a predator, and disease may all act in a **density-dependent** way. On the other hand, are **density-independent** factors more important, such as the harshness of the winter?

These are the sorts of questions we can ask about populations. The answers are not to be found at the population level, however, but in terms of individuals and how they interact with one another. This unit is about the important questions, and some examples of the answers which appear when somebody takes the trouble to study a population in detail.

26.2 Population size, structure and change

The first step in describing a population is to estimate its size. This is relatively easy with plants, and animals which do not move about. The density (number of individuals per unit area) can be estimated if numbers are counted in small, representative sample areas. The size of the entire population can then be estimated if the area of the habitat is known. Animals which can move are more tricky. They can hide, run away and mix in with the rest of the population after being counted. They have to be tracked, trapped or marked, in all manner of ways, depending on the species. Even humans present problems. It takes an official census to get a really accurate idea of how many people live in your own town.

Organisms which reproduce asexually present another problem: what is an individual? Is it an independent living body, or is it a genetic individual, all the cells derived from one fertilized egg? This is no problem in vertebrate animals, and many invertebrates. A grey wagtail is a grey wagtail, however you look at it. The problem is much more serious in organisms which reproduce asexually, especially plants. If I am gloomily trying to estimate the number of daisy plants in my lawn, do I simply count the number of rosettes of leaves (easy) or try to work out which ones came originally from the same seed (very tricky)? The 'working unit' type of individual is called a **ramet**. The genetic individual is called the **genet**. The choice of which to study depends on the point of the research. An agricultural scientist investigating weeds competing with a crop may well be interested in ramets. Genets will be more important in evolutionary biology.

Density: Number of individuals per unit area.
Density-dependent factors: Factors which vary according to the population density.
Density-independent factors: Factors which are not affected by population density.
Ramet: One way of defining an individual – a more or less independent living body.
Genet: Another way of defining an individual – all the living material derived from one fertilized egg.

Populations are not collections of identical individuals. Many have a mixture of ages (Fig. 26.1). Others may be even-aged, but individuals still vary in size or stage of growth (Fig. 26.2). Describing population structure is a major refinement, but it still does not help to answer the big question. Why is a population the size it is?

Four things can change population size. A population can only grow through

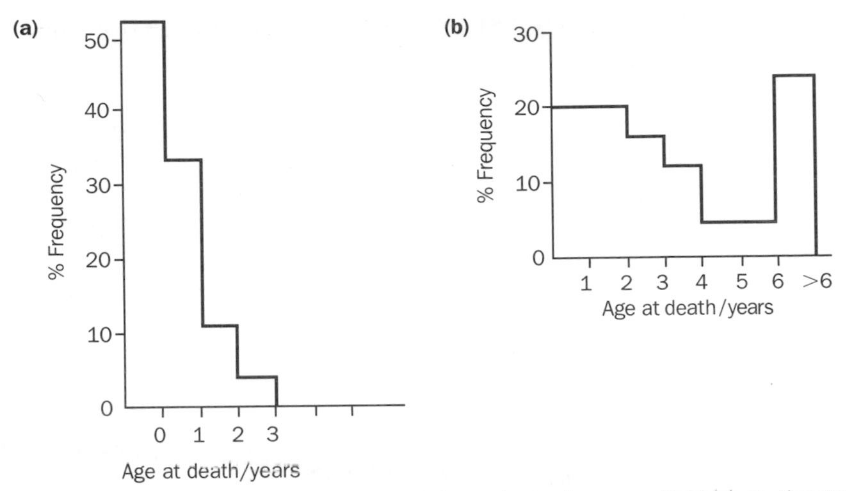

Fig. 26.1 Life spans of shoots in two populations of a tundra moss, *Polytrichum alpestre*. (a) when another moss species was present and (b) in pure stand.

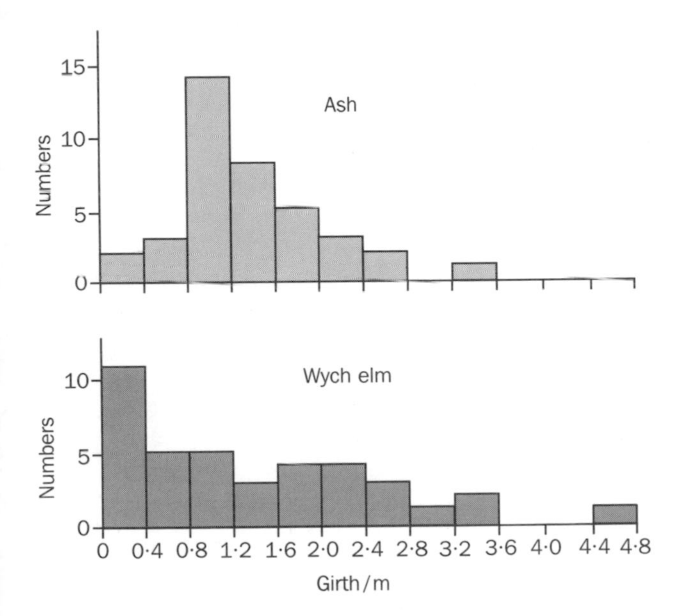

Fig. 26.2 Size structure of ash (*Fraxinus excelsior*) and wych elm (*Ulmus glabra*) populations in an English woodland.

Fig. 26.3 Factors affecting population size.

birth or immigration. It can only shrink through emigration or death. Population size is the result of a balance between these four processes (Fig. 26.3). This basic truth is a touchstone in population ecology. Much effort goes into describing how and when these factors operate in a given population (Box 26.1). It is even harder to work out what controls the rates of birth, death, immigration and emigration.

Analysing population dynamics

Raw data on the timings of birth and death in populations can be analysed to produce life tables and fecundity schedules. They show how the risk of dying and the chance of reproducing change as organisms go through their lives.

The details of life tables vary according to the life cycle of the organism. Table 26.1 shows the ideas as simply as possible. The references at the end of the unit will introduce you to the complexities of real life tables.

We follow a cohort of individuals, that is all the individuals born into the population during a narrow window of time, perhaps a single breeding season. These individuals are monitored throughout their lives. We must be able to identify them, perhaps by tagging an animal or accurately mapping a plant's position. At regular intervals, we record how many have died. We also record how many offspring have been born to our cohort. A life table is drawn up, as in Table 26.1.

Table 26.1 Life table (imaginary example)

Age/ years x	Number surviving at age x a_x	Proportion of original cohort surviving to age x l_x	Proportion of original cohort which dies between x and x+1 d_x	Mortality rate (the chance of an individual which survives to x dying before x+1 q_x	k value (see text) k_x
0	200	1.000	0.600	0.60	0.40
1	80	0.400	0.280	0.70	0.52
2	24	0.120	0.095	0.79	0.68
3	5	0.025	0.025	1.00	—
4	0	0.000	—	—	—

The proportion of the cohort surviving (l_x) allows us to plot a survivorship curve: l_x against age (graph (a)). It is more usual to plot $\log_{10}(l_x)$ against age, because this makes clear how risk of death changes with age. When death risk is the same at every stage of life, the graph is a descending straight line (dotted line on graph (b)). The graph for our example shows that risk of death is greater, later in life. This is confirmed by a graph of mortality rate ($q_x=d_x/l_x$) against age (graph (c)).

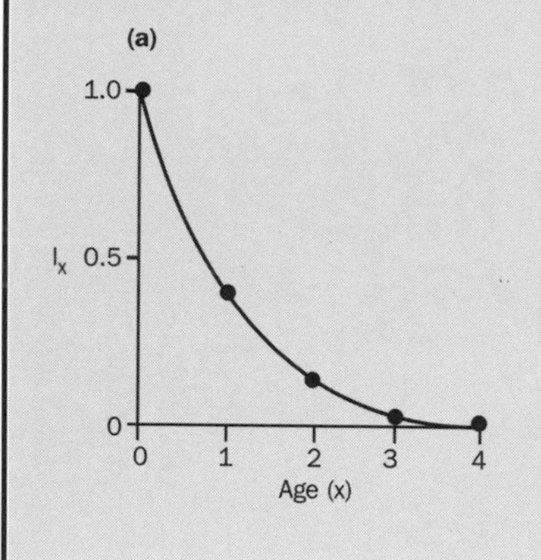

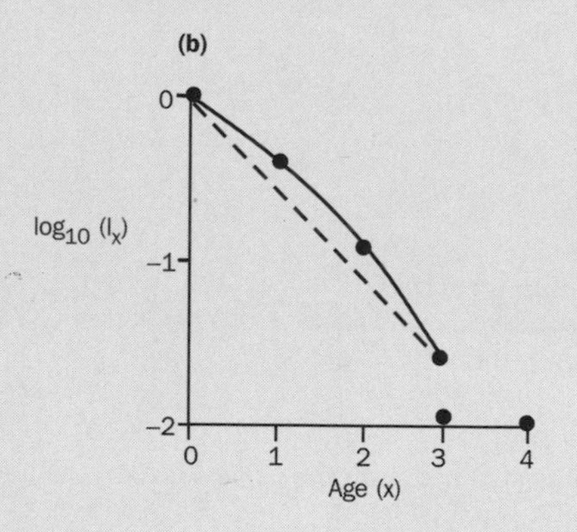

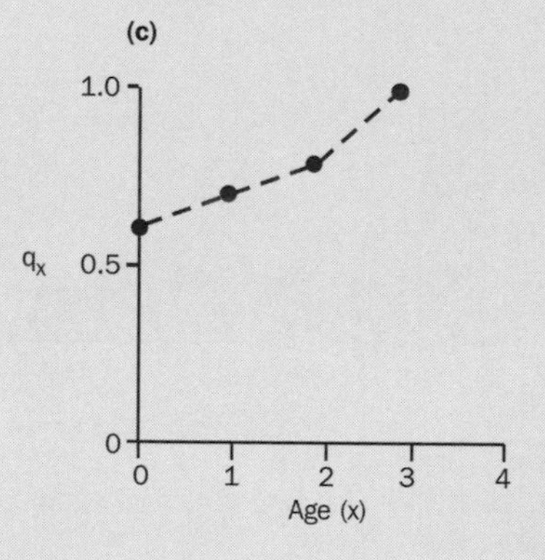

Birth

When in their lives do individuals reproduce? What controls the number of offspring? Most populations are held to one type of life cycle by their genes and/or their environments. **Semelparous** organisms are those which reproduce just once. For them, life is a period of resource collecting which ends in a burst of reproduction, then death. Annual plants are semelparous. Some perennials are as well, for example the foxglove and teasel. There are semelparous animals, too, for example the mayflies,

Semelparous organisms: Organisms which can reproduce only once.

Box 26.1

Mortality rate is a useful statistic but it cannot be summed. If we regrouped the data into two-year intervals, q_x at age 2 would be 0.975 (try it and see), yet we cannot get this figure by adding up existing q values.

The k value, or killing power, (k_x) is another measure of the risk of dying at a particular age. It is calculated as $\log_{10}(a_x)/\log_{10}(a_{x+1})$. It can be summed. This proves useful when trying to discover whether death at one particular stage of the life cycle determines changes in population size (Box 26.4).

While recording survivorship, we also record number of offspring born at different ages. This information is the basis of a fecundity schedule, as in Table 26.2.

Table 26.2 A fecundity schedule, compiled from data on reproduction

Age /years x	Number surviving at age x a_x	Total offspring between x and x+1	Average offspring per individual alive at x B_x
0	200	0	0
1	80	400	5.0
2	24	156	6.5
3	5	11	2.2
4	0	—	—

A graph of B_x against x shows how the chance of reproducing is not constant throughout life.

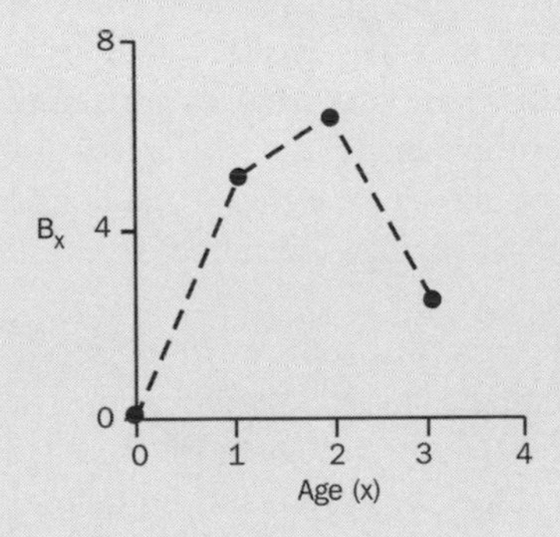

the Atlantic salmon and the common octopus. **Iteroparous** organisms may reproduce again and again, indefinitely. Cats, swans and pine trees are familiar examples.

Semelparous organisms can invest more of their resources in reproduction. They do not have to save any for further growth. This suggests that their populations could grow faster than similar iteroparous species. At first sight this seems a crushing advantage. However, iteroparity is an advantage when being big is important. Individuals which have reproduced and continue to grow will be larger than they were in

Iteroparous organisms: Organisms which are capable of reproducing more than once.

their first season. Large individuals may survive harsh winters better, or compete for resources more effectively. Then, less reproduction early in life can mean much more reproduction later. Delaying a single burst of reproduction has similar benefits, but there is an increased risk of dying without leaving any offspring at all.

The benefits of being big are so general, that the existence of short-lived, semelparous organisms needs some explaining. Sometimes they are exploiting a window of opportunity which does not last for long. Foxgloves are typically plants of woodland clearings. A tree falls, allowing light to the forest floor. Foxgloves germinate, grow and reproduce in spectacular fashion in less than two years. By this time, shrubs and young trees will be beginning to starve them of light, so there is little point in keeping back resources for the future. Seeds are dispersed and lie dormant in the soil until another clearing appears. Arable fields are rich in annual weeds. All vegetation above ground is destroyed by ploughing each year, so there is no incentive to hold back on reproduction. Many groups of insects are tied to semelparity by their specialized life cycles. For example, the eggs of mayflies (order Ephemeroptera) hatch as nymphs, specialized for feeding in fresh water. These eventually give rise to very short-lived adults, adapted for dispersal by flying, and reproduction. This is a one way process. The adult cannot revert to the nymph stage, so any resources not used for reproduction are wasted when the adult dies.

A semelparous plant. *Megacarpaea polyandra* grows as a low rosette of leaves for many years, going out in one glorious burst of reproduction. Native to the Himalayas, this one is doing its stuff in a Scottish garden.

Size and resource capture are plainly important in understanding the differences in patterns of reproduction between populations. They also determine the differences in reproductive success within a single population. This is seen particularly clearly in plants. As density increases, the resources available to the population are shared out more ways. Each individual gets a smaller share. It may be that some individuals get too little to survive, and die (see Death). However, most plants have a plastic, easily modified growth form. Very small or very large individuals may still be able to reproduce. The higher the density, the smaller the individuals. If the individuals are smaller, they can produce fewer seeds. So, as density increases, the average individual will produce fewer offspring. This effect is known as density-dependent **fecundity** (Box 26.2).

Density is a measure of the crowding of the population *as a whole*. This does not mean that each individual has the same experience. Individuals suffer shortage of resources due to local crowding. Some will be more crowded than others. Some will compete more effectively for resources. As a result, some individuals will produce far more offspring than others. This variation is essential if natural selection is to occur (Unit 2). When looking at density effects, we average across all this biologically important variation. Density-dependent fecundity is not the whole story. It is simply an application of one of life's sadder facts: the more ways you cut the cake, the smaller the pieces.

Density-dependent fecundity certainly operates in animal as well as plant populations. However, it tends to operate over a smaller density range in animals. This is because most animal bodies are far less plastic than plants. They show much less variation in adult size.

Much variation in reproductive success is not related to density. For example, swifts in Britain raise more young, on average, in warm dry summers when the insects they eat are more abundant. (Note, though, that it is the number of young which are successfully raised which varies here, not the number of eggs laid.) Density-independent fecundity may be important in determining changes in population size from year to year, but it cannot regulate population size, as we shall see (Section 26.3).

Fecundity: The number of offspring an individual produces.

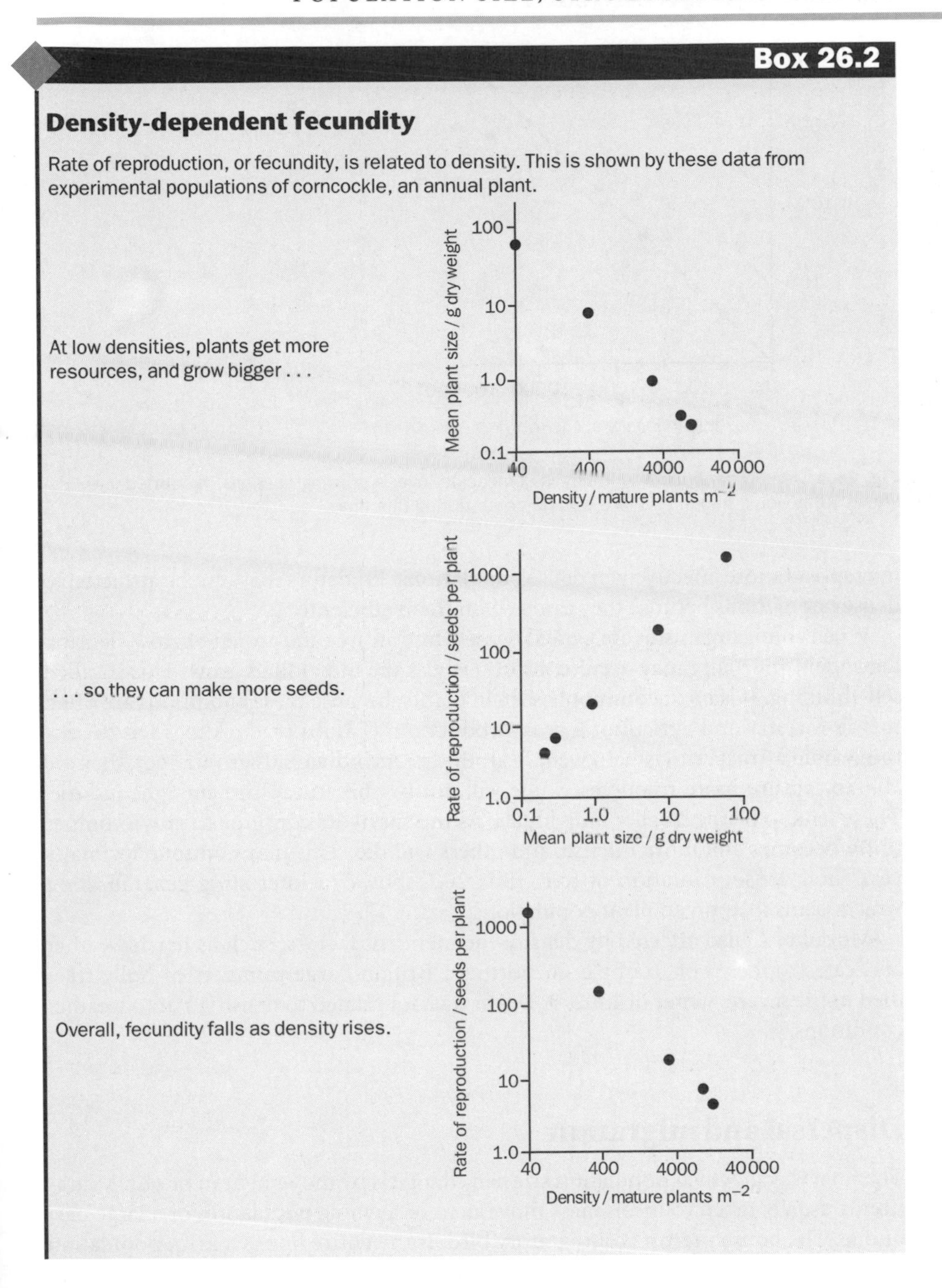

Death

Shortage of resources can lead to death. This may be direct, due to starvation. Alternatively, weakened individuals may be more prone to disease or predation. Very many populations show density-dependent mortality (Fig. 26.4). As density increases, the risk of an individual dying also increases. In denser populations, individuals tend to capture fewer resources. Some may get too little to survive. Disease

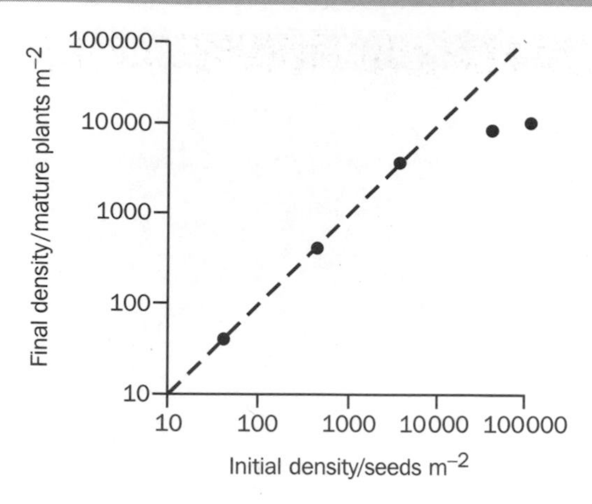

Fig. 26.4 Density-dependent mortality in corncockle, over a growing season. The dotted line shows what would happen if there was no death during this time.

may spread more effectively in dense populations. Predators may also be attracted to dense populations because they can feed on them efficiently.

When young organisms are packed densely into an area and are left to grow, density-dependent mortality may occur continuously as the individuals grow. This is called self-thinning. It is most commonly seen in plants, because these conditions are often met in forestry and agricultural crop production. (Lambs or ducklings left to self-thin would attract criticism on welfare grounds!) As individuals grow larger, they are able to capture more resources. Some will not be able to capture enough, and die. These tend to be the smaller individuals. As the survivors continue to grow, competition becomes still more intense, and others will die. This may continue for many years in a dense plantation of trees. Box 26.3 shows an interesting generalization which seems to apply to plant populations.

Mortality is also affected by density-independent factors, such as harsh weather and catastrophes. For example, in northern Britain, large numbers of holly trees died in the severe winter of 1962–3. Death was not related to density, but to weather conditions.

Dispersal and migration

The final way in which population size can change is by movement in or out. Migration is usually taken to mean mass movement of a whole population or a large part of one. The house martin is a migratory bird. If the entire house martin population of England uproots itself and flies to southern Africa, the population does not necessarily change in size, it simply moves. Dispersal of individuals away from their parents can affect population size, however.

It is clear that individuals do move between and within populations. Pigeons and ladybirds fly, badgers and lizards run, thistle seeds and mushroom spores blow on the wind, cherry seeds are carried in the guts of animals that feed on them, whilst white clover plants move through grassland by growth, rooting as they go. The reasons why dispersal might go on seem clear enough. Taken to an extreme, if individuals could not move away from their parents at all, their would be intense local crowding,

Box 26.3

Self-thinning: a general rule?

As individuals in a dense population grow, the smaller individuals get too little light, or other resources. Some of them die: self-thinning.

Data on self-thinning in plant populations are usually plotted in the following way. A population at a given time is plotted as one point on the graph.

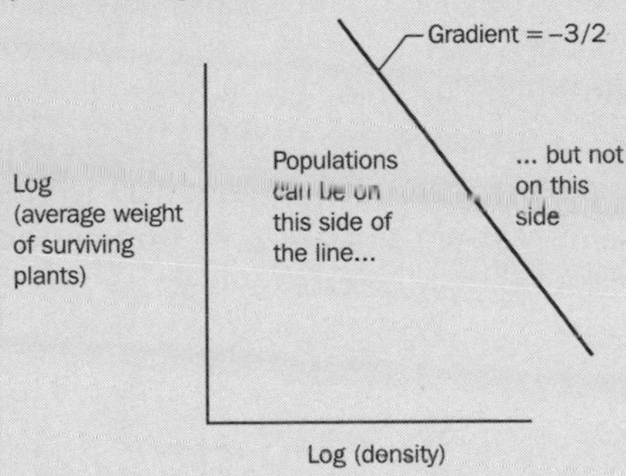

As plants grow, the population moves upwards on these axes. If self-thinning takes place, it also moves to the left. For a species in a given environment, there seems to be a limit to how large plants can be at a particular density, as follows.

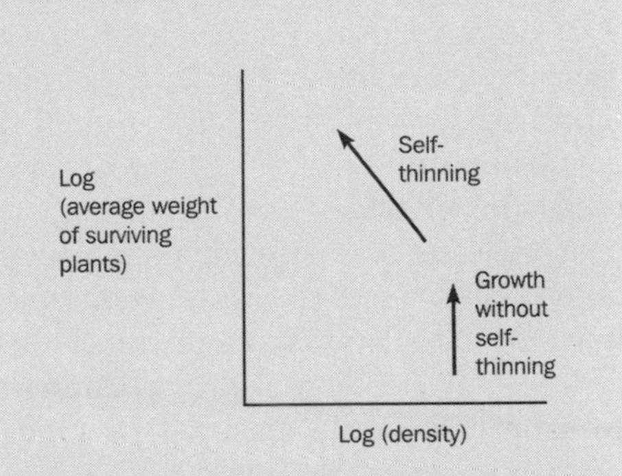

This is no surprise: at very high density, there would not be enough room for giant individuals, let alone enough resources to go round. What is more surprising is that the gradient of the line seems to be almost the same for a wide range of plants, from weeds to trees: $-3/2$ (that is, -1.5). This is a bit of a mystery, but probably has something to do with how plant mass (related to volume, in m^3) is packed onto ground area (in m^2).

Those of us who believed that self-thinning populations might routinely track along this line, may have been over enthusiastic. It is best thought of as an outer limit.

with all that follows from that. Dispersal gives at least some chance of finding a better place to live, with enough resources for survival. It is particularly important in unstable, short-lived habitats: there is no future in staying put. Early successional species (Section 27.3) colonize a habitat early on as natural woodland develops. They grow and reproduce quickly, but are eventually shaded out by larger, longer lived trees with shade-tolerant seedlings. Early successional species typically disperse their seeds more widely than later successional species, since by the time the parent has reproduced, the habitat is only suitable for the later species. The birches are typical early successional trees. Their seeds have papery wings which allow them to travel long distances on the wind. The dipterocarps of rain forest in south east Asia are at the other extreme. These late successional trees have seeds which are typically dispersed only a matter of metres from the parent.

There is also a genetic incentive for dispersal. Inbreeding can lead to less successful reproduction. There is an increased risk of offspring being homozygous for harmful recessive alleles (Unit 12). Dispersal tends to reduce inbreeding, since groups of relatives are scattered through the wider population.

Dispersal happens, but does it affect population size in a big way? It is hard to give a general answer, because this is a difficult area to investigate. In one classic study, it turned out that 57% of the breeding birds in a great tit population were immigrants, that is they hatched in another population. However, does immigration equal emigration? If it always did, the population size would not be affected. On the other hand, successful breeding populations might support less successful ones by overall emigration. The critical experiment is to cut off one natural population from the neighbouring ones, and see what happens. This has been done too rarely.

Is dispersal density-dependent? It would make sense, and there is anecdotal evidence. For example, migrating swarms of ladybirds, searching for food and turning to cannibalism, were common in England in the summer of 1976, following a population explosion when food became scarce. Lemmings, Arctic rodents whose populations fluctuate considerably, are famous for mass dispersal when density is high. Conclusive experimental evidence for density-dependent emigration is rare, but exists for some species. For example, the chance of an adult Colorado beetle leaving its population in summer increases with density.

It would also seem likely that immigration might also be density-dependent in some species. Lower density populations might attract immigrants because competition is less intense. On the other hand, there may be safety in numbers within a denser population. There is evidence of an optimum flock size in some birds, presumably reflecting a trade-off between these conflicting factors (Section 29.4). However, a flock is not the same as a population.

26.3 Fluctuation and regulation

Populations fluctuate in size from year to year. What causes these fluctuations? Could there be mechanisms which regulate population size, tending to push numbers back up or down if they move away from a particular level?

Density-dependent processes provide a mechanism which could regulate the size of any population. This is why they are so interesting. If population size rises, mortality also rises but fecundity falls. Together, they bring population size back down again. If the population gets smaller, fecundity rises and mortality falls, leading to an increase in population size. Either way, if the population size moves away from some set level, a chain of events tends to bring it back towards that level. This is an example of homeostasis (Unit 15), not within a cell or body, but within a population. There is no need to suggest that this is good for the population as a whole, although it might be. Nor are individuals doing it deliberately, on their own or in a group. It is a natural consequence of selfish individuals living, reproducing and dying whilst competing for limited resources.

It can happen, but does it? It could be that, in practice, factors acting in a density-independent way keep the population at a relatively low level. If this were the case, the population would show little evidence of density-dependent mortality. The regulatory system would still exist, even though it would not be the cause of the population changes which are seen. This situation can crop up in any homeostatic system. For example, the temperature in my refrigerator is normally kept at around 2°C. If it gets too warm, the compressor starts and the cooling system removes heat. If it gets too cool the compressor cuts out. However, if my small son keeps opening the door and putting in hot water bottles, the temperature may never get low enough for the compressor to cut out. The homeostatic mechanism is there, but a Martian recently landed in my kitchen would see little evidence of it by simply observing the 'fridge. It would need to experiment with the system, raising and lowering the temperature to see how the 'fridge responds. This is what we are doing when we control population density in experiments to obtain graphs like those in Fig. 26.4 and Box 26.2.

The next step is to see how wild populations fluctuate and what controls this. Do density-dependent factors *regulate* population size? Do density-independent factors *determine* the year-to-year changes, in practice? It is quite possible for both to be true in the same population. In a winter moth population in an English woodland, the key factor determining changes in population size from one year to the next was loss of eggs and young larvae in winter and early spring. This was mainly due to starvation of larvae which hatched before the oak leaves which they eat had emerged. This proved to be density-independent. Only one factor was strongly density-dependent. This was predation of pupae, yet this explained little of the year-to-year variation. So the population has a regulatory mechanism, but this is not what determines population size in a given year. Box 26.4 shows how to use life table data in key factor analysis.

Box 26.4

Key factor analysis

What determines year to year changes in population size? Are these factors density-dependent? We can investigate these questions using k values from life tables (Box 26.1).

A cohort is followed through its life cycle. A k value is estimated for each stage. (This could be year 1, year 2, year 3; or egg, pupa, larva, adult; or seed, seedling, adult before flowering, flowering adult; and so on.) With some cunning mathematics, it is even possible to calculate a k value for reduced reproduction. All the k values are added up to give a total. This is then repeated for other cohorts, usually in the following years.

An artificially simple example shows how the data are plotted:

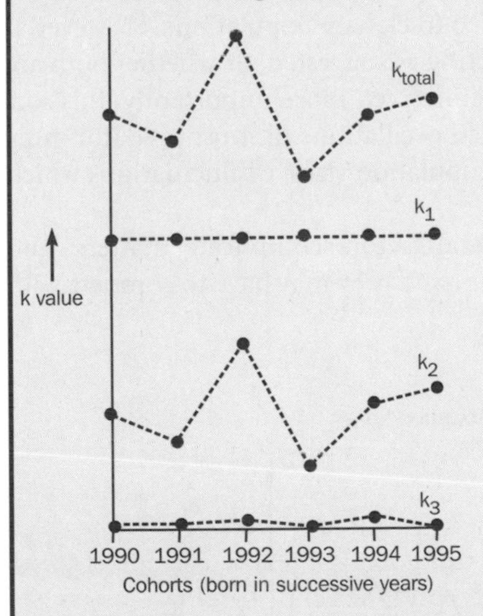

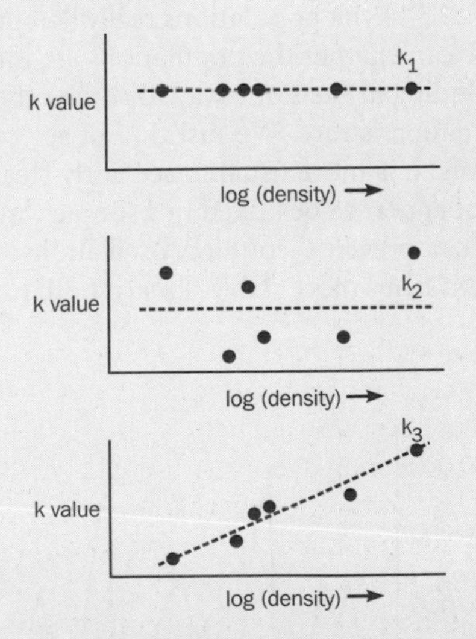

The left-hand graph shows that k_1 explains most of the death risk, but is constant from year to year. It explains none of the variation. k_2, however, closely follows the changes in k_{total}. It is the key factor which determines population size. k_3 explains very little of the death risk.

These factors can also be tested for density-dependence. In the right-hand graphs, k values for each cohort are plotted against the density of that cohort. It is clear that only k_3 is related to density. k_3 has the potential to regulate population size.

26.4 Population changes in interacting species

Competition between individuals of one species has a major effect on population dynamics. However, in most habitats, individuals of other species are out to get the same resources. Competition between the species could have important effects on both populations. This is vital in understanding how communities are made up, and so is treated as part of community ecology (Section 27.4).

Other populations interact by feeding. Predators are animals which catch and eat other animals. Herbivores are animals which feed on plants. Parasites live in or on another organism's body for most or all of their lives, and feed on it (Boxes 31.10 and 31.11, for example). Parasitoids are similar, but are insects in which only the larva feeds in this way.

Here is a simple argument about predators and their prey. It could equally well apply to herbivores and plants, or to parasites/parasitoids and their hosts. Consider one predator species feeding on one prey species. As the prey population grows, there is more food for predators. The predator population will also grow, with some time delay. However, as the predator population soars, many more of the prey species will be eaten, and its population will shrink. This will be followed by a decline in predator numbers, as they run short of food. As predators become scarcer, the prey population grows again, and so on. This very straightforward, over simple view of predator–prey dynamics predicts coupled oscillations: cycles of population change in which predator numbers follow prey numbers. Sometimes, this actually occurs in nature. A very famous and misleading example is the coupled oscillations in numbers of snowshoe hares (prey) and lynx (predator) in the Canadian Arctic (Fig. 26.5). Lynx populations really do appear to track hare populations. However, it is unclear whether the populations are interacting as suggested, or whether both are responding in the same way to some other factor. Even more importantly, this sort of situation is rare. We just do not see coupled oscillations in most predator–prey systems. It is more usual to see fairly steady population sizes, or fluctuations which do not appear to be linked in a simple way.

What prevents coupled oscillations? Several factors complicate matters, and make systems more stable. Firstly, both predator and prey may have to compete with

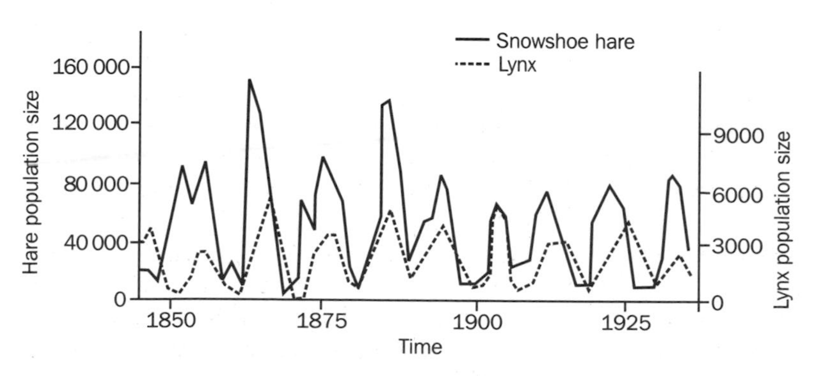

Fig. 26.5 Oscillations in populations of snowshoe hares and lynx.

other species. This will tend to damp down the dramatic increases predicted by the simple model. Secondly, many predators feed on more than one species. Well-adapted predators forage optimally (Section 29.4). There is a trade-off between the benefits of foraging (food captured) and the costs of foraging (energy used, time spent which could have been used for sex, and so on). An optimal forager maximizes the profit. Some prey may be better as food than others, but the predator might only bother with it when their density is above a certain level. Below that level, it may be more profitable to switch to a more common species, even if it has less food value. Prey switching also stabilizes populations, since the prey suffer more when they are at high density. Thirdly, many prey species are eaten by more than one predator, again blurring the simple relationship. Finally, our simple model did not take into account the patchiness of the environment. Predators may not act in the same way on all prey individuals. A tasty grass living on the clifftop is grazed by sheep wherever they can get at it. However, the sheep cannot, or will not, graze under the gorse bushes. Any grass plants growing there are safe. Parasitoids are typically clustered: several in some hosts, but none in many others. In each case, part of the population is bearing the brunt of the predation, leaving another part unscathed. This too tends to stabilize both predator and prey populations: neither can suffer or gain too much.

Going beyond these simple ideas would mean a great deal of detailed evidence and mathematical models. Both are fascinating, but this is not the place to go into them. Instead, we return to the idea that the life histories of different species are evolutionary solutions to the basic dilemma of dividing limited resources.

26.5 Trade-offs and strategies

The perfect, selfish individual would reproduce a lot. It would start to reproduce very early in life and continue to do so frequently up to a ripe old age. Each time, it would produce a huge number of offspring, all very large, packed with resources to help them become established, and all well cared for. This appealing life style does not exist. Nothing even approaches it. Each of these desirable characteristics requires resources, and resources are limited. Reproducing early in life, or a great deal at one time, makes an individual less likely to survive to reproduce again, as already discussed (Section 26.2: Birth). The individual can go for 'a lot, later' or 'less, now', but not 'a lot, now'. Only a certain amount of resources are available for the offspring at any one time. These resources can be packaged in different ways. A few offspring can each receive a lot, or many offspring can each be given a little. For example, coconut palms produce huge seeds (the coconuts themselves) but only a few at a time. Orchids produce millions of seeds, but each is minute. A female cod may lay millions of eggs in her lifetime, most of which are fertilized. No further resources are given, and most die shortly after being laid. A human will rarely produce more than 15 fertilized eggs in her lifetime (usually much less in most societies). Each is larger than the cod's eggs, but the big difference is that vast resources are put into each one after fertilization, both before birth and afterwards. The chances of them surviving are much higher. Codfish and people are opposite extremes. Neither extreme is better in itself. Each suits some habitats and ways of life.

r and K selection

The idea of 'r and K' selection is an attempt to link these strategies to particular types of habitat. In a stable, long-lived habitat, it is argued that competition is likely to be important because populations have had time to grow large. Density-dependent mortality is likely to be important. Under these conditions a population experiences K selection. A K strategy includes long life (a safe bet if the habitat is stable), delayed reproduction, for similar reasons, and small numbers of large offspring which will have the best chance in the struggle for survival. Further, since the habitat is constant, there is no pressing need for many offspring to 'explore' new habitats.

r selection takes place in unstable, short-lived habitats. Competition will be less intense, it is suggested, and density-independent mortality caused by environmental fluctuations will be more significant. r strategists will be short-lived and reproduce early, since the habitat may change. They produce many small offspring, because mortality is density-dependent. The proportion which dies does not depend on density. The more that start out in life, the more survivors there will be. Dispersal is important, too, if the habitat cannot be relied upon. More offspring can reach more new habitats. Please do not be put off by the strange way in which letters are used to name these types of selection! They come out of mathematical models of population growth. r represents the rate at which a population is capable of growing: r-selected populations can grow quickly at first. K symbolizes the maximum population size. K-selected species, it is argued, can build up large, persistent populations.

Without being too critical, it is easy to see examples which seem to fit the theory. Cereal fields are an obvious place to look for r-selected plants, because the vegetation is destroyed every year by ploughing. As predicted, most cornfield weeds are annuals, with early reproduction. Their seeds are small, at least compared with most trees, so an individual can produce many seeds, in relation to its size. Undisturbed forests are obvious K habitats. Their dominant trees, for example oak and beech in much of England, are usually long-lived, with delayed reproduction and big seeds.

Like most simple ideas, r and K selection has had a fair amount of scientific mud thrown at it. Some has stuck. For a start, most organisms are not clearly r or K selected. These are the extremes of a continuous scale. It is often possible to say that one species is *relatively* r selected, whilst a related species is *relatively* K selected. However, some species will not fit neatly into the scheme. They combine r and K characteristics, sometimes in an extreme fashion. Many of our western European orchids are plants of undisturbed grassland. Grazing and/or mowing prevent woody plants surviving in these habitats. Herbaceous perennials like orchids can survive because they produce new shoots each year from underground organs. Most orchids show delayed reproduction, a K characteristic. After seed dispersal, the plant grows underground for several years, gaining nutrients from a symbiotic fungus. Only then does a photosynthetic shoot appear, and this does not reproduce until it is big enough. Some grassland orchids are long-lived, with significant numbers of individuals living for 30 years or perhaps much longer. Again, this is towards the K end of the spectrum. Yet the orchids have some of the tiniest seeds, and produce vast numbers. These are r characteristics. The symbiotic fungus allows seedlings to survive, despite having very little food reserve in the seed.

The concept of r and K selection may be flawed, but it is a reminder of an important principle. Natural selection acts on life history characteristics when selfish individuals live together in populations.

Summary

- ◆ A population is all the individuals of a species in a habitat.
- ◆ Two definitions of 'individual' are used. A ramet is a working unit. A genet is a genetic individual.
- ◆ Populations vary in size, age structure and size structure.
- ◆ Population size is affected by four things: birth, death, emigration and immigration.
- ◆ A factor is described as 'density-dependent' if its rate depends on population density.
- ◆ Rates of birth, death and dispersal are frequently density-dependent.
- ◆ These factors are also affected by things which are independent of density, such as weather conditions.
- ◆ Population sizes fluctuate.
- ◆ Density-dependent processes can regulate the size of any population.
- ◆ In many populations, density-independent processes explain much of the year-to-year fluctuations in size. The regulatory systems are still there.
- ◆ Populations are affected by populations of other species. These include competitors, predators, herbivores, prey, parasites and parasitoids.
- ◆ Natural selection leads to evolution of distinct life-history strategies.

Exercises

26.1. How would you expect the age structure of the human population of an affluent European country (Switzerland, say) to differ from that of a poor, developing country (Burma, for example)?

26.2. Look at the survivorship curve in Box 26.1. Sketch the graph of $\log_{10}(l_x)$ you would expect for another population in which there was a greater risk of dying *earlier* in life.

26.3. Throughout this unit I have been harping on about density and density-dependence. Why is this so important in population ecology?

26.4. *Vulpia membranacea* is a grass native to sand dunes in Western Europe. It is a semelparous annual plant which germinates in the autumn, grows through the winter and dies after flowering in the spring. Even in flower it is not a large plant, typically a few hundred millimetres tall and having just a few shoots. Its seeds are large for a grass, and it produces relatively few of them. This is linked to its high rate of survival from the seed to the mature adult stage.
Can *Vulpia* be easily classified as r or K selected? Explain your reasons.

Further reading

Beeby, A. *Applying Ecology* (London: Chapman & Hall, 1993). This book includes a treatment of population management.

Begon, M., Harper, J.L. and Townsend, C.R. *Ecology: Individuals, Populations and Communities* (2nd ed.) (Oxford: Blackwell Science, 1990). Not the first big textbook of ecology, but the first which I felt reflected the range of questions modern ecologists are asking. Population ecology is covered well, without losing sight of other levels of organization, such as communities and ecosystems.

Community Ecology

Contents

Connections

► Communities are made up of populations, and you should have some knowledge of population ecology (Unit 26) before studying this unit. Ecology can also be studied on the larger scale of the ecosystem (Unit 28).

27.1 Communities

A **community** is all the individuals of all the species living in an area. It is a group of populations. Usually, communities of one broad type of organism are studied. For example, my large intestine has a bacterial community; my skin has another. A pasture has a plant community, dominated by various species of grass. It also has an animal community, dominated by cattle, perhaps, but there will be other animals present. Rabbits, snails and slugs may graze, and the soil will contain large numbers of invertebrates. It is perfectly all right to study communities of even smaller groupings, for example the moss community of a moorland area, the fish community of a coral reef or the bird community of a forest.

The first step in studying a community must be to describe it. Normally this involves making a list of species, with an indication of how much of each species is present. There are several measures of abundance. Density is the number of individuals per unit area. Cover is the percentage of the ground area covered by a particular species. It is only appropriate for plants, or animals like mussels which do not move around. Frequency is a measure of relative abundance, the percentage of unit samples which contained the species. The unit sample could be anything appropriate, for example a 0.25 m^2 sample of ground area in a plant community, a sticky yellow sheet hung in a tree for 24 hours in the case of some sorts of insects, a five minute trawl with a fine mesh net for planktonic algae, and so on.

Communities are not museum exhibits, however. They are made up of populations of living, reproducing, dying individuals. It is not surprising that they change through time. Communities change on two scales. **Community dynamics** are the small-scale, relatively short-term changes that go on within a single community. Some species become more common, others become rarer, but the overall nature of the community does not change. It is still a grassland, or a coniferous forest.

Community: All the individuals of all the species living in an area.

Community dynamics: Small-scale, short-term changes in community structure.

Succession is the larger scale, longer term changes. One type of community replaces another. If a field is fenced off and farmer, livestock and tractor keep away, grassland becomes woodland.

The big questions in community ecology are these: How do communities get to be the way they are? How can some species live together but not others? What are the assembly rules for communities? Why are some communities more diverse than others? These questions are all related. They are also very difficult to answer. After a brief discussion of community dynamics and succession, the bulk of this unit is a tour of the various ways people have tackled these questions.

27.2 Community dynamics

Communities change. Have a picnic in a grassy field, look around, then return to repeat the experience a few years later. The plant community may look the same, but there will be differences. Perhaps some plant species have become more common, at the expense of others. Every population will have been changing, even if its overall density has not altered. Births and deaths will have occurred; new individuals may be in different places; old individuals may have survived and moved, slowly but surely, by growth. Community dynamics are these small-scale, relatively short-term changes in the interacting populations which make up a community.

Little is known about the dynamics of most communities. These are not obvious changes. Careful records, repeated for years, are needed to discover them. Fig. 27.1 shows a rare and beautiful example: the changes in abundance of three plant species in a tiny sample area of grassland, recorded over 44 years. In this small patch, a species seems to dominate for a few years. Then it is largely replaced by another. The whole process is presumably going on throughout the habitat, but out of step: different species dominate in different places. Too few studies show up these fine-scale changes. Even fewer give clues as to what causes them.

Succession: Larger scale, longer term changes in community structure.

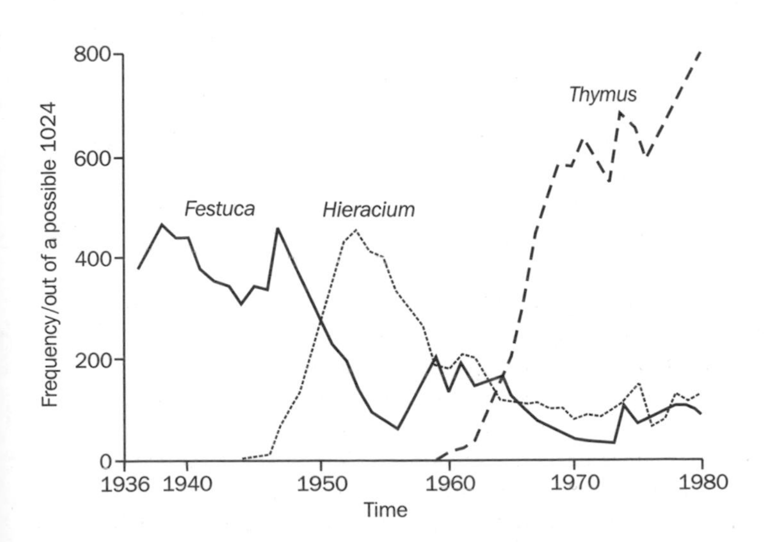

Fig. 27.1 Changes in the abundance of three plant species in an enclosed plot of grassland.

27.3 Succession

If community dynamics are changes going on within a community, succession is the replacement of one community with another. Whole groups of species are replaced by others. The appearance of the habitat may change drastically. Studies of succession have usually centred on plant communities. This is because plant and animal communities do not interact in a symmetrical way. The plant community provides food and homes for members of the animal community, but not the other way round. Animal succession may broadly follow plant succession. However, this does not mean that animals have no influence on plant succession. Feeding, and seed dispersal by animals are likely to have an effect.

Sometimes, succession is quick enough to study directly. For example, the succession of bacteria and fungi in a compost heap will be finished within a very few years of setting up the heap. Many other successions probably take well over 100 years. A few unselfish biologists have set up long-term studies, trusting their successors to carry them on. It is easier to study situations where succession began at different times in different places. Two famous examples are succession behind a retreating glacier in Alaska, and in abandoned fields in the eastern USA. In each case, the succession begins with a plant community based on low-growing, herbaceous (non-woody) plants. In each case it ends up dominated by trees (Box 27.1).

Box 27.1

Two successions

Where glaciers once stood

When glaciers retreat, they leave behind a nutrient-poor mixture of clay and stones: glacial till. Succession on glacial till has been studied at Glacier Bay, Alaska, where there has been a dramatic, well-documented retreat of ice since the 18th century.

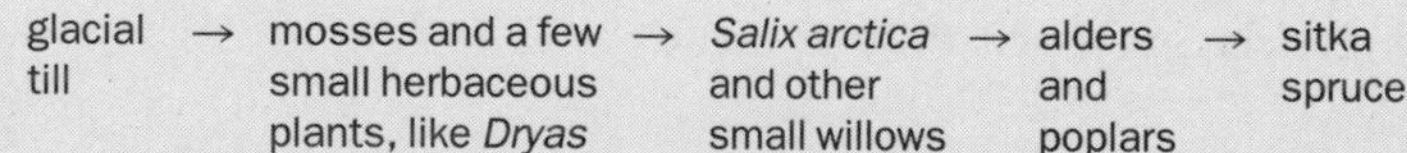

As succession progresses, the soil becomes deeper and more acidic. The amount of nitrogen in the soil increases until sitka spruce becomes established. It appears that early successional plants make soil conditions suitable for later successional species to invade (facilitation). Later successional plants exclude earlier species through competition (inhibition).

How the West was won, and where it got us

European settlers in North America colonized the east coast first. During the 19th century, there was a westward movement. Families and communities uprooted themselves; agriculture spread to new areas. The result: abandoned fields in the eastern USA. Succession in old fields has been studied in various places by comparing the vegetation of fields abandoned at different times. The general pattern is as follows (details of species vary from place to place).

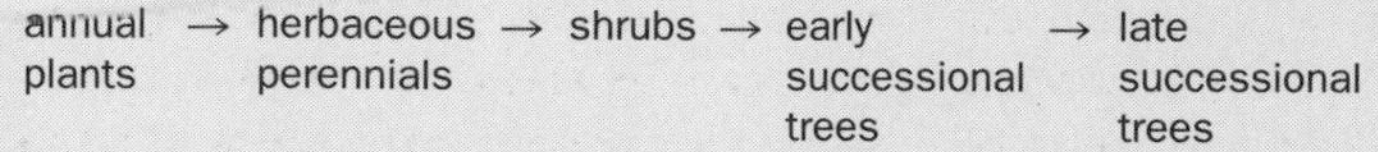

Inhibition is clearly taking place. It could simply be that the faster growing herbaceous plants are getting a share of the resources while the woody plants are still small. However, facilitation may have a role too. Experiments suggest that, in some areas, the build up of nitrogen early on encourages invasion by later species.

The unpromising beginnings of succession behind a glacier.

The final community in a succession is called the **climax community**. There has been a bitter and rather unnecessary argument over whether or not climax communities really exist. The climax has sometimes been defined in a rigid way: the composition of the community must not change *at all*. It certainly would be surprising to find a community like this. We know that small-scale dynamic changes go on in long-lived communities. Occasional disturbance by, say, storms or forest fires, would also keep pushing the community away from equilibrium so we might never see the final, stable state. It is more useful to define climax loosely as the broad type of community at the end of a succession: 'woodland dominated by oak and alder', for example.

Succession happens because species are different. Some disperse better than others. Some grow and reproduce quicker than others. Some are better competitors in a quick scramble for resources, others can tolerate a long, drawn-out contest. All organisms will affect their environment in some way, perhaps making it more or less suitable for other species.

Early successional species are likely to be good at dispersal: otherwise, they would not be there yet. Organisms may disperse through space or time. Animals walk, swim or run from one place to another, while seeds and spores are blown, catapulted or carried by animals. Many plants have seed banks buried in the soil, a dormant army from the past, waiting to invade when conditions are right. Looking at it another way, early successional species need to be good dispersers because, as succession goes on, they need to find new habitats. Quick growth and reproduction are important for the same reason. Annual plants are most commonly found in early successional habitats.

When trees enter a habitat, competition for light takes on a new dimension. Woody stems mean that plants can be tall, and the tall plant gets the light. Herbaceous plants which survived in open, early successional habitats, may be shaded out as woodland develops. Other herbaceous species which can tolerate shade may take their place under the trees. In a forest clearing, there will be a scramble for light as tree cover begins to develop. Early successional trees like birches and aspens have to succeed in this situation. They tend to grow upwards quickly, producing an open crown with several layers. A crown like this catches light from all around. It is also cheap to produce, as less wood is needed. Late successional trees such as oak and beech tend to be slower growing. They have to be able to tolerate shade until, eventually, they overtop the early successional species and shade them out. Their crowns tend to be a dense leafy shell around a supporting structure of branches inside the crown. They are adapted for a long, competitive struggle with their neighbours (Section 26.5).

So far, everything points to succession working by **inhibition**. Later species make life impossible for earlier ones. However, there are situations where earlier species set up the conditions which later species need: **facilitation**. For example, in the succession behind a glacier described in Box 27.1, the soil is thin and lacking nitrogen at first. Two early successional plants, alder and *Dryas*, have nitrogen-fixing bacteria in root nodules (Box 18.3). They increase soil nitrogen to a point at which spruces can grow. Dead organic matter also accumulates. This helps buffer the soil against extremes of water content, allowing a wider range of plants to survive.

Climax community: The final community in a succession.

Inhibition: Late successional species make conditions unsuitable for early successional species.

Facilitation: Early successional species make conditions suitable for later successional species.

27.4 Explaining community structure

Studies of succession give some clues about why one type of community replaces another. They tell us nothing about why a community is the way it is. They do not tell us why some species are present while other, often very similar species are not. The rest of this unit deals with various ways of tackling this question. The concept of the niche comes first. This is necessary to make sense of some of the ideas.

Niches

A **niche** is the set of environmental conditions in which a species can live and reproduce. Salmon can only live within a certain temperature range. Too cold: the water freezes and they die. Too hot: they also die. They can only survive when dissolved oxygen is above a certain level. A particular type of river bed is needed for egg-laying. Only certain invertebrate animals are suitable as food. There must be a migration route from the river where they breed to the sea where they feed and grow. These are probably only a few of the environmental variables involved. The combination of all these conditions which suit the salmon is called its niche. It is not the same as a habitat. A **habitat** is a place. Some habitats fall within the salmon's niche, others do not. It can only live in those habitats which do.

Some species could survive in a habitat if there were no competitors, but cannot survive if better competitors are there too. This is the great ecological lesson of gardening. Loads of exotic plant species will grow in a flower bed if you space out the individuals and remove the weeds. Let the weeds grow, and many species die out. The niche of a species when competitors are absent is called its **fundamental niche**. When competitors are present it can survive under a smaller range of conditions: the **realized niche.**

The competitive exclusion principle

No two species are the same. (If they were, they would not be different species.) This implies that if two species are competing for the same limited resources, some difference will make one species a better competitor. The difference may be small, but it will be there. The better competitor will survive; the other will die out. This is the competitive exclusion principle. It explains why very similar species are often found in different habitats, not the same one. For example, a study of three species of *Sphagnum* (bog mosses) shows a clear example of competitive exclusion. They naturally grow in slightly different habitats. All three species could grow in each habitat if the other species were weeded out. This means that all habitats were within the fundamental niche of each species. However, competition between them leads to only one species surviving in each habitat. The species occupy their realized niches. The realized niche is not necessarily the conditions under which the species grows best. It is the conditions under which it grows better than the competitors.

The competitive exclusion principle suggests that communities will be sets of species which do not compete with one another. This would lead to very simple communities, especially in plants, which all need more or less the same things. So why do we see species-rich communities all around us?

Niche: The set of environmental conditions in which a species can live and reproduce.

Habitat: The place where an organism lives, or, the type of place where a species lives.

Fundamental niche: The niche when competitors are not present.

Realized niche: The niche when competitors are present.

Studies of competition between species

Experiments on the effects of density, linked to mathematical models, have helped our understanding of population biology. Communities are made up of interacting populations, so it seems sensible to extend these experiments to mixtures of species. In the most rigorous experiments, a wide range of densities of one species is combined with a wide range of densities of another species, in all possible combinations. This generates a lot of rather unwieldy data, which are best handled using mathematical models (see references at the end of the unit). The upshot of this sort of work is as follows: several types of model seem to fit experimental data quite well. They predict that even when there is competition between the species, stable coexistence is possible, especially when competition within each population is more intense than competition between the two populations. (They also predict that when competitive interactions are rather different, competitive exclusion is possible, or even unpredictable, chaotic behaviour.) This suggests that the competitive exclusion principle is only partially correct. Coexisting species may be competing to some extent, but there must still be differences in the way they capture resources.

The habitat is not uniform

Habitats may be made up of several slightly different microhabitats. These might be nothing more than bumps in the ground where the soil is drier and dips where it is wetter. Competitive exclusion may go on in each microhabitat. If different species compete better under different conditions, several similar species may coexist in the habitat as a whole. This is the case in the earlier *Sphagnum* example, where the three 'habitats' could be found very close together in the same bog. The three species exclude one another locally, but coexist on a wider scale.

Their niches are different

A **guild** is a group of species using the same types of resources in a similar way. Very often, competition within a guild is limited or non-existent. In many plant communities, different species take up water and minerals at different depths in the soil: some have deeper roots than others. Five closely related bird species, the blue tit, great tit, coal tit, marsh tit and willow tit, all feed on insects in an English wood. However, they concentrate on different tree species, different heights above the ground and so different insect species. Two bumble bee species in the Rocky Mountains usually feed on nectar from different plant species. In each case, there is a niche difference.

Sometimes there is evidence that the difference is only in the realized niches. For example, in the bumble bees, either species visited both types of flower if the other species was removed. When together, however, each species is more successful than the other at collecting nectar from one type of flower, so the difference is set up. This could be seen as simple competitive exclusion, but in practice, the bees are likely to forage where it is most efficient, avoiding competition.

There is little or no evidence of competition between the species in other cases, such as the five tits. Here, it is possible that the differences are in the fundamental niche, although the only way to know for sure is to keep out all but one species from experimental areas, and look for signs of that species broadening its niche.

Guild: A group of species using the same types of resources in a similar way.

Different resources are limiting

An individual may need several resources. Not all of them limit its growth at any one time (see also Section 10.5). For example, most soils contain enough molybdenum to satisfy all the plants living there. Add more, and you see no effect. It only begins to limit growth at extremely low levels. This can also be true for resources needed in larger amounts: light, phosphate, carbon dioxide and so on. The only way to find out is to add more. If growth increases, it is limiting.

If two species are limited by different resources, there is no reason why they should not coexist. This was clearly shown by some experiments on single-celled algae. They were grown under experimental conditions, where limiting resources were one of two minerals. Mixtures of two species were given different ratios of the two nutrients. At some nutrient ratios, both species were limited by the same resource. Then, one species drove the other to extinction. However, at intermediate ratios when the two species were limited by different nutrients, they coexisted.

Predators and herbivores promote diversity

Predators and herbivores damage and kill the organisms they feed on. They are certainly selective about what they feed on. Common sense, backed up by some experimental evidence, suggests that an efficient predator might switch from one prey species to another if the first becomes scarce. Predators and herbivores may be promoting coexistence and diversity by feeding on abundant species which might otherwise dominate communities. There is no question of this being intentional. It is simply a side-effect of predators looking after their own interests by feeding efficiently.

Even a predator which does not switch between prey species may promote coexistence. If it feeds on several species, keeping their densities low, it may reduce competition and make competitive exclusion less likely.

Experiments do support this prediction. The grasslands on chalk in southern England are particularly species rich. It was suspected that grazing by rabbits might be promoting this diversity. Sample areas were fenced to keep rabbits out. Over a few years, diversity declined. The areas became dominated by a one species of grass. A similar effect was seen when epidemics of myxomatosis, a viral disease, swept through rabbit populations.

Another famous experiment involved starfish feeding on the animal community of a rocky shore in Washington state, USA. The community was quite diverse, with several mollusc and barnacle species as well as one starfish species. The starfish were excluded from a sample area, and the community began to change. Within a couple of years the community was dominated by one mollusc, a mussel, with small numbers of one barnacle.

Communities may not reach equilibrium

Competitive exclusion is the outcome of a process which takes time. It may take several generations before the population of a poor competitor falls to zero. The system is said to have come to equilibrium once population sizes stop changing. What if the system never reaches equilibrium? Perhaps some factor keeps stepping in to push it away from an end point which is never reached.

Cornfields experience massive disturbance, year after year. As a result, their plant and animal communities have a low diversity. Intermediate levels of disturbance may lead to maximum diversity.

Disturbance may prevent a community reaching equilibrium. The disturbance could be storm, fire, slash-and-burn agriculture, or a plague of locusts eating everything in sight, for example. If disturbance happens in different places at different times, the community may become a mosaic of patches at different stages of succession.

Intermediate levels of disturbance may lead to greatest diversity. Too frequent, too intense disturbance means that only fast-living r-selected species can survive (Section 26.5). Too little disturbance, and the community becomes dominated by a few K-selected species. At intermediate levels, the community will be as patchy as possible, with a mixture of early and late successional species. Several studies support this idea. For example, Fig. 27.2 shows the diversity of corals in increasingly disturbed areas of an Australian reef. Another non-equilibrium idea applies to communities where the limiting resource is a space to live. As soon as an individual dies,

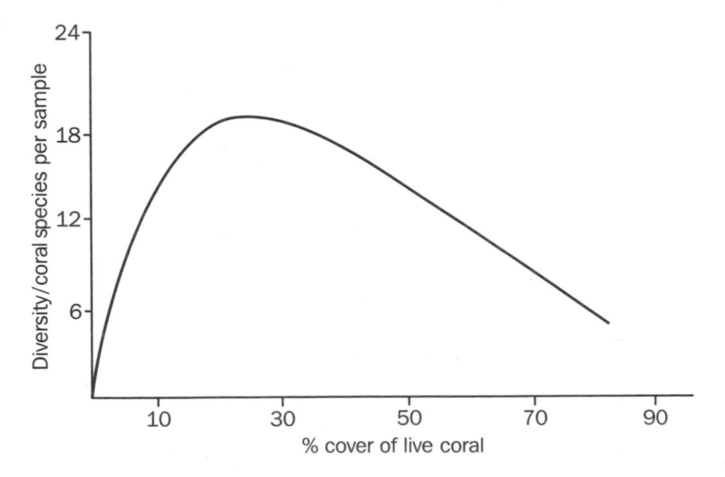

Fig. 27.2 Effect of disturbance on diversity in Australian coral reefs (dead coral is a sign of disturbance).

its place is taken by a lucky young individual. A diverse fish community on an Australian reef has been interpreted in this way. Many species have similar niches. Fish larvae seem to colonize vacant spots at random. The first to get there wins the space, regardless of species. Competition for space is essentially a lottery which any species can win sometimes, but none can win all the time. The competitive exclusion principle actually still applies here. There must be some species differences which would make some better colonizers, perhaps better dispersal. The point is that if the differences are small, the competitive advantage will be small. Only minor, random events would be needed to keep the community away from equilibrium.

Other ideas relate to highly divided environments. Examples might be intestines, the habitat of parasitic worms, or flowers where thrips, tiny insects, feed and reproduce. These communities are split between tiny 'islands'. There is nowhere for a tapeworm to live in between two animals guts. Competition only occurs between individuals in the same island. If only a few live in each island, there is a chance that some islands will only contain one species. If there are lots of very small islands, the chance becomes very high. So even if one species is a better competitor, the other will survive because a proportion of the population escapes competition. The odds are shortened even more if one species is a better competitor but the other is better at dispersing between islands. Stable coexistence is then a serious possibility.

All these ideas about how species can coexist should not be seen as alternatives. They may all be true sometimes, perhaps often. They are different facets of the complex solution to a very simple question.

27.5 Biological control

Our species has a history of moving other species around the world, for agriculture, for ornament or by accident. Some of these species find themselves in the unnaturally simple communities of agricultural fields, plantations or glasshouses. Others 'escape', and become part of natural communities, changing them in the process. Often, their natural enemies have not travelled with them. This may be a good thing for a crop species. For example, the potato has been carried from South America to many parts of the world, but the fungus which causes potato blight has not always travelled with it. However, it can also help an introduced species to succeed in a natural community. For example, prickly pear cacti (*Opuntia* species) were introduced to Australian gardens in the 19th Century. No native animals fed on them, they invaded natural

Recently is has been fashionable to study community complexity. Complexity itself is quite a complex idea. It includes not only the number of species in a community but also how many of these species interact (for example through competition, predator–prey interactions, mutualisms), and how strong these interactions are. Many studies have looked for a link between complexity and stability. Stability is how well the community recovers from an event that changes it.

I owe it to you to summarize these ideas and discoveries in a dispassionate way, and I can not. I am uneasy about the whole idea of complexity. When I look at communities I do not see complexity. I see species richness and coexistence, but not all the interactions between species. I can guess at them, or get evidence of them by experiments, but I can't see them. I am not really interested in explaining complexity, or using it to explain other things. My prejudice is that it is not a useful or interesting thing to be measuring.

I am also sure that I am wrong. Several biologists whose work I respect and admire have been involved in this field. Others have written about it approvingly. However, I cannot convince you of something I do not believe in, so go elsewhere and hear the story from believers. My only defence – you can't expect a passionate biologist to be dispassionate; at least, not *all* the time.

plant communities, and out-competed many native species. The cactus was controlled by introducing one of its natural enemies from South America. Larvae of the moth *Cactoblastis cactorum* feed on the cactus, limiting its success. This is an example of **biological control**.

In other cases, a plant and a pest might be introduced without the pest's natural enemies. A classic example of this situation comes from the citrus groves of California, where a scale-insect pest, *Icerya purchasi*, became a problem in the mid 19th Century. It was brought under control by introducing two of its natural enemies. A fly larva (*Cryptochaetum* species) is a parasite of the scale. A beetle, *Rodolia cardinalis*, eats it. Together, they have largely controlled the pest to the present day.

Some biological control programmes involve simply increasing the numbers of an existing enemy. Aphids are pests of a wide range of glasshouse crops in Britain and elsewhere. They have a wide range of predators and parasites, but conditions in glasshouses may not allow big populations of these animals to develop. For example, glasshouses may be emptied and sterilized between crops, to control disease-causing fungi and other pests. Aphids can reproduce asexually, and recover from this population crash faster than their enemies. One solution is to use insecticides. The alternative is to introduce natural enemies early in the growth of the crop, before the aphid population is too large. Predatory lacewing, ladybird and midge larvae, and parasitic wasp larvae have been used in this way.

Developing a biological control programme is a tricky business. Many attempts fail. Sometimes closely related species are mixed up. The enemies of the wrong species are introduced. Alternatively, the enemy may establish, and settle to a stable population size without seriously affecting the population of the host species. The enemy may even escape, and become a pest itself. For example, the cane toad was introduced to Australia from the West Indies, to control insect pests of sugar cane. It is a non-specific predator, and has spread across Queensland, eating native invertebrates in natural communities. These problems can only be avoided by good identification, and careful experiments before release. The skills of the taxonomist and ecologist are in demand.

Biological control: The addition of a species to a community, once or repeatedly, to control the abundance of another species.

Summary

- A community is all the individuals of all the species living in an area.
- Communities change through time.
- Some of these changes are small-scale and short-term, within a particular community: community dynamics.
- Others changes are on a larger scale. One community replaces another: succession.
- Early successional species may make conditions suitable for late successional species: facilitation.
- Late successional species may exclude earlier ones through competition: inhibition.
- A niche is the set of environmental conditions in which a species can survive and reproduce.
- The fundamental niche is the niche when competitors are not present.
- The realized niche is the niche when competitors are present.

- The competitive exclusion principle states that two different species cannot share the same niche. The better competitor will exclude the other.
- Species may coexist by occupying different niches.
- If the habitat is not uniform, different species may compete better in different microhabitats.
- Two species may coexist because they are limited by different resources.
- Predators and herbivores promote diversity by feeding on more common species and/or reducing competition throughout the community.
- Disturbance can increase diversity. It may prevent competitive exclusion from running its course.
- Highly divided habitats increase the chance that some individuals will escape competition.
- Human activity changes communities in many ways.
- Biological control is the addition of a natural enemy of a pest species to a community.

Exercises

27.1. In a study of the root systems of several common plants of the North American prairies, it was found that they occupied different levels in the soil. They appear to occupy different niches. How could you find out whether the differences are in the fundamental or realized niches?

27.2. A small, woody plant species requires acidic soil, but cannot survive in full shade. Some species with these characteristics are found behind retreating Alaskan glaciers. Would you expect them to be found in early, middle or late successional communities? Explain.

27.3. *Littorina littorea* is a small mollusc which feeds on algae between low- and high-water marks on rocky shores. Like other herbivores, it shows distinct preferences for certain algal species. The effect of varying *Littorina* density on the diversity of various algal communities was investigated. In some, increasing density led to an increase in diversity. In others, the reverse was true. Attempt to explain these observations.

27.4. A mussel, *Mytilus californianus*, coexists with a brown alga, *Postelsia palmaeformis*, on part of the Pacific coast of North America. *Mytilus* is longer lived than *Postelsia*, which cannot recolonize where *Mytilus* becomes established. *Mytilus* out-competes *Postelsia*. Suggest how they might be able to coexist, despite this.

Further reading

Beeby, A. *Applying Ecology*. (London: Chapman & Hall, 1993). This book includes material on conservation of communities.

Begon, M., Harper, J.L. and Townsend, C.R. *Ecology: Individuals, Populations and Communities* (2nd ed.) (Oxford: Blackwell Science, 1990). Not the first big textbook of ecology, but the first which I felt reflected the range of questions modern ecologists are asking.

The Ecology of Ecosystems

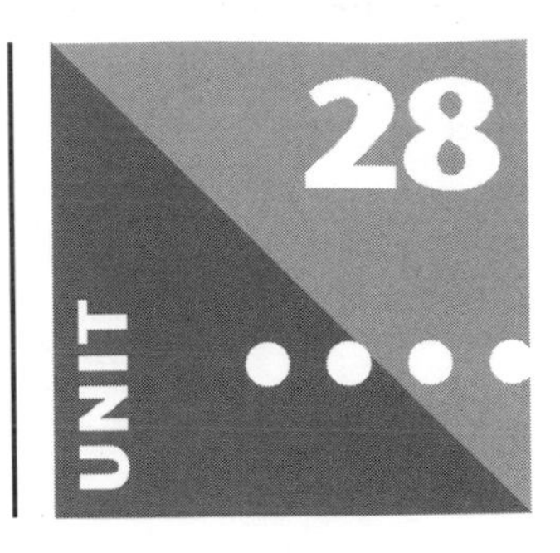

Connections

▶ This unit is about the study of ecology on the biggest scale. It can mostly be understood on its own, although you will get a more rounded view of the science of ecology by reading it with Units 26 and 27. A basic understanding of homeostasis (Unit 15) is assumed.

Contents

28.1 Ecosystems

Life inhabits a thin layer smeared around the Earth's surface: the **biosphere**. From the depths of the ocean to the stratosphere, it is only a few miles thick. Here at the surface of the Earth, the signs of life are everywhere. Even if we try not to see the organisms themselves, the physical environment has been shaped by living things. The oxygen in the air was made by photosynthesis. There was little or no oxygen gas on the early Earth before the first photosynthetic cells appeared. Soils are built up by organisms. They are complex, highly structured mixtures of dead organic matter, mineral particles, plant roots, bacteria, animals, water and air. All around are structures made by animals: nests, worm casts, railway lines.

Living things interact with the physical world. The study of **ecosystems** recognizes this. An ecosystem is all the living things in a habitat, plus all the non-living things. A woodland ecosystem is not just the trees, the other plants, the animals and microorganisms living around them, but also the soil, the rock beneath, the weather, the dead organic matter lying around, the air including any pollution, and the chainsaws which invade from time to time. The study of ecosystems centres on how energy and materials flow between living things, and between life and the physical environment.

The first step is a framework for looking at movement of energy and materials: the food chain.

Biosphere: The zone at the Earth's surface where life is found.

Ecosystem: All the living things in a habitat, plus all the non-living things as well.

28.2 Food chains

Photosynthetic organisms (mainly plants) get the energy they need from light. They get the chemical raw materials for growth from the physical environment, as carbon dioxide, water, nitrate, and so on. Everything else gets energy and raw materials by feeding on something else. A food chain is a group of organisms linked by feeding. The first is photosynthetic, generally a plant. The next feeds on the plant. This in turn is fed on by the next, and so on. Energy and materials flow through food chains.

The levels in a food chain are called trophic levels ('trophic' means 'to do with feeding'). Consider the food chain

oak → winter moth → blue tit → sparrowhawk

This has four trophic levels. The oak is the producer, since it produces its own food molecules in photosynthesis. The other three are consumers. They consume ready-made food molecules. The winter moth is described as a primary consumer, the blue tit is a secondary consumer, the sparrowhawk a tertiary consumer. Some food chains are longer, some are shorter. Fig. 28.1 shows more examples of food chains.

A food chain does not tell you everything that eats or is eaten by these organisms. This is the job of a food web (Fig. 28.2). A food chain is one of many paths through a food web.

These food chains are all based on herbivores grazing plants. Ecosystems have a second group of food chains based on the pool of dead organic matter. Fungi, bacteria and some animals such as earthworms feed on this. They are in turn eaten by animals.

Energy and raw materials enter the food chain through producers. Primary con-

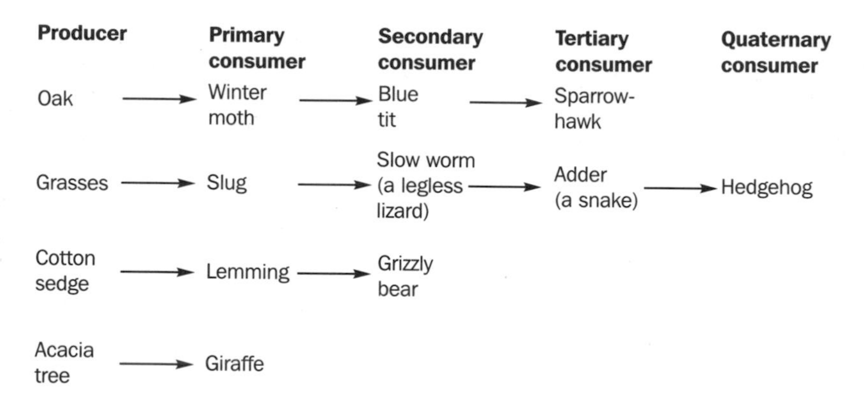

Fig. 28.1 Food chains.

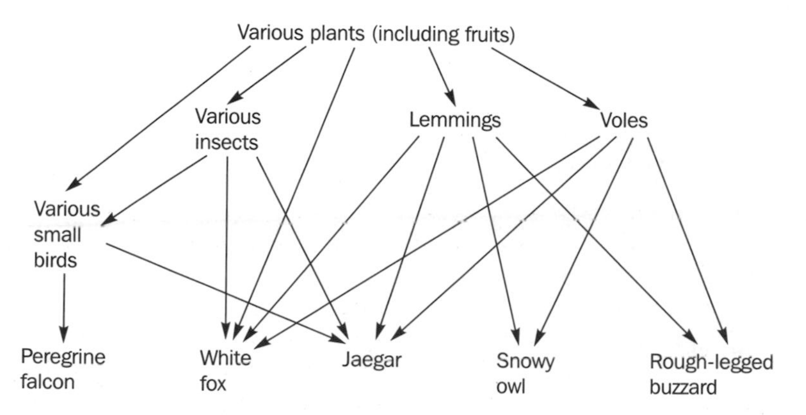

Fig. 28.2 Part of a tundra food web.

sumers can only get what they need from producers. Secondary consumers can only get it from primary consumers. Energy enters the chain as light. It is passed along the chain as energy-storing molecules. At every stage, energy is lost to the environment as heat. Food chains convert light energy to heat. There is no way back. Life goes on because the sun constantly bathes the Earth in light.

Chemical raw materials are quite different. There is not a constant rain of carbon, nitrogen, phosphorus and so on, hitting the Earth from space. There are only so many carbon atoms on Earth, and only those in the form of carbon dioxide can enter food chains based on photosynthesis. Some food chains are based on bacteria which use simple organic molecules from the environment as an energy source. Even these rely, in the end, on photosynthetic organisms to produce these molecules. If all the carbon dioxide in the atmosphere was used up, life would cease. It carries on because carbon cycles through a number of pools or stores in the ecosystem, and CO_2 is constantly regenerated. Geochemical cycles like this are known for all the elements important for life. Some parts of the cycles go on within a single ecosystem. Others operate over the whole biosphere.

Energy enters food chains as light. Life goes on because the sun bathes the Earth in light, day after day.

28.3 Geochemical cycles

I have a lot of socks. They can be in any one of several places. At the moment, I am wearing one pair, three pairs are on the washing line, two are in the laundry basket, none are in the washing machine, and three pairs are in the chest of drawers. I have also discovered one pair (very smelly) in my old shoes in the porch. Adding up the socks in these various pools, I seem to have ten pairs. They go through a sock cycle: from feet to laundry basket to washing machine to washing line to chest of drawers and back to feet again, with the occasional detour to old shoes. Socks flow between the pools in well-defined ways. They do not necessarily follow each other round the cycle like trains on a track. Some stay in a pool for much longer than others. It makes sense to look at geochemical cycles in the same way. Atoms of the element in question exist in several pools. Biological, chemical and physical processes set up flows between these pools. The carbon cycle is an easy example.

A great deal of carbon is in the form of carbon dioxide. Some is in the atmosphere. Some is dissolved in the oceans and other water bodies. Some rocks, notably limestone, contain carbon. Fossil fuels like coal and oil are another geological carbon pool. Lots of carbon is locked up in living things and detritus, dead organic matter. The flows between these pools are simple to describe (Fig. 28.3), though much harder to put numbers on.

Carbon leaves the CO_2 pool in the biosphere by photosynthesis. It moves through food chains by feeding. Carbon returns to CO_2 by respiration at every stage of the food chain. Lots of carbon may be locked up in these pools, for example in tree trunks. Only a few ecosystems, such as peat bogs, seem to steadily accumulate carbon, without limit. Under special conditions, dead organic matter may be fossilized as coal or incorporated into limestone as it forms.

Humans have a place in the carbon cycle as ordinary consumers. However, we affect the cycle in two other ways. Firstly, when we burn fossil fuels we shift a lot of carbon into the CO_2 pool. Secondly, human activity tends to replace ecosystems which store a lot of carbon by ones which store less. For example, forests with their huge stores of carbon in wood, tend to be replaced by grassland, cereal fields and the

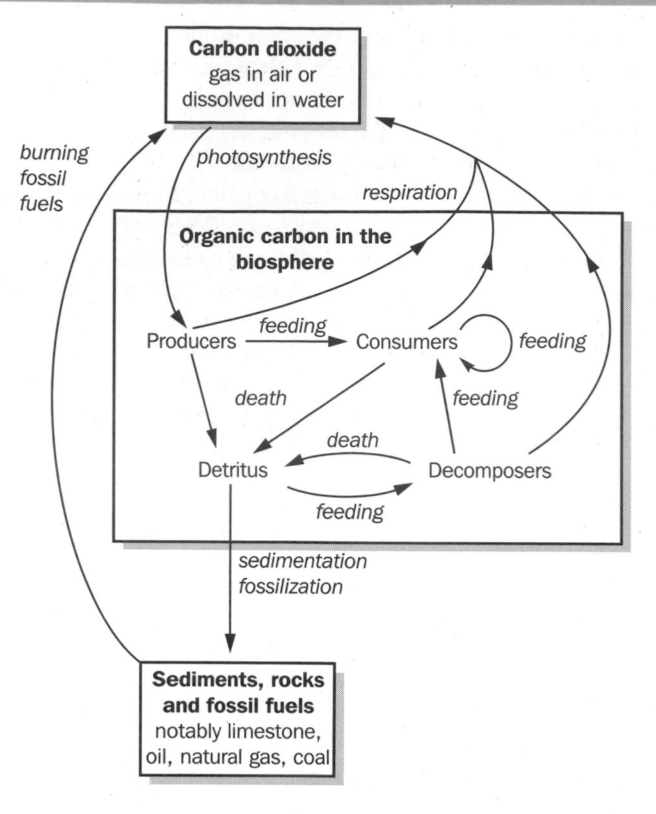

Fig. 28.3 The carbon cycle.

cities which they feed. The balance of carbon is shifted to CO_2 as the vegetation of cleared forest is burned or left to rot. Overall, then, people tend to shift carbon into the CO_2 pool. This is very bad news.

Carbon dioxide is one of several gases that absorb infra-red radiation ('heat radiation'). This tends to warm the atmosphere. If the concentration of CO_2 in the

A significant flow of carbon into the carbon dioxide pool.

air rises, the atmosphere may get warmer. This is the greenhouse effect. Since weather and climate are driven by heat in the atmosphere, this could change weather patterns across the world.

There is evidence of increasing CO_2 levels in the air over the last few decades. There is also evidence that the atmosphere has been getting warmer. This does not mean that it is caused by human activity, and it is clear that climates have always fluctuated. However, the longer it goes on, the more likely it becomes that people are causing global warming. At this point, a detailed understanding of how the carbon cycle behaves is needed. Does an increase in CO_2 lead to an increase in photosynthesis by plankton in the sea? If so, the system may be able to absorb a great deal of carbon with little effect on the atmosphere. This would be especially important if the carbon

Box 28.1

The nitrogen cycle

The pool of simple nitrogen compounds in the soil allows cycling within the ecosystem. The relatively small flows to and from N_2 gas in the air allow global cycling.

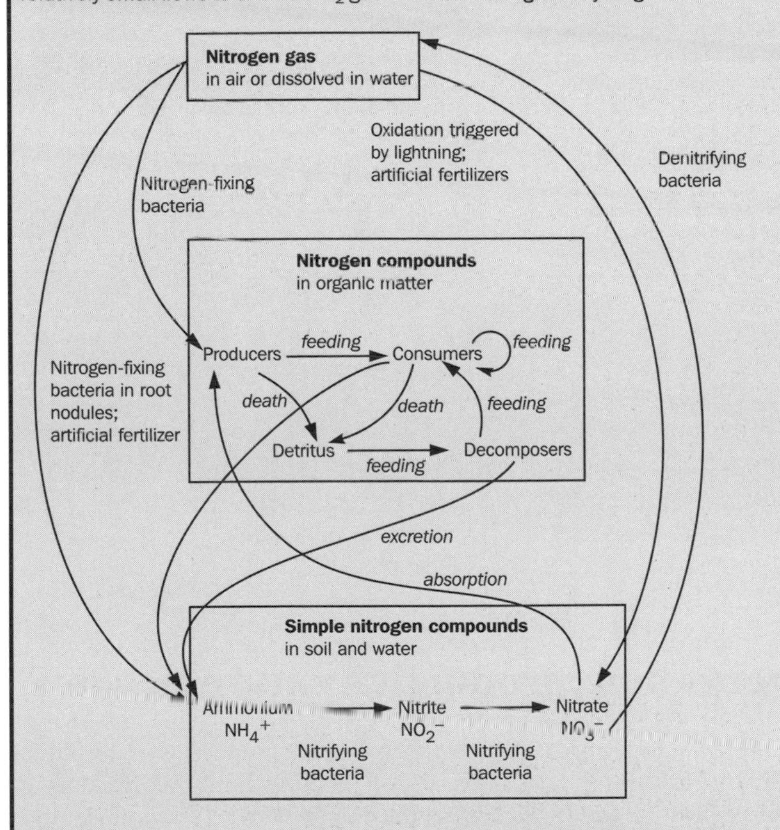

Nitrogen-fixing bacteria have an enzyme complex which allows them to use atmospheric nitrogen (Box 18.3).

Nitrifying bacteria oxidize inorganic nitrogen compounds, using them as an energy source.

Denitrifying bacteria live in anaerobic environments. They use nitrogen as an oxidizing agent in respiration, instead of oxygen.

became trapped indefinitely in ocean sediments. Could photosynthesis in the sea be encouraged by dumping limiting minerals, like iron, into the sea? If these checks and balances within the biosphere do exist, how far can they be pushed? These are exciting questions for scientists, as well as urgent problems for human society.

Carbon cycling involves CO_2 entering the air or water. Air and water move about freely, so carbon is carried between ecosystems around the world quite readily. In some geochemical cycles, minerals in the soil are the pools which allow cycling within ecosystems. Soil tends to stay in one place, so the global links are less strong. This is seen in the nitrogen cycle (Box 28.1) where nitrogen gas in the air is the global carrier of nitrogen. In the phosphorus cycle (Box 28.2), the huge pool of phosphate in rocks provides a steady input to the biosphere. The global cycle itself is unimaginably slow, relying on rock formation, the cycling of the Earth's crust and weathering of rocks to move significant amounts of phosphate from the sea to the land.

Box 28.2

The phosphorus cycle

Phosphorus exists in the biosphere as phosphate ions and phosphoryl groups (Box 3.2) in organic molecules. There is no significant pool of phosphorus in the atmosphere. Cycling takes place within ecosystems in the usual ways, but global cycling is very slow. Water carries phosphate from the land to the sea. More becomes available as rocks are weathered. Rock formation in the ocean sediments, and movements of the Earth's crust complete the cycle.

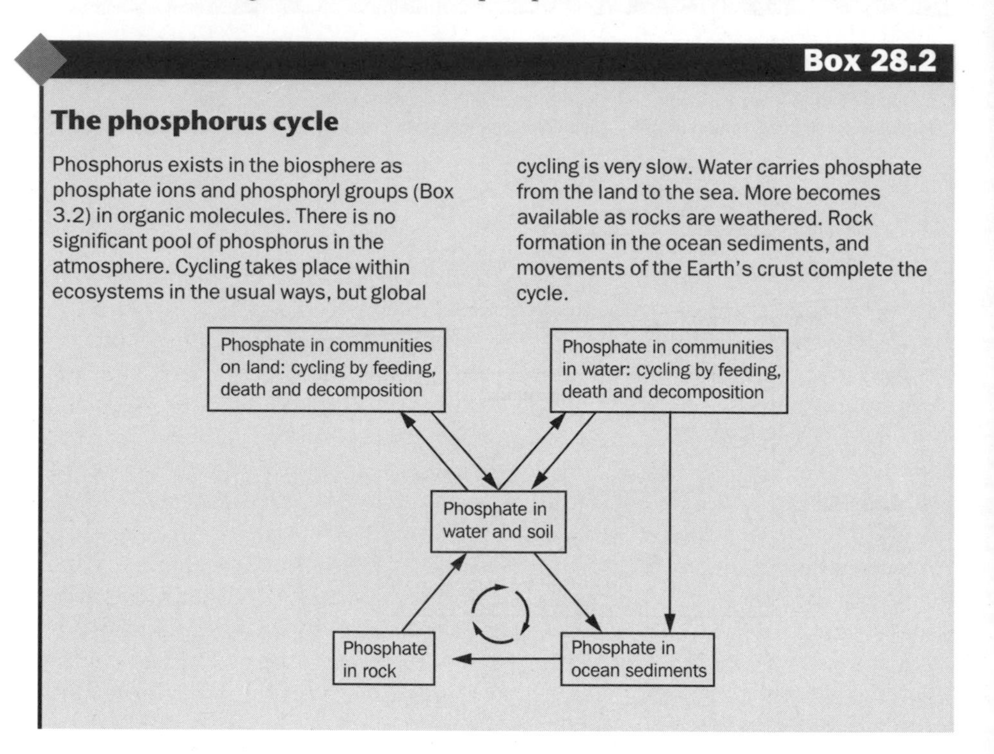

28.4 Energy flow, productivity and efficiency

Energy flows through food chains. Light energy enters at the producer level, by photosynthesis (Unit 10). It leaves as heat at every level, because almost all reactions in cells release some heat energy. In particular, the reactions of respiration (Unit 9) release a lot of heat. Again, it helps to see energy in terms of pools and flows (Fig. 28.4). Each trophic level of the grazing food chain and the detritus food chain is a pool, as well as the pool of detritus. Feeding, heat loss and death cause the flows.

Productivity: The rate at which energy enters a trophic level.

Deeper understanding of food chains requires putting numbers to the pools and flows. **Productivity** is a measure of how fast energy enters a trophic level. It is measured in units of energy (kilojoules) per unit area (square metres) per unit time

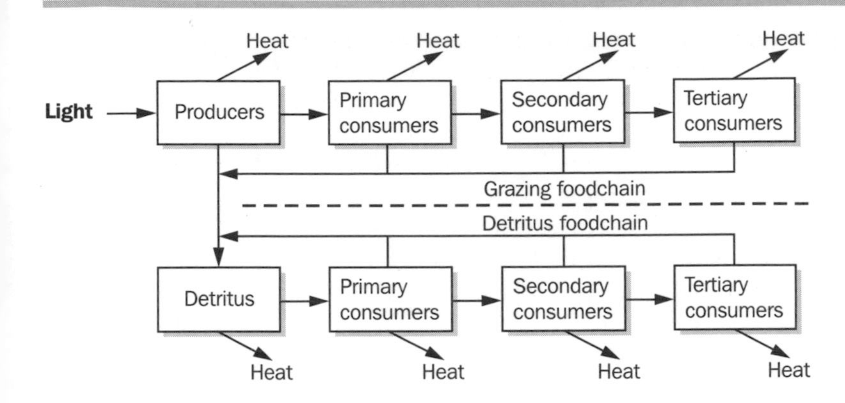

Fig. 28.4 Energy flow in an ecosystem.

(year): kJ m^{-2} yr^{-1}. Gross primary productivity is the rate at which energy enters the producers. Net primary productivity (NPP) is gross primary productivity minus the rate of energy loss as heat from the producers. In other words, NPP is the rate at which energy becomes available to the next trophic level, the primary consumers. Estimates of NPP range from as little as 100 kJ m^{-2} yr^{-1} for some arid areas, to 100 000 kJ m^{-2} yr^{-1} for the most productive agricultural ecosystems in the tropics. The NPP of a British grassland or forest ecosystem might be of the order of 10 000 kJ m^{-2} yr^{-1}. Productivity is high when conditions are good for plant growth. On land, productivity is decreased by shortage of light, water or minerals, by very high or low temperatures, or by an inefficient layout of leaves which allows some light to miss the leaves altogether. In the sea, productivity is mainly due to photosynthetic algae in the plankton, the floating organisms in the surface layers of the sea. Too deep, and there is little or no light. Minerals frequently limit marine productivity, often nitrogen, phosphorus or iron. Productivity tends to be higher where minerals enter the surface layers. Water near the sea bed picks up minerals from sediments, so upwellings of deep sea water near the edges of continents lead to high productivity. Rivers also carry minerals from the land to the sea.

The productivity of any other trophic level can also be estimated. In the grazing food chain, the productivity of each trophic level is much less than the productivity of the trophic level it feeds on. This is because some energy is lost as heat, some production is never eaten but goes into the detritus pool, and some of what is eaten passes through the gut and into the detritus pool as faeces. The percentage of the productivity of one trophic level which enters the next trophic level is called the **transfer efficiency**. Transfer efficiencies usually turn out to be somewhere between 1% and 50%. Beyond that, it is unwise to generalize!

Efficiency is a term which sounds positive. Employers want efficient workers; I want my motorbike to be efficient, to avoid spending too much on petrol. A high transfer efficiency does not imply that the food chain is somehow working well, or that the organisms are cooperating to transfer energy. It is simply a statement of the way things are.

The loss of energy along a food chain limits its length. There comes a point when there is just too little energy in a trophic level to support another. There is no clear rule about how long a chain can be. However, very large, very productive, stable ecosystems can in general support longer chains than small, unproductive or highly unstable ones. Long food chains are more likely to be found in rain forest in Brazil, than in a Scottish barley field. The idea that the amount of 'stuff' in a trophic level

Transfer efficiency: The percentage of the productivity of one trophic level which enters the next trophic level.

Box 28.3

Ecological pyramids

The ecological pyramid is a way of making the point that energy and materials are progressively lost as they move along food chains. It is a bar chart on its side, with the bars arranged centrally. Each bar represents a trophic level. The length of the bar represents some measurable feature of the trophic level.

Pyramids of energy

These are the most usual. The length of the bar represents the productivity of one trophic level in an ecosystem. Productivity always declines as we go along a food chain, so these diagrams really are pyramid-shaped. In this example, each transfer efficiency is 10%.

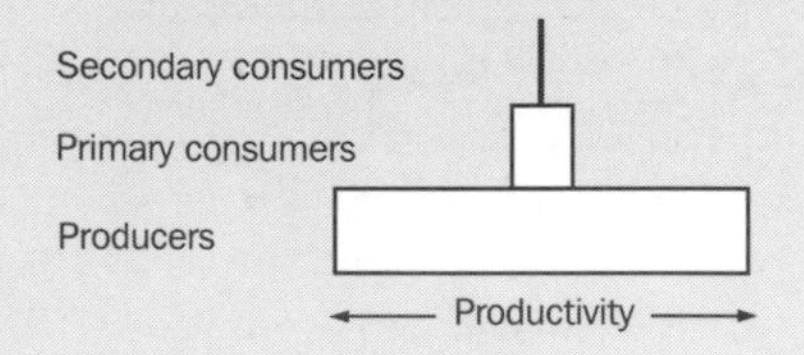

Two other types of pyramid are occasionally drawn.

Pyramids of biomass

These show the mass of organic matter in each trophic level at a particular time.

Pyramids of numbers

These show the number of individuals in each trophic level. They need not be pyramid-shaped. For example, if many tiny herbivores are feeding on enormous producers (insects on trees?) the 'pyramid' will have a bulge.

decreases as you go along a food chain is the basis of the old-fashioned idea of ecological pyramids. I find them an obscure way of presenting very simple data, but if you insist, have a look at Box 28.3.

Patterns of energy flow

Analysis of energy flow in different ecosystems can show up interesting differences. Some ecosystems seem to be accumulating energy, others losing it. The key to this is the ratio of primary productivity to total heat loss. If the value of the ratio is exactly one, productivity must equal heat loss. The amount of energy locked up in the system stays constant. If the ratio is greater than one, more energy enters than leaves. It is being stored, either in living organisms as they grow or in an accumulating pool of dead matter. A plantation of young trees would have a high ratio, which would fall as the trees became mature. If the ratio is less than one, there is an overall loss of energy. This must be because the mass of living and/or dead matter is decreasing. A woodland in which all the trees are old and dying, where grazing animals prevent new tree seedlings establishing, would probably fall into this category.

Energy fixed by producers can go down either the grazing food chain or the detritus food chain. The majority probably enters the detritus chain in most ecosystems, but there is some variation. In most forests and streams, only a trivial proportion of production is grazed. Most plant material becomes food only after it has died. Grazing tends to be more important in grasslands, and most significant in the plankton communities of the sea, where perhaps half of the primary productivity is grazed.

28.5 Gaia

Humans influence the world in a big way. As well as creating and destroying ecosystems, people affect geochemical cycles acting on a global scale. For example, activities as different as cooking, bus travel and clearing forest to make way for crops, add carbon dioxide to the air. It would be naive to think that all that extra CO_2 stays there. It is perfectly possible that the more CO_2 enters the atmosphere, the faster photosynthesis takes it out again: possible, but in no way certain. Could the carbon cycle or other cycles be regulated in a homeostatic way? Could it be that homeostasis on a global scale maintains conditions in the biosphere suitable for life? This is the essence of the **Gaia** hypothesis. Emotive slogans such as 'the Earth is alive' have sometimes been linked to the hypothesis. Attractive to some and off-putting to others, they distract attention from the main, generally unanswered question: Is the Gaia hypothesis correct?

Homeostasis is the ways in which some variable is buffered against change (Unit 15). If the variable increases, a chain of events is set off which returns the variable to its original value. If it decreases, a chain of events leads to an increase. To suggest that this happens on a global scale should not imply that there is deliberate control. There is no need to invoke direct intervention by a creator God, a great Earth Mother, or some sort of happily cooperating world-alliance of living things. In Section 26.4 we saw that population size is regulated in a homeostatic way, simply through individuals looking after their own interests in a world where resources are limited. Environmental variables might be regulated simply by supply and demand. If the level of CO_2 in the air, nitrate in the soil, phosphate in the sea or whatever, increases, biological processes that use it may speed up. The level will then fall again.

Even if true, this does not mean that we can stop worrying about our effects on geochemical cycles. Homeostatic systems can break down in two ways. Firstly, if pushed away from equilibrium too strongly, the system cannot cope. If I leave the oven door open, it will never get up to temperature, let alone be kept at that level by homeostasis. Similarly, even if photosynthesis does mop up extra CO_2, there may come a point when it can mop up no more. Secondly, the system itself may become damaged. If I cut the electric cable leading to the oven, or smash the thermostat, it can no longer regulate its own temperature. James Lovelock, originator of the Gaia hypothesis, believes that the main carrier of sulphur through the biosphere is dimethyl sulphide, released to the air by some seaweeds. He suggests that large-scale farming of other seaweeds could disrupt the sulphur cycle by displacing these species, perhaps preventing a homeostatic system from working.

Gaia is still hypothesis, not established fact. Its great importance is to focus attention on the need to understand geochemical cycles in a detailed, quantitative way. At a time when growing human populations affect ecosystems and geochemical cycles in ways which are far from subtle, this cannot be a bad thing.

Gaia: The hypothesis that homeostasis on a global scale maintains conditions which happen to be suitable for life.

Summary

- Living things interact with the non-living things around them.
- An ecosystem is all the living things in a habitat, plus all the non-living things.
- A food chain is a group of organisms linked by feeding.
- Every food chain begins with a producer, usually a plant. The other organisms in the chain are called consumers.
- Producers take in energy from light, and simple molecules from the environment.
- Primary production is the rate at which energy enters the producers of an ecosystem. It varies wildly from place to place, depending on how suitable the conditions are for photosynthesis.

◆ Up to half the primary production is grazed. The rest dies and enters the detritus food chain.

◆ Energy flows though food chains, from producers to consumers. It is lost forever as heat.

◆ The length of a food chain is limited by the loss of energy at each stage.

◆ Chemical elements are passed along food chains, but are not made or destroyed. They cycle between various pools in organisms and the environment.

◆ Some geochemical cycles involve free flow of materials throughout the biosphere. Others are dominated by cycling within ecosystems.

◆ The Gaia hypothesis suggests that some geochemical cycles may involve homeostatic control. This may keep conditions in the biosphere suitable for life.

Exercises

28.1. **(i)** Global cycling of nitrogen is far more rapid than that of phosphorus. Why?
(ii) Nitrogen and phosphorus compounds are both useful as fertilizers. Any country with appropriate technology and a good energy supply can manufacture its own nitrogen fertilizers, but phosphorus fertilizers can only be made economically in some places. Explain.

28.2. Among non-scientists, I sometimes hear an argument that goes like this. 'Photosynthesis takes carbon dioxide out of the atmosphere. A lot of photosynthesis goes on in tropical rain forests. If rain forests are cut down, more of the CO_2 produced by respiration will remain in the atmosphere. This contributes to the greenhouse effect, so cutting down tropical rain forests is a bad thing.' To what extent is this argument correct?

28.3. Look at the food chains of Fig. 28.1. Even if we knew the amount of energy entering each producer each year, we have no way of predicting the rate at which energy enters, say, hedgehogs or grizzly bears. What else would we need to know in order to estimate these things?

28.4. In an imaginary ecosystem, net primary productivity is 20 000 kJ m^{-2} yr^{-1}. The transfer efficiency at each level is 15%. Calculate the net productivity of the secondary consumers.

Further reading

Aber, J.D. and Melillo, J.M. *Terrestrial Ecosystems.* (Philadelphia: Saunders, 1991). A modern introduction to the quantitative study of ecosystems.

Barnes, R.S.K. and Hughes, R.N. *An Introduction to Marine Ecology* (2nd ed.) (Oxford: Blackwell Science, 1988). This book deals with a range of marine ecosystems, but extends beyond this level, to population biology and life histories.

Beeby, A. *Applying Ecology.* (London: Chapman & Hall, 1993). This book includes conservation and restoration of ecosystems.

Begon, M., Harper, J.L. and Townsend, C.R. *Ecology: Individuals, Populations and Communities* (2nd ed.) (Oxford: Blackwell Science, 1990). Not the first big textbook of ecology, but the first which I felt reflected the range of questions modern ecologists are asking.

Moss, B. *Ecology of Fresh Waters: Man and Medium* (2nd ed.) (Oxford: Blackwell Science, 1988). A thorough introduction to an important group of ecosystems.

Behaviour

Connections

▶ This unit should make sense even if you know little about other areas of Biology. Animal behaviour poses some difficult evolutionary problems, however. You should have a clear understanding of natural selection (Unit 2, especially Sections 2.4 and 2.5) before reading this unit.

Contents

29.1 What is behaviour?

Behaviour is what animals do. When we study behaviour, we aim to find out what animals are doing, and why.

It is spring. We are standing quietly outside a lambing shed, leaning against the makeshift wall and looking in. Thirty heavily-pregnant ewes are in there, being given shelter and extra food in the last stage of pregnancy. At first sight they do not appear to be doing much. Watch them for a while, and see what goes on. Some stand up. Others lie down. Some eat the hay provided for them. Occasionally one urinates. All these activities are behaviours. Why do they sometimes lie and sometimes stand? How much of the time do they lie down, and why? Does this change as pregnancy goes on? There are more subtle interactions taking place. From time to time, one ewe comes up to another and nudges her. Sometimes, the second ewe holds her ground. More often, she gives way. What are we seeing here? Are the ewes competing for a good position, or is one simply asserting her dominance over the other? Is there a social hierarchy among these sheep? If so, why? How did the top ewe get to that position, while others settled to the bottom of the social scale?

A bird has perched on the hazel bush beside the shed. Even without looking, I recognize it as a blackbird by its song. How did it develop the ability to sing this characteristic song? Did it learn from its parents, or are there genetic instructions? Now it stops singing. It moves its head about, then flies off with a harsh alarm call. The alarm call has an apparent benefit to other blackbirds, but less obviously to the individual that made it. How could this behaviour have evolved?

We have been asking four types of question while watching animals behave. What triggers behaviour? How does the ability to behave in that way develop? What is its value to the animal? How did the behaviour evolve? In these simple observations and questions lies the science of animal behaviour.

Most people think sheep are boring old munchers. Perhaps they are, but their behaviour repays study (see text).

This unit deals first with what triggers behaviour, then how behaviour develops. The value and evolution of behaviour are important and difficult questions. They are examined by looking at four areas of behaviour as examples: searching for food, living in groups, fighting, and alarm calls.

29.2 What triggers behaviour?

Many behaviours are clearly a direct response to a stimulus. If I stand at the back door and clap my hands, the collared doves eating our broad beans fly off. They cannot see me. The stimulus must be the sudden loud noise. A stimulus which triggers a behaviour is called a releaser. A famous example shows that releasers may be very simple. Male sticklebacks defend a territory against rival males in spring. In experiments, they would attack models only if they had a red underside, like real males. This was the releaser: the shape of the model mattered very little.

More often, we cannot identify a single reliable releaser for a behaviour. Animals respond in a complex way to a range of stimuli, as well as the internal motivations which drive them to eat, drink, sleep and reproduce.

29.3 How do behaviours develop?

Some behaviours are instinctive. Animals are born able to do them correctly, at the right times. My baby son is expert at sucking milk. This involves coordinated movements of the tongue and jaw. Two other automatic responses (reflex actions) are needed to make this work. The first is a response to a touch on one cheek: he turns towards it and opens his mouth. The second is a response to liquid in the throat: he swallows. Like other babies, he could do this at birth, without learning. It is an instinctive behaviour. Other behaviours are learned, bowling a cricket ball for example. Instinctive and learned behaviours are extremes. Very often, instinct and learning interact. The song of the blackbird is an example of this.

29.4 Areas of behaviour

Searching for food

It is often clear that a behaviour is connected with foraging. A bee flies from flower to flower, collecting nectar and pollen. A black-headed gull probes the soil of a wet field for invertebrates. From time to time it catches and eats something. How is searching organized? Do they do it efficiently?

Most behaviours have costs as well as benefits. Foraging uses up more energy than staying at home sleeping, and it is more dangerous, but the animal that stays at home starves. Sometimes foraging is not worthwhile. If a bee uses more energy finding and collecting nectar from a flower than it gets from the nectar, it has lost energy. There was no point in foraging. In general, the longer it takes to find a flower, the more energy it must get from it if the behaviour is to be worthwhile. There is clear evidence that bees only forage when it is worthwhile.

It seems likely that animals optimize the balance between gains and losses when foraging, maximizing the energy profit. Studies of foraging behaviour attempt to test this idea. Nutritional value, abundance and distribution of different types of food are all factors in the equation.

It is not hard to see how optimal foraging could evolve. The energy gained from foraging can be used for other things, especially reproduction and caring for offspring. The greater the energy profit, the more offspring an animal can leave. Optimal foraging means higher fitness (as defined in Section 2.2).

Living in groups

Black-headed gulls and dunlin feed in flocks; gannets nest in dense colonies; herrings swim around in shoals; antelope form herds; prairie dogs share burrows. Communal living is widespread. It has one big advantage. The more animals there are, the sooner one of them will spot a predator approaching the group. The whole group can then escape. Alternatively, each individual can get away with spending less time looking out for predators, leaving more time for feeding and other activities. At first sight this might seem like a group selection argument (Section 2.2). Could group-forming behaviour be undermined by 'selfish' individuals which rely on others watching for them, and feed all the time? However, watching for predators probably benefits the individual directly. If an animal spots a predator itself, it escapes ahead of the others, and stands even less chance of being caught. There is no evolutionary problem here.

Group living also has disadvantages. Predators may be attracted by large concentrations of prey. Parasites and diseases may spread more quickly in dense populations. If animals are feeding in a group, there may be less food per individual, and competition may become more intense (Unit 26).

These issues have been examined in many species by comparing how much time animals spend on different behaviours in different-sized groups. For example, one would expect individuals in a bigger flock to spend less time watching for predators, but more time fighting over food or space. The remaining time could be spent feeding. Very many studies support this idea. Many suggest that feeding flocks have an optimum size. In a flock of this size, the balance between feeding and fighting leaves the maximum time for feeding. Changes to the food supply or the level of predation may change the optimum size.

Optimum-sized flocks may not be common, however. An optimal flock would be the most attractive flock for an individual to join. The flock would grow. In doing so, it would stop being optimal.

Fighting

Animals fight over limited resources. When stags fight over mates, and robins skirmish at the edges of their territories, they are competing for something they both need.

Research on fighting in various species leads to some general conclusions. Firstly, fighting has costs as well as benefits. It uses time and energy, and carries a risk of injury or death, as well as bringing the chance of getting valuable resources. Fighting is more likely if the benefits are great: benefits must exceed costs.

Secondly, fights are usually decided on either fighting ability or on how much the resource is worth to each animal. In some fights, the better fighter wins. This is true when red deer stags fight over females. In others, the individual which has most to gain from the fight usually wins. This makes sense. The risks are the same for both individuals, but the benefits are greater for one. That individual might be prepared to fight harder. Territories are a good example of this sort of resource. If I am a robin, my territory is much more valuable to me than anyone else's territory. I know my way around it, where the food is, where I can hide, and so on. I have invested time and energy in learning all about it. It means more to me than to others who might want it. I am likely to win contests in my own territory.

Thirdly, fights rarely go as far as serious injury. Most animals have ways of predicting the outcome of the fight before it escalates, and backing down if they are likely to lose. This has clear benefits to both individuals if predictions are accurate.

Box 29.1

The hawk and dove game

Simple models based on game theory shed light on the evolution of fighting behaviour.

Imagine a population in which there are two alternative fighting strategies, each controlled by genes. All animals have the same fighting ability. They simply make decisions about when to fight and when to back off in different ways. The two strategies are called 'hawk' and 'dove'. The names are simply labels, and have more to do with how we talk about politicians' attitudes to war than with the real animals with those names. There is an attacker and an opponent in each fight.

If a dove attacks a hawk, the dove will soon back off again, with no benefit but no injury. If a dove attacks another dove, they end up sharing the resource.

If a hawk attacks a dove, the dove will back off before getting hurt: the hawk gets the resource. If a hawk attacks another hawk, the fight escalates. Both lose more through injury than they gain.

This table shows how these conflicts affect the attacker's fitness. It is called a pay-off matrix. The numbers are chosen simply to illustrate the strategies.

Attacker \ Opponent	hawk	dove
hawk	−1	+1
dove	0	$+\frac{1}{2}$

+ means a gain in fitness
− means a drop in fitness

The matrix can tell us if there is an evolutionarily stable strategy (ESS; Section 2.2). An ESS must be able to invade a population which operates the other strategy, but cannot be invaded itself.

In a population of only hawks, all fights have a payoff of −1. Imagine a mutant individual, which follows the dove strategy in that population. All its fights would have a payoff of 0. The dove would have a higher fitness than the hawks: its genes would spread. This means that 'hawk' is not an ESS.

In a pure dove population, all fights would have a payoff of $+\frac{1}{2}$. A mutant hawk would have payoffs of +1. It would have a higher fitness. In other words, neither strategy is an ESS. Each can invade the other.

However, a mathematical argument shows that mixed strategies can be stable. Individuals sometimes behave as hawks, sometimes as doves. This can be stable even if the decision to be a hawk or a dove is taken at random. Strategies which take account of what the opponent is like, or who the resource is most valuable to, can lead to stability very simply.

Imagine a third strategy, called 'bourgeois'. Bourgeois strategists take into account who already owns the resource. Resources such as territories are usually worth more to their owners than to attackers. A bourgeois will behave like a hawk if it owns the resource, but like a dove if it does not. The pay-off matrix might be as follows:

Attacker \ Opponent	hawk	dove	bourgeois
hawk	−1	+1	0
dove	0	$+\frac{1}{2}$	$+\frac{1}{4}$
bourgeois	$-\frac{1}{2}$	$+\frac{3}{4}$	$+\frac{1}{2}$

The four types of fight in the top left corner are just as in the first matrix. The other five types of fight involve bourgeois strategists. Their pay-offs are calculated as follows.

Imagine many fights where a bourgeois attacks a dove. On average, the bourgeois will own the resource half the time, and will behave like a hawk. The other half of the time it will not own the resource, so will act like a dove. Half the time the pay-off will be the same as a hawk–dove fight, +1. The rest of the time it will be as a dove–dove fight, $+\frac{1}{2}$. The average pay-off will be $(\frac{1}{2} + 1) \div 2 = +\frac{3}{4}$. You should be able to confirm the other four pay-offs for yourself.

Is 'bourgeois' an ESS? It can invade a pure hawk population because the bourgeois–hawk payoff (−1/2) is greater than the hawk–hawk payoff (−1). In the same way it can invade a pure dove population ($+\frac{3}{4}$ is greater than $+\frac{1}{2}$). A bourgeois population cannot be invaded by hawks (0 is less than $+\frac{1}{2}$) or doves ($+\frac{1}{4}$ is less than $+\frac{1}{2}$). So 'bourgeois' is an ESS.

This simple model helps us to understand why territory holders usually win contests. Not only does backing down to a territory holder increase fitness, but it is a stable strategy too.

The winner gets the same resources, but minimizes the cost of acquiring it. The loser still gains nothing, but loses less. Predictions are based on assessing the opponent's fighting ability, either directly or by indirect signals such as body size.

The final point relates to the evolution of fighting strategies. It is not enough for a strategy to be a good one: it must also be a stable one (Section 2.2). A population of individuals which operates an evolutionarily stable strategy cannot be invaded by individuals with an alternative strategy. These ideas are based on models taken from game theory. Box 29.1 shows a simple example.

Alarm calls

The screeching chatter of a blackbird as it flies away is a familiar alarm call. Very many animal species have them. It is easy to see how alarm calls could benefit the population as a whole, because other individuals are warned, and can escape. There is evidence to support this in some species (although not in blackbirds!). However, this cannot explain how the behaviour evolved (Section 2.2). It is hard to see how alarm calls could benefit the individual directly. They may even carry a cost, by attracting predators. However, if an individual's offspring are nearby, an alarm call is just another example of parental care. The offspring carry the parent's genes. A gene which leads to this behaviour could well be favoured.

In some social animals such as the prairie dog, an American rodent, alarm calls are made whether or not offspring are present. In these inbred societies of close relatives, kin selection seems to be at work (Section 2.3).

An alternative explanation of alarm calls is that they are directed at the predator. 'I know you are there, so don't bother' could sometimes be the meaning.

29.5 Play?

We tend to pick out events from the continuous stream of animal behaviour, and analyse their significance in terms of survival, security and sex. It is very hard to see the direct benefits of some behaviours. Why do jackdaws fly about wildly on windy days? Why do people play cards?

Why do caged polar bears sometimes pace back and forth, shaking their heads in a way that is not seen in the wild? This sort of stereotyped behaviour in zoo animals can sometimes be alleviated by making their environments more complex, or giving them toys. It is as if looking after the animals' physical needs is not enough. They need to be given the opportunity to behave in certain ways. Do some behaviours develop, or keep in trim, mental processes which are used in other contexts? Do some behaviours build up relationships within social groups? Do animals play?

Summary

- The science of animal behaviour involves studying what animals do, and why.
- We identify particular behaviours to study in the constant stream of animal behaviour.
- We can ask several types of question about a behaviour. What triggers it? How does it develop? What is the value of it to the animal? How did it evolve?

- Some behaviours are a simple response to a stimulus (a releaser).
- Other behaviours are the result of interaction between external stimuli and internal motivations.
- Some behaviours are instinctive. Animals can do them without having to learn. Other behaviours are learned. Often, both instinct and learning are involved.
- Foraging has costs and benefits. Optimal foraging means searching for food in a way which maximizes the net energy gain.
- Group living means that predators will probably be spotted sooner. There is a cost in terms of increased competition.
- Animals fight over limited resources.
- Fighting is more likely if the resource at stake is very valuable.
- Sometimes fights are won by the better fighter. Sometimes they are won by the animal to which the resource is most valuable.
- Individuals assess their opponents in the early stages of a fight. This may lead to one backing down before the fight escalates.
- Alarm calls benefit the whole population. They probably evolved because they particularly benefit offspring or relatives.
- For a behaviour to evolve, it must be evolutionarily stable.

Exercises

29.1. The first step in almost any study of animal behaviour is to watch the animals carefully, with no preconceptions about what they are doing. The next is to make an ethogram. This is a list of all the behaviours you have observed, along with a precise description of each, to show how you have defined each behaviour. Find a convenient animal, and do just this. Depending on where you live, it might be a pet, a farm animal, something living under stones in your garden, or an animal in the local zoo. There is no 'right' answer to this exercise, but you can find an example of an ethogram in the Answers section.

29.2. A feeding flock of a species of bird has an optimal size. Predict how optimal flock size will change if

(i) more food is available;
(ii) there are more predators in the area.

29.3. Konrad Lorenz, a pioneer in the study of behaviour, described how two doves placed in the same cage fought until one was killed. Explain this observation, given that fights in the wild rarely lead to serious injury.

Further reading

Krebs, J.R. and Davies, N.B. *Behavioural Ecology: an Evolutionary Approach* (3rd ed.) (Oxford: Blackwell Science, 1993). An excellent undergraduate textbook which focuses on the functions and evolution of behaviour.

Ridley, M. *Animal Behaviour* (2nd ed.) (Oxford: Blackwell Science, 1995). A short, wide-ranging and accessible introduction: an excellent starting point.

PART 7

Applied Topics

Disease

Connections

▶ This unit draws on principles from almost every unit of this book. An understanding of the ways in which organisms protect themselves (Unit 23) is particularly important.

Contents

30.1 Disease – from within and without

Nothing works properly all the time. Cars break down; computer programs crash; customs officials can never completely prevent smuggling; the best legal systems sometimes lock up the innocent and let the guilty go free. Wherever there is an organized system, there is an opportunity for it to go wrong. Organisms are no exception. Life goes hand in hand with disease.

Disease is not simply something that hits the organism from outside. External and internal factors influence whether or not an organism goes down with a disease.

Pathogens, disease-causing organisms, are the most obvious external factors. Viruses are nucleic acids which parasitize cells (Boxes 30.1–30.6). Other pathogens include bacteria (Boxes 30.7–30.9), protoctists (Boxes 30.10, 30.12, 30.13), animals (Box 30.11) and fungi (see Box 30.13). Prion (Box 30.19) is a protein which seems to be capable of transmitting disease on its own.

Even diseases caused by pathogens are influenced by internal factors. Organisms are well defended (Unit 23). The nature of the defences determines the course of the disease, and whether or not it occurs at all. In some cases, the body's attempts to control a pathogen cause more harm than the pathogen itself. Hepatitis A patients suffer symptoms because the immune system destroys liver cells infected by the virus. Tuberculosis involves another disastrous response to a pathogen (Box 30.9). Some diseases result from a specific inability to combat infection. The HIV virus, which causes AIDS, infects helper T-lymphocytes, compromising the immune system. AIDS patients may die of infections which are rarely seen in other people. In the autoimmune diseases (Box 30.20) the immune system attacks the body's own cells.

At first sight, genetic diseases seem to come purely from within (Box 30.14). However, their ultimate cause is mutation, which is affected by external factors such as radiation and mutagenic chemicals. Cancers (Boxes 30.15–30.18) are caused by accumulated mutations in certain genes. Exposure to mutagens increases the risk of cancer.

Other diseases may be caused by dietary deficiency (Box 21.1 and 30.22), or by a failure of homeostasis (Box 30.21).

Symptoms are not related to the molecular basis of disease in a simple way. A single chemical defect can lead to diverse symptoms (Box 30.21). At the same time, a single symptom can have many causes (Box 30.22).

The rest of this unit illustrates the biological basis of disease, through a series of boxes. Many, but not all the examples are taken from human disease.

Box 30.1

Viruses

Viruses are parasites of cells. They are not made up of cells. Instead, they are nucleic acid hijackers which take control of cells for their own ends. The machinery of the cell is used to make viral proteins, using the information coded in the viral nucleic acid, as well as to replicate the nucleic acid itself. The result is many more viruses which can infect other cells.

All viruses have:

- **Nucleic acid:** either DNA or RNA. There may be one or several pieces. Their genomes may have anything from three to over 200 genes.
- **A protein coat (capsid):** This is a shell of one or more types of protein, in a regular arrangement. It encloses the nucleic acid. The protein coat is only present when the virus is outside host cells.

Some also have:

- **A membrane, the envelope:** This surrounds the capsid, but is not made by the virus itself. It is a modified fragment of the plasma membrane of the last cell invaded by the virus.

Viruses do not have:

- organelles
- cytoplasm
- enzymes (with a few very special exceptions)
- any sort of metabolism.

Viruses cause disease because they kill cells. If successful, they destroy the cell when it breaks open to release new viruses. If the virus is unsuccessful, programmed cell death may be triggered at an early stage of infection (Section 25.5). Symptoms depend on which type of cell is infected.

Box 30.2 shows how RNA viruses overcome the problem of having no DNA. Boxes 30.3–30.6 illustrate the structure and 'lives' of a range of viruses.

Box 30.2

RNA viruses

A cell transcribes a gene many times to make a pool of mRNA molecules. Each one is translated many more times, making a larger pool of proteins. RNA viruses have a problem: they might miss out on the first stage of this amplification process. They also have a problem with replication: cells replicate DNA, not RNA. There are four ways out of this situation.

In each case, Ⓢ represents a sense strand, Ⓐ a complimentary antisense strand (Section 7.2). Notice that some viruses have double-stranded RNA.

1. Ⓢ RNA ⟶ Ⓐ RNA ⟶ Ⓢ RNA ⟶ new viruses
 └⟶ protein synthesis

Both steps need an RNA-directed RNA polymerase (a replicase). Cells do not have such things. However, the virus has a gene for this enzyme, which is expressed when the viral RNA enters the cell. The same process leads to amplification and replication. Rhinovirus, which causes the common cold, is an example.

2. Ⓐ RNA ⟶ Ⓢ RNA ⟶ Ⓐ RNA ⟶ new viruses
 └⟶ protein synthesis

Replicase is needed in these viruses as well. The viral RNA is an antisense strand, so its genes cannot be expressed until a sense strand has been copied. The virus carries a replicase molecule inside the capsid. Rabies virus is a well-understood example.

3. (S/A) RNA ⟶ Ⓢ RNA ⟶ Ⓐ RNA ⟶ new viruses
 └⟶ protein synthesis

Reovirus works like this. It infects epithelial cells of the mammalian gut and airways, but does not cause any recognizable disease. Reovirus includes a replicase molecule. Only the antisense strand of the double-stranded RNA is transcribed.

4. Ⓢ RNA ⟶ Ⓐ DNA ⟶ (S/A) DNA ⟶ Ⓢ RNA ⟶ new viruses
 └⟶ protein synthesis

The whole process of making double-stranded DNA from a single-stranded RNA template is carried out by reverse transcriptase. This enzyme is coded for by a viral gene, and a molecule is included within the capsid. The double-stranded DNA becomes part of a host-cell chromosome before it is transcribed. When the cell divides, the viral DNA is passed on to the daughter cells. Viruses which do this are called retroviruses. They include the HIV virus which causes AIDS by infecting helper T-lymphocytes.

Box 30.3

Adenovirus

- A double-stranded DNA virus with about 30 genes.
- The capsid is an icosahedron (a regular polyhedron with 20 identical faces). This is close to being a sphere. It has a low surface area/volume ratio, so a given amount of protein can hold a lot of DNA. The regular structure is made up of repeated protein units. Having a single type of capsid protein makes the most of a limited number of genes.
- It has no envelope. Special protein spikes in the capsid insert into the host cell plasma membrane. They disrupt it locally, allowing the virus to slip in.
- Adenovirus commonly infects the nose and throat. Infections of the smaller airways in the lung may be serious in young children.
- It is transmitted in mucus droplets in the air.
- Adenovirus hides from the immune system. Infected cells cannot pass viral antigens to T-lymphocytes (Section 23.5). Normally, class 1 MHC protein would sit in the plasma membrane, presenting antigens to T-cells. A viral protein prevents newly formed MHC from reaching the membrane.

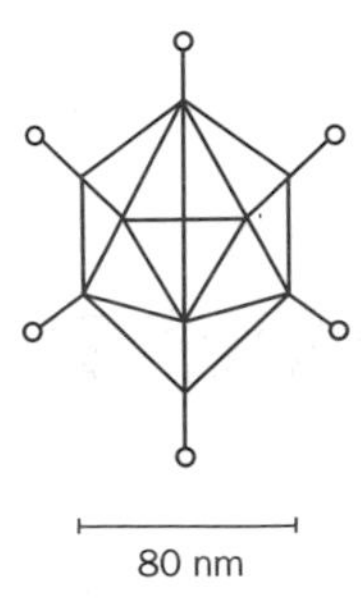

Box 30.4

Herpes simplex virus

- A double-stranded DNA virus infecting humans.
- The capsid has an icosahedral structure (Box 30.3).
- The virus has an envelope. It fuses with the host cell's plasma membrane, allowing the capsid and DNA in.
- Two different strains cause cold sores and genital herpes.
- Transmission is by physical contact: mouth-to-mouth for cold sores, or sexual for genital herpes.
- The virus infects the host for a long time, but usually hides from the immune system and is inactive (latency). It hides in sensory neurones and only occasionally breaks out to multiply in other tissues. This explains why cold sores disappear and reappear.

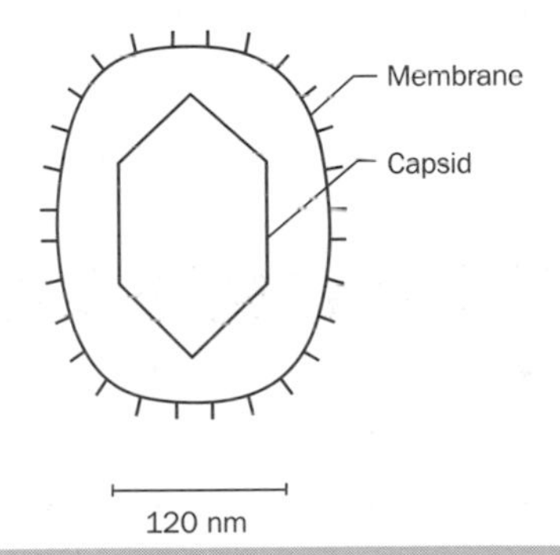

Box 30.5

Tobacco mosaic virus (TMV)

- A single-stranded RNA virus, with six genes.
- It is rod-shaped. Like the icosahedron, this is a regular structure which can be made up of many identical proteins.
- A single strand of RNA runs the length of the virus. Proteins are arranged around it, forming a spiral shell.
- TMV can self-assemble. Purified TMV RNA and capsid proteins can bind together to form infective viruses, even in the test-tube.
- TMV infects a wide range of flowering plants. Symptoms vary with host species and include stunted growth, mottling and distortion of leaves. Infections spread throughout the plant. It can be transmitted by aphids and other insects which feed on phloem sap.

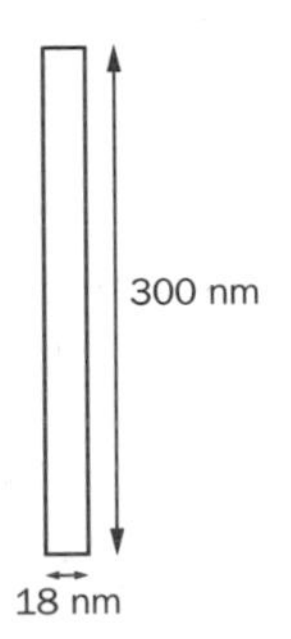

Box 30.6

'Phage T4

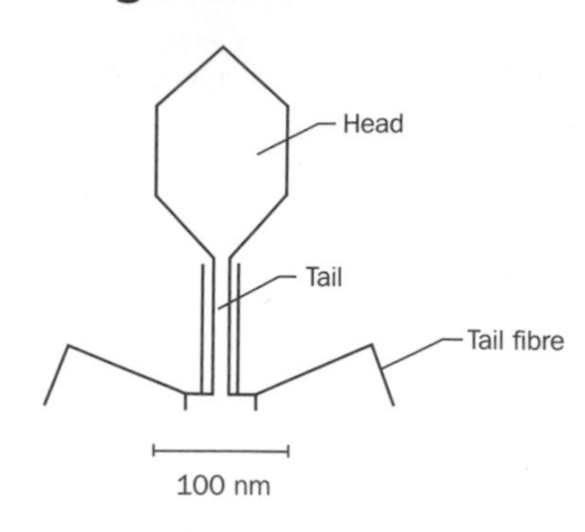

- Bacteriophages ('phages for short) are viruses which infect bacteria. T4, like most others, is a big, double-stranded DNA virus. It has about 150 genes. It infects *E. coli*, the mammalian gut bacterium.
- The capsid has a complex structure. It consists of an icosahedral head; a tail made up of two hollow tubes, one inside the other; and six tail fibres. The DNA is in the head.
- The tips of the tail fibres bind to the bacterial wall.
- The outer tube of the tail contracts, forcing the rigid inner tube through the wall. DNA can then enter the cell.
- This complex virus cannot self-assemble. Many of its genes code for proteins which control construction.

Box 30.7

Clostridial infections

Many bacteria cause disease because they secrete toxins. Toxins are proteins or glycoproteins which affect the working of the body. Bacteria in the genus *Clostridium* show that the body can be harmed without bacteria invading living tissue.

Clostridium botulinum

This bacterium makes a toxin which blocks neuromuscular junctions, causing paralysis and death. The bacterium need not even enter the gut. Small amounts of toxin in contaminated food are enough to cause symptoms. In the early years of the food-canning industry, canned fish products were sometimes responsible.

Clostridium perfringens

This species enters the gut in contaminated meat. Once there, it produces a toxin which kills epithelial cells: clostridial food poisoning.

Clostridium tetani

This species is an anaerobe which can grow among the dead tissue of a deep wound, having entered as spores in soil. Unlike the other two species, it multiplies in the wound. As it does so, it secretes a neurotoxin which results in muscle spasm, paralysis and death. The disease is called tetanus.

It is interesting to note that *C. perfringens* can live in a similar way, in wounded muscle. This causes gas gangrene.

Box 30.8

Streptococcus pyogenes

This bacterium infects epithelial surfaces, and the spaces between cells in the underlying tissues. Some strains infect the throat, causing sore throats and tonsillitis. Others infect the skin, causing open sores. It can spread alarmingly quickly.

It holds on to the extracellular matrix, using polysaccharides on its cell surface which bind to fibronectin (Section 4.5: The cytoskeleton). Protein chains extend out from the cell surface, interfering with the complement system (Section 23.5) and resisting phagocytosis.

It releases a group of toxins and other proteins. The immune system recognizes one of the toxins, and causes local swelling which would normally stop infection spreading (Section 23.4). However, *S. pyogenes* can cut through these defences. Some toxins kill white blood cells. Hyaluronidase digests the GAGs of the extracellular matrix. Another toxin, streptokinase, leads to breakdown of fibrin clots in the intercellular spaces. It mimics tissue plasminogen activator, which activates a fibrin-digesting enzyme. (This toxin is used to treat heart-attack patients.)

Two other problems are caused by the body's response to the bacterium: acute glomerulonephritis and rheumatic fever. Acute glomerulonephritis occurs when antigen/antibody/complement complexes get stuck in the glomerulus (Section 22.4). This leads to inflammation, and the glomerulus fails. If this happens to many nephrons, the kidney as a whole fails. Following streptococcal infections in a few people, small and persistent swellings appear in heart tissue, joints and skin. They contain macrophages and lymphocytes. This is rheumatic fever. The immune system is presumably responding to some interaction between a streptococcal antigen and molecules in the body's own cells. The details are not understood.

Box 30.9

Tuberculosis

Tuberculosis in humans is caused by two species of bacteria, *Mycobacterium tuberculosis* and *M. bovis*. *M. tuberculosis* is spread from person to person by tiny airborne mucus droplets which are breathed in and deposited in the alveoli. *M. bovis* is contracted from infected milk.

M. tuberculosis infects the alveoli of the lungs. It makes no toxins, and does not kill cells directly. During the early stages of infection, there is little damage to lung tissue.

Macrophages engulf the bacteria by phagocytosis. A bacterial product prevents lysosomes fusing with phagosomes, protecting the bacteria. They can even reproduce inside the macrophages.

Some macrophages clump together to form a large mass up to 2 mm across with many nuclei, called a granuloma. The granuloma causes damage in two main ways. Firstly, neighbouring cells are somehow affected by it, and are destroyed by the body's defences. Secondly, granulomas are slowly replaced by scar tissue. This then contracts, distorting the adjacent airways.

Granulomatous diseases like tuberculosis, leprosy and various fungal infections clearly show the interaction between internal and external factors in causing disease. The body's response causes more damage than the pathogen itself.

Box 30.10

Leishmaniasis

Leishmania species are parasitic protoctists. Like many parasites, they infect more than one host species at different stages of the life cycle, mammals and sandflies (a group of blood-sucking insects). The sandfly carries the parasite between mammals: it is a vector. This parasite reproduces in the vector; some do not.

Different species infect macrophages in different parts of the body. Three species cause human disease: *L. donovani* (spleen and liver), *L. tropica* (skin) and *L. braziliense* (epithelia lining mouth and nose).

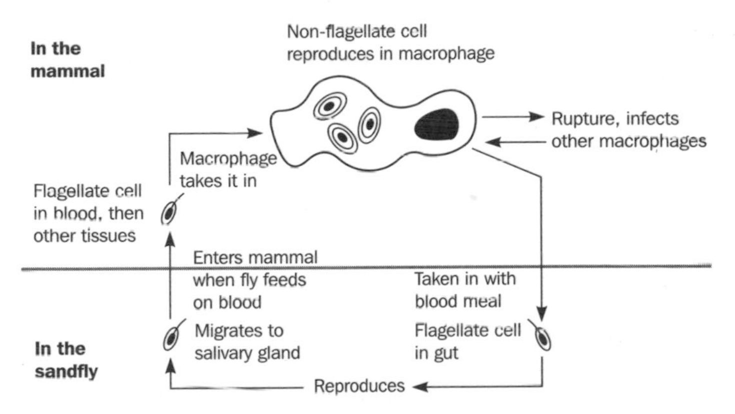

Box 30.11

Schistosomiasis

Schistosoma mansoni is a parasitic animal (Phylum Platyhelminthes, Class Trematoda), related to the liver fluke. It has two hosts: humans and a tropical snail, as well as free-living, aquatic forms.

Like many invertebrates, it has several larval stages.

Unlike the protoctist *Leishmania*, it does not invade cells. Patients affected by *Schistosoma* are weak, but in spite of progressive liver damage, the disease is rarely fatal.

Most familiar parasitic animals belong to the Platyhelminthes, Annelida, Nematoda and Arthropoda (Class Insecta). Some, like this, live within the body. Others, including fleas, live externally. Plants also have animal parasites, mostly nematodes.

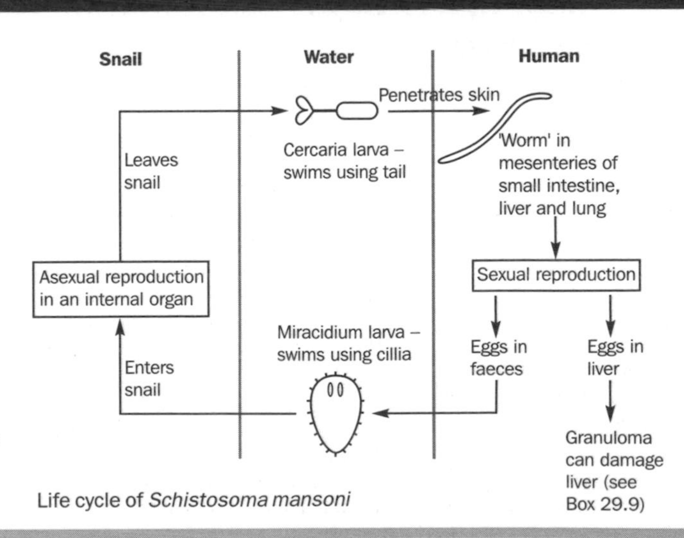

Life cycle of *Schistosoma mansoni*

Box 30.12

Club root

- This plant disease is caused by a slime mould (Kingdom Protoctista, Phylum Eumycetozoa), *Plasmodiophora brassicae*.
- Several species in the family Brassicaceae are affected, including cabbage, cauliflower and wallflower.
- The slime mould invades the roots. Enlarged plant cells contain plasmodia: masses of slime mould cytoplasm containing many nuclei.
- The roots have a thickened, warty appearance. Plant growth is poor.
- Spores are released into the soil when the root dies. The soil acts as a reservoir of infection.

Box 30.13

Potato blight

- Potato blight is caused by an oomycete, *Phytophthora infestans*.
- The hyphae grow through the leaves, stems and tubers of the potato plant. Fine extensions enter the cells and digest them.
- Branches of the mycelium grow out through the stomata. They produce spores which can infect other plants, or other parts of the same plant. They are carried by wind or in splashing rain drops. One type can swim.
- There is only one host. The fungus survives the winter along with the plant, as infected tubers.

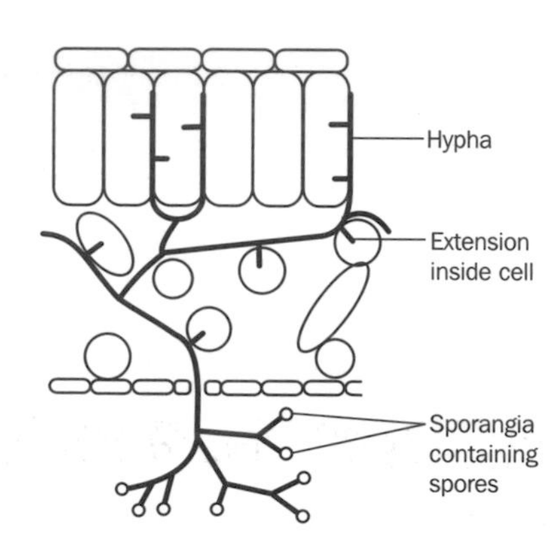

- Most of the important parasites of plants are fungi (basidiomycetes and ascomycetes) or oomycetes. Most have this general life style: hyphae feed in plant organs, producing spores on the outside. The genetic details of their life cycles may be very different, but they infect their hosts in broadly similar ways.

Box 30.14

Genetic disease

Some diseases are caused by mutant genes inherited from the parents. Most of the serious genetic diseases are caused by recessive alleles. The disease appears in the children of two heterozygotes. All manner of proteins can be affected by mutations in the genes which code for them, as these human examples show.

Enzymes

Mutant forms of the enzymes which convert glycogen to glucose cause a group of diseases.

In McArdle's disease, the enzyme deficiency prevents the muscle getting enough energy during strenuous exercise, leading to muscle cramps. The other three conditions lead to far too much glycogen being stored in the liver, which is enlarged. Blood sugar level drops too low between meals, and the body relies more on fat metabolism. Von Gierke's disease is the most serious, with gross enlargement of the liver; infection may trigger a fatal drop in blood sugar level.

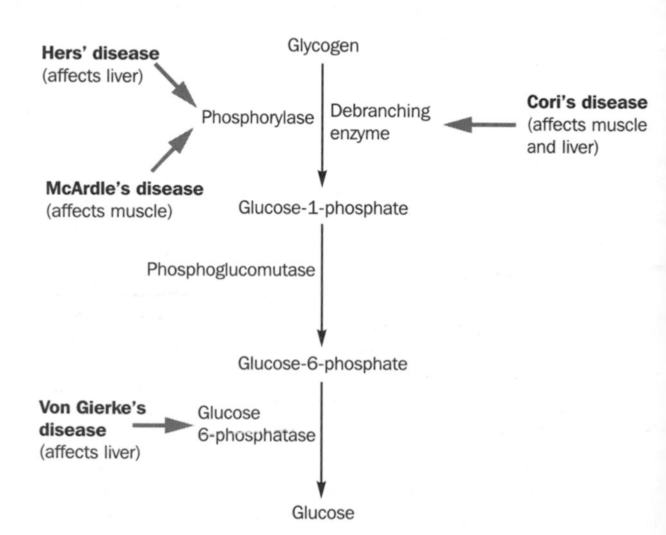

Box 30.14 (cont)

Structural protein

Muscular dystrophy is caused by a mutation in the gene for dystrophin. This cytoskeleton protein is found in skeletal muscle cells. It binds actin filaments, and probably binds them to a membrane protein. The effect is serious: muscle cells die.

Carrier protein

In cystic fibrosis, there is a defective chloride channel protein in the plasma membrane of epithelial cells which secrete mucus. The disruption to membrane transport leads to over-production of mucus. This clogs the airways in the lungs, bungs up the pancreatic duct, and makes absorption of digested food less efficient.

Proteins involved in cell signalling and control of cell division

Mutations in these proteins are some of the causes of cancer (Box 30.16).

Box 30.15

Tumours and cancer

A **tumour** is a mass of a body's own tissue which grows faster than normal. **Benign tumours** stay in the region where they began. They do not invade surrounding tissues, although they may push against them as they grow. Their cells do not go off and set up new tumours. Benign tumours are not usually life-threatening.

Malignant tumours can invade adjoining tissues. They can also metastasize, that is set up new tumours elsewhere in the body. They are life-threatening because they can invade vital organs, even if they began in a relatively harmless position. To have a malignant tumour is to have cancer. Most of our knowledge of cancer comes from mammals: humans and mice in particular.

We can look for causes of cancer at two levels, inside the cell (Box 30.16) and outside the cell (Box 30.17). External events lead to internal changes.

Box 30.16

Causes of cancer I: inside the cell

Cancer cells are special because:

- they can divide again and again, without the checks and balances seen in normal tissue growth (Section 25.5);
- they do not differentiate in a normal way, and the whole cell population keeps on dividing;
- they can bind to, digest and cross a basal lamina to invade other tissues.

Cancers are genetic diseases

Cancer cells have mutations in genes involved in regulating division and differentiation. All sorts of mutations are known. Several mutations are needed, in different genes. Cells collect mutations as they live and divide. Only when the cell has a certain collection of mutations can it start to form a tumour. Other mutations are required if the tumour is to become malignant. This all takes time, so most cancers appear in middle to old age.

Each tumour starts with its own collection of mutations. Further mutations often occur within the tumour, so it becomes a mosaic of different mutant cell types.

At least three groups of genes can be involved in cancer.

Oncogenes are cancer-causing mutations with a dominant effect. The normal forms of these genes mostly code for proteins with a role in cell signalling. Some code for receptors, others for components of signalling pathways within the cell. The cells lose the ability to respond to some type of signal from neighbouring cells. Mutations in *ras* genes coding for Ras proteins (Section 16.4) behave in this way. As a rule, having two or more oncogenes makes for faster tumour development.

Tumour suppressor genes are cancer-causing mutations with a recessive effect. The normal forms of these genes tend to code for proteins which control cell division. For example, *p53* is a mammalian gene coding for a protein which inhibits cell division when a cell is exposed to radiation which could damage DNA. Mutations in *p53* are common in many types of cancer. Most cancers of the colon show mutation and/or loss of the two copies of *p53*.

Mutations in **genes affecting DNA repair** do not cause cancer directly. However, they make mutations in other genes more likely. As a result, they speed up the mutation-collecting process which leads to cancer.

Some cancers run in families

Very simply, if you inherit a mutant form of one of these genes, your cells have a head start in collecting mutations. Cancer is more likely, earlier in your life. Hereditary retinoblastoma, a rare cancer which affects children's retinas, is caused by inheriting one mutant copy of a tumour suppressor gene.

Box 30.17

Causes of cancer II: outside the cell

Mutagens

Cancers are caused by mutations. Anything which causes mutation makes cancer more likely. Mutagens (Section 7.5) are also carcinogens. It can be hard to link exposure to a mutagen to a particular type of cancer. There is a long time delay, and only a chance of cancer developing.

Viruses

Some cancers are linked to viral infection. It is, again, not easy to prove the connection. It is especially hard in humans where the most obvious experiments are unethical. In the best understood cases, it seems that the virus has picked up an oncogene from a host cell, at some time in its history. If the oncogene is passed on to another cell, the cell is a step along the way to being cancerous.

DNA viruses can do this. So can retroviruses: RNA viruses in which DNA is made from an RNA template on infection.

Cancers linked to viruses include cervical cancer in women. Two strains of the wart virus are probably involved. These are not the strains that cause warts and verrucas on the hands and feet! Cervical cancer behaves like a sexually transmitted disease.

Viruses may be major risk factors for certain cancers, but they are not the only causes.

Box 30.18

Plants get tumours, but not cancer

Abnormal tissue growths do appear on plants. Galls are highly organized. They are formed around the larvae of gall wasps, and provide a home for them. It seems likely that the larvae are genetically engineering the plant cells, slipping in genes which disrupt normal development. The bacterium *Agrobacterium tumefaciens* causes a disease called crown gall. Swellings containing the bacterium develop on stems. Here, the genetic engineering is well understood. Several genes are introduced into the plant cells, by a plasmid.

Galls and crown gall do not metastasize, however. Plants do not seem to get cancer, and even these benign tumours arise from interactions with other organisms, not accumulated mutations.

Why not? We do not really know, but there are some promising leads. Firstly, plant cells have walls, so cannot wander around the plant. This means that plant tumours cannot metastasize. Secondly, changes to the controls on cell division seem to have very little effect on plant development. Experiments on *Arabidopsis* (a plant we met in Box 25.2) show that transgenic plants with too much, or too little cell division, still form normal organs. The positional information governing organ development seems to be robust enough to survive big changes in cell number. The modular body plan of plants is also flexible enough to accommodate over-large organs.

Box 30.19

Prion diseases

Many species of mammals suffer from spongiform encephalopathies. These brain diseases involve death of some brain cells, progressive loss of brain function, and eventually death. The brain is found to contain many small cavities: it is spongy. Well-known examples are scrapie in sheep, bovine spongiform encephalopathy (BSE) in cattle, and Creutzfeldt-Jacob disease (CJD) in humans.

Some cases seem to crop up by themselves, but these diseases can also be transmitted from one individual to another. Transmission by eating affected brain, spinal cord and perhaps other organs (cannibalism!), is well established.

What is being transmitted? It is certainly not a parasite, bacterium or virus. Much evidence points towards it being a unique pathogenic protein: a prion, or PrP. This protein (according to the hypothesis) comes in two forms. The normal form, in healthy brains, is called PrP^{C}. Infected brains have PrP^{Sc}.The two forms have the same amino acid sequence, so mutation is not involved. However, they have different tertiary structures (Section 3.5).

The most surprising part of all this is that exposing a normal brain to PrP^{Sc} leads to the existing PrP^{C} being converted to PrP^{Sc}. It is an infection which spreads by recruitment, rather than reproduction. As I write (early 1997), the details of how conversion takes place and how symptoms are caused remain to be worked out. Some workers still doubt that the infectious agent is a protein on its own.

These diseases can cross between species. For example, it is clear that BSE has crossed from cattle to mice and macaques in experiments, to domestic cats and to various mammals in zoos, by contaminated food. It seems likely that a new variant of CJD is caused in the same way. Interestingly, prions from different species can infect different ranges of host, yet keep some of their distinctive characteristics in the new host.

By the time you read this, we may be closer to an explanation of these effects.

Box 30.20

Autoimmune disease

The mammalian immune system does not normally recognize the body's own antigens (Section 23.5). Sometimes, however, this self-tolerance breaks down: autoimmunity. Autoimmunity has been implicated in several diseases.

- In myasthenia gravis, acetylcholine receptors in the neuromuscular junction are attacked. This results in muscle weakness, especially in the muscles which move the eye, and in the neck.
- Systemic lupus erythematosus is a sinister, often fatal condition. The immune system attacks many tissues at once.

The causes of autoimmune diseases are poorly understood. Sometimes there may be a similarity between a foreign antigen and a self antigen. Lymphocyte clones which become common in response to a foreign molecule may then respond to the self molecule as well. Some T-cell clones which have a low affinity for a self molecule may escape detection before birth. They only respond when the immune system is challenged by infection. Alternatively, the normal controls on the immune system may be flawed. Interestingly, susceptibility to autoimmune disease has an inherited component.

Box 30.21

Diabetes: one cause, many symptoms

Diabetes mellitus is caused by a failure of homeostasis. Glucose levels in the blood are poorly controlled, and fluctuate according to how much is being used and absorbed. **Juvenile onset diabetes** is caused by loss of the β cells in the pancreas, which make insulin. This may be due to viral infection, an autoimmune response (Box 30.20) or other unknown factors. The body produces little or no insulin. There are several effects:

- Cells take up and use less glucose than normal.
- Blood glucose levels remain higher than normal for long periods after eating.
- Cells which are being prevented from taking up glucose still need an energy supply. Fat reserves are mobilized, and the overall pattern of lipid metabolism is greatly changed. Levels of keto acids increase, which can cause a drop in blood pH, potentially fatal. Patients may also eat a lot and/or lose weight.
- The extra glucose in the blood and tissue fluid draws water out of cells by osmosis.
- Excess glucose in kidney filtrate reduces water reabsorption in the nephron (Section 22.4). The body dehydrates. Uncontrolled diabetics often drink excessive amounts of water.

Box 30.22 **Anaemia: one symptom, many causes**, can be found over page.

Box 30.22

Anaemia: one symptom, many causes

Anaemia is a reduction in the ability of blood to carry oxygen. There are fewer erythrocytes (red blood cells), and so a reduction in haemoglobin concentration.

Erythrocytes are made by division of stem cells in the bone marrow (Box 25.1). In humans, they normally die after about three months. Anaemia happens when erythrocytes are lost faster than they are made. Human anaemias have very many causes. Each one either reduces production or increases loss.

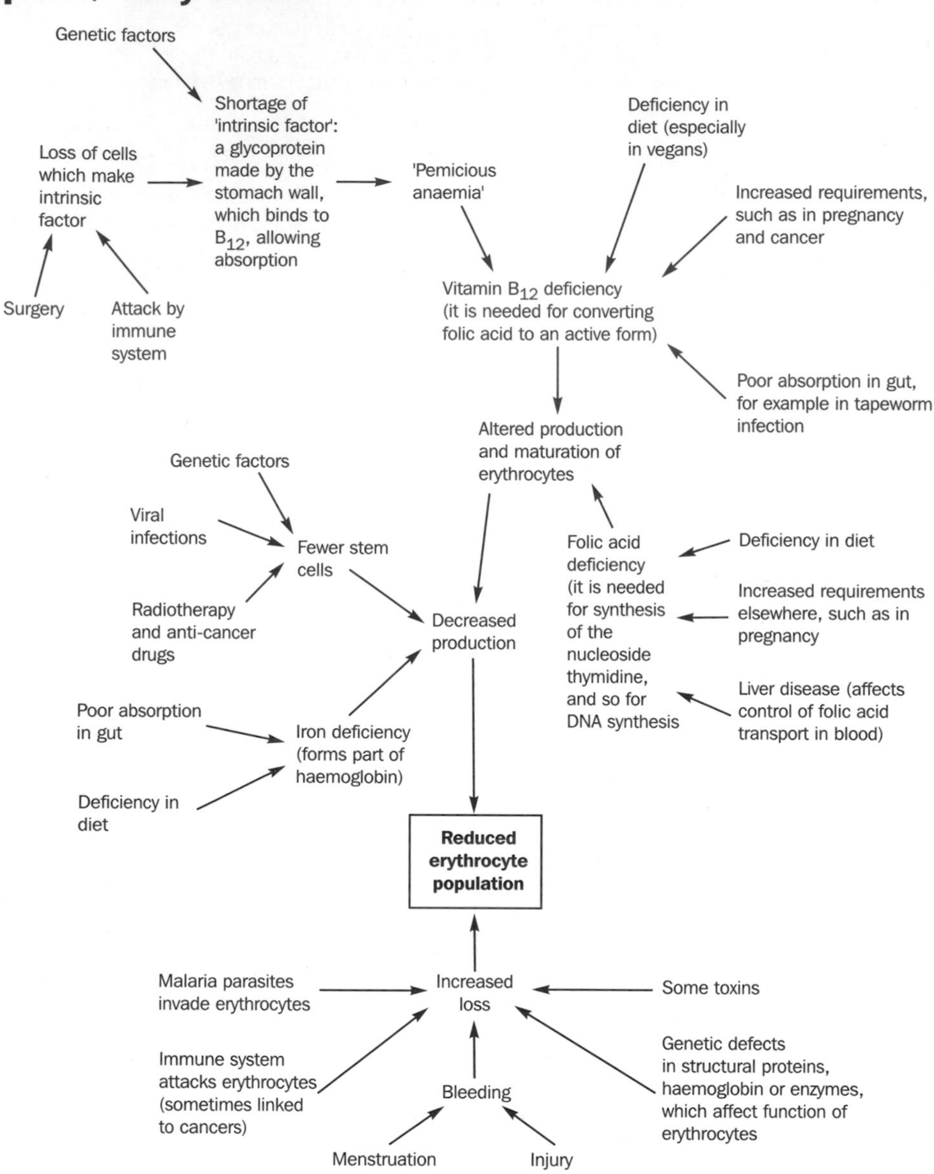

Summary

- Disease is caused by internal and external factors.
- Pathogens are disease-causing organisms.
- Some species and individuals can resist a pathogen better than others.
- Pathogens include bacteria, **protoctists**, animals and viruses.

- Viruses are nucleic acids which parasitize cells. They have no cell structure or metabolism of their own. They cause disease by killing cells.
- Bacteria produce toxins which affect the working of the body.
- Some bacteria invade tissues. In other cases, the toxin is absorbed.
- Sometimes, the immune response to a pathogen causes more damage than the pathogen itself.
- Many pathogens of plants are fungi or oomycetes (protoctists). Their hyphae grow through plant tissues, feeding as they go.
- Genetic diseases involve mutant alleles inherited from the parents.
- A tumour is a mass of a body's own tissue which grows faster than normal.
- A malignant tumour is one which can set up new tumours at a distance and in other tissues: cancer.
- Cancer is caused by accumulated mutations in genes concerned with cell signalling, control of cell division and DNA repair.
- ◆ Prion is a poorly understood infectious protein.
- ◆ In autoimmune diseases, the immune system attacks the body's own cells.
- ◆ A single molecular problem may have many different symptoms. Paradoxically, a single symptom may have many different causes.

Exercises

30.1. The mutation rates of many RNA viruses are enormously high. In some, the error rate in each generation is so high that the genome is on the verge of being scrambled.

(i) Suggest why these high mutation rates occur.
(ii) Can you see any advantage to a virus in a high mutation rate?

30.2. The flow chart in Box 30.22 shows many ultimate causes of a reduced erythrocyte population. Classify these causes as internal or external.

30.3. Box 30.21 describes juvenile onset diabetes mellitus, caused by loss of the β cells in the pancreas. Maturity onset diabetes has a similar set of symptoms, but the pancreas releases insulin normally. What could be the problem?

30.4. (i) Viruses must enter cells, wholly or partly, in order to cause disease, but most pathogenic bacteria do not. Explain.
(ii) One bacterium mentioned in this unit does enter cells. Which, and how?

Further reading

None of these books addresses the subject head on, but I do not think that advanced text-books of pathology, written for medics, would be quite the thing either. Between them, you should find more information on many areas of interest.

Cox, F.E.G. *Modern Parasitology*. (Oxford: Blackwell Science, 1993). A student textbook which makes a link with the specialist literature.

Guyton, A.C. and Hall, J.E. *Textbook of Medical Physiology* (9th ed.) (Philadelphia: Saunders,

1996). This book touches on the basis of many human diseases, as do most textbooks of medical physiology.

Guyton, A.C. *Human Physiology and Mechanisms of Disease* (5th ed.) (Philadelphia: Saunders, 1992). Effectively a shorter, boiled-down version.)

Jones, D.G. *Plant Pathology*. (Milton Keynes: Open University Press, 1987). A general introduction, straightforward and concise.

Postgate, J. *Microbes and Man* (3rd ed.) (Cambridge: Cambridge University Press, 1992). A classic, highly readable introduction to the world of microbes. Their relevance to disease and biotechnology are included.

Prescott, L.M., Harley, J.P. and Klein, D.A. *Microbiology* (3rd ed.) (Dubuque IA: Wm C. Brown, 1996). A comprehensive textbook including good sections on medical and industrial microbiology.

Singleton, P. *Bacteria in Biotechnology and Medicine* (3rd ed.) (Chichester: Wiley, 1995). A concise introduction which parallels the coverage of this topic in bigger textbooks of general microbiology.

Stryer, L. *Biochemistry* (4th ed.)(New York: Freeman, 1995). The molecular biology of viruses is introduced very well.

Biotechnology

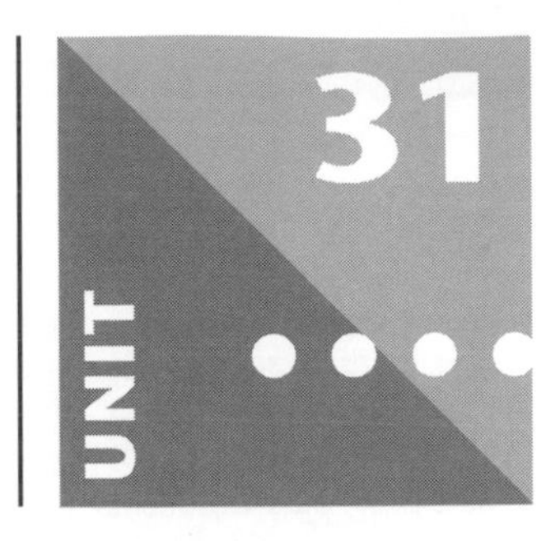

Connections

► This unit is about applications of Biology, not basic principles. Important biotechnologies involve microbes, aerobic and anaerobic respiration (Unit 9), enzymes (Unit 5), antibodies (Unit 23) and DNA (Units 3 and 7). You should understand these areas thoroughly before studying this unit.

Contents

31.1 Introducing biotechnology

Biotechnology is the use of biological systems in industry. The term is used to include applications of microbes, enzymes, antibodies and recombinant DNA technology, but not the use of organisms in agriculture.

This unit covers some important techniques and examples of applications, as a series of boxes. Uses of microbes are examined first (Boxes 31.1 to 31.9). These range from processes which have traditionally been carried out using 'low technology', such as brewing and yoghurt-making, to modern applications, for example making antibiotics. Next, enzyme technologies are considered (Boxes 31.10 to 31.12). These may use isolated enzymes, or entire microbial cells. However, the cells are used as vehicles for a single enzyme, not as complete living units. Tissue culture (Box 31.13) and techniques using antibodies (Box 31.14) are discussed next. Finally, this unit introduces recombinant DNA technology, techniques and applications. Here, pure and applied science are developing together (Boxes 31.15 to 31.29). New techniques lead to new discoveries and new uses. In this area, it is neither easy, nor very useful to draw a line between biotechnology and biology.

Box 31.1

Brewing

Brewing is the production of beer, not a general name for fermentation industries. The aim is a drink with desirable alcohol content, carbon dioxide content and flavour.

The ingredients of beer and their use

Malt: The substrate for fermentation. Barley (rarely wheat) grains are germinated to the stage at which amylase has been produced but before significant starch has been broken down. At this point it is dried. Gibberellins are sometimes used to control germination. Malt may be roasted to a greater or lesser extent to affect the beer's character, for example stouts and porters have a proportion of very dark roasted malt. Unmalted grains or refined sugars are sometimes added to the malt.

Water: Mineral content is critical for beer flavour. Modern brewers often add minerals to the natural water supply to create the desired levels. Malt is ground then steeped in water. Amylase

(Box 31.1 continues over page)

Box 31.1 *(cont)*

in the malt breaks down the starch to maltose. It is heated to stop the process at a desired point, and filtered. The filtrate is called the wort.

Hops: The flower heads of a vine, *Humulus lupulus*. They contain a soluble resin which gives a bitter flavour, and have some antibacterial properties. They have been used in beer only since the 16th century; other bitter herbs were used in the past. Hops are added to the wort after boiling.

Yeast: There are two species: *Saccharomyces cerevisiae* – 'top fermenting' or 'ale' yeast, and *S. carlsbergensis* – 'bottom fermenting' or 'lager' yeast. Each species has much variation: almost all breweries use selected strains. One local style of beer in Belgium still uses yeasts found naturally on the grain.

The yeast is added to the wort and allowed to ferment (Section 9.4):

$$\text{sugars} \rightarrow \rightarrow \rightarrow \text{ethanol} + \text{carbon dioxide}$$

S. cerevisiae:

- requires warmer temperatures (20°C);
- ferments more rapidly so rapid gas evolution carries yeast to top of vat;
- produces wide range of molecules which give complexity of flavour (notably fruity and spicy elements);
- used for ales: older type of beer with centres of production in England and Belgium.

S. carlsbergensis:

- works best at lower temperatures (12–15°C);
- ferments more slowly: yeast stays at bottom;
- yeast contributes little to flavour of beer, which tends to be more simple;
- used for lagers (origin in the Czech Republic, now world wide).

This stage of fermentation is carried out in open vats. Carbon dioxide escapes.

Different strains of yeast, of either species, have different alcohol tolerances. Very strong (10% alcohol) or weak (under 3%) beers can be made using suitable strains of either species.

After brewing

'Real' beers:

- **Ales** Fermented beer, along with any yeast in suspension, is put into wooden or aluminium casks, rarely bottles. Yeast continues to ferment, producing dissolved carbon dioxide. This gives bubbles and helps to drive the beer from the cask. A hole in the cask is stoppered with different types of plug to regulate carbon dioxide loss. Flavour continues to develop until the beer is drawn from the cask.
- **Lagers** Fermented beer, along with some yeast, is placed in large barrels under cool conditions for many weeks (lagering), before drinking or bottling.

Other beers: keg, canned, and most bottled beers. The fermented beer is heated to kill all yeast, then filtered and put into sealed aluminium kegs, cans or bottles. The stabilized product keeps better, but does not develop further character and has to be driven from the keg by compressed carbon dioxide or nitrogen.

I have presented brewing in some detail, to show some of the subtleties of any microbial process. There is a hidden message: respect the microbe, respect the brewer, respect the product. And yes, I do like beer.

Box 31.2

Vinegar production

All types of vinegar are dilute solutions of ethanoic (acetic) acid. Vinegar is used as a preservative in pickles and chutneys, as well as for flavouring food. Vinegar production requires acetic acid bacteria, a group of Gram positive, aerobic, rod-shaped bacteria. They can use ethanol as a substrate in respiration:

$$\text{ethanol} \rightarrow \text{ethanal} \rightarrow \text{ethanoic acid} \rightarrow \text{acetyl–CoA} \rightarrow \text{Krebs cycle}$$

Some oxidize ethanol all the way to CO_2, via the Krebs cycle (Section 9.2). Others, such as the genus *Gluconobacter*, lack some enzymes of the Krebs cycle, so they release ethanoic acid as a waste product. These are used to make vinegar.

Vinegar can be made from wine, cider or diluted distilled spirit.

Bacteria meet alcohol in three main ways:

- The Orleans process is traditional, and still used for good wine vinegar. Wine is put in wooden vats carrying bacteria. A culture forms over the surface. Oxygen enters slowly by diffusion, so the process takes weeks.
- Substrate is circulated through tanks which contain bacterial cultures on wood shavings. Air is bubbled through, and the process takes days.
- Pure cultures in modern fermenters (Box 31.6) are increasingly used.

Box 31.3

Yoghurt

Yoghurt is a preserved milk product. It has a distinctive taste and a thicker texture than milk.

Lactic acid bacteria are a group of Gram positive bacteria. They lack the respiratory chain (Section 9.2: Oxidative phosphorylation) so respire anaerobically, producing lactate as a waste product (Section 9.4). Two species are used in yoghurt making: *Lactobacillus bulgaricus* and *Streptococcus thermophilus*. They form a simple community, and each uses waste products released by the other. *L. bulgaricus* digests proteins to shorter peptides which *S. thermophilus* can use. *S. thermophilus* produces methanoic acid which *L. bulgaricus* uses.

As lactate is made, pH falls. This gives a sour taste and prevents growth of most other bacteria. The milk protein caseinogen precipitates as insoluble casein under acidic conditions, so the milk thickens. Other products such as ethanal influence the taste of the yoghurt. Products marketed as 'bio' yoghurt are made with cultures containing two extra species. The result is a creamier flavour.

Box 31.4

Cheese

Cheese is another preserved milk product. The details depend on the type of cheese, but the process has two main stages: curdling and ripening.

Curdling

The soluble milk protein caseinogen precipitates as casein. This is caused by lactic acid bacteria producing acidic conditions, or by the enzyme chymosin (rennin). Chymosin helps make milk protein digestible in the stomachs of young mammals. It is extracted from calves' stomachs after slaughter, or made by transgenic bacteria (Box 31.9) in fermenters (Box 31.6).

Curdling leads to separation into curds

(the insoluble proteins and fats) and whey (the rest of the milk). The curds are used to make cheese.

Ripening

Ripening develops the flavour and character of a cheese. It involves proteolysis and lipolysis.

Proteolysis:

proteins → amino acids → ammonia, amines, organic acids

The amount of proteolysis varies from 30% in cheddar to almost 100% in Camembert.

Lipolysis:

fats → fatty acids + glycerol

Ripening is done by lactic acid bacteria which grow through the body of hard cheeses, or by fungi on the surface of soft cheeses. In a few cases, specific flavours are brought about by growth of a mould through the cheese. These 'blue' cheeses include stilton and roquefort.

Box 31.5

Sewage treatment

Sewage is what we flush down the lavatory and tip down the sink. It is rich in organic molecules: they are potential food for microbes. The aim of sewage treatment is to feed the microbes at the sewage works rather than in a river, where their increased growth could use too much oxygen, endangering animals.

Methane-producing communities also exist in rubbish heaps, and occasionally the methane is used commercially. They are widely used in small 'biogas fermenters', where dung or plant material is decomposed, producing a domestic gas supply.

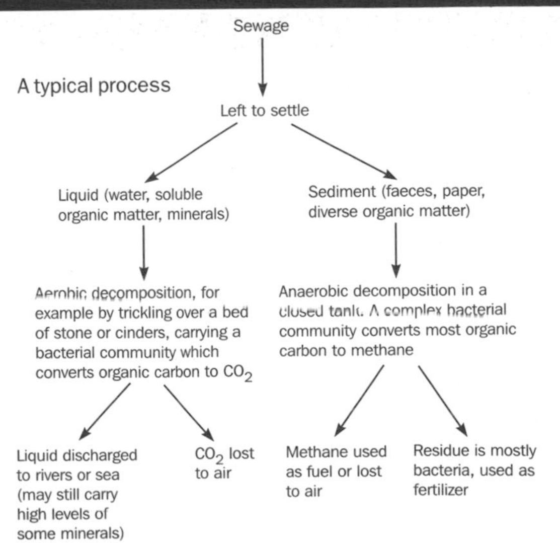
A typical process
Sewage
Left to settle
Liquid (water, soluble organic matter, minerals)
Sediment (faeces, paper, diverse organic matter)
Aerobic decomposition, for example by trickling over a bed of stone or cinders, carrying a bacterial community which converts organic carbon to CO_2
Anaerobic decomposition in a closed tank. A complex bacterial community converts most organic carbon to methane
Liquid discharged to rivers or sea (may still carry high levels of some minerals)
CO_2 lost to air
Methane used as fuel or lost to air
Residue is mostly bacteria, used as fertilizer

Box 31.6

Fermenter technology

Fermenters allow growth of pure microbial cultures in controlled, sterile conditions. Air is bubbled through, at least for aerobic microbes. The contents must be stirred to keep the microbes in suspension. Temperature, pH and perhaps the concentrations of particular chemicals are monitored and kept constant.

The diagram shows a standard type of fermenter in which stirring and aeration are powered separately. It is used for batch processes. The tank is filled, the culture allowed to grow for some time, then the tank is emptied and the product extracted. Fermenters like this could have a volume of anything from a few cubic metres to hundreds of cubic metres. Other designs in which mixing is driven by air flow are also in use.

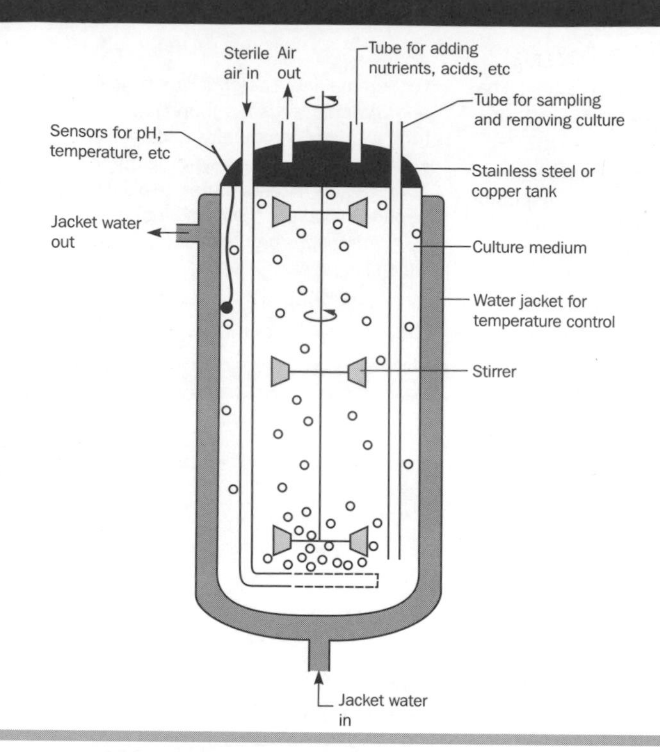

Box 31.7

Antibiotics

Antibiotics are substances made by microbes which are toxic to other microbe species. It is not clear how much value they have as chemical weapons in the struggle for resources. They are produced industrially to help combat bacterial disease.

Most antibiotics used as drugs are made by microbes grown in batch fermenters (Box 31.6). For example, penicillins are made by the moulds *Penicillium notatum* and *P. chrysogenum*.

Different penicillins have different R groups. Some are added by the microbe itself, using organic molecules present in the culture medium. Others can be added chemically, after the drug has been extracted and purified.

The structure of any penicillin

Box 31.8

Single cell protein (SCP)

Some industrial processes make a continuous supply of energy-rich waste products. These can be turned into food very efficiently by feeding it to appropriate microbes in fermenters. The microbes themselves are then used as a high-protein animal feed, rarely in meat substitutes for human consumption. Yeasts and filamentous fungi are commonly used, but prokaryotic SCP has been made.

More surprisingly, high-energy substrates which already have a use can prove economic. However, a well-publicized SCP process in the UK, which used methanol as an energy supply, was abandoned on economic grounds.

Some substrates which have been used:

- starchy/sugary wastes from sugar refining, fruit, cereal and potato processing, paper-making;
- whey from cheese-making, containing proteins and lactose;
- methanol;
- methane (natural gas);
- hydrocarbons from oil refining.

Toxins are a big worry, especially when petrochemicals are used. Also, while cattle may not be too fussy, people need a lot of persuading before they will eat an unfamiliar microbial product.

Box 31.9

Microbes and mining

The sulphur-oxidizing bacteria are an interesting group. They are chemotrophs, and get their energy by oxidizing the sulphur in hydrogen sulphide or metal sulphides. The sulphur ends up, in effect, as sulphuric acid. The metal ends up as an ion in solution. This can be a problem in coal and metal mines.

When bacteria, water and oxygen meet iron pyrite (FeS_2), a common mineral, the water becomes very acidic. Water flowing from mines can cause serious pollution. However, sulphur-oxidizing bacteria can also be put to use, extracting metals from low grade ores which would normally just be dumped beside mines.

The most important copper ore is chalcopyrite ($CuFeS_2$). *Thiobacillus ferrooxidans* can oxidize both sulphides and Fe^{2+} ions. When using chalcopyrite, it releases Cu^{2+} into solution.

Microbes are used to extract other metals, including uranium, cobalt, lead and zinc.

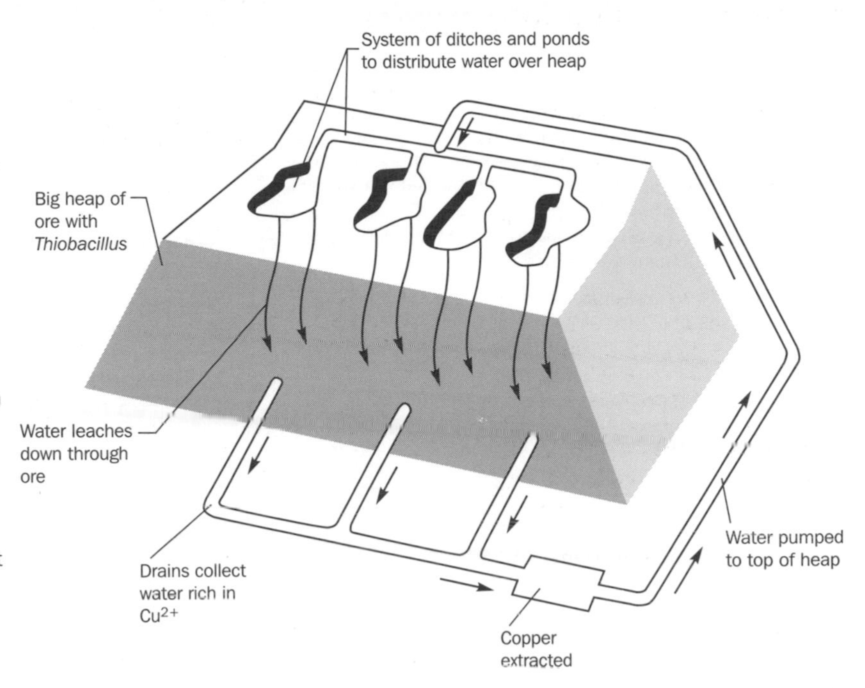

Box 31.10

Why use enzymes in industry?

Enzymes are more or less specific catalysts. They are reusable (if you can recover them!). They operate at relatively low temperatures, and atmospheric pressure, unlike many non-enzyme processes in industry. They are biodegradable. Most are not harmful, so can safely be used in the food industry. All this makes them attractive in processes which involve biological molecules.

Most industrial enzymes are ones which are secreted by the microbes that make them. It is simpler to extract them from the culture medium in a fermenter than from the cells themselves.

Some industrial enzymes need to work at temperatures which would denature most enzymes. Microbes from warmer places, like hot springs, can be useful here. Some enzymes used in biological washing powders have been designed for extra stability. Specific amino acid residues in an existing type of enzyme are changed, to make disulphide bridges at key points. This is done by changing the gene which codes for the enzyme: site-directed mutagenesis.

Box 31.11

Immobilizing enzymes

It makes economic sense to recover and reuse an enzyme. This is tricky, if not impossible, unless the enzyme is fixed on or in something else: immobilized. Enzymes can be immobilized by:

- trapping them in a gel (silica gel, polyacrylamide or polysaccharide). Small substrate molecules can get in, but big enzyme molecules cannot pass through the mesh.
- adsorbing them onto a surface (for example charcoal or collagen).
- covalently bonding them to a cellulose mesh.

Immobilization makes continuous processes feasible. A constant stream of substrate passes slowly through a bed of immobilized enzymes. Whole microbial cells can be immobilized too. Entire enzyme systems can then be used, but there is no contamination because the microbes are trapped.

Box 31.12

Some applications of enzymes

Here are two very successful and well known examples.

High fructose syrups

The traditional sweetener in the food industry has been sucrose from sugar beet or sugar cane. High fructose syrup is a new competitor. It can be made from starch extracted from cereals, grown cheaply in temperate parts of the world. It contains roughly equal amounts of glucose and fructose. Four different enzymes are required. (Chemical details of starch are given in Section 3.4.)

Starch → Amylopectin (branched) → **αamylase** (bacterial) → Dextrins → **Pullulanase** (fungal) **Glucoamylase** (bacterial) → Glucose

Starch → Amylose (unbranched) → **Glucoamylase** (bacterial) → Glucose

Glucose → **Glucose isomerase** (bacterial) → Fructose

The final step is needed because fructose tastes sweeter than glucose. The aim is to get the maximum sweetness from each gram of starch.

Glucose isomerase is used immobilized, still in the bacteria which make it. The other three enzymes are secreted by cells, so can be produced in fermenters and purified.

Washing powders

If I splatter my shirt with curry or red wine, the stains I am left with are a complex mixture of biological molecules. 'Biological' washing powders include proteases and lipases to digest them. They must be resistant to high temperatures if they are to avoid being denatured during washing. Bacteria from hot springs have provided some useful enzymes. Artificially stable enzymes can be made, using recombinant DNA technology to add disulphide bridges which help the enzyme retain its shape.

Box 31.13

Tissue culture

Plant cells

Plant cells can be cultured on agar plates containing nutrients. The big advantage is that many individual plants can be produced more quickly than by conventional propagation. Also, many clones can be kept alive in a small space.

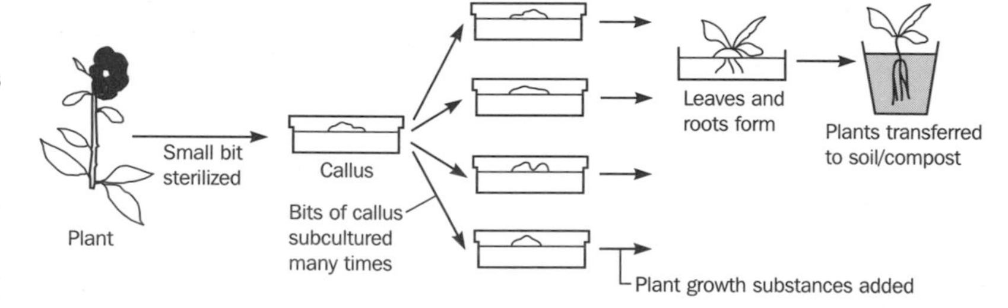

Many plant organs will form undifferentiated callus cells when injured. They will divide freely in culture. Particular combinations of plant growth substances, such as auxins and cytokinins, bring about differentiation to form shoots and roots.

Single shoot tips can also be used as tiny cuttings, rooted in culture medium. Some advantages of plant tissue culture are:

- new crop plant clones can be propagated quickly;
- some (but not all) ornamental plants which are slow-growing or hard to propagate can be produced efficiently;
- callus cells are convenient for gene-transfer experiments;
- transgenic plants can be kept secure during culture and propagated rapidly.

Animal cells

Many animal cell types can be cultured on the surface of a sterile plastic dish. Culture solution is washed over them. Its uses include studying cell function, hormone action and the effects of drugs. The economic benefits are real, but often indirect, unlike plant tissue culture.

Box 31.14

Making and using antibodies

Antibodies are proteins which bind other molecules (antigens), especially proteins and glycoproteins, in a strong and specific way. This makes them useful in detecting and locating proteins. Antibodies are made by B-lymphocytes (Section 23.5). There are two ways of manufacturing antibodies to a particular antigen.

The simpler, traditional way

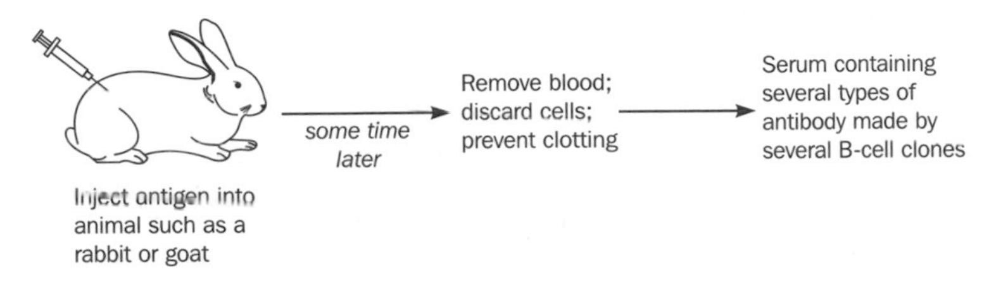
Inject antigen into animal such as a rabbit or goat
some time later
Remove blood; discard cells; prevent clotting
Serum containing several types of antibody made by several B-cell clones

However, a mixture of several antibodies may not be specific enough for some uses. Also, this provides just one batch of antibodies. If more are needed, the process must be repeated, and a different mixture of antibodies will be produced.

Monoclonal antibodies

A more tricky procedure, which relies on cell culture (Box 31.13).

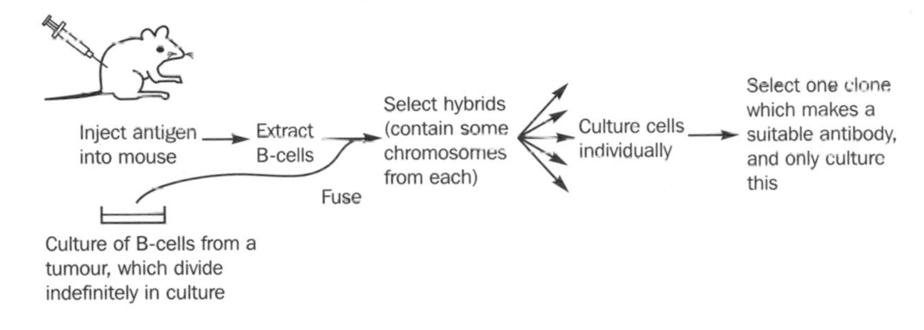
Inject antigen into mouse
Extract B-cells
Fuse
Culture of B-cells from a tumour, which divide indefinitely in culture
Select hybrids (contain some chromosomes from each)
Culture cells individually
Select one clone which makes a suitable antibody, and only culture this

This gives a long-term supply of a single type of antibody.

Finding the antibodies

Even when antibodies have bound to the antigen, there is still a problem: we need to be able to see where they are, or how many there are. Various techniques suit particular jobs.

- **Immunofluorescence:** Antibodies are bound to a fluorescent dye. This is useful for finding the antigen using the light microscope. Only the cells or organelles containing the antigen fluoresce.
- **Immunogold electron microscopy** works like immunofluorescence, but under the electron microscope. Antibodies are linked to tiny gold particles, which absorb electrons. They show up as black dots.
- **Enzyme-linked antibodies:** Antibodies are linked to an enzyme which makes an insoluble, coloured product. This accumulates around the antigen, where the antibody is bound. It can be used for locating molecules in cells and tissues, and has applications in pathology. Home pregnancy tests use an enzyme-linked antibody to human chorionic gonadotropin, a hormone made by the young embryo. The colour only appears when the antibody binds the hormone, and the coloured product is made.
- In the last three methods, the enzyme or label is usually attached to a **secondary antibody.** This binds to any other antibody, not to the antigen itself. The specific antibody is added first. It binds the antigen, but does not show up. Then the secondary antibody is added. It binds to the first antibody, making it easy to detect. This makes production simpler: only one type of antibody needs to be modified, whatever is being investigated. It also gives some amplification, since several secondary antibodies bind to one original antibody.
- **Radioimmunoassay:** Antibodies carry a radioactive label. The radioactivity of the sample is a measure of the number of antibodies binding to it. This can be a super-sensitive way of measuring very low concentrations of biological molecules. It has revolutionized the study of hormones.

Box 31.15

Cutting and joining DNA

Recombinant DNA technology ('genetic engineering') requires many techniques for handling DNA. Cutting and joining DNA are two of the most basic.

Cutting

DNA is cut by restriction endonucleases ('restriction enzymes'). They occur naturally in bacteria. Restriction enzymes recognize particular sequences in DNA and cut it at that point. Different enzymes recognize different sequences. However, they all recognize short 'palindromic' sequences: the base sequence in one strand, in the 5′–3′ direction, is the same as the sequence on the opposite strand, in the 5′–3′ direction. Also, they cut DNA at the same point *in the sequence*, not necessarily the same point in the DNA. In most cases, this leaves the end of one strand unpaired: a sticky end. Sticky ends made by one enzyme are all the same, and can pair up with each other in any combination. This means that DNA can be recombined in new ways.

Sequence	Restriction enzyme	After cutting
5'–GAATTC–3' 3'–CTTAAG–5'	*Eco*RI →	–G + AATTC– –CTTAA G– sticky ends
5'–GGATCC–3' 3'–CCTAGG–5'	*Bam*HI →	–G + GATCC– –CCTAG G–
5'–GGCC–3' 3'–CCGG–5'	*Hae*III →	–GG + CC– –CC GG– (no sticky ends this time)

Joining

DNA ligase joins the sugar phosphate backbone of DNA. In nature, this happens in DNA replication, at the ends of Okazaki fragments (Box 7.1). The enzyme is used to link pieces of DNA whose sticky ends have paired up.

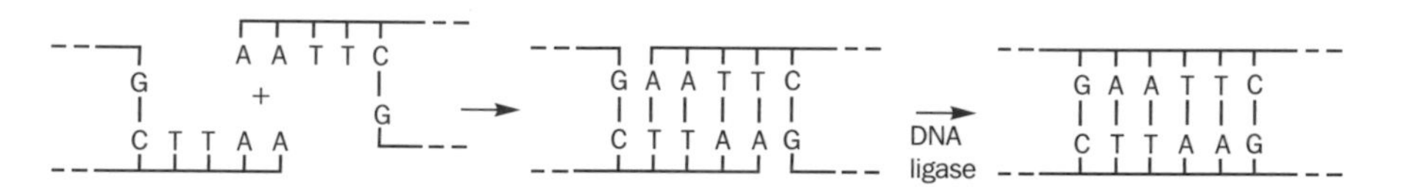

Box 31.16

Separating DNA

Gel electrophoresis is used to separate DNA fragments according to length. Sometimes we really need to know how long a piece of DNA is. More often, we need to isolate a particular DNA fragment from a mixture. Fragments are separated, then the right bit can be selected using a probe (Box 31.17).

In electrophoresis, a DNA suspension is put in a well at one end of a strip of gel. Electrodes are put at each end. DNA has a negative charge (remember the phosphoryl groups, Section 3.8?). It is attracted to the positive electrode. Shorter fragments move through the pores in the gel more easily, so they travel further. The fragments are separated into bands, according to length. All the bands can be stained, or particular ones picked out using a probe.

Different gels have different pore sizes. The small pores of polyacrylamide give good separation of small fragments, but large pieces hardly move. Agarose gels are more suitable for these.

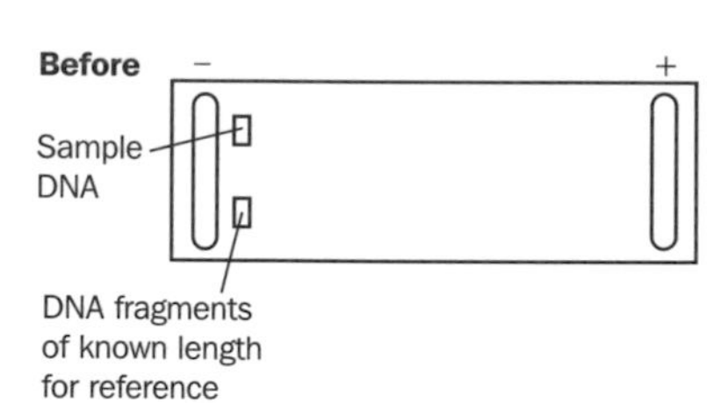

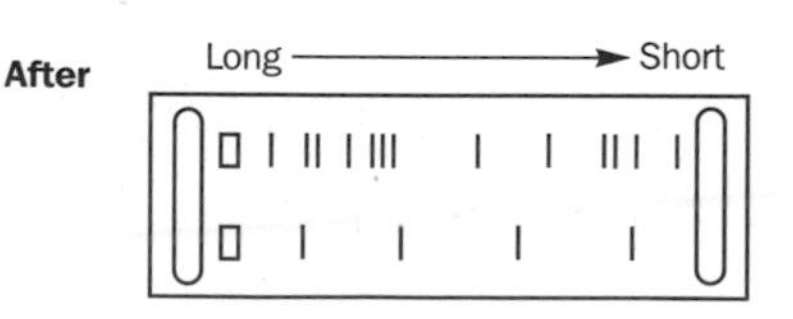

Box 31.17

Probes and hybridization

Once DNA fragments have been separated by electrophoresis (Box 31.16) we often need to know which fragments contain a particular sequence.

A probe is a synthetic DNA fragment. It is single-stranded, and complimentary to the sequence it is meant to detect. If the two strands of the sample DNA are separated, the probe will attach to the target sequence by complimentary base pairing (DNA hybridization). The probe is made from radioactively labelled nucleotides, so can be detected by photographic film. The complete technique is called Southern blotting.

DNA fragments separated by electrophoresis.

Buffer solution soaks up through the gel. DNA binds to nitrocellulose film.

DNA strands separated using alkali; film flooded with solution of radioactive probe; excess rinsed off.

Gel left on photographic film; radioactive fragments show up.

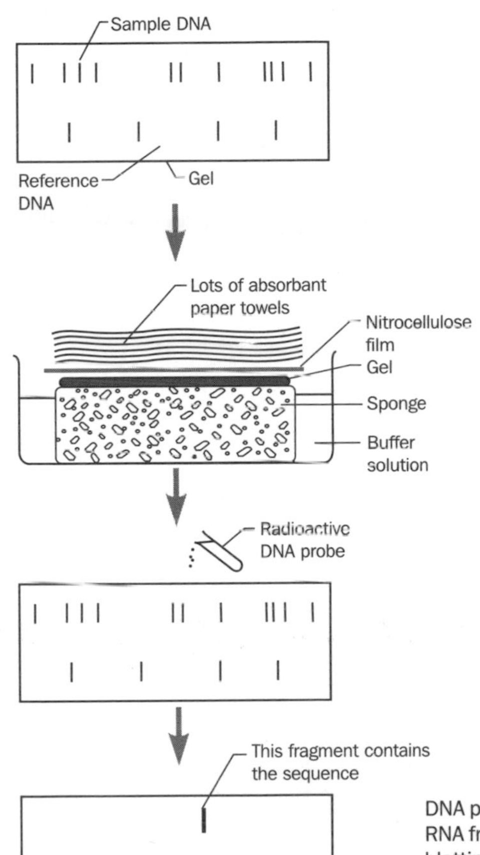

DNA probes can also be hybridized with RNA fragments. This is called Northern blotting. (A molecular biologists' joke: Southern blotting is named after its inventor, not the point of the compass.)

Box 31.18

Getting DNA into cells

There are two related problems: how to make a foreign DNA fragment stable enough such that it will survive inside a cell, and how to smuggle that DNA into the cell.

Plasmids: DNA loops in bacteria, small 'optional extra' chromosomes. They are passed from cell to cell in nature. A recombinant plasmid, partly plasmid and partly foreign DNA, is an excellent way of stabilizing and smuggling foreign DNA into bacteria (Boxes 31.19, 31.23, 31.24). They can also be introduced into animal cells by electroporation (see below). One plasmid, T_i, is used to carry DNA into plant cells. It is made by a bacterium (*Agrobacterium tumefaciens*) which causes tumour-like swellings on plant stems.The plasmid enters the plant cells and changes the course of cell division and development. Recombinant T_i plasmids are often used in plant genetic engineering.

Viruses: DNA viruses (Boxes 30.3, 30.4) which infect bacteria or animal cells are useful vectors. Recombinant viral DNA is packaged in viral coat proteins. It will then

(Box 31.18 continues over page)

Box 31.18 *(cont)*

infect the target cell. It is important to use genetically altered strains of viruses that will not take control of the cell and kill it.

Electroporation: Used to increase uptake of plasmids by bacteria or animal cells. The cells are put in a solution containing the plasmids. An intense electric field is applied in short pulses. Big holes open up in the plasma membrane for a short time. Plasmids may enter through the holes.

The 'shotgun' method: An extreme solution to the problem of plant cells having walls. Tiny metal fragments covered in foreign DNA are blasted at cells. They rip through the walls and some end up inside.

Microinjection: We often need a mammal's entire body to be made up of transgenic cells. Foreign genes are injected straight into the nucleus of the fertilized egg, before it starts to divide.

Liposomes: Artificial phospholipid vesicles which can fuse with membranes. They may prove useful in delivering DNA to animal cells.

Box 31.19

Cloning DNA in bacteria

Sometimes we want to introduce new genes into bacteria. Sometimes we simply want to use bacteria as plasmid factories, to clone DNA which will later be put into another type of cell. Either way, the basic technique is as shown here:

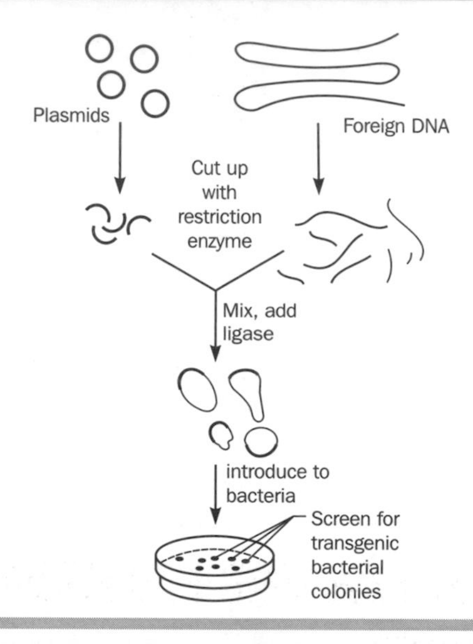

There are three big questions:

1. How do we get the foreign DNA? (Box 31.20)
2. How do we persuade the foreign DNA to enter the host cell's chromosome? (Box 31.21)
3. How can we identify transgenic bacteria? (Box 31.22)

Box 31.20

Sources of DNA

Finding the gene to be cloned is a needle-in-a-haystack problem. Even prokaryotes have thousands of genes; mammals have around 100 000. There are three possible strategies. Two involve screening 'libraries' of DNA fragments for the right bit. The third avoids the search by making the gene from scratch.

Genomic DNA libraries

DNA is extracted from an organism and digested by a restriction enzyme. Every section of DNA, gene or non-coding sequence, is on one fragment or another. Lots of cells will have been used, each with a full compliment of DNA. The restriction enzyme always cuts in the same place, so there will be many copies of each fragment. The fragments are incorporated into plasmids or viruses (Box 31.18) which are then introduced into a population of bacteria. Each recombinant bacterium will have a tiny part of the foreign genome. As it forms a colony, it will copy that single piece of the genome each time the cell divides. A library like this can be kept going indefinitely. The colonies are screened to find the one carrying the gene of interest (Box 31.21).

cDNA libraries

Genomic DNA libraries can be huge. A slimmer alternative is to make a library of just the genes active in a particular type of cell. The active genes are the ones making RNA. First, the mRNA is extracted from cells. Then reverse transcriptase (a viral enzyme, Box 30.2) is used to make a DNA strand which is complementary to each mRNA. This is called cDNA. DNA polymerase is then used to make a second strand. The cDNA molecules are cloned in plasmids or viruses as before.

Synthetic genes

Sometimes we know the amino acid sequence of the polypeptide made by a gene. It is easy to work back to a DNA sequence which would produce this polypeptide. Polynucleotides can be synthesized to order, one nucleotide at a time. In this way, small synthetic genes can be made.

Box 31.21

Getting foreign DNA into a chromosome

It is not enough to introduce a piece of DNA to a cell. It must become part of the chromosome, so it is replicated before the cell divides.

Similar DNA sequences pair up. This happens naturally at prophase I of meiosis in eukaryotes (Section 11.2). Then, the DNA sequences may break and rejoin in a new way: recombination. However, a small piece of foreign DNA may pair with a very similar sequence in the host cell chromosome at other times, especially in prokaryotes. This can be used to slip foreign DNA into a chromosome.

For example, we want to put a gene (G) into a chromosome at point Y. Sequences X and Z are either side of Y. We construct a recombinant plasmid with G sandwiched between X and Z. In the cell, the two copies of X and Z may pair. Recombination may then slip G into the chromosome.

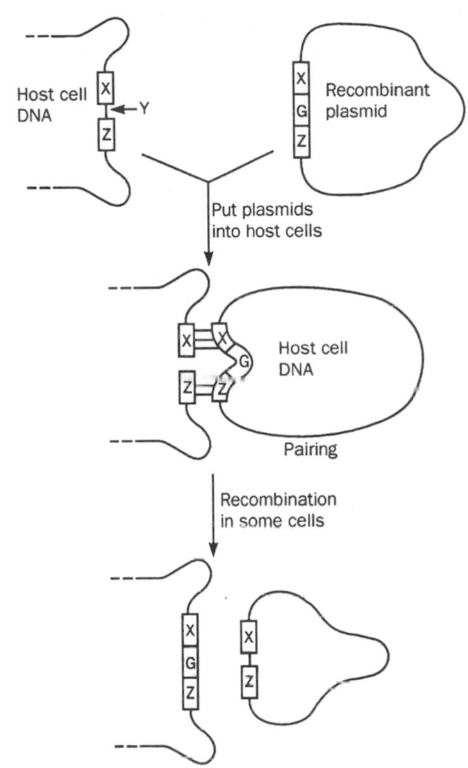

Box 31.22

Screening a library for a gene

A DNA library is simply lots of colonies of transgenic bacteria, growing in a series of Petri dishes. They are kept going by transferring bacteria from dish to dish on paper discs.

The same idea can be used to sample all the colonies without destroying them.

After this, we can go back to the plate, scrape off the colony which has the gene, and culture it alone. When many identical colonies have grown, the plasmid carrying the gene can be isolated. The result is large amounts of a short piece of DNA which includes the target gene.

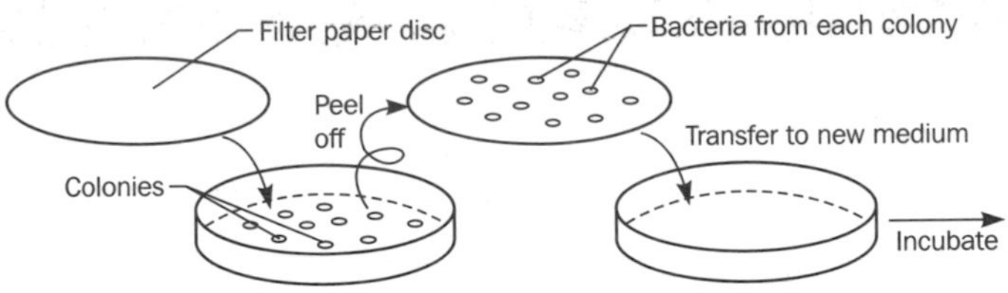

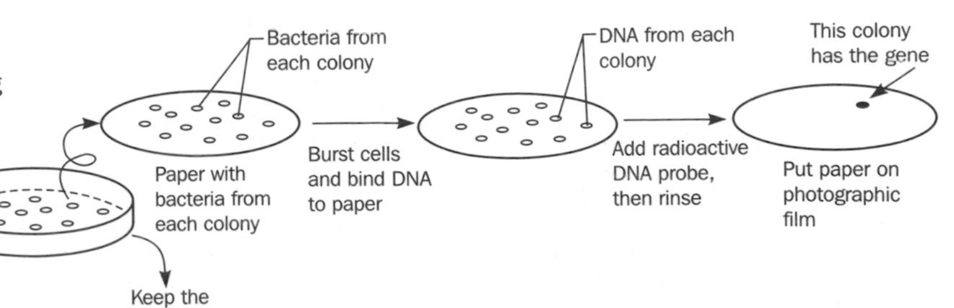

Box 31.23

Genetic engineering: case history I

Aim: to make a strain of the bacterium *E. coli* which makes human growth hormone. This can be used to treat dwarfs.

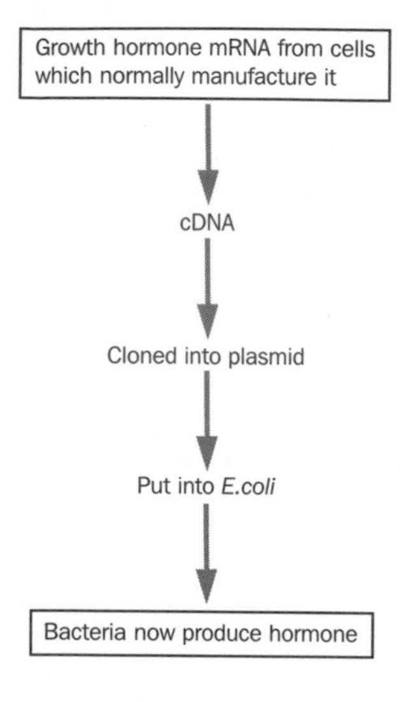

Box 31.24

Genetic engineering: case history II

Aim: to make a harmless strain of *Vibrio cholerae* (the cholera bacterium) for use as a vaccine (Box 23.3). It should not make the toxin which causes disease symptoms, but should still have antigens which the immune system can recognize.

Strategy: the toxin gene is replaced by a mercury resistance gene, which allows us to select the recombinant cells.

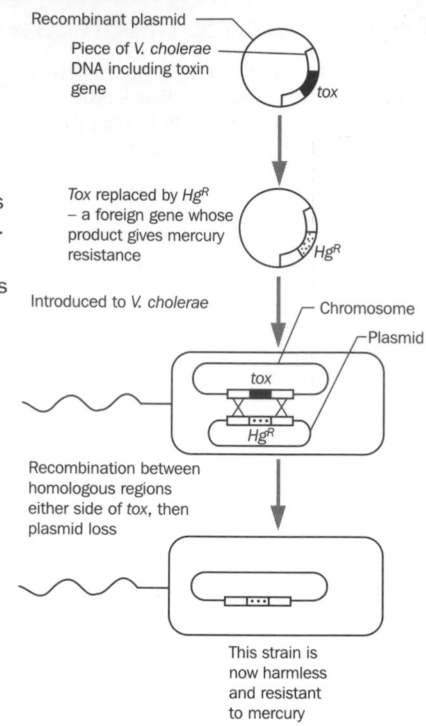

Box 31.25

Genetic engineering: case history III

Aim: production of human tissue plasminogen activator (*tPA*) in mouse milk. This protein activates an enzyme which digests fibrin clots (Box 23.1).

Strategy: the *tPA* gene is put into the middle of a gene coding for a milk protein. The human *tPA* gene will only be expressed in the mammary gland.

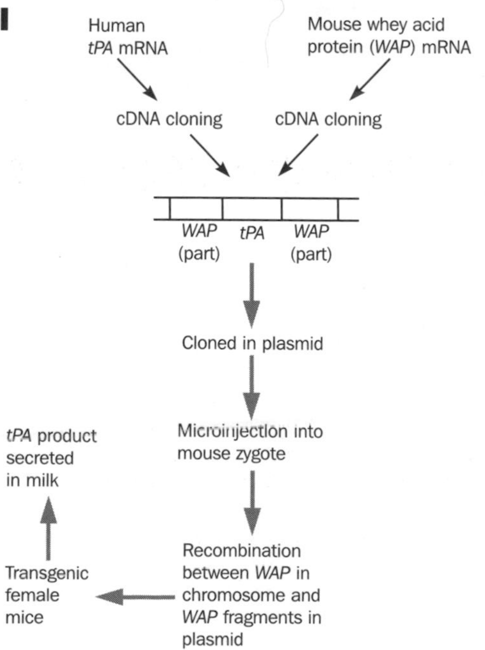

Box 31.26

Knocking out a gene to study its function

A wonderful technique, gene targeting, replaces a working mouse gene with a useless version. This lets us study the functions of mammalian genes in a new and direct way: knock out the gene, and see what happens to the mouse.

Gene targeting relies on homologous recombination (Box 31.21). This is not easy in eukaryotic cells. For 99.9% of the time, the entire piece of foreign DNA inserts somewhere else by another mechanism. Cunning use of drug resistance and sensitivity genes allows us to screen for these cells.

For example, we want to replace a functional allele, **A** with a mutant allele, **a**. **a** is simply **A** with a very large section deleted.

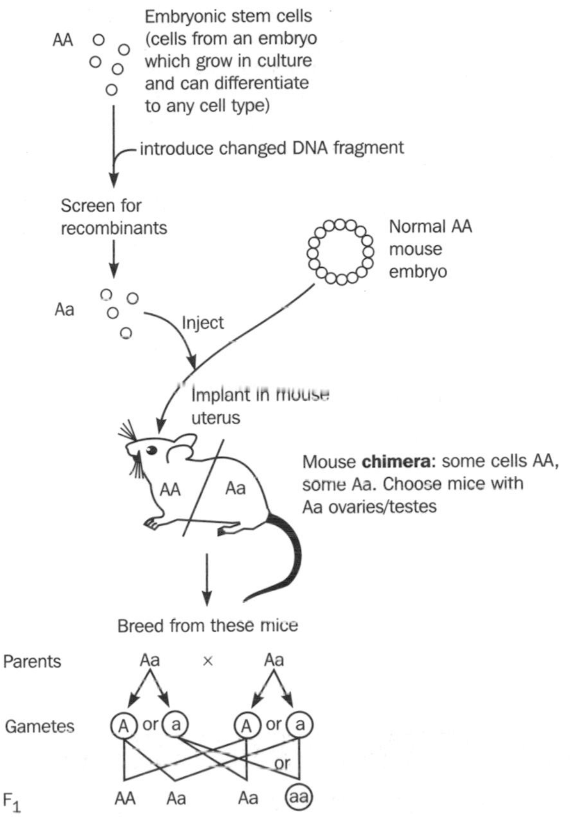
AA
Embryonic stem cells (cells from an embryo which grow in culture and can differentiate to any cell type)
introduce changed DNA fragment
Screen for recombinants
Normal AA mouse embryo
Aa
Inject
Implant in mouse uterus
AA
Aa
Mouse chimera: some cells AA, some Aa. Choose mice with Aa ovaries/testes
Breed from these mice
Parents
Aa × Aa
Gametes
A or a
A or a
or
F1
AA Aa Aa aa
25% of these mice are homozygous for the mutation: study these!

Box 31.27

Gene therapy

It may soon be possible to treat some diseases caused by a single mutant allele, by gene therapy. The idea is that the mutant allele is replaced by a normal version.

Somatic cell therapy involves introducing them to the body cells which need them. This would have to be repeated throughout life, and the 'repair' would not be passed on to any children.

Germ line therapy would involve replacing the gene in the fertilized egg, by homologous recombination (Box 31.21). The entire embryo would then carry the normal gene: a permanent repair.

Somatic cell therapy is being tried for a few diseases, notably cystic fibrosis (CF). CF patients have an abnormal chloride channel in epithelia which secrete mucus. They secrete too much thick mucus, which clogs the lungs and makes digestion inefficient. There have been attempts to introduce the normal allele to the epithelia of the lungs and nose, using viruses or liposomes. Success has been limited so far.

Germ line therapy will probably be technically possible quite soon. We have some tricky ethical questions to face before then.

Box 31.28

Cloning rare DNA fragments in the test-tube

A technique called **polymerase chain reaction (PCR)** does just this. It relies on short, synthetic DNA strands (primers) which bind to the start of each strand in a target sequence.

The cycle is repeated 20 to 30 times. The target fragment is copied each time, but other fragments are ignored. This is because the primer has not bound to them, and DNA polymerase can only link nucleotides to an existing chain (Box 7.1).

PCR can be used to amplify DNA from tiny tissue samples for genetic fingerprinting (Box 31.29). It can also be used to identify new members of gene families in cDNA preparations. This is because a primer which binds to part of one gene will also bind to the same sequence in a related gene.

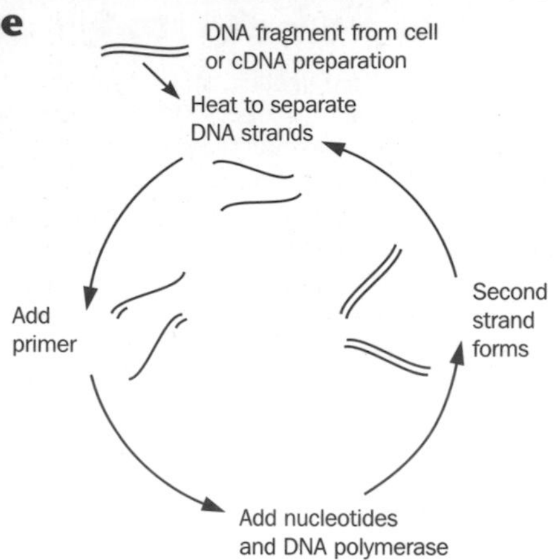

Box 31.29

DNA profiling

Tiny tissue samples, from drops of blood or semen to hair follicles, can be linked to one individual by this sensitive technique.

Satellite sequences are non-coding DNA sequences which are repeated many times, one after the other. The number of repeats is very variable. Each satellite has many different copy numbers which are inherited like alleles of any normal gene. Like any normal gene, a person will have two 'copies', in this case two sets of repeated sequences. For example, one satellite might be inherited as follows:

Copy number

Parents: Fred 13,27; Jill 21,31

Gametes: (13) or (27); (21) or (31)

F_1: 13,21 13,31 27,21 27,31

So if Jill's son William has copy numbers of 17 and 21, he could not be Fred's son (this can prove useful in paternity testing).

In DNA profiling, a DNA sample is taken. Primers are added which bind to several different satellites, and these are amplfied by PCR (Box 31.28). Once there is enough DNA, the cloned fragments are separated by electrophoresis (Box 31.16). There will be two bands for each type of satellite investigated. Copy numbers are so variable that, if several satellites are investigated, one can almost prove that a sample came from a particular individual.

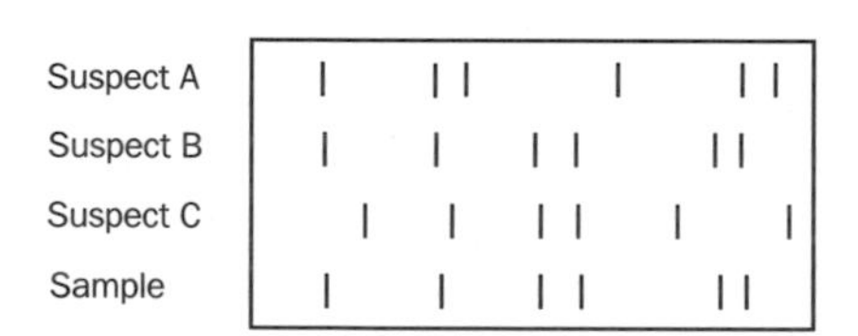

Example gel

The sample may have come from B, but could not have come from A or C.

The term 'DNA fingerprinting' refers to a similar technique. More specific probes are used, which only bind to one satellite sequence. Each sample will give only two bands on the gel.

Summary

- Biotechnology is the use of biological systems in industry.
- It includes uses of microbes, enzymes, antibodies and recombinant DNA (genetic engineering).
- Pure and applied science progress together, especially in molecular genetics.

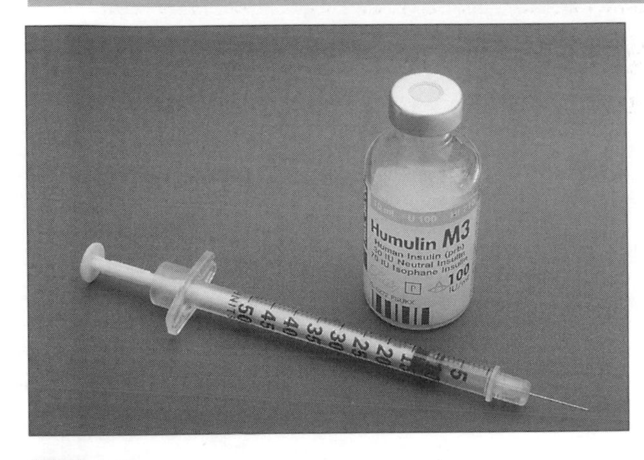

Human insulin for diabetics was among the first genetically engineered products available commercially.

Exercises

31.1. Do the following applications of microorganisms involve aerobic or anaerobic processes?

(i) brewing; **(ii)** extracting copper from low-grade ore; **(iii)** vinegar making;
(iv) sewage treatment; **(v)** yoghurt-making.

31.2. **(i)** When introducing a foreign gene into a plasmid, it is important that both plasmid and foreign DNA are cut by the same restriction enzyme. Why?
(ii) When making a genomic DNA library, the restriction enzyme might cut the DNA at a site within the target gene. How could this problem be overcome?

31.3. Which human cells would you use to make a cDNA library if you wanted to clone genes for the following?

(i) insulin; **(ii)** follicle stimulating hormone; **(iii)** sucrase; **(iv)** haemoglobin;
(v) immunoglobulin; **(vi)** Na^+/K^+ ATPase.

31.4. Which techniques involving antibodies could be used to:

(i) determine the distribution of ribosome proteins within the nucleus;
(ii) estimate the concentration of luteinizing hormone in a blood sample;
(iii) find out where an auxin receptor protein is found in a section of plant stem?

Further reading

Brown, T.A. *Gene Cloning* (3rd ed.) (London: Chapman & Hall, 1995). A thorough introduction for undergraduates. Techniques and theory are well integrated. Full of clear diagrams.

Postgate, J. *Microbes and Man* (3rd ed.) (Cambridge: Cambridge University Press, 1992). A classic, highly readable introduction to the world of microbes. Their relevance to disease and biotechnology are included.

Prescott, L.M., Harley, J.P. and Klein, D.A. *Microbiology* (3rd ed.) (Dubuque IA: Wm C. Brown, 1996). A comprehensive textbook including good sections on medical and industrial microbiology.

Singleton, P. *Bacteria in Biotechnology and Medicine* (3rd ed.) (Chichester: Wiley, 1995). A concise introduction which parallels the coverage of this topic in bigger textbooks of general microbiology.

Smith, C.A. and Wood, E.J. *Molecular Biology and Biotechnology*. (London: Chapman & Hall, 1991). A clear, straightforward introduction.

Answers to Exercises

Unit 1

1.1. (i) Kingdom Plantae, phylum Angiospermophyta, class Dicotyledonae.
(ii) *Digitalis purpurea* L. (Don't forget to underline, or print in italics, the name but not the authority.)
1.2. (i) Kingdom Animalia, phylum Chordata, class Reptilia.
(ii) *Bufo calamita* Laurenti.
1.3. This character is seen in two groups, section Gladioloides and subsection Grandibracteata. Subsection Grandibracteata shares four characteristics with other groups, but not with section Gladioloides: these are characters 4, 5, 6 and 7. If we placed subsection Grandibracteata next to section Gladioloides, we would have to assume that all these four characters had evolved twice. Moreover, they would have evolved *together* twice. This is far less likely than the original assumption: that character 3 evolved twice. A taxonomist will take other characteristics into account too, as well as 'overall similarity'. Not everything is shown in the cladogram.
1.4. I can't help you with this one. I know what I can find outside my own back door, but wherever you live, you are likely to find something else. This is one of the wonderful things about Biology: celebrate variation and diversity!

Unit 2

2.1. (i) The story assumes that because a plant grows deeper roots in order to reach water, it can pass this characteristic on to its offspring. This is called 'inheritance of acquired characteristics'. It would involve phenotypic characters somehow changing the DNA. We know of no mechanism which could do this.
(ii) The ancestors of this species invaded the dry habitat. The population was genetically variable. Some individuals had a combination of alleles which led to roots which grew deeper than others. These individuals gained more water in summer. On average, they survived and reproduced more than the others. These alleles became more common, over generations, leading to a deep-rooted population. Any new mutations which increased this deep-rooting behaviour would have been favoured in the same way.
2.2. Natural selection requires genetic variation, and variation in survival and/or reproductive success between individuals. Genetic variation is probably as great as ever, but several factors reduce variation in survival and reproduction. Improvements in medicine reduce the selection pressure against genes which make an individual more susceptible to some disease. The organization of most human societies allows individuals with conditions such as blindness to survive and reproduce, in a way which is not seen in other ape species. In developed, and many developing countries, availability of contraception, coupled with economic circumstances give individuals a degree of choice over their own fertility. This may limit natural selection, but does not prevent it. We have little idea about which genes are being favoured in modern human populations.
2.3. (i) Mole rat colonies are highly inbred. Individuals within a colony share very many genes, so helping the colony as a whole has a similar effect to helping oneself, in evolutionary terms. Specialized castes, which work for the benefit of the colony as a whole, may develop in these conditions.
(ii) It might act as a genetic 'safety valve' preventing excessive inbreeding by occasionally giving the colony's genes a chance to combine with another colony's genes.
2.4. This is typical of the effects of sexual selection. Males are able to father far more offspring than females can bear – if they can get the mates! As a result, males compete for females: females select males and often use plumage as one of the indicators of male quality.
2.5. I can't help you with this one! However, I bet you will not be able to find everything on your list in any other species.
2.6. (i) You and your father, your father and grandfather, your grandfather and great-grandfather each have a relatedness of 0.5.
$0.5 \times 0.5 \times 0.5 = 0.125$
(ii) You and your mother, and she and her sister each have a relatedness of 0.5.
$0.5 \times 0.5 = 0.25$
(iii) She and you are each related only through your mother: in each case, the relatedness is 0.5.
$0.5 \times 0.5 = 0.25$

Unit 3

3.1. Phosphoryl, hydroxyl (3), carbonyl, carboxylate. The molecule, by the way, is 3-deoxyarabinoheptulosonate-7-phosphate, an intermediate in the synthesis of some amino acids from carbohydrates (I didn't expect you to work that out . . . .).
3.2. (i) It is an amino acid. Its side chain identifies it as cysteine. Like other amino acids, it is a component of polypeptides and proteins. Its special significance is that it forms disulphide bridges: covalent links between peptide chains.
(ii) High pH (alkaline). The carboxyl group has dissociated, and the amino group has not picked up an H^+ ion.
3.3. It is a monosaccharide sugar. It is a hexose, and is in the pyranose ring form. It is shown in the β form. (The molecule is galactose, important to mammals as part of the disaccharide lactose.)
3.4. It is a monounsaturated fatty acid. (It happens to be called palmitoleic acid.)
3.5. Amylose, DNA, catalase.
3.6.

	DNA	**RNA**
Which pentose?	deoxyribose	ribose
Which bases?	A T G C	A U G C
Single- or double-stranded?	normally double	normally single
Roles in cells	information store	various, connected with the expression of genetic information

Unit 4

4.1. One: vacuole, endoplasmic reticulum, peroxisome, lysosome.
Two: nucleus, plastid, mitochondrion.
None: ribosome
4.2.

	Prokaryotic	**Eukaryotic**
Wall present?	yes	sometimes
Plasma membrane present?	yes	yes
Membrane-bound organelles present?	few or none	many
Is DNA in a nucleus?	no	yes
Chromosomes: shape and number	single closed loop	several open chains
How do flagella move?	rotate	beat
Relative size of ribosome	smaller	larger
Relative size of cell	smaller	larger

4.3. Ribosomes are needed for protein synthesis. The ribosomes making insulin will be bound to **endoplasmic reticulum**, forming **rough ER**, because insulin is to be secreted. There will be well-developed **Golgi apparatus** for packaging insulin into **vesicles** for secretion. All these processes require energy from respiration, so the cell will have many **mitochondria**.
4.4 (i) tight junctions; **(ii)** gap junctions; **(iii)** plasmodesmata.

Unit 5

5.1 It is easy to spot that **(iii)**, **(v)**, **(vi)** and **(viii)** are enzymes, even if you know nothing about them: their names end in *-ase*. The others are not, but some knowledge is needed here, since a few enzyme names do not follow the rule, for example pepsin and lysozyme.
5.2. In each case, enzymes are losing their activity. All the yeast cells' enzymes are denatured by boiling. Heavy metal ions are reversible inhibitors of many enzymes. Imbalances in blood pH will lead to imbalances in pH of cell contents: enzymes are denatured by abnormally acidic or alkaline conditions.
5.3. Oxidation of methanol is catalysed by an enzyme. Ethanol is closely related to methanol, and can bind to the active site of the same enzyme. It acts as a competitive inhibitor of this reaction. Methanol is broken down more slowly, and is excreted by the kidneys before too much of the toxic product builds up.
5.4. Firstly, the control point is at the start of the pathway. This means that intermediates need not build up if the pathway is inhibited. Secondly, high $NADP^+$ concentration is a sign of low NADPH concentration, because a single pool of NADP shifts between the two

forms. NADPH is a useful product of the pathway: if it gets low, the pathway must speed up. It also turns out that this first reaction is normally far from equilibrium, another important characteristic of a control step.

Unit 6

6.1. (i) active transport **(ii)** endocytosis
(iii) mass flow **(iv)** diffusion.
6.2. Starting at the top right, with the proton pumps of the respiratory chain, and working clockwise:
proton pumps: simple carriers, active
ATPase: simple carrier, passive as illustrated, but if supplied with ATP in experiments, it can actively pump protons out of the matrix
brown fat proton carrier: simple carrier, passive
ATP/ADP carrier: antiport, passive
shuttle 2: a pair of simple, passive carriers, it would appear
shuttle 1: no molecules cross the membrane here
fatty acid transport: unclear
carriers of pyruvate and Krebs cycle intermediates: passive symport with H^+ or passive antiport with OH^-
movement of dissolved gases: requires no carriers or channels
porin: forms a giant, non-specific channel.
6.3. (i) $\psi_w = \psi_s + \psi_p = -1000\text{ kPa} + 600\text{ kPa} = -400\text{ kPa}$
(ii) In each case, water moves from higher to lower water potential. The cell has water potential −400 kPa. Remember that, for example, −400 is *higher* than −500.
A: water leaves cell B: water leaves cell
C: no net movement of water D: water enters cell
pure water: water enters cell.
6.4. Other vesicles are constantly fusing with the plasma membrane. They bring newly synthesized proteins to the plasma membrane; they release molecules which are secreted by the cell. If cell surface area stays constant, the rate at which new membrane arrives must equal the rate at which it is budded off.
6.5. The protein is an LDL receptor in the plasma membrane. It binds to LDL, allowing it to enter by receptor-mediated endocytosis.

Unit 7

7.1. (i) mRNA: GGCCAUGCUCGCAGGGACCUCGUGCCC-AACUAAAUAGCCG
(ii) Using the three-letter codes for amino acids (Box 3.6): met-leu-ala-gly-thr-ser-ile-pro-thr-lys. Remember that translation begins at an AUG codon, which defines the reading frame. Translation of this mRNA is ended by a UAG codon.
(iii) A mutation has led to a UGC codon being replaced by a stop codon.
(iv) An insertion of one nucleotide in the DNA responsible for the fourth codon has led to frameshift, changing all the subsequent amino acids. This results in the tenth codon becoming UAA, ending translation.
7.2. mRNA provides the information which controls the amino acid sequence of the polypeptide being made. rRNAs form part of each ribosomal subunit. tRNAs match the correct amino acid to each codon.

7.3.

	Replication	Transcription	Translation
Does it happen to whole chromosomes or single genes?	whole chromosomes	single genes	single genes
Where does it happen in a eukaryotic cell?	nucleus	nucleus	cytoplasm
What is made?	DNA	RNA	polypeptide
Chemical machinery involved	DNA polymerase and associated enzymes	RNA polymerase	ribosome and tRNAs
Site where it begins	initiation sites	promoter	start codon (AUG)

7.4. (i) The longer strand is DNA. The mRNA is shorter, because introns have been edited out. The exons are paired up, while the bits of DNA which loop out are the introns. This is one of the first pieces of evidence for the existence of introns.
(ii) DNA and RNA would be the same length. Introns are rare in prokaryotes.

Unit 8

8.1. The only significant input is the potential energy of molecules in its food. This is converted to various energy stores within the body (still potential). Heat is lost from each reaction involved. Some energy is used for transport across membranes. It may be transferred to the potential energy of a concentration gradient; in the end, much of this will end up as heat. Some energy will be converted to kinetic energy, for example when the greyhound runs, but again this is rapidly converted to heat. So, a greyhound converts potential energy to heat.
8.2. It is gaining an electron, so it is being reduced.
8.3. NADPH is becoming oxidized. We know that oxidation and reduction always go together, so 1,3-diphosphoglycerate must be being reduced.
8.4. It must require an energy input, because it is coupled to an energy-releasing reaction.
8.5. Probably not. An energy-requiring reaction which needed relatively little energy would still have to be coupled to this standard reaction. Most of the large energy output would be wasted as heat. Think about money again. If we could only have one unit of currency, would you prefer a low value coin or a very high value note? If you bought a carrot, you would get no change from the note: a few small coins would be more efficient.

Unit 9

9.1. (i) No direct effect. The respiratory chain sets up the proton gradient, but does not rely on it being there.
(ii) DNP inhibits ATP synthesis. It is powered by the proton gradient: if the gradient breaks down, ATP synthesis cannot continue.
(iii) Glycolysis and the Krebs cycle operate faster than normal. This is because the cell contains more and more ADP, but less and less ATP. These are normally signs that a higher rate of respiration is needed. There is a knock-on effect on oxidative phosphorylation: because NADH is being formed more rapidly, electron transport speeds up.
The energy stored in the proton gradient is lost as heat, instead of being stored in ATP.
9.2. NADH cannot be oxidized. $FADH_2$ can still be oxidized, since it passes its electrons directly to cytochrome reductase. However, more and more NAD ends up as NADH, at the expense of NAD^+. The shortage of NAD^+ inhibits glycolysis and the Krebs cycle. The overall effect is much like lack of oxygen.
9.3. Glycolysis produces ATP and NADH. In anaerobic respiration which produces lactate or ethanol, the NADH is recycled, without harnessing its stored energy. Only the ATP produced directly can be used. In this alternative system, energy is transferred from NADH to extra ATP, because the terminal electron acceptor allows oxidative phosphorylation to carry on. Overall, more ATP is made for every molecule of glucose used.
9.4. Firstly, children are growing: this requires energy. Secondly, they have a higher surface area/volume ratio than adults, and may lose heat to the environment more rapidly. Children need to respire more, for each kilogram of body tissue, in order to maintain body temperature.
9.5. At first, respiration was entirely aerobic. The substrates were carbohydrates. As oxygen demands rose, or levels became depleted, the seedling carried out more and more anaerobic respiration, as well as aerobic respiration.

Unit 10

10.1. (i) Oxygen is only produced during non-cyclic photophosphorylation. Electrons are transported from water to NADP, by way of both photosystems.
(ii) $NADP^+$ is the natural Hill reagent. Electron transport is only possible when it is present. In an isolated chloroplast, all the NADP may end up as NADPH, or may leak out. Then, some other Hill reagent is necessary if oxygen is to be produced.
10.2. (i) Glycerol-3-phosphate.
(ii) Yes. Much of the carbon which has been fixed and reduced to triose phosphates is recycled to ribulose bisphosphate, keeping the cycle going.
10.3. Twice. Once in the early morning, once at dusk.
10.4. (i) Carbon fixation by Rubisco takes place in the bundle sheath of C_4 plants. The carbon pumping system means that CO_2 concentration here is higher than it would be in a C_3 leaf, at the same external concentration. External CO_2 concentration would have to fall much lower for a C_4 plant to reach its compensation point.
(ii) NADPH is used to reduce fixed carbon in the mesophyll, and is reformed in the bundle sheath. The system pumps reducing power as well as carbon. The C_3 carbon reduction cycle, taking place in the bundle sheath, requires NADPH and ATP. Much of the NADPH is supplied in this way, but ATP must be generated by light reactions in the bundle sheath itself. Cyclic photophosphorylation produces ATP but not NADPH. Interestingly, it does not produce oxygen, making conditions in the bundle sheath even less favourable for photorespiration.
(iii) During the day, malate is decarboxylated, so CO_2 levels remain high. Once the malate is used up, late in the day, CO_2 levels fall, making conditions more suitable for photorespiration.

Unit 11

11.1. (i) 16 (a diploid root cell has 32, so a haploid pollen nucleus must have half this number).
(ii) 2, because two homologous chromosomes each carry the gene. One copy came from each of the plant's parents. They may be identical, but they may be different alleles.
(iii) 4, because each chromosome has replicated.

11.2.

Feature	Mitosis	Meiosis
Number of divisions	1	2
Number of daughter cells	2	4
Is parent cell haploid or diploid?	either	diploid
What happens to chromosome number?	stays the same	halves
Are daughter cells genetically identical to one another?	yes	no
Are chromosomes replicated before division?	yes	yes
Do homologues pair up?	no	yes
Are chiasmata formed?	no	yes

11.3. The forms with fewer chromosomes will release their variation more slowly, all else being equal. Random assortment of chromosomes will have less of a mixing effect, when there are fewer chromosomes. Interestingly, the frequency of these forms is correlated with how exposed is the shore. The reasons are not clear.

11.4. (i) The species with higher chromosome numbers probably evolved as autopolyploids. Notice that $2 \times 28 = 56$, and $3 \times 28 = 84$. The higher chromosome numbers may have arisen independently. Alternatively, each may have arisen once, giving polyploid species which later gave rise to more new species.
(ii) Perhaps the forms with 82 chromosomes are aneuploids, in which two types of chromosome are linked to form a single longer one. In a diploid cell, this would decrease the chromosome number by two.
All this is idle speculation. You and I need to know much, much more about the situation before we could come to firm conclusions. As far as I know, much of this work has not yet been done.

Unit 12

12.2. (i)

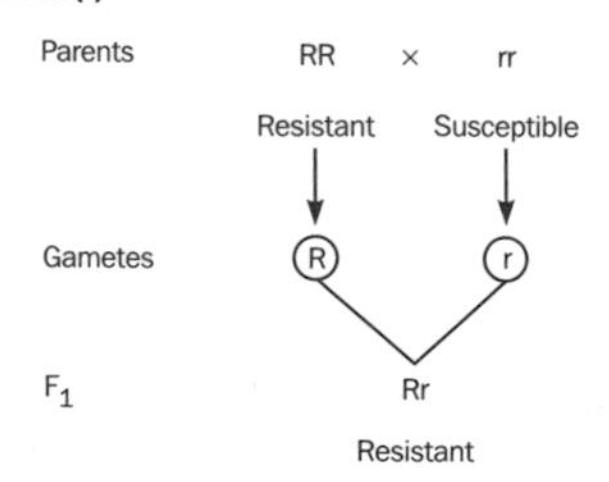

All offspring are resistant to warfarin.
(ii)

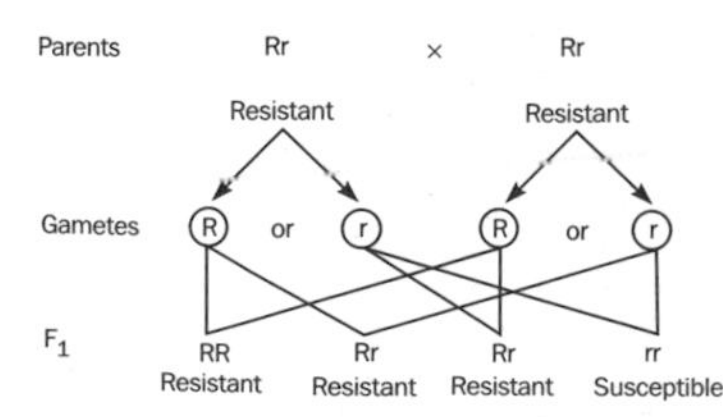

About three quarters of the offspring are likely to be resistant.
(iii) Mate the rat with a number of susceptible females. We know they must have the genotype **rr**. If any of the offspring of these crosses is susceptible, the male must be a heterozygote, since these offspring must inherit an **r** allele from each parent. If there are no susceptible offspring, the male is probably homozygous.

12.2 The bull was heterozygous for this gene. The lethal allele is recessive, so the bull had the normal phenotype. The lethal allele was passed on to some of its offspring, which were also heterozygotes since their mothers were normal. However, when these heterozygotes mated with the bull, some calves inherited the lethal allele from both parents. They were homozygotes and died. This is why the offspring of incestuous relationships have a higher risk of genetic disorders.

12.3 (i) male EeDd, female eedd; **(ii)** male EEDD, female eedd; **(iii)** male EEDd, female eedd; **(iv)** male EeDD, female eedd.

12.4. (i) The allele is recessive. This must be true because the individuals with this condition are children of non-sufferers: each parent must have been heterozygous. The gene is carried on the **X** chromosome, although the evidence here is not conclusive. Every sufferer in this family tree is male. The numbers are not great, however, and this could happen just by chance.
(ii) Any symbol will do, although the convention is to use a lower case letter for the recessive allele and the same letter, but upper case, for the dominant allele. I will use **D** for the normal allele and **d** for the deafness allele. It helps to show which sex chromosomes are present when writing the genotype.
A: X^dY B: X^DX^d C: X^DY D: X^DX^d
E: X^DX^d F: either X^DX^d or X^DX^D.

12.5.

11 — 2·1 — 5·2 — 3·6
C B E D A

Unit 13

13.1. Internal fertilization means that the egg and zygote have a protected, controlled environment. This is most important in land animals, where the egg can be kept moist even if conditions outside the body are dry. Birds, insects and mammals are all classic groups of land animals, and all have internal fertilization.

13.2. Seminiferous tubule, duct within testis, epididymis, vas deferens, urethra, vagina, cervix, uterus, oviduct.

13.3. (i) Birds' eggs are bigger than fish eggs because they store more food. This means that the young bird can grow much bigger than the young fish, before it needs to feed. The cost to the bird is in terms of numbers. A limited amount of food can be divided into many small eggs, or a few big ones.
(ii) Bird's eggs are bigger than mammals' eggs because the bird packages all the food needed for embryo development into the egg. The mammal gives most of the food to the embryo through the placenta. The large investment in each mammalian embryo means that relatively few eggs are made, even though they are small.

13.4. (i) The gonadotropins stimulate follicle development and ovulation. If too little of the hormones is released, egg development and release may not be triggered. In particular, if the LH surge is not great enough, ovulation will not take place.
(ii) Chorionic gonadotropin is a hormone made by the young embryo (Box 13.5). Its natural role is to mimic LH, keeping the corpus luteum active for longer, and so inhibiting menstruation. However, it will also stimulate ovulation if given at the right time in the cycle, mimicking the LH surge.

13.5. Oestrogen and progesterone inhibit secretion of FSH and LH. This in turn means that follicles do not start to develop, and ovulation does not take place. No eggs – no pregnancy.

13.6. If a gene which led to more female offspring existed, breeders could certainly select cows with this gene, and breed from them. The argument of Section 13.9 does not apply here, because the breeder ensures that genes favouring males are selected against. If a female-biased population was established, but then left to breed at random, selection would then favour male-biasing genes, and the ratio would move back towards 1:1. All this assumes that there is genetic variation in the sex ratio of offspring, from which to select. In cattle, and probably most other mammals, that variation is simply not there. The genetic system of sex determination (Section 12.4) appears to be very robust.

Unit 14

14.1. (i) flowering plant, conifer, *Selaginella*, fern, moss, *Ulva*.
(ii) flowering plant, conifer, *Selaginella*, fern, moss, *Ulva*, *Spirogyra*, *Chlamydomonas*.
(iii) flowering plant, conifer, *Selaginella*, fern, moss, *Ulva*, *Spirogyra*, *Chlamydomonas*.
(iv) flowering plant, conifer.
(v) Selaginella, fern, moss, Ulva, Spirogyra, Chlamydomonas.
(vi) flowering plant, conifer.
(vii) flowering plant.
(viii) moss, Ulva, Spirogyra, Chlamydomonas.
(ix) flowering plant, conifer, *Selaginella*.

14.2. (i) and **(iv)** would be expected to produce nectar. In **(ii)**, the insect is not visiting with the aim of feeding, and in the others no animals need be attracted.

14.3. Specialized nectar-feeding animals such as butterflies, bees and hummingbirds need a supply of nectar throughout their lives. In colder regions, there tend to be few flowers around in the winter. This is not a problem if the animal is dormant (bees) or in another stage of the life cycle (many butterflies). It *is* a big problem for birds, however.

14.4 (i) fruit; **(ii)** seed; **(iii)** embryo; **(iv)** endosperm; **(v)** testa.

Unit 15

15.1. (iii) Concentration of the hormone adrenaline in the blood must vary widely, if it is to be any use as a chemical signal. **(v)** Rate of heat production by the body is one of the factors which is varied in order to maintain body temperature.
Although slimmers may not realize this, the amount of fat stored in fat cells *is* under homeostatic control, although not tightly (Section 21.6).

15.2. The pancreas would no longer secrete insulin. When blood glucose rose above the set point, it would rise higher, and stay high for longer. It would decline in time, because it is used by cells throughout the body, and is excreted in the urine if it is above a certain level. This is one of the causes of diabetes mellitus. See Box 30.21 for details of the knock-on effects of high blood glucose.

15.3. (i) The set point has just been raised. Body temperature has not changed, but is now below the set point. The hypothalamus attempts to raise body temperature in all the normal ways.
(ii) The set point has been lowered, towards its normal level. Now temperature is above the set point, so sweating and dilatation of skin arterioles help the body to lose heat.

Unit 16

16.1. These are hydrophobic hormones, which can cross the plasma membrane and bind directly to receptors in the nucleus. There is no need for a second messenger, and no need for a G protein to couple the receptor to the system which releases the second messenger.
16.2 High thyroxine levels inhibit secretion of thyrotropin levels by the pituitary. This leads to a fall in thyroxine release.
16.3. In most cases there is a general lack of hormone secretion by the anterior pituitary. Lack of growth stimulating hormone, follicle stimulating hormone, luteinizing hormone and thyrotropin lead to the symptoms mentioned. In the other cases, there is some specific defect in the system which secretes growth stimulating hormone.
16.4. If hormones remained in the blood, they would continue to be detected by the target cells, long after the signal they carry has any relevance. For a signal to be a signal, it must only be there some of the time.

Unit 17

17.1. The Na^+/K^+ ATPase stops working. The membrane is leaky enough to allow slow rates of diffusion. Sodium and potassium gradients very slowly break down by diffusion and are not built up again by active transport. The overall effect: the resting potential very slowly becomes less negative.
17.2. (i) Colour vision depends on the cones, which are not stimulated by dim light. Only the rods are effective.
(ii) Groups of rods are linked to a single sensory neurone, unlike cones which have their own neurones. This means that the rods give poor resolution of fine detail, which is required for reading small print.
17.3 The main cause of fatigue is that the presynaptic terminal runs out of neurotransmitter. The synaptic vesicles store large amounts, but very rapid impulses lead to it being released faster than it is made. Also, some neurotransmitter receptors may become inactivated by repeated stimulation.

Unit 18

18.1. Heating the trunk will kill all living cells. Water is carried in xylem vessels and tracheids. They are dead anyway, so are not affected. Sugars are carried in the phloem, made up of living cells.
18.2. Most importantly, the water inside vessels and tracheids is under tension: the pressure inside is very low. Lignin helps the cell wall resist collapsing inwards. These cells also have a role in supporting the plant, especially in woody stems.
18.3. (i) Water pressures in the xylem are negative, that is below atmospheric pressure. This suggests that the water is under tension. Water pressure is lower (more negative) higher up the tree. This is evidence for a pressure gradient up the tree: water moves from higher pressure at the bottom to lower pressure at the top.
(ii) Transpiration is usually greater in the middle of the day than in the early morning (brighter light, higher temperature). More water loss will lower water potential in the leaves still further. The pressure gradient up the tree will be steeper; pressure at any intermediate point will be lower than before.
18.4. Carbohydrates reach the flower in the phloem, in the form of sucrose. Sucrose is a disaccharide, consisting of glucose linked to fructose. If a sucrase enzyme is present, nectar will contain these three sugars in various proportions.
18.5. (i) source: leaves and stems; sink: tubers
(ii) source: roots; sink: flowers.

Unit 19

19.1. They are small and/or thin. No cell is far from the surface. They rely on diffusion to carry molecules through the body.
19.2. The curve is shifted to the right if the blood has a lower oxygen affinity. This means that the blood of the active fish will unload more of its oxygen in the tissues, supporting the higher respiration rate needed for fast swimming. If oxygen levels in the water are high, uptake in the gills will probably not be affected by the shift.
19.3. (i) If carbon monoxide (CO) is bound to a haemoglobin molecule, oxygen cannot also be bound. The blood loses part, or most, of its capacity to carry oxygen. Vital organs get too little oxygen, and fail. Patients who survive severe CO poisoning may have brain damage if the oxygen levels fell too low in the brain.
(ii) O_2 and CO are competing for access to haemoglobin. CO has the advantage of binding much more slowly, but a high O_2/CO ratio will still increase the amount of bound oxygen, and decrease the bound carbon monoxide, some of which may then be breathed out.
(iii) Inhaling air with a high CO_2 content will increase the level of CO_2 in the blood. This is the main factor which stimulates rate and depth of breathing. Deeper, more rapid breathing will clear any carbon monoxide from the lungs more effectively. This in turn means that there will be a steeper CO concentration gradient from blood to air in the lungs, encouraging CO to leave haemoglobin.
19.4. (i) Increase, by increasing heart rate.
(ii) Decrease, by decreasing heart rate.
(iii) Decrease, by decreasing return of blood to the heart, so stretching the walls less and decreasing stroke volume.
19.5. (i) Faster. The SAN is only the pacemaker because it sets off impulses faster than any other part of the heart. An ectopic pacemaker must be even faster if it is to beat the SAN. However, if impulses are blocked from passing to the ventricles, a second pacemaker will develop below the block, and will probably be slower: the two sides of the block are now independent.
(ii) The impulse will reach different parts of the heart in the wrong sequence, disrupting the normal pattern of contraction. The heart is likely to become a less efficient pump.

Unit 20

20.1. Nose, throat, past the epiglottis, trachea (including larynx), bronchus, bronchioles, terminal bronchiole, atrium, alveolus, cell of alveolar wall, endothelial cell around capillary.
20.2. (i) Breathing will be much more difficult than normal, because there is more resistance to air flow in the airways.
(ii) The lung will lose much of its capacity for gas exchange, because the total surface area of alveoli has decreased. Exercise becomes difficult or impossible.
(iii) Loss of alveolar walls means loss of capillaries. The right side of the heart has to create a higher pressure, maximizing blood flow to those that are left. The extra effort leads to an increase in muscle mass on the right side, and often to heart failure.
My great uncle, a heavy smoker, suffered from emphysema before his early death. The memory of this previously vigorous man unable to move from his chair, oxygen mask always at hand, still haunts me.
20.3. An injury has allowed air to enter the pleural cavity. This breaks the fluid seal between the two pleural membranes. There is no longer anything to hold the lung out against the chest wall. Air can enter in one of two ways. If the outer pleural membrane is punctured, it comes in from the outside the body. If the inner pleural membrane and underlying lung tissue is damaged, it enters from the lung itself.
20.4. At rest, her breathing rate was enough to keep the concentration of CO_2 in her blood at its normal level. When she hyperventilated, gas exchange took place more rapidly. CO_2 levels in her blood fell, while O_2 levels rose. When she returned to natural breathing, the low CO_2 levels are detected in the respiratory centre of the medulla. This results in a slower than normal ventilation rate, until CO_2 levels have climbed back to normal.
20.5. In water, gill lamellae are separate, and water flows between them. The skin also carries out gas exchange, but it is unimportant because the gills have such a large area, and such a rich blood supply.
In air, the gill lamellae tend to stick together, greatly reducing surface area for gas exchange. The skin is unaffected, so its contribution becomes a greater proportion of the total. Air contains more oxygen than water, so the skin can absorb more oxygen than it would have done in water.

Unit 21

21.1.

Substance	Digested to ...	Enzyme(s) needed	Value to the body
Proteins	amino acids	proteases, peptidases	making proteins, also as an energy supply if abundant
Fats (triglycerides)	fatty acids, glycerol	lipases	energy supply
Cholesterol	absorbed directly	–	stiffens membranes, waterproofs skin, may be converted to other steroids (mammals can make their own, but insects need it in the diet)
Starch	glucose	amylase, maltase, dextrinases	energy supply
Lactose	glucose, galactose	lactase	energy supply
Sucrose	glucose, fructose	sucrase	energy supply
Cellulose	glucose	cellulase	energy supply

21.2. Blood leaving the gut is very variable in composition, depending on what has been eaten. This system allows the liver to make some modifications before the blood rejoins the general circulation. For example, some excess glucose can be removed, and liver cells may begin to take up and detoxify various toxins before they have been diluted in the circulation.
21.3. One possibility is grossly to overeat, use all the nitrogen-containing molecules in the food, then get rid of most of the carbohydrate. Aphids do this. Phloem sap moves up the stylet and into the gut, under pressure. Some sugars and most or all of the amino acids are removed. Slightly modified phloem sap then passes out of the gut, still rich in sugars (some ants exploit this 'honeydew' as part of their diet).
A second possibility is to develop a relationship with nitrogen-fixing bacteria, which can use nitrogen gas to make organic nitrogen compounds. They are found in the guts of various wood-eating invertebrates.

21.4. Leptin is released when the fat stores in fat cells increase. It is not released if they are shrinking. Lack of leptin leads to hunger. When blood sugar levels are low for a long period, fat stores will be broken down. The leptin system can certainly explain the first observation. Carbohydrate and lipid metabolism are closely linked, since the body is in the business of managing its energy reserves as a whole. It is certainly feasible that artificially low fatty acid levels in an experimental animal could lead to fat breakdown in the fat cells, explaining observation (c). However, it is not easy to see how the second observation could be explained. The leptin feedback loop may not be the whole story.

Unit 22

22.1. (a) Since the frog lives in fresh water, salts will not diffuse in, but water will tend to enter the body by osmosis. The frog loses this water as dilute urine.
(b) Living in fresh water, the frog's problem is to get salts, and retain them against a concentration gradient. Active uptake across the body surface is one part of the solution.
22.2. There is an energy cost in reabsorbing useful molecules from the filtrate. However, it is a fail-safe system. As well as excreting waste products, the mammalian kidney can excrete soluble toxins which have entered the body in the diet. This is especially important for herbivores, which eat plants that defend themselves chemically. If all excretion involved active secretion, a carrier would be needed for each toxin. A new toxin could not be lost. In a filtration–reabsorption system, unfamiliar molecules are excreted.
22.3.

Component of blood	Filtered out?	Reabsorbed?
Amino acids	yes	yes
Creatinine	yes	no
Glucose	yes	yes
Platelets	no	–
Proteins	no	–
Red blood cells	no	–
Salts	yes	partly
Small, soluble toxic molecules	yes	no
Urea	yes	no
Uric acid	yes	no
Water	yes	partly
White blood cells	no	–

22.4. It would be harmful. The salt in the sea water would have to be excreted. Even if he makes his urine as concentrated as possible, the water he has gained is not enough to do this: he must lose extra water in getting rid of the salt. Sea water dehydrates the body.
22.5. (i) The patient will always produce lots of dilute urine. (This condition is called diabetes insipidus, and can also be caused by a failure to respond to ADH.)
(ii) There will be a general failure of reabsorption: urine volume will be high, and it will contain glucose and amino acids. (This is called Fanconi syndrome: it can have other causes.)
(iii) Excess sodium, and so water will be reabsorbed, leading to a low urine volume. The other side of the issue is an increase in blood pressure. Interestingly, the raised blood pressure in turn increases filtration in the glomerulus, tending to reverse the change in urine production. The lasting symptom is high blood pressure.

Unit 23

23.1. If helper T-cells are killed, the body is less able to mount an organized immune response to any antigen, including HIV antigens. Rapid mutation in genes coding for HIV antigens means that a patient's lymphocyte clones which are effective against the virus now, may not be effective in the future.
23.2. Blood of this type does not contain antigens of either of the two most important blood group systems. If there is no time to check the patients blood group, O Rh– blood has the lowest chance of a harmful immune response.
23.3. Blood clots more slowly than usual, because platelet activation is one of the factors which triggers clotting. This is more of a problem when capillaries are damaged, because when larger vessels are damaged collagen is exposed, which also sets off the process.
23.4. Swelling is a hypersensitive reaction, which extends beyond the infected or irritated area, preventing the spread of infection.
23.5. Producing a defence chemical only when needed is economical of energy and materials. However, they only work if the infection is detected. If it is present all the time, this is not a problem.

Unit 24

24.1. (i) Hunting for food and other resources, finding a mate, escaping or hiding from predators, following favourable conditions in a changing environment.
(ii) Sponges, corals, barnacles, hydrozoans, some molluscs (such as mussels), for example. These animals have good defences against predators, such as the shells of molluscs and barnacles, and the stinging cells of corals and hydrozoans. They are all animals which live in water, which tends to allow food to come to them: many are filter feeders (Unit 20). It also allows external fertilization: sperm can swim through the water. Interestingly, many animals which live their adult lives in one spot have a juvenile stage which can move (the planktonic larvae of barnacles, for example). This allows the genetic advantages of wider outbreeding, and gives the offspring a chance to find a better place to live.
24.2. Actin, myosin, tubulin (the protein which forms microtubules), dynein, cross-linking proteins in the eukaryotic flagellum, the filament and motor of the prokaryotic flagellum, collagen (the fibrous protein in bone, also in cartilage), channels and receptors in the membranes of muscle cells, troponin, tropomyosin. The moral of the story? Big biological events have small, molecular causes.
24.3. Plant cells have walls; animal cells do not. Systems involving actin and myosin can change the shape of an animal cell, which can lead to body movement if repeated in many cells. The cell wall is far more rigid, and cannot be distorted in this way. However, the cell wall can resist high pressures without bursting, and can be stretched or distorted by the large forces involved. Turgor changes can drive plant cell movements, but would simply burst animal cells.
24.4. Soft, light wood uses less energy and materials to make. This means that a tree can grow taller, given the same amount of energy. However, the wood is weaker. A trunk made of softer wood cannot support such a large system of branches and leaves. There is a trade-off here. As a crude generalization, a gap or clearance in a forest is first colonized by fast-growing trees with softer, lighter wood. They exploit the increased light levels quickly, and grow upwards rapidly as they compete with one another. Later on, slower growing trees with stronger wood and shade-tolerant seedlings grow up beneath them. In the end, their strong trunks support taller, denser leaf canopies which shade out the earlier species. One is built for speed, the other for the long haul.
24.6. I cannot help you here: it all depends on the animal you are studying.
24.7. (i) The spider will not be able to extend its legs. They end up fully flexed, beneath its body.
(ii) Vertebrates have an internal skeleton, and their blood is contained in blood vessels. The mechanical components of a hydraulic system simply are not there.

Unit 25

25.1. (i) Regulative development.
(ii) Regulative and mosaic development refer to the origin of positional information, not what it leads to. Cells respond to positional information by becoming locked in to a particular fate: they become determined. As embryo development continues, cells lose the ability to form whole embryo, even before they become obviously differentiated.
25.2. (i) P_{fr} **(ii)** P_r **(iii)** P_r **(iv)** P_r **(v)** P_{fr}.
25.3 Length of the dark period is what matters here. A critical day length of 14 hours is really a critical night length of 10 hours. So treatments **(iii)**, **(iv)** and **(v)** will initiate flowering.

Unit 26

26.1. On average, people in affluent countries live longer than those in the poorest countries. Also, it is a reasonable generalization that as countries become developed the birth rate falls. These two factors both lead to an age structure which is more heavily biased towards younger individuals in poorer countries. These are generalizations, of course, and each country will have individual circumstances.
26.2. The curve will be concave, as opposed to the convex curve shown in Box 26.1.
26.3. An individual in a population is affected by other members of the population, and by other factors, both living and non-living. Density gives us an idea of how intensely an individual will experience the effects of other members of its own population. This is useful for two reasons. Firstly, density-dependence allows a population to regulate its own size (there is no suggestion that the individuals are trying to do this, or that it is desirable for the individual or the population: it just happens). Having seen how the population affects itself, we can go on to see how other factors interact with this. Secondly, the graphs which demonstrate density-dependent effects lead to mathematical models which predict how population size changes over time. This is very attractive. However, density-independent factors must also be taken into account, and it is much less easy to model their effects.
26.4. In short, no. r characteristics include semelparity, short life span and small size. However, large seed size is generally thought of as a K characteristic.

Unit 27

27.1. Grow the plants in controlled experimental conditions. Compare the plants growing together, with each species growing on its own. If the difference is the fundamental niche, the roots will be distributed in the same way, whether in mixture or alone.
27.2. Mid-successional communities. Early on, the soil is too alkaline. The earlier communities facilitate these species, by reducing soil pH. Late in succession, forest

trees create too much shade: they are inhibiting these species.

27.3. In the first situation (seen in rock pools) the species which the mollusc favours are the dominant competitors. An increase in grazing pressure releases other species from competition. In the second situation (seen on rocks uncovered at low tide) the favoured species are already subordinate to other species. Increasing grazing simply makes rare species rarer still.

27.4. The habitat is subject to disturbance by waves, clearing gaps in the mussel beds. *Postelsia* can colonize more quickly, so occupies these gaps for a while, before being excluded by the re-establishing mussels.

Unit 28

28.1. (i) Nitrogen gas is the global carrier of nitrogen: air moves around the globe relatively quickly. Global cycling of phosphorus relies on movements of the Earth's crust: unimaginably slow.

(ii) Nitrogen gas is available everywhere. Minerals rich in phosphates are the basis of most phosphorus fertilizers, and they are not found everywhere.

28.2 Rain forests are very productive, so photosynthesis there does remove a great deal of CO_2 from the atmosphere. However, a rain forest is an ecosystem, not a plant community. All that plant material is going somewhere. Some is grazed, some is stored in wood for many years but ultimately becomes food for decomposers. Unless fossilized or preserved plant material is accumulating continuously, all the carbon which enters must also leave through respiration. A mature forest is probably doing very little to change the concentration of carbon dioxide in the atmosphere. However, a lot of carbon is locked up in the biomass of the forest. Replacing the forest with almost any other ecosystem will mean that less carbon is locked up on each square metre of land. The balance ends up in the atmosphere, through decomposition or burning. This is the real problem: a one-off shot of carbon into the atmosphere when the forest is destroyed. (There are other, even more potent reasons for conserving forests, of course.)

28.3. Firstly, we would need an estimate of the transfer efficiency of each step. Secondly, we would need to know the major features of the food webs in which these chains are found. Thirdly, we would need to know the relative importance of the organisms in these chains in the diets of the animals which eat them. For example, the adder is probably not a big part of the hedgehog's diet in most habitats, most of the time.

28.4. Net productivity of producers =
$20\,000$ kJ $m^{-2}\,yr^{-1}$.
Net productivity of primary consumers =
$0.15 \times 20\,000$ kJ $m^{-2}\,yr^{-1} = 3000$ kJ $m^{-2}\,yr^{-1}$.
Net productivity of secondary consumers =
0.15×3000 kJ $m^{-2}\,yr^{-1} = 450$ kJ $m^{-2}\,yr^{-1}$.

Unit 29

29.1. To set you on track with your 'own' species, here is part of an ethogram for captive bears. It was based on observations of Kodiak, Asiatic black and polar bears at Zoo Atlanta, USA. The full ethogram included 13 behaviours, grouped as passive, active and abnormal. Notice how closely each behaviour is defined. Some, like Maintenance, are really groups of related behaviours.

Passive

- *Rest:* Bear sits or lies with head down or eyes visibly closed.
- *Passive/alert:* Bear lies, sits or stands with head up and eyes open
- *Maintenance:* Bear grooms self with mouth and/or paws, scratches, defaecates, urinates, shakes, rubs its body on an object.

Active

- *Object investigation:* Bear peers closely at, mouths or manipulates non-food items such as dead branches and logs, rocks, leaves, exhibit walls and so on.
- *Locomote:* Bear walks or runs quadrupedally or bipedally

Abnormal

- *Masturbate:* Bear rubs genital area with paw or on substrate, scratches genitals repetitively, leans over and mouths penis, etc.
- *Self-stimulation:* Bear clutches its own limbs, bites its own limbs, rocks back and forth repetitively, swings head repetitively.

29.2. (i) If more food is available, birds will be able to meet get the food they need in less time. This releases time which can be spent in other activities. If the group size stays the same, fighting may increase. If the group size falls, the extra time is available for scanning for predators, since fewer birds must each spend longer scanning. Prediction: optimal flock size decreases.

(ii) If more predators are about, more scanning may be required. If flock size stays the same, each bird might scan more and so feed less. If flock size increases, more birds are scanning, so each individual may not have to change its time allocation. Prediction: optimal flock size increases.

29.3. Fights end when the loser backs out. In birds, this is likely to involve flying away. This is not possible in a small cage. Fighting behaviours which avoid injury in the wild, may not be effective in captivity.

Unit 30

30.1. (i) Double-stranded DNA has a back-up information store: the antisense strand. Mismatched bases are a clue that mutation has happened, making repair a viable option. This is not the case for single-stranded RNA. Also, cells have evolved systems for replicating DNA with high fidelity. These systems will not help an invading RNA virus.

(ii) High mutation rates mean that viral antigens change over time, even within the course of an infection. The virus can, perhaps, keep one step ahead of the immune system.

30.2.

Internal	External
genetic factors	surgery
attack by immune system	dietary deficiency of iron, vitamin B_{12} or folic acid
menstruation	poor uptake of iron or vitamin B_{12}
liver disease?	increased vitamin requirements
	injury
	toxins
	liver disease?

30.3. Target cells are failing to respond to the insulin which is released, probably because there are too few receptors. (An increased rate of insulin destruction would be an intelligent, but incorrect, alternative answer.)

30.4. (i) It is not easy for relatively large particles like viruses, let alone bacteria, to enter or leave a cell, but viruses rely on the host cell for transcription, translation and replication of nucleic acids. This can only happen if the virus, or at least its DNA or RNA enters the cell. There is no such pressing need for most bacterial pathogens to enter cells.

(ii) Mycobacterium, which causes tuberculosis, is taken into macrophages by phagocytosis, but survives and reproduces there.

Unit 31

31.1. (i) Anaerobic; **(ii)** aerobic; **(iii)** aerobic; **(iv)** both; **(v)** anaerobic.

31.2. (i) The enzyme will leave the same sticky ends on each fragment, wherever it comes from. These will be able to pair up in new combinations, as well as the original ones. If different enzymes were used, plasmid DNA would only be able to pair with other plasmid DNA.

(ii) If the target gene is not present in the library, try again with different restriction enzymes: until one is found which does not recognize a site within the gene

31.3. (i) β cells of the pancreas; **(ii)** cells of the anterior pituitary; **(iii)** epithelial cells lining the small intestine; **(iv)** immature erythrocytes; **(v)** B lymphocytes; **(vi)** any cell.

31.4. (i) Immunogold electron microscopy; **(ii)** radioimmunoassay; **(iii)** immunofluorescence.

Glossary

Active site: The part of an enzyme which binds the substrate, and where the reaction takes place.

Active transport: Transport which requires an energy input.

Adaptations: Characteristics of organisms which suit them to an environment or way of life.

Aerobic respiration: Respiration requiring oxygen.

Allele: One version of a gene.

Altruism: Behaviour which benefits others, rather than oneself.

Anaerobic respiration: Respiration not requiring oxygen.

Aneuploidy: Cells having one or a few extra or too few chromosomes.

Antheridia: Structures in gametophytes which make male gametes.

Antibodies: Proteins which bind to antigens, made by B lymphocytes and released to the bloodstream.

Antigen: Any foreign molecule that the immune system can detect and respond to.

Antiports: Carriers which transport two different molecules in opposite directions.

Apical meristems: The growing points at the tips of roots and shoots where cell division occurs.

Apoplast: The network of cell walls in a plant tissue.

Archegonia: Structures in gametophytes which make female gametes.

Arteries: Blood vessels which carry blood away from the heart.

Asexual reproduction: Reproduction which makes clones of a single parent.

ATP: Adenosine triphosphate; a nucleotide which acts as an energy currency in cells.

Autotrophic nutrition: Making food by building up complex molecules from simple substances.

Auxiliary pigments: Light-absorbing molecules other than chlorophyll which take part in photosynthesis.

Basal metabolic rate: Metabolic rate at rest.

Batesian mimicry: Harmless species look like a harmful one.

Biological control: The addition of a species to a community, once or repeatedly, to control the abundance of another species.

Biosphere: The zone at the Earth's surface where life is found.

Blastula: An early embryo which is a hollow sphere of cells.

Carriers: Protein 'machines' which transport molecules or ions across membranes.

Catalysts: Substances that increase the rate of a chemical reaction without being used up.

Central nervous system: The brain and spinal cord.

Channels: Passive protein pores in membranes.

Charged molecule: A molecule which has extra, or too few electrons.

Chiasma: Two strands of DNA visibly crossing where crossing over has happened.

Chromatids: Identical copies of a chromosome, formed by DNA replication before mitosis or meiosis.

Chromosome: A long DNA molecule.

Cladistics: A method of making classifications, based on cladograms.

Cladogram: A tree-like diagram, showing when related groups diverged from one another.

Climax community: The final community in a succession.

Closed circulatory system: Blood stays inside blood vessels.

Codon: A sequence of three nucleotides which codes for one amino acid.

Community dynamics: Small-scale, short-term changes in community structure.

Community: All the individuals of all the species living in an area.

Compensation point: The environmental conditions under which rates of photosynthesis and respiration are equal.

Composite structure: Made up of fibres runing through a matrix.

Crossing over: Exchange of sections of DNA between homologous chromosomes.

Cyclic photophosphorylation: Light reactions which require light absorption only by PSI. Only ATP is made. Water is not split.

Denature: To inactivate an enzyme by changing its shape.

Density: Number of individuals per unit area.

Density-dependent factors: Factors which vary according to the population density.

Density-independent factors: Factors which are not affected by population density.

Development: The process by which a body takes on its mature form.

ΔG: The free energy change for a reaction.

Differentiation: The process by which a cell takes on a specialized, mature form.

Diffusion: The net movement of molecules from a region of higher concentration to a region of lower concentration.

Digestion: Physical and chemical breakdown of food.

Diploid: Having two sets of chromosomes.

DNA replication: The making of an identical copy of a DNA molecule.

Dominant allele: An allele which has an effect on the phenotype in homozygotes and heterozygotes.

Double circulation: Blood passes through the heart twice per circuit of the body.

Ecosystem: All the living things in a habitat, plus all the non-living things as well.

Egestion: The removal of material from the gut. Egestion is not a form of excretion.

Egg (= ovum): The female gamete of an animal.

Electron micrograph: An image made using an electron microscope, which can resolve much smaller structures than the light microscope.

Endocrine cells: Cells which make hormones.

Endoderm, mesoderm, ectoderm: The inner, middle and outer layers of an animal's body wall.

Entropy: The amount of disorder in a system.

Enzymes: Catalytic proteins made by cells.

Evolution: Inherited changes in a population, accumulated over a number of generations.

Excretion: The removal of waste products from cells and organisms.

Exon: A section of a eukaryotic gene which codes for part of a polypeptide.

Extinction: The death of an entire species.

Facilitation: Early successional species make conditions suitable for later successional species.

Fecundity: The number of offspring an individual produces.

Fertilization: The fusion of gametes to make a zygote; two haploid cells join, forming a diploid cell.

Fitness: A measure of how many offspring a particular variant leaves.

Force: A push or a pull.

Fruit: A structure which surrounds one or more seeds, formed from an ovary.

Fundamental niche: The niche when competitors are not present.

Gaia: The hypothesis that homeostasis on a global scale maintains conditions which happen to be suitable for life.

Gametes: The haploid cells which join in fertilization.

Gametophyte: A haploid plant body which makes gametes.

Gas exchange: Taking in gases used by the body and getting rid of waste gases to the environment.

Gastrula: An early embryo with a newly formed gut.

Gene: An inherited instruction for cell development; a section of a chromosome which codes for one polypeptide, and so for some characteristic.

Genet: Another way of defining an individual – all the living material derived from one fertilized egg.

Genetic drift: Random changes in allele frequencies in a population.

Genotype: The combination of alleles an organism has.

Gills: Organs for gas exchange between animals' bodies and water.

Glycolysis: The pathway which converts glucose to pyruvate.

Growth: An increase in body size.

Guild: A group of species using the same types of resources in a similar way.

Habitat: The place where an organism lives, or, the type of place where a species lives.

Haploid: Having one set of chromosomes.

Heritability: A measure of how much of the variation in a characteristic is controlled by genes.

Heterospory: The practice of making two types of spore.

Heterotrophic nutrition: The use of complex organic molecules as food.

Heterozygote: An individual which has two different alleles at one locus.

Homeostasis: The ways in which variables are kept at a constant level within a cell or body.

Homologous chromosomes (=homologues): Two

chromosomes of the same type in one cell, one inherited from each parent.

Homozygote: An individual which has two identical alleles at one locus.

Hormones: Chemical signals made in one cell type, carried in the blood, and detected by another cell type.

Hydrophilic molecules: Attract water molecules.

Hydrophobic molecules: Do not attract water, and clump together.

Hypertonic: Having a higher solute concentration than the surroundings.

Hypotonic: Having a lower solute concentration than the surroundings.

Immune system: A system which recognizes foreign molecules and cells, and makes them harmless.

***In vitro* fertilization:** Fertilization under laboratory conditions, outside the body.

Inhibition: Late successional species make conditions unsuitable for early successional species.

Intron: A section of DNA which does not code for polypeptide sequence, between the exons.

Isotonic: Having the same solute concentration as the surroundings.

Iteroparous organisms: Organisms which are capable of reproducing more than once.

Krebs cycle: A cycle of reactions which oxidizes pyruvate to carbon dioxide.

Limiting factor: If a factor limits the rate of some process, increasing the level of the factor will increase the rate of the process.

Lipoprotein: A cluster of lipid molecules bound to a protein.

Locus: The place on a chromosome where a particular gene is found.

Lymphokines: Signalling molecules made by white blood cells.

Mass flow: Movement of the whole body of a liquid or gas.

Megaspore: A large female spore.

Meiosis: A diploid nucleus divides to make four haploid nuclei.

Metabolic pathway: A sequence of reactions in a cell which together make a useful product.

Metabolic rate: Respiration rate of an entire animal.

Microspore: A small male spore.

Mitosis: A nucleus divides to make two identical clones.

Molecule: A group of atoms linked by covalent bonds.

Mullerian mimicry: A group of harmful species look similar.

Mutation: Random changes to a cell's DNA.

Nephron: The working unit of the kidney, which makes urine.

Neurotransmitters: Molecules which carry information across synapses.

Niche: The set of environmental conditions in which a species can live and reproduce.

Non-cyclic photophosphorylation: Light reactions which require light absorption by both photosystems. ATP and NADPH are made. Water is used, oxygen is made.

Open circulatory system: Blood bathes all body cells.

Organelle: Compartment or other structure inside a cell.

Osmoconformers: Animals which cannot control the salt concentration of their body fluids.

Osmoregulators: Animals which can control the salt concentration of their body fluids.

Osmosis: The net movement of water across a partially permeable membrane, from a region of lower solute concentration to a region of higher solute concentration.

Oxidation: Loss of electrons.

Oxidative phosphorylation: Transfer of energy from NADH and $FADH_2$ to ATP in the respiratory chain.

Oxygen dissociation curve: A graph of oxygen saturation of haemoglobin against partial pressure of oxygen.

Paracrines: Local signalling molecules.

Partial pressure: The component of total pressure exerted by one gas in a mixture.

Peristalsis: The movement of material through a tubualr organ, by waves of muscular contraction.

Phenetic classification: Classification based on organisms' characteristics.

Phenotype: An organism's characteristics.

Phloem: A plant tissue specialized for transporting sucrose and amino acids around the plant.

Photosynthesis: Building up carbohydrates, using carbon dioxide, water and light energy, and making oxygen as a waste product.

Photosystems: Highly organized arrays of chlorophyll molecules, other pigments and proteins, in the thylakoid membrane.

Phyletic classification: Classification based on the evolutionary history of a group.

Phylogeny: The evolutionary history of a group.

Plasma membrane: The membrane surrounding a cell.

Polar body: Tiny cell made during egg formation, containing an unwanted nucleus.

Polar molecule: A molecule which has small local charge, but no overall charge.

Pollination: The transfer of pollen from anther to stigma.

Polymorphism: The situation in which more than one version of a gene exists in a population.

Polyploidy: Cells having whole extra sets of chromosomes.

Population: Every member of a species in a habitat.

Positional information: Chemical signals which tell cells where they are in an embryo.

Primary oocyte: An egg-forming cell in the first stage of meiosis.

Productivity: The rate at which energy enters a trophic level.

Ramet: One way of defining an individual – a more or less independent living body.

Reading frames: The three ways an mRNA sequence can be divided into codons.

Realized niche: The niche when competitors are present.

Recessive allele: An allele which does not affect the phenotype in heterozygotes.

Red Queen hypothesis: The idea that interactions between species may drive ongoing evolution in both species.

Reduction: Gain of electrons.

Relatedness: The probability that two individuals will share a particular gene as a result of sharing ancestors.

Respiration: The breakdown, in cells, of energy-rich molecules, with the release of energy. **NB:** Respiration **is not** breathing or gas exchange.

RNA splicing: Removes the non-coding introns from the mRNA.

Seed: A tiny sporophyte, usually with food stores, in a tough outer coat.

Semelparous organisms: Organisms which can reproduce only once.

Sex-linked genes: Genes carried on the X chromosome.

Sexual reproduction: Reproduction involving fertilization. Offspring are variable.

Single circulation: Blood passes through the heart once per circuit of the body.

Skeleton: A system of structures which supports a body.

Speciation: The formation of new species

Species: Groups of populations which can interbreed in nature to produce fertile offspring.

Sperm (= spermatozoa): The male gametes of animals.

Spermatids: Haploid cells which are differentiating as sperm.

Spermatocytes: Sperm-forming cells in the process of meiosis.

Spermatogonia: Diploid cells in the testis which divide by meiosis, in order to make sperm.

Sporangia: Structures in sporophytes which make spores.

Sporophyte: A diploid plant body which makes spores.

Stomata (singular, stoma): Holes in the leaf epidermis surrounded by pairs of **guard cells**.

Substrate: A molecule which takes part in a reaction.

Succession: Larger scale, longer term changes in community structure.

Symplast: The network of cell contents in a plant tissue, linked by plasmodesmata.

Symports: Carriers which transport two different molecules in the same direction.

Synapses: Tiny gaps between adjacent neurones.

T-cell antigen receptors: Proteins which bind to antigens, made by T-lymphocytes and attached to their plasma membranes.

Taxonomy: The science of classification.

The founder effect: Small populations, set up as offshoots of larger ones, may by chance have very different allele frequencies.

Tracheae and tracheoles: Air filled pipes in insects' bodies.

Transcription: The formation of an mRNA strand using the DNA of one gene as a template.

Transfer efficiency: The percentage of the productivity of one trophic level which enters the next trophic level.

Translation: The making of a polypeptide, following the information coded for by mRNA.

Transpiration: The uptake, transport and loss of water by plants.

Tropism: Growth response to the direction of a stimulus.

Veins: Blood vessels which carry blood to the heart.

Ventilation: The ways in which water or air are brought to and from gills and lungs.

Xylem: A plant tissue specialized for water transport.

Zygote: The diploid cell formed in fertilization.

Bibliography

Aber, J.D. and Melillo, J.M. (1991) *Terrestrial Ecosystems* (Philadelphia: Saunders).

Alberts, B., Bray, D., Lewis, J., Raff, M., Roberts, K. and Watson, J.D. (1994) *Molecular Biology of the Cell* (3rd ed.) (New York: Garland).

Atkins, P. (1994) *Physical Chemistry* (5th ed.) (Oxford: Oxford University Press).

Barnes, R.S.K. and Hughes, R.N. (1988) *An Introduction to Marine Ecology* (2nd ed.) (Oxford: Blackwell Science).

Barnes, R.S.K., Calow, P. and Olive, P.J.W. (1993) *The Invertebrates: a New Synthesis* (2nd ed.) (Oxford: Blackwell Science).

Beeby, A. (1993) *Applying Ecology* (London: Chapman & Hall).

Begon, M., Harper, J.L. and Townsend, C.R. (1990) *Ecology: Individuals, Populations and Communities* (2nd ed.) (Oxford: Blackwell Science).

Berenbaum, M. (1995) *Bugs in the System: Insects and their Impact on Human Affairs.* (Reading MA: Addison-Wesley).

Browder, L.W., Erickson, C.A. and Jeffery, W.R. (1991) *Developmental Biology* (3rd ed.) (Philadelphia: Saunders).

Brown, T.A. (1995) *Gene Cloning* (3rd ed.) (London: Chapman & Hall).

Bulmer, M. (1994) *Theoretical Evolutionary Ecology* (Sunderland MA: Sinauer).

Clark, M.S. and Wall, W.J. (1996) *Chromosomes* (London: Chapman & Hall).

Cowen, R. (1995) *History of Life* (2nd ed.) (Oxford: Blackwell Science).

Cox, F.E.G. (1993) *Modern Parasitology* (Oxford: Blackwell Science).

Craigmyle, M.B.L. (1986) *A Colour of Histology* (2nd ed.) (London: Wolfe Medical Publications).

Dawkins, R. (1989) *The Selfish Gene* (2nd ed.) (Oxford: Oxford University Press).

Dennett, D.C. (1995) *Darwin's Dangerous Idea* (London: Penguin).

Goodwin, B. (ed.) (1991) *Development* (London: Hodder & Stoughton).

Gould, S.J. (1977) *Ever Since Darwin* (London: Penguin).

Gould, S.J. (1983) *Hens' Teeth and Horses' Toes* (London: Penguin).

Gow, N.A.R. and Gadd, G.M. (1995) *The Growing Fungus* (London: Chapman & Hall).

Gullan, P.J. and Cranston, P.S. (1994) *The Insects: an Outline of Entomology* (London: Chapman & Hall).

Guyton, A.C. (1992) *Human Physilogy and Mechanisms of Disease* (5th ed.) (Philadelphia: Saunders).

Guyton, A.C. and Hall, J.E. (1996) *Textbook of Medical Physiology* (9th ed.) (Philadelphia: Saunders).

Hall, D.O. and Rao, K.K. (1994) *Photosynthesis* (5th ed.) (Cambridge: Cambridge University Press).

Hartl, D.L. (1996) *Essential Genetics* (Sudbury MA: Jones & Bartlett).

Ingold, C.T. and Hudson, H.J. (1993) *The Biology of Fungi* (6th ed.) (London: Chapman & Hall).

Ingrouille, M. (1992) *Diversity and Evolution of Land Plants* (London: Chapman & Hall).

Johnson, M.H. and Everitt, B.J. (1988) *Essential Reproduction* (3rd ed.) (Oxford: Blackwell Science).

Jones, D.G. (1987) *Plant Pathology* (Milton Keynes: Open University Press).

Jones, S. (1993) *The Language of the Genes* (London: Harper-Collins).

Krebs, J.R. and Davis, N.B. (1993) *Behavioural Ecology: an Evolutionary Approach* (3rd ed.) (Oxford: Blackwell Science).

Lewin, B. (1994) *Genes V* (Oxford: Oxford University Press).

Lewin, R. (1997) *Patterns in Evolution: the New Molecular View* (New York: Scientific American Library).

Lewis, R. and Evans, W. (1997) *Chemistry* (Basingstoke: Macmillan Press).

Margulis, L., Schwartz, K.V. and Dolan, M. (1994) *The Illustrated Five Kingdoms* (New York: Harper Collins).

Maynard Smith, J. (1989) *Evolutionary Genetics* (Oxford: Oxford University Press).

McNeill, Alexander R. (1990) *Animals* (Cambridge: Cambridge University Press).

McNeill, Alexander R. (1992) *Exploring Biomechanics* (New York: Scientific American Library).

Moss, B. (1988) *Ecology of Fresh Waters: Man and Medium* (2nd ed.) (Oxford: Blackwell Science).

Pechenick, J.A. (1996) *Biology of the Invertebrates* (3rd ed.) (Dubuque IA: Wm C. Brown).

Postgate, J. (1992) *Microbes and Man* (3rd ed.) (Cambridge: Cambridge University Press).

Prescott, L.M., Harley, J.P. and Klein, D.A. (1996) *Microbiology* (3rd ed.) (Dubuque IA: Wm C. Brown).

Proctor, M.C.P., Lack and Yeo, P.F. (1992) *The Natural History of Pollination* (London: Collins).

Ridley, M. (1995) *Animal Behaviour* (2nd ed.) (Oxford: Blackwell Science).

Ridley, M. (1996) *Evolution* (2nd ed.) (Oxford: Blackwell Science).

Ridge, I. (1992) *Plant Physiology* (London: Hodder & Stoughton).

Roitt, I., Brostoff, J. and Male, D. (1996) *Immunology* (4th ed.) (London: Mosby).

Rudall, P. (1992) *Anatomy of Flowering Plants* (2nd ed.) (Cambridge: Cambridge University Press).

Salisbury, F.B. and Ross, C.W. (1992) *Plant Physiology* (4th ed.) (Belmont CA: Wadsworth).

Schmidt-Nielsen, K. (1990) *Animal Physiology: Adaptation and Environment* (4th ed.) (Cambridge: Cambridge University Press).

Singleton, P. (1995) *Bacteria in Biotechnology and Medicine* (3rd ed.) (Chichester: Wiley).

Smith, C.A. and Wood, E.J. (1991a) *Biological Molecules* (London: Chapman & Hall).

Smith C.A. and Wood, E.J. (1991b) *Molecular Biology and Biotechnology* (London: Chapman & Hall).

South, G.R. and Whittick, A. (1987) *Introduction to Phycology* (Oxford: Blackwell Science).

Stryer, L. (1995) *Biochemistry* (4th ed.) (New York: Freeman).

Wilson, E.O. (1992) *The Diversity of Life* (London: Penguin).

Withers, P.C. (1992) *Comparative Animal Physiology* (Fort Worth: Saunders).

Young, J.Z. (1950) *The Life of Vertebrates* (Oxford; Oxford University Press).

Index

Page references in **bold** type refer to major entries

H

I

Q

R

T